Lecture Notes in Computer Science 4800

Commenced Publication in 1973
Founding and Former Series Editors:
Gerhard Goos, Juris Hartmanis, and Jan van Leeuwen

Arnon Avron   Nachum Dershowitz
Alexander Rabinovich (Eds.)

Pillars
of Computer Science

Essays Dedicated to Boris (Boaz) Trakhtenbrot
on the Occasion of His 85th Birthday

 Springer

Volume Editors

Arnon Avron
Nachum Dershowitz
Alexander Rabinovich
Tel Aviv University
School of Computer Science
Ramat Aviv, Tel Aviv, 69978 Israel
E-mail: {aa; nachumd; rabinoa}@post.tau.ac.il

Library of Congress Control Number: 2008920893

CR Subject Classification (1998): F.1, F.2.1-2, F.4.1, F.3, D.2.4, D.2-3, I.2.2

LNCS Sublibrary: SL 1 – Theoretical Computer Science and General Issues

ISSN   0302-9743
ISBN-10   3-540-78126-9 Springer Berlin Heidelberg New York
ISBN-13   978-3-540-78126-4 Springer Berlin Heidelberg New York

Springer is a part of Springer Science+Business Media

springer.com

© Springer-Verlag Berlin Heidelberg 2008

Typesetting: Camera-ready by author, data conversion by Scientific Publishing Services, Chennai, India
Printed on acid-free paper   SPIN: 12227471   06/3180   5 4 3 2 1 0

Dedicated to

Boris (Boaz) Trakhtenbrot

in honor of his eighty-fifth birthday,
with deep admiration and affection.

A mighty man of valor ... his name was Boaz

אִישׁ גִּבּוֹר חַיִל ... וּשְׁמוֹ בֹּעַז

(Book of Ruth, 2:1)

Wishing him many,
many happy returns.

Boris Abramovich Trakhtenbrot
(b. 1921)

Preface

The Person

Boris Abramovich Trakhtenbrot[1] (Борис Абрамович Трахтенброт)– his Hebrew given name is Boaz (בועז) – is universally admired as a founding father and long-standing pillar of the discipline of computer science. He is the field's preeminent distinguished researcher and a most illustrious trailblazer and disseminator. He is unmatched in combining farsighted vision, unfaltering commitment, masterful command of the field, technical virtuosity, æsthetic expression, eloquent clarity, and creative vigor with humility and devotion to students and colleagues.

For over half a century, Trakhtenbrot has been making seminal contributions to virtually all of the central aspects of theoretical computer science, inaugurating numerous new areas of investigation. He has displayed an almost prophetic ability to foresee directions that are destined to take center stage, a decade or more before anyone else takes notice. He has never been tempted to slow down or limit his research to areas of endeavor in which he has already earned recognition and honor. Rather, he continues to probe the limits and position himself at the vanguard of a rapidly developing field, while remaining, as always, unassuming and open-minded.

Trakhtenbrot is a grand visionary who pioneered many fascinating directions and innovative concepts. Even when working on his doctoral dissertation (in the late 1940s), while every logician was thinking about infinite structures, he proved that the set of first-order formulæ valid on finite structures is not recursively enumerable. This result, which bears his name, precludes the possibility of any completeness theorem for first-order predicate calculus on finite structures. As such, it was the first important result in "finite model theory", and heralded a field that rose dramatically in popularity over the subsequent decades, as more and more researchers realized its centrality and multitudinous applications.

This was just an early instance of very many ideas of genius that Trakhtenbrot brought to the field. He was the first to introduce the use of monadic second-order logic as a specification formalism for the infinite behavior of finite automata. This logic turned out to be very fundamental; various temporal logics are just "sugared" fragments of the monadic logic. The classic theorem that finite automata and weak monadic second-order logic are expressively equivalent was established by Trakhtenbrot, and independently by Büchi and Elgot, in 1958. Trakhtenbrot also initiated the study of topological aspects of ω-languages

---

[1] Boris/Boaz   Abramovich/Avramovich/Avraamovich   Trakhtenbrot/Trahtenbrot/Trachtenbrot/Trajtenbrot.

and operators and provided a characterization of operators computable by finite automata. Furthermore, he provided solutions to special cases of the Church synthesis problem, which was later solved by Büchi and Landweber. The equivalence between the monadic logic and automata and the solvability of the Church problem have provided the necessary underlying mathematical framework for the development of formalisms for the description of interactive systems and their desired properties, for the algorithmic verification and the automatic synthesis of correct implementations given logical specifications, and for the advanced algorithmic techniques that are now embodied in industrial tools for verification and validation.

Trakhtenbrot was among the very first to consider time and space efficiency of algorithms (using what he called "signalizing functions") and speak about abstract complexity measures, at a time when most others cast doubt on the very notion. His justly famous and truly elegant "Gap Theorem" and his development – with his student, Janis Barzdins – of the "crossing sequence" method were groundbreaking in this regard. His paper on "auto-reducibility" provided a turning point in abstract complexity. In the USSR, these works quickly became very influential, and, in the US, complexity took over as the central preoccupation of theoretical computer scientists.

Early on, Trakhtenbrot recognized that the classic conceptual view of computation as a sequential process does not suffice to capture the operation of modern computers. Computer networks, reactive systems, and concurrent computation are all not describable in traditional terms. Accordingly, many of his more recent works deal with various aspects of concurrency, including data flow networks, Petri nets, partial-order versus branching-time equivalence, bi-simulation, real-time automata, and hybrid systems. His operative style remains classic Trakhtenbrot: patient in-depth survey of existing literature, uncompromising evaluation and critical comparison of existing approaches, followed by his own extraordinary and prescient contributions.

The list of topics upon which Trakhtrenbrot has made a lasting impression is breathtaking in its scope: decidability problems in logic, finite automata theory, the connection between automata and monadic second-order logic, complexity of algorithms, abstract complexity, algorithmic logic, probabilistic computation, program verification, the lambda calculus and foundations of programming languages, programming semantics, semantics and methodology for concurrency, networks, hybrid systems, and much more. Despite this prolificacy of subjects, the entire body of his work demonstrates the same unique melding of supreme mathematical prowess, with profound depth and thoroughness.

A roll call of Trakhtenbrot's students reads like the "Who's Who" of computer science in the USSR. (See his academic genealogy in the first chapter of this volume.) Trakhtenbrot was instrumental in the building of the computer science department in Novosibirsk, he collaborated with computer designers in the Soviet Union, and he helped in the establishment of a department of theoretical informatics in Jena.[2] The Latvian school of computer science flourished under the tutelage of his students, Brazdins and Rūsiņš Freivalds. In 1980, he emigrated from the Soviet Union and joined Tel Aviv University's School of Mathematical Sciences. There he was instrumental in the growth phase of its computer science department, now a School of Computer Science in its own right, a leading academic center in the Mideast. Though nominally long-retired, he remains vitally active.

Trakhtenbrot is a master pedagogue and expositor. He consistently sets aside time and effort for writing surveys and textbooks. His book, *Algorithms and Automatic Computing Machines*, first written in Russian in 1957, was translated into English and a dozen other languages, and is universally recognized as the first important text in the field. Two major contributions to computer-science education were his 1965 book, *Introduction to the Theory of Finite Automata*, and his 1973 book on *Finite Automata (Behavior and Synthesis)*, both widely translated. A whole generation of computer scientists was shaped by his books. Moreover, he played the key rôle in the dissemination of Soviet computer science research in the West, writing surveys on such topics as Soviet approaches to brute force search (*perebor*). (See his publication list in the first chapter of this volume.)

On several occasions, Trakhtenbrot has treated his readers to glimpses of his life under totalitarian rule. (See Chapter 2 of this volume for a detailed scientific autobiography.) While in the USSR, he was barred from attending most of the international congresses to which he had been invited. He suffered under the last stages of the Stalin era, plagued as it was with persecution and victimization of "idealists", "cosmopolitans", etc. Philosophers and logicians of the like of Russel, Carnap, and Tarski (= Tajtelbaum) were taboo, and anyone who respected their ideas was suspect, especially Jews like Trakhtenbrot.[3] His contributions are astounding under any measure; how much more so when consideration is given to the fact that he worked under the most adverse conditions: persecution, lack of support, almost no access to foreign meetings, and so on. His undaunted spirit should serve as an inspiration to the rest of the world.

Celebrating his Birthdays

Zeroeth Birthday

Boris Abramovich Trakhtenbrot was born on 20 February 1921 (Gregorian), according to official records, in Brichevo, a small North Bessarabian shtetl (presently in Moldova).

---

[2] The Friedrich Schiller University in Jena bestowed a degree of doctor *honoris causa* on Trakhtenbrot in October 1997.

[3] Boaz and other "idealists" did receive the support and encouragement of enlightened people like Andrei Kolmogorov, Alexey Lyapunov, Piotr Novikov (his advisor), Andrey Markov, and Sophia Yanovskaya, all great scientists whom Trakhenbrot always mentions with deep affection and gratitude.

Sixtieth Birthday

In the summer of 1979, Zdzislaw Pawlak proposed "to publish a collection of contributions by outstanding scientists in the field of theoretical computer science and foundations of mathematics in order to honor the 60th anniversary of Professor B. A. Trachtenbrot from Novosibirsk." He, along with Calvin Elgot, Erwin Engeler, Maurice Nivat, and Dana Scott were to edit the festschrift. However, with the untimely death of Cal Elgot, and Boaz's impending immigration to Israel with his family,[4] the project had to be abandoned. Boaz arrived in Israel on 26 December 1980.

Seventieth Birthday

In June 1991, Zvi Galil, Albert Meyer, Amir Pnueli, and Amiram Yehudai organized "An International Symposium on Theoretical Computer Science in honor of Boris A. Trachtenbrot on the occasion of his Retirement and Seventieth Birthday". The event took place in Tel Aviv and brought together many of the world's foremost scientists, including: Samson Abramsky, Georgy Adelson-Velsky, Arnon Avron, Val Breazu-Tannen, Manfred Broy, Bob Constable, Nachum Dershowitz, Zvi Galil, Rob van Glabbeek, Yuri Gurevich, Leonid Levin, Jean-Jacques Lévy, Gordon Plotkin, Amir Pnueli, Vaughan Pratt, Alex Rabinovich, Wolfgang Reisig, Vladimir Sazanov, Eli Shamir, Michael Taitslin, and Klaus Wagner. At that time, Albert Meyer eloquently highlighted his enormous debt to Trachtenbrot, the scientist, and appreciation of Trachtenbrot, the person, a debt and appreciation that countless other scientists share.

On this occasion, very many well-wishers who could not attend sent their blessings by other means:

- Bob McNaughton wrote, "We were colleagues in research... since our work in logic and automata theory in the 1960's was so close. Since very early in my own career (dating back to 1950) and continuing to the present I have always had reason to admire your contributions."
- Vitali Milman: "The name of Trachtenbrot I heard first time in the 60s when I was still a student in Kharkov. We studied his book on "Automata" and considered him to be a "father" of Russian computer science. I was very proud when in '80 it was said he will join our department, and was very happy that he spent the last decade working with us."
- Robin Milner said, "You are one of the founders of our subject. Every time a new decade of computer scientists come along, they re-invent the subject. I hope you stay with us a long time, to make sure that we re-invent it properly."

---

[4] We would be remiss if we did not take this opportunity to acknowledge the crucial rôle of Berta Isakovna (née Rabinovich), Boaz's wife of many years, who has lovingly, selflessly, and steadfastly supported Boaz "through fire and water". Berta was also a motherly figure for his many students, whom she always welcomed warmly and for whom she invariably prepared the most delicious meals.

– John Shepherdson wrote to Albert Meyer, saying, "Please tell Boris that, for over 40 years I have enjoyed reading his highly significant and beautifully written papers. I would like to thank him for the help he has given me in correspondence.... I regard him not only as one of the most significant and elegant logicians of the generation after Goedel, Church and Kleene, but also as a very warm, friendly and helpful human being."

– Dana Scott: "My heartfelt thanks for all the scientific contributions you have made over your career.... Your discoveries and insights, and your encouragement and stimulation of others have been exceptional and invaluable. You have lots of admirers."

– Jerzy Tiuryn: "You are one of the unquestionable fathers of theoretical computer science in the USSR."

For details of that event, see the report by Val Breazu-Tannen in *SIGACT News* (vol. 22, no. 4, Fall 1991, pp. 27–32). The hope was that the talks at this meeting would lead to a festschrift, but – despite concerted efforts at the time – that hope never materialized. Still, Trakhtenbrot's colleagues and former students from Latvia did publish a volume, "Dedicated to Professor B. A. Trakhtenbrot, father of Baltic Computer Science, on the occasion of his 70th birthday," entitled, *Baltic Computer Science*, in this same series (*Lecture Notes in Computer Science*, vol. 502, Springer-Verlag, May 1991).

Eightieth Birthday

In July 2001, in honor of his eightieth birthday and his "very important contribution to Formal Languages and Automata", Trakhtenbrot was invited to give a keynote address on "Automata, Circuits, and Hybrids: Facets of Continuous Time" at the joint session of the *International EATCS Colloquium on Automata, Languages and Programming (ICALP)* and of the *ACM Symposium on Theory of Computing (SIGACT)*, held on the island of Crete.

Eighty-Fifth Birthday

On Friday, 28 April 2006, the School of Computer Science at Tel Aviv University held a "Computation Day Celebrating Boaz (Boris) Trakhtenbrot's Eighty-Fifth Birthday". At a gala birthday party in Jaffa, some of the messages we received were read, including those from Samson Abramsky, Leonid Levin ("Allow me, from afar, to express my great admiration of deep insight and technical power that was always so characteristic of your work, and that have been combined with human decency and fairness that was not always easy to find in the environment in which you spent a big part of your life."), Maurice Nivat, Robin Milner, Grisha Mints ("Your name is known to every [even beginning] logician."), Gordon Plotkin, Dana Scott, Wolfgang Thomas ("I would like to express my feelings of deep thanks for your guidance of our research community, and also of amazement about your unfailing energy which pushed us over many decades."), and Igor Zaslavsky together with his Armenian colleagues ("We, the Armenian mathematical logicians and computer scientists, congratulate you with profound reverence on your 85th anniversary. ... Your human nobleness and total devotion to science always served as an inspiring example for us.").

The scientific program included the following lectures:

1. "Introductory Remarks", Dany Leviatan (Rector, Tel Aviv University).
2. "Provably Unbreakable Encryption in the Limited Access Model", Michael Rabin (Hebrew University and Harvard University).
3. "Verification of Software and Hardware", Zohar Manna (Stanford University).
4. "Why Sets? On Foundations of Mathematics and Computer Science", Yuri Gurevich (Microsoft, Redmond).
5. "Models of Bounded Complexity in Describing Decidable Classes in Predicate Logics with Time", Anatol Slissenko (Université de Paris 12).
6. "Linear Recurrences for Graph Polynomials", Janos Makowsky (Technion).
7. "Unusual Methods for Executing Scenarios", David Harel (Weizmann Institute).
8. "The Church Synthesis Problem with Parameters", Alex Rabinovich (Tel Aviv University).
9. "Concluding Remarks", Boaz Trakhtenbrot (Tel Aviv University).

This Volume

As a follow-up to that 85th birthday event, we asked Boaz's students and colleagues (those who did not attend, as well as those who did) to contribute to a volume in his honor. This book is the result of that effort. More precisely, it is the culmination of an effort that has been brewing for almost thirty years, sometimes on a high flame, other times on low.

The collection of articles in this book begins with historical overviews by Trakhtenbrot and Albert Meyer. These are followed by 34 technical contributions (each of which was reviewed by one or two readers), which cover a broad range of topics:[5]

- **Foundations:** Papers by Arnon Avron; by Andreas Blass and Yuri Gurevich; and by Udi Boker and Nachum Dershowitz.
- **Mathematical logic:** Papers by Sergei Artemov; by Matthias Baaz and Richard Zach; by Moti Gitik and Menachem Magidor; and by Grigori Mints.
- **Logics for computer science:** Papers by Johan van Benthem and Daisuke Ikegami; by Dov Gabbay, by Daniel Lehmann; by Alexander Rabinovich (student of Boaz) and Amit Shomrat; and by Moshe Vardi.
- **Mathematics for computer science:** Papers by Jan Bergstra, Yoram Hirshfeld, and John Tucker; by Leonid Levin; by Johann Makowsky and Eldar Fischer; and by Boris Plotkin and Tatjana Plotkin.

---

[5] Our assignment of authors to categories is of necessity somewhat haphazard, because many of the contributions themselves span more than one topic.

- **Automata, formal languages, and logic:** Papers by Michael Dekhtyar (student) and Alexander Dikovsky; by Rūsiņš Freivalds (student); by Michael Kaminski and Tony Tan; and by Wolfgang Thomas.
- **Asynchronous computation:** Papers by Irina Lomazova (student); by Antoni Mazurkiewicz; and by Wolfgang Reisig.
- **Semantics of programming languages:** Papers by David Harel, Shahar Maoz, and Itai Segall; by Masahito Hasegawa, Martin Hofmann, and Gordon Plotkin; and by Vladimir Sazonov (student).
- **Verification:** Papers by Michael Dekhtyar (student), Alexander Dikovsky, and Mars Valiev (student); by Daniel Leivant; by Oded Maler, Dejan Nickovic, and Amir Pnueli; by César Sánchez, Matteo Slanina, Henny Sipma, and Zohar Manna; and by Valery Nepomniaschy (student).
- **Software engineering:** Papers by Mikhail Auguston ("grand-student" of Boaz) and Mark Trakhtenbrot (son of Boaz); and by Janis Barzdins (student), Audris Kalnins (grand-student), Edgars Rencis, and Sergejs Rikacovs.

We offer this modest volume to Boaz in honor of his birthday and in recognition of his grand contributions to the field.

5 December 2007   Arnon Avron
25 Kislev 5768   Nachum Dershowitz
א' חנוכה ה'תשס"ח   Alexander Rabinovich
Tel Aviv

Organization

Reviewers

The editors were aided by the following individuals, to whom they are most grateful:

Gul Agha	Rajesh Karmani
Matthias Baaz	Menachem Kojman
Johan van Benthem	Sumit Nein
Andreas Blass	Wojciech Penczek
Stan Burris	Wolfgang Reisig
Bruno Courcelle	Shmuel Tyshberowitz
Yuri Gurevich	Moshe Vardi
Yoram Hirshfeld	Greta Yorsh
Tirza Hirst	Anna Zamansky

About the Cover

The cover illustration is from *Architectura civil, recta y obliqua, considerada y dibuxada en el Templo de Jerusalem* (Viglevani [= Vigevano], 1678) by Juan Caramuel y Lobkowitz (b. Madrid, 1606). It is a rendition of the Jachin (right) and Boaz (left) ornamental bronze columns of King Solomon's Temple in Jerusalem, cast by Hiram of Tyre (I Kings 7:13–22), and is reproduced courtesy of Antiquariat Turszynski, Herzogstr. 66, 80803 München, Germany.

Table of Contents

From Logic to Theoretical Computer Science – An Update[*]

Boris A. Trakhtenbrot

School of Computer Science, Tel Aviv University, Ramat Aviv 69978, Israel

1 Foreword

In October 1997, whilst touching up this text, exactly 50 years had past since I was accepted for graduate studies under P.S. Novikov. I started then to study and do research in logic and computability, which developed, as time will show, into research in Theoretical Computer Science (TCS).

After my emigration (in December 1980) from the Soviet Union (SU), I was encouraged by colleagues to experience the genre of memoirs. That is how [67,68] appeared, and more recently [72], conceived as contributions to the history of TCS in the SU. The present paper is intended as a more intimate perspective on my research and teaching experience. It is mainly an account of how my interests shifted from classical logic and computability to TCS, notably to Automata and Computational Complexity. Part of these reminiscences, recounting especially the scientific, ideological and human environment of those years (roughly 1945–67), were presented earlier at a Symposium (June 1991) on the occasion of my retirement. Occasionally, I will quote from [67,68,72], or will refer to them.

Before starting the main narrative I would like to recall some important circumstances which characterized those years.

First of all, the postwar period was a time of ground-breaking scientific developments in Computability, Information Theory, and Computers. That is widely known and needs no comment. The subjects were young and so were their founders. It is amazing that at that time the giants, Church, Kleene, Turing, and von Neumann, were only in their thirties and forties!

Now, about the specific background in the Soviet Union.

The genealogical tree of TCS in the SU contains three major branches, rooted in A.N. Kolmogorov (1903–87), A.A. Markov (1903–79) and P.S. Novikov (1901–75). In those troublesome times, these famous mathematicians also had the reputation of men with high moral and democratic principles. Their scientific interests, authority and philosophies influenced the development of mathematical logic, computability for several generations, and subsequently TCS in the SU.

Whereas Markov and Kolmogorov contributed directly to TCS, Novikov's involvement occurred through his strong influence on his disciples and collaborators. The most prominent of them – A.A. Lyapunov (1911–1973) – became a

---

[*] This chapter is an expanded and updated version of "From logic to theoretical computer science", which appeared in *People and Ideas in Theoretical Computer Science*, Cristian S. Calude, ed., Springer-Verlag, Singapore, pp. 314–342, 1998.

A. Avron et al. (Eds.): Trakhtenbrot/Festschrift, LNCS 4800, pp. 1–38, 2008.

Fig. 1. Three generations of scientists (from left to right): A. Slissenko, A.A. Markov, Jr., B.A. Kushner, B.A. Trakhtenbrot (from the archive of the Markov family)

widely recognized leader of "Theoretical Cybernetics" – the rubric which covered at that time most of what is considered today to belong to TCS.

As a matter of fact, for many offspring of those three branches, including myself, the perception of TCS was as of some kind of applied logic, whose conceptual sources belong to the theoretical core of mathematical logic. The affiliation with Logic was evident at the All-Union Mathematical Congress (Moscow, 1956), where Theoretical Cybernetics was included in the section on mathematical logic. Other examples: books on Automata [17,13] appeared in the series *Mathematical Logic and Foundations of Mathematics*; also, my first papers on Computational Complexity were published in Anatoly I. Maltsev's (1909-67) journal *Algebra and Logic*.

The early steps in TCS coincided with attacks of the official establishment on various scientific trends and their developers. In particular, Cybernetics was labeled a "pseudo-science", and Mathematical Logic – a "bourgeois idealistic distortion". That was the last stage of the Stalin era with persecution and victimization of "idealists", "cosmopolitans", etc. The survival and the long overdue recognition of Mathematical Logic and Cybernetics is in many respects indebted to Lyapunov, Markov, Novikov, Kolmogorov and S.A. Yanovskaya (1896–1966). But even after that, academic controversies often prompted such bureaucratic repression as the prevention of publications and the denial of degrees. Difficulties with publications also happened because of the exactingness and self-criticism of the authors and/or their mentors, or because the community was far from prepared to appreciate them. I told about that in [67] and [68].

Above, the emphasis was on the Soviet side; now, some remarks on the international context in which research in TCS was conducted in the SU.

The chronology of events reveals that quite a number of ideas and results in TCS appeared in the SU parallel to, independent of, and sometimes prior to, similar developments in the West. This parallelism is easy to explain by the fact that these were natural ideas occurring at the right time. In particular, that is how comprehensive theories of Automata and of Computational Complexity emerged in the 50s and 60s; I will elaborate on this subject in the next sections. But for a variety of reasons, even in those cases where identical or similar results were obtained independently, the initial motivation, the assessment of the results and their impact on the development and developers of TCS did not necessarily coincide. In particular, in the SU specific interest in complexity theory was aroused by discussions on the essence of brute force algorithms (*perebor* – in Russian). However, despite this difference in emphasis from the motivating concerns of the American researchers, after a few years these approaches virtually converged.

In the past, the priority of Russian and Soviet science was constantly propounded in Soviet official circles and media. This unrestrained boasting was cause for ironic comments in the West and for self-irony at home. But, as a matter of fact, the West was often unaware of developments in the SU, and some of them went almost entirely unnoticed. To some extent this was a consequence of the isolation imposed by language barriers and sociopolitical forces. In particular, travels abroad were a rare privilege, especially to the "capitalist" countries. My first trip abroad, for example, took place in 1967, but visits to the West became possible only in 1981 after my emigration to Israel.

Against this unfavorable background it is worth mentioning also the encouraging events and phenomena, which eased the isolation.

The International Mathematical Congress in Moscow (1966) was attended by the founders of our subject, namely, Church, Kleene, Curry, Tarski and other celebrities. It was an unforgettable and moving experience to have first-hand contact with these legendary characters. Later, Andrey P. Ershov (1931–1988) managed to organize a series of International Symposia on "Theoretical Programming", attended also by people from the West, and among them F. Bauer, E. Dijkstra, E. Engeler, C.A.R. Hoare, D. Knuth, J. McCarthy, R. Milner, M. Nivat, D. Park, M. Paterson, J. Schwartz, and A. van Wijngaarden. For many years, A. Meyer used to regularly send me proceedings of the main TCS symposia, a way to somehow compensate for the meetings my colleagues and myself were prevented from attending. This was part of our unusual and long-term contact by correspondence, which – after my emigration – switched to direct collaboration. All this reinforced our sense of belonging to the international TCS community.

2 Early Days

2.1 Brichevo

I was born in Brichevo, a village in Northern Bessarabia (now Moldova). Though my birth place has nothing to do with my career or with other events I am going to write about, let me begin with the following quotation:

> Brichevka, a Jewish agricultural settlement, founded in 1836. According to the general (1897) census of the population – 1644 inhabitants, 140 houses... (from vol. 5 of *The Jewish Encyclopedia*, St. Petersburg, 1912; translated from Russian).

Among the first settlers were Eli and Sarah Helman, the grandparents of my maternal grandfather. World War II brought about the collapse of Brichevo (or Brichevka). The great majority of the population did not manage to flee and were deported to the notorious Transnistria camps; only a small number survived and they dispersed over countries and continents. For years I used "Brichevo" as a reliable password: easy for me to remember, apparently impossible for outsiders to guess, and still a way to retain the memory of a vanished community.

After completing elementary school in Brichevo I attended high school in the neighboring towns of Belts and Soroka, where I was fortunate to have very good teachers of mathematics. My success in learning, and especially in mathematics, was echoed by the benevolence of the teachers and the indulgence of my fellow pupils. The latter was even more important to me, since it to some degree compensated for the discomfort and awkwardness caused by my poor vision.

2.2 Kishinev

In 1940 I enrolled in the Faculty of Physics and Mathematics of the newly-established Moldavian Pedagogical Institute in Kishinev. The curriculum covered a standard spectrum of teachers' training topics. In particular, mathematical courses presented basics in Calculus, Linear Algebra and Algebra of Polynomials, Analytical Geometry, Projective Geometry, Foundations of Geometry (including Lobachevski Geometry), Elements of Set Theory and Number Theory.

On June 22, 1941, Kishinev (in particular the close neighbourhood of our campus) was bombed by German air forces. In early July, I managed to escape from the burning city. Because of vision problems, I was released from military service and, after many mishaps, arrived as a refugee in Chkalov (now Orenburg) on the Ural River. Here, I enrolled in the local pedagogic institute. A year later we moved to Buguruslan in the Chkalov region, to where the Kishinev Institute was evacuated to in order to train personnel for the forthcoming return home as soon as our region would be liberated. Almost all the lecturers were former high school teachers – skilled people whose interests lay in the pedagogic aspects of mathematics and physics. (There were no recipients of academic degrees among them, but one of the instructors in the Chkalov institute bore the impressive name Platon Filosofov). Nikolai S. Titov, a former Ph.D. student of the Moscow University, who happened to flee to Buguruslan, lectured on Set Theory. I was deeply impressed by the beauty and novelty of this theory. Unfortunately, this was only a transient episode in those hard and anxious days. Actually, during the war years 1941–1944, my studies were irregular, being combined with employment in a felt boot factory, a storehouse and, finally, in the Kuybyshev-Buguruslan Gas Trust.

In August 1944 the institute was evacuated to Kishinev and I returned to my native regions for a position in the Belts college to train elementary school teachers. Only a year later did I take my final examinations and qualify as a high school mathematics teacher. That was my mathematical and professional background in September 1945 when (already at the age of $24\frac{1}{2}$) I decided to take a chance and seriously study mathematics.

2.3   Chernovtsy

I enrolled at the University of Chernovtsy (Ukraine) to achieve the equivalent of a master's degree in mathematics. In that first postwar year the university was involved in the difficult process of restoration. Since my prior education covered only some vague mathematical-pedagogical curriculum with examinations partially passed without having attended lectures, I did not know much to start off with. But there were only a few students and the enrollment policy of the administration was quite liberal. There were also only a few academic staff in our Faculty of Physics and Mathematics and soon I became associated with Alexander A. Bobrov, a prominent character on the general background. A.A. (b. 1912), who completed his Ph.D. thesis in 1938 under Kolmogorov, gave an original course in Probabilities. The distinguishing quality was not so much in the content of the course as in his style (completely new to me) of teaching and of involving the audience. A.A. did not seem to be strongly committed to his previously prepared lectures; during class he would try to examine new ideas and to improvise alternative proofs. As such trials did not always succeed, he would not hesitate to there and then loudly criticize himself and appeal to the audience for collaboration. This challenging style was even more striking in a seminar he held on Hausdorff's famous book on Set Theory, with the participation of both students and academic staff. Due to the "Bobrovian" atmosphere dominating the seminar, I started to relish the idea of research in this fascinating area. A.A. also helped me secure a job in the newly founded departmental scientific library. My primary task was to take stock of the heaps of books and journals extracted earlier from basements and temporary shelters, and to organize them into some bibliographical service. I remember reverently holding volumes of the *Journal fuer reine und angewandte Mathematik* with authentic papers and pictures of Weierstrass and other celebrities. As I later understood, the mathematical library was exclusively complete, and, as a matter of fact, disposed of all the important journals before WWII. As there was only a handful of graduate students it soon turned out that my library was not in much demand – in truth, for days there were no visitors at all; so most of the time I shared the roles of supplier and user of the library services. Through self study I mastered a significant amount of literature and reached some scientific maturity. I soon identified *Fundamenta Mathematicae* to be the journal closest to my interests in Descriptive Set Theory. All the volumes, starting with the first issue dated 1921, were on my table and I would greedily peruse them.

After considering some esoteric species of ordered sets, I turned to the study of delta-sigma operations, a topic promoted by Andrei N. Kolmogorov and also tackled in *Fundamenta*. At this stage, Bobrov decided that it was the right time to bring me together with the appropriate experts and why not with Kolmogorov himself! In the winter of 1946 Kolmogorov was expected to visit Boris V. Gnedenko (1912–1995) at the Lvov university. So far so good, except that at the last moment Kolmogorov canceled his visit. Gnedenko did his best to compensate for that annoying failure. He showed me exclusive consideration, invited me to lunch at his home and attentively inquired about all my circumstances. It was the first time that I had talked to a full professor and I felt somehow shy in his presence and in the splendour of his dwelling. B.V. listened to me patiently and, I guess, was impressed not as much by my achievements (which were quite modest, and after all, beyond the field of his main interests) but by my enthusiastic affection for Descriptive Set Theory. Anyway, he explained to me that for the time-being Kolmogorov had other research preferences and it would be very useful to contact Piotr S. Novikov and Alexei A. Lyapunov who, unlike Kolmogorov and other descendants of the famous Lusin set-theoretical school, were mostly still active in the field.

During this period I met Berta I. Rabinovich, who was to become my wife.

In the summer of 1946, I visited Moscow for the first time. Because it was vacation time and since no prior appointments had been set up, it was very difficult to get hold of people. Nevertheless, I managed to see Kolmogorov for a short while at the university and to give him my notes on delta-sigma operations. He was in a great hurry, so we agreed to meet again in a couple of weeks on my way back home; unfortunately this did not work out. Novikov was also unreachable, being somewhere in the countryside. I was more fortunate with A.A. Lyapunov, in whose house I spent a wonderful evening of scientific discussions alternated with tea-drinking with the whole family. A.A. easily came to know my case and presented me with a deeper picture of the Moscow set-theoretical community with a stress on the current research done by Novikov and by himself. He offered to inform Novikov in detail about my case and suggested that I visit Moscow at a more appropriate time for further discussions.

My second trip to Moscow was scheduled for May 1947 on the very eve of my graduation from the Chernovtsy University, when, beyond pure mathematics, the question of my forthcoming (if any) Ph.D. studies was on the agenda. All in all I had to stay in Moscow for at least a couple of weeks and that required appropriate logistics – a very nontrivial task at that time, in particular, because of the food rationing system and the troublesome train connections. Alas, at the first connection of the Lvov railroad station, local pickpockets managed to cut out the pocket with all my money. Despite this most regrettable incident, the trip ultimately turned out to be quite successful. The meetings with Novikov were very instructive and warm. And again, as in the case of the Lyapunovs, the atmosphere in the Novikov family was friendly and hospitable. Occasionally, Novikov's wife, Ludmila V. Keldysh (1904–1976), a prominent researcher in set theory in her own right, as well as A.A. Lyapunov, would also participate in the conversations. Counterbalancing my interests and efforts towards Descriptive Set Theory, Novikov called my attention to new developments I was not aware of in provincial Chernovtsy. He pointed to the path leading from a handful of hard set theoretical problems to modern concepts of mathematical logic and computability theory. He also offered his support and guidance should I agree to follow this path. I accepted Novikov's generous proposal, although with a sense of regret about my past dreams about Descriptive Set Theory.

Novikov held a permanent position at the Steklov Mathematical Institute of the USSR Academy. At that time, departments of mathematical logic did not yet exist in the USSR, but Novikov, together with Sofia A. Yanovskaya, had just started a research seminar "Mathematical Logic and Philosophical problems of Mathematics" in Moscow University, unofficially called The Bolshoy (great) Seminar. So, it was agreed that wherever other options might arise, Novikov would undertake my supervision and would do his best to overcome bureaucratic barriers.

3   Ph.D. Studies

In October 1947, I began my Ph.D. studies at the Kiev Mathematical Institute of the Ukrainian Academy of Sciences. The director of the institute, Mikhail A. Lavrentiev (1900–81), approved my petition to specialize in mathematical logic under P.S. Novikov and agreed to grant me long-term scientific visits to Moscow where I would stay with my advisor.

In Moscow, the Bolshoy seminar was then the main medium in which research and concomitant activities in that area were conducted. In particular, it was the forum where mathematical logicians from the first post-war generation (mostly students of P.S. Novikov, S.A. Yanovskaya and A.N. Kolmogorov) joined the community, reported on their ongoing research, and gained primary approval of their theses; and that is also what happened to me.

The atmosphere dominating the meetings of the seminar was democratic and informal. Everybody, including the students, felt and behaved at ease without strong regulations and formal respect for rank. I was happy to acquire these habits and later to promote them at my own seminars.

Actually the seminar was the successor of the first seminar in the USSR for mathematical logic, which was founded by Ivan I. Zhegalkin (1869–1947). After Zhegalkin's death it became affiliated with the Department of History of Mathematical Sciences of the Moscow University, whose founder and head was Yanovskaya. Its exceptional role in the development of mathematical logic in the USSR is a topic of its own and I will touch on it only very briefly.

The seminar usually engaged in a very broad spectrum of subjects from mathematical logic and its applications as well as from foundations and philosophy of mathematics. Here are some of the topics pursued by the senior participants:

Novikov – consistency of set-theoretical principles; Yanovskaya – philosophy of mathematics and Marx's manuscripts; Dmitri A. Bochvar (a prominent chemist in his main research area) – logic and set-theoretical paradoxes; Victor I. Shestakov (professor of physics) – application of logic to the synthesis and analysis of circuits.

Among the junior participants of the seminar I kept in close contact with the three Alexanders:

Alexander A. Zykov (1922–), also a Ph.D. student of Novikov, was at that time investigating the spectra of first order formulas. A.A. called my attention to Zhegalkin's decidability problem, which became the main topic of my Ph.D. thesis. He also initiated the correspondence, with me sending lengthy letters to Kiev with scientific Moscow news. This epistolary communication, followed later by correspondence with Kuznetsov and Sergey V. Yablonski (1925–1998). was a precious support in that remote time.

Alexander V. Kuznetsov (1927–87) was the secretary of the Bolshoy Seminar and conducted regular and accurate records of all meetings, discussions and problems. For years he was an invaluable source of information. For health reasons, A.V. did not even complete high school studies. As an autodidact in extremely difficult conditions, he became one of the most prominent soviet logicians. I had the good fortune to stay and to collaborate with him.

Alexander S. Esenin-Volpin (1924–) (the son of the famous Russian poet Sergey A. Esenin) was a Ph.D. student in topology under Pavel A. Alexandrov, but he early on became involved in logic and foundations of mathematics. A.S. became most widely known as an active fighter for human rights, and already in the late forties the KGB was keeping an eye on him. In the summer of 1949, we met in Chernovtsy, where he had secured a position after defending his thesis. Shortly thereafter he disappeared from Chernovtsy and we later learned that he had been deported to Karaganda (Kazakhstan). A couple of years later, I received a letter from him through his mother. I anxiously opened the letter, fearful of what I was about to learn. The very beginning of the letter was characteristic of Esenin-Volpin's eccentric character – "Dear Boris, let f be a function...".

During the years of my Ph.D. studies (1947–50), I actively (though not regularly) participated in the seminar meetings. Also the results which made up my thesis, "The decidability problem for finite classes and finiteness definitions in set theory" were discussed there. S. A. Yanovskaya offered the official support of the department in the future defense at the Kiev Institute of Mathematics; the other referees were A. N. Kolmogorov, A. A. Lyapunov and B. V. Gnedenko.

My thesis [43] included the finite version of Church's Theorem about the undecidability of first order logic: the problem of whether a first order formula is valid in all finite models is, like the general validity problem, undecidable, but in a technically different way. The novelty was in the formalization of the algorithm concept. Namely, I realized that, in addition to the process of formal inference, the effective process of (finite) model checking could also be used as a universal approach to the formalization of the algorithm concept. This observation anticipated my future concern with constructive processes on finite models.

Other results of the thesis which are seemingly less known, deal with the connection between deductive incompleteness and recursive inseparability.

In 1949 I proved the existence of pairs of recursively enumerable sets which are not separable by recursive sets. I subsequently learned that P.S. Novikov had already proved this, but, as usual, had not taken the trouble to publish what he considered to be quite a simple fact. (Note, that in 1951, Kleene who independently discovered this fact, published it as "a symmetric form of Godel's theorem".) In the thesis I showed that the recursive inseparability phenomenon implies that the means of any reasonably defined set theory are not enough to answer the question of whether two different finiteness definitions are equivalent. This incompleteness result was also announced in my short note [42] presented by A. N. Kolmogorov to the *Doklady*, but after Novikov's cool reaction to "inseparability", I refrained from explicitly mentioning that I had used these very techniques. Clearly, A. N. had forgotten that these techniques were in fact developed in the full text of my thesis and he later proposed the problem to his student Vladimir A. Uspenski (1930–) . Here is a quotation from *History of Mathematics* [40, p. 446]: "A. N. Kolmogorov pointed to the possible connection between the deductive incompleteness of some formal systems and the concept of recursive inseparability (investigated also by Trakhtenbrot). V. A. Uspenski established (1953) results, which confirm this idea...."

Those early years were a period of fierce struggle for the legitimacy and survival of mathematical logic in the USSR. Therefore the broad scope of the agendas on the Bolshoy seminar was beneficial not only for the scientific contacts between representatives of different trends, but also, in the face of ideological attacks, to consolidate an effective defense line and to avoid isolation and discredit of mathematical logic. For us, the junior participants of the seminar, it was also a time when we watched the tactics our mentors adopted to face or to prevent ideological attacks. Their polemics were not free of abundant quotations from official sources, controlled self-criticism and violent attacks on real and imaginary rivals.

It was disturbing then (and even more painful now) to read S.A. Yanovskaya's notorious prefaces to the 1947–1948 translations of Hilbert and Ackermann's *Principles of Mathematical Logic* and Tarski's *Introduction to Logic and the Methodology of Deductive Sciences* in which Russell was blamed as a warmonger and Tarski, as a militant bourgeois. Alas, such were the rules of the game and S.A. was not alone in that game. I remember the hostile criticism of Tarski's book by A.N. Kolmogorov (apparently at a meeting of the Moscow Mathematical Society): "Translating Tarski was a mistake, but translating Hilbert was the correct decision," he concluded. This was an attempt to grant some satisfaction to the attacking philosophers in order to at least save the translation of Hilbert-Ackermann's book. I should also mention that S.A. was vulnerable – she was Jewish – a fact of which I was unaware for a long time. I learned about it in the summer of 1949 during Novikov's visit to Kiev. He told me then with indignation about official pressure on him "to dissociate from S.A. and other cosmopolitans".

However difficult the situation was, we – the students of that time – were not directly involved in the battle which we considered to be only a confrontation of titans. As it turned out this impression was wrong.

4   Toward TCS

In December 1950 after the defense of my thesis, I moved to Penza, about 700 km. SE of Moscow, for a position at the Belinski Pedagogical Institute.

At the beginning it was difficult for me to appropriately pattern my behaviour to the provincial atmosphere so different from the informal, democratic surroundings of the places I came from. These circumstances unfavorably influenced my relationships with some of the staff and students (in particular because of the constant pressure and quest for high marks). Because of this, though I like teaching, at the beginning, I did not derive satisfaction from it.[1] The situation was aggravated after a talk on mathematical logic I delivered to my fellow mathematicians. The aim of the talk entitled "The method of symbolic calculi in mathematics", was to explain the need and the use of exact definitions for the intuitive concepts "algorithm" and "deductive system". I was then accused of being "an idealist of Carnap-species". In that era of Stalin paranoia such accusations were extremely dangerous. At diverse stages of the ensuing developments, P.S. Novikov and A.A. Lyapunov (Steklov Mathematical Institute) and to some degree A.N. Kolmogorov and Alexander G. Kurosh (1908–1971) (Moscow Mathematical Society) were all involved in my defense, and S.A. Yanovskaya put my case on the agenda of the Bolshoy seminar. This story is told in [72].

My health was undermined by permanent tension, fear and overwork (often more than 20 hours teaching weekly). It goes without saying that for about two years I was unable to dedicate enough time to research. It was in those circumstances that only the selfless care and support of my wife Berta saved me from collapse. I should also mention the beneficial and calming effect of the charming middle-Russian landscape which surrounded our dwelling. Cycling and skiing in the nearby forest compensated somewhat for our squalid housing. (Actually, until our move to Novosibirsk in 1961, we shared a communal flat, without water and heating facilities, with another family.)

But despite all those troubles I remember this period mainly for its happy ending. In the summer of 1992, forty years after this story took place, Berta and I revisited those regions. The visit to Penza was especially nostalgic. Most of the participants of those events had already passed away. Only the recollections and of course the beautiful landscape remained.

---

[1] Of course, I also had good students and one of them, Ilya Plamennov (1924–), was admitted through my recommendation to Ph.D. studies at the Moscow University. Later he became involved in classified research and was awarded the most prestigious Lenin Prize (1962).

Returning to the "Idealism" affair, the supportive messages I received from Moscow stressed the urgent need for a lucid exposition of the fundamentals of symbolic calculi and algorithms for a broad mathematical community. They insisted on the preparation of a survey paper on the topic, which "should be based on the positions of Marxism-Leninism and contain criticism of the foreign scientists-idealists". There was also an appeal to me to undertake this work which would demonstrate my philosophical ideological loyalty. Nevertheless I did not feel competent to engage in work which covered both a mathematical subject and official philosophical demands. These demands were permanently growing and changing; they could bewilder people far more experienced than myself. So it seemed reasonable to postpone the project until more favorable circumstances would allow separation of logic from official philosophy. Indeed, such a change in attitude took place gradually, in particular due to the growing and exciting awareness of computers.

In 1956 the journal *Mathematics in School* published my tutorial paper "Algorithms and automated problem solving". Its later revisions and extensions appeared as books which circulated widely in the USSR and abroad [50]. (Throughout the years I was flattered to learn from many people, including prominent logicians and computer scientists, that this tutorial monograph was their own first reading on the topic as students and it greatly impressed them.)

Meanwhile, I started a series of special courses and seminars over and above the official curriculum, for a group of strong students. These studies covered topics in logic, set theory and cybernetics, and were enthusiastically supported by the participants. Most of them were later employed in the Penza Computer Industry where Bashir I. Rameev, the designer of the "Ural" computers, was a prominent figure. Later, several moved with me to Novosibirsk. They all continued to attend the seminar after graduating from their studies. We would gather somewhere in the institute after a full day of work in Rameev's laboratories (the opposite end of town), inspired and happy to find ourselves together. Here is a typical scene – a late winter's evening, frosty and snowy, and we are closing our meeting. It is time to disperse into the lonely darkness, and Valentina Molchanova, a most devoted participant of our seminar, has still to cross the frozen river on her long walk home.

The publication of my tutorial on algorithms and the above-mentioned work with students increased my pedagogical visibility to such a degree that I was instructed by the Education Ministry, to compile the program of a course "Algorithms and Computers" for the pedagogical institutes. Moreover, the Ministry organized an all-Russian workshop in Penza, dedicated to this topic, with the participation of P.S. Novikov, A.I. Maltsev, and other important guests from Moscow.

In Penza there was a lack of scientific literature, not to mention normal contacts with well established scientific bodies. This obvious disadvantage was partially compensated by sporadic trips to Moscow for scientific contacts (and food supply), as well by correspondence with Kuznetsov, Sergey V. Yablonsky and Lyapunov.

I continued the work on recursive nonseparability and incompleteness of formal theories [44,48], started in the Ph.D. thesis. At the same time, I was attracted by Post's problem of whether all undecidable axiomatic systems are of the same degree of undecidability. This super-problem in Computability and Logic, with a specific flavour of descriptive set theory, was for a long time on the agenda of the Bolshoy Seminar. It inspired also my work on classification of recursive operators and reducibilities [45]. Later, A. V. Kuznetsov joined me and we extended the investigation to partial recursive operators in the Baire space [18]. These issues, reflected our growing interest in relativized algorithms (algorithms with oracles) and in set-descriptive aspects of computable operators, I worked then on a survey on this subject, but the (uncompleted) manuscript was never published. Nevertheless, the accumulated experience helped me later in the work on relativized computational complexity.

In 1956 Post's problem was solved independently by Albert A. Muchnik (1934–) – a young student of P. S. Novikov – and by the American Richard Friedberg. Their solutions were very similar and involved the invention of the priority method of computability theory. At that point it became clear to me that I had exhausted my efforts and ambitions in this area, and, that I am willing to switch to what nowadays would be classified as "Theoretical Computer Science". From the early 50's this research was enthusiastically promoted by A. A. Lyapunov and S. V. Yablonski under the general rubric "Theoretical Cybernetics"; it covered switching theory, minimization of boolean functions, coding, automata, program schemes, etc. Their seminars at the Moscow University attracted many students and scholars, and soon became important centers of research in these new and exciting topics. I was happy to join the cybernetics community through correspondence and trips to Moscow. The general atmosphere within this fresh and energetic community was very friendly, and I benefited much from it. Many "theoretical cybernetists" started with a background in Mathematical Logic, Computability and Descriptive Set Theory and were considerably influenced by these traditions. So, no wonder that, despite my new research interests in Switching and Automata Theory, I considered myself (as did many others) to be a logician. My formal "conversion" to Cybernetics happened on Jan. 9, 1960 when Sergey L. Sobolev (1908–1989) invited me to move to the Novosibirsk Akademgorodok and to join the cybernetics department of the new Mathematical Institute.

Topics in combinational complexity were largely developed by the Yablonski school, which attributed exceptional significance to asymptotic laws governing synthesis of optimal control systems. The impetus for these works was provided by Shannon's seminal work on synthesis of circuits. However, the results of S. V. Yablonski, Oleg B. Lupanov and their followers surpassed all that was done in the West at that time, as can be seen from Lupanov's survey [19]. But focusing on asymptotic evaluations caused the oversight of other problems for which estimates up to a constant factor are still important.

A *perebor* algorithm, or *perebor* for short, is Russian for what is called in English a "brute force" or "exhaustive search" method. Work on the synthesis and minimization of boolean functions led to the realization of the role of *perebor* as a trivial optimization algorithm, followed by Yablonski's hypothesis of its non-elimination. In 1959 he published a theorem which he considered proof of the hypothesis [77]. However the interpretation of the problem given in his results was not universally convincing – a presage of future controversies in the TCS community. I told this story in detail in [67], and will touch it briefly in the next section.

In the winter of 1954, I was asked to translate into Russian a paper by A. Burks and J. Wright, two authors I didn't know earlier. Unexpectedly, this episode strongly influenced my "Cybernetical" tastes and provided the impetus to research in Automata Theory. A curious detail is that in [7], the authors don't even mention the term "automaton", and focus on Logical Nets as a mathematical model of physical circuits. Afterwards. "Logical Nets" would also appear in the titles of my papers in Automata Theory, even though the emphasis was not so much on circuitry, as on operators, languages and logical specifications.

The use of propositional logic, promoted independently by V.I. Shestakov and C. Shannon, turned out to be fruitful for combinational synthesis, because it suffices to precisely specify the behaviour of memoryless circuits. However, for the expression of temporal constraints one needs other, appropriate, specification tools, which would allow to handle synthesis at two stages: At the first, behavioral stage, an automaton is deemed constructed once we have finite tables defining its next-state and output functions, or, equivalently, its canonical equations. This serves as raw material for the next stage, namely for structural synthesis, in which the actual structure (circuit) of the automaton is designed. (Note, that in [20] Kleene does not yet clearly differentiate between the stages of behavioural and structural synthesis.) After some exercises in structural synthesis I focused on behavioral synthesis and began to collaborate with Nathan E. Kobrinsky (1910–85), who at that time held a position in the Penza Polytechnical Institute. Our book "Introduction to the Theory of Finite Automata" [17] was conceived as a concord of pragmatics (N.E.'s contribution) and theory (summary of my results). The basic text was written in 1958, but the book was typeset in 1961, and distributed only in early 1962, when both of us had already left Penza.

5   Automata

5.1   Languages and Operators

The concept of a finite automaton has been in use since the 1930s to describe the growing automata now known as Turing machines. Paradoxically, though finite automata are conceptually simpler than Turing machines, they were not systematically studied until the Fifties, if we discount the early work of McCulloch and Pitts. A considerable part of the collection "Automata Studies [20] was already devoted to finite automata. Its prompt translation into Russian, marked the beginning of heightened interest by Soviet researchers in this field. In particular, the translation included a valuable appendix of Yuri T. Medvedev (one of the translators), which simplified and improved Kleene's results, and anticipated some of Rabin and Scott's techniques for nondeterministic automata.

As in the West, the initial period was characterized by absence of uniformity, confusion in terminology, and repetition of basically the same investigations with some slight variants. The subject appeared extremely attractive to many Soviet mathematicians, due to a fascination with automata terminology with which people associated their special personal expectations and interests. Automata professionals who came from other fields readily transferred their experience and expertise from algebra, mathematical logic, and even physiology to the theory of finite automata, or developed finite-automata exercises into approaches to other problems.

Kleene's regular expressions made evident that automata can be regarded as certain special algebraic systems, and that it is possible to study them from an algebraic point of view. The principal exponents of these ideas in the SU were Victor M. Glushkov (1923–1982) and his disciples, especially Alexander A. Letichevski, Vladimir N. Red'ko, Vladimir G. Bodnarchuk. They advocated also the use of regular expressions as a primary specification language for the synthesis of automata. Later, adherents of this trend in the SU and abroad developed a rich algebraic oriented theory of languages and automata (see [34]).

Counterbalancing this "algebra of languages" philosophy, I followed a "logic of operators" view on the subject, suggested by A. Burks and J. Wright. In [7] they focused on the input-output behaviour of logical nets, i.e. on operators that convert input words in output words of the same length, and infinite input sequences into infinite output sequences.[2] Apparently, they were the first to study infinite behaviour of automata with output, and to (implicitly) characterize input-output operators in terms of retrospection and memory. Furthermore, they considered Logical Nets as the basic form of interaction between input-output agents.

To summarize, Burks and Wright suggested the following ideas I adopted and developed in my further work on the subject:

1. Priority of semantical considerations over (premature) decisions concerning specification formalisms.
2. Relevance of infinite behaviour; hence, ω-sequences as an alternative to finite words.
3. The basic role of operators as an alternative to languages.

Accordingly to those ideas, I focused on two set-theoretical approaches to the characterization of favorite operators and ω-languages (i.e. sets of ω-sequences). The first is in terms of memory; hence, operators and languages with finite memory. The second one, follows the spirit of Descriptive Set Theory (DST), and selects operators and ω languages by appropriate metrical properties and

---

[2] Compare this with D. Scott's argumentation in [38]: "The author (along with many other people) has come recently to the conclusion that the functions computed by the various machines are more important – or at least more basic – than the sets accepted by these devices. The sets are still interesting and useful, but the functions are needed to understand the sets. In fact by putting the functions first, the relationship between various classes of sets becomes much clearer. This is already done in recursive function theory and we shall see that the same plan carriers over the general theory."

set-theoretical operations. (Note, that the set of all ω-sequences over a given alphabet can be handled as a metrical space with suitably chosen metrics.)

My first reaction on the work of Burks and Wright was [49], submitted in 1956 even before the collection *Automata Studies* was available. A footnote added in proof mentions: "the author learned about Moore's paper in [20], whose Russian translation is under print".

The paper [49] deals with operators, and distinguishes between properties related to retrospection, which is nothing but a strong form of continuity, and those related to finite memory. In [55] a class of finite-memory ω-languages is defined which is proved to contain exactly those ω-languages, that are definable in second order monadic arithmetic. Independently Büchi found for them a characterization in terms of the famous "Büchi automata".

In the paper [51], I started my main subject – synthesis of automata, developed later in the books [17] and [63].

6   Experiments and Formal Specifications

Usually, verbal descriptions are not appropriate for the specification of input-output automata. Here are two alternative approaches:

1. *Specification by examples.* This amounts to assembling a table which indicates for each input word x, belonging to some given set M, the corresponding output word z. Further, the synthesis of the automaton is conceived as an interpolation, based on that table. This approach was very popular among soviet practitioners, and suggested the idea of algorithms for automata-identification. Such an algorithm should comprise effective instructions as to:
   (a) What questions of the type "what is the output of the black box for input x?" should be asked?
   (b) How should the answers to these questions be used to ask other questions?
   (c) How to construct an automaton which is consistent with the results of the experiment?
   In his theory of experiments [20], Moore proved that the behavior of an automaton with k states can be identified (restored) by a multiple experiment of length $2k - 1$. Independently, I established in [49] the same result, and used it in [17] to identify automata, with an a priori upper bound of memory. I conjectured also in [49] that the restorability degree of "almost" all automata is of order $\log k$, i.e. essentially smaller than $2k - 1$. This conjecture was proved by Barzdins and Korshunov [63]. Barzdin also developed frequency identification algorithms [63], which produce correct results with a guaranteed frequency, even when there is no a priori upper bound of the memory. The complexity estimation for such algorithms relies on the proof of the $\log k$ conjecture. Later Barzdin and his collective in Riga significantly developed these ideas into a comprehensible theory of inductive learning.

2. *Formal Specifications.* The second approach, initiated by S.C. Kleene in [20], amounts to designing special specification formalisms, which suitably use logical connectives. However the use of only propositional connectives runs into difficulties, because they cannot express temporal relationships.

Actually, Kleene's paper in [20] contains already some hints as to the advisability and possibility of using formulas of the predicate calculus as temporal specifications. Moreover, Church [10] attributes to Kleene the following:

> *Characterization Problem:* Characterize regular events directly in terms of their expression in a formalized language of ordinary kind, such as the usual formulations of first or second order arithmetic.

6.1   Towards Logical Specifications

The years 1956–61 marked a turning point in the field, and Church reported about that at the 1962-International Mathematical Congress. Here is a quotation from [10]: "This is a summary of recent work in the application of mathematical logic to finite automata, and especially of mathematical logic beyond the propositional calculus."

Church's lecture provides a meticulous chronology of events (dated when possible up to months) and a benevolent comparison of his and his student J. Friedman's results with work done by Büchi, Elgot and myself. Nevertheless, in the surveyed period (1956–62) the flow of events was at times too fast and thus omission prone. That is why his conclusion: "all overlaps to some extent, though more in point of view and method than in specific content" needs some reexamination. Actually, the reference to Büchi's paper [6], as well as the discussion of my papers [51,53], were added only "in proof" to the revised edition of the lecture (1964). My other Russian papers [54,55] were still unknown to Church at that time.

Independently, I, myself [51], and somewhat later A. Church [9] developed languages based on the second order logic of monadic predicates with natural argument. Subsequently, another variant was published by R. Büchi [5].

In those works the following restrictions were assumed:

- Trakhtenbrot (1958) [51]: restricted first order quantification;
- Church (1959) [9]: no second order quantification;
- Büchi (1960) [5]: restriction to predicates that are true only on a finite set of natural numbers.

All these languages are particular cases of a single language, widely known now as S1S – Second Order Monadic Logic with One Successor, in which all the restrictions above are removed.

Various arguments can be given in favor of choosing one language or another, or developing a new language. Nevertheless, two requirements seem to be quite natural: The first one (expressiveness) represents the interest of the client, making easier for him the formulation of his intention. The second requirement reflects the viewpoint of the designer; there must be an (fairly simple (?)) algorithm for the synthesis problem in the language.

These two requirements are contradictory. The more comprehensive and expressive the language, the more universal and so more complex is the algorithm. Moreover, if the language is too comprehensive the required algorithm may not exist at all. It turned out that the choice of S1S supports the demand of expressiveness and still guarantees a synthesis algorithm. Indeed, one can show, that all other known specification formalisms can be embedded naturally into S1S. However, this process is in general irreversible.

6.2 Synthesis

Church's lecture focuses on four problems, namely: 1) simplification; 2) synthesis; 3) decision; 4) Kleene's Characterization Problem.[3]

Problem 2, better known as the Church-synthesis problem, amounts roughly to the following: Given a S1S-formula $A(x, y)$:

a) Does there exist an automaton M with input x and output y, whose behaviour satisfies $A(x, y)$?
b) If yes, construct such an automaton.

By solutions are understood algorithms that provide the correct answers and/or constructions.

Problem 4 presumes the invention of a logical formalism L (actually – a rich sublanguage of S1S), which expresses exactly the operators (or events) definable by finite automata, and is equipped with two translation algorithms: (i) from formulas to automata (Kleene-synthesis) and (ii) from automata to formulas (Kleene-analysis).

In accordance with the above classification, [17] deals with Kleene-synthesis and Kleene-analysis.

Actually, in [17] we used the following three formalisms to specify input-output operators:

1. At the highest level – formulas of S1S.
2. At the intermediate level – finite input-output automata represented by their canonical equations.
3. At the lower level – logical nets.

Note that in [17] regular expressions are not considered!

Correspondingly, we dealt there with both behavioral synthesis (from 1 to 2) and with structured synthesis (from 2 to 3).

Büchi was the first to use automata theory for logic and proved [6] that S1S is decidable. These achievements notwithstanding, the general Church-synthesis problem for specifications in full S1S remained open, not counting a few special classes of S1S-formulas, for which the problem was solved by Church and myself (see [8] and [54]). The game theoretic interpretation of Church-synthesis is due to McNaughton [22]. R. Büchi and L. Landweber used this interpretation to solve

---

[3] Of course there is also the problem of *efficiency*: estimate and improve the complexity of the algorithms and/or the succinctness of the results they provide.

the general Church-synthesis problem. Note, that the original proof in [22] was erroneous. Unfortunately, I did not detect this error, which was reproduced in the Russian edition of [63], and corrected later by L. Landweber in the English translation.

Part 1 of the book [63] constitutes a revised version of my lectures at Novosibirsk University during the spring semester of 1966. It summarizes the results of Church, Büchi-Landweber, McNaughton and myself, as explained above. Part 2, written by Barzdin, covers his results on automaton identification.

6.3   About the Trinity

The choice of the three formalisms in [17] is the result of two decisions. The first identifies three levels of specifications; one can refer to them respectively as the declarative, executable and interactive levels. The second chooses for each of these levels a favorite formalism. In [17] those were, respectively, S1S-Formulas, Automata and Logical Nets; these three are collectively called "The Trinity" in [70]. The first decision is more fundamental, and is recognizable also in computational paradigms beyond finite automata. The second decision is flexible even for finite automata; for example, the Trinity does not include Regular Expression (in [17], they are not even mentioned!) After Pnueli's seminal work, Linear Temporal Logics (LTL) became very popular as a declarative formalism. But note that the various versions of LTL are in fact just the friendly syntactical sugar of S1S-fragments, and that the most extended one, called ETL, has the same expressive power as the whole S1S. In this sense, one can argue that S1S is the genuine temporal logic, and that the Trinity has a basic status. Moreover, recent computational paradigms are likely to revive interest in the original Trinity and its appropriate metamorphoses.

7   Complexity

7.1   Entering the Field

In 1960, I moved to the Akademgorodok, the Academic Center near Novosibirsk, where, through the initiative and guidance of Lyapunov, the Department of Theoretical Cybernetics was established within the Mathematical Institute.

I continued to work on automata theory, which I had begun at Penza, at first focusing mainly on the relationship between automata and logic, but also doing some work in structural synthesis [46,52,57]. At that time automata theory was quite popular, and that is what brought me my first Ph.D. students in Novosibirsk: M. Kratko, Y. Barzdin, V. Nepomnyashchy.

However, this initial interest was increasingly set aside in favor of computational complexity, an exciting fusion of combinatorial methods, inherited from switching theory, with the conceptual arsenal of the theory of algorithms. These ideas had occurred to me earlier in 1955 when I coined the term "signalizing function" which is now commonly known as "computational complexity measure".

(But note that "signalizing" persisted for a long time in Russian complexity papers and in translations from Russian, puzzling English-speaking readers.) In [47] the question was about arithmetic functions f specified by recursive schemes R. I considered there the signalizing function that for a given scheme R and nonnegative x, returns the maximal integer used in the computation of $f(x)$ according to R. As it turned out, G.S. Tseytin (1936–), then a student of A.A. Markov at Leningrad University, began in 1956 to study time complexity of Markov's normal algorithms. He proved nontrivial lower and upper bounds for some concrete tasks, and discovered the existence of arbitrarily complex 0-1 valued functions (Rabin's 1960 results became available in the SU in 1963). Unfortunately, these seminal results were not published by Tseytin; later, they were reported briefly (and without proofs) by S.A. Yanovskaya in the survey [78].

Because of my former background, my interest in switching theory, automata, etc. never did mean a break with Mathematical Logic and Computability. In fact, the Sixties marked a return to those topics via research in complexity of computations.

I profited from the arrival of Janis M. Barzdins (1937–) and Rusins V. Freivalds (1942–) in Novosibirsk as my postgraduate students. These two, both graduates of the Latvian University in Riga, engaged actively and enthusiastically in the subject. Alexey V. Gladkiy (1928–) and his group in mathematical linguistics also became interested in complexity problems, concerning grammars and formal languages. Soon other people joined us, mainly students of the Novosibirsk University. My seminar "Algorithms and Automata" was the forum for the new complexity subjects, and often hosted visitors from other places. This is how research in computational complexity started in Novosibirsk; a new young generation arose, and I had the good fortune to work with these people over a lengthy period.

Subsequently, I joined forces with A.V. Gladkiy in a new department of our Mathematical Institute, officially called the Department of Automata Theory and Mathematical Linguistics. Its staff in different periods included our former students Mikhail L. Dekhtyar, (1946–), Mars K. Valiev (1943–), Vladimir Yu. Sazonov (1948–), Aleksey D. Korshunov (1936–), Alexander Ya. Dikovski (1945–), Miroslav I. Kratko (1936–) and Valeriy N. Agafonov (1943–1997).

The basic computer model we used was the Turing machine with a variety of complexity measures; for example, besides time and space, also the number of times the head of the machine changes its direction. Along with deterministic machines we considered also nondeterministic machines, machines with oracles, and probabilistic machines.

It is not surprising that we were attracted by the same problems as our colleagues in the West, notably – J. Hartmanis and R. Stearns. Independently and in parallel we worked out a series of similar concepts and techniques: complexity measures, crossing sequences, diagonalization, gaps, speed-up, relative complexity, to cite the most important ones.

Blum's machine-independent approach to complexity was new for us, and it aroused keen interest in our seminar. But, when later, at a meeting with Tseytin,

I began telling him about Blum's work, he interrupted me almost at once and proceeded to set forth many basic definitions and theorems. As it turned out, he had realized it for some time already, but had never discussed the subject in public!

My "gap" theorem [61] was stimulated by Blum's theory. It illustrated a set of pathological time-bounding functions which need to be avoided in developing complexity theory. Meyer and McCreight's "Honesty Theorem" [21] showed how this can be done through the use of appropriate "honest" functions.

In 1967, I published a set of lecture notes [61] for a course "Complexity of Algorithms and Computations" that I had given in Novosibirsk. The notes contained an exposition of results of Blum and Hartmanis-Stearns, based on their published papers, as well as results of our Novosibirsk group: my "gap" theorem, Barzdin's crossing sequences techniques [2], and my other results reported on our seminar [58,59,60].

I sent a copy of these notes to M. Blum (by then at Berkeley). Further I am quoting Albert Meyer [24]:

> Blum passed on a copy of the Trakhtenbrot notes to me around 1970 when I was at MIT since I knew of a graduate student who was interested in translating them. His work was not very satisfactory, but then Filloti came to MIT to work as a post-doc with me and did a respectable job. By this time the notes began to seem dated to me (about five years old in 1972!) and I decided that they needed to be revised and updated. This youthful misjudgment doomed the project since I was too impatient and perfectionist to complete the revision myself, and the final editing of the translation was never completed.

In the academic year 1970–71, V.N. Agafonov continued my 1967-course, and published the lecture notes [1] as Part 2 of "Complexity of Algorithms and Computations". But, unlike Part 1, which focused on complexity of computations (measured by functions) Part 2 was dedicated to descriptive complexity of algorithms (measured by numbers). It contained a valuable exposition of the literature around bounded Kolmogorov complexity and pseudo-randomness, including contributions of Barzdin and of Valery himself.

7.2   Towards Applications

In the SU it was fully in the tradition of Algorithm Theory to handle applications of two kinds: (i) Proving or disproving decidability for concrete problems, (ii) Algorithmic interpretation of mathematical concepts (for example – along the line of constructive analysis in the Markov School). So, it seemed natural to look for similar applications in the complexity setting.

The attitude of the "classical" cybernetics people (notably, Yablonski) to the introduction of the theory of algorithms into complexity affairs was quite negative. The main argument they used was that the theory of algorithms is essentially a theory of diagonalization, and is therefore alien to the complexity area that requires combinatorial constructive solutions. And indeed, except some simple lower bounds supported by techniques of crossing sequences, all our early results rested on the same kind of "diagonalization" with priorities, as in classical computability theory.

But whereas in Algebra and Logic there were already known natural examples for undecidability phenomena which were earlier analyzed in the classical theory, no natural examples of provable complexity phenomena were known. This asymmetry was echoed by those who scoffed at the emptiness of the diagonal techniques with respect to applications of complexity theory. In particular, they distrusted the potential role of algorithm based complexity in the explanation of *perebor* phenomena, and insisted on this view even after Kolmogorov's new approach to complexity of finite objects.

In the summer of 1963, during a visit by A.N. Kolmogorov to the Novosibirsk University, I learned more about his new approach to complexity and the development of the concepts of information and randomness by means of the theory of algorithms. In the early cybernetics period, it was already clear that the essence of problems of minimization of boolean functions was not in the particular models of switching circuits under consideration. Any other natural class of 'schemes', and ultimately any natural coding of finite objects (say, finite texts) could be expected to exhibit similar phenomena, and, in particular, those related to *perebor*. But, unlike former pure combinatorial approaches, the discovery by Kolmogorov (1965), and independently by Solomonoff (1964) and Chaitin (1966), of optimal coding for finite objects occurred in the framework of algorithm and recursive function theory. (Note that another related approach was developed by A.A. Markov (1964) and V. Kuzmin (1965).)

7.3 Algorithms and Randomness

I became interested in the correlation between these two paradigms back in the Fifties, when P.S. Novikov, called my attention to algorithmic simulation of randomness in the spirit of von Mises-Church strategies. Ever since, I have returned to this topic at different times and for various reasons, including the controversies around *perebor*. Since many algorithmic problems encounter essential difficulties (non existence of algorithms or non existence of feasible ones), the natural tendency is to use devices that may produce errors in certain cases. The only requirements are that the probability or frequency of the errors does not exceed some acceptable level and that the procedures are feasible. In the framework of this general idea, two approaches seemed to deserve attention: probabilistic algorithms and frequential algorithms.

In the academic year 1969–70, I gave a course "Algorithms and Randomness" which covered these two approaches, as well as algorithmic modeling of of Mizes-Church randomness.

The essential features of a frequential algorithm M are generally as follows.

1. M is deterministic, but each time it is applied, it inputs a whole suitable sequence of inputs instead of an individual one, and then inputs the corresponding sequence of outputs.

2. The frequency of the correct outputs must exceed a given level.

The idea of frequency computations is easily generalized to frequency enumerations, frequency reductions, etc.

I learned about a particular such model from a survey by McNaughton (1961), and soon realized that as in the probabilistic case, it is impossible to compute functions that are not computable in the usual sense [56].

Hence, the following questions [65]:

1. Is it possible to compute some functions by means of probabilistic or frequential algorithms with less computational complexity than that of deterministic algorithms?
2. What reasonable sorts of problems (not necessarily computation of functions) can be solved more efficiently by probabilistic or frequential algorithms than by deterministic ones? Do problems exist that are solvable by probabilistic or frequential algorithms but not by deterministic algorithms?

These problems were investigated in depth by Barzdin [3], Freivald [12] and their students.

7.4 Relativized Complexity

Computations with oracles are a well established topic in the Theory of Algorithms, especially since Post's classical results and the solution of his famous problem by Muchnik and Friedberg. So it seemed to me quite natural to look how such issues might be carried to the complexity setting [62]. At this point I should mention that Meyer's confession, about the translation of my lecture notes, points only on a transient episode in our long-time contacts. Let me quote Albert again [24]:

> Repeatedly and independently our choices of scientific subareas, even particular problems, and in one instance even the solution to a problem, were the same. The similarity of our tastes and techniques was so striking that it seemed at times there was a clairvoyant connection between us. Our relationship first came about through informal channels – communications and drafts circulated among researchers, lecture notes, etc. These various links compensated for the language barrier and the scarcity of Soviet representation at international conferences. Through these means there developed the unusual experience of discovering an intellectual counterpart, tackling identical research topics, despite residing on the opposite side of the globe.... Today ... we find ourselves collaborating firsthand in an entirely different area of Theoretical Computer Science than complexity theory to which we were led by independent decisions reflecting our shared theoretical tastes.

As to computations with oracles, we both were attracted by the question:to what extent can be simplified a computation by bringing in an oracle, and how accurately can the reduction of complexity be controlled depending on the choice of the oracle? This was the start point for a series of works of our students (mainly M. Dekhtyar, M. Valiev in Novosibirsk and N. Lynch at MIT) with similar results of two types: about oracles which do help (including the estimation of the help) and oracles which cannot help. The further development of the subject by Meyer and M. Fischer ended with a genuine complexity-theoretic analog to the famous Friedberg-Muchnik theorem. It reflects the intuitive idea that problems might take the same long time to solve but for different reasons! Namely:

> *There exist nontrivial pairs of (decidable!) sets, such that neither member of a pair helps the other be computed more quickly.*

Independently of Meyer and Fisher, and using actually the same techniques, I obtained an improvement of this theorem. That happened in the frame of my efforts to use relative algorithms and complexity in order to formalize intuitions about mutual independence of tasks and about *perebor*.

7.5 Formalizing Intuitions

Autoreducibility. When handling relativized computations it is sometimes reasonable to analyze the effect of restricted access to the oracle. In particular, this is the case with the algorithmic definition of "collectives", i.e. of random sequences in the sense of von Mises-Church. This definition relies on the use of "selection strategies", which are relative algorithms with restricted access to oracles. A similar situation arises with the intuition about mutual independence of individual instances which make up a general problem [64]. Consider, for example, a first order theory T. It may well happen that there is no algorithm, which, for an arbitrary given formula $\mathcal{A}$, decides whether $\mathcal{A}$ is provable or not in T. However, there is a trivial procedure W which reduces the question about $\mathcal{A}$ to similar questions for other formulas; W just inquires about the status of the formula $(\neg(\neg\mathcal{A}))$. The procedure W is an example of what may be called *autoreduction*. Now, assume that the problem is decidable for the theory T, and hence the correct answers can be computed directly (without autoreduction). It still might happen that one cannot manage without very complex computations, whereas the autoreduction above is simple.

A guess strategy is a machine M with oracle, satisfying the condition: for every oracle G and natural number n, the machine M, having been started with n as input, never addresses the oracle with the question "$n \in G$?" (although it may put any question "$\nu \in G$?" for $\nu \neq n$). A set G is called *autoreducible* if it possesses an autoreduction, i.e. a guess strategy which, having been supplied with the oracle G, computes the value $G(n)$ for every n. Otherwise G is *non-autoreducible*, which should indicate that the individual queries "$n \in G$?" are mutual independent.

It turned out that:

- (i) The class of non-autoreducible sequences is essentially broader than the class of random sequences.

– (ii) There are effectively solvable mass problems M of arbitrary complexity with the following property: autoreductions of M are not essentially less complex than their unconditional computations.

Understanding *Perebor*. Disputes about *perebor*, stirred by Yablonski's paper [77], had a certain influence on the development, and developers of complexity theory in the Soviet Union. By and large, reflections on *perebor* spurred my interest in computational complexity and influenced my choice of special topics, concerning the role of sparse sets, immunity, oracles, frequency algorithms, probabilistic algorithms, etc. I told this story in detail in [67]; below I reproduce a small fragment from [67].

The development of computational complexity created a favorable background for alternative approaches to the *perebor* topics: the inevitability of *perebor* should mean the nonexistence of algorithms that are essentially more efficient. My first attempt was to explain the plausibility of *perebor* phenomena related to the "frequential Yablonski-effect"; it was based on space complexity considerations. Already at this stage it became clear that space complexity was too rough and that time complexity was to be used. Meanwhile I began to feel that another interpretation of *perebor* was worth considering, namely, that the essence of *perebor* seemed to be in the complexity of interaction with a "checking mechanism", as opposed to the checking itself. This could be be formalized in terms of oracle machines or reduction algorithms as follows. Given a total function f that maps binary strings into binary strings, consider Turing machines, to compute f, that are equipped with the oracle G that delivers (at no cost!) the correct answers to queries "$f(x) = y$?" (x, y may vary, but f is always the same function). Among them is a suitable machine $M_{perebor}$ that computes $f(x)$ by subsequently addressing the oracle with the queries

$$f(x) = B(0)?, f(x) = B(1)?, \ldots , f(x) = B(i)? \ldots$$

where $B(i)$ is the i-th binary string in lexicographical order. Hence, in the computation of the string $f(x)$ the number of steps spent by $M_{perebor}$ is that represented by the string $f(x)$. I conjectured in 1966 that for a broad spectrum of functions f, no oracle machine M can perform the computation essentially faster. As for the "graph predicates" $G(x, y) =$def $f(x) = y$, it was conjectured that they would not be too difficult to compute. From this viewpoint, the inevitability of *perebor* could be explained in terms of the computational complexity of the reduction process. The conjecture was proved by M. I. Dekhtyar in his master's thesis (1969) for different versions of what "essentially faster" should mean. Using modern terminology, one can say that Dekhtyar's construction implicitly provides the proof of the relativized version of the NP $\neq$ P conjecture. For the first time, this version was explicitly announced by Baker, Gill and Soloway (1975) together with the relativized version of the NP $=$ P conjecture. Their intention was to give some evidence to the possibility that neither NP $=$ P nor NP $\neq$ P is provable in common formalized systems. As to my conjecture, it had nothing to do with the ambitious hopes to prove the independence of the

NP = P conjecture. As a matter of fact, I then believed (and to some extent do so even now) that the essence of *perebor* can be explained through the complexity of relative computations based on searching through the sequence of all binary strings. Hence, being confident that the true problem was being considered (and not its relativization!), I had no stimulus to look for models in which *perebor* could be eliminated.

To the Perebor account [67] it is worth adding the following quotations from my correspondence with Mike Sipser (Feb. 1992):

S.   You write that Yablonski was aware of *perebor* in the early 50's, and that he even conjectured that *perebor* is inevitable for some problems in 1953–54. But the earliest published work of Yablonski that you cite is 1959. Is there a written publication which documents Yablonski's awareness of these issues at the earlier time? This seems to be an important issue, at least from the point of establishing who was the first to consider the problem of eliminating brute force search. Right now the earliest document I have is Godel's 1956 letter to Von-Neumann.

T.   I cannot remember about any publication before 1959 which documents Yablonski's awareness of these issues but I strongly testify and confirm that (a quotation follows from my paper [59]): *"Already in 1954 Yablonski conjectured that the solution of this problem is in essence impossible without complex algorithms of the kind of perebor searching through all the versions...."* He persistently advocated this conjecture on public meetings (seminars and symposia).

S.   Second, is it even clear that Yablonski really understands what we presently mean by eliminating brute force search? He claimed to have proven that it could not be eliminated in some cases back in 1959. So there must be some confusion.

T.   That is indeed the main point I am discussing in Section 1 of my *perebor* paper [67]. The conclusion there is that there is no direct connection between Yablonski's result and what we presently mean by eliminating *perebor*. Hence the long year controversy with Yablonski.

S.   I'd appreciate your thoughts on how to handle Yablonski's contribution to the subject.

T.   I would mention three circumstances:
   – 1. In Yablonski's conjecture the notion of *perebor* was a bit vague and did not anticipate any specific formalization of the idea of *complexity*. Nevertheless (and may be just due to this fact) it stimulated the investigation of *different approaches* to such a formalization, at least in the USSR.
   – 2. Yablonski pointed from the very beginning on very attractive candidates for the status of problems which need essentially *perebor*. See Sect. 1 of [67], where synthesis of circuits is considered in this context.

- 3. Finally, he made the point that for his candidates the disaster caused by *perebor* might be avoided through the use of probabilistic methods.

... Let me mention that as an alternative to Yablonski's approach I advocated the idea of complexity of computations with oracles. In these terms, I formulated a conjecture which presently could be interpreted as the relativised version of P not equal NP. This conjecture was proved by my student M. Dekhtyar [11].

Turning Points. The controversies around *perebor* were exacerbated by the emergence of the new approach to complexity of algorithms and computations. And it was precisely this approach which was relevant for the genuine advance in the investigation of *perebor* in the seminal works of Leonid Levin in the SU and the Americans, Steven Cook and Richard Karp.

The discovery of NP-complete problems gave evidence to the importance of the Theory of Computational Complexity. Soon another prominent result strengthened this perception. In 1972, A. Meyer and Stockmeyer (see [23]) found the first genuine natural examples of inherently complex computable problems. This discovery was particularly important for me because the example came from the area of automata theory and logic in which I had been involved for a long time. Clearly, for the adherents of the algorithmic approach to complexity, including myself, these developments confirmed the correctness of their views on the subject and the worthwhileness of their own efforts in the past. However the time had also come for new research decisions.

In the 70s, certain trends began to develop, which ultimately resulted in fundamental changes. Part of my group (Agafonov, Lomazova, Sazonov, Valiev) and other participants of our seminar became increasingly interested and involved in the investigation of the theory of programming. On the other side, people previously engaged in complexity started to lose interest in this subject.

Against this background, our relationship improved with the Department of Theoretical Programming, headed in the Computer Center by Andrei P. Ershov (1931–1988) – one of the most prominent leaders of programming in the SU. Many of his collaborators and students participated in our seminar. Quoting V.E. Kotov: "In years of stagnation Ershov managed to create around himself a healthy political and humane situation, completely different than that outside; it made our life and work easier." I benefited from his liberal, benevolent, position; then it was very important. Unfortunately, it did not work against the devising of a hostile official mathematical establishment (I told this story in [75]).

In 1979, I came to the difficult decision about emigration to Israel. We departed in December 1980 with the traumatic prospect of separation (in those times most likely for ever) with relatives, friends, colleagues, students. Most of the remaining staff members left the Mathematics Institute, and our department fell apart.

Since January 1, 1981, I am affiliated with Tel Aviv University. But this is another story!

8   Epilogue

For a long time, I was not actively involved in automata and computational complexity, being absorbed in other topics. During that period both areas underwent impressive development, which is beyond the subject of this account.

My entry into the field happened at an early stage, when formation of concepts and asking the right questions had high priority, at least as far as solving well established problems. This is also reflected in my exposition above, in which the emphasis was rather on the conceptual framework in the area. Some of those concepts and models occurred in very specific contexts, or were driven by curiosity rather than by visible applications. Did they anticipate problems beyond their first motivation? I would like to conclude with some remarks about this.

The first is connected to the new and very attractive paradigms of Timed Automata and Hybrid Systems (HS). Nowadays, despite significant achievements, the area is still dominated by an explosion of models, concepts and ad hoc notation, a reminder of the situation in Automata Theory in the Fifties. However, I believe that the classical conceptual framework can still help to elucidate the intuitions underlying the new paradigms, and to avoid reinvention of existing ideas [31].

One way to do so is to start with two separate and orthogonal extensions of the basic model of a finite automaton M. The first one is by interconnecting M with an oracle N, which is also an automaton, but, in general, with an infinite set of states. Typically, think about a logical net over components M and N, with subsequent hiding of N. Whatever M can do while using N is called its relativization with respect to this oracle. The other extension is with continuous time (instead of discrete time, as in the classical case) but without oracles.

For each of these extensions, considered apart, it becomes easier to clarify how (if any) and to what extent the heritage of classical Automata Theory can be adapted.

Appropriate combinations of the two might facilitate the adaptation of classical heritage, whenever it makes sense.

The next remark is about a resurgence of interest in autoreducibility and frequency computations.

It was instructive to learn that the idea of restricting access to oracles, now underlies several concepts, which are in fact randomized and/or time bounded versions of autoreducibility: coherence, checkability, self-reducibility, etc. Most of these concepts were identified independently of (though later than) my original autoreducibility, and have occupied a special place in connection with program checking and secure protocols (see [4] for details and references).

On the other hand, the idea of frequency computation was extended to bounded query computations and parallel learning. Also, interesting relationships were discovered between autoreducibility, frequency computations and various other concepts.

My final remark is about the continuous conceptual succession since my youthful exercises in Descriptive Set Theory (DST), which should be clear from the previous exposition. In particular, it is quite evident that computational complexity is inspired by computability. But the succession can be traced back even to DST; just keep in mind the ideas which lead from classification of sets and functions to classification of what is computable, and ultimately to hierarchies within computational complexity.

9   Addendum – November 2007

These sketchy notes address three additional research topics I have pursued since the 1970s. Theory of Programming is the last project I was involved in before leaving the SU. I developed it further while in Tel Aviv University through collaboration with Albert Meyer and Joseph Halpern [76] at MIT. Research on the next two topics, Concurrency and Continuous-Time Paradigms, started in Tel Aviv, where I managed to attract to these subjects some of my colleagues and students. In particular, Alex Rabinovich became my main collaborator and coauthor.

These notes are mainly compiled from non-technical parts of texts adopted from previous publications. I am aware of the imperfection of such undertakings. Unfortunately, at this moment, I have no possibility to provide a more comprehensive and lucid account and to give tribute to those who shared the efforts with me.

9.1   Theory of Programming

Introductory Remarks. In the 70s a broad spectrum of topics was on the agenda of the SU community and, in particular of our group. Investigation (and even the primary definitions) of program schemes started long ago in the SU with the works of Lyapunov, Yanov, Ershov and others. Other developments, such as comparative schematology, program logic, as well as verification and specification of programs came mainly from the West.

On the other hand, our interests were significantly inspired by Algol-68, the Scott-Strachey theory of denotational semantics [39] and the importance of lambda-calculus for the theory of programming (see the essay [69]).

Functional languages. The thesis that lambda-calculus underlies programming languages. was first extensively argued by Peter Landin (the author of ISWIM) in his 1965–1966 papers. As a major innovation to lambda-calculus, ISWIM includes the additional binding mechanisms through the let and letrec constructs, which permit the statement of declarations (definitions) in a convenient programming style. Scott's language LCF [37] is ISWIM enriched with fixed point operators and conditionals; PCF is that part to which an appropriate arithmetical signature is added. Whereas lambda-calculus is usually recognized as a sequential language, Scott raised the idea to enrich PCF with parallelism facilities, say with the parallel function OR. Hence, the distinction between sequential and parallel functions and the comparative power of parallel functions suggested themselves.

Imperative languages. However, this was a purely functional approach not assuming the imperative features of real programming languages like FORTRAN, Algol, and Pascal. These languages allow constructs that are alien to the spirit of lambda-calculus, e.g. assignments and **goto**'s. Such constructs, directly inherited from the von Neumann computer architecture, were nicknamed "dirty features" by adepts of pure functional programming. Especially **goto**'s were long ago recognized as a troublesome control mechanism by the pioneers of structured programming (Dijkstra's "**goto**'s considered harmful"). Note that John Backus, the designer of FORTRAN, later joined the criticism of "dirty features". He called for the liberation of programming from the von Neumann style and outlined a new functional language. Hence, the question: to what extent and in what form can Landin's thesis be adopted to imperative languages? That is what I became interested in.

Presented below is a summary of research performed by our group.

Work on Program Schemes, Program Logic, and Specification and Verification of Programs

- V. N. Agafonov:
  - Syntactical analysis in compilation.
  - Typology (semantics) of programming languages and verification of programs.
  - Specification of programs.
- I. A. Lomazova:
  - Inductive conditions in Hoare's logic for programs with loops (independent of Apt, Bergstra, Tucker).
  - Semantics and complete Hoare-type algorithmic logic for programs with **goto**.
- B. A. Trakhtenbrot:
  - Universality of classes of program schemes.
  - Recursive program schemes and computable functionals.
  - Relaxation rules and completeness of algorithmic logic.
- M. K. Valiev:
  - Axiomatization and decision complexity for variants of PDL (independent of Halpern).

Work on LCF-PCF

- V. Yu. Sazonov [35,36]: All the results below were obtained independently (and even prior) to G. Plotkin's "LCF as a programming language" [26] and related works on semantics of type-free lambda calculus by M. Hyland and C. Wadsworth.

  - Precise characterization of the expressive power of the programming part of Scott's language LCF (currently known as PCF). This was given in terms of computational strategies – a precursor of currently widely used game semantics.

- Exact correspondence between operational and denotational semantics in terms of computational strategies (in both typed and type-free settings).
- Characterization of degrees of parallelism in computations.

Work on Algol-Like Languages [66,76,14]

My view on the essence of Algol-like languages was crystallized in discussions with V. Sazonov. The never-published technical report [66] is my first account on the subject. As I learnt later, J. C. Reynolds promoted similar ideas on the essence of what he called "Idealized Algol" [32,33].

The main idea is to face the impediments of the original Algol through the design of an appropriate Algol-like language (Idealized Algol) that can be explained in the lambda-calculus core and supported with denotational semantics. Here are some of the principles that characterize this class of languages:

- Exclude goto's but preserve assignments.
- The language is fully typed.
- Higher order procedures of all finite types are allowed.
- There is a clear distinction between locations and storable values (integers, say).
- Blocks with local storage, and sharing (aliasing) are allowed.

Actually, these languages preserve as much as possible of the rich expressive power of the original Algol, and have sufficient structure to yield a rich algebra and proof theory. Moreover, the denotational semantics of a program is provided by a two-step process:

1. A purely syntactic translation of the program into an ISWIM expression. This step provides the "true" ISWIM syntax for the program – a worthy alternative for the Algol-jargon which came down through history.
2. Assignment of semantics to ISWIM in the standard way, assuming an adequate choice of domains, which is consistent with the underlying intuition.

Thus, programs simply inherit their semantics from the ISWIM-terms into which they are translated. In this way, procedures are entirely explained at a purely functional level – independent of the interpretation of program constructs – by continuous models for lambda-calculus. However, the usual (complete partial order) models are not adequate to model local storage allocation. New domains of store models are offered to solve this problem and partial correctness theory over store models is developed.

9.2 Concurrency

Introductory Remarks. Before the emergence of modern theory of concurrency, logical nets presented the main model of concurrency in automata theory (see Sect. 5). The interaction of components in a logical net is often called synchronous (we would prefer "simultaneous") interaction. Our concern below is about asynchronous nets and we call them simply *nets*.

Nets are widely used in the theory of concurrency. One evident reason for that is the convenience of visualizing the communication structure of systems, as it comes to light for example in Petri nets or in data flow nets. On the other hand, as emphasized by Pratt, nets seem to cover almost any situation which involves "sharing" or communication. For an engineer it could mean sharing of component terminals by connecting them electrically; a mathematician can consider sharing of variables in a system of equations.

Modularity reflects the Frege Principle: any two expressions $expr_1$ and $expr_2$ that have the same meaning (semantics) can be replaced by each other in every appropriate context without changing the meaning of the overall expression. In a conventional syntax with signature Σ (call it "textual" as opposed to the graphical syntax of nets), a complex piece of syntax $expr$ may be uniquely decomposed into simpler sub-pieces: $expr = op(expr_1, \ldots, expr_k)$, where op is in Σ. Typically, a denotational semantics is formulated in such a compositional style and hence supports modularity. However, often one starts with an operational semantics that lacks a compositional structure. Then a standard way to prove modularity is to discover a compositional semantics which is equivalent to the operational one.

In net models of concurrency, a syntax is provided by some specific class NN of labeled graphs called nets, and a semantics is usually defined globally either in an operational style through appropriate firing (enabling) rules, as in the case of Petri nets, or as a solution of a system of equations. However, the notion of context makes sense for nets and therefore modularity of nets may be defined and investigated.

For a long time, despite the rich and suggestive information of the Petri pictorial approach, the question of compositionality or modularity had not been raised. For elementary Petri nets modularity was established by Antoni Mazurkiewicz. Namely, he discovered a compositional semantics which is equivalent to the original "token game" semantics. This compositional approach to Petri nets provided the initial stimulus to our work.

Challenges that we Addressed:

- Unification of concepts and approaches for the numerous models and semantics proposed for nets [27].
- The "causal (called also true) vs. interleaving semantics" dilemma for Petri nets.
- Modularity issues for nets (especially, for data flow nets).
- The close subject of compositional proof systems for nets ([71]).

Below we consider in more detail our work on causality and modularity issues. Petri nets and data flow nets are the fundamental paradigms we focused on.

Discerning Causality. The question here is: to what extent is the causal (partial order) semantics relevant for concurrent computational problems?

It appears that usually one starts with a common interleaving semantics sem_1 for the system under consideration. Yet, at a later stage, a causal semantics sem_2

is chosen which is consistent with sem_1 in a natural sense. In [16], we studied possible choices of causal semantics for C/E Petri nets. Modularity arguments are used to show that there is a unique such semantics satisfying some simple and necessary modularity and consistency conditions. Moreover, this is true also for other formalisms, such as Milner's CCS and Hoare's CSP. However, in other situations the proper choice of causal semantics still has to be justified by extra conditions. This is illustrated for P/T nets, i.e. Petri nets with bounded capacities (a model that reflects distribution of resources).

Data Flow Nets [28,29]. Unlike other types of nets, here the main concern is the I(nput)/O(utput) behavior of a system (under appropriate topological restrictions). Kahn's Principle states that for special deterministic agents (Kahn automata) the I/O behavior of a net can be obtained from the I/O behaviors of its components as the solution of an appropriate system of equations. This implies I/O compositionality (and hence also modularity) for such nets. In an attempt to generalize Kahn's result to the nondeterministic agent merge, Brock and Ackermann observed that I/O modularity fails. This is the so called Brock-Ackermann anomaly. Hence, two problems are raised by the pioneering works of Kahn and of Brock-Ackermann:

1. Why do data flow nets behave compositionally for Kahn-automata but not for Brock-Ackermann automata?
2. What are the fundamental limits for the applicability of Kahn's Principle?

We carried out an extensive development of dataflow semantics which provides very precise answers to these questions. We showed that modularity may fail even for components with functional I/O behavior. We gave a characterization of functional agents ("smoothness") which is both sufficient and necessary to support modularity and Kahn's Principle. Moreover, the class SMOOTH of these automata is the unique largest I/O-modular class of automata with functional behavior. Any extension of the class of Kahn automata by a component with "ambiguous" I/O-behavior – not necessarily merge – spoils modularity [29,28]. Analyzing the possible deviations from Kahn's Principle we identified two kinds of anomalies. The "meagerness" anomaly may occur even for functional agents and not only for ambiguous ones, as sometimes is misunderstood the original Brock-Ackermann example. On the other hand, the "ambiguity"-anomaly is indeed rooted in the semantics of nondeterminism. These results were credited as an essential progress towards an "ultimate Kahn's Principle".

Nets of Relations. A conceptual and technical novelty we started is the idea to consider semantics of nets of relations [30]. We identified observable relations and nets of observable relations as appropriate tools for the investigation of data flow nets over nondeterministic agents. We showed that the main source of the Brock-Ackermann anomaly is in the semantics of nets of relations. If one considers nets over a subclass of observable relations, it may happen that the semantics over such nets is modular.

9.3   Continuous-Time Paradigms

In Sect. 5 we briefly discussed the status of, and the relationship between, three basic formalism of automata theory: automata, logic and circuits. These are collectively called there "The Trinity". As emphasized in the position paper [70], more recent developments in the theory of real-time systems put forward the task of lifting the classical trinity to continuous time. Some of the work in this direction is summarized in a special issue of Fundamenta Informaticae ([73]). Below I quote the contents of, and the preface to this issue.

The Papers in [73]:

[HR]   Hirshfeld, Y., and Rabinovich, A.: Logics for Real Time: Decidability and Complexity [15].

[S]   Slissenko, A.: A Logic Framework for Verification of Timed Algorithms [41].

[T]   Trakhtenbrot, B. A.: Understanding Automata Theory in the Continuous Time Setting [74].

[PRD]   Pardo, D., Rabinovich, A., and Trakhtenbrot, B. A.: Synchronous Circuits over Continuous Time: Feedback Reliability and Completeness [25].

From the Preface of the Editor: In the last 15 years, research in Computer Science has involved many paradigms in which continuous time appears whether in a pure way or in cooperation with discrete time. In particular, this is evident in subjects concerning Automata, Circuits and Logic. This issue of FI consists of four papers dedicated to such subjects. The papers will be referred to below as [HR], [S], [PRT], [T] (the initials of the authors). [HR] and [S] are about continuous-time logics, whereas in [T] and [PRT] the focus is on automata and circuits (logical nets).

In [HR] the concern is about Monadic Logics of Order (MLO) and their relationship to Temporal Logics. For discrete time, decidability of Second Order Monadic Logic (SOML) follows from the connection between SOML and automata theory. It is well known that SOML covers numerous temporal logics (TL). Ultimately, these TL may be considered as syntactically sugared versions of SOML fragments. In [HR] further facts of this kind are established for continuous time. However, in this case, instead of traditional automata-theoretic techniques one needs to use properly general theorems from logic.

The logic framework developed in [S] is based on First Order Timed Logic (FOTL), that allows functions and predicates with more than one argument; moreover, it allows also some arithmetic. This makes the logic expressive enough to represent, more or less directly, continuous-time properties of distributed algorithms. But, on the other hand, it makes the logic undecidable. The fundamental observation that, nevertheless, permits the efficient use this logic for verification is as follows: the underlying theories of continuous time (e.g. the theory of real addition, Tarski algebra, etc.) are decidable or have much better complexity than the corresponding theories of discrete time. Interesting decidable classes of the verification problem are based on appropriate properties of FOTL.

The companion papers [T] and [PRT] draw their initial motivation from the literature on Hybrid Automata, Circuits and related control problems. Concrete problems about circuits (feedback reliability, completeness) and control (sample-and-hold architectures in continuous time) are also the subject of [T] and [PRT]. Yet, a more general contribution is the development of a conceptual framework that allows one to highlight the genuine distinctions and similarities between the discrete-time and continuous-time tracks.

There is a growing feeling in the community that the literature on these subjects, as well on the related logics, is plagued by a Babel of models, constructs and formalisms with an amazing discord of terminology and notation. Further models and formalisms are engendered, and it is not clear where to stop. Hence, appeals like:

> look back to sort out what has been accomplished and what needs to be done... by surveying logic-based and automata-based real-time formalisms and putting them into a perspective. (R. Alur and T. Henzinger).

> ... isolate the right concepts, ... formulate the right models, and discard many others, that do not capture the reality we want to understand.... (J. Hartmanis).

[HR], [T] and [PRT] are strongly committed to the analysis in depth of the various continuous-time paradigms and to their robust conceptual integration in mainstream Automata Theory and Logic. These papers come to the following conclusions about some misconceptions in the previously suggested models and logics:

1. The standard continuous-time model for logic was ignored as a yardstick; instead, different kinds of sequences of continuous-time bits were used. This may have been an attempt to pursue the connection with automata theory since automata were traditionally associated with sequences. This was the main cause in the rejection of the classical model. It complicated the subsequent research. The choice of the temporal logic became an arbitrary decision [HR].
2. Input/output behavior of automata was ignored in favor of generating devices. Functions (in particular, input/output behavior of automata) are more fundamental than sets (say, languages accepted by automata). Accordingly, circuits and feedback should be defined explicitly as generic concepts. Surprisingly, it has been left unobserved in the literature that some flaws in the conceptual decisions concerning continuous time are identifiable already at the level of discrete-automata theory [T].

References

1. Agafonov, V.N.: Complexity of algorithms and computations (part 2). Lecture Notes, p. 146. Novosibirsk State University (1975)
2. Barzdin, J.M.: Complexity of the recognition of the symmetry predicate in Turing machines. Problemy Kibernetiki 15, 245–248 (1965)
3. Ja Barzdin, M.: On computability on probabilistic machines. DAN SSSR 189, 699–702 (1969)
4. Beigel, R., Feigenbaum, J.: On being incoherent without being very hard. Computational Complexity 2, 1–17 (1992)
5. Büchi, J.R.: Weak second-order arithmetic and finite automata. Z. Math. Logik Grundlagen Math. 6, 66–92 (1960)
6. Büchi, J.R.: On a decision method in restricted second order arithmetic. In: Proc. of the 1960 Intl. Cong. on Logic, Philosophy and Methodology of Sciences, pp. 1–11. Stanford Univ. Press (1962)
7. Burks, A., Wright, J.: Theory of logical nets. Proc. IRE 41(4) (1953)
8. Church, A.: Applications of recursive arithmetic to the problem of circuit synthesis. In: Summaries of the Summer Institute of Symbolic Logic, vol. 1, pp. 3–50. Cornell Univ., Ithaca, NY (1957)
9. Church, A.: Application of recursive arithmetic to the theory of computers and automata, notes, summer conference course. In: Advanced Theory of the Logical Design of Digital Computers, pp. 1–68. University of Michigan (1959)
10. Church, A.: Logic, arithmetic and automata. In: Proceedings of Intl. Congress of Mathematicians, pp. 23–35 (1962)
11. Dekhtyar, M.I.: The impossibility of eliminating complete search in computing functions from their graphs. In: DAN SSSR 189, pp. 748–751 (1969)
12. Freivald, R.V.: Fast probabilistic algorithms. LNCS, vol. 74, pp. 57–69 (1979)
13. Glushkov, V.M.: Synthesis of digital automata, Fizmatgiz, Moscow (1962)
14. Halpern, J., Meyer, A., Trakhtenbrot, B.A.: The semantics of local storage, or what makes the free list free (preliminary report). In: Conference Record of the XI ACM Symposium on Principles of Programming Languages (POPL), pp. 245–257 (1984)
15. Hirshfeld, Y., Rabinovich., A.: Logics for real time: decidability and complexity. Fundamenta Informatica 62(1), 1–28 (2004)
16. Hirshfeld, Y., Rabinovich, A., Trakhtenbrot, B.A.: Discerning causality in interleaving behavior. In: Proceedings of Logic at Botic 1989, Pereslavl Zalessky, USSR (July 1989)
17. Kobrinski, N.E., Trakhtenbrot, B.A.: Introduction to the Theory of Finite Automata, Fizmatgis, Moscow, pp. 1–404 (1962), English translation. In: Studies in Logic and the Foundations of Mathematics, North-Holland(1965)
18. Kuznetsov, A.V., Trakhtenbrot, B.A.: Investigation of partial recursive operators by techniques of Baire spaces. Doklady AN SSR 105 6, 896–900 (1955)
19. Lupanov, O.B.: An approach to systems synthesis – a local coding principle. Problems of Cybernetics 14, 31–110 (1965)
20. McCarthy, J., Shannon, C. (eds.): Automata Studies, Princeton (1956)
21. McCreight, E.M., Meyer, A.R.: Classes of computable functions defined by bounds on computation. In: Proc. of 1st STOC, pp. 79–88 (1969)
22. McNaughton, R.: Finite-state infinite games. In: Project MAC Rep., September 1965, MIT, Cambridge (1965)
23. Meyer, A.R.: Weak monadic second order theory of successor is not elementary recursive. In: Proj. MAC, MIT, Cambridge (1973)

24. Meyer, A.R.: unpublished memo; see the second chapter of this collection
25. Pardo, D., Rabinovich, A., Trakhtenbrot, B.A.: Synchronous circuits over continuous time: feedback reliability and completeness. Fundamenta Informatica 62(1), 123–137 (2004)
26. Plotkin, G.: LCF considered as a programming language. Theoretical Comp. Science 5, 223–257 (1977)
27. Rabinovich, A., Trakhtenbrot, B.A.: Behavior structures and nets of processes. Fundamenta Informaticae 11, 357–403 (1988)
28. Rabinovich, A., Trakhtenbrot, B.A.: Nets of processes and data-flow. In: de Bakker, J.W., de Roever, W.-P., Rozenberg, G. (eds.) Linear Time, Branching Time and Partial Order in Logics and Models for Concurrency. LNCS, vol. 354, pp. 574–602. Springer, Heidelberg (1989)
29. Rabinovich, A., Trakhtenbrot, B.A.: Nets and data-flow interpreters. In: Proceedings of LICS (1989)
30. Rabinovich, A., Trakhtenbrot, B.A.: Communication among relations. In: Proceedings of the 17th Colloquium on Automata, Languages and Programming, Warwick, England (1990)
31. Rabinovich, A., Trakhtenbrot, B.A.: From finite automata toward hybrid systems. In: Chlebus, B.S., Czaja, L. (eds.) FCT 1997. LNCS, vol. 1279, pp. 411–422. Springer, Heidelberg (1997)
32. Reynolds, J.C.: Idealized Algol and its specification logic, Technical Report, Syracuse University pp. 1-81 (1981)
33. Reynolds, J.C.: The essence of Algol. In: de Bakker, van Vliet (eds.) International Symposium on on Algorithmic Languages, North-Holland, pp. 345–372 (1981)
34. Rozenberg, G., Salomaa, A. (eds.): Handbook of Formal Languages. pp. I–III, Springer, Berlin (1997)
35. Sazonov, V.Yu., Expressibility of functions in Scott's LCF language, Algebra i Logika 15, 308–320 (in Russian); 192–206 in English edition, 1976.
36. Sazonov, V.Y.: Functionals computable in series and in parallel. Siberian Math. Journal 17(3), 648–672 (1976) (in Russion); 498–516 in English edition
37. Scott, D.A.: Type-theoretical alternative to CUCH, OWHY, ISWIM. Theoretical Computer Science 121, 411–440 (1993) Reprint of a manuscript, Oxford University (1969)
38. Scott, D.: Some definitional suggestions in automata theory.J. of Computer and Syst. Sci., 187–212 (1967)
39. Scott, D., Strachey, C.: Toward a mathematical semantics of computer languages. In: Proceedings of a Symposium on Computer and Automata, New York (1971)
40. Shtokalo, I.Z. (ed.): History of Mathematics, Kiev (in Russian, 1970)
41. Slissenko, A.: A logic framework for verification of timed algorithms. Fundamenta Informaticae 62(1) (August 2004)
42. Trakhtenbrot, B.A.: The impossibility of an algorithm for the decidability problem on finite classes. Doklady AN SSR 70(4), 569–572 (1950)
43. Trakhtenbrot, B.A.: Decidability problems for finite classes and definitions of finite sets. Ph.D. Thesis, Math. Inst. of the Ukrainian Academy of Sciences, Kiev (1950)
44. Trakhtenbrot, B.A.: On recursive separability. Doklady AN SSR 88(6), 953–956 (1953)
45. Trakhtenbrot, B.A.: Tabular representation of recursive operators. Doklady AN SSR 101(4), 417–420 (1955)
46. Trakhtenbrot, B.A.: The synthesis of non-repetitive schemas. Doklady AN SSR 103(6), 973–976 (1955)

47. Trakhtenbrot, B.A.: Signalizing functions and tabular operators. Uchionnye Zapiski Penzenskogo Pedinstituta (Transactions of the Penza Pedagogoical Institute) 4, 75–87 (1956)
48. Trakhtenbrot, B.A.: On the definition of finite set and the deductive incompleteness of set theory. Izvestia AN SSR 20, 569–582 (1956)
49. Trakhtenbrot, B.A.: On operators, realizable by logical nets. Doklady AN SSR 112(6), 1005–1006 (1957)
50. Trakhtenbrot, B.A.: Algorithms and Computing Machines, Gostechizdat (1957) second edition by Fizmatgiz, 1960; English translation in the series: Topics in Mathematics, D.C. Heath and Company, Boston, pp. 1–101 (1963)
51. Trakhtenbrot, B.A.: The synthesis of logical nets whose operators are described in terms of monadic predicates. Doklady AN SSR 118(4), 646–649 (1958)
52. Trakhtenbrot, B.A.: The asymptotic estimate of the logical nets with memory. Doklady AN SSR 127(2), 281–284 (1959)
53. Trakhtenbrot, B.A.: Some constructions in the monadic predicate calculus. Doklady AN SSR 138(2), 320–321 (1961)
54. Trakhtenbrot, B.A.: Finite automata and the monadic predicate calculus. Doklady AN SSR 140(2), 326–329 (1961)
55. Trakhtenbrot, B.A.: Finite automata and the monadic predicate calculus. Siberian Math. Journal 3(1), 103–131 (1962)
56. Trakhtenbrot, B.A.: On the frequency computation of recursive functions. Algebra i Logika, Novosibirsk 1(1), 25–32 (1963)
57. Trakhtenbrot, B.A.: On the complexity of schemas that realize many-parametric families of operators. Problemy Kibernetiki 12, 99–112 (1964)
58. Trakhtenbrot, B.A.: Turing Computations with Logarithmic Delay. Algebra i Logika, Novosibirsk 3(4), 33–48 (1964)
59. Trakhtenbrot, B.A.: Optimal computations and the frequency phenomena of Yablonski. Algebra i Logika, Novosibirsk 4(5), 79–93 (1965)
60. Trakhtenbrot, B.A.: On normalized signalizing functions for Turing computations. Algebra i Logika, Novosibirsk 5(6), 61–70 (1966)
61. Trakhtenbrot, B.A.: The Complexity of Algorithms and Computations, Lecture Notes, ed. by Novosibirsk University, pp. 1–258 (1967)
62. Trakhtenbrot, B.A.: On the complexity of the mutual-reduction algorithms in the construction of Novikov and Boone. Algebra i Logika 8, 50–71 (1969)
63. Trakhtenbrot, B.A., Barzdin, J.M.: Finite Automata (Behavior and Synthesis), Nauka, Moscow, pp. 1–400 (1970) English translation in: Fundamental Studies in Computer Science 1, North-Holland (1973)
64. Trakhtenbrot, B.: On autoreducibility. Doklady AN SSR 192(6), 1224–1227 (1970)
65. Trakhtenbrot, B.A.: Notes on the complexity of probabilistic machine computations. In: Theory of Algorithms and Mathematical Logic, ed. by the Computing Center of the Academy of Sciences, pp. 159–176 (1974)
66. Trakhtenbrot, B.A.: On denotational semantics and axiomatization of partial correctness for languages with procedures as parameters and with aliasing (extended abstract), Technical Report, Tel Aviv University, p. 20 (August 1981)
67. Trakhtenbrot, B.A.: A survey of Russian approaches to perebor (brute-force search) algorithms. Annals of the History of Computing 6(4), 384–400 (1984)
68. Trakhtenbrot, B.A.: Selected Developments in Soviet Mathematical Cybernetics, Monograph Series, sponsored by Delphic Associates, Washington, XIV + 122 pages (1985)

69. Trakhtenbrot, B.A.: Comparing the Church and Turing approaches: two prophetical messages. In: The Turing Universal Machine – A Half Century Survey, pp. 603–630. Oxford University Press, Oxford (1988)
70. Trakhtenbrot, B.A.: Origins and metamorphoses of the Trinity: logic, nets, automata. In: Proc. of the 10th IEEE Symposium on LICS, San Diego (1995)
71. Trakhtenbrot, B.A.: On the power of compositional proofs. Fundamenta Informaticae 30(1), 83–95 (1997)
72. Trakhtenbrot, B.A.: In memory of S.A. Yanovskaya (1896–1966) on the centenary of her birth. Modern Logic 7(2), 160–187 (1997)
73. Trakhtenbrot, B.A.(ed.): Special Issue on Continuous-Time Paradigms in Logic and Automata. Fundamenta Informaticae 62(1) (August 2004)
74. Trakhtenbrot, B.A.: Understanding Basic Automata Theory in the Continuous Time Setting. Fundamenta Informaticae 62(1), 69–121 (2004)
75. Trakhtenbrot, B.A.: In memory of Andrei P. Ershov. In: Ershov, a Scientist and Human Being (in Russian), Publishing House of the Russian Academy of Sciences in Novosibirsk (2006)
76. Trakhtenbrot, B.A., Halpern, J., Meyer, A.R.: From denotational to operational and axiomatic semantics for Algol-like languages (an overview). In: Clarke, E., Kozen, D. (eds.) Logic of Programs 1983. LNCS, vol. 164, pp. 474–500. Springer, Heidelberg (1984)
77. Yablonski, S.V.: Algorithmic difficulties in the synthesis of minimal contact networks (in Russian), Problems of Cybernetics, vol. 2, Moscow (1959)
78. Yanovskaya, S.: Mathematical logic and fundamentals of mathematics. In: Mathematics in the USSR for 40 Years, Moscow, Fizmatgiz, pp. 13–120 (1959)

Reminiscences*

Albert R. Meyer

Massachusetts Institute of Technology, Cambridge, MA

In his early memoirs, Trakhtenbrot told several stories. The greater part is an intellectual history of Soviet research in theoretical computer science from the 1950's to the late 70's. A second story – of academic and political disputes that shaped the course of Soviet research in the area – is briefly indicated. Finally, there is a laconic suggestion of the scientific life of a gifted, prolific mathematician and scholar.

Of course the emphasis in each story is on the Soviet side, but the international context in which research in theoretical computer science has been conducted for several decades is very apparent. The parallel between Soviet and American research is especially visible to me personally because, long before I had the pleasure of meeting and later collaborating with Trakhtenbrot, I was first delighted and then warmly thrilled to watch through the medium of research papers and notes the thinking of a scholarly soul mate. Repeatedly and independently, Trakhtenbrot's and my choices of scientific sub-areas, even particular problems, and in one instance even the solution to a problem, were the same. The similarity of our tastes and techniques was so striking that it seemed at times there was a clairvoyant connection between us.

This personal story offers an alternative, more intimate perspective on the nature of Soviet/Western research interaction in the area of theoretical computer science, as well as some additional biographical information about the author of these memoirs.

Trakhtenbrot probably became most widely known in America because of his tutorial monograph on *Algorithms and Automatic Computing Machines*. I did not become aware of Trakhtenbrot for another half a dozen years, and actually realized only much later that I had studied this book as a graduate student in 1963 when it first became available in translation. I remembered it well as an exceptionally clear and elegant introduction to the basic ideas of computability theory. It was another decade before I learned first hand from the author something of the circumstances under which it was written: the new Ph.D. Trakhtenbrot, who arrived in 1950 in the University at Penza, was certainly not of proletarian background – a Jew who spoke eight languages, whose research was decidedly abstract and "pure", and who, if his present manner may accurately be extrapolated back over fifty years, must have seemed to the casual observer an easy fit to the stereotype of an absent-minded professor. Whispered accusations of bourgeois idealism were heard: in that era of Stalinist paranoia they were gravely threatening. The book was written to demonstrate that this apparently unworldly scholar could produce an object at least of pedagogical value

---

* This contribution is based on a draft written already in 1985.

A. Avron et al. (Eds.): Trakhtenbrot/Festschrift, LNCS 4800, pp. 39–45, 2008.
© Springer-Verlag Berlin Heidelberg 2008

to the Socialist state. It succeeded admirably not only in increasing its author's professional visibility, but possibly in keeping him out of prison.

Trakhtenbrot is also fortunate to have found a basic theorem of mathematical logic which is now named after him. Trakhtenbrot's Theorem is a finitary variant of the undecidablity of first-order logic: the problem of whether a first-order formula is valid in all finite models is, like the general validity problem, undecidable, but in a technically different way (co-r.e. as opposed to r.e.). Although this result is accepted as a core result of classical logic, it already reveals Trakhtenbrot's concerns with constructive processes on finite structures. Not bourgeois idealism at all, really.

It was through my own exposure to the pioneering research on computational complexity theory by Hartmanis & Stearns at GE research in Schenectedy and Manuel Blum at MIT in the late 60's that I learned about Trakhtenbrot. There were only a few published papers and no books documenting this exciting new area. Trakhtenbrot had written a set of lecture notes for a course on complexity theory he gave in Novosibirsk and had sent a copy of these notes to Blum (by then at Berkeley). The notes contained a valuable exposition of the results of Blum and Hartmanis-Stearns – based on their published papers in American journals – as well as new results by Trakhtenbrot: his "gap" theorem and his automaton-theoretic analysis using "crossing sequences" of the complexity of transferring information on a linear storage tape. The concern with problems of *perebor* which led Trakhtenbrot and his group to these interests are outlined in his memoir and differed slightly in emphasis from the motivating concerns of the American researchers, but within a couple of years after publication of the basic results in the West, the approaches of Trakhtenbrot's and the American groups had virtually converged. Indeed, a principal result of Borodin's Ph.D under Hartmanis at Cornell in 1969 was his independent version of Trakhtenbrot's "gap theorem". Likewise, the method of crossing sequences was developed independently by the Israeli Rabin (then at Harvard), Hartmanis at GE and Hennie at MIT. The roots of the crossing sequence technique lie directly in the classic papers of 1959 by Rabin and Scott and by Shepherdson on finite automata; these papers were well known to the world research community, so there is no mystery at the independent duplication of the results. Still, I noted at the time that Trakhtenbrot (and Hennie's) development of the "crossing sequence" technique went an extra elegant step beyond Hartmanis'. This was my first hint of the flair and penetrating quality of Trakhtenbrot's style.

Trakhtenbrot's "gap theorem" showed similar quality. Its technical details are immaterial here, but its general nature reveals Trakhtenbrot's refined mathematical aesthetics. There is an easily satisfied side-condition which was needed in the proofs by Hartmanis-Stearns and Blum of their fundamental results establishing the existence of inherently complex problems. I remember myself as a student pointing out after an early lecture by Hartmanis that he had neglected to mention the need for the side-condition in his presentation, and he acknowledged that it was technically necessary but didn't seem worth highlighting. The scientific significance of the results was not impaired by the side-condition. But mathematically, it is intellectually wasteful if not exasperating to use a hypothesis that is not strictly necessary. The "gap theorem" demonstrated that the side-condition was absolutely necessary, and dramatically illustrated a set of pathological time-bounding functions which needed to be avoided in developing complexity theory.

In fact, the gap theorem was the principal stimulus for joint research carried out by my first student, Ed McCreight, and myself. Our "Honesty Theorem" showed how the pathological "dishonest" time-bounding functions of the "gap theorem" could always be replaced by "honest" functions satisfying the side-condition. This method of proof involved one of the most elaborate applications up to that time of the priority method of recursion complexity theory.

It's worth interjecting here an early instance of the astonishing parallels between Soviet and Western research in logic and the theory of computation. The fundamental problem of Post about whether all undecidable axiomatic systems were of the same degree of undecidability was solved in 1956, a dozen years after its formulation. This was achieved, independently, by Mucnik in the Soviet Union and Freidberg, a Harvard undergraduate, within a few months of each other. Their solutions were very similar and involved the invention of the priority method of computability theory.

At the time of this work with McCreight, we had not seen any Soviet writings. Perhaps Hartmanis, a Latvian emigré, was a Russian reader who saw Trakhtenbrot's notes, or perhaps Blum's Romanian emigré student Filloti, who was fluent in Russian, read them. In any case, we learned of the results of Trakhtenbrot's notes by word-of-mouth from Borodin, Hartmanis, and Blum.

Blum passed on a copy of the Trakhtenbrot notes to me around 1970, when I was at MIT, since I knew of a graduate student who was interested in translating them. His work was not very satisfactory, but then Filloti came to MIT to work as a post-doc with me, and did a respectable job. By this time the notes began to seem dated to me (about five years old in 1972) and I decided that they needed to be revised and updated. This youthful misjudgment doomed the project, since I was too impatient and too much of a perfectionist to complete the revision myself, and the final editing of the translation was never completed.

Some very personal issues arise at this point in my story which are nevertheless appropriate to Trakhtenbrot's account. Other neurotic attitudes of mine also undermined the project of publishing Trakhtenbrot's notes. I was at the time enmeshed in a pained marriage with two small children – a marriage which ended in divorce a few years later – and the depression I suffered during those years, which I carefully concealed from my professional colleagues, carried over into doubts about my own research and the significance of the field of complexity theory in general. These doubts, and perhaps other invidious feelings about the productivity of other researchers who seemed more prolific and less depressed than myself, left me inhibited at the final stages of several publishing projects, of which the Trakhtenbrot notes were, I regret to report, just one instance.

As the reader will learn, two of the principal characters on the Soviet side, the prodigies Tseitin and Levin, suffered in one way or another severe inhibitions about communicating their results, and Trakhtenbrot indicates that Tseitin suffered doubts similar to mine about the value of his own outstanding contributions. Whether this is coincidental or a typical reflection of the personality factors that lead young mathematicians into their vocation I remain unsure, but the psychological parallels here are as striking as the intellectual ones.

It amazes me in retrospect how I was actually able to make good scientific use of my doubts to suggest new questions to test the adequacy of our theories. One aspect of complexity theory which had dissatisfied me from the beginning was that in the theory "complexity" was essentially synonymous with "time-consuming to calculate". This definition was inconsistent with certain intuitive ideas about complexity; in particular, it seemed that there ought to be computational problems that were time-consuming, but were nevertheless intuitively not complex in that they merely involved a very simple computation for their solution, though the simple computational procedure might need a very large number of repetitions to produce its result. Thus, there ought to be distinctions among equally time-consuming computational problems reflecting the intuitive idea that two problems might take the same long length of time to solve but for different reasons.

There is no obvious formulation of what a "reason" for complexity in a problem might be, but the famous solution to Post's problem by Friedberg/Mucnik mentioned above offered what seemed to me a straightforward formulation of a precise conjecture which captured the idea. If one could construct two equally time-consuming problems with the property that, even given the ability to compute solutions to instances of one problem in no time at all, the other problem remained as time-consuming to solve as without this instantaneous ability, and vice-versa given the other problem, then it seemed legitimate to say the problems were difficult for different reasons. Once the conjecture was formulated, the construction of such sets became a matter of making suitable modifications of the "classical" priority methods of recursion theory. Working with my expert colleague Fischer, it took only a few weeks to polish the details, leaving one purely æsthetic flaw which seemed amateurish: the construction of the two sets involved defining two of each kind of bell and whistle needed in the construction, one for each set. This seemed wasteful and clogged up the reasoning with subscripts $i = 1, 2$.

Complexity theory was still a new, small field in 1972, and I was hungry for the excitement and reassurance that hearing new results in my area provided. The American Mathematical Society at that time offered a computer search service of journal abstracts to which one could subscribe and specify a rather sophisticated protocol to generate titles, abstracts, and even reprints of articles depending on the degree of match with the subscribers interest profile, and I was an enthusiastic subscriber. The service brought me many of the abstracts published in the *Doklady* by the characters in Trakhtenbrot's memoir.

One day, an abstract from Trakhtenbrot himself turned up of which I needed to read only the title, "On auto-reducibility". Here was the simple repair for the æsthetic flaw: don't construct two sets, neither of which helps the other be computed more quickly; construct one set such that answering membership questions about one part of the set does not help in computing solutions to membership questions about any other parts – a set that is not auto-reducible. Trakhtenbrot had thus come to look for similar results to ours, had obtained them with similar methods, and had added the final elegant touch we had missed. This was the event which confirmed my growing admiration for his ability. It also again reassured me that these abstract, speculative, positively obscure preoccupations of mine were not utterly narcissistic – or at least not *uniquely* narcissistic, because they were shared by an older, experienced researcher whose abilities had been certified by an altogether different establishment. I loved Trakhtenbrot for that reassurance. And given my secret depression and self-doubts, it helped to have a beloved mentor whom I hadn't met and who worked half a world away – little risk of rejection or disappointment that way.

Trakhtenbrot has emphasized that Westerners were too frequently unaware of the contributions of their Soviet counterparts, but as the story above suggests, this was not my own experience at the time. The failure of the AMS abstracting service, which I found so valuable, was an event that surprised and disappointed me at the time, but which helps explain Trakhtenbrot's impression of what he worried was a parochial neglect of Eastern research by Western researchers in theoretical computer science. The service, I was told, failed for lack of subscribers; I was one of the few who actually made use of it. Today, as a senior scientist in the now much larger and better known area of theoretical computer science, the failure of the AMS service seems much more understandable: there are far too many interesting results being discovered and far too many ingenious and significant papers to keep up with. The problem is not to find papers to read but to avoid being overwhelmed by them. I would not be a subscriber today.

The story of the parallels between Trakhtenbrot's and my research areas has too many more chapters to spell out much further. Suffice it to say that we found ourselves happily and fruitfully collaborating firsthand in an entirely different area of theoretical computer science than complexity theory to which we were led by independent decisions reflecting our shared theoretical tastes.

But there is one more personal epiphany which adds some perspective on Trakhtenbrot's story of the academic disputes between the Moscow establishment personified by Yablonskii versus Kolmogorov and Trakhtenbrot's own Novosibirsk group. The doubts that plagued me about the significance of complexity theory – with an emotional force which undoubtedly sprang from intimate aspects of my personal life – also were reinforced by external criticism from logicians and computer scientists alike. The problem was that, however provocative the theorems of complexity sounded, in the final analysis all the early results rested on the same kind of "diagonalization with priorities" which formed the core of classical computability theory. But unlike the classical theory for which natural instances of the kind of undecidability phenomena analyzed in the theory were well-known elsewhere in Logic and Algebra, no instances of provable complexity phenomena were known. For example, I remember presenting my results on sets that were complex-for-different-reasons at a 1971 logicians' meeting.

During the question period after my talk I received one cool question from an eminent logician: "Do your results give any information about the complexity of deciding propositional tautologies?" The lay reader should read this question as "What does what you're doing have to do with the price of eggs?" I had nothing to say about eggs.

These criticisms were echoed by those of Yablonskii, who scoffed at the emptiness of the diagonal technique. But let there be no misinterpretation of Trakhtenbrot's softly toned account of the disputes with Yablonskii. The man was clever enough to fasten on a weak point of complexity theory at that stage, but his actions in attacking the proponents of the theory to the extent of destroying careers and denying students their degrees cannot be accounted for out of sincere intellectual doubt; this was a villainous careerism which the Soviet system seems to have fostered. My feelings were hurt that the mainstream community of logicians and computability theorists were initially cool to my interests, but my career was never in jeopardy, and if I could not find support – moral or financial – among them, there were other communities of engineers and computer scientists with positions and grants. With Yablonskii in centralized charge in Moscow of higher degree granting, promotions, and even scheduling of research meetings, working in an area he opposed proved to be a perilous professional choice for my Soviet counterparts.

Shortly after my admission of ignorance about the economic principles of egg pricing, an American, Cook, at Toronto (and independently, though without comparable recognition, Levin in Moscow) discovered them. The precise computational complexity of deciding propositional tautologies remains open, but it is now understood to be the central problem of theoretical computer science, known, through the development of a rich theory, to be equivalent in complexity to hundreds of other apparently unrelated problems.

The excitement of Cook's discovery and its elaboration a few months later by Karp reawakened my interest in research at a time when I was actually taking some first steps towards leaving a scientific career altogether. In 1972, jointly with a very talented student, named Stockmeyer, I found the first genuinely natural examples of inherently complex computable problems. I knew there were only a small handful of people who understood the field deeply enough to appreciate immediately the significance of our examples, so it was with pride and anticipation that I sent the earliest draft of our results to Trakhtenbrot in Novosibirsk, where they were indeed received with immediate celebration. Sending these results first to Trakhtenbrot was doubly appropriate, because, though I did not know it at the time, Trakhtenbrot was one of the seminal researchers in the area of automata theory and logic from which Stockmeyer's and my first example came.

We come now to the question of to what degree these individual anecdotes represent a pattern of East/West scientific collaboration in the theoretical computer science. Trakhtenbrot – as noted above – was concerned that Western researchers were too unaware of Eastern research and may therefore have mistakenly underestimated their ability and the potential that would exist, were the stifling climate under Yablonskii's stewardship to abate, for dramatic contributions to be made. But while I agree that the names and contributions of several Soviet scholars, perhaps especially Barzdin, were not as prominent in the West as they may have deserved, nevertheless when I review the major scientific discoveries in the area in the 70s and 80s, I find major Soviet contributions widely and quickly recognized in the West. Thus, the ingenious tree manipulation algorithm of Adelson-Velsky was widely taught in undergraduate computer science courses throughout the world, the theoretically efficient algorithm for linear programming of V'jugin, Nemirovsky and Khachian was the focus of more than one entire scientific colloquium in the West after it was noticed (after, I should add, an uncertain delay) in the Soviet literature in 1979, and a major advance in the study of combinational complexity by a student of Kolmogorov named Razborov captured the imagination of the American research community.

So, I do not find great underestimation or neglect of Soviet activity, though the collegial connections with the West are rarely very close, as one might expect given the obstacles to travel and communication imposed. So, the names of Eastern researchers are not especially audible in informal conversation in Western scientific circles.

As Trakhtenbrot once said, "There can be no question but that in terms of sheer magnitude of pioneering efforts, the work of Western computer scientists exceeds that of their Soviet colleagues." The history of parallels confirms the abilities and contributions of the Soviet research community, but, on the other hand, can be read as showing that the West did not particularly need the Eastern contributions, since they would undoubtedly have been forthcoming anyway from corresponding Western work.

Nevertheless, Trakhtenbrot has called attention to potentially outstanding results such as those of Barzdins and Tseitin, which may have been unduly neglected in the West. It is in the last analysis impressive – and a testament to the vitality of theoretical ideas – that valuable contributions were continually made by Soviet researchers within an academic bureaucracy that would have overwhelmed a Western researcher.

Boris A. Trakhtenbrot*:
Academic Genealogy and Publications

Arnon Avron, Nachum Dershowitz, and Alexander Rabinovich

1 Trakhtenbrot's Genealogy

According to the Mathematics Genealogy Project (at http://genealogy.math.ndsu.nodak.edu), Trakhtenbrot's ancestral dag is rooted in Otto Mencke (who received his degree in 1666, but whose advisor is not listed):

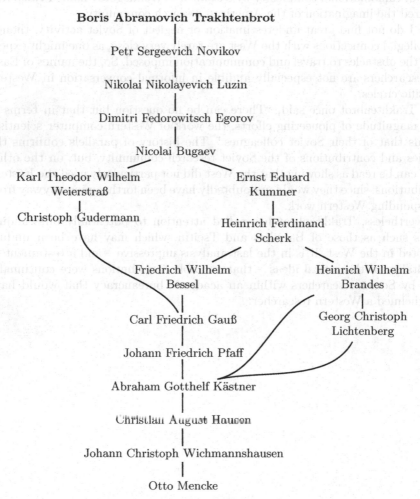

Boris Abramovich Trakhtenbrot

Petr Sergeevich Novikov

Nikolai Nikolayevich Luzin

Dimitri Fedorowitsch Egorov

Nicolai Bugaev

Karl Theodor Wilhelm Weierstraß   Ernst Eduard Kummer

Christoph Gudermann   Heinrich Ferdinand Scherk

Friedrich Wilhelm Bessel   Heinrich Wilhelm Brandes

Carl Friedrich Gauß   Georg Christoph Lichtenberg

Johann Friedrich Pfaff

Abraham Gotthelf Kästner

Christian August Hausen

Johann Christoph Wichmannshausen

Otto Mencke

---

* Sometimes transcribed as Trahtenbrot, Trachtenbrot, Trajtenbrot. His patronymic "A." is variously spelled: Abramovich, Avraamovich, Avramovich.

A. Avron et al. (Eds.): Trakhtenbrot/Festschrift, LNCS 4800, pp. 46–57, 2008.
© Springer-Verlag Berlin Heidelberg 2008

The next section enumerates Trakhtenbrot's 16 doctoral students, the titles of their dissertations, and their own academic progeny. It is followed by a bibliography of Trakhtenbrot's hundred-plus publications, including 7 books.

2   Trakhtenbrot's Progeny

There follows a list of Trakhtenbrot's 16 doctoral students, their 56 students, and their 13 students, in turn, for a grand total of 85 known academic descendants – to date.

1. Miroslav Kratko, *Unsolvability of Completeness Problem in Automata Theory*, Institut of Mathematics, Siberian Branch of the USSR Academy of Sciences, Novosibirsk (1964)
   (a) Oleg Revin (1974)
   (b) Valery Pavlenko (1979)
   (c) Miroslav Pritula (1983)
   (d) Yuri Yatsyshyn (1988)
   (e) Igor Yanovich (1989)
   (f) Andrei Nazarok (1989)
   (g) Andrei Khruzin (1990)

2. Nikolai Beljakin, *About the External Memory of Turing Machines*, Institut of Mathematics, Siberian Branch of the USSR Academy of Sciences, Novosibirsk (co-supervised with Petr Novikov) (1964)
   (a) Valery A. Ganov (1975)
   (b) Elena G. Nikiforova (1987)
   (c) Alla N. Gamova (1992)
   (d) Leonid N. Pobedin (1993)
   (e) Ruslana V. Ganova (1998)
   (f) Evgeniya V. Gailit (2004)

3. Janis Barzdins, *Universality Problems in the Theory of Growing Automata*, Institute of Mathematics, Siberian Branch of the USSR Academy of Sciences, Novosibirsk (1965)
   (a) Audris Kalnins (1971)
      i. Edgars Celms (2007)
      ii. Valdis Vitolins (2007)
   (b) Evalds Ikaunieks (1973)
   (c) Janis Kalnins (1974)
   (d) Jānis Bičevskis (1979)
      i. Guntis Arnicans
      ii. Girts Karnitis
      iii. Viesturs Vezis
      iv. Jānis Zuters
   (e) Karlis Podnieks (1979)
   (f) Mihails Augustons (1983)
   (g) Andrejs Auzins (1988)

Fig. 1. On the occasion of the "Logic at Botik '89" symposium in Pereslavl-Zalesski (from left to right): V. Yu. Sazonov, J. M. Barzdins, R. V. Freivalds, A. Ja. Dikovsky, B. A. Trakhtenbrot, M. K. Valiev, M. I. Kratko, V. A. Nepomnyashchy

  (h) Maris Treimanis (1988)
  (i) Juris Strods (1988)
  (j) Alvis Brazma (1989)
  (k) Juris Borzovs (1989)
          i. Guntis Urtāns (unofficial advisor) (2003)
          ii. Baiba Apine (2003)
          iii. Mārtiņš Gils (2005)
          iv. Darja Šmite (2007)
  (l) Ilona Etmane (1990)
  (m) Karlis Cerans (1992)
  (n) Ugis Sarkans (1998)
  (o) Girts Linde (2004)

4. Valery Nepomnyaschy, *Investigation of Bases and Sufficient Classes of Recursive Functions*, Institute of Mathematics, Siberian Branch of the USSR Academy of Sciences, Novosibirsk (1967)
   (a) Kostovsky Valery (1986)
   (b) Sergey Vorobyov (1987)
   (c) Nikolay Shilov (1987)
          i. Natalia Garanina (2004)

Fig. 2. On the banks of the Volga, after the (followup) "Logical Foundations of Computer Science" symposium, held near Tver in July 1992 (from left to right): M. N. Sokolovskiy, M. I. Dekhtyar, M. K. Valiev, B. A. Trakhtenbrot, V. Yu. Sazonov, I. A. Lomazova, V. N. Agafonov

- (d) Alexander Sulimov (1991)
- (e) Alexander Ustimenko (1997)
- (f) Igor Anureev (1998)
- (g) Tatiana Churina (2000)
- (h) Vitaly Kozura (2004)
- (i) Elena Okunishnikova (2004)
- (j) Aleksey Promsky (2004)

5. Alexei Korshunov, *An Investigation of Weight, Diameter and Ranks of Automata with a Big Memory*, Institute of Mathematics, Siberian Branch of the USSR Academy of Sciences, Novosibirsk (1967)
   - (a) Miron N. Sokolovskiy (1979)
6. Mars K. Valiev, *On Complexity of Word Problem for Finitely Presented Groups*, Institute of Mathematics, Siberian Branch of the USSR Academy of Sciences, Novosibirsk (1969)

7. Valery Agafonov, *On Algorithms, Frequency and Probability*, Institute of Mathematics, Novosibirsk (1969)
8. Djavkathodja Hodjaev, *On Complexity of Computations on Turing Machines with Oracles*, Institute of Cybernetics, Tashkent (1970)
9. Zoya Litvintseva, *Complexity of Some Algorithmic Problems for Groups and Semigroups*, Novosibirsk State University (1970)

10. Rūsiņš Freivalds, *Completeness up to Coding for Systems of Multiple-Valued Functions*, Institute of Mathematics, Siberian Branch of the USSR Academy of Sciences, Novosibirsk (1972)
    (a) Efim Kinber (1974)
    (b) Agnis Andzans (1983)
        i. Līga Rāmana (2004)
        ii. Ilze France (2005)
    (c) Maris Alberts (1989)
    (d) Daina Taimina (1990)
    (e) Janis Kaneps (1991)
    (f) Juris Viksna (1994)
    (g) Andris Ambainis (1997)
    (h) Juris Smotrovs (1999)
    (i) Dainis Geidmanis (1999)
    (j) Marats Golovkins (2003)
    (k) Maksims Kravcevs (2006)
    (l) Arnolds Kikusts (2007)

11. Anatoli Vaisser, *Problems of the Complexity and Stability of Probabilistic Algorithms*, Tomsk (Vladimir P. Tarasenko served as *de jure* advisor[1]) (1976)
12. Vladimir Sazonov, *On the Semantics of Applicative Algorithmic Languages*, Novosibirsk State University (1976)
    (a) Alexei Lisitsa (1997)

13. Michael Dekhtyar, *On the Complexity of Relativized Computations*, Moscow State University (1977)
    (a) Sergej Dudakov (2000)

14. Irina Lomazova, *Semantics and Algorithmic Logic for Programs with goto*, Institute of Mathematics of the USSR Academy of Sciences (Andrei P. Ershov served as *de jure* advisor[1]) (1981)
    (a) Vladimir Bashkin (2003)

15. Alex Barel, *On Collective Behavior of Automata*, Tel Aviv University (co-supervised with Ilya Piatetski-Shapiro) (1984)
16. Alexander Rabinovich, *Nets and Processes*, Tel Aviv University (1989)
    (a) Greta Yorsh (co-supervised with Shmuel Sagiv) (2007)

---

[1] Tarasenko and Ershov genrously volunteered to serve as official supervisors, so as to circumvent bureaucratic repression on the part of the official mathematical establishment, which would have endangered the careers of Trakhtenbrot's students.

3   Trakhtenbrot's Publications

Trakhtenbrot's books have been very influential and were translated into a wide range of languages, including English, French, German, and Japanese. His 96 additional publications span the period from 1950 until the present and have left indelible marks all over the field of theoretical computer science.

3.1   Books

1. *Algoritmy i mashinnoe reshenie zadach* (in Russian) [*Algorithms and Automatic Computing Machines* = Алгорифмы и машинное решение задач], Gostekhizdat, 1st ed., Moscow, 95 pages. Issue 26 of the series *Popular Lectures in Mathematics*, 1957.

    (a) *Wieso konnen Automaten Rechnen: Eine Einführung in die logisch-mathematischen Grundlagen programmgesteuerter Rechenautomaten* (German, translated from Russian), Deutscher Verlag der Wissenschaften, Berlin, 1959.

    (b) *Los Algorirmos y la Resolucion Automatica de Problemas* (Spanish, translated from Russian), Editorial Mir, Moscow, 1977.

    (c) (Japanese translation), Shigeo Kawano, Tokio-Toshio, 1959.

    (d) *Algorytmy i automatyczne rozwiazyvanie zadan*, Panstwowe Wydawnictwo Naukowe, Warszawa (Polish), 1961.

    (e) *Algorithme et Resolution de Problemes par des Machines* (French, translated from Russian), Editions Mir, Moscow; appears as part 2 of I. Yaglom, B. Trakhtenbrot, H. Ventsel, A. Solodovnikov, *Nouvelles Orientations des Mathematiques*, Editions Mir, Moscow, 1975.

2. *Algoritmy i mashinnoe reshenie zadach* (in Russian) [*Algorithms and Automatic Computing Machines* = Алгорифмы и машинное решение задач], 2nd ed., Fizmatgiz, Moscow, 120 pages, 1960.

    (a) *Algoritmy a strojove reseni ulokh* (Czech, translated from Russian), Nakladatelstvi Ceskoslovenske akademie ved, Praha, 1963.

    (b) *Algorithmes et machines a calculer* (French, translated by A. Chauvin from Russian), Dunod, Paris, 1963.

    (c) *Algorytmy i mashinno resavane na zadachi* (Bulgarian, translated from Russian), Dyrzhavno Izdatelstvo "Tekhnika", Sofia, 1963.

    (d) *Algorithms and Automatic Computing Machines* (translated and adapted from Russian by J. Kristian, J. D. McCawley, and S. A. Schmitt), in the series *Topics in Mathematics*, D. C. Heath and Company, Boston, 1963.

    (e) Japanese translation, Tosho, Tokyo, 1964.

    (f) *Algoritmi e Macchine Calcolatrici Automatiche* (Italian, translated from English), Progresso TecnicoEditoriale, Milano, 1964.

    (g) *Algoritmalar ve Otomatik Hesap Makinalari* (Turkish), Turk Matematik Dernegi Yaynlari, Istanbul, 1964.

(h) *Introduction a la Teoria Matematica de las Computadoras y de la Programacion* (Spanish, translated from Russian), Siglo Veintiuno Editores SA, Mexico, 1967.

3. N. E. Kobrinski and B. A. Trakhtenbrot, *Vvedenie v Teoriiu Konechnykh Avtomatov* (in Russian) [*Introduction to the Theory of Finite Automata* = Введение в теорию конечных автомато], Fizmatgiz, Moscow, 404 pages, 1962.

(a) *Introduction to the Theory of Finite Automata* (translated from Russian; translation edited by J. C. Shepherdson), in: Studies in Logic and the Foundations of Mathematics, North-Holland Publishing Company, Amsterdam, 1965.

(b) *Uvod do Teorie Konecnych Automatu* (Czech, translated from Russian), Nakladatelstvi Technicke Literatury, Praha, 1967.

(c) *Einfuhrung in die Theorie endlicher Automaten* (German, translated from Russian), Akademie-Verlag, Berlin, 1967.

4. *Slozhnost' Algoritmov i Vychislenii* (in Russian) [*The Complexity of Algorithms and Computations* = Сложность алгорифмов и вычислений], Lecture Notes, edited by Novosibirsk University, 258 pages, 1967.

5. B. A. Trakhtenbrot and Ya. M. Barzdin, *Konechnye Avtomaty (Povedenie i Sintez)* (in Russian) [*Finite Automata (Behavior and Synthesis)* = Конечные автоматы (поведение и синтез)], Nauka, Moscow, 400 pages, 1970.

(a) *Finite Automata: Behavior and Synthesis* (translated from Russian by D. Louvish; translation edited by E. Shamir and L. H. Landweber) in: Series Fundamental Studies in Computer Science, North-Holland, Amsterdam, and Elsevier, New York, 1973.

6. *Algoritmy i Vychislitelnye Avtomaty* (in Russian) [*Algorithms and Computational Automata* = Алгорифмы и вычислителные автоматы], Sovetskoe Radio, Moscow, 200 pages, 1974.

(a) *Algorithmen und Rechenautomaten* (German, translated from Russian), Deutcher Verlag der Wissenschaften, Berlin, 1977.

(b) *Sto SU Algoritmi: Algoritmi i racunski automati. Skolska Knjiga* (Horvat-Serbian, translated from Russian), Zagreb, 1978.

(c) *Algoritmusok es absztrak Automatak* (Hungarian, translated from Russian), Muszaki Konivkiado, Budapest, Mir Konivkiado, Moszkva, 1978.

7. *Selected Developments in Soviet Mathematical Cybernetics: Finite automata, combinational complexity, algorithmic complexity*, in: Monograph Series on Soviet Union, Delphic Associates, Falls Church, VA, xiv + 122 pages, 1985.

References

1. The impossibility of an algorithm for the decidability problem on finite classes. Doklady AN SSR 70(4), 569–572 (1950)
2. Decidability problems for finite classes and definitions of finite sets, Ph.D. Thesis, Math. Inst. of the Ukrainian Academy of Sciences, Kiev (1950)
3. On recursive separability. Doklady AN SSR 88(6), 953–956 (1953)
4. Tabular representation of recursive operators. Doklady AN SSR 101(4), 417–420 (1955)
5. Modelling functions on finite set. Uchionnye Zapiski Pennenskogo Pedinstituta (Transactions of the Pensa Pedagogical Institute), (2), 61–78 (1955)
6. The synthesis of non-repetitive schemas. Doklady AN SSR 103(6), 973–976 (1955)
7. Kuznetsov, A.V., Trakhtenbrot, B.A.: Investigation of partial recursive operators by techniques of Baire spaces. Doklady AN SSR 105(6), 896–900 (1955)
8. On the definition of finite set and the deductive incompleteness of set theory. Izvestia AN SSR 20, 569–582 (1956)
9. Signalizing functions and tabular operators. Uchionnye Zapiski Penzenskogo Pedinstituta (Transactions of the Penza Pedagogoical Institute) (4), 75–87 (1956)
10. On effective operators and their properties connected to continuity. In: Proceedings of the Third All-Union Mathematical Congress, Moscow, vol. 2, pp. 147–148 (1956)
11. Algorithms and automatic solution of problems. Mathematics in the School, (4–5) (1956)
12. Descriptive classifications in recursive arithmetic. In: Proceedings of the Third All-Union Mathematical Congress, Moscow, vol. 1, p. 185 (1956)
13. Applying some topological invariants to the synthesis of two-polar switching schemes. In: Proceedings of the Third All-Union Mathematical Congress, Moscow, vol. 1, p. 136 (1956)
14. On operators, realizable by logical nets. Doklady AN SSR 112(6), 1005–1006 (1957)
15. The synthesis of logical nets whose operators are described in terms of monadic predicates. Doklady AN SSR 118(4), 646–649 (1958)
16. On the theory of non-repetitive schemes. Trudy Mathem. Instituta im. Steklova (Transactions of the Steklov Mathem. Institute) 51, 226–269 (1958)
17. The asymptotic estimate of the logical nets with memory. Doklady AN SSR 127(2), 281–284 (1959)
18. Kobrinski, N.E., Trakhtenbrot, B.A.: Fundamentals of the theory of logical nets. In: Computing Techniques and Their Application, Moscow, pp. 248–268 (1959)
19. Sketching of a general theory of logical nets. In: Logical Investigations, The Institute of Philosophy of the Academy of Sciences, Moscow, pp. 352–378 (1959)
20. Some constructions in the monadic predicate calculus. Doklady AN SSR 138(2), 320–321 (1961)
21. Finite automata and the monadic predicate calculus. Doklady AN SSR 140(2), 326–329 (1961)
22. Finite automata and the monadic predicate calculus. Siberian Mathematical Journal 3(1), 103–131 (1962)
23. Investigations on the synthesis of finite automata, Doctoral Thesis, Math. Inst. of the Siberian Branch of the USSR Academy of Sciences, Novosibirsk (1962)
24. On the frequency computation of recursive functions. Algebra i Logika 1(1), 25–32 (1963)
25. Finite automata. In: Proceedings of the Fourth All-Union Math. Congress, 1961, Nauka, Moscow, vol. 2, pp. 93–101 (1964)

26. On an estimate for the weight of a finite tree. Siberian Mathematical Journal 5(1), 186–191 (1964)
27. On the complexity of schemas that realize many-parametric families of operators. Problemy Kibernetiki 12, 99–112 (1964)
28. Turing computations with logarithmic delay. Algebra i Logika 3(4), 33–48 (1964)
29. Automata theory. Automation of Electronic Industries Encyclopedia 1 (1964)
30. Optimal computations and the frequency phenomena of Yablonski. Algebra i Logika 4(5), 79–93 (1965)
31. On normalized signalizing functions for Turing computation. Algebra i Logika 5(6), 61–70 (1966)
32. On complexity of computations. In: Abstracts of the International Mathematical Congress, Section 1, Moscow, p. 26 (1966)
33. Barzdin, Y.M., Trakhtenbrot, B.A.: On the current situation in the behavioral theory ofautomata. In: Abstracts of the Conference on Automata Theory and Artificial Intelligence, Tashkent, Computing Center of the USSR Academy of Sciences, pp. 10–11 (1968)
34. Trakhtenbrot, B.A., Valiev, M.: On the complexity of the mutual reduction of the identity-problem for finite definable groups and the decidability of enumerable sets. In: Abstracts of the IX All-Union Symposium on Algebra, Gomel, pp. 41–42 (1968)
35. On the complexity of the mutual-reduction algorithms in the construction of Novikov and Boone. Algebra i Logika 8, 50–71 (1969)
36. The theory of algorithms in the USSR. In: History of Mathematics in our Country, Kiev, pp. 409–431 (1970)
37. On autoreducibility. Doklady AN SSR 192(6), 1224–1227 (1970)
38. Some applications of the theory of complexity of computations. Inform. Bulletin of the Cybernetics Scientific Council, Academy of Sciences (5), 61–74 (1970)
39. Complexity of computations on discrete automata. Studien zur Algebra und ihre Anwendungen, p. 152. Akademie Verlag, Berlin (1972)
40. On autoreducible and non-autoreducible predicates and sets. Investigations in the Theory of Algorithms and Mathematical Logic, Computer Center, Academy of Sci. 1, 211–235 (1973)
41. Frequency computations, Trudy Matem. Instituta im. Steklova Transactions of the Steklov Mathematical Institute 133, 221–232 (1973)
42. Formalization of some notions in terms of computation complexity. In: Proceedings of the 1st International Congress for Logic,Methodology and Philosophy of Science, Studies in Logic and Foundations of Mathematics, vol. 74, pp. 205–214 (1973)
43. On universal classes of program-schemes, International Symposium on Theoretical Programming, Novosibirsk, Lecture Notes in Computer Science, vol. 5, Springer-Verlag, Berlin, pp. 144–151 (1974)
44. Notes on the complexity of probabilistic machine computations. In: The Computing Center of the Academy of Sciences (ed.) Theory of Algorithms and Mathematical Logic, pp. 159–176 (1974)
45. Alexey Andreevich Lyapunov (in Russian), Mathematics in the School (3) (1974) (a) German translation in: Mathematik in der Schule, Jg 13, no. 4, pp. 203–209, Berlin (1975) (b) Alexey Andreevich Lyapunov. In: Essays on the History of Informatics in Russia, Nauchno-Izdatel'skiy Tsentr Sibirskogo Otdeleniya Rossiyskoy Akademii Nauk, Novosibirsk, pp. 470–480 (1998)
46. Abstraktnaya teoriya avtomatov (Abstract automata theory; in Russian). Encyclopedia Kibernetika 1, 11–12 Kiev (1974)
47. Avtomatov teorija (Theory of automata; in Russian), Encyclopedia of Cybernetics, vol. 1, pp. 57–60, Kiev (1974)

48. Mery slozhnosti v teorii avtomatov (Complexity measures in automata theory; in Russian). Encyclopedia of Cybernetics 1, 588–590 (1974)

49. Povedenie avtomatov (Behavior of automata; in Russian). Encyclopedia of Cybernetics 2, 166–169 (1974)

50. Yazyk logicheskiy dlya zadaniya avtomatov (Logical language for specifications of automata; in Russian). Encyclopedia of Cybernetics, Kiev, vol. 1, pp. 595–596 (1974)

51. On problems solvable by successive trials. In: Becvar, J. (ed.) MFCS 1975. LNCS, vol. 32, pp. 125–138. Springer, Heidelberg (1975)

52. Recursive program schemas and computable functionals. In: Mazurkiewicz, A. (ed.) MFCS 1976. LNCS, vol. 45, pp. 137–151. Springer, Heidelberg (1976)

53. Frequency algorithms and computations. In: Gruska, J. (ed.) MFCS 1977. LNCS, vol. 53, pp. 148–161. Springer, Heidelberg (1977)

54. On the semantics of algorithmic languages. In: Proceedings of the Symposium on Development Perspectives in System and Theoretical Programming, Novosibirsk, pp. 77–84 (1978)

55. Semantics and logic of algorithm languages. Semiotika i Informatika, Moscow 13, 47–85 (1979)

56. On the completeness of algorithmic logic. Kibernetika, Kiev (2), 6–11 (1979)

57. Relaxation rules of algorithmic logic. In: Schlender, B., Frielinghaus, W. (eds.) GI-Fachtagung 1974. LNCS, vol. 7, pp. 453–462. Springer, Heidelberg (1974)

58. Vospitanie matematiko-logicheskoy kultury uchashchikhs'a (On mathematical-logical education; in Russian), In: Olimpiada, Algebra, Kombinatorika, Nauka, Novosibirsk, pp. 26–52 (1979)

59. Some reflections on the connection between computer science and the theory of algorithms. In: Knuth, D.E., Ershov, A.P. (eds.) Algorithms in Modern Mathematics and Computer Science. LNCS, vol. 122, pp. 461–462. Springer, Heidelberg (1981)

60. On denotational semantics and axiomatization of partial correctness for languages with procedures as parameters and with aliasing (Extended abstract), Technical Report, Tel Aviv University, 20 pages (August 1981)

61. Trakhtenbrot, B.A., Halpern, J., Meyer, A.: From denotational to operational and axiomatic semantics for Algol-like languages (An overview). In: Clarke, E., Kozen, D. (eds.) Logic of Programs 1983. LNCS, vol. 164, pp. 474–500. Springer, Heidelberg (1984)

62. Halpern, J., Meyer, A., Trakhtenbrot, B.A.: The semantics of local storage, or what makes the free list free (preliminary report). In: Conference Record of the XI ACM Symposium on Principles of Programming Languages (POPL), pp. 245–257 (1984)

63. A survey on Russian approaches to perebor (brute force search algorithms). Annals of the History of Computing 6(4), 384–400 (1984)

64. On logical relations in program semantics. In: Proceedings of the Conference on Mathematical Logic and its Applications, Dedicated to the 80th Anniversary of Kurt Gödel, Druzhba, Bulgaria, pp. 213–229. Plenum Press, New York (1987)

65. Comparing the Church and Turing approaches: Two prophetical messages. In: The Universal Turing Machine – A Half-Century Survey, pp. 603–630. Oxford University Press, Oxford (1988) (a) 2nd edn. Springer-Verlag, Wien, pp. 557–582 (1995)

66. Editor's foreword, Special Issue on Concurrency, Fundamenta Informaticae, vol. xi, pp. 327–329. North-Holland, Amsterdam (1988)

67. Rabinovich, A., Trakhtenbrot, B.A.: Behavior structures and nets of processes, Fundamenta Informaticae, vol. xi, pp. 357–403. North-Holland, Amsterdam (1988)

68. Hirshfeld, Y., Rabinovich, A., Trakhtenbrot, B.A.: Discerning causality in interleaving behavior. In: Meyer, A.R., Taitslin, M.A. (eds.) Logic at Botik 1989. LNCS, vol. 363, pp. 146–162. Springer, Heidelberg (1989)
69. Understanding nets. In: Kreczmar, A., Mirkowska, G. (eds.) MFCS 1989. LNCS, vol. 379, pp. 133–134. Springer, Heidelberg (1989)
70. Rabinovich, A., Trakhtenbrot, B.A.: Nets of processes and data flow. In: de Bakker, J.W., de Roever, W.-P., Rozenberg, G. (eds.) Linear Time, Branching Time and Partial Order in Logics and Models for Concurrency. LNCS, vol. 354, pp. 574–602. Springer, Heidelberg (1989)
71. Rabinovich, A., Trakhtenbrot, B.A.: Nets and data flow interpreters. In: Proceedings of Fourth Annual Symposium on Logic in Computer Science (LICS), Asilomar, CA, pp. 164–174 (June 1989)
72. Rabinovich, A., Trakhtenbrot, B.A.: Communication among relations. In: Paterson, M. (ed.) ICALP 1990. LNCS, vol. 443, pp. 294–307. Springer, Heidelberg (1990)
73. Mazurkiewicz, A., Rabinovich, A., Trakhtenbrot, B.A.: Connectedness and synchronization. In: Bjørner, D., Kotov, V. (eds.) Images of Programming. IFIP Series, North-Holland, Amsterdam (1991) Mazurkiewicz, A., Rabinovich, A., Trakhtenbrot, B.A.: Connectedness and synchronization. Theoretical Computer Science 90(1), 171–184 (1991)
74. Rabinovich, A., Trakhtenbrot, B.A.: On nets, algebras and modularity. In: Ito, T., Meyer, A.R. (eds.) TACS 1991. LNCS, vol. 526, pp. 176–203. Springer, Heidelberg (1991)
75. Compositional proofs for networks of processes. Fundamenta Informaticae 20, 231–275 (1994)
76. Origins and metamorphoses of the Trinity: Logic, nets, automata. In: Proceedings of the 10th IEEE Symposium on Logic in Computer Science, San Diego, pp. 506–507 (1995)
77. On the power of compositional proofs. Fundamenta Informaticae 30(1), 83–95 (1997)
78. In memory of S.A. Yanovskaya (in Russian), Researches in History of Mathematics, series 2, vol. 2(37), Russian Academy of Sciences, The Institute for History of Natural Sciences and Technology, pp. 109–127 (1997) (a) In memory of S.A. Yanovaskaya (1896–1966) on the centenary of her birth, Modern Logic 7(2), 160–187 (1997)
79. Rabinovich, A.M., Trakhtenbrot, B.A.: From finite automata toward hybrid systems. In: Chlebus, B.S., Czaja, L. (eds.) FCT 1997. LNCS, vol. 1279, pp. 411–422. Springer, Heidelberg (1997)
80. Automata and hybrid systems (seven-lecture mini-course), Technical Report no. 153, UPMAIL, Uppsala University, pp. 1–89 (1998)
81. From logic to theoretical computer science. In: Calude, C.S. (ed.) People and Ideas in Theoretical Computer Science, pp. 314–342. Springer, Heidelberg (1998)
82. Automata and their interaction: Definitional suggestions. In: Ciobanu, G., Păun, G. (eds.) FCT 1999. LNCS, vol. 1684, pp. 54–89. Springer, Heidelberg (1999)
83. Automata, circuits and hybrids: Facets of continuous time. In: Orejas, F., Spirakis, P.G., van Leeuwen, J. (eds.) ICALP 2001. LNCS, vol. 2076, pp. 4–23. Springer, Heidelberg (2001) Automata, circuits and hybrids: Facets of continuous time (Extended abstract). In: Proceedings of the 33rd Annual ACM Symposium on Theory of Computing, Hersonissos, Crete, Greece, pp. 754–755 (July 2001)
84. Automata, logic, circuits: The impact of continuous time, Abstracts of the International Conference Mathematical Logic, Algebra and Set Theory, Dedicated to the 100th Anniversary of P. S. Novikov, p. 48 (August 2001)

85. Remembering Alexey Andreevich. In: Proceedings of the Conference Dedicated to the 90th Anniversary of A. A. Lyapunov, Joint Institute of Informatiques, Novosibirsk, Siberian Branch of the Russian Academy of Sciences (2001)
86. Editor's preface, Special Issue on Continuous Time Paradigms in Logic and Automata. Fundamenta Informaticae 62(1) v–vii (2004)
87. Understanding basic automata theory in the continuous time setting. Fundamenta Informaticae, 62(1), 69–121 (2004)
88. Pardo, D., Rabinovich, A., Trakhtenbrot, B.A.: Synchronous circuits over continuous time: Feedback reliability and completeness. Fundamenta Informaticae 62(1), 123–137 (2004)
89. In memory of Andrey Petrovich Ershov (in Russian). In: Marchuk, A.G. (ed.) Andrey Petrovich Ershov, Scientist and Man, pp. 343–351 Publishing House of Siberian Branch of the Russian Academy of Sciences (2006)
90. Avron, A., et al. (eds.): From logic to theoretical computer science. LNCS, vol. 4800. Springer, Heidelberg

Symmetric Logic of Proofs

Sergei Artemov

CUNY Graduate Center, 365 Fifth Ave., New York, NY 10016, USA
SArtemov@gc.cuny.edu

Dedicated to B.A. Trakhtenbrot on the occasion of his 85th birthday.

Abstract. The Logic of Proofs LP captures the invariant propositional properties of proof predicates *t is a proof of F* with a set of operations on proofs sufficient for realizing the whole modal logic S4 and hence the intuitionistic logic IPC. Some intuitive properties of proofs, however, are not invariant and hence not present in LP. For example, the choice function '+' in LP, which is specified by the condition $s{:}F \lor t{:}F \to (s{+}t){:}F$, is not necessarily symmetric. In this paper, we introduce an extension of the Logic of Proofs, SLP, which incorporates natural properties of the standard proof predicate in Peano Arithmetic:

t is a code of a derivation containing F,

including the symmetry of Choice. We show that SLP produces Brouwer-Heyting-Kolmogorov proofs with a rich structure, which can be useful for applications in epistemic logic and other areas.

1 Introduction

In [15], Gödel used the modal logic S4 to axiomatize classical provability and provide the formal provability semantics to the intuitionistic propositional logic IPC by reducing IPC to S4. The question of the provability semantics of S4 itself was left open and found its resolution in [1,2] via the Logic of Proofs LP[1] which provided a complete axiomatization of the proof predicate

t is a proof of F

in a propositional language with a sufficiently rich system of operations on proofs. On the other hand, LP can realize every S4 derivation by recovering corresponding proof terms at every occurrence of the modality (the Realization Theorem from [1,2], cf. Theorem 1). The combination of these two features renders LP a bridge between intuitionistic logic and the realm of formal mathematical proofs in the style of Brouwer-Heyting-Kolmogorov:

$$\text{IPC} \hookrightarrow \text{S4} \hookrightarrow \text{LP} \hookrightarrow \textit{Gödelian proof predicates.}$$

---

[1] The first (incomplete) sketch of the logic of proofs was given by Gödel in one of his lectures of 1938 [16]. This lecture was published only in 1995; by that time, the complete system LP for the Logic of Proofs had already been discovered and shown to provide a desired provability semantics for intuitionistic logic (cf. [1,2]).

A. Avron et al. (Eds.): Trakhtenbrot/Festschrift, LNCS 4800, pp. 58–71, 2008.
© Springer-Verlag Berlin Heidelberg 2008

In this diagram, IPC $\hookrightarrow$ S4 denotes Gödel's faithful embedding of IPC into S4 [15], S4 $\hookrightarrow$ LP signifies the Realization Theorem which faithfully embeds S4 into LP [1,2], and LP $\hookrightarrow$ *Gödelian proof predicates* refers to the arithmetical soundness (and completeness) theorem ([1,2]) for the Logic of Proofs. We refer the reader to [2] and surveys [6,7] for detailed discussion of these matters.

2   LP Basics

The Logic of Proofs has three basic operations on proofs: *Application* '$\cdot$' (binary), *Choice* '$+$' (binary), and *Proof Checker* '!'(unary). *Proof polynomials* are terms built by these operations from *proof variables* $x, y, z, \ldots$ and *proof constants* $a, b, c, \ldots$. The formulas of LP are defined by the grammar

$$A = S \mid A \to A \mid A \wedge A \mid A \vee A \mid \neg A \mid t{:}A$$

where t stands for any proof polynomial and S for any sentence letter. As usual, we shorten $s \cdot t$ to st when convenient. The binding priority from strong to weak is $!, \cdot, +, :, \neg, \wedge, \vee, \to$. In particular, $t(u+v){:}F \to tu{:}F \vee tv{:}F$ denotes

$$\{[t\cdot(u+v)]{:}F\} \to \{[(t\cdot u){:}F] \vee [(t\cdot v){:}F]\}.$$

The postulates of the Logic of Proofs LP are

1. a fixed set of axioms for classical propositional logic with *Modus Ponens* as its only rule of inference, e.g., the set from [18];
2. $s{:}(F \to G) \to (t{:}F \to (s\cdot t){:}G)$ (*Application*);
3. $t{:}F \to !t{:}(t{:}F)$ (*Proof Checker*);
4. $s{:}F \to (s+t){:}F, \quad t{:}F \to (s+t){:}F$ (*Choice*);
5. $t{:}F \to F$ (*Reflection*);
6. *Constant Specification Rule: If c is a proof constant and A is an axiom from 1-5, then infer c:A.*

LP is closed under substitutions of proof polynomials for proof variables and formulas for propositional variables, and enjoys the deduction theorem.

Constant Specification CS is a set of formulas $\{c_1{:}A_1, c_2{:}A_2, \ldots\}$ where each A_i is an axiom and each c_i is a proof constant. Each derivation in LP naturally generates a (finite) constant specification CS introduced in this derivation by the Constant Specification Rule. By LP_{CS}, we mean a subsystem of LP where the Constant Specification Rule is only allowed to produce formulas from a given constant specification CS. If CS contains all formulas $c{:}A$ where A is an axiom and c is a proof constant, then LP_{CS} is LP itself.

The principal feature of the Logic of Proofs is the Internalization Principle[2] which states that

      whenever $\vdash F$, there is a proof polynomial p such that $\vdash p{:}F$,

---

[2] In his brief sketch of the logic of proofs in [16], Gödel cited the Internalization Principle as one of its features.

which is nothing more than the explicit version of the modal Necesitation Rule

$$\frac{\vdash F}{\vdash \Box F.}$$

The following Theorem 1 discloses a connection between the Logic of Proofs LP and Gödel's provability logic S4; this result has been crucial for providing intuition-istic logic with the intended Brouwer-Heyting-Kolmogorov semantics of proofs. This theorem shows that the provability modality in S4 can indeed be read as

$$\Box F \quad = \quad \textit{there is proof of } F$$

using the language of Skolem-style operations on proofs rather than quantifiers. For example, a sentence

$$\Box F \;\rightarrow\; \Box G$$

in the logic of formal provability[3] reads as

$$\exists x \,(\text{'}x \text{ is a proof of } F\text{'}) \;\rightarrow\; \exists y \,(\text{'}y \text{ is a proof of } G\text{'}),$$

whereas, by the Realization Theorem, this reads as an LP sentence

$$x{:}F \;\rightarrow\; t(x){:}G,$$

for an appropriate proof polynomial $t(x)$.

The usual Skolemization does not work for quantifiers on proofs, and a totally new technique has been invented here.

Theorem 1. [Realization Theorem for LP ([1,2])] *There is an algorithm which, given S4 $\vdash F$, substitutes each occurrence of the modality in F by an appropriate proof polynomial such that the resulting formula F^r is derivable in LP. Moreover, such a realization can be made in a way that respects Skolem-style reading of $\Box X$ as 'there is a proof of X': each negative occurrence of $\Box$ can be realized by a proof variable, and each positive occurrence of $\Box$ is realized by a proof polynomial depending of those variables.*

The size of realizing proof polynomials can be limited by a quadratic function in the length of a cut-free derivation of F in S4 ([12]). A semantical proof of the Realization theorem which is not based on cut-elimination in S4 was suggested in [14]. R. Kuznets in [12] showed that S4 cannot be realized in LP without using self-referential proof assertions of the sort $t{:}F(t)$.

Corollary 1. S4 *is the forgetful projection of* LP.

Proof. It is straightforward that the forgetful projection of LP is S4-compliant; it suffices to notice that the forgetful projections of all axioms of LP are provable in S4 and the rules of LP are S4-sound. By Theorem 1, every theorem of S4 is a forgetful projection of some LP-theorem. □

---

[3] This approach led to the well-known Provability Logic GL (cf. [11,27]).

3   Symmetric Provability Interpretation

The intended provability semantics of LP is given by interpreting $t:F$ as the arithmetical proof predicate

$$t \text{ is a proof of } F,$$

where 'proofs' are understood in a multi-conclusional way, i.e., a proof can yield more than one theorem (think of a Hilbert-style proof sequence that proves all formulas occurring in this sequence).

A body of work in this area shows that proof realization of modality necessarily requires the multi-conclusion reading of proof predicates. What follows is a light informal argument which hints at what is going on. Imagine a variant $\mathcal{LP}$ of the logic of proofs capable of realizing S4. Then $\mathcal{LP}$ should be able to realize the modality in an easy S4 theorem $\Box A$ where A is a propositional modal-free axiom (e.g., $P \rightarrow (Q \rightarrow P)$ for propositional letters P and Q). Such a realization should be of the form $t:A$ for some proof term t and $\mathcal{LP} \vdash t:A$. Assume that $\mathcal{LP}$ is closed under the substitution of propositional formulas for propositional letters; all provability logics have this feature since they describe schemas valid under all arithmetical interpretations, and this property survives substitution. Let A' be a substitutional instance of A, syntactically different from the latter. By the substitution closure, $\mathcal{LP} \vdash t:A'$. Hence $\mathcal{LP} \vdash t:A \wedge t:A'$ and t represents a proof of two different theorems.

Moreover, the logic of single-conclusion proofs contains some principles which are inconsistent with modal logic. For example (cf. [2]), the principle $\neg (x:\top \wedge x: (\top \wedge \top))$ is valid for single-conclusion proofs whereas its natural modal language presentation via the 'forgetful projection,' $\neg(\Box\top \wedge \Box(\top \wedge \top))$, is false in any modal logic. The logic of single-conclusion proofs has been axiomatized in [10] (without operations) and in [20,21] (with the operations Application and Proof Checker[4]). For further progress in this direction cf. [22].

In the context of the Logic of Proofs, one has to consider the whole class of proof predicates and axiomatize only invariant properties, i.e., those that hold for all proof predicates (from a given class). There is a good reason for this. The language of the Logic of Proofs is rather expressive and captures individual properties of proofs which should not count as general logic laws. For example, the formula $x:(\top \wedge \top) \rightarrow x:\top$ claims that any (multi-conclusion) proof of the conjunction $\top \wedge \top$ should prove $\top$ as well. Apparently, this 'principle' is not invariant since by changing an axiom system, one can obtain two proof predicates in which this formula both holds and does not hold, respectively.

The soundness and completeness theorems from [1,2] state that the Logic of Proofs LP captures exactly all invariant properties of multi-conclusion proof predicates with natural computable operations on proofs corresponding to Application, Proof Checker, and Choice[5]. We refer the reader to [6,7,3] for more details.

---

[4] The operation Choice '+' is incompatible with single-conclusion proof semantics

[5] E. Goris suggested in [17] a natural interpretation of LP in Bounded Arithmetic where all of these operations are PTIME-computable.

In the rest of this section, we introduce a symmetric version of the standard provability semantics for the Logic of Proofs which differs slightly from the one given in [2], Sect. 6 and is more convenient for revealing the natural structure of proof operations.

Consider the proof predicate $PROOF(x, y)$ for Peano Arithmetic PA which is a natural arithmetical formalization of the usual Hilbert-style definition of proofs:

x is a number of a proof sequence which contains a formula with a number y .

An interpretation of the language of LP maps propositional letters to sentences of PA, and proof variables and constants to Gödel numbers of Hilbert-style proofs in PA. This is the only difference between the symmetric semantics and the standard provability semantics from [2], Sect. 6, where proof variables and constants are mapped to arbitrary natural numbers, and not necessarily the codes of PA-derivations. Note that each proof sequence in PA is a complete proof of each sentence occurring within it.

For the proof predicate $PROOF(x, y)$, $s \cdot t$ can be interpreted as the operation which concatenates the codes of proofs corresponding to s and t and adds to the right all formulas G such that for some F, $F \to G$ and F occur in s and t respectively.

Here is a more formal description. Let s', t', ... denote arithmetical interpretations of s, t, etc., i.e., Gödel numbers of Hilbert-style proofs in PA. Let also $*$ denote the concatenation on Gödel numbers of finite sequences. Then we define

$$(s \cdot t)' = s' * t' * \ulcorner G_1 \urcorner * \ldots * \ulcorner G_n \urcorner \tag{1}$$

where $\ulcorner G_1 \urcorner, \ldots, \ulcorner G_n \urcorner$ are Gödel numbers of single-formula sequences for each G_i for which there exists F such that $F \to G_i$ and F are in the proofs with the numbers s' and t' respectively. For example, if $s' = \ulcorner F \to G \urcorner$ and $t' = \ulcorner F \urcorner$, then

$$(s \cdot t)' = \ulcorner F \to G \urcorner * \ulcorner F \urcorner * \ulcorner G \urcorner, \qquad (t \cdot s)' = \ulcorner F \urcorner * \ulcorner F \to G \urcorner.$$

The Choice operation $s + t$ can be interpreted as the concatenation of proof sequences corresponding to s and t respectively:

$$(s + t)' = s' * t'. \tag{2}$$

The Proof Checker is a primitive recursive operation that takes a proof t and, for each F such that $t : F$ holds, produces a proof of all such $t : F$'s; such an operation can be traced back to the proof of Gödel's Second Incompleteness Theorem (cf. [26]).

We call $PROOF(x, y)$ with operations (1), (2), and Proof Checker as above the *symmetric arithmetical semantics of the Logic of Proofs*.

We can see that within the standard semantics, the Choice function is symmetric:

$$(s + t) : F \leftrightarrow (t + s) : F;$$

moreover, it satisfies the principles

$$(s+t){:}F \leftrightarrow s{:}F \vee t{:}F,$$

$$u(s+t){:}F \leftrightarrow (us+ut){:}F,$$

$$(s+t)u{:}F \leftrightarrow (su+tu){:}F.$$

None of these principles is derivable in LP. The point is that these principles are not invariant! For example, one can easily devise an interpretation where $(s+t){:}F \rightarrow s{:}F \vee t{:}F$ does not hold. Indeed, interpret $s+t$ and $s \cdot t$ as follows:

$$(s \cdot t)^{\natural} = s^{\natural} * t^{\natural} * \ulcorner G_1 \urcorner * \ldots * \ulcorner G_n \urcorner,$$

as in (1), and

$$(s+t)^{\natural} = (s \cdot t)^{\natural}.$$

It is clear that $s{:}F \vee t{:}F \rightarrow (s+t){:}F$ holds, since $(s+t)^{\natural}$ contains the concatenation of $s^{\natural}$ and $t^{\natural}$. However, $(s+t)^{\natural}$ and $(t+s)^{\natural}$ can very well prove different sets of theorems, cf. an example about $(s \cdot t)'$ and $(t \cdot s)'$ above. In this case, neither $(s+t){:}F \leftrightarrow (t+s){:}F$ nor $(s+t){:}F \leftrightarrow s{:}F \vee t{:}F$ holds.

Under yet another interpretation:

$$(s \cdot t)^{\flat} = s^{\flat} * t^{\flat} * \ulcorner G_1 \urcorner * \ldots * \ulcorner G_n \urcorner,$$

as in (1), and

$$(s+t)^{\flat} = (s \cdot t)^{\flat} * (t \cdot s)^{\flat}.$$

Choice '+' becomes symmetric,

$$(s+t){:}F \leftrightarrow (t+s){:}F,$$

but $(s+t){:}F \rightarrow s{:}F \vee t{:}F$ generally does not hold.

4 Justification and Epistemic Semantics

A formal justification semantics for LP was offered by Mkrtychev in [25]. This helped to establish the decidability of LP ([25]), find complexity bounds ([23,24]), and establish the disjunctive property of LP ([19]).

A Mkrtychev model is a pair $M = (\mathcal{A}, \Vdash)$, where

- $\Vdash$ is the usual truth evaluation of propositional letters;
- $\mathcal{A}$ is an *admissible evidence* predicate $\mathcal{A}(t, F)$ defined on pairs *(term, formula)*. The intuition behind $\mathcal{A}$ is that $\mathcal{A}(t, F)$ means

$$t \text{ is an admissible evidence for } F.$$

The admissible evidence predicate respects operations on proofs, i.e., $\mathcal{A}$ satisfies the natural closure conditions copied from the axioms of LP:

Application: $\mathcal{A}(s, F \rightarrow G)$ and $\mathcal{A}(t, F)$ implies $\mathcal{A}(s{\cdot}t, G)$;
Proof Checker: $\mathcal{A}(t, F)$ implies $\mathcal{A}(!t, t{:}F)$;
Choice: $\mathcal{A}(s, F)$ or $\mathcal{A}(t, F)$ implies $\mathcal{A}(s+t, F)$.

Given a model, the truth relation ⊪ is extended to all formulas by stipulating that

- ⊪ respects classical Boolean connectives;
- ⊪ t:F iff '⊪ F and $\mathcal{A}(t, F)$.'

For a given constant specification CS, a model $M = (\mathcal{A}, \Vdash)$ is a CS-model iff $\mathcal{A}(c, B)$ holds for any c:$B \in CS$. It is an easy exercise to check, by induction on derivations in LP_{CS}, the soundness of LP with respect to Mkrtychev semantics:

If $\mathsf{LP}_{CS} \vdash F$, *then* F *holds in each* CS-model.

In particular, all formulas from CS are true in all CS-models.

We present here a different (from that shown in [25]) proof of the completeness theorem for LP with respect to Mkrtychev models by the standard maximal completeness sets construction. Let W be the collection of all maximal consistent sets over LP_{CS}. For each $\Gamma \in W$, we define the truth relation ⊪$_\Gamma$ on propositional letters as

$$\Vdash_\Gamma p \quad iff \quad p \in \Gamma$$

and the admissible evidence predicate as

$$\mathcal{A}_\Gamma(t, F) \quad iff \quad t{:}F \in \Gamma .$$

The aforementioned closure conditions on $\mathcal{A}_\Gamma$ are obviously met. Let us check *Choice*. Suppose $\mathcal{A}_\Gamma(s, F)$ holds. Then s:$F \in \Gamma$. Since s:$F \to (s{+}t)$:$F \in \Gamma$ (as an LP-axiom) and Γ is maximal consistent, $(s{+}t)$:$F \in \Gamma$, too. Hence $\mathcal{A}_\Gamma(s{+}t, F)$. Moreover, c:$B \in \Gamma$ for all c:B from CS, by maximality and consistency of Γ. Therefore, for each Γ, $M = (\mathcal{A}_\Gamma, \Vdash_\Gamma)$ is an LP_{CS}-model.

The next step is to establish the Truth Lemma: *for each formula* F,

$$\Vdash_\Gamma F \quad iff \quad F \in \Gamma.$$

Induction on F. The base case is given by the definition of the model. The Boolean cases are straightforward. Consider the case when F is t:G. If t:$G \in \Gamma$, then $G \in \Gamma$ as well, since t:$G \to G \in \Gamma$ and Γ is deductively closed. By the induction hypothesis, ⊪$_\Gamma G$. Moreover, $\mathcal{A}_\Gamma(t, G)$ also holds, by the definition of $\mathcal{A}_\Gamma$. Hence ⊪$_\Gamma t$:G.

Now let t:$G \notin \Gamma$. Then $\mathcal{A}_\Gamma(t, G)$ does not hold, by the definition. Hence ⊮$_\Gamma t$:G.

To finish the completeness theorem, it suffices to note that if $\mathsf{LP}_{CS} \nvdash F$, then $\{\neg F\}$ is a consistent set. By the standard Lindenbaum construction, find its maximal consistent extension Γ. Since $\neg F \in \Gamma$, $F \notin \Gamma$. By the Truth Lemma, for $\mathcal{A}_\Gamma$ and ⊪$_\Gamma$, ⊮$_\Gamma F$.

There are several useful refinements of Mkrtychev semantics known:

1. Exact Evidence Model ([25]): *For every model* $M = (\mathcal{A}, \Vdash)$, *there is a model* $M' = (\mathcal{A}', \Vdash')$ *such that*
   - M *and* M' *are equivalent, i.e., for each* F, ⊪F *iff* ⊪$'F$;
   - $\mathcal{A}'$ *is exact, i.e.,* $\mathcal{A}'(t, F)$ *iff* ⊪$'t$:F.

2. Minimal Model ([19]): *For each constant specification CS, there is an exact evidence model $M = (\mathcal{A}, \Vdash)$ such that for each $t{:}F$, $M \Vdash t{:}F$ iff $\mathsf{LP}_{CS} \vdash t{:}F$.*

The minimal model theorem yields the Disjunctive Property ([19]):

$$\mathsf{LP}_{CS} \vdash s{:}F \vee t{:}G \text{ iff } (\mathsf{LP}_{CS} \vdash s{:}F \text{ or } \mathsf{LP}_{CS} \vdash t{:}G).$$

An epistemic Kripke-style semantics for LP was offered by Fitting [14,13]. A Fitting model may be regarded as a Kripke model, each node of which is a Mkrtychev model with a monotone admissible evidence function:

$$\text{if } uRv \text{ then } \mathcal{A}_u \subseteq \mathcal{A}_v.$$

The new condition which specifies the truth of proof assertions at a given node is as follows

$$u \Vdash t{:}F \quad \text{iff} \quad \mathcal{A}_u(t, F) \text{ holds and } v \Vdash F \text{ for every } v \text{ with } uRv.$$

Proper modifications of Fitting semantics can accommodate multiple modalities and proof assertions and are playing a key role connecting the Logic of Proofs with epistemic modal logics ([4,5,8,9]).

5 Choice Function '+' in LP

The operation '+' which is called *Choice, Union, Sum,* or *Plus* indeed performs something which can be described as *choice*. The behavior of '+' is governed by the logical principle

$$s{:}F \vee t{:}F \rightarrow (s+t){:}F,$$

which states that '+' takes two proofs s and t, at least one of which is indeed a proof of F, and produces the output $s + t$, which is a proof of F. 'Under the hood,' this operation chooses a proof of F between s and t.

The following theorem was established in [19]; it showed that the Choice operation in LP is weakly symmetric and weakly equivalent to a disjunction.

Theorem 2. [19] *For any constant specification CS, the following are equivalent:*

1. $\mathsf{LP}_{CS} \vdash (s+t){:}F$
2. $\mathsf{LP}_{CS} \vdash (t+s){:}F$
3. $\mathsf{LP}_{CS} \vdash s{:}F \vee t{:}F.$

Theorem 3 below shows that none of these properties hold in LP in a strong sense, i.e., internally.

Theorem 3. *Let x, y be proof variables and P a propositional letter. Then*

1. $\mathsf{LP} \nvdash (x+y){:}P \rightarrow (y+x){:}P$
2. $\mathsf{LP} \nvdash (x+y){:}P \rightarrow x{:}P \vee y{:}P$

3. $\mathsf{LP} \nvdash x(y+z){:}P \to (xy+xz){:}P$
4. $\mathsf{LP} \nvdash (y+z)x{:}P \to (yx+zx){:}P.$

Proof. In principle, these statements could be proven by proper use of the arithmetical counterexamples. However, since not all details about the provability semantics of LP were given here, we present a proof based on Mrktychev models.

To establish (1), consider a Mkrtychev model $M = (\mathcal{A}, \Vdash)$ where $\Vdash P$ and $\mathcal{A}(t, F)$ holds iff t is different from x, y, and $y+x$. To verify that M is a legitimate model, it suffices to check the closure properties of $\mathcal{A}$. The only relevant case is *Choice* in a configuration when $\mathcal{A}(y+x, G)$ does not hold. In this situation, neither $\mathcal{A}(y, G)$ nor $\mathcal{A}(x, G)$ holds, which does not constitute a violation of the closure property of $\mathcal{A}$. It is easy to see that in this model, $\Vdash (x+y){:}P$ since both $\mathcal{A}(x+y, P)$ and $\Vdash P$ hold. On the other hand, $\nVdash (y+x){:}P$ since $\mathcal{A}(y+x, P)$ does not hold. Overall,

$$\nVdash (x+y){:}P \to (y+x){:}P.$$

Item (2) immediately follows from (1) since $\mathsf{LP} \vdash x{:}P \vee y{:}P \to (y+x){:}P.$

Item (3). Similar to 1. Consider a Mkrtychev model $M = (\mathcal{A}, \Vdash)$ where $\Vdash P$ and $\mathcal{A}(t, F)$ holds iff t is different from any subterm of $xy+xz$, i.e., $t \notin \{xy+xz, xy, xz, x, y, z\}$. The closure property holds since in all applicable clauses when $\mathcal{A}$ can be false in the conclusion, the assumptions are also false, by the definition of $\mathcal{A}$. In this model, $x(y+z){:}P$ is true since $x(y+z)$ is not a subterm of $xy+xz$, but $(xy+xz){:}P$ is false.

Item(4) can be treated similarly to (3).   □

6   Symmetric Logic of Proofs

As we have already discussed, the Logic of Proofs LP has two prominent features which determine its foundational significance: it has a natural provability semantics[6], and it suffices for realizing S4, hence IPC, thus making LP a kind of Brouwer-Heyting-Kolmogorov semantics for intuitionistic logic.

In this chapter, we introduce the *Symmetric Logic of Proofs*, SLP, extending LP itself by postulating a property borrowed from the symmetric arithmetical interpretation.

SLP has all the postulates of LP and one additional principle:

Symmetry Principle:

$$t(u+v){:}F \leftrightarrow tu{:}F \vee tv{:}F, \tag{3}$$

$$(u+v)t{:}F \leftrightarrow ut{:}F \vee vt{:}F. \tag{4}$$

---

[6] Moreover, LP provides a complete axiomatization of the class of all multi-conclusion proof predicates.

Notational conventions. We assume that this principle also covers the case when there is no multiplication by t at all, i.e.

$$(u+v){:}F \leftrightarrow u{:}F \vee v{:}F. \tag{5}$$

With this convention, the Symmetry Principle subsumes the Choice Axiom of LP and hence may be considered a generalization of the latter.

Clearly,

$$\mathsf{LP} \subset \mathsf{SLP}.$$

Theorem 4. SLP *is sound with respect to the symmetric provability interpretation.*

Proof. We have to establish soundness of the Symmetry Principle.

Let us start with $tu{:}F \vee tv{:}F \to t(u+v){:}F$. Fix the (symmetric arithmetical) interpretation t', u', v', F' of s, u, v, and F respectively and suppose that $(tu{:}F)'$ holds. By the definition of the symmetric arithmetical interpretation, at least one of the following cases holds:

1. $F' \in t'$;
2. $F' \in u'$;
3. there is X such that $X \to F' \in t'$ and $X \in u'$.

In cases (1) and (2), $F' \in [t(u+v)]'$, by the definition of the symmetric arithmetical interpretation. In case (3), $X \in u'{*}v'$, hence $X \in (u+v)'$, hence $F' \in [t(u+v)]'$ as well. The case when $[tv{:}F]'$ holds is symmetric.

Let us examine $t(u+v){:}F \to tu{:}F \vee tv{:}F$. Suppose $F' \in [t(u+v)]'$. By the definition of the symmetric arithmetical interpretation, at least one of the following cases holds:

1. $F' \in t'$;
2. $F' \in (u+v)'$, i.e., $F' \in u' {*} v'$;
3. there is X such that $X \to F' \in t'$ and $X \in u' {*} v'$.

In case (1), both $(tu{:}F)'$ and $(tv{:}F)'$ hold. In case (2), either $F' \in u'$, and then $(tu{:}F)'$, or $F' \in v'$, and then $(tv{:}F)'$. In case (3), either $X \in u'$, hence $(tu{:}F)'$; or $X \in v'$, hence $(tv{:}F)'$. In any case, $(tu{:}F \vee tv{:}F)'$ holds.

The cases $ut{:}F \vee vt{:}F \leftrightarrow (u+v)t{:}F$ and $u{:}F \vee v{:}F \leftrightarrow (u+v){:}F$ are treated similarly. $\qquad\square$

Theorem 5. SLP *is closed under substitution and enjoys the Internalization Property.*

Proof. Trivial from the definition of SLP and the fact that no new rules of inference were added to SLP as compared to LP. $\qquad\square$

Theorem 6. SLP *enjoys the Realization Theorem with respect to* S4.

Proof. Indeed, suppose F is derivable in S4. By the Realization Theorem for LP (Theorem 1), there is a realization of F, F^r by proof polynomials in the basis $\{\cdot, +, !\}$ which is derivable in LP. Since SLP extends LP, $\mathsf{SLP} \vdash F^r$. $\qquad\square$

Theorem 7. S4 *is the forgetful projection of* SLP.

Proof. It suffices to note that the Symmetry Principle has forgetful projections of the sort $\Box X \leftrightarrow \Box X \vee \Box X$, trivially provable in S4. By Theorem 6, every theorem of S4 is a forgetful projection of some SLP-theorem. $\Box$

These results, along with the provability semantics for SLP, show that the latter gives a Brouwer-Heyting-Kolmogorov-style semantics for S4 and IPC as well.

$$\text{IPC} \;\hookrightarrow\; \text{S4} \;\hookrightarrow\; \text{SLP} \;\hookrightarrow\; \textit{Gödelian proof predicates.}$$

Theorem 8. *Let CS be a constant specification. Let $s \sim t$ mean that for any formula F, $s{:}F \leftrightarrow t{:}F$ is provable in* SLP *with this CS. Then the following holds:*

1. $s + t \;\sim\; t + s$ *(commutativity of Choice);*
2. $s + (t + u) \;\sim\; (s + t) + u$ *(associativity of Choice);*
3. $s + s \;\sim\; s$ *(idempotency of Choice);*
4. $t(u + v) \;\sim\; tu + tv$ *(left distributivity);*
5. $(u + v)t \;\sim\; ut + vt$ *(right distributivity).*

Proof. All the derivations below are in SLP.

1. $(s+t){:}F \;\to\; s{:}F \vee t{:}F \;\to\; t{:}F \vee s{:}F \;\to\; (t+s){:}F$
2. $[s+(t+u)]{:}F \;\leftrightarrow\; s{:}F \vee t{:}F \vee u{:}F \;\leftrightarrow\; [(s+t)+u]{:}F$
3. $(s+s){:}F \;\leftrightarrow\; s{:}F \vee s{:}F \;\leftrightarrow\; s{:}F$
4. $t(u+v){:}F \;\leftrightarrow\; tu{:}F \vee tv{:}F \;\leftrightarrow\; tu + tv{:}F$
5. $(u+v)t{:}F \;\leftrightarrow\; ut{:}F \vee vt{:}F \;\leftrightarrow\; ut + vt{:}F$ $\Box$

Mkrtychev models for SLP are the usual Mkrtychev LP-models with admissible evidence predicates which respect the Symmetry Principle:

$$\mathcal{A}(t(u+v), F) \quad \textit{iff} \quad \text{`}\mathcal{A}(tu, F) \textit{ or } \mathcal{A}(tv, F)\text{'} \tag{6}$$

$$\mathcal{A}((u+v)t, F) \quad \textit{iff} \quad \text{`}\mathcal{A}(ut, F) \textit{ or } \mathcal{A}(vt, F)\text{'} \tag{7}$$

$$\mathcal{A}(u+v, F) \quad \textit{iff} \quad \text{`}\mathcal{A}(u, F) \textit{ or } \mathcal{A}(v, F)\text{.'} \tag{8}$$

Theorem 9. *For each constant specification CS,* SLP *is sound and complete for* SLP *Mkrtychev models.*

Proof. Soundness of SLP is straightforward. We have only to check that Symmetry holds in each SLP-model. Let us consider (3). Let $\Vdash t(u+v){:}F$. Then $\Vdash F$ and $\mathcal{A}(t(u+v), F)$. By (6), $\mathcal{A}(tu, F)$ holds or $\mathcal{A}(tv, F)$ holds, hence $\Vdash tu{:}F$ or $\Vdash tv{:}F$. In either case, $\Vdash tu{:}F \vee tv{:}F$. The remaining clauses for soundness are checked in the same manner.

Completeness can be established by a maximal consistent set construction as in Sect. 4. One need only check that the canonical model (the set of maximal consistent sets with $\mathcal{A}$ and $\Vdash$ as in Sect. 4) is indeed an SLP-model. For this, is suffices to check that for each maximal consistent set Γ, conditions (6), (7), and (8) hold. Let us check (6).

Suppose $\mathcal{A}(t(u+v), F)$ holds for this Γ. This means that $t(u+v){:}F \in \Gamma$. Since $t(u+v){:}F \to tu{:}F \vee tv{:}F \in \Gamma$, by maximality of Γ, either $tu{:}F \in \Gamma$ or $tv{:}F \in \Gamma$, hence '$\mathcal{A}(tu, F)$ or $\mathcal{A}(tv, F)$' holds for this Γ.

Now let '$\mathcal{A}(tu, F)$ or $\mathcal{A}(tv, F)$' hold in Γ. Then either $tu{:}F \in \Gamma$ or $tv{:}F \in \Gamma$. By the Symmetry Principle (3), since Γ is deductively closed, $t(u+v){:}F \in \Gamma$, which yields that $\mathcal{A}(t(u+v), F)$ holds for this Γ.

The remaining clauses are checked similarly.   □

Fitting models for SLP are obtained from those for LP by adding conditions (6), (7), and (8), respectively. The following theorem can be easily established along the lines of Theorem 9:

Theorem 10. *For each constant specification CS, SLP is sound and complete for SLP Fitting models.*

7   Discussion

Note that Application in SLP is neither associative nor commutative since neither of these properties hold for the symmetric arithmetical interpretation (Sect. 3).

A natural attempt to add more ring structure to SLP by introducing a constant 0 for the empty derivation does not work well with the symmetric provability interpretation. In particular, $t + 0 = t$, but $t \cdot 0$ is t rather than 0 here[7]. This could be fixed to $t \cdot 0 = 0$ by bending the provability semantics a little. However, under any modification of semantics, the identity $\neg 0{:}\top$ holds, which spoils the connection to modal logic. Indeed, the forgetful projection of the latter formula is $\neg\Box\top$, which is false in any normal modal logic. This observation alone should not discourage us from considering the addition of 0 to a version of SLP, though. The language of proof polynomials should then be extended by a special constant 0 which cannot be used as 'c' in the Constant Specification Rule (Sect. 2). A new axiom schema $\neg 0{:}F$ should also be added.

Note that the relation '$\sim$' from Theorem 8 is an equivalence relation on proof polynomials. However, '$\sim$' is not a congruence, e.g., $s \sim t$ does not generally yield $!s \sim !t$. For example, $x \sim x + x$ (Theorem 8.3), but not $!x \sim !(x + x)$. Indeed, in the symmetric provability interpretation, $!x$ should contain proofs of $x{:}F$ for all $F \in x$, whereas $!(x + x)$ contains proofs of $(x + x){:}F$ for all such F's. This observation still leaves an opportunity for '$\sim$' to be a congruence in some !-free variant of SLP. It is easy to check that in SLP, $s \sim t$ and $u \sim v$ yield $s + u \sim t + v$. It is not known whether the same holds for '$\cdot$', i.e., $su \sim tv$ as well. Such a rule for '$\cdot$' is sound for the standard symmetric provability interpretation, which suggests that a proper version of this rule either holds in, or can be safely added to, SLP. Answers to these questions could lead to an adequate notion of the *equality of proofs* in Justification Logic in general. We leave this, however, for future studies.

---

[7] This observation is due to V.N. Krupski.

Other natural steps in this direction would be to clarify the questions of decidability and complexity for SLP, to check the disjunctive property, and to describe adequate Gentzen and tableaux proof systems.

We conjecture that SLP is arithmetically complete with respect to the class of proof predicates for which the Symmetry Principle holds. An intriguing question remains open about the logic of proofs for the standard symmetric provability interpretation.

Acknowledgements

This paper was inspired by discussions with Joan and Yiannis Moschovakis at the 6th Panhellenic Logic Symposium, Volos, Greece, 5-8 July 2007. The author is very grateful to Mel Fitting, Evan Goris, Vladimir Krupski, Roman Kuznets, and Elena Nogina whose advice helped with this paper. Many thanks to Karen Kletter for editing this text.

References

1. Artemov, S.: Operational modal logic. Technical Report MSI 95-29, Cornell University (1995)
2. Artemov, S.: Explicit provability and constructive semantics. Bulletin of Symbolic Logic 7(1), 1–36 (2001)
3. Artemov, S.: Operations on proofs that can be specified by means of modal logic. In: Advances in Modal Logic., vol. 2, pp. 59–72. CSLI Publications, Stanford University (2001)
4. Artemov, S.: Evidence-based common knowledge. Technical Report TR-2004018, CUNY Ph.D. Program in Computer Science (2005)
5. Artemov, S.: Justified common knowledge. Theoretical Computer Science 357(1–3), 4–22 (2006)
6. Artemov, S.: On two models of provability. In: Gabbay, M.Z.D.M., Goncharov, S.S. (eds.) Mathematical Problems from Applied Logic II, pp. 1–52. Springer, New York (2007)
7. Artemov, S., Beklemishev, L.: Provability logic. In: Gabbay, D., Guenthner, F. (eds.) Handbook of Philosophical Logic, 2nd edn., vol. 13, pp. 229–403. Kluwer, Dordrecht (2004)
8. Artemov, S., Nogina, E.: Introducing justification into epistemic logic. Journal of Logic and Computation 15(6), 1059–1073 (2005)
9. Artemov, S., Nogina, E.: On epistemic logic with justification. In: van der Meyden, R. (ed.) Theoretical Aspects of Rationality and Knowledge. Proceedings of the Tenth Conference (TARK 2005), Singapore, June 10–12, 2005, pp. 279–294. National University of Singapore (2005)
10. Artemov, S., Strassen, T.: Functionality in the basic logic of proofs. Technical Report IAM 93-004, Department of Computer Science, University of Bern, Switzerland (1993)
11. Boolos, G.: The Logic of Provability. Cambridge University Press, Cambridge (1993)
12. Brezhnev, V., Kuznets, R.: Making knowledge explicit: How hard it is. Theoretical Computer Science 357(1–3), 23–34 (2006)

13. Fitting, M.: A semantics for the logic of proofs. Technical Report TR-2003012, CUNY Ph.D. Program in Computer Science (2003)
14. Fitting, M.: The logic of proofs, semantically. Annals of Pure and Applied Logic 132(1), 1–25 (2005)
15. Gödel, K.: Eine Interpretation des intuitionistischen Aussagenkalkuls. Ergebnisse Math. Kolloq. 4, 39–40 (1933) English translation In: Feferman, S. et al. (eds.) Kurt Gödel Collected Works, vol. 1, pp 301–303. Oxford University Press, Oxford, Clarendon Press, New York (1986)
16. Gödel, K.: Vortrag bei Zilsel. In: Feferman, S. (ed.) Kurt Gödel Collected Works, vol. III, pp. 86–113. Oxford University Press, Oxford (1995)
17. Goris, E.: Logic of proofs for bounded arithmetic. In: Grigoriev, D., Harrison, J., Hirsch, E.A. (eds.) CSR 2006. LNCS, vol. 3967, pp. 191–201. Springer, Heidelberg (2006)
18. Kleene, S.: Introduction to Metamathematics. Van Norstrand (1952)
19. Krupski, N.V.: On the complexity of the reflected logic of proofs. Theoretical Computer Science 357(1), 136–142 (2006)
20. Krupski, V.N.: Operational logic of proofs with functionality condition on proof predicate. In: Adian, S., Nerode, A. (eds.) LFCS 1997. LNCS, vol. 1234, pp. 167–177. Springer, Heidelberg (1997)
21. Krupski, V.N.: The single-conclusion proof logic and inference rules specification. Annals of Pure and Applied Logic 113(1–3), 181–206 (2001)
22. Krupski, V.N.: Referential logic of proofs. Theoretical Computer Science 357(1), 143–166 (2006)
23. Kuznets, R.: On the complexity of explicit modal logics. In: Clote, P.G., Schwichtenberg, H. (eds.) CSL 2000. LNCS, vol. 1862, pp. 371–383. Springer, Heidelberg (2000)
24. Milnikel, R.: Derivability in certain subsystems of the Logic of Proofs is Π_2^p-complete. Annals of Pure and Applied Logic 145(3), 223–239 (2007)
25. Mkrtychev, A.: Models for the logic of proofs. In: Adian, S., Nerode, A. (eds.) LFCS 1997. LNCS, vol. 1234, pp. 266–275. Springer, Heidelberg (1997)
26. Smoryński, C.: The incompleteness theorems. In: Barwise, J. (ed.) Handbook of Mathematical Logic, pp. 821–865. North Holland, Amsterdam (1977)
27. Solovay, R.M.: Provability interpretations of modal logic. Israel Journal of Mathematics 28, 33–71 (1976)

Synthesis of Monitors for
Real-Time Analysis of Reactive Systems

Mikhail Auguston[1] and Mark Trakhtenbrot[2]

[1] Naval Postgraduate School, Monterey, CA, USA
[2] Holon Institute of Technology, Holon, Israel
auguston@cs.nps.navy.mil,markt@hit.ac.il

*To Boaz – with love and respect, and with thanks for the great privilege to
learn so much from the "first hands". – M.T.*

Abstract. In model-driven development of reactive systems, statecharts are
widely used for formal description of their behavior, providing a sound basis for
verification, testing and code generation. The paper presents an approach for
dynamic analysis of reactive systems via run-time monitoring of code generated
from statechart-based models. The core of the approach is automatic creation
of monitoring statecharts from formulas that specify system's behavioral
properties (including real-time constraints) in a proposed assertion language.
Such monitors are then translated into code together with the system model, and
executed concurrently with the system code. The approach allows for a realistic
analysis of reactive systems (and in particular of their real-time aspects), as
monitoring is supported in system's actual operating environment. This
especially relates to design-oriented models that include mapping of abstract
model elements into those of the underlying operating system. This way, the
natural restrictions inherent to simulation and verification are overcome.

1 Introduction

Development of reliable reactive systems is a significant challenge, especially due to
their complex behavior. There has been a great deal of research on the development of
formal methods for specification, design, analysis and verification of reactive
systems.

For precise specification of system behavioral properties, various types of temporal
logic are widely used. These include linear temporal logic [17], which offers special
temporal operators for reasoning about past and future properties of behavioral
sequences, and MTL [6], which supports expression of real-time constraints through
definition of durations for future temporal operators. Some specification formalisms
suggest various kinds of syntax sugar that make the specification task more user friendly
for designers who are not logicians. For example, with the LA language in [21], temporal
properties look as a combination of stylized English with C-like expressions.

In [3], the temporal logic details are hidden "behind the scenes", and instead, patterns
are used that allow to specify common properties (such as existence, absence, response,
precedence, etc.) and scope in which the property should hold. This approach is used, for

A. Avron et al. (Eds.): Trakhtenbrot/Festschrift, LNCS 4800, pp. 72–86, 2007.

example, in the Statemate ModelCertifier verification tool [20] that offers a rich library of pre-defined property patterns, where each pattern looks as a parameterized natural language sentence. Paper [7] introduces a language for pattern definition as a way to create extendable sets of property patterns. Sugar [4], that evolved from the temporal logic CTL, provides several layers for property specification and verification; in particular, extended regular expressions are used to describe execution sequences on which temporal properties are checked.

On the other hand, model-based system development has become the way to design, implement and validate reactive systems. Statecharts, first introduced in [11], have become a standard for behavior design in popular model-based methodologies such as structured and object-oriented design [8]. Various tools (e.g., Statemate [13], [16]), Rhapsody [10], BetterState [23]) support the creation of executable models using statecharts, and their analysis through simulation, execution of automatically generated code, and, in Statemate, verification. Ongoing research on model-based testing covers, among other issues, test generation from statechart models [5].

A powerful method of dynamic analysis is run-time monitoring of system execution. While monitoring relates to testing (where a system is examined on selected test scenarios), it is common to consider it as run-time verification whereas each test execution is checked against a formally defined system property. A number of tools have been developed for monitoring various types of programs (including real-time systems); see, for example [1], [2], [21]. The relevant assertion languages allow for expressing a wide range of properties in terms of events that occur in the running code, and for defining tool reactions when a violation is found or when the run was successful. An important problem here is the gap between the system specification, which usually refers to high-level objects, and monitors, which refer to implementation-level events (such as function calls, etc.). Some issues related to derivation of monitors from system specification are considered in [19].

Model-based development leads to narrowing of this gap, as monitoring can be performed on the model (rather than the implementation) level. Statemate [13,16] supports the use of the so-called testbench (watchdog) statechart. Such a chart is not part of the system model; rather, it is used aside of the model to play either the role of a driver (acting as an environment and producing system inputs) or a monitor (watching the system for proper behavior or abnormalities). To perform its role, the testbench is executed in parallel with the model. Testbench charts are used in a variety of Statemate tools for model analysis: simulation, model checking, and with generated code.

Violation of the monitored property can be expressed and observed as entering an error state in the monitor chart. For example, Fig. 1 shows a simple statechart for monitoring of the following property: "Processing of a request must be accomplished within 5 seconds and before receiving the next request".

An important feature of monitor statecharts is that they have access to all elements in the system model. In other words, visibility from the monitor is supported both for observable elements (events, conditions and data items) that belong to system's interface with the environment, and for internal elements such as states or events used for internal communication between system components. This allows for both black-box and more detailed white-box monitoring, and makes localization of design problems easier.

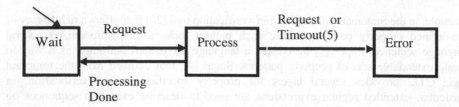

Fig. 1. A simple monitor chart

2 Dynamic Analysis with Synthesized Monitors

This paper presents an approach to dynamic analysis of reactive systems modeled with statecharts.

The analysis is based on run-time monitoring of code generated from the system model. The code is checked against the system specification describing the required and forbidden behaviors; these are expressed in a proposed assertion language described below in Section 5. The main idea underlying this approach is the automatic synthesis of monitors directly from system specification. This is achieved through translation of the specification into an equivalent testbench statechart(s). This step is followed by generating C code from the system model and from the created monitor, and by their simultaneous execution. Appropriate diagnostics is produced during the execution and/or upon its completion.

An important advantage of our approach is that it allows for a realistic analysis of reactive systems, as monitoring is supported in system's actual operating environment. In particular, analysis of *real-time* aspects in system behavior becomes feasible, as opposed to usual analysis tools that deal with *simulated time* (this is detailed in the next section).

Note that even though such realistic analysis of generated code is the primary goal of this work, having the intermediate stage at which monitor charts are synthesized (instead of direct translation of assertions into code) has an added value. Namely, the obtained charts can be also used by both Statemate simulation and model checker as monitor testbenches.

3 Underlying Semantics

Check of real-time constraints is of special interest in run-time monitoring. For systems derived from statechart-based models, it directly relates to the underlying time model.

The monitor synthesis approach suggested in this paper adopts an operational semantics of statecharts, as described in [14]. This semantics is the basis for simulation, code generation and model checking tools in the Statemate working environment for modeling and analysis of reactive systems [13].

Over years, a number of different versions for statechart semantics were suggested, starting from [15,18] and including more recent [12] (for UML statecharts). The main focus of these papers is on "what is in a step", i.e. how to accurately define system's basic reaction to external stimuli.

Statemate semantics [14] was the first to address also the time-related constructs in statecharts: timeouts and scheduled actions. It considers two time models: synchronous (for clock-driven systems, like pure hardware) and asynchronous (for event-driven systems, like typical software).

In the synchronous model, duration of all steps is assumed to be the same, regardless of how "heavy" the executed actions are. In fact, this model doesn't distinguish between concepts of step and time: at each clock cycle, a step is performed.

As for the asynchronous model, it clearly makes such a distinction: steps are considered to take zero time, and the system executes a chain of steps until stabilization (a super-step); only then the clock is advanced and new inputs are accepted.

The above time models are based on the assumption that the system is fast enough to complete its reactions to external stimuli before the next stimulus arrives.

4   Real-Time Monitoring vs. Other Analysis Methods

Existing analysis tools (such as the Statemate model checker called ModelCertifier [20]) follow this abstraction, and simulate the time passage accordingly.

However, in reality every single action takes some real time. Hence, when verification of a system model (based on simulated time) concludes that the system fulfils its timing constraints, this may turn out to be wrong in the real world.

There are also other limitations in addressing time-related properties by tools such as ModelCertifier. As noticed in [20], "choosing time model has a strong impact on complexity and expressiveness for the further verification process". Analysis with synchronous time model is easier, while "there are semantic aspects coming with the asynchronous model which are sometimes very difficult for the verification process".

We consider the run-time monitoring of code generated from the system model as a way to overcome the various limitations of the verification process:

(1) Generated code for the system and its monitor are executed in real time. Even though such code is often considered as having only prototype quality, it is fast enough and allows for meaningful checks of time constraints. In particular, this way it is possible to check whether the system indeed reacts faster than inputs are arriving to it. Such checks are beyond the scope of simulation and model checking tools.

(2) There is no restriction related to the state space of the tested model, and execution of compiled code (for the model and the monitor) is fast. The reason is that every test execution of generated code explores states of one particular scenario. On the other hand, model checking has to explore (in the worst case) the entire state space of the system. Hence it becomes slow (and in fact unfeasible) for large real-world models.

(3) Our approach allows monitoring of code generated from models augmented by attributes that reflect the various implementation oriented decisions. Such attributes are used to define how the various abstract features of the model are mapped into specific elements of the target real-time operating system. They may describe division of the system into tasks of various types (periodic, etc.), mapping of model elements into events of the target RTOS, etc. For example, code generator described in [22]

automatically translates statechart models augmented with design attributes into a highly optimized production quality code for the OSEK operating system widely used in the automotive industry for embedded microcontroller development. With our approach, code generated with this and similar tools can be executed and monitored in its realistic hardware-in-the-loop operating environment. This kind of analysis is impossible with model checking.

(4) Model checking requires that all data be properly restricted, to guarantee that a finite state model is analyzed. This requirement is problematic for input data, if there is not enough information about the system environment. No such restrictions are relevant for monitoring, and moreover, monitored code derived from the system model can be connected to real sources of input data.

5 Assertion Language

To specify and monitor real-time properties of reactive systems, we use an assertion language that integrates a number of powerful features found in temporal logic and in the FORMAN language (introduced in [1], [2], and used in a number of tools).

(1) *Boolean expressions* can refer to any element in the system model, and express properties of system configurations. For example: *in(S) and (x>5)* means that currently the system is in state *S* and *x* is greater than 5.

(2) *Regular expressions* allow for description of state and event sequences. Consider for example, the expression:

Open (Read | Write)* Close

It describes executions in which all *Read* and *Write* operations are executed, in any order, strictly after *Open* but before *Close*.

(3) *Temporal formulas* express order properties fulfilled by system execution sequences. They are built using unrestricted future temporal operators *NEXT, ALWAYS, EVENTUALLY, UNTIL* and their past counterparts: *PREVIOUS, ALWAYS_WAS, SOMETIME_WAS, SINCE*. Following [17], we consider formulas for the following types of properties:

Safety: *ALWAYS (P)*
Guarantee: *EVENTUALLY (P)*
Obligation: Boolean combination of safety and guarantee
Response: *ALWAYS (EVENTUALLY (P))*
Persistence: *EVENTUALLY (ALWAYS (P))*
Reactivity. Boolean combination of response and persistence.

According to [17], any temporal formula is equivalent to a reactivity formula; the other five types of formulas are allowed for more flexibility.

(4) *Real-time constraints* are expressed with a version of the above operators obtained by attaching appropriate time characteristics. For example, *ALWAYS (10) P* means that *P* is continuously true in the 10 time units interval that starts at the current moment, while *SOMETIME_WAS (10) P* denotes that *P* was true at least once during

the last 10 time units. With this extension, *P* in the above formulas is now allowed to be a restricted (future or past) formula. Note that we don't allow an unrestricted temporal operator to be nested within a restricted one.

(5) *Actions* define what should be done when a property violation is found, or when the property holds for the checked run. Typically, this includes sending an appropriate message. In general, any user-defined function can be used here to provide a meaningful report that may include, for example, interesting statistics and other profiling information (frequency of occurrence for a certain event, total time spent by the system in a certain state, etc.). For this, actions can use the appropriate attributes of the referred objects (e.g., the time at which a certain interval was entered).

The examples in Section 6 illustrate the use of this assertion language. Since the language is based on constructs described elsewhere (see [1], [16] and [17]), detailed description of its syntax and semantics is omitted here.

Nevertheless, there is a delicate semantic issue that should be mentioned here. System specification usually expresses properties of infinite execution sequences (as a reactive system performs an ongoing interaction with its environment). Correspondingly, the traditional semantics of temporal operators is also defined for infinite execution sequences. However, monitoring usually deals with finite (truncated) runs, and this requires a proper definition of the semantics for cases when there is a doubt as to what would have been the property formula value if the execution had not been stopped. Paper [9] studies several ways of reasoning with temporal logic on truncated executions. We follow the so called neutral view discussed in [9]; this is illustrated by the following example. Consider the assertions:

ALWAYS (*P* → EVENTUALLY (10) *Q*)
ALWAYS (*P* → ALWAYS (10) *Q*)

and suppose that the run is completed (truncated) 4 seconds after the last occurrence of event *P* (we assume that each of the properties held for all earlier occurrences of *P*). If there was no *Q* after the last *P*, then the first assertion is considered to be false for this run (even though continuation of the run could reveal that *Q* does occur in 10 seconds after *P*, as required). On the contrary, if *Q* held continuously after the last *P* and until the end of the run, then the second assertion is considered to be true. In general, it is the user's responsibility to make the on-satisfy and on-failure actions detailed enough, so that he can better understand the monitoring results (e.g. whether a real violation was found, or it is in doubt due to the state at which the execution was truncated).

6 Examples

To illustrate our approach, we consider the Early Warning System (EWS) example from [16]. We present its verbal description followed by the statechart presenting the behavioral design of the system. We then give examples of assertions and, for one of them, show its translation into a monitor statechart according to our translation scheme.

The EWS receives a signal from an external source. When the sensor is connected, the EWS performs signal sampling every 5 seconds; it processes the sampled signal and checks whether the resulting value is within a specified range. If the value is out of range, the system issues a warning message on the operator display. If the operator does not respond to this warning within a given time interval (15 seconds), the system prints a fault message and stops monitoring the signal. The range limits are set by the operator. The system is ready to start monitoring the signal only after the range limits are set. The limits can be redefined after an out-of-range situation has been detected, or after the operator has deliberately stopped the monitoring.

Fig. 2 shows a statechart describing the EWS, similar to the one in [16]. The main part of EWS behavior is detailed in the state *ON*. It contains two *AND*-components that represent the EWS controller (upper component) and the sensor (lower component) acting concurrently. Events *DO_SET_UP, EXECUTE,* and *RESET* represent the commands that can be issued by the operator. Timing requirements are represented by delays that trigger the corresponding transitions. The *AND*-components can communicate; for example, see event *CONNECT_OFF* sent from the controller to the sensor.

Following are four examples of assertions that reflect some of the above requirements for EWS:

1) *ALWAYS (EXECUTE → SOMETIME_WAS (DO_SET_UP))*
Monitoring of signal should be preceded by setting range limits.
2) *ALWAYS (OUT_OF_RANGE →*
        EVENTUALLY (15) (RESET or started(PRINT_ALARM))

This assertion requires that in the out-of-range situation, within 15 seconds either the operator responds or a fault message is printed. In fact, it is too weak as it allows

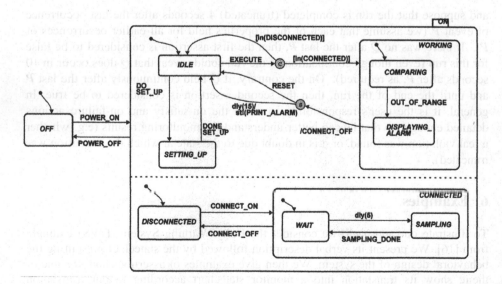

Fig. 2. Statechart for Early Warning System

printing of alarm even when less than 15 time units elapsed without a reset. The next assertion addresses this requirement properly.

3) *ALWAYS (*
   ALWAYS_WAS (15) (in(DISPLAY_ALARM) & not RESET)
   → *started(PRINT_ALARM))*

If during 15 time units the system displays an alarm but there is no operator respond, then a fault message is printed.

4) *ALWAYS (SAMPLING_DONE* →
   ALWAYS (5) in(WIAT) or EVENTUALLY (5) CONNECT_OFF)

After signal sampling is finished, there is a 5-second pause before the next sampling, unless the sensor is disconnected.

Note that the first assertion is violated for the given statechart; this happens in the following scenario: *POWER_ON; CONNECT_ON; EXECUTE.*

As for the other assertions, they are valid as long as the system remains in its *ON* state (i.e., *POWER_OFF* doesn't occur); otherwise they can be violated. Consider, for example the second of the assertions. Fig. 3 shows how it is translated into a monitor statechart (here, as in any other monitor chart, *F* denotes an accepting state and *D* denotes a rejecting state). Now consider a 15-seconds interval that follows occurrence of *OUT_OF_RANGE*, such that:

- there is no *RESET* in this interval
- event *POWER_OFF* occurs 7 seconds after occurrence of *OUT_OF_RANGE*.
- the system remains in state *OFF* for the following 8 seconds.

In this scenario, the assertion is violated (no reset and no message printed in 15 seconds after *OUT_OF_RANGE*), and the monitor enters its rejecting state *D*.

7 Implementation Outline

Statemate boolean expressions obtained from basic predicates (like *in(DISPLAY_ ALARM))*, guarding conditions, and event occurrences are directly visible from monitor statechart; in this sense, their monitoring is trivial. In monitors such expressions can be used just as transition triggers, similar to the example in Fig.1.

In the rest of this section, we present an outline of the translation scheme for restricted and unrestricted temporal formulas allowed by our assertion language (see Section 5 above). Though not fully formalized here, the presentation clearly shows the technique used for synthesis of monitors from assertions.

Let *P, S* be basic boolean formulas that do not contain any temporal operators, and let *Q* denote any formula.

The general idea behind the presented translation patterns is as follows:

- *P* → *Q* means that *P* is used as a trigger to start monitoring of formula *Q* ; for each occurrence of *P*, a new thread of *Q* monitoring is started.

- absence of the trigger *(P→ ...)* means that start of execution is the only trigger event.

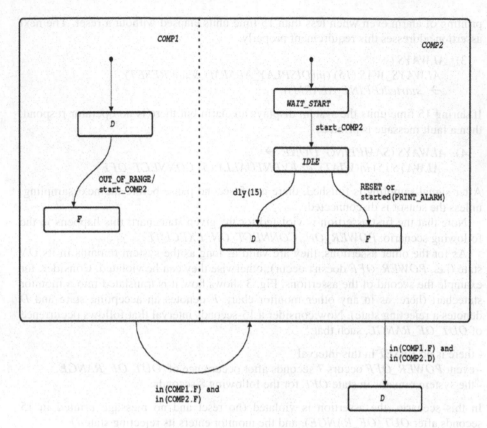

Fig. 3. Monitor chart for the assertion
ALWAYS (OUT_OF_RANGE → EVENTUALLY (15) (RESET or started(PRINT_ALARM)))

7.1 Translation of Restricted Operators

If a formula includes only restricted future temporal operators, like in

$FRM \equiv P \rightarrow TL\_Operator\ (N1)\ TL\_Operator\ (N2)\ ....\ TL\_Operator\ (Nk\ )\ S$

then its value becomes known after (i.e. it needs to be monitored during), at most,
$t(FRM) = N1 + N2 + ... + Nk$ time units from the triggering event P. For example:

$P \rightarrow ALWAYS\ (5)\ EVENTUALLY\ (10)\ S$

is monitored during, at most, 15 time units from the moment when event *P* was
triggered. Namely, for each time point within the 5-seconds interval that follows
occurrence of *P*, a new monitoring thread is started that checks whether S occurs
during the next 10 time units.

As an illustration, Fig. 4 schematically shows the translation pattern for *FRM* ≡
EVENTUALLY (N) P, where *P* itself is either a basic or a restricted future formula.

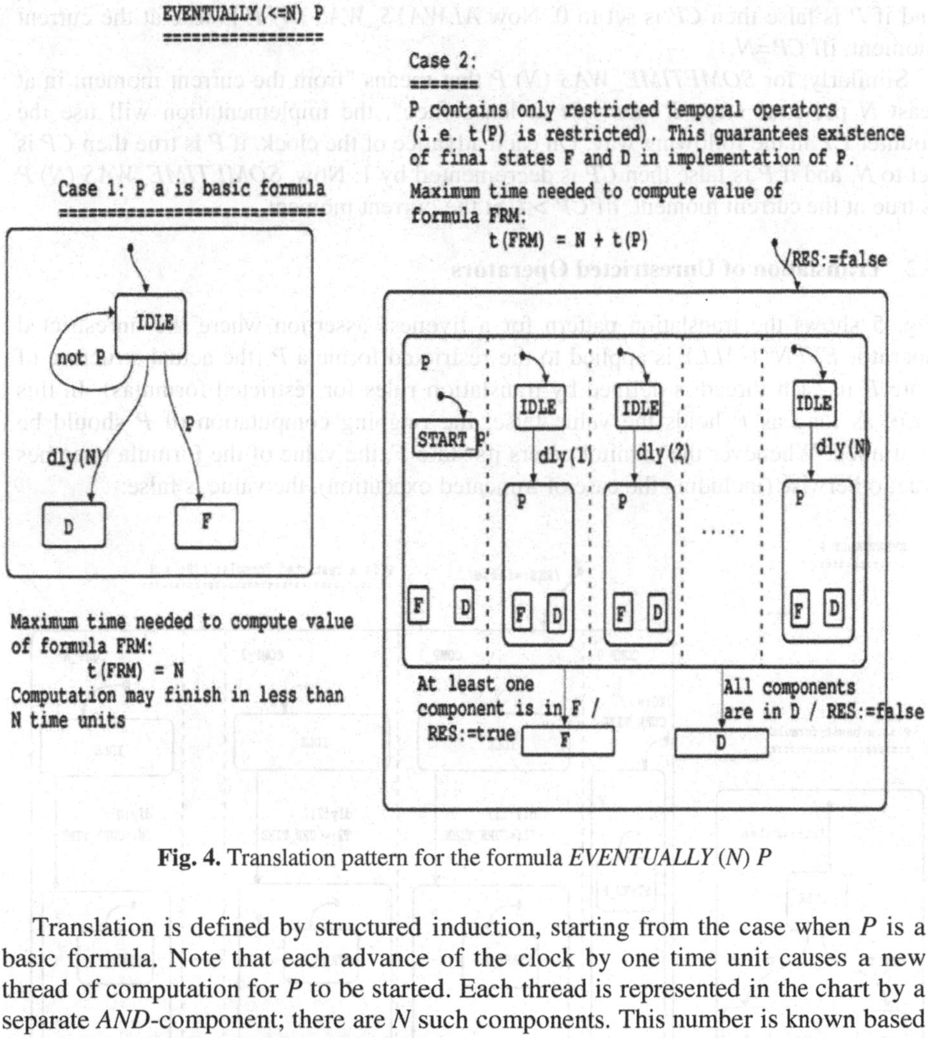

EVENTUALLY(<=N) P
==================

Case 2:
=======
P contains only restricted temporal operators
(i.e. t(P) is restricted). This guarantees existence
of final states F and D in implementation of P.

Case 1: P a is basic formula
=============================

Maximum time needed to compute value of
formula FRM:
$$t(FRM) = N + t(P)$$

Maximum time needed to compute value
of formula FRM:
$$t(FRM) = N$$
Computation may finish in less than
N time units

At least one
component is in F /
RES:=true

All components
are in D / RES:=false

Fig. 4. Translation pattern for the formula EVENTUALLY (N) P

Translation is defined by structured induction, starting from the case when P is a basic formula. Note that each advance of the clock by one time unit causes a new thread of computation for P to be started. Each thread is represented in the chart by a separate AND-component; there are N such components. This number is known based on an analysis of the translated formula.

Note that every restricted future formula is translated into a chart containing two designated states: accepting state F, and rejecting state D; there are no transitions exiting from F and D in such a chart. The value of the formula is true when computation ends in F, and false when it ends in D. If execution of the monitored system is truncated before completion of the formula computation, then (in the spirit of the neutral view as defined in [9]) the value is decided to be true for the ALWAYS-formula and false for the EVENTUALLY-formula.

For restricted past formulas, only a finite segment of the execution should be monitored in order to decide whether the formula is true or false. Consider, for example, ALWAYS_WAS (N) P that means "during N time units preceding the current moment, P was continuously true". The implementation uses a counter CP associated with the formula; on each advance of the clock, if P is true then CP is incremented, and if *P* is false then *CP* is set to 0. Now *ALWAYS_WAS (N) P* is true at the current moment, iff *CP=N*.

Similarly, for *SOMETIME_WAS (N) P* that means "from the current moment in at least *N* previous steps *P* was true at least once", the implementation will use the counter *CP* in the following way: On each advance of the clock, if *P* is true then *CP* is set to *N*, and if *P* is false then *CP* is decremented by 1. Now, *SOMETIME_WAS (N) P* is true at the current moment, iff *CP > 0* at the current moment.

7.2 Translation of Unrestricted Operators

Fig. 5 shows the translation pattern for a liveness assertion where the unrestricted operator *EVENTUALLY* is applied to the restricted formula *P* (the actual structure of state *P* in each thread is defined by translation rules for restricted formulas). In this case, as long as *P* holds the value false, the ongoing computation of *P* should be continued. Whenever the monitor enters its state *F*, the value of the formula becomes true; otherwise (including the case of truncated execution), the value is false.

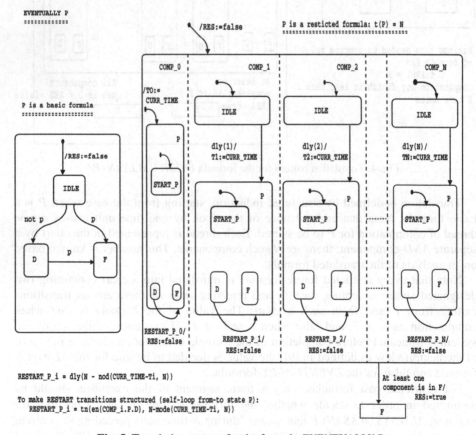

Fig. 5. Translation pattern for the formula *EVENTUALLY P*

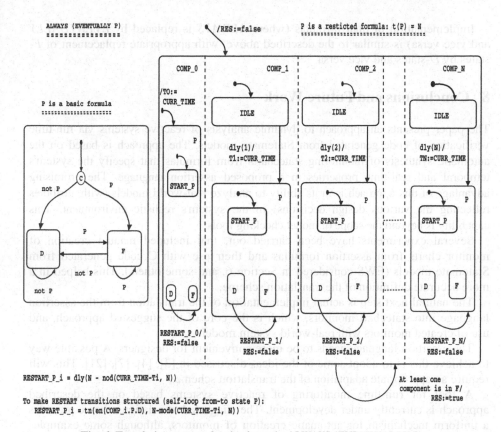

Fig. 6. Translation pattern for the formula *ALWAYS (EVENTUALLY P)*

Note that since obtaining a value of *P* may require up to *t(P)* time units, there are *t(P)* threads computing *P*. When a cycle of *P* computation is completed with the value false (the component reaches its state *D*), it is restarted again.

Also note the delays: *RESTART_P_i* is defined in such a way that with each advance of the clock by one time unit, a new cycle of *P* computation is started. Restarting *P* immediately upon its completion in state *D* would have caused a violation of such synchronization in case that a certain cycle takes less time than *t(P)*. This, in turn, could lead to wrong computation of the entire formula. Note that for proper evaluation of *RESTART_P_i*, each thread "remembers" the time at which it starts computation of *P*. To make the statechart more structured (so that the restarting transition will exit from state *P* itself, and not from its sub-state *D*), a proper modification in definition of event *RESTART_P_i* should be done; see details in Fig.5).

Fig. 6 shows the translation pattern for property *ALWAYS (EVENTUALLY (P))*. Here, computation of *EVENTUALLY (P)* is restarted whenever it gets the value true, i.e., when the chart in Fig. 5 enters state *F* (at the top level of the hierarchy). In other words, such implementation can be obtained by redirecting the transition from *F* back to the *AND*-state.

Implementation of dual formulas (where *ALWAYS* is replaced by *EVENTUALLY* and vice versa) is similar to the described above, with appropriate replacement of *F*-states by *D*-states and vice versa.

8   Conclusions and Future Work

The paper presents an approach to dynamic analysis of reactive systems via run-time verification of code generated from Statemate models. The approach is based on the automatic synthesis of monitoring statecharts from formulas that specify the system's temporal and real-time properties in a proposed assertion language. The promising advantage of this approach is in its ability to analyze real-world models (with attributes reflecting the various design decisions) in the system's realistic environment. This capability is beyond the scope of model checking tools.

Several experiments have been carried out, that included manual creation of monitor charts from assertion formulas and their use with C code generated from Statemate models (EWS considered in Section 6, and some others). This helped in a more accurate definition of the translation scheme.

The natural next step is actual implementation of the translation from the assertion language into statechart monitors, which is the core of the suggested approach, and use of created monitors with real-world system models.

The assertion language needs to be more convenient for designers. A possible way to achieve this is to adopt some of the ideas discussed in [3], [4], [7], [21]. This will require an appropriate adaptation of the translation scheme.

A tool for run-time monitoring of reactive systems based on the described approach is currently under development. The suggested translation scheme provides a uniform mechanism for automatic creation of monitors, although some examples show that, in certain cases, more compact and optimized monitors can be produced. Further research is needed to define a more efficient translation scheme, both for synchronous and asynchronous time models.

Finally, an interesting challenge is to check a similar approach with a UML-based design paradigm that uses an object-oriented version of statecharts for behavior description. Here an additional advantage could be in monitoring of systems where objects are created dynamically such that their amount is not limited in advance (model checking analysis of such systems is clearly problematic).

Acknowledgments

This work has been supported in part by the U.S. Office of Naval Research Grant # N00014-01-1-0746.

References

1. Auguston, M.: Program Behavior Model Based on Event Grammar and its Application for Debugging Automation. In: 2nd Int'l Workshop on Automated and Algorithmic Debugging, AADEBUG 1995, pp. 277–291 (May 1995)

2. Auguston, M., Gates, A., Lujan, M.: Defining a Program Behavior Model for Dynamic Analyzers. In: 9th International Conference on Software Engineering and Knowledge Engineering, SEKE 1997, pp. 257–262 (June 1997)
3. Avrunin, G.S., Corbett, J.C., Dwyer, M.B.: Property Specification Patterns for Finite-State Verification. In: 2nd Workshop on Formal Methods in Software Practice, pp. 7–15 (March 1998)
4. Beer, I., et al.: The Temporal Logic Sugar. In: Berry, G., Comon, H., Finkel, A. (eds.) CAV 2001. LNCS, vol. 2102, pp. 363–367. Springer, Heidelberg (2001)
5. Bogdanov, K., Holcombe, M., Singh, H.: Automated Test Set Generation for Statecharts. In: Hutter, D., Traverso, P. (eds.) FM-Trends 1998. LNCS, vol. 1641, pp. 107–121. Springer, Heidelberg (1999)
6. Chang, E.S., Manna, Z., Pnueli, A.: Compositional Verification of Real-time Systems. In: Proceedings of the 9th IEEE Symposium Logic in Computer Science (LICS 1994), pp. 458–465. IEEE Computer Society Press, Los Alamitos (1994)
7. Corbett, J.C., Dwyer, M.B., Hatcliff, J.R.: A Language Framework for Expressing Checkable Properties of Dynamic Software. In: Havelund, K., Penix, J., Visser, W. (eds.) SPIN 2000. LNCS, vol. 1885, pp. 205–223. Springer, Heidelberg (2000)
8. Douglass, B.P., Harel, D., Trakhtenbrot, M.: Statecharts in Use: Structured Analysis and Object-Orientation. In: Rozenberg, G. (ed.) EEF School 1996. LNCS, vol. 1494, pp. 368–394. Springer, Heidelberg (1998)
9. Eisner, C., et al.: Reasoning with Temporal Logic on Truncated Paths. In: Hunt Jr., W.A., Somenzi, F. (eds.) CAV 2003. LNCS, vol. 2725, pp. 27–39. Springer, Heidelberg (2003)
10. Gery, E., Harel, D., Palatchi, E.: Rhapsody: A Complete Lifecycle Model-Based Development System. In: Butler, M., Petre, L., Sere, K. (eds.) IFM 2002. LNCS, vol. 2335, pp. 1–10. Springer, Heidelberg (2002)
11. Harel, D.: Statecharts: A Visual Formalism for Complex Systems. Science of Computer Programming 8, 231–274 (1987)
12. Harel, D., Kugler, H.: The Rhapsody Semantics of Statecharts (or, On the Executable Core of the UML). In: Ehrig, H., et al. (eds.) INT 2004. LNCS, vol. 3147, pp. 325–354. Springer, Heidelberg (2004)
13. Harel, D., et al.: STATEMATE: A Working Environment for the Development of Complex Reactive Systems. IEEE Trans. on Software Engineering 16(4), 403–414 (1990)
14. Harel, D., Naamad, A.: The STATEMATE Semantics of Statecharts. ACM Trans. on Software Engineering Method 5(4), 293–333 (1996)
15. Harel, D., et al.: On the Formal Semantics of Statecharts. In: Proc. 2nd IEEE Symp. on Logic in Computer Science, Ithaca, NY, pp. 54–64 (1987)
16. Harel, D., Politi, M.: Modeling Reactive Systems with Statecharts: The STATEMATE Approach. McGraw-Hill, New York (1998)
17. Manna, Z., Pnueli, A.: The Temporal Logic of Reactive and Concurrent Systems. Springer, Heidelberg (1991)
18. Pnueli, A., Shalev, M.: What is in a Step: On the Semantics of Statecharts. In: Ito, T., Meyer, A.R. (eds.) TACS 1991. LNCS, vol. 526, pp. 244–264. Springer, Heidelberg (1991)
19. Richardson, D., Leif Aha, S., Owen O'Malley, T.: Specification-based Test Oracles for Reactive Systems. In: Proc. Fourteens Intl. Conf. on Software Engineering, Melbourne, pp. 105–118 (1992)
20. Bienmüller, T., Damm, W., Wittke, H.: The STATEMATE Verification Environment - Making It Real. In: Emerson, E.A., Sistla, A.P. (eds.) CAV 2000. LNCS, vol. 1855, pp. 561–567. Springer, Heidelberg (2000)

21. Strichman, O., Goldring, R.: The 'Logic Assurance (LA)' System - A Tool for Testing and Controlling Real-Time Systems. In: Proc. 8th Israeli Conference on Computer Systems and Software Engineering, pp. 47–56 (1997)
22. Thanne, M., Yerushalmi, R.: Experience with an Advanced Design Flow with OSEK Compliant Code Generation for Automotive ECU's. Dedicated Systems Magazine, Special Issue on Development Methodologies & Tools, pp. 6–11 (2001)
23. Wind River Systems, Inc. BetterState, http://www.windriver.com/products/betterstate/index.html

A Framework for Formalizing Set Theories Based on the Use of Static Set Terms

Arnon Avron

School of Computer Science
Tel Aviv University, Tel Aviv 69978, Israel
aa@math.tau.ac.il

To Boaz Trakhtenbrot: a scientific father, a friend, and a great man.

Abstract. We present a new unified framework for formalizations of axiomatic set theories of different strength, from rudimentary set theory to full ZF. It allows the use of set terms, but provides a *static* check of their validity. Like the inconsistent "ideal calculus" for set theory, it is essentially based on just two set-theoretical principles: extensionality and comprehension (to which we add $\in$-induction and optionally the axiom of choice). Comprehension is formulated as: $x \in \{x \mid \varphi\} \leftrightarrow \varphi$, where $\{x \mid \varphi\}$ is a legal set term of the theory. In order for $\{x \mid \varphi\}$ to be legal, φ should be *safe* with respect to $\{x\}$, where safety is a relation between formulas and finite sets of variables. The various systems we consider differ from each other mainly with respect to the safety relations they employ. These relations are all defined purely syntactically (using an induction on the logical structure of formulas). The basic one is based on the safety relation which implicitly underlies commercial query languages for relational database systems (like SQL).

Our framework makes it possible to reduce all extensions by definitions to abbreviations. Hence it is very convenient for mechanical manipulations and for interactive theorem proving. It also provides a unified treatment of comprehension axioms and of absoluteness properties of formulas.

1 Introduction

The goal of this paper is to develop a unified, user-friendly framework for formalizations of axiomatic set theories of different strength, from rudimentary set theory to full ZF. The work in a formal system that is constructed within such a framework should be very close to the way work in set theories is practically done in reality. In particular, it should be possible to employ in a natural way all the usual set notations and constructs as found in textbooks on naive or axiomatic set theory (and *only* such notations).

Our starting point is what is known as the "ideal calculus" for naive set theory (see [10], Sect. III.1). This very simple calculus is based on just two set-theoretical principles: extensionality and full comprehension. It thus exactly reflects our initial, immediate intuitions concerning sets (before becoming aware

A. Avron et al. (Eds.): Trakhtenbrot/Festschrift, LNCS 4800, pp. 87–106, 2008.
© Springer-Verlag Berlin Heidelberg 2008

of the inconsistencies they involve). Now in its most transparent formal presentation, the ideal calculus employs set terms of the form $\{x \mid \varphi\}$, where x is a variable and φ is *any* formula in which x occurs free. Then the comprehension principle is most succinctly formulated as follows:

$$x \in \{x \mid \varphi\} \leftrightarrow \varphi$$

Unfortunately, it is well known that this principle leads to paradoxes (like Russel's paradox). Hence all set theories that are believed to be consistent impose constraints on the use of this principle. In all textbooks the choice of these constraints is guided by semantic intuitions (like the limitation of size doctrine [10,16]), especially the question: what operations on sets are "safe". Since it is one of our main purposes to remain as close to the "ideal calculus" as possible, on one hand, and we aim at computerized systems, on the other, we shall translate the various semantic principles into *syntactic* constraints on the logical form of formulas. Given a set theory S, we shall call a formula $\varphi(x)$ (which may have free variables other than x) S-safe with respect to x if $\{x \mid \varphi\}$ is a valid term of S (which intuitively means that according to the principles accepted by S, the set denoted by this term exists for all values of the other parameters). Thus "safety" will basically be here a relation between formulas and variables. (Actually, in order to define it syntactically we shall need to generalize it to a relation between formulas and finite sets of variables.) The various systems we consider differ from each other only with respect to the safety relations they employ.

Another problem solved in our framework is that official formalizations of axiomatic set theories in almost all textbooks are based on some standard first-order languages. In such languages terms are variables, constants, and sometimes function applications (like $x \cap y$). What is usually *not* available in the *official* languages of these formalizations is the use of set terms of the form described above ($\{x \mid \varphi\}$). As a result, already the formulation of the axioms is quite cumbersome, and even the formalization of elementary proofs becomes something practically incomprehensible. In contrast, *all* modern texts in all areas of mathematics (including set theory itself) use such terms extensively. For the purpose of mechanizing real mathematical practice and for automated or interactive theorem proving, it is therefore important to have formalizations of ZF and related systems which allow the use of such terms. Now, set terms *are* used in all textbooks on first-order set theories, as well as in several computerized systems. However, whenever they are intended to denote *sets* (rather than *classes*) they are introduced (at least partially) in a *dynamic* way, based for example on the "extension by definitions" procedure (see [20], Sect. 4.6): In order to be able to introduce some set term for a *set* (as well as a new operation on *sets*) it is necessary first to justify this introduction by *proving* a corresponding existence *theorem*. (The same is basically true in case set terms are *officially* used to denote "classes", as in [18], Sect. I.4.) The very useful complete separation we have in first-order logic between the (easy) check whether a given expression is a well-formed term or formula, and the (difficult) check whether it is a theorem, is thus lost. By analogy to programs: texts in such dynamic languages can only be "interpreted", but not "compiled". In contrast, a crucial feature of our framework is that although it makes extensive use of set terms, the languages used in it are all *static*: the task of verifying that a given term or formula is well-formed is decidable, easily mechanizable, and completely separated from any task connected with proving theorems (like finding proofs or checking validity of given ones). Expanding the language is allowed only through *explicit* definitions (i.e. new valid expressions of an extended language will just be abbreviations for expressions in the original language). This feature has the same obvious advantages that static type-checking has over dynamic type-checking.[1]

Two other important features of the framework we propose are:

- It provides a unified treatment of two important subjects of set theory: axiomatization and absoluteness (the latter is a crucial issue in independence proofs and in the study of models of set theories – see e.g. [17]). In the usual approaches these subjects are completely separated. Absoluteness is investigated mainly from a syntactic point of view, axiomatizations – from a semantic one. Here both are given the same syntactic treatment. In fact, the basis of the framework is its formulation of rudimentary set theory, in which only terms for absolute sets are allowed. The other set theories are obtained from it by small changes in the definitions of the safety relations.[2]

- Most of our systems (including the one which is equivalent to ZF) have the remarkable property that every set or function that is implicitly definable in them already has a term in the corresponding language denoting it. More precisely: if $\varphi(x, y_1, \ldots, y_n)$ is a formula such that $\forall y_1, \ldots, y_n \exists! x \varphi$ is provable, then there is a term $t(y_1, \ldots, y_n)$ such that $\varphi(y_1, \ldots, y_n, t(y_1, \ldots, y_n))$ is provable. Hence, there is no need for the procedure of extension by definitions, and introduction of new symbols is reduced to using *abbreviations*.

---

[1] The closest attempt I am aware of to develop a language for sets that employs static set terms can be found Sect. 5.1 of [7]. However, the construction there is rather complicated, and far remoted from actual mathematical practice. (The terms have the form: $\{t_{n+1} : x_0 C_0 t_0, x_1 C_1 t_1, \ldots, x_n C_n t_n \mid \varphi\}$, where each C_i is either $\in$ or $\subseteq$, φ is a formula, and $t_1, \ldots, t_n$ are terms such that $Fv(t_i) \cap \{x_1, \ldots, x_n\} \subseteq \{x_0, \ldots, x_{i-1}\}$). Moreover: the use of these terms does not have the two important features described below, and cannot serve as a basis for a framework of the type developed here.

[2] It should perhaps be noted that the idea that existence of sets $\{x \mid \varphi\}$ might be connected with absoluteness properties of φ occurs also (though with a very different formalization) in Ackermann's set theory [1], which turned out to be equivalent (once one adds regularity) to ZF [19]. The connections (if any) between Ackermann's approach and the present one are yet to be determined, and will be investigated in the future. (I am grateful to an anonymous referee for bringing Ackermann's set theory to my attention).

2 A Description of the General Framework

2.1 Languages

Officially, every set theory S has in our formal framework its own language $L(S)$. $L(S)$ is determined by the *safety relation* $\succ_S$ on which S is based. The sets of terms and formulas of $L(S)$ and $\succ_S$ are usually defined by a simultaneous recursion. For every S the clauses for $L(S)$ in this recursive definition are the following (where $Fv(exp)$ denotes the set of free variables of exp):

- Every variable is a term.
- The constant ω is a term.
- If x is a variable, and φ is a formula such that $\varphi \succ_S \{x\}$, then $\{x \mid \varphi\}$ is a term (and $Fv(\{x \mid \varphi\}) = Fv(\varphi) - \{x\}$).
- If t and s are terms then $t = s$ and $t \in s$ are atomic formulas.
- If φ and ψ are formulas, and x is a variable, then $\neg\varphi$, $(\varphi \wedge \psi)$, $(\varphi \vee \psi)$, and $\exists x \varphi$ are formulas.

Note. We have included the constant ω in all our languages in order to be able to have in all of them closed terms for denoting constant sets (see e.g. the definition of $\emptyset$ in Sect. 3.1). However, in most of our systems nothing is assumed about ω and its interpretation. Only in systems that include the infinity axiom we put the constant ω (which is available anyway) to the further use of denoting the set whose existence is guaranteed by this axiom.

2.2 Logic

Basically, the logic we will use in most of our systems is the usual first-order logic with equality. One should note however the following differences/additions:

1. Our languages provide much richer classes of terms than those allowed in orthodox first-order systems. In particular: a variable can be bound in them within a term. The notion of a term being free for substitution is generalized accordingly (also for substitutions within terms!). As usual this amounts to avoiding the capture of free variables within the scope of an operator which binds them. Otherwise the rules/axioms concerning the quantifiers and terms remain unchanged (for example: $\varphi[x \mapsto t] \to \exists x \varphi$ is valid for *every* term t which is free for x in φ).
2. The rule of α-conversion (change of bound variables) is included in the logic.
3. The substitution of equals for equals is allowed within any context (under the usual conditions concerning bound variables).
4. In analogy to the previous rule concerning identity of terms, we assume similar rule(s) allowing the substitution of a formula for an equivalent formula in any context in which the substitution makes sense. In particular, the following schema is valid whenever $\{x \mid \varphi\}$ and $\{x \mid \psi\}$ are legal terms:

$$\forall x(\varphi \leftrightarrow \psi) \to \{x \mid \varphi\} = \{x \mid \psi\}$$

2.3   Axioms

The main part of all our systems consists of the following axioms and axiom schemes (our version of the ideal calculus, augmented with the assumption that we are dealing with the cumulative universe):

Extensionality:

$$- \forall y(y = \{x \mid x \in y\})$$

Comprehension Schema:

$$- \forall x(x \in \{x \mid \varphi\} \leftrightarrow \varphi)$$

The Regularity Schema ($\in$-induction):

$$- (\forall x(\forall y(y \in x \rightarrow \varphi[x \mapsto y]) \rightarrow \varphi)) \rightarrow \forall x\varphi$$

Notes:

1. Thus the main parts of the various set theories we shall consider will differ only with respect to the power of their comprehension scheme. This, in turn, again depends only on the safety relation used by each. Hence also the differences in strength between the systems will mainly be due to the differences between their safety relations.
2. It is easy to see (see [4]) that our assumptions concerning the underlying logic and the comprehension schema together imply that the above formulation of the extensionality axiom is equivalent to the more usual one:

$$\forall z(z \in x \leftrightarrow z \in y) \rightarrow x = y$$

3. The first two axioms immediately entail the following two principles (where t is an arbitrary valid term):

   $- \{x \mid x \in t\} = t$ (provided $x \notin Fv(t)$)
   $- t \in \{x \mid \varphi\} \leftrightarrow \varphi[x \mapsto t]$ (provided t is free for x in φ)

   These principles are counterparts of the reduction rules (η) and (β) (respectively) from the λ-calculus. Like their counterparts, they are designed to be used as simplification rules (at least in the solution of elementary problems).

The Axiom of Choice. The full set theory ZFC has one more axiom that does not fit into the formal framework described above: AC (the axiom of choice). It seems that the most natural way to incorporate it into our framework is by further extending the set of terms, using Hilbert's ε symbol, together with its usual characterizing axiom (which is equivalent to the axiom of global choice):

$$\exists x\varphi \rightarrow \varphi[x \mapsto \varepsilon x\varphi]$$

It should be noted that this move is not in line with our stated goal of employing only standard notations used in textbooks, but some price should be paid for including the axiom of choice in a system.

2.4 Safety Relations

As emphasized above, the core of each of our systems is the safety relation it employs. Now the idea of using such relations is due to the similarity (noted first in [4]) between issues of safety and domain independence in database theory ([2,24]), and issues of set-existence and absoluteness in set theory. This similarity allows us to apply in the context of set theories the purely syntactic approach to safety of formulas that has been developed in database theory.

From a logical point of view, a database of scheme $D = \{P_1, \ldots, P_n\}$ is just a given set of *finite* interpretations of the predicate symbols $P_1, \ldots, P_n$. A query language for such a database is an ordinary first-order language with equality, the signature of which includes $\{P_1, \ldots, P_n\}$. Ideally, every formula ψ of a query language can serve as a query. If ψ has free variables then the answer to ψ is the set of tuples which satisfy it in some intended structure, where the interpretations of $P_1, \ldots, P_n$ is given by the database. If ψ is closed then the answer to the query is either "yes" or "no" (which can be interpreted as $\{\emptyset\}$ and $\emptyset$, respectively). However, an answer to a query should be finite and computable, even if the intended domain is infinite. Hence only "safe" formulas, the answers to which always have these properties, should be used as queries. In fact, an even stronger property of formulas is usually taken to be crucial. Safe queries should be *domain independent* ([24,2]) in the following sense:

Definition 1. [3] *Let σ be a signature which has no function symbols, and whose set of predicate symbols includes $D = \{P_1, \ldots, P_n\}$. A query $\varphi(x_1, \ldots, x_n)$ in σ is called D-d.i. (D-domain-independent) if whenever S_1 and S_2 are structures for σ such that S_1 is a substructure of S_2, and the interpretations of $\{P_1, \ldots, P_n\}$ in S_1 and S_2 are identical, then for all $a_1 \in S_2, \ldots, a_n \in S_2$:*

$$S_2 \models \varphi(a_1, \ldots, a_n) \quad \leftrightarrow \quad a_1 \in S_1 \wedge \ldots \wedge a_n \in S_1 \wedge S_1 \models \varphi(a_1, \ldots, a_n)$$

Thus a domain-independent query is a query the answer to which depends only on the information included in the database, and on the objects which are mentioned in the query. Practical database query languages are designed so that only d.i. queries can be formulated in them. Unfortunately, it easily follows from *Trakhtenbrot's Theorem* (see [9]) that it is undecidable which formulas are d.i. (or "safe" in any other reasonable notion of safety of queries, like "finite and computable"). Therefore all commercial query languages (like SQL) allow to use as queries only formulas from some syntactically defined class of d.i. formulas. Many explicit proposals of decidable, syntactically defined classes of safe formulas have been made in the literature. Perhaps the simplest among them is the following class $SS(D)$ ("syntactically safe" formulas for a database scheme D) from [24] (originally designed for languages in which every term is either a variable or a constant):[4]

---

[3] This is a slight generalization of the usual definition ([24]), which applies only to free Herbrand structures which are generated by adding to σ some new set of constants.

[4] What we present below is both a generalization and a simplification of Ullman's original definition.

1. $P_i(t_1, \ldots, t_{n_i}) \in SS(D)$ in case P_i (of arity n_i) is in D.
2. $x = c$ and $c = x$ are in $SS(D)$ (where x is a variable and c is a constant).
3. $\varphi \vee \psi \in SS(D)$ if $\varphi \in SS(D)$, $\psi \in SS(D)$, and $Fv(\varphi) = Fv(\psi)$.
4. $\exists x \varphi \in SS(D)$ if $\varphi \in SS(D)$.
5. If $\varphi = \varphi_1 \wedge \varphi_2 \wedge \ldots \wedge \varphi_k$ then $\varphi \in SS(D)$ if the following conditions are met:
   (a) For each $1 \leq i \leq k$, either φ_i is atomic, or φ_i is in $SS(D)$, or φ_i is a negation of a formula of either type.
   (b) Every free variable x of φ is limited in φ. This means that there exists $1 \leq i \leq k$ such that x is free in φ_i, and either $\varphi_i \in SS(D)$, or there exists y which is already limited in φ, and $\varphi_i \in \{x = y, y = x\}$.

There is one clause in this definition which is somewhat strange: the last one, which treats conjunction. The reason why this clause does not simply tell us (like in the case of disjunction) when a conjunction of *two* formulas is in $SS(D)$, is the desire to take into account the fact that once the value of y (say) is known, the formula $x = y$ becomes safe. In order to replace this problematic clause by a more concise one (which at the same time is more general) the formula *property* of d.i. was turned in [4] into the following *relation* between a formula φ and finite subsets of $Fv(\varphi)$:

Definition 2. *Let σ be as in Definition 1. A formula $\varphi(x_1, \ldots, x_n, y_1, \ldots, y_k)$ in σ is D-d.i. with respect to $\{x_1, \ldots, x_n\}$ if whenever S_1 and S_2 are structures as in Definition 1, then for all $a_1 \in S_2, \ldots, a_n \in S_2$ and $b_1 \in S_1, \ldots, b_k \in S_1$:*

$$S_2 \models \varphi(\overrightarrow{a}, \overrightarrow{b}) \quad \leftrightarrow \quad a_1 \in S_1 \wedge \ldots \wedge a_n \in S_1 \wedge S_1 \models \varphi(\overrightarrow{a}, \overrightarrow{b})$$

Obviously, a formula φ is D-d.i. iff it is D-d.i. with respect to $Fv(\varphi)$. On the other hand the formula $x = y$ is only partially D-d.i.: it is D-d.i. with respect to $\{x\}$ and $\{y\}$, but not with respect to $\{x, y\}$.

A particularly important observation is that a formula φ is D-d.i. with respect to $\emptyset$, if whenever S_1 and S_2 are structures as in Definition 1, then for all $b_1, \ldots, b_k \in S_1$, $S_2 \models \varphi(\overrightarrow{b}) \leftrightarrow S_1 \models \varphi(\overrightarrow{b})$. Such formulas may be called D-absolute. Obviously, this notion of D-absoluteness is closely related to the set-theoretical notion of absoluteness. However, as it is, it is not really a generalization of the notion used in set theory. In addition to $=$, the language of set theory has only one binary predicate symbol: $\in$. Now the notion of $\{\in\}$-absoluteness is useless (since if the interpretations of $\in$ in two standard models S_1 and S_2 of ZF are identical, then S_1 and S_2 are identical). The notion of $\emptyset$-absoluteness, in contrast, *is* identical to the most general notion of absoluteness as defined e.g. in [17] (p. 117), but that notion is of little use in set theory. Thus Δ_0-formulas are not $\emptyset$-absolute. Indeed, in order for Δ_0-formulas to be absolute for structures S_1 and S_2 (where S_1 is a substructure of S_2), we should assume that S_1 is a *transitive* substructure of S_2. This means that if b is an element of S_1, and $S_2 \models a \in b$, then a belongs to S_1, and $S_1 \models a \in b$. In other words: the formula $x \in y$ should be d.i. with respect to $\{x\}$ (but not with respect to $\{y\}$). In [4] and [6] this observation was used for developing a general framework for domain independence and absoluteness, and it was shown that this framework has deep applications in computability theory.

The similarity between d.i. and absoluteness is also the crucial observation on which the present framework for set-theories is based. However, in order to exploit this similarity here we do not need the full general framework developed in [4,6]. It suffices to introduce the following general, abstract notion of a safety relation (which is based on Ullman's notion of syntactic safety, but its use is not confined to database theory):

Definition 3. *A relation* $\succ$ *between formulas* φ *and subsets of* $Fv(\varphi)$ *is a safety relation if it satisfies the following conditions:*

1. *If* $\varphi \succ X$ *then* $X \subseteq Fv(\varphi)$.
2. *If* $\varphi \succ X$ *and* $Z \subseteq X$, *then* $\varphi \succ Z$.
3. *If* $\varphi \succ \{x_1, \ldots, x_n\}$ *and* $v_1, \ldots v_n$ *are* n *distinct variables not occurring in* φ, *then* $\varphi[x_1 \mapsto v_1, \ldots, x_n \mapsto v_n]$.
4. $\varphi \succ \emptyset$ *if* φ *is atomic.*
5. $t = x \succ \{x\}$ *and* $x = t \succ \{x\}$ *if* $x \notin Fv(t)$.
6. $\neg\varphi \succ \emptyset$ *if* $\varphi \succ \emptyset$.
7. $\varphi \vee \psi \succ X$ *if* $\varphi \succ X$ *and* $\psi \succ X$.
8. $\varphi \wedge \psi \succ X \cup Y$ *if* $\varphi \succ X$, $\psi \succ Y$, *and* $Y \cap Fv(\varphi) = \emptyset$.
9. $\exists y\varphi \succ X - \{y\}$ *if* $y \in X$ *and* $\varphi \succ X$.

Note. Recall that we are taking $\wedge, \vee, \neg$ and $\exists$ as our primitives. Moreover: we take $\neg(\varphi \rightarrow \psi)$ as an abbreviation for $\varphi \wedge \neg\psi$, and $\forall x_1, \ldots, x_k\varphi$ as an abbreviation for $\neg\exists x_1, \ldots, x_k\neg\varphi$. This entails the following important property of "bounded quantification": If $\succ$ is a safety relation, $\varphi \succ \{x_1, \ldots, x_n\}$, and $\psi \succ \emptyset$, then $\exists x_1 \ldots x_n(\varphi \wedge \psi) \succ \emptyset$ and $\forall x_1 \ldots x_n(\varphi \rightarrow \psi) \succ \emptyset$. The latter can easily be generalized, and the generalization can be used for an alternative definition of safety relations in case the negation connective may be used only before atomic formulas, and the negation of φ, $\overline{\varphi}$, is inductively defined for complex formulas (a common procedure in proof theory): strengthen condition 4 above to $\varphi \succ \emptyset$ if φ is a literal, and replace condition 6 by: $\forall x_1 \ldots x_n\varphi \succ \emptyset$ if $\overline{\varphi} \succ \{x_1, \ldots, x_n\}$.

Examples

- For first order languages with equality, having no function symbols and no predicate symbols other than those in D, partial D-d.i. (Definition 2) is a safety relation. A syntactic counterpart $\succ$ directly corresponding to $SS(D)$ is inductively defined by using the clauses of Definition 3 and the assumption that $\varphi \succ Fv(\varphi)$ for every atomic formula φ of the form $P_i(t_1, \ldots, t_{n_i})$.
- Let L be the language of PA (Peano's Arithmetic), and let $\mathcal{N}$ be the standard model of PA. Define a relation $\succ_{\mathcal{N}}$ on L by: $\varphi(x_1, \ldots, x_n, y_1, \ldots, y_l) \succ_{\mathcal{N}} \{x_1, \ldots, x_n\}$ if the set $\{\langle k_1, \ldots, k_n \rangle \in \mathcal{N}^n \mid \varphi(k_1, \ldots, k_n, m_1, \ldots, m_l)\}$ is finite and computable (as a function of $m_1, \ldots, m_l$) for all $m_1, \ldots, m_l$ in $\mathcal{N}$.[5]

---

[5] In the case $l = 0$ an intentional meaning of "computable" is meant, but we shall not get into details here. See [4,6] for more details.

Then $\succ_N$ is a safety relation, and $\varphi \succ_N \emptyset$ iff φ defines a decidable predicate. A useful syntactic approximation $\succ_b$ of $\succ_N$ can in this case inductively be defined by using the clauses of Definition 3 and the assumption that $x < t \succ_b x$ if $x \notin Fv(t)$. The set $\{\varphi \mid \varphi \succ_b \emptyset\}$ is a straightforward extension of Smullyan's set of Σ_0 formulas (see [23], P. 41), which can serve as a basis for the usual arithmetical hierarchy. It is interesting to note that a succinct inductive definition of $\succ_b$ can be given which is almost identical to that of the basic safety relation $\succ_{RST}$ of set theory (see Definition 5). The only difference is that the condition $x \in t \succ_b x$ in Definition 5 should be replaced by $x < t \succ_b x$.

Next we describe the way safety relations are used in our framework for set theories. The basic idea is that φ should be safe for $\{x\}$ in a set theory S iff the collection $\{x \mid \varphi\}$ is accepted as a set by S. This leads to the following definition:

Definition 4. *Let L be a language which has $\in$ among its binary predicate symbols. An $\in$-safety relation for L is a safety relation $\succ$ for L which satisfies the following condition:*

- $x \in t \succ \{x\}$ *if x is a variable such that $x \notin Fv(t)$.*

All the safety relations used in our framework are $\in$-safety relations.

3   The Rudimentary Set Theory *RST*

Our basic system is the one which corresponds to the minimal $\in$-safety relation:

Definition 5. *The relation $\succ_{RST}$ is inductively defined as follows:*

1. $\varphi \succ_{RST} \emptyset$ *if φ is atomic.*
2. $\varphi \succ_{RST} \{x\}$ *if $\varphi \in \{x = t, t = x, x \in t\}$, and $x \notin Fv(t)$.*
3. $\neg\varphi \succ_{RST} \emptyset$ *if $\varphi \succ_{RST} \emptyset$.*
4. $\varphi \vee \psi \succ_{RST} X$ *if $\varphi \succ_{RST} X$ and $\psi \succ_{RST} X$.*
5. $\varphi \wedge \psi \succ_{RST} X \cup Y$ *if $\varphi \succ_{RST} X$, $\psi \succ_{RST} Y$, and $Y \cap Fv(\varphi) = \emptyset$.*
6. $\exists y\varphi \succ_{RST} X - \{y\}$ *if $y \in X$ and $\varphi \succ_{RST} X$.*

It is easy to see that $\succ_{RST}$ is indeed an $\in$-safety relation. We denote by *RST* (Rudimentary Set Theory) the set theory it induces (within the framework described above). The following theorem about *RST* can easily be proved:

Theorem 1. *Given an expression E and a finite set X of variables, it is decidable in polynomial time whether E is a valid term of RST, whether it is a valid formula of RST, and if the latter holds, whether $E \succ_{RST} X$.*

Note. The last theorem is of a crucial importance from implementability point of view, and it obtains also for all the extensions of *RST* discussed (explicitly or implicitly) below. In order to ensure it, we did not include in the definition of safety relations the natural condition that if $\varphi \succ X$ and ψ is (logically) equivalent to φ (where $Fv(\varphi) = Fv(\psi)$) then also $\psi \succ X$. However, we obviously do have that if $\vdash_{RST} \varphi \leftrightarrow \psi$ then $\vdash_{RST} x \in \{x \mid \varphi\} \leftrightarrow \psi$, and so $\vdash_{RST} \exists Z \forall x.x \in Z \leftrightarrow \psi$.

3.1   The Power of *RST*

In the language of *RST* we can introduce as *abbreviations* (rather than as extensions by definitions) most of the standard notations for sets used in mathematics. Again, all these abbreviations should be used in a purely static way: no justifying propositions and proofs are needed. Here are some examples:

- $\emptyset =_{Df} \{x \mid x \in \omega \wedge x \neq x\}$.
- $\{t_1, \ldots, t_n\} =_{Df} \{x \mid x = t_1 \vee \ldots \vee x = t_n\}$ (where x is new).
- $\langle t, s \rangle =_{Df} \{\{t\}, \{t, s\}\}$.
- $\langle t_1, \ldots, t_n \rangle$ is $\emptyset$ if $n = 0$, t_1 if $n = 1$, $\langle \langle t_1, \ldots, t_{n-1} \rangle, t_n \rangle$ if $n \geq 2$.
- $\{x \in t \mid \varphi\} =_{Df} \{x \mid x \in t \wedge \varphi\}$, provided $\varphi \succ_{RST} \emptyset$. (where $x \notin Fv(t)$).
- $\{t \mid x \in s\} =_{Df} \{y \mid \exists x. x \in s \wedge y = t\}$ (where y is new, and $x \notin Fv(s)$).
- $s \times t =_{Df} \{x \mid \exists a \exists b. a \in s \wedge b \in t \wedge x = \langle a, b \rangle\}$ (where x, a and b are new).
- $\{\langle x_1, \ldots, x_n \rangle \mid \varphi\} =_{Df} \{z \mid \exists x_1 \ldots \exists x_n. \varphi \wedge z = \langle x_1, \ldots, x_n \rangle\}$, provided $\varphi \succ_{RST} \{x_1, \ldots, x_n\}$, and $z \notin Fv(\varphi)$.
- $s \cap t =_{Df} \{x \mid x \in s \wedge x \in t\}$ (where x is new).
- $s \cup t =_{Df} \{x \mid x \in s \vee x \in t\}$ (where x is new).
- $s - t =_{Df} \{x \mid x \in s \wedge x \notin t\}$ (where x is new).
- $S(x) =_{Df} x \cup \{x\}$
- $\bigcup t =_{Df} \{x \mid \exists y. y \in t \wedge x \in y\}$ (where x and y are new).
- $\bigcap t =_{Df} \{x \mid \exists y(y \in t \wedge x \in y) \wedge \forall y(y \in t \rightarrow x \in y)\}$ (where x, y are new).

It is straightforward to check that in all these abbreviations the right hand side is a valid term of *RST* (provided that the terms/formulas occurring in it are valid terms/well-formed formulas of *RST*). We explain $s \times t$ by way of example: since a and b are new, $a \in s \succ_{RST} \{a\}$, and $b \in t \succ_{RST} \{b\}$. Since $b \notin Fv(a \in s)$, this implies that $a \in s \wedge b \in t \succ_{RST} \{a, b\}$. Similarly, $a \in s \wedge b \in t \wedge x = \langle a, b \rangle \succ_{RST} \{a, b, x\}$. It follows that $\exists a \exists b. a \in s \wedge b \in t \wedge x = \langle a, b \rangle \succ_{RST} \{x\}$. Hence our term for $s \times t$ (which is the most natural one) is a valid term of *RST*.

Lemma 1. *There is a formula $OP(z, x, y)$ in the basic language of RST (i.e.: without set terms) such that:*

1. $\vdash_{RST} OP(z, x, y) \leftrightarrow z = \langle x, y \rangle$
2. $OP(z, x, y) \succ_{RST} \{x, y\}$.

Proof: Let $Pa(z, x, y) \equiv_{Df} x \in z \wedge y \in z \wedge \forall w(w \in z \rightarrow w = x \vee w = y)$. Then $Pa(z, x, y) \succ_{RST} \{x, y\}$, and $\vdash_{RST} Pa(z, x, y) \leftrightarrow z = \{x, y\}$. Let $OP(z, x, y)$ be the formula $\exists u \exists v(Pa(z, u, v) \wedge Pa(u, x, x) \wedge Pa(v, x, y))$. $\square$

With the help of OP we can define all the standard basic operations related to relations and functions. For example:

- $Dom(s) =_{Df} \{x \mid \exists z \exists y(z \in s \wedge OP(z, x, y))\}$
- $Rng(s) =_{Df} \{y \mid \exists z \exists x(z \in s \wedge OP(z, x, y))\}$
- $t \upharpoonright s =_{Df} \{x \in t \mid \exists z \exists y OP(x, y, z) \wedge y \in s\}$

In RST we can also introduce as abbreviations the terms used in the λ-calculus for handling explicitly defined functions which are sets (except that our terms for functions should specify the domains of these functions, which should also be explicitly definable sets). Moreover: the reduction rules of the λ-calculus for these terms are easy theorems of RST. Thus the notation for λ-set and function application are introduced as follows:

- $\lambda x \in s.t =_{Df} \{\langle x, t\rangle \mid x \in s\}$ (where $x \notin Fv(s)$)
- $f(t) =_{Df} \bigcup Rng(f \upharpoonright \{t\})$

(Note that $f(t)$ is defined for *every* f and t, but when f denotes a function F, and t denotes an element a in F's domain, then $f(t)$ indeed denotes the value of F at a.) We can easily check now that rules β and η obtain in RST:

- $\vdash_{RST} u \in s \rightarrow (\lambda x \in s.t)u = t[x \mapsto u]$ (if u is free for x in t).
- $\vdash_{RST} u \notin s \rightarrow (\lambda x \in s.t)u = \emptyset$ (if u is free for x in t).
- $\vdash_{RST} \lambda x \in s.t(x) = t \upharpoonright s$ (in case $x \notin Fv(t)$).

Exact characterizations of the operations that are explicitly definable in RST, and of the strength of RST, are given in the following theorems and corollary (the proofs of which will be given in [6]).

Theorem 2

1. If F is an n-ary rudimentary function[6] then there exists a formula φ s. t.:
   (a) $Fv(\varphi) = \{y, x_1, \ldots, x_n\}$
   (b) $\varphi \succ_{RST} \{y\}$
   (c) $F(x_1, \ldots, x_n) = \{y \mid \varphi\}$.
2. If φ is a formula such that:
   (a) $Fv(\varphi) = \{y_1, \ldots, y_k, x_1, \ldots, x_n\}$
   (b) $\varphi \succ_{RST} \{y_1, \ldots, y_k\}$
   then there exists a rudimentary function F such that:

$$F(x_1, \ldots, x_n) = \{\langle y_1, \ldots, y_k\rangle \mid \varphi\}.$$

Corollary 1. If $Fv(\varphi) = \{x_1, \ldots, x_n\}$, and $\varphi \succ_{RST} \emptyset$ then φ defines a rudimentary predicate P. Conversely, if P is a rudimentary predicate then there is a formula φ such that $\varphi \succ_{RST} \emptyset$ and φ defines P.

Theorem 3. RST is equivalent to the system obtained from Gandy's "Basic Set Theory" BST ([12]) by the addition of the $\in$ $-induction$ schema.

---

[6] The class of rudimentary set functions was introduced independently by Gandy ([12]) and Jensen ([15]). See also [8], Sect. IV.1.

3.2 Generalized Absoluteness

For simplicity of presentation, we assume the cumulative universe V of ZF, and formulate our definitions accordingly. It is easy to see that V is a model of RST (with the obvious interpretations of RST's terms).

Definition 6. *Let $\mathcal{M}$ be a transitive model of RST. Define the relativization to $\mathcal{M}$ of the terms and formulas of RST recursively as follows:*

- $t_{\mathcal{M}} = t$ *if t is a variable or a constant.*
- $\{x \mid \varphi\}_{\mathcal{M}} = \{x \mid x \in \mathcal{M} \wedge \varphi_{\mathcal{M}}\}$.
- $(t = s)_{\mathcal{M}} = (t_{\mathcal{M}} = s_{\mathcal{M}})$ $(t \in s)_{\mathcal{M}} = (t_{\mathcal{M}} \in s_{\mathcal{M}})$.
- $(\neg\varphi)_{\mathcal{M}} = \neg\varphi_{\mathcal{M}}$ $(\varphi \vee \psi)_{\mathcal{M}} = \varphi_{\mathcal{M}} \vee \psi_{\mathcal{M}}$. $(\varphi \wedge \psi)_{\mathcal{M}} = \varphi_{\mathcal{M}} \wedge \psi_{\mathcal{M}}$.
- $(\exists x \varphi)_{\mathcal{M}} = \exists x (x \in \mathcal{M} \wedge \varphi_{\mathcal{M}})$.

Definition 7. *Let T be an extension of RST such that $V \models T$.*

1. *Let t be a term, and let $Fv(t) = \{y_1, \ldots, y_n\}$. We say that t is T-absolute if the following is true (in V) for every transitive model $\mathcal{M}$ of T:*

$$\forall y_1 \ldots \forall y_n. y_1 \in \mathcal{M} \wedge \ldots \wedge y_n \in \mathcal{M} \rightarrow t_{\mathcal{M}} = t$$

2. *Let φ be a formula, and let $Fv(\varphi) = \{y_1, \ldots, y_n, x_1, \ldots, x_k\}$. We say that φ is T-absolute for $\{x_1, \ldots, x_k\}$ if $\{\langle x_1, \ldots, x_k \rangle \mid \varphi\}$ is a set for all values of the parameters $y_1, \ldots, y_n$, and the following is true (in V) for every transitive model $\mathcal{M}$ of RST:*

$$\forall y_1 \ldots \forall y_n. y_1 \in \mathcal{M} \wedge \ldots \wedge y_n \in \mathcal{M} \rightarrow [\varphi \leftrightarrow (x_1 \in \mathcal{M} \wedge \ldots \wedge x_k \in \mathcal{M} \wedge \varphi_{\mathcal{M}})]$$

Thus a term is T-absolute if it has the same interpretation in all transitive models of T which contains the values of its parameters, while a formula is T-absolute for $\{x_1, \ldots, x_k\}$ if it has the same extension (which should be a set) in all transitive models of T which contains the values of its other parameters. In particular: φ is T-absolute for $\emptyset$ iff it is absolute relative to T in the usual sense of set theory (see e.g. [17]), while φ is T-absolute for $Fv(\varphi)$ iff it is domain-independent in the sense of database theory (see Definition 1) for transitive models of T.

Theorem 4

1. *Any valid term t of RST is RST-absolute.*
2. *If $\varphi \succ_{RST} X$ then φ is RST-absolute for X.*

The proof is by a simultaneous induction on the complexity of t and φ.

4   Stronger Set Theories

The definability of $\{t, s\}$ and of $\bigcup t$ in the language of RST means that the axioms of pairing and union are provable in RST. We turn now to the question how to deal with the other comprehension axioms of ZF within the proposed framework. We start first with the axioms that remain valid if we limit ourselves to hereditarily finite sets. We show that the addition of each of them to RST corresponds to adding to the definition of $\succ_{RST}$ a certain syntactic condition.

4.1   Basic ZF: The Full Separation and Replacement Schemes

Theorem 5. *Let T be an extension of RST, based on some safety relation $\succ_T$ which extends $\succ_{RST}$.*

1. *If $\succ_T$ satisfies the condition:*
   (Sep) $\varphi \succ_T \emptyset$ *for every formula φ*
   then the axiom schema of separation is derivable in T.
2. *If $\succ_T$ satisfies the condition:*
   (Rep) $\exists y\varphi \wedge \forall y(\varphi \to \psi) \succ_T X$ *if $\psi \succ X$, and $X \cap Fv(\varphi) = \emptyset$.*
   then the axiom schema of replacement is derivable in T.

Proof: In the presence of condition (Sep), $\{x \mid x \in z \wedge \varphi\}$ is a valid term for every φ, and this implies the separation schema.

Suppose now that $\succ_T$ Satisfies (Rep). The proof that the replacement schema is derivable in T is more difficult than in the previous case, because unlike the other comprehension axioms of ZF, the official formulation of replacement has the form of a *conditional*:

$$(\forall y \exists v \forall x(\varphi \Leftrightarrow x = v)) \Rightarrow (\exists Z \forall x.x \in Z \Leftrightarrow (\exists y.y \in w \wedge \varphi))$$

where $v, w, Z \notin Fv(\varphi)$. To prove this in T, let A be the formula $\forall x(\varphi \Leftrightarrow x = v)$. Reasoning in T, assume $\forall y \exists v A$ (this is the left hand side of the implication we want to prove). This and the definition of the formula A logically imply $(\exists v A \wedge \forall v(A \to x = v)) \Leftrightarrow \varphi$. But by (Rep), $\exists v A \wedge \forall v(A \to x = v) \succ_T \{x\}$. Hence $\exists y.y \in w \wedge (\exists v A \wedge \forall v(A \to x = v)) \succ_T \{x\}$. Thus the comprehension axiom of T implies: $\exists Z \forall x.x \in Z \Leftrightarrow (\exists y.y \in w \wedge (\exists v A \wedge \forall v(A \to x = v)))$. This and the above conclusion of $\forall y \exists v A$ together entail $\exists Z \forall x.x \in Z \Leftrightarrow (\exists y.y \in w \wedge \varphi)$. $\square$

Definition 8

1. *The safety relation $\succ_{BZF}$ is obtained from $\succ_{RST}$ by replacing clauses 1 and 3 of its definition with (Sep) and (Rep).*
2. *The system BZF is defined like RST, using $\succ_{BZF}$ instead of $\succ_{RST}$.*

Note. Any formula φ is logically equivalent to $\exists y\varphi \wedge \forall y(\varphi \to \exists x.x = \omega)$, where y is a dummy variable. Hence (Sep) is superfluous in the presence of (Rep) (This corresponds to the well-known fact that separation is derivable from replacement). In particular, to get BZF it suffices to add to $\succ_{RST}$ only (Rep).

Theorem 6. *Let BZF^* be the system in the pure first-order fragment of the language of BZF (i.e. with no set terms) which is obtained from BZF by replacing its comprehension axiom with the following safe comprehension schema:*

$$(SCn) \qquad \exists Z(\forall x.x \in Z \Leftrightarrow \varphi)$$

where φ is in the language of BZF^*, $\varphi \succ_{BZF} \{x\}$, and $Z \notin Fv(\varphi)$. Let ZF^{--} be ZF without the powerset axiom and the infinity axiom. Then BZF, BZF^*, and ZF^{--} are all equivalent.[7]

---

[7] Note again (see the note in Sect. 2.1) that although ZF^{--} can talk about ω, as far as this theory is concerned, ω could be any set whatsoever.

Proof: Obviously, every theorem of BZF^* is also a theorem of BZF. That every theorem of ZF^{--} is a theorem of BZF^* can be shown exactly like in the proof of Theorem 5.

To complete the cycle, it remains to show that BZF is a conservative extension of ZF^{--}. For this we define recursively for every formula φ of BZF a translation $\varphi^{(I)}$ into the language of ZF^{--} such that $Fv(\varphi^{(I)}) = Fv(\varphi)$:

-   If φ is an atomic formula in the language of ZF^{--} then $\varphi^{(I)} = \varphi$.
-   Suppose φ is an atomic formula which contains a set term. Let $t = \{x \mid \psi\}$ (where $\psi \succ_{BZF} x$) be a maximal set term of φ. Define:

$$\varphi^{(I)} = \exists Z(\forall x(x \in Z \Leftrightarrow \psi^{(I)}) \wedge (\varphi[t \mapsto Z])^{(I)})$$

    where Z is a new variable, and $\varphi[t \mapsto Z]$ is the formula obtained from φ by replacing every occurrence of t in φ by Z.
-   Let $(\varphi \wedge \psi)^{(I)} = (\varphi)^{(I)} \wedge (\psi)^{(I)}$, $(\exists x \varphi)^{(I)} = \exists x(\varphi)^{(I)}$ etc.

Next, we show how to express the safety relation $\succ_{BZF}$ within the language of ZF^{--}. From Lemma 1 it easily follows that there is a formula $B_n(x_1, \ldots, x_n, z)$ in the language of ZF^{--} such that $B_n(x_1, \ldots, x_n, z) \succ_{RST} \{x_1, \ldots, x_n\}$ and $B_n(x_1, \ldots, x_n, z)$ is equivalent in RST to $\langle x_1, \ldots, x_n \rangle \in z$. Let $set_{x_1,\ldots,x_n}\varphi$ be $(\varphi \to \varphi)$ for $n = 0$, $\exists Z \forall x_1 \ldots \forall x_n(B_n(x_1, \ldots, x_n, Z) \Leftrightarrow \varphi)$ for $n > 0$ (where $Z \notin Fv(\varphi)$) [8]. Let $Set_{x_1,\ldots,x_n}\varphi$ be the universal closure of $set_{x_1,\ldots,x_n}\varphi$. Note that $Set_x\varphi$ formalizes the application to φ of the comprehension principle. We show by induction on the structure of a formula φ of BZF that if $\varphi \succ_{BZF} \{x_1, \ldots, x_n\}$ then $Set_{x_1,\ldots,x_n}\varphi^{(I)}$ is a theorem of ZF^{--}.

1. The case $n = 0$ is trivial

2. (a) If t is a variable or a constant of BZF then

    -   $set_x x = t$ and $set_x t = x$ follow from the pairing axiom.

    -   $set_x x \in t$ is a logically valid formula.

    (b) If $t = \{y \mid \psi\}$ (where $\psi \succ_{BZF} y$) and $\varphi = p(x, t)$, where $p(x, t)$ is in $\{x = t, t = x, x \in t\}$, and $x \notin Fv(t)$ $(= Fv(\psi) - \{y\})$, then $\varphi^{(I)}$ is $\exists Z(\forall y(y \in Z \Leftrightarrow \psi^{(I)}) \wedge p(x, Z))$. By induction hypothesis for ψ we have $\vdash_{ZF^{--}} Set_y\psi^{(I)}$. This means that $\vdash_{ZF^{--}} \exists Z(\forall y(y \in Z \Leftrightarrow \psi^{(I)}))$, and so $\vdash_{ZF^{--}} \exists! Z(\forall y(y \in Z \Leftrightarrow \psi^{(I)}))$. By part (a) also $\vdash_{ZF^{--}} Set_x p(x, Z)$. Now it is easy to show that $(\exists! ZA \wedge \forall Z set_x B) \to set_x \exists Z(A \wedge B)$ is logically valid in case $x \notin Fv(A)$. This implies that $\vdash_{ZF^{--}} Set_x\varphi^{(I)}$.

3. $set_{x_1,\ldots,x_n}(\varphi \vee \psi)^{(I)}$ follows from $set_{x_1,\ldots,x_n}\varphi^{(I)}$ and $set_{x_1,\ldots,x_n}\psi^{(I)}$ by the axioms of union and pairing.

---

[8] This is a generalization of the notation $Set_x\varphi$ from [20], P. 240.

4. To simplify notation, assume that $Fv(\varphi) = \{x, z\}$, $Fv(\psi) = \{x, y, z\}$, and that $\varphi \succ_{BZF} \{x\}$, $\psi \succ_{BZF} \{y\}$ (and so $\varphi \wedge \psi \succ_{BZF} \{x, y\}$). By induction hypothesis, $\vdash_{ZF^{--}} Set_x\varphi^{(I)}$, and $\vdash_{ZF^{--}} Set_y\psi^{(I)}$. Reasoning in ZF^{--}, this means that there are sets $Z(z)$ and $W(x,z)$ such that $x \in Z(z) \Leftrightarrow \varphi^{(I)}$ and $y \in W(x, z) \Leftrightarrow \psi^{(I)}$. It follows that

$$\{\langle x, y \rangle \mid (\varphi \wedge \psi)^{(I)}\} = \bigcup_{x \in Z(z)} \{x\} \times W(x, z)$$

$Set_{x,y}(\varphi \wedge \psi)^{(I)}$ follows therefore by the axioms of replacement and union, and the fact that the existence of Cartesian products is provable in ZF^{--}.

5. Deriving $Set_{X-\{y\}}\exists y\varphi^{(I)}$ in ZF^{--} from $Set_X\varphi^{(I)}$ is left to the reader.

6. Assume that $\vdash_{ZF^{--}} set_{x_1,\ldots,x_n}\psi^{(I)}$, and $\{x_1, \ldots, x_n\} \cap Fv(\varphi) = \emptyset$. We show that $\vdash_{ZF^{--}} set_{x_1,\ldots,x_n}(\exists y\varphi \wedge \forall y(\varphi \to \psi))^{(I)}$. This is immediate from the fact that if $\{x_1, \ldots, x_n\} \cap Fv(\varphi) = \emptyset$ then $\exists y\forall x_1 \ldots x_n((\exists y\varphi \wedge \forall y(\varphi \to \psi)) \to \psi)$ is logically valid [9], together with the following lemma:

Lemma: Assume that $\{y_1, \ldots, y_k\} \cap Fv(\varphi) = \emptyset$, $\vdash_{ZF^{--}} set_{x_1,\ldots,x_n}\psi$ and $\exists y_1, \ldots, y_k\forall x_1, \ldots, x_n(\varphi \to \psi)$ is logically valid. Then $\vdash_{ZF^{--}} set_{x_1,\ldots,x_n}\varphi$.

Proof of the Lemma: $\exists y_1, \ldots, y_k\forall x_1, \ldots, x_n(\varphi \to \psi)$ logically implies the formula $\exists y_1, \ldots, y_k\forall x_1, \ldots, x_n(\varphi \leftrightarrow .\psi \wedge \varphi)$. It is easy however to see that if $\{y_1 \ldots y_k\} \cap Fv(\varphi) = \emptyset$ then $Set_{x_1,\ldots,x_n}\varphi$ logically follows in first order logic from $Set_{x_1,\ldots,x_n}\phi$ and $\exists y_1 \ldots y_k\forall x_1 \ldots x_n(\varphi \leftrightarrow \phi)$. Hence we only need to prove that $set_{x_1,\ldots,x_n}(\psi \wedge \varphi)$ follows in ZF^{--} from $set_{x_1,\ldots,x_n}\psi$. This is immediate from the axiom of subsets.

Now we show that if $\vdash_{BZF} \varphi$ then $\vdash_{ZF^{--}} \varphi^{(I)}$. Since obviously $\varphi^{(I)} = \varphi$ in case φ is in the language of ZF^{--}, this will end the proof of the theorem. Now the inference rules are identical in the two systems, and our translation preserves applications of these rules. It suffices therefore to show that the translations of the comprehension axioms of BZF are theorems of ZF^{--}. Well, if $\varphi \succ_{BZF} \{x\}$ then the translation of the φ-instance of this schema is $\forall x(\exists Z(\forall x(x \in Z \leftrightarrow \varphi^{(I)}) \wedge x \in Z) \leftrightarrow \varphi^{(I)})$. It is easy to see that this formula follows in ZF^{--} from $Set_x\varphi^{(I)}$. The latter formula, in turn, is provable in ZF^{--} by what we have proved above (since $\varphi \succ_{BZF} \{x\}$).

This completes the proof of Theorem 6.   □

As noted at the end of the introduction, in mathematical practice new symbols for relations and functions are regularly introduced in the course of developing a theory. This practice is formally based on the "extensions by definitions" procedure (see e.g. [20], Sect. 4.6). Now, while new relation symbols are introduced just as abbreviations for (usually) longer formulas, new function symbols are introduced in a dynamic way: once $\forall y_1, \ldots, y_n\exists! x\varphi$ is proved (where

---

[9] For the proof of the validity of this formula show that it follows from $\exists y_1 \ldots y_k\varphi$ as well as from $\neg\exists y_1 \ldots y_k\varphi$.

$Fv(\varphi) = \{y_1, \ldots, y_n, x\}$) then a new n-ary function symbol F_φ can conservatively be introduced, together with a new axiom: $\forall y_1, \ldots, y_n(\varphi[x \mapsto F_\varphi(y_1, \ldots, y_n)])$. Now a particularly remarkable property of BZF and its extensions is that this dynamic procedure is not needed for them. The required terms are available in advance, and every new function symbol we might wish to use may be introduced statically, as an abbreviation for an already existing term (in particular: any set which has an implicit definition in some extension of BZF has an explicit definition in that extension, using a set term):

Theorem 7. *For any formula φ of BZF such that $Fv(\varphi) = \{y_1, \ldots, y_n, x\}$), there exists a term t_φ of BZF such that $Fv(t_\varphi) = \{y_1, \ldots, y_n\}$, and*

$$\vdash_{BZF} \forall y_1, \ldots, y_n \exists! x \varphi \rightarrow \forall y_1, \ldots, y_n(\varphi[x \mapsto t_\varphi])$$

Proof: Define $\iota x \varphi = \{z \mid \exists x \varphi \wedge \forall x(\varphi \rightarrow z \in x)\}$ (where z is a new variable, not occurring in φ). This is a valid term of BZF by the new clause in the definition of $\succ_{BZF}$. Now it can easily be proved that

$$\vdash_{BZF} \forall y_1, \ldots, y_n(\exists! x \varphi \rightarrow \forall x(\varphi \leftrightarrow x = \iota x \varphi))$$

It follows that $\iota x \varphi$ is a term t_φ as required.   □

Corollary 2. *Every instance of the replacement schema (in the language of BZF) is derivable in BZF.*[10]

Proof: From the last theorem (and the definition of $\succ_{BZF}$) it follows that

$$\vdash_{BZF} \forall y \exists! x \varphi \rightarrow \forall x(\exists y.y \in w \wedge \varphi \leftrightarrow x \in \{x \mid \exists y.y \in w \wedge x = \iota x \varphi\}).$$

Note. $\iota x \varphi$ intuitively denotes the unique x such that φ, in case such exists. However, our $\iota x \varphi$ is always meaningful, and denotes $\emptyset$ if there is no set that satisfies φ, and the intersection of all the sets which satisfy φ in case there is more than one such set.

4.2   The Powerset Axiom

Theorem 8

1. Let T be an extension of BZF, based on some safety relation $\succ_T$ which extends $\succ_{BZF}$. If $\succ_T$ satisfies the condition:

   (Pow) $\forall y(y \in x \rightarrow \varphi) \succ (X - \{y\}) \cup \{x\}$ if $\varphi \succ X$, $y \in X$, and $x \notin Fv(\varphi)$.

   Then the powerset axiom is derivable in T.
2. Let $\succ_{BZFP}$ be the safety relation obtained from $\succ_{BZF}$ by adding condition (Pow) to its definition, and let the system $BZFP$ be defined like RST, using $\succ_{BZFP}$ instead of $\succ_{RST}$. Then $BZFP$ is equivalent to $ZF - Inf$ (ZF without the infinity axiom).

---

[10] This corollary provides a direct, short proof that BZF is an extension of ZF^{--}.

Proof: For the first part, note that in the presence of condition (Pow) the powerset axiom immediately follows from the facts that $y \in z \succ_{RST} y$, and that $P(z) = \{x \mid \forall y(y \in x \rightarrow y \in z)\}$. The proof of the second part is similar to that of Theorem 6. □

Another method (which may look more natural and is the one used in [5]) to add the power of the powerset axiom to the systems described above, is to extend the language by taking $\subseteq$ as an extra primitive binary relation symbol. A definition of a system which is equivalent to $ZF - Inf$ can then be obtained from the definition of BZF by making the following two changes:

- Replace $\succ_{BZF}$ with $\succ_{ZF-I}$, where $\succ_{ZF-I}$ is defined like $\succ_{BZF}$, but with one extra condition:

  $x \subseteq t \succ_{ZF-I} \{x\}$ if x is a variable, t is a term, and $x \notin Fv(t)$.

- Add the usual definition of $\subseteq$ in terms of $\in$ as an extra *axiom*:

$$\forall x \forall y(x \subseteq y \leftrightarrow \forall z(z \in x \rightarrow z \in y))$$

Alternatively, since $\subseteq$ is now taken as primitive, it might be more natural to use it as such in our axioms. This means that instead of adding the above axiom, it might be preferable to replace the single extensionality axiom of BZF with the following three extensionality axioms:

(Ex1) $x \subseteq y \wedge y \subseteq x \rightarrow x = y$

(Ex2) $z \in x \wedge x \subseteq y \rightarrow z \in y$

(Ex3) $x \subseteq y \vee \exists z(z \in x \wedge z \notin y)$

4.3   The Axiom of Infinity

Finally we turn to the axiom of infinity — the only axiom that necessarily takes us out of the realm of (hereditarily) finite sets. As long as we take FOL (First-Order Logic) as the underlying logic, it seems impossible to incorporate it into our systems by just imposing new simple syntactic conditions on the safety relation. Instead the easiest and most natural way to add its power to the systems discussed so far, is to add to them *Peano's Axioms* as new axioms:

- $\emptyset \in \omega$
- $\forall x(x \in \omega \rightarrow S(x) \in \omega)$ (where $S(x)$ is defined like in Sect. 3.1)
- $\varphi[x \mapsto \emptyset] \wedge \forall x(\varphi \rightarrow \varphi[x \mapsto S(x)]) \rightarrow \forall x(x \in \omega \rightarrow \varphi)$

Note that because we are assuming the $\in$-induction schema, the above induction schema can actually be replaced by the following single axiom:

$$(\emptyset \in y \wedge \forall x(x \in y \rightarrow S(x) \in y)) \rightarrow \omega \subseteq y$$

Theorem 9. ([5]) *Let ZF^+ be the system obtained from RST by adding (Rep) and (Pow) to the definition of the safety relation, and the above Peano's axioms to the set of axioms. Then ZF^+ is equivalent to ZF.*

5   Using Transitive Closure Logic

Introducing the infinity axioms into a system is a major step that from a computational and proof-theoretical point of view takes us to a completely different level. As is clear from the form we gave to this introduction, it incorporates inductive reasoning into the systems. In order to introduce such reasoning already on the *logical* level, and to keep as far as possible the uniformity of our framework, it is most natural to use as the underlying logic a logic which is stronger than FOL, but still reasonably manageable from a computational point of view. Now in [3] it was argued that languages and logics with transitive closure operation TC provide the best framework for the formalization of mathematics. Following this suggestion seems particularly suitable in the present context, since with TC the difference between set theories which assume infinity, and set theories which are valid also in the universe of hereditarily finite sets, can again be reduced to differences in the underlying syntactic safety relations.

Definition 9. ([14,22]) Let L be a (first-order) language. The language L_{TC} is obtained from L by adding the following clause to the definition of a formula: If φ is a formula, x, y are distinct variables, and t, s are terms, then $(TC_{x,y}\varphi)(t, s)$ is a formula (in which every occurrence of x and y in φ is bound). The intended meaning of $(TC_{x,y}\varphi)(t, s)$ is the following "infinite disjunction": (where $w_1, w_2, \ldots$, are all new):

$$\varphi[x \mapsto s, y \mapsto t] \vee \exists w_1(\varphi[x \mapsto s, y \mapsto w_1] \wedge \varphi[x \mapsto w_1, y \mapsto t]) \vee$$
$$\vee \exists w_1 \exists w_2 (\varphi[x \mapsto s, y \mapsto w_1] \wedge \varphi[x \mapsto w_1, y \mapsto w_2] \wedge \varphi[x \mapsto w_2, y \mapsto t]) \vee \ldots$$

The most important relevant facts shown in [3] concerning TC are:

1. If L contains a constant 0 and a (symbol for) a pairing function, then all types of finitary inductive definitions of relations and functions (as defined by Feferman in [11]) are available in L_{TC}.
2. Let V_0 be the smallest set including 0 and closed under the operation of pairing. Let U be the smallest set of first-order terms in a language with a constant for 0 and a function symbol for pairing. Let $\mathcal{PTC}^+$ be the smallest set of formulas which includes all formulas of the form $t = s$ for $t, s \in U$, and is closed under $\vee, \wedge$ and TC. Then a subset S of V_0 is recursively enumerable iff there exists a formula $\varphi(x)$ of $\mathcal{PTC}^+$ such that $S = \{x \in V_0 \mid \varphi(x)\}$.
3. By generalizing a particular case which has been used by Gentzen in [13], mathematical induction can be presented as a logical rule of languages with TC. Indeed, Using a Gentzen-type format, a general form of this principle can be formulated as follows:

$$\frac{\Gamma, \psi, \varphi \Rightarrow \Delta, \psi[x \mapsto y]}{\Gamma, \psi[x \mapsto s], (TC_{x,y}\varphi)(s, t) \Rightarrow \Delta, \psi[x \mapsto t]}$$

where x and y are not free in Γ, Δ, and y is not free in ψ.

Now if we are interested in set theories which are valid under the assumption that all sets are (hereditarily) finite, then the comprehension axiom remains valid if TC is included in the language, and the following clause is added to the definition of a safety relation in the extended language:

(TC-fin) $(TC_{x,y}\varphi)(x,y) \succ X$ if $\varphi \succ X$, and $\{x,y\} \subseteq X$.

On the other hand, for set theories which assume the existence of infinite sets the following stronger principle should be adopted:

(TC-inf) $(TC_{x,y}\varphi)(x,y) \succ X$ if $\varphi \succ X$, and $\{x,y\} \cap X \neq \emptyset$.

Let PST (for "Predicative Set Theory") be the extension of RST which has TC in its language, and is based on the safety relation $\succ_{PST}$ obtained from $\succ_{RST}$ by adding (TC-inf) as a new clause. Then the infinity axiom is derivable in PST, since one can introduce there the set $\mathcal{N}$ of natural numbers as follows:[11]

$$\mathcal{N} = \{x \mid x = \emptyset \vee \exists y.y = \emptyset \wedge (TC_{x,y}(x = S(y)))(x,y)\}$$

It is not difficult to see that PST still has the properties of RST described in Theorems 4 and 5.

Note. The set of valid formulas of TC-logic is not r.e. (or even arithmetical). Hence no sound and complete formal system for it is possible. It follows that PST and its extensions cannot be fully formalized, and so appropriate formal approximations (yet to be determined) of the underlying *logic* should be used.

References

1. Ackermann, W.: Zur Axiomatik der Mengenlehre. Mathematische Annalen 131, 336–345 (1956)
2. Abiteboul, S., Hull, R., Vianu, V.: Foundations of Databases. Addison-Wesley, Reading (1995)
3. Avron, A.: Transitive closure and the mechanization of mathematics. In: Kamareddine, F. (ed.) Thirty Five Years of Automating Mathematics, pp. 149–171. Kluwer Academic Publishers, Dordrecht (2003)
4. Avron, A.: Safety signatures for first-order languages and their applications. In: Hendricks, et al. (eds.) First-Order Logic Revisited, pp. 37–58. Logos Verlag, Berlin (2004)

---

[11] We could then adopt $\mathcal{N} = \omega$ as an axiom. Alternatively, we could omit the constant ω from the language, add the clause $x \neq x \succ_{PST} \{x\}$ to the definition of $\succ_{PST}$, define $\emptyset$ in PST as $\{x \mid x \neq x\}$, and then use the name ω rather than $\mathcal{N}$. Note also that if we start with BZF (or one of its extensions) as the basis, then $\emptyset$ (and so $\mathcal{N}$) may be defined in the language without the constant ω by the closed term:

$$\emptyset =_{DF} \{x \mid (\exists y.y = y) \wedge \forall y(y = y \to x \in y)\}.$$

5. Avron, A.: Formalizing set theory as it is actually used. In: Asperti, A., Bancerek, G., Trybulec, A. (eds.) MKM 2004. LNCS, vol. 3119, pp. 32–43. Springer, Heidelberg (2004)
6. Avron, A.: Constructibility and decidability versus domain independence and absluteness. Theoretical Computer Science (2007), doi:10.1016/j.tcs.2007.12.008
7. Cantone, D., Omodeo, E., Policriti, A.: Set Theory for Computing. Springer, Heidelberg (2001)
8. Devlin, K.J.: Constructibility. Perspectives in Mathematical Logic. Springer, Heidelberg (1984)
9. Di Paola, R.A.: The recursive unsolvability of the decision problem for the class of definite formulas. J. ACM 16(2), 324–327 (1969)
10. Fraenkel, A., Bar-Hillel, Y., Levy, A.: Foundations of Set Theory. North-Holland, Amsterdam (1973)
11. Feferman, S.: Finitary inductively presented logics. In: Logic Colloquium 1988, pp. 191–220. North-Holland, Amsterdam (1989)
12. Gandy, R.O.: Set-theoretic functions for elementary syntax. In: Axiomatic Set Theory, Part 2, pp. 103–126. AMS, Providence, Rhode Island (1974)
13. Gentzen, G.: Neue fassung des widerspruchsfreiheitsbeweises für die reine zahlentheorie. Forschungen zur Logik, N.S. (4), 19–44 (1938)
14. Immerman, N.: Languages which capture complexity classes. In: 15th Symposium on Theory of Computing, Association for Computing Machinery, pp. 347–354 (1983)
15. Jensen, R.B.: The fine structure of the constructible hierarchy. Annals of Mathematical Logic 4, 229–308 (1972)
16. Hallett, M.: Cantorian Set Theory and Limitation of Size. Clarendon Press, Oxford (1984)
17. Kunen, K.: Set Theory, An Introduction to Independence Proofs. North-Holland, Amsterdam (1980)
18. Levy, A.: Basic Set Theory. Springer, Heidelberg (1979)
19. Reinhardt, W.R.: Ackermann's set theory Equals ZF. Annals of Mathematical Logic 2, 189–249 (1970)
20. Shoenfield, J.R.: Mathematical Logic. Addison-Wesley, Reading (1967)
21. Shoenfield, J.R.: Axioms of set theory. In: Barwise, J. (ed.) Handbook of Mathematical Logic, North-Holland, Amsterdam (1977)
22. Shapiro, S.: Foundations Without Foundationalism: A Case for Second-order Logic. Oxford University Press, Oxford (1991)
23. Smullyan, R.M.: The Incompleteness Theorems. Oxford University Press, Oxford (1992)
24. Ullman, J.D.: Principles of Database and Knowledge-Base Systems. Computer Science Press (1988)

Effective Finite-Valued Approximations of General Propositional Logics

Matthias Baaz[1] and Richard Zach[2,*]

[1] Technische Universität Wien, Institut für Diskrete Mathematik und
Geometrie E104, A–1040 Vienna, Austria
baaz@logic.at
[2] University of Calgary, Department of Philosophy,
Calgary, Alberta T2N 1N4, Canada
rzach@ucalgary.ca

Dedicated to Professor Trakhtenbrot on the occasion of his 85th birthday.

Abstract. Propositional logics in general, considered as a set of sentences, can be undecidable even if they have "nice" representations, e.g., are given by a calculus. Even decidable propositional logics can be computationally complex (e.g., already intuitionistic logic is PSPACE-complete). On the other hand, finite-valued logics are computationally relatively simple—at worst NP. Moreover, finite-valued semantics are simple, and general methods for theorem proving exist. This raises the question to what extent and under what circumstances propositional logics represented in various ways can be approximated by finite-valued logics. It is shown that the minimal m-valued logic for which a given calculus is strongly sound can be calculated. It is also investigated under which conditions propositional logics can be characterized as the intersection of (effectively given) sequences of finite-valued logics.

1 Introduction

The question of what to do when faced with a new logical calculus is an old problem of mathematical logic. Often, at least at first, no semantics are available. For example, intuitionistic propositional logic was constructed by Heyting only as a calculus; semantics for it were proposed much later. Linear logic was in a similar situation in the early 1990s. The lack of semantical methods makes it difficult to answer questions such as: Are statements of a certain form (un)derivable? Are the axioms independent? Is the calculus consistent? For logics closed under substitution, many-valued methods have often proved valuable since they were first used for proving underivabilities by Bernays [5] in 1926 (and later by others, e.g., McKinsey and Wajsberg; see also [17, § 25]). The method is very simple. Suppose you find a many-valued logic in which the axioms of a given calculus are tautologies, the rules are sound, but the formula in question is not a tautology: then the formula cannot be derivable.

* Research supported by the Natural Sciences and Engineering Research Council of Canada.

A. Avron et al. (Eds.): Trakhtenbrot/Festschrift, LNCS 4800, pp. 107–129, 2008.

Example 1. Intuitionistic propositional logic is axiomatized by the following calculus **IPC**:

1. Axioms:

$$a_1 \quad A \supset A \wedge A$$
$$a_2 \quad A \wedge B \supset B \wedge A$$
$$a_3 \quad (A \supset B) \supset (A \wedge C \supset B \wedge C)$$
$$a_4 \quad (A \supset B) \wedge (B \supset C) \supset (A \supset C)$$
$$a_5 \quad B \supset (A \supset B)$$
$$a_6 \quad A \wedge (A \supset B) \supset B$$
$$a_7 \quad A \supset A \vee B$$
$$a_8 \quad A \vee B \supset B \vee A$$
$$a_9 \quad (A \supset C) \wedge (B \supset C) \supset (A \vee B \supset C)$$
$$a_{10} \quad \neg A \supset (A \supset B)$$
$$a_{11} \quad (A \supset B) \wedge (A \supset \neg B) \supset \neg A$$
$$a_{12} \quad A \supset (B \supset A \wedge B)$$

2. Rules (in usual notation):

$$\frac{A \quad A \supset B}{B} \text{ MP}$$

Now consider the two-valued logic with classical truth tables, except that $\neg$ maps both truth values to "true". Then every axiom except a_{10} is a tautology and modus ponens preserves truth. Hence a_{10} is independent of the other axioms.

To use this method to answer underivability question in general it is necessary to find many-valued matrices for which the given calculus is sound. It is also necessary, of course, that the matrix has as few tautologies as possible in order to be useful. We are interested in how far this method can be automatized.

Such "optimal" approximations of a given calculus may also have applications in computer science. In the field of artificial intelligence many new (propositional) logics have been introduced. They are usually better suited to model the problems dealt with in AI than traditional (classical, intuitionistic, or modal) logics, but many have two significant drawbacks: First, they are either given solely semantically or solely by a calculus. For practical purposes, a proof theory is necessary; otherwise computer representation of and automated search for proofs/truths in these logics is not feasible. Although satisfiability in many-valued propositional logics is (as in classical logic) NP-complete [16], this is still (probably) much better than many other important logics.

On the other hand, it is evident from the work of Carnielli [6] and Hähnle [12] on tableaux, and Rousseau, Takahashi, and Baaz et al. [2] on sequents, that finite-valued logics are, from the perspective of proof *and* model theory, very close to classical logic. Therefore, many-valued logic is a very suitable candidate if one looks for approximations, in some sense, of given complex logics.

What is needed are methods for obtaining finite-valued approximations of the propositional logics at hand. It turns out, however, that a shift of emphasis is in order here. While it is the *logic* we are actually interested in, we always are given only a *representation* of the logic. Hence, we have to concentrate on approximations of the representation, and not of the logic per se.

What is a representation of a logic? The first type of representation that comes to mind is a calculus. Hilbert-type calculi are the simplest conceptually and the oldest historically. We will investigate the relationship between such calculi on the one hand and many-valued logics or effectively enumerated sequences of many-valued logics on the other hand. The latter notion has received considerable attention in the literature in the form of the following two problems: Given a calculus $\mathbf{C}$,

1. find a minimal (finite) matrix for which $\mathbf{C}$ is sound (relevant for non-derivability and independence proofs), and
2. find a sequence of finite-valued logics, preferably effectively enumerable, whose intersection equals the theorems of $\mathbf{C}$, and its converse, given a sequence of finite-valued logics, find a calculus for its intersection (exemplified by Jaśkowski's sequence for intuitionistic propositional calculus, and by Dummett's extension axiomatizing the intersection of the sequence of Gödel logics, respectively).

For (1), of course, the best case would be a finite-valued logic $\mathbf{M}$ whose tautologies *coincide* with the theorems of $\mathbf{C}$. $\mathbf{C}$ then provides an axiomatization of $\mathbf{M}$. This of course is not always possible, at least for *finite*-valued logics. Lindenbaum [15, Satz 3] has shown that any logic (in our sense, a set of formulas closed under substitution) can be characterized by an *infinite*-valued logic. For a discussion of related questions see also Rescher [17, § 24].

In the following we study these questions in a general setting. Consider a propositional Hilbert-type calculus $\mathbf{C}$. It is (weakly) sound for a given m-valued logic if all its theorems are tautologies. Unfortunately, it turns out that it is undecidable if a calculus is sound for a given m-valued logic. However, for natural stronger soundness conditions this question is decidable; a finite-valued logic for which $\mathbf{C}$ satisfies such soundness conditions is called a *cover* for $\mathbf{C}$. The optimal (i.e., minimal under set inclusion of the tautologies) m-valued cover for $\mathbf{C}$ can be computed. The next question is, can we find an approximating sequence of m-valued logics in the sense of (2)? It is shown that this is impossible for undecidable calculi $\mathbf{C}$, and possible for all decidable logics closed under substitution. This leads us to the investigation of the *many-valued closure* $\mathrm{MC}(\mathbf{C})$ of $\mathbf{C}$, i.e., the set of formulas which are true in all covers of $\mathbf{C}$. In other words, if some formula can be shown to be underivable in $\mathbf{C}$ by a Bernays-style many-valued argument, it is not in the many-valued closure. Using this concept we can classify calculi according to their many-valued behaviour, or according to how good they can be dealt with by many-valued methods. In the best case $\mathrm{MC}(\mathbf{C})$ equals the theorems of $\mathbf{C}$ (This can be the case only if $\mathbf{C}$ is decidable). We give a sufficient condition for this being the case. Otherwise $\mathrm{MC}(\mathbf{C})$ is a proper superset of the theorems of $\mathbf{C}$.

Axiomatizations $\mathbf{C}$ and $\mathbf{C}'$ of the same logic may have different many-valued closures $\mathrm{MC}(\mathbf{C})$ and $\mathrm{MC}(\mathbf{C}')$ while being model-theoretically indistinguishable.

Hence, the many-valued closure can be used to distinguish between **C** and **C'** with regard to their proof-theoretic properties.

Finally, we investigate some of these questions for other representations of logics, namely for decision procedures and (effectively enumerated) finite Kripke models. In these cases approximating sequences of many-valued logics whose intersection equals the given logics can always be given.

Some of our results were previously reported in [4], of which this paper is a substantially revised and expanded version.

2   Propositional Logics

Definition 2. A *propositional language* $\mathcal{L}$ consists of the following:

1. propositional variables: $X_1, X_2, X_3, \ldots$
2. propositional connectives of arity n_j: $\square_1^{n_1}, \square_2^{n_2}, \ldots, \square_r^{n_r}$. If $n_j = 0$, then $\square_j$ is called a *propositional constant*.
3. Auxiliary symbols: (,), and , (comma).

Formulas and subformulas are defined as usual. We denote the set of formulas over a language $\mathcal{L}$ by $\mathrm{Frm}(\mathcal{L})$. By $\mathrm{Var}(A)$ we mean the set of propositional variables occurring in A. A *substitution* σ is a mapping of variables to formulas, and if F is a formula, $F\sigma$ is the result of simultaneously replacing each variable X in F by $\sigma(X)$.

Definition 3. The *depth* $\mathrm{dp}(A)$ of a formula A is defined as follows: $\mathrm{dp}(A) = 0$ if A is a variable or a 0-place connective (constant). If $A = \square(A_1, \ldots, A_n)$, then let $\mathrm{dp}(A) = \max\{\mathrm{dp}(A_1), \ldots, \mathrm{dp}(A_n)\} + 1$.

Definition 4. A *propositional Hilbert-type calculus* **C** in the language $\mathcal{L}$ is given by

1. a finite set $A(\mathbf{C}) \subseteq \mathrm{Frm}(\mathcal{L})$ of axioms.
2. a finite set $R(\mathbf{C})$ of rules of the form

$$\frac{A_1 \quad \ldots \quad A_n}{C} \; r$$

where $C, A_1, \ldots, A_n \in \mathrm{Frm}(\mathcal{L})$

A formula F is a *theorem* of **L** if there is a derivation of F in **C**, i.e., a finite sequence

$$F_1, F_2, \ldots, F_s = F$$

of formulas s.t. for each F_i there is a substitution σ so that either

1. $F_i = A\sigma$ where A is an axiom in $A(\mathbf{C})$, or
2. there are $F_{k_1}, \ldots, F_{k_n}$ with $k_j < i$ and a rule $r \in R(\mathbf{C})$ with premises $A_1, \ldots, A_n$ and conclusion C, s.t. $F_{k_j} = A_j\sigma$ and $F_i = C\sigma$.

If F is a theorem of $\mathbf{C}$ we write $\mathbf{C} \vdash F$. The set of theorems of $\mathbf{C}$ is denoted by $\mathrm{Thm}(\mathbf{C})$.

Remark 5. The above notion of a propositional rule is the one usually used in axiomatizations of propositional logic. It is, however, by no means the only possible notion. For instance, Schütte's rules

$$\frac{A(\top) \quad A(\bot)}{A(X)} \qquad \frac{C \leftrightarrow D}{A(C) \leftrightarrow A(D)}$$

where X is a propositional variable, and A, C, and D are formulas, does not fit under the above definition. And not only do they not fit this definition, the proof-theoretic behaviour of such rules is indeed significantly different from other "ordinary" rules. For instance, the rule on the left allows the derivation of all tautologies with n variables in number of steps linear in n; with a Hilbert-type calculus falling under the definition, this is not possible [3].

Remark 6. Many logics are more naturally axiomatized using sequent calculi, in which structure (sequences of formulas, sequent arrows) are used in addition to formulas. Many sequent calculi can easily be encoded in Hilbert-type calculi in an extended language, or even straightforwardly translated into Hilbert calculi in the same language, using constructions sketched below:

1. Sequences of formulas can be coded using a binary operator $\cdot$. The sequent arrow can simply be coded as a binary operator $\rightarrow$. For empty sequences, a constant Λ is used. We have the following rules, to assure associativity of $\cdot$:

$$\frac{X \cdot \big((U \cdot (V \cdot W)) \cdot Y\big) \rightarrow Z}{X \cdot \big(((U \cdot V) \cdot W) \cdot Y\big) \rightarrow Z} \qquad \frac{\big(X \cdot (U \cdot (V \cdot W))\big) \cdot Y \rightarrow Z}{\big(X \cdot ((U \cdot V) \cdot W)\big) \cdot Y \rightarrow Z}$$

   as well as the respective rules without X, without Y, without both X and Y, with the rules upside-down, and also for the right side of the sequent (20 rules total).

2. The usual sequent rules can be coded using the above constructions, e.g., the $\wedge$-Right rule of **LJ** would become:

$$\frac{U \rightarrow V \cdot X \quad U \rightarrow V \cdot Y}{U \rightarrow V \cdot (X \wedge Y)}$$

3. If the language of the logic in question contains constants and connectives which "behave like" the Λ and $\cdot$ on the left or right of a sequent, and a conditional which behaves like the sequent arrow, then no additional connectives are necessary. For instance, instead of $\cdot$, Λ on the left, use $\wedge$, $\top$; on the right, use $\vee$, $\bot$, and use $\supset$ instead of $\rightarrow$. Addition of the rule

$$\frac{\top \supset X}{X}$$

   would then result in a calculus which proves exactly the formulas F for which the sequent $\rightarrow F$ is provable in the original sequent calculus.

4. Some sequent rules require restrictions on the form of the side formulas in a rule, e.g., the □-right rule in modal logics:

$$\frac{\Box\Pi \to A}{\Box\Pi \to \Box A}$$

It is not immediately possible to accommodate such a rule in the translation. However, in some cases it can be replaced with another rule which can. E.g., in **S4**, it can be replaced by

$$\frac{\Pi \to A}{\Box\Pi \to \Box A}$$

which can in turn be accommodated using rules such as

$$\frac{X \supset Y}{\Box X \supset \Box Y} \quad \frac{U \wedge \Box(X \wedge Y) \supset V}{U \wedge (\Box X \wedge \Box Y) \supset V} \quad \frac{U \wedge \Box\Box Y \supset V}{U \wedge \Box Y \supset V}$$

(in the version with standard connectives serving as · and sequent arrow).

Definition 7. A propositional Hilbert-type calculus is called *strictly analytic* iff for every rule

$$\frac{A_1 \ldots A_n}{C} r$$

it holds that $\mathrm{Var}(A_i) \subseteq \mathrm{Var}(C)$ and $\mathrm{dp}(A_i\sigma) \le \mathrm{dp}(C\sigma)$ for every substitution σ.

This notion of strict analyticity is orthogonal to the one employed in the context of sequent calculi, where "analytic" is usually taken to mean that the rules have the subformula property (the formulas in the premises are subformulas of those in the conclusion). A strictly analytic calculus in our sense need not satisfy this. On the other hand, Hilbert calculi resulting from sequent calculi using the coding above need not be strictly analytic in our sense, even if the sequent calculus has the subformula property. For instance, the contraction rule does not satisfy the condition on the depth of substitution instances of the premises and conclusion. The standard notion of analyticity does not entail decidability, since for instance cut-free propositional linear logic **LL** is analytic but **LL** is undecidable [14]. Our notion of strict analyticity does entail decidability, since the depth of the conclusion of a rule in a proof is always greater or equal to the depth of the premises, and so the number of formulas that can appear in a proof of a given formula is finite.

Definition 8. A *propositional logic* **L** in the language $\mathcal{L}$ is a subset of $\mathrm{Frm}(\mathcal{L})$ closed under substitution.

Every propositional calculus **C** defines a propositional logic, namely $\mathrm{Thm}(\mathbf{C})$, since $\mathrm{Thm}(\mathbf{C})$ is closed under substitution. Not every propositional logic, however, is axiomatizable, let alone finitely axiomatizable by a Hilbert calculus. For instance, the logic

$$\{\Box^k(\top) \mid k \text{ is the Gödel number of a}$$
$$\text{true sentence of arithmetic}\}$$

is not axiomatizable, whereas the logic

$$\{\Box^k(\top) \mid k \text{ is prime}\}$$

is certainly axiomatizable (it is even decidable), but not by a Hilbert calculus using only $\Box$ and $\top$. (It is easily seen that any Hilbert calculus for $\Box$ and $\top$ has either only a finite number of theorems or yields arithmetic progressions of $\Box$'s.)

Definition 9. A *propositional finite-valued logic* $\mathbf{M}$ is given by a finite set of truth values $V(\mathbf{M})$, the set of *designated truth values* $V^+(\mathbf{M}) \subseteq V(\mathbf{M})$, and a set of truth functions $\widetilde{\Box}_j \colon V(\mathbf{M})^{n_j} \to V(\mathbf{M})$ for all connectives $\Box_j \in \mathcal{L}$ with arity n_j.

Definition 10. A *valuation* v is a mapping from the set of propositional variables into $V(\mathbf{M})$. A valuation v can be extended in the standard way to a function from formulas to truth values. v *satisfies* a formula F, in symbols: $v \models_{\mathbf{M}} F$, if $v(F) \in V^+(\mathbf{M})$. In that case, v is called a *model* of F, otherwise a *countermodel*. A formula F is a *tautology* of $\mathbf{M}$ iff it is satisfied by every valuation. Then we write $\mathbf{M} \models F$. We denote the set of tautologies of $\mathbf{M}$ by $\mathrm{Taut}(\mathbf{M})$.

Example 11. The sequence of m-valued Gödel logics $\mathbf{G}_m$ is given by $V(\mathbf{G}_m) = \{0, 1, \ldots, m-1\}$, the designated values $V^+(\mathbf{G}_m) = \{0\}$, and the following truth functions:

$$\widetilde{\neg}_{\mathbf{G}_m}(v) = \begin{cases} 0 & \text{for } v = m-1 \\ m-1 & \text{for } v \neq m-1 \end{cases}$$

$$\widetilde{\vee}_{\mathbf{G}_m}(v, w) = \min(a, b)$$
$$\widetilde{\wedge}_{\mathbf{G}_m}(v, w) = \max(a, b)$$

$$\widetilde{\supset}_{\mathbf{G}_m}(v, w) = \begin{cases} 0 & \text{for } v \geq w \\ w & \text{for } v < w \end{cases}$$

In the remaining sections, we will concentrate on the relations between propositional logics $\mathbf{L}$ represented in some way (e.g., by a calculus), and finite-valued logics $\mathbf{M}$. The objective is to find many-valued logics $\mathbf{M}$, or effectively enumerated sequences thereof, which, in a sense, approximate the the logic $\mathbf{L}$.

The following well-known product construction is useful for characterizing the "intersection" of many-valued logics.

Definition 12. Let $\mathbf{M}$ and $\mathbf{M}'$ be m and m'-valued logics, respectively. Then $\mathbf{M} \times \mathbf{M}'$ is the mm'-valued logic where $V(\mathbf{M} \times \mathbf{M}') = V(\mathbf{M}) \times V(\mathbf{M}')$, $V^+(\mathbf{M} \times \mathbf{M}') = V^+(\mathbf{M}) \times V^+(\mathbf{M}')$, and truth functions are defined component-wise. I.e., if $\Box$ is an n-ary connective, then

$$\widetilde{\Box}_{\mathbf{M} \times \mathbf{M}'}(w_1, \ldots, w_n) = \langle \widetilde{\Box}_{\mathbf{M}}(w_1, \ldots, w_n), \widetilde{\Box}_{\mathbf{M}'}(w_1, \ldots, w_n) \rangle.$$

For convenience, we define the following: Let v and v' be valuations of $\mathbf{M}$ and $\mathbf{M}'$, respectively. $v \times v'$ is the valuation of $\mathbf{M} \times \mathbf{M}'$ defined by: $(v \times v')(X) = \langle v(X), v'(X) \rangle$. If $v^\times$ is a valuation of $\mathbf{M} \times \mathbf{M}'$, then the valuations $\pi_1 v^\times$ and $\pi_2 v^\times$ of $\mathbf{M}$ and $\mathbf{M}'$, respectively, are defined by $\pi_1 v^\times(X) = v$ and $\pi_2 v^\times(X) = v'$ iff $v^\times(X) = \langle v, v' \rangle$.

Lemma 13. $\mathrm{Taut}(\mathbf{M} \times \mathbf{M}') = \mathrm{Taut}(\mathbf{M}) \cap \mathrm{Taut}(\mathbf{M}')$

Proof. Let A be a tautology of $\mathbf{M} \times \mathbf{M}'$ and v and v' be valuations of $\mathbf{M}$ and $\mathbf{M}'$, respectively. Since $v \times v' \models_{\mathbf{M} \times \mathbf{M}'} A$, we have $v \models_{\mathbf{M}} A$ and $v' \models_{\mathbf{M}'} A$ by the definition of $\times$. Conversely, let A be a tautology of both $\mathbf{M}$ and $\mathbf{M}'$, and let $v^\times$ be a valuation of $\mathbf{M} \times \mathbf{M}'$. Since $\pi_1 v^\times \models_{\mathbf{M}} A$ and $\pi_2 v^\times \models_{\mathbf{M}'} A$, it follows that $v^\times \models_{\mathbf{M} \times \mathbf{M}'} A$. □

The definition and lemma are easily generalized to the case of finite products $\prod_i \mathbf{M}_i$ by induction.

The construction of Lindenbaum [15, Satz 3] shows that every propositional logic can be characterized as the set of tautologies of an infinite-valued logic. $\mathbf{M}(\mathbf{L})$ is defined as follows: the set of truth values $V(\mathbf{M}(\mathbf{L})) = \mathrm{Frm}(\mathcal{L})$, and the set of designated values $V^+(\mathbf{M}(\mathbf{L})) = \mathbf{L}$. The truth functions are given by

$$\widetilde{\Box}(F_1, \ldots, F_n) = \Box(F_1, \ldots, F_n)$$

Since we are interested in finite-valued logics, the following constructions will be useful.

Definition 14. Let $\mathrm{Frm}_{i,j}(\mathcal{L})$ be the set of formulas of depth $\leq i$ containing only the variables $X_1, \ldots, X_j$. The finite-valued logic $\mathbf{M}_{i,j}(\mathbf{L})$ is defined as follows: The set of truth values of $\mathbf{M}_{i,j}(\mathbf{L})$ is $V = \mathrm{Frm}_{i,j}(\mathcal{L}) \cup \{\top\}$; the designated values $V^+ = (\mathrm{Frm}_{i,j}(\mathcal{L}) \cap \mathbf{L}) \cup \{\top\}$. The truth tables for $\mathbf{M}_{i,j}(\mathbf{L})$ are given by:

$$\widetilde{\Box}(v_1, \ldots, v_n) = \\ = \begin{cases} \Box(F_1, \ldots, F_n) & \text{if } v_j = F_j \text{ for } 1 \leq j \leq n \\ & \text{and } \Box(F_1, \ldots, F_n) \in \mathrm{Frm}_{i,j}(\mathcal{L}) \\ \top & \text{otherwise} \end{cases}$$

Proposition 15. *Let v be a valuation in $\mathbf{M}_{i,j}(\mathbf{L})$. If $v(X) \notin \mathrm{Frm}_{i,j}(\mathcal{L})$ for some $X \in \mathrm{Var}(A)$, then $v(A) = \top$. Otherwise, v can also be seen as a substitution σ_v assigning the formula $v(X) \in \mathrm{Frm}_{i,j}(\mathcal{L})$ to the variable X. Then $v(A) = A$ if $\mathrm{dp}(A\sigma_v) \leq i$ and $= \top$ otherwise.*

If $A \in \mathrm{Frm}_{i,j}(\mathcal{L})$, then $A \in \mathrm{Taut}(\mathbf{M}_{i,j}(\mathbf{L}))$ iff $A \in \mathbf{L}$; otherwise $A \in \mathrm{Taut}(\mathbf{M}_{i,j}(\mathbf{L}))$. In particular, $\mathbf{L} \subseteq \mathrm{Taut}(\mathbf{M}_{i,j}(\mathbf{L}))$.

Proof. By induction on the depth of A. □

When looking for a logic with as small a number of truth values as possible which falsifies a given formula we can use the following construction.

Proposition 16. *Let $\mathbf{M}$ be any many-valued logic, and $A_1, \ldots, A_n$ be formulas not valid in $\mathbf{M}$. Then there is a finite-valued logic $\mathbf{M}' = \Phi(\mathbf{M}, A_1, \ldots, A_n)$ s.t.*

1. *$A_1, \ldots, A_n$ are not valid in $\mathbf{M}'$,*
2. *$\mathrm{Taut}(\mathbf{M}) \subseteq \mathrm{Taut}(\mathbf{M}')$, and*

3. $|V(\mathbf{M'})| \leq \xi(A_1, \ldots, A_n)$, where $\xi(A_1, \ldots, A_n) = \prod_{i=1}^{n} \xi(A_i)$ and $\xi(A_i)$ is the number of subformulas of A_i $+1$.

Proof. We first prove the proposition for $n = 1$. Let v be the valuation in $\mathbf{M}$ making A_1 false, and let $B_1, \ldots, B_r$ $(\xi(A_1) = r + 1)$ be all subformulas of A_1. Every B_i has a truth value t_i in v. Let $\mathbf{M'}$ be as follows: $V(\mathbf{M'}) = \{t_1, \ldots, t_r, \top\}$, $V^+(\mathbf{M'}) = V^+(\mathbf{M}) \cap V(\mathbf{M'}) \cup \{\top\}$. If $\Box \in \mathcal{L}$, define $\widetilde{\Box}$ by

$$\widetilde{\Box}(v_1, \ldots, v_n) = \begin{cases} t_i & \text{if } B_i \equiv \Box(B_{j_1}, \ldots, B_{j_n}) \\ & \text{and } v_1 = t_{j_1}, \ldots, v_n = t_{j_n} \\ \top & \text{otherwise} \end{cases}$$

(1) Since t_r was undesignated in $\mathbf{M}$, it is also undesignated in $\mathbf{M'}$. But v is also a truth value assignment in $\mathbf{M'}$, hence $\mathbf{M'} \not\models A_1$.

(2) Let C be a tautology of $\mathbf{M}$, and let w be a valuation in $\mathbf{M'}$. If no subformula of C evaluates to $\top$ under w, then w is also a valuation in $\mathbf{M}$, and C takes the same truth value in $\mathbf{M'}$ as in $\mathbf{M}$ w.r.t. w, which is designated also in $\mathbf{M'}$. Otherwise, C evaluates to $\top$, which is designated in $\mathbf{M'}$. So C is a tautology in $\mathbf{M'}$.

(3) Obvious.

For $n > 1$, the proposition follows by taking $\Phi(\mathbf{M}, A_1, \ldots, A_n) = \prod_{i=1}^{n} \Phi(\mathbf{M}, A_i)$ □

3   Many-Valued Covers for Propositional Calculi

A very natural way of representing logics is via calculi. In the context of our study, one important question is under what conditions it is possible to find, given a calculus $\mathbf{C}$, a finite-valued logic $\mathbf{M}$ which approximates as well as possible the set of theorems $\mathrm{Thm}(\mathbf{C})$. In the optimal case, of course, we would like to have $\mathrm{Taut}(\mathbf{M}) = \mathrm{Thm}(\mathbf{C})$. This is, however, not always possible. In fact, it is in general not even possible to decide, given a calculus $\mathbf{C}$ and a finite-valued logic $\mathbf{M}$, if $\mathbf{M}$ is sound for $\mathbf{C}$. In some circumstances, however, general results can be obtained. We begin with some definitions.

Definition 17. A calculus $\mathbf{C}$ is *weakly sound* for an m-valued logic $\mathbf{M}$ provided $\mathrm{Thm}(\mathbf{C}) \subseteq \mathrm{Taut}(\mathbf{M})$.

Definition 18. A calculus $\mathbf{C}$ is *t-sound* for an m-valued logic $\mathbf{M}$ if

(∗) All axioms $A \in A(\mathbf{C})$ are tautologies of $\mathbf{M}$, and for every rule $r \in R(\mathbf{C})$ and substitution σ: if for every premise A of r, $A\sigma$ is a tautology, then the corresponding instance $C\sigma$ of the conclusion of r is a tautology as well.

Definition 19. A calculus $\mathbf{C}$ is *strongly sound* for an m-valued logic $\mathbf{M}$ if

(∗∗) All axioms $A \in A(\mathbf{C})$ are tautologies of $\mathbf{M}$, and for every rule $r \in R(\mathbf{C})$: if a valuation satisfies the premises of r, it also satisfies the conclusion.

$\mathbf{M}$ is then called a *cover* for $\mathbf{C}$.

We would like to stress the distinction between these three notions of soundness. soundness. The notion of weak soundness is the familiar property of a calculus to produce only valid formulas (in this case: tautologies of **M**) as theorems. This "plain" soundness is what we actually would like to investigate in terms of approximations. More precisely, when looking for a finite-valued logic that approximates a given calculus, we are content if we find a logic for which **C** is weakly sound. This is unfortunately not possible in general.

Proposition 20. *It is undecidable if a calculus **C** is weakly sound for a given m-valued logic **M**.*

Proof. Let **C** be an undecidable propositional calculus, let F be a formula, and let C and, for each $X_i \in \text{Var}(F)$, C_i be new propositional constants (0-ary connective) not occurring in **C**. Let $\sigma \colon X_i \mapsto C_i$ be a substitution. Clearly, $\mathbf{C} \vdash F$ iff $\mathbf{C} \vdash F\sigma$. Now let **C′** be **C** with the additional rule

$$\frac{F\sigma}{C}$$

and let **M** be an m-valued logic which assigns a non-designated value to C and otherwise interprets every connective as a constant function with a designated value. Then every formula except a variable of the original language is a tautology, and C is not. **M** is then weakly sound for **C** over the original language, but weakly sound for **C′** iff C is not derivable. Moreover, $\mathbf{C'} \vdash C$ iff $\mathbf{C'} \vdash F\sigma$, i.e., iff $\mathbf{C} \vdash F$. If it were decidable whether **M** is weakly sound for **C′** it would then also be decidable if $\mathbf{C} \vdash F$, contrary to the assumption that **C** is undecidable. □

On the other hand, it *is* obviously decidable if **C** is strongly sound for a given matrix **M**.

Proposition 21. *It is decidable if a given propositional calculus is strongly sound for a given m-valued logic.*

Proof. (∗∗) can be tested by the usual truth-table method. □

It is also decidable if **C** is t-sound for a matrix **M**, although this is less obvious:

Proposition 22. *It is decidable if a given propositional calculus is t-sound for a given m-valued logic **M**.*

Proof. Let r be a rule with premises $A_1, \ldots, A_n$ and conclusion C containing the variables $X_1, \ldots, X_k$, and σ a substitution. If $A_1\sigma, \ldots, A_n\sigma$ are tautologies, but C is not, then (∗) is violated and r is not weakly sound. Given σ, this is clearly decidable. We have to show that there are only a finite number of substitutions σ which we have to test.

Let $Y_1, \ldots, Y_l$ be the variables occurring in $X_1\sigma, \ldots, X_k\sigma$. We show first that it suffices to consider σ with $l = m$. For if v is a valuation in which $C\sigma$ is false, then at most m of $Y_1, \ldots, Y_l$ have different truth values. Let τ be a substitution so that $\tau(Y_i) = Y_j$ where j is the least index such that $v(Y_j) = v(Y_i)$. Then (1)

$v(C\sigma\tau) = v(C\sigma)$ and hence $C\sigma\tau$ is not a tautology; (2) $A_i\sigma\tau$ is still a tautology; (3) there are at most m distinct variables occurring in $A_1\sigma\tau$, ..., $A_n\sigma\tau$, $C\sigma\tau$.

Now every $B_i = X_i\sigma$ defines an m-valued function of m arguments. There are m^{m^m} such functions. Whether $A_i\sigma$ is a tautology only depends on the function defined by B_i, but it is not prima facie clear which functions can be expressed in $\mathbf{M}$. Nevertheless, we can give a bound on the depth of formulas B_i that have to be considered. Suppose σ is a substitution of the required form with $B_i = X_i\sigma$ of minimal depth and suppose that the depth of B_i is greater than $m' = m^{m^m}$. Now consider a sequence of formulas C_1, ..., $C_{m'+1}$ with $C_1 = B_i$ and each C_j a subformula of C_{j-1}. Each C_j also expresses an m-valued function of m arguments. Since there are only m' different such functions, there are $j < j'$ so that C_j and $C_{j'}$ define the same function. The formula obtained from B_i by replacing C_j by $C_{j'}$ expresses the same function. Since this can be done for every sequence of C_j's of length $> m'$ we eventually obtain a formula which expresses the same function as B_i but of depth $\leq m'$, contrary to the assumption that it was of minimal depth.   □

Now, if $\mathbf{C}$ is strongly sound for $\mathbf{M}$, it is also t-sound; and if it is t-sound, it is also weakly sound. The converses, however, are false:

Example 23. Let $\mathcal{L}$ be the language consisting of a unary connective $\square$ and a binary connective $\lhd$, and let $\mathbf{C}$ be the calculus consisting of the sole axiom $X \lhd \square X$ and the rules

$$\frac{X \lhd Y \quad Y \lhd Z}{X \lhd Z}\ r_1 \qquad \frac{X \lhd X}{Y}\ r_2$$

It is easy to see that the only derivable formulas in $\mathbf{C}$ using only rule r_1 are substitution instances of $\square^\ell X \lhd \square^k X$ with $\ell < k$. In particular, no substitution instance of the premise of r_2, $X \lhd X$, is derivable. It follows that rule r_2 can never be applied. We now show that if $\mathbf{C}$ is strongly sound for an m-valued matrix $\mathbf{M}$, Taut$(\mathbf{M}) = \text{Frm}(\mathcal{L})$, i.e., $\mathbf{M}$ is trivial. Suppose $\mathbf{M}$ is given by the set of truth values $V = \{1, \ldots, m\}$. Since $X \lhd \square X$ must be a tautology, $\widetilde{\lhd}(i, \widetilde{\square}(i)) \in V^+$ for $i = 1$, ..., m. Since $\mathbf{C}$ is strongly sound for rule r_1, and by induction, $\widetilde{\lhd}(i, \widetilde{\square}^k(i)) \in V^+$ for all k. Since V is finite, there are i and k such that $i = \widetilde{\square}^k(i)$. Then $\widetilde{\lhd}(i, i) \in V^+$. Since $\mathbf{C}$ is strongly sound for r_2, we have $V = V^+$. However, $\mathbf{C}$ is weakly sound for non-trivial matrices, e.g., $\mathbf{M}'$ with $V' = \{1, \ldots, k\}$, $V^+ = \{k\}$, $\widetilde{\square}(i) = i + 1$ for $i < k$ and $= k$ otherwise, and $\widetilde{\lhd}(i, j) = k$ if $i < j$ or $j = k$ and $= 1$ otherwise. $\mathbf{C}$ is, however, also not t-sound for this matrix.

Example 24. Consider the calculus with propositional constants T, F, and binary connective $\neq$, the axiom $T \neq F$ and the rules

$$\frac{Y \neq X}{X \neq Y}\ r_1 \qquad \frac{X \neq T \quad X \neq F}{Y}\ r_2$$

and the matrix with $V = \{0, 1, 2\}$, $V^+ = \{2\}$, $\widetilde{T} = 2$, $\widetilde{F} = 0$, and $\widetilde{\neq}(i, j) = 2$ if $i \neq j$ and $= 0$ otherwise. Clearly, the only derivable formulas are $T \neq F$ and

$F \neq T$, which are also tautologies. The calculus is not strongly sound, since for $v(X) = 1$, $v(Y) = 0$ the premises of r_2 are designated, but the conclusion is not. It is, however, t-sound: only a substitution σ with $v(X\sigma) = 1$ for all valuations v would turn both premises of r_2 into tautologies, and there can be no such formulas. Hence, we have an example of a calculus t-sound but not strongly sound for a matrix.

Example 25. The **IPC** is strongly sound for the m-valued Gödel logics $\mathbf{G}_m$. For instance, take axiom a_5: $B \supset (A \supset B)$. This is a tautology in $\mathbf{G}_m$, for assume we assign some truth values a and b to A and B, respectively. We have two cases: If $a \leq b$, then $(A \supset B)$ takes the value $m-1$. Whatever b is, it certainly is $\leq m-1$, hence $B \supset A \supset B$ takes the designated value $m - 1$. Otherwise, $A \supset B$ takes the value b, and again (since $b \leq b$), $B \supset A \supset B$ takes the value $m - 1$.

Modus ponens passes the test: Assume A and $A \supset B$ both take the value $m - 1$. This means that $a \leq b$. But $a = m - 1$, hence $b = m - 1$.

Now consider the following extension $\mathbf{G}_m^\top$ of $\mathbf{G}_m$: $V(\mathbf{G}_m^\top) = V(\mathbf{G}_m) \cup \{\top\}$, $V^+(\mathbf{G}_m^\top) = \{m - 1, \top\}$, and the truth functions are given by:

$$\widetilde{\Box}_{\mathbf{G}_m^\top}(\bar{v}) = \begin{cases} \top & \text{if } \top \in \bar{v} \\ \widetilde{\Box}_{\mathbf{G}_m}(\bar{v}) & \text{otherwise} \end{cases}$$

for $\Box \in \{\neg, \supset, \wedge, \vee\}$. **IPC** is not strongly sound for $\mathbf{G}_m^\top$, since a valuation with $v(X) = \top$, $v(Y) = 0$ would satisfy the premises of rule MP, X and $X \supset Y$, but not the conclusion Y. However, a calculus in which the conclusion of each rule contains all variables occurring in the premises, is strongly sound (such as a calculus obtained from **LJ** using the construction outlined in Remark 6).

Example 26. Consider the following calculus **K**:

$$X \leftrightarrow \bigcirc X \qquad \frac{X \leftrightarrow Y}{X \leftrightarrow \bigcirc Y} \; r_1 \qquad \frac{X \leftrightarrow X}{Y} \; r_2$$

It is easy to see that the corresponding logic consists of all instances of $X \leftrightarrow \bigcirc^k X$ where $k \geq 1$. This calculus is only strongly sound for the m-valued logic having all formulas as its tautologies. But if we leave out r_2, we can give a sequence of many-valued logics $\mathbf{M}_i$, for each of which **K** is strongly sound: Take for $V(\mathbf{M}_n) = \{0, \ldots, n - 1\}$, $V^+(\mathbf{M}_n) = \{0\}$, with the following truth functions:

$$\widetilde{\bigcirc} v = \begin{cases} v + 1 & \text{if } v < n - 1 \\ n - 1 & \text{otherwise} \end{cases}$$

$$v \widetilde{\leftrightarrow} w = \begin{cases} 0 & \text{if } v < w \text{ or } v = n - 1 \\ 1 & \text{otherwise} \end{cases}$$

Obviously, $\mathbf{M}_n$ is a cover for **K**. On the other hand, $\mathrm{Taut}(\mathbf{M}_n) \neq \mathrm{Frm}(\mathcal{L})$, e.g., any formula of the form $\bigcirc(A)$ takes a (non-designated) value > 0 (for $n > 1$). In fact, every formula of the form $\bigcirc^k X \leftrightarrow X$ is falsified in some $\mathbf{M}_n$.

4   Optimal Covers

By Proposition 21 it is decidable if a given m-valued logic $\mathbf{M}$ is a cover of $\mathbf{C}$. Since we can enumerate all m-valued logics, we can also find all covers of $\mathbf{C}$. Moreover, comparing two many-valued logics as to their sets of tautologies is decidable, as the next theorem will show. Using this result, we see that we can always generate optimal covers for $\mathbf{C}$.

Definition 27. For two many-valued logics $\mathbf{M}_1$ and $\mathbf{M}_2$, we write $\mathbf{M}_1 \trianglelefteq \mathbf{M}_2$ iff $\mathrm{Taut}(\mathbf{M}_1) \subseteq \mathrm{Taut}(\mathbf{M}_2)$.

$\mathbf{M}_1$ is *better* than $\mathbf{M}_2$, $\mathbf{M}_1 \lhd \mathbf{M}_2$, iff $\mathbf{M}_1 \trianglelefteq \mathbf{M}_2$ and $\mathrm{Taut}(\mathbf{M}_1) \neq \mathrm{Taut}(\mathbf{M}_2)$.

Theorem 28. *Let two logics $\mathbf{M}_1$ and $\mathbf{M}_2$, m_1-valued and m_2-valued respectively, be given. It is decidable whether $\mathbf{M}_1 \lhd \mathbf{M}_2$.*

Proof. It suffices to show the decidability of the following property: There is a formula A, s.t. (*) $\mathbf{M}_2 \models A$ but $\mathbf{M}_1 \not\models A$. If this is the case, write $\mathbf{M}_1 \lhd^* \mathbf{M}_2$. $\mathbf{M}_1 \lhd \mathbf{M}_2$ iff $\mathbf{M}_1 \lhd^* \mathbf{M}_2$ and not $\mathbf{M}_2 \lhd^* \mathbf{M}_1$.

We show this by giving an upper bound on the depth of a minimal formula A satisfying the above property. Since the set of formulas of $\mathcal{L}$ is enumerable, bounded search will produce such a formula iff it exists. Note that the property (*) is decidable by enumerating all assignments. In the following, let $m = \max(m_1, m_2)$.

Let A be a formula that satisfies (*), i.e., there is a valuation v s.t. $v \not\models_{\mathbf{M}_1} A$. W.l.o.g. we can assume that A contains at most m different variables: if it contained more, some of them must be evaluated to the same truth value in the counterexample v for $\mathbf{M}_1 \not\models A$. Unifying these variables leaves (*) intact.

Let $B = \{B_1, B_2, \ldots\}$ be the set of all subformulas of A. Every formula B_j defines an m-valued truth function $f(B_j)$ of m variables where the values of the variables which actually occur in B_j determine the value of $f(B_j)$ via the matrix of $\mathbf{M}_2$. On the other hand, every B_j evaluates to a single truth value $t(B_j)$ in the countermodel v.

Consider the formula A' constructed from A as follows: Let B_i be a subformula of A and B_j be a proper subformula of B_i (and hence, a proper subformula of A). If $f(B_i) = f(B_j)$ and $t(B_i) = t(B_j)$, replace B_i in A with B_j. A' is shorter than A, and it still satisfies (*). By iterating this construction until no two subformulas have the desired property we obtain a formula A^*. This procedure terminates, since A' is shorter than A; it preserves (*), since A' remains a tautology under $\mathbf{M}_2$ (we replace subformulas behaving in exactly the same way under all valuations) and the countermodel v is also a countermodel for A'.

The depth of A^* is bounded above by $m^{m^m+1} - 1$. This is seen as follows: If the depth of A^* is d, then there is a sequence $A^* = B'_0, B'_1, \ldots, B'_d$ of subformulas of A^* where B'_k is an immediate subformula of B'_{k-1}. Every such B'_k defines a truth function $f(B'_k)$ of m variables in $\mathbf{M}_2$ and a truth valued $t(B'_k)$ in $\mathbf{M}_1$ via v. There are m^{m^m} m-ary truth functions of m truth values. The number of distinct truth function-truth value pairs then is m^{m^m+1}. If $d \geq m^{m^m+1}$, then two of the B'_k, say B'_i and B'_j where B'_j is a subformula of B'_i define the same truth function and the same truth value. But then B'_i could be replaced by B'_j, contradicting the way A^* is defined. $\qquad\Box$

Corollary 29. *It is decidable if two many-valued logics define the same set of tautologies. The relation $\trianglelefteq$ is decidable.*

Proof. $\mathrm{Taut}(\mathbf{M}_1) = \mathrm{Taut}(\mathbf{M}_2)$ iff neither $\mathbf{M}_1 \lhd^* \mathbf{M}_2$ nor $\mathbf{M}_2 \lhd^* \mathbf{M}_1$. $\qquad\Box$

Let $\simeq$ be the equivalence relation on m-valued logics defined by: $\mathbf{M}_1 \simeq \mathbf{M}_2$ iff $\mathrm{Taut}(M_1) = \mathrm{Taut}(M_2)$, and let MVL_m be the set of all m-valued logics over $\mathcal{L}$ with truth value set $\{1, \ldots, m\}$. By $\mathcal{M}_m$ we denote the set of all sets $\mathrm{Taut}(\mathbf{M})$ of tautologies of m-valued logics $\mathbf{M}$. The partial order $\langle \mathcal{M}_m, \subseteq \rangle$ is isomorphic to $\langle \mathrm{MVL}_m/ \simeq, \trianglelefteq / \simeq \rangle$.

Proposition 30. *The optimal (i.e., minimal under $\lhd$) m-valued covers of $\mathbf{C}$ are computable.*

Proof. Consider the set $C_m(\mathbf{C})$ of m-valued covers of $\mathbf{C}$. Since $C_m(\mathbf{C})$ is finite and partially ordered by $\trianglelefteq$, $C_m(\mathbf{C})$ contains minimal elements. The relation $\trianglelefteq$ is decidable, hence the minimal covers can be computed. $\qquad\Box$

Example 31. By Example 25, **IPC** is strongly sound for $\mathbf{G}_3$. The best 3-valued approximation of **IPC** is the 3-valued Gödel logic. In fact, it is the only 3-valued approximation of *any* sound calculus $\mathbf{C}$ (containing modus ponens) for **IPL** which has less tautologies than classical logic **CL**. This can be seen as follows: Consider the fragment containing $\perp$ and $\supset$ ($\neg B$ is usually defined as $B \supset \perp$). Let $\mathbf{M}$ be some 3-valued strongly sound approximation of $\mathbf{C}$. By Gödel's double-negation translation, B is a classical tautology iff $\neg\neg B$ is true intuitionistically. Hence, whenever $\mathbf{M} \models \neg\neg X \supset X$, then $\mathrm{Taut}(\mathbf{M}) \supseteq \mathbf{CL}$. Let 0 denote the value of $\perp$ in $\mathbf{M}$, and let $1 \in V^+(\mathbf{M})$. We distinguish cases:

1. $0 \in V^+(\mathbf{M})$: Then $\mathrm{Taut}(\mathbf{M}) = \mathrm{Frm}(\mathcal{L})$, since $\perp \supset X$ is true intuitionistically, and by modus ponens: $\perp, \perp \supset X/X$.
2. $0 \notin V^+(\mathbf{M})$: Let u be the third truth value.
   (a) $u \in V^+(\mathbf{M})$: Consider $A \equiv ((X \supset \perp) \supset \perp) \supset X$. If $v(X)$ is u or 1, then, since everything implies something true, A is true (Note that we have $Y, Y \supset (X \supset Y) \vdash X \supset Y$). If $v(X) = 0$, then (since $0 \supset 0$ is true, but $u \supset 0$ and $1 \supset 0$ are both false), A is true as well. So $\mathrm{Taut}(\mathbf{M}) \supseteq \mathbf{CL}$.
   (b) $u \notin V^+(\mathbf{M})$, i.e., $V^+(\mathbf{M}) = \{1\}$: Consider the truth table for implication. Since $B \supset B$, $\perp \supset B$, and something true is implied by everything, the upper right triangle is 1. We have the following table:

$$\begin{array}{c|ccc} \supset & 0 & u & 1 \\ \hline 0 & 1 & 1 & 1 \\ u & v_1 & 1 & 1 \\ 1 & v_0 & v_2 & 1 \end{array}$$

Clearly, v_0 cannot be 1. If $v_0 = u$, we have, by $((X \supset X) \supset \perp) \supset Y$, that $v_1 = 1$. In this case, $\mathbf{M} \models A$ and hence $\mathrm{Taut}(\mathbf{M}) \supseteq \mathbf{CL}$. So assume $v_0 = 0$.

    i. $v_1 = 1$: $\mathbf{M} \models A$ (Note that only the case of $((u \supset 0) \supset 0) \supset u$ has to be checked).

    ii. $v_1 = u$: $\mathbf{M} \not\models A$.

    iii. $v_1 = 0$: With $v_2 = 0$, $\mathbf{M}$ would be incorrect ($u \supset (1 \supset u)$ is false). If $v_2 = 1$, again $\mathbf{M} \models A$. The case of $v_2 = u$ is the Gödel logic, where A is not a tautology.

Note that it is in general impossible to effectively construct a $\trianglelefteq$-minimal m-valued logic $\mathbf{M}$ with $\mathbf{L} \subseteq \mathrm{Taut}(\mathbf{M})$ if $\mathbf{L}$ is given independently of a calculus, because, e.g., it is undecidable whether $\mathbf{L}$ is empty or not: e.g., take

$$\mathbf{L} = \begin{cases} \{\Box^k(\top)\} & \text{if } k \text{ is the least solution of } D(x) = 0 \\ \emptyset & \text{otherwise} \end{cases}$$

where $D(x) = 0$ is the Diophantine representation of some undecidable set.

5   Effective Sequential Approximations

In the previous section we have shown that it is always possible to obtain the best m-valued covers of a given calculus, but there is no way to tell *how good* these covers are. In this section, we investigate the relation between sequences of many-valued logics and the set of theorems of a calculus $\mathbf{C}$. Such sequences are called *sequential approximations* of $\mathbf{C}$ if they verify all theorems and refute all non-theorems of $\mathbf{C}$, and *effective* sequential approximations if they are effectively enumerable. This is also a question about the limitations of Bernays's method. On the negative side, an immediate result says that calculi for undecidable logics do not have effective sequential approximations. If, however, a propositional logic is decidable, it also has an effective sequential approximation (independent of a calculus). Moreover, any calculus has a uniquely defined *many-valued closure*, whether it is decidable or not. This is the set of all sentences which cannot be proved underivable using a Bernays-style many-valued argument. If a calculus has an effective sequential approximation, then the set of its theorems equals its many-valued closure. If it does not, then its closure is a proper superset. Different calculi for one and the same logic may have different many-valued closures according to their degree of analyticity.

Definition 32. Let $\mathbf{L}$ be a propositional logic and let $\mathbf{A} = \langle \mathbf{M}_1, \mathbf{M}_2, \mathbf{M}_3, \ldots, \rangle$ be a sequence of many-valued logics s.t. (1) $\mathbf{M}_i \trianglelefteq \mathbf{M}_j$ iff $i \geq j$.

    $\mathbf{A}$ is called a *sequential approximation* of $\mathbf{L}$ iff $\mathbf{L} = \bigcap_{j \in \omega} \mathrm{Taut}(\mathbf{M}_j)$. If in addition $\mathbf{A}$ is effectively enumerated, then $\mathbf{A}$ is an *effective sequential approximation*.

    If $\mathbf{L}$ is given by the calculus $\mathbf{C}$, and each M_j is a cover of $\mathbf{C}$, then $\mathbf{A}$ is a *strong (effective) sequential approximation* of $\mathbf{C}$ (if $\mathbf{A}$ is effectively enumerated).

    We say $\mathbf{C}$ is *effectively approximable*, if there is such a strong effective sequential approximation of $\mathbf{C}$.

Condition (1) above is technically not necessary. Approximating sequences of logics in the literature (see next example), however, satisfy this condition. Furthermore, with the emphasis on "approximation," it seems more natural that the sequence gets successively "better."

Example 33. Consider the sequence $\mathbf{G} = \langle \mathbf{G}_i \rangle_{i \geq 2}$ of Gödel logics and intuitionistic propositional logic $\mathbf{IPC}$. $\mathrm{Taut}(\mathbf{G}_i) \supset \mathrm{Thm}(\mathbf{IPC})$, since $\mathbf{G}_i$ is a cover for $\mathbf{IPC}$. Furthermore, $\mathbf{G}_{i+1} \lhd \mathbf{G}_i$. This has been pointed out by [10], for a detailed proof see [11, Theorem 10.1.2]. It is, however, not a sequential approximation of $\mathbf{IPC}$: The formula $(A \supset B) \vee (B \supset A)$, while not a theorem of $\mathbf{IPL}$, is a tautology of all $\mathbf{G}_i$. In fact, $\bigcap_{i \geq 2} \mathrm{Taut}(\mathbf{G}_i)$ is the set of tautologies of the infinite-valued Gödel logic $\mathbf{G}_\aleph$, which is axiomatized by the rules of $\mathbf{IPC}$ plus the above formula. This has been shown in [8] (see also [11, Section 10.1]). Hence, $\mathbf{G}$ is a strong effective sequential approximation of $\mathbf{G}_\aleph = \mathbf{IPC} + (A \supset B) \vee (B \supset A)$.

Jaśkowski [13] gave an effective strong sequential approximation of $\mathbf{IPC}$. That $\mathbf{IPC}$ is approximable is also a consequence of Theorem 48, with the proof adapted to Kripke semantics for intuitionistic propositional logic, since $\mathbf{IPL}$ has the finite model property [9, Ch. 4, Theorem 4(a)].

The natural question to ask is: Which logics have (effective) sequential approximations; which calculi are approximable?

First of all, any propositional logic has a sequential approximation, although it need not have an effective approximation.

Proposition 34. *Every propositional logic $\mathbf{L}$ has a sequential approximation.*

Proof. A sequential approximation of $\mathbf{L}$ a is given by $\mathbf{M}_i = \mathbf{M}_{i,i}(\mathbf{L})$ (see Definition 14). Any formula $F \notin \mathbf{L}$ is in $V(\mathbf{M}_k)$ for $k = \max\{\mathrm{dp}(F), j\}$ where j is the maximum index of variables occurring in F. By Proposition 15, F is falsified in $\mathbf{M}_k$. Also, $\mathrm{Taut}(\mathbf{M}_i) \supseteq \mathbf{L}$, and $\mathbf{M}_i \lhd \mathbf{M}_{i+1}$.   □

Corollary 35. *If $\mathbf{L}$ is decidable, it has an effective sequential approximation.*

Proof. Using a decision procedure for $\mathbf{L}$, we can effectively enumerate the $\mathbf{M}_{i,i}(\mathbf{L})$.   □

Proposition 36. *If $\mathbf{L}$ has an effectively sequential approximation, then $\mathrm{Frm}(\mathcal{L}) \setminus \mathbf{L}$ is effectively enumerable.*

Proof. Suppose there is an effectively enumerated sequence $\mathbf{A} = \langle \mathbf{M}_1, \mathbf{M}_2, \ldots \rangle$ s.t. $\bigcap_{j \geq 1} \mathrm{Taut}(\mathbf{M}_j) = \mathbf{L}$. If $F \notin \mathbf{L}$ then there would be an index i s.t. F is false in $\mathbf{M}_i$. But this would yield a semi-decision procedure for non-members of $\mathbf{L}$: Try for each j whether F is false in $\mathbf{M}_j$. If $F \notin \mathbf{L}$, this will be established at $j = i$.   □

Corollary 37. *If $\mathbf{C}$ is undecidable, then it is not effectively approximable.*

Proof. $\mathrm{Thm}(\mathbf{C})$ is effectively enumerable. If $\mathbf{C}$ were approximable, it would have an effective sequential approximation, and this contradicts the assumption that the non-theorems of $\mathbf{C}$ are not effectively enumerable.   □

Example 38. This shows that a result similar to that for **IPC** cannot be obtained for full propositional linear logic.

If **C** is not effectively approximable (e.g., if it is undecidable), then the intersection of all covers for **C** is a proper superset of Thm(**C**). This intersection has interesting properties.

Definition 39. The *many-valued closure* MC(**C**) of a calculus **C** is the set of formulas which are true in every many-valued cover for **C**.

Proposition 40. MC(**C**) *is unique and has an effective sequential approximation.*

Proof. MC(**C**) is unique, since it obviously equals $\bigcap_{\mathbf{M} \in S}$ Taut(**M**) where S is the set of all covers for **C**. It is also effectively approximable, an approximating sequence is given by

$$\mathbf{M}_1 = \mathbf{M}'_1$$
$$\mathbf{M}_i = \mathbf{M}_{i-1} \times \mathbf{M}'_i$$

where $\mathbf{M}'_i$ is an effective enumeration of S.   □

Since MC(**C**) is defined via the many-valued logics for which **C** is strongly sound, it need not be the case that MC(**C**) = Thm(**C**) even if **C** is decidable. (An example is given below.) On the other hand, it also need not be trivial (i.e., equal to Frm($\mathcal{L}$)) even for undecidable **C**. For instance, take the Hilbert-style calculus for linear logic given in [1,19], and the 2-valued logic which interprets the linear connectives classically and the exponentials as the identity. All axioms are then tautologies and the rules (modus ponens, adjunction) are strongly sound, but the matrix is clearly non-trivial.

Corollary 41. *If* **C** *is strictly analytic, then* MC(**C**) = Thm(**C**).

Proof. We have to show that for every $F \notin$ Thm(**C**) there is a finite-valued logic **M** which is strongly sound for **C** and where $F \notin$ Taut(**M**). Let $X_1, \ldots,$ X_j be all the variables occurring in F and the axioms and rules of **C**. Then set $\mathbf{M} = \mathbf{M}_{\mathrm{dp}(F),j}(\mathrm{Thm}(\mathbf{C}))$.

By Proposition 15, $F \notin$ Taut(**M**) and all axioms of **C** are in Taut(**M**). Now consider a valuation v in **M** and suppose $v(A_i) \in V^+$ for all premises A_i of a rule of **C**. We have two cases: if $v(X) = \top$ for some variable X appearing in a premise A_i, then, since **C** is strictly analytic, X also appears in the conclusion C and hence $v(C) = \top$. Otherwise, let σ_v be the substitution corresponding to v. If $v(A_i) = \top$ for some i, this means that $\mathrm{dp}(A_i\sigma) > \mathrm{dp}(F)$. By strict analyticity, $\mathrm{dp}(C\sigma) \geq \mathrm{dp}(A_i\sigma) > \mathrm{dp}(F)$ and hence $v(C) = \top$. Otherwise, $v(A_i) = A_i\sigma$ for all premises A_i. Since $v(A_i)$ is designated, $A_i\sigma \in$ Thm(**C**), hence $C\sigma \in$ Thm(**C**). Then either $v(C) = C\sigma$ or $v(C) = \top$, and both are in V^+.   □

Example 42. The last corollary can be used to uniformly obtain semantics for strictly analytic Hilbert calculi. Strict analyticity of the calculus is a necessary condition, as Example 23 shows. The calculus given there is decidable, though not strictly analytic, and has only trivial covers. Its set of theorems nevertheless has an effective sequential approximation, i.e., it is the intersection of an infinite sequence of finite-valued matrices which are weakly sound for **C**. For this it is sufficient to give, for each formula A s.t. $\mathbf{C} \nvdash A$, a matrix **M** weakly sound for **C** with $A \notin \mathrm{Taut}(\mathbf{M})$. Let the depth of A be k, and let

$$V_0 = \mathrm{Var}(A) \cup \{\dagger\}$$
$$V_{i+1} = V_i \cup \{B \lhd C \mid B, C \in V_i\} \cup \{\Box B \mid B \in V_i\}$$

Then set $V = V_k$, $V^+ = \{B \lhd C \mid B \lhd C \in V, C \equiv \Box^l B\} \cup \{\dagger\}$. The truth functions are defined as follows:

$$\widetilde{\Box}(B) = \begin{cases} \dagger & \text{if } B \in V_k \text{ but } B \notin V_{k-1}, \text{ or } B = \dagger \\ \Box B & \text{otherwise} \end{cases}$$

$$\widetilde{\lhd}(B, C) = \begin{cases} \dagger & \text{if } C \in V_k \text{ but } C \notin V_{k-1}, \text{ or } B = \dagger \\ \dagger \lhd C & \text{else if } B \in V_k \text{ but } B \notin V_{k-1} \\ B \lhd C & \text{otherwise} \end{cases}$$

The axiom $X \lhd \Box X$ is a tautology. For if $v(X) = \dagger$, then $v(\Box X) = \dagger$ and hence $v(X \lhd \Box X) = \dagger \in V^+$. If $v(X) \in V_k$ but $\notin V_{k-1}$, then $v(\Box X) = \dagger$ and $v(X \lhd \Box X) = \dagger$. Otherwise $v(\Box X) = \Box B$ for $B = v(A)$ and $v(X \lhd \Box X) = \dagger$ (if $\Box B \in V_k$) or $= B \lhd \Box B \in V^+$.

If $v(X) = \dagger$, then $v(X \lhd Z) = \dagger$. Otherwise $v(X \lhd Z) \in V^+$ only if $v(X) = B$ and $v(Y) = \Box^l B$ and $B \notin V_k$. Then, in order for $v(Y \lhd Z)$ to be $\in V^+$, either $v(Y \lhd Z) = \dagger$, in which case $v(Z) \in V_k$ but $\notin V_{k-1}$, and hence $v(Y \lhd Z) = \dagger$, or $v(Y \lhd Z) = \Box^l B \lhd \Box^{l'} B$ with $l < l'$, in which case $v(Y \lhd Z) = B \lhd \Box^{l'} B$.

However, $A \notin Taut(\mathbf{M})$. For it is easy to see by induction that in the valuation with $v(X) = X$ for all variables $X \in \mathrm{Var}(A)$, $v(B) = B$ as long as $B \in V_k$ and so $v(A)$ can only be designated if $A \equiv B \lhd \Box B$ for some B, but all such formulas are derivable in **C**.

However, there are substitution instances of $X \lhd X$, viz., for any σ with $X\sigma$ of depth $> k$, for which $(X \lhd X)\sigma$ is a tautology. Even though **C** is weakly sound for **M**, it is not t-sound.

So far we have concentrated on approximations of logics given via calculi. However, propositional logics are also often defined via their semantics. The most important example of such logics are modal logics, where logics can be characterized using families of Kripke structures. If these Kripke structures satisfy certain properties, they also yield sequential approximations of the corresponding logics. Unsurprisingly, for this it is necessary that the modal logics have the *finite model property*, i.e., they can be characterized by a family of finite Kripke structures. The sequential approximations obtained by our method are only effective, however, if the Kripke structures are effectively enumerable.

Definition 43. A *modal logic* **L** has as its language $\mathcal{L}$ the usual propositional connectives plus two unary *modal operators:* $\square$ (necessary) and $\lozenge$ (possible). A *Kripke model* for $\mathcal{L}$ is a triple $\langle W, R, P \rangle$, where

1. W is any set: the set of *worlds*,
2. $R \subseteq W^2$ is a binary relation on W: the *accessibility relation*,
3. P is a mapping from the propositional variables to subsets of W.

A modal logic **L** is characterized by a class of Kripke models for **L**.

This is called the *standard semantics* for modal logics (see [7, Ch. 3]). The semantics of formulas in standard models is defined as follows:

Definition 44. Let **L** be a modal logic, $\mathcal{K}_\mathbf{L}$ be its characterizing class of Kripke models. Let $K = \langle W, R, P \rangle \in \mathcal{K}_\mathbf{L}$ be a Kripke model and A be a modal formula.

If $\alpha \in W$ is a possible world, then we say A is *true in* α, $\alpha \models_\mathbf{L} A$, iff the following holds:

1. A is a variable: $\alpha \in P(X)$
2. $A \equiv \neg B$: not $\alpha \models_\mathbf{L} B$
3. $A \equiv B \wedge C$: $\alpha \models_\mathbf{L} B$ and $\alpha \models_\mathbf{L} C$
4. $A \equiv B \vee C$: $\alpha \models_\mathbf{L} B$ or $\alpha \models_\mathbf{L} C$
5. $A \equiv \square B$: for all $\beta \in W$ s.t. $\alpha \, R \, \beta$ it holds that $\beta \models_\mathbf{L} B$
6. $A \equiv \lozenge B$: there is a $\beta \in W$ s.t. $\alpha \, R \, \beta$ and $\beta \models_\mathbf{L} B$

We say A is *true in* K, $K \models_\mathbf{L} A$, iff for all $\alpha \in W$ we have $\alpha \models_\mathbf{L} A$. A is *valid in* **L**, $\mathbf{L} \models A$, iff A is true in every Kripke model $K \in \mathcal{K}_\mathbf{L}$. By Taut(**L**) we denote the set of all formulas valid in **L**.

Many of the modal logics in the literature have the *finite model property (fmp)*: for every A s.t. $\mathbf{L} \not\models A$, there is a finite Kripke model $K = \langle W, R, P \rangle \in \mathcal{K}$ (i.e., W is finite), s.t. $K \not\models_\mathbf{L} A$ (where **L** is characterized by $\mathcal{K}$). We would like to exploit the fmp to construct sequential approximations. This can be done as follows:

Definition 45. Let $K = \langle W, R, P \rangle$ be a finite Kripke model. We define the many-valued logic $\mathbf{M}_K$ as follows:

1. $V(\mathbf{M}_K) = \{0, 1\}^W$, the set of 0-1-sequences with indices from W.
2. $V^+(\mathbf{M}_K) = \{1\}^W$, the singleton of the sequence constantly equal to 1.
3. $\widetilde{\neg}_{\mathbf{M}_K}, \widetilde{\vee}_{\mathbf{M}_K}, \widetilde{\wedge}_{\mathbf{M}_K}, \widetilde{\supset}_{\mathbf{M}_K}$ are defined componentwise from the classical truth functions
4. $\widetilde{\square}_{\mathbf{M}_K}$ is defined as follows:

$$\widetilde{\square}_{\mathbf{M}_K}(\langle w_\alpha \rangle_{\alpha \in W})_\beta = \begin{cases} 1 & \text{if for all } \gamma \text{ s.t. } \beta \, R \, \gamma, \, w_\gamma = 1 \\ 0 & \text{otherwise} \end{cases}$$

5. $\widetilde{\lozenge}_{\mathbf{M}_K}$ is defined as follows:

$$\widetilde{\lozenge}_{\mathbf{M}_K}(\langle w_\alpha \rangle_{\alpha \in W})_\beta = \begin{cases} 1 & \text{if there is a } \gamma \text{ s.t. } \beta \, R \, \gamma \text{ and } w_\gamma = 1 \\ 0 & \text{otherwise} \end{cases}$$

Furthermore, v_K is the valuation defined by $v_K(X)_\alpha = 1$ iff $\alpha \in P(X)$ and $= 0$ otherwise.

Lemma 46. *Let* **L** *and* K *be as in Definition 45. Then the following hold:*

1. *Every valid formula of* **L** *is a tautology of* $\mathbf{M}_K$.
2. *If* $K \not\models_\mathbf{L} A$ *then* $v_K \not\models_{\mathbf{M}_K} A$.

Proof. Let B be a modal formula, and $K' = \langle W, R, P' \rangle$. We prove by induction that $v_{K'}(B)_\alpha = 1$ iff $\alpha \models_\mathbf{L} B$:

B is a variable: $P'(B) = W$ iff $v_K(B)_\alpha = 1$ for all $\alpha \in W$ by definition of v_K.

$B \equiv \neg C$: By the definition of $\widetilde{\neg}_{\mathbf{M}_K}$, $v_K(B)_\alpha = 1$ iff $v_K(C)_\alpha = 0$. By induction hypothesis, this is the case iff $\alpha \not\models_\mathbf{L} C$. This in turn is equivalent to $\alpha \models_K B$. Similarly if B is of the form $C \wedge D$, $C \vee D$, and $C \supset D$.

$B \equiv \Box C$: $v_K(B)_\alpha = 1$ iff for all β with $\alpha\ R\ \beta$ we have $v_K(C)_\beta = 1$. By induction hypothesis this is equivalent to $\beta \models_\mathbf{L} C$. But by the definition of $\Box$ this obtains iff $\alpha \models_\mathbf{L} B$. Similarly for $\Diamond$.

(1) Every valuation v of $\mathbf{M}_K$ defines a function P_v via $P_v(X) = \{\alpha \mid v(X)_\alpha = 1\}$. Obviously, $v = v_{P_v}$. If $\mathbf{L} \models B$, then $\langle W, R, P_v \rangle \models_\mathbf{L} B$. By the preceding argument then $v(B)_\alpha = 1$ for all $\alpha \in W$. Hence, B takes the designated value under every valuation.

(2) Suppose A is not true in K. This is the case only if there is a world α at which it is not true. Consequently, $v_K(A)_\alpha = 0$ and A takes a non-designated truth value under v_K. $\Box$

The above method can be used to construct many-valued logics from Kripke structures for not only modal logics, but also for intuitionistic logic. Kripke semantics for **IPL** are defined analogously, with the exception that $\alpha \models A \supset B$ iff $\beta \models A \supset B$ for all $\beta \in W$ s.t. $\alpha\ R\ \beta$. **IPL** is then characterized by the class of all finite trees [9, Ch. 4, Thm. 4(a)]. Note, however, that for intuitionistic Kripke semantics the form of the *assignments* P is restricted: If $w_1 \in P(X)$ and $w_1\ R\ w_2$ then also $w_2 \in P(X)$ [9, Ch. 4, Def. 8]. Hence, the set of truth values has to be restricted in a similar way. Usually, satisfaction for intuitionistic Kripke semantics is defined by satisfaction in the *initial* world. This means that every sequence where the first entry equals 1 should be designated. By the above restriction, the only such sequence is the constant 1-sequence.

Example 47. The Kripke tree with three worlds

yields a five-valued logic $\mathbf{T}_3$, with $V(\mathbf{T}_3) = \{000, 001, 010, 011, 111\}$, $V^+(\mathbf{T}_3) = \{111\}$, the truth table for implication

$\supset$	000	001	010	011	111
000	111	111	111	111	111
001	010	111	010	111	111
010	001	001	111	111	111
011	000	001	010	111	111
111	000	001	010	011	111

$\perp$ is the constant 000, $\neg A$ is defined by $A \supset \perp$, and $\vee$ and $\wedge$ are given by the componentwise classical operations.

The Kripke chain with four worlds corresponds directly to the five-valued Gödel logic $\mathbf{G}_5$. It is well know that $(X \supset Y) \vee (Y \supset X)$ is a tautology in all $\mathbf{G}_m$. Since $\mathbf{T}_3$ falsifies this formula (take 001 for X and 010 for Y), we know that $\mathbf{G}_5$ is not the best five-valued approximation of **IPL**.

Furthermore, let

$$O_5 = \bigwedge_{1 \leq i < j \leq 5} (X_i \supset X_j) \vee (X_j \supset X_i) \text{ and}$$

$$F_5 = \bigvee_{1 \leq i < j \leq 5} (X_i \supset X_j).$$

O_5 assures that the truth values assumed by $X_1, \ldots, X_5$ are linearly ordered by implication. Since neither $010 \supset 001$ nor $001 \supset 010$ is true, we see that there are only four truth values which can be assigned to $X_1, \ldots, X_5$ making O_5 true. Consequently, $O_5 \supset F_5$ is valid in $\mathbf{T}_3$. On the other hand, F_5 is false in $\mathbf{G}_5$.

Theorem 48. *Let* $\mathbf{L}$ *be a modal logic characterized by a set of finite Kripke models* $\mathcal{K} = \{K_1, K_2, \ldots\}$. *A sequential approximation of* $\mathbf{L}$ *is given by* $\langle \mathbf{M}_1, \mathbf{M}_2, \ldots \rangle$ *where* $\mathbf{M}_1 = \mathbf{M}_{K_1}$, *and* $\mathbf{M}_{i+1} = \mathbf{M}_i \times \mathbf{M}_{K_{i+1}}$. *This approximation is effective if* $\mathcal{K}$ *is effectively enumerable.*

Proof. (1) Taut($\mathbf{M}_i$) $\supseteq$ Taut($\mathbf{L}$): By induction on i: For $i = 1$ this is Lemma 46 (1). For $i > 1$ the statement follows from Lemma 13, since Taut($\mathbf{M}_{i-1}$) $\supseteq$ Taut($\mathbf{L}$) by induction hypothesis, and Taut($\mathbf{M}_{K_i}$) $\supseteq$ Taut($\mathbf{L}$) again by Lemma 46 (1).

(2) $\mathbf{M}_i \trianglelefteq \mathbf{M}_{i+1}$ from $A \cap B \subseteq A$ and Lemma 13.

(3) Taut($\mathbf{L}$) = $\bigcap_{i \geq 1}$ Taut($\mathbf{M}_i$). The $\subseteq$-direction follows immediately from (1). Furthermore, by Lemma 46 (2), no non-tautology of $\mathbf{L}$ can be a member of all Taut($\mathbf{M}_i$), whence $\supseteq$ holds. $\qquad\square$

Remark 49. Finitely axiomatizable modal logics with the fmp always have an effective sequential approximation, since it is then decidable if a given finite Kripke structure satisfies the axioms. Urquhart [20] has shown that this is not true if the assumption is weakened to recursive axiomatizability, by giving an example of an undecidable recursively axiomatizable modal logic with the fmp. Since this logic cannot have an effective sequential approximation, its characterizing family of finite Kripke models is not effectively enumerable. The preceding theorem thus also shows that the many-valued closure of a calculus for a modal logic with the fmp equals the logic itself, provided that the calculus contains modus ponens and necessitation as the only rules. (All standard axiomatizations are of this form.)

6   Conclusion

Our brief discussion unfortunate must leave many interesting questions open, and suggests further questions which might be topics for future research. The main open problem is of course whether the approach used here can be extended to the case of first-order logic. There are two distinct questions: The first is how to check if a given finite-valued matrix is a cover for a first-order calculus. Is this decidable? One might expect that it is at least for "standard" formulations of first-order rules, e.g., where the rules involving quantifiers are monadic in the sense that they only involve one variable per rule. The second question is whether the relationship $\lhd^*$ is decidable for n-valued first order logics. Another problem, especially in view of possible applications in computer science, is the complexity of the computation of optimal covers. One would expect that it is tractable at least for some reasonable classes of calculi which are syntactically characterizable.

We have shown that for strictly analytic calculi, the many-valued closure coincides with the set of theorems, i.e., that they are effectively approximable by their finite-valued covers. Is it possible to extend this result to a wider class of calculi, in particular, what can be said about calculi in which modus ponens is the only rule of inference (so-called Frege systems)? For calculi which are not effectively approximable, it would still be interesting to characterize the many-valued closure. For instance, we have seen that the many-valued closure of linear logic is not equal to linear logic (since linear logic is undecidable) but also not trivial (since all classical non-tautologies are falsified in a 2-valued cover). What is the many-valued closure of linear logic? For those (classes of) logics for which we have shown that sequential approximations are possible, our methods of proof also do not yield optimal solutions. For instance, for modal logics with the finite model property we have shown that all non-valid formulas can be falsified in the many-value logic obtained by coding the corresponding Kripke countermodel. But there may be logics with fewer truth-values which also falsify these formulas. A related question is to what extent our results on approximability still hold if we restrict attention to many-valued logics in which only one truth-value is designated. The standard examples of sequences of finite-valued logics approximating, e.g., Łukasiewicz or intuitionistic logic are of this form, but it need not be the case that every approximable logic can be approximated by logics with only one designated value.

References

1. Avron, A.: The semantics and proof theory of linear logic. Theoret. Comput. Sci. 57, 161–184 (1988)
2. Baaz, M., Fermüller, C.G., Salzer, G., Zach, R.: Labeled calculi and finite-valued logics. Studia Logica 61, 7–33 (1998)
3. Baaz, M., Zach, R.: Short proofs of tautologies using the schema of equivalence. In: Meinke, K., Börger, E., Gurevich, Y. (eds.) CSL 1993. LNCS, vol. 832, pp. 33–35. Springer, Heidelberg (1994)

4. Baaz, M., Zach, R.: Approximating propositional calculi by finite-valued logics. In: 24th International Symposium on Multiple-valued Logic. ISMVL 1994 Proceedings, pp. 257–263. IEEE Press, Los Alamitos (1994)
5. Bernays, P.: Axiomatische Untersuchungen des Aussagenkalküls der "Principia Mathematica". Math. Z. 25, 305–320 (1926)
6. Carnielli, W.A.: Systematization of finite many-valued logics through the method of tableaux. J. Symbolic Logic 52, 473–493 (1987)
7. Chellas, B.F.: Modal Logic: An Introduction. Cambridge University Press, Cambridge (1980)
8. Dummett, M.: A propositional calculus with denumerable matrix. J. Symbolic Logic 24, 97–106 (1959)
9. Gabbay, D.M.: Semantical Investigations in Heyting's Intuitionistic Logic. In: Synthese Library, vol. 148, Reidel, Dordrecht (1981)
10. Gödel, K.: Zum intuitionistischen Aussagenkalkül. Anz. Akad. Wiss. Wien 69, 65–66 (1932)
11. Gottwald, S.: A Treatise on Many-valued Logics. Research Studies Press, Baldock (2001)
12. Hähnle, R.: Automated Deduction in Multiple-Valued Logics. Oxford University Press, Oxford (1993)
13. Jaśkowski, S.: Recherches sur la système de la logique intuitioniste. In: Actes du Congrès International de Philosophie Scientifique 1936, Paris, vol. 6, pp. 58–61 (1936)
14. Lincoln, P.D., Mitchell, J., Scedrov, A., Shankar, N.: Decision problems for propositional linear logic. In: Proceedings 31st IEEE Symposium on Foundations of Computer Science. FOCS 1990, pp. 662–671. IEEE Press, Los Alamitos (1990)
15. Łukasiewicz, J., Tarski, A.: Untersuchungen über den Aussagenkalkül. Comptes rendus des séances de la Société des Sciences et des Lettres de Varsovie Cl III 23, 30–50 (1930), English translation in [18, 38–59]
16. Mundici, D.: Satisfiability in many-valued sentential logic is NP-complete. Theoret. Comput. Sci. 52, 145–153 (1987)
17. Rescher, N.: Many-valued Logic. McGraw-Hill, New York (1969)
18. Tarski, A.: Logic, Semantics, Metamathematics, 2nd edn. Hackett, Indianapolis (1983)
19. Troelstra, A.S.: Lectures on Linear Logic. CSLI Lecture Notes, vol. 29. CSLI, Standford, CA (1992)
20. Urquhart, A.: Decidability and the finite model property. J. Philos. Logic 10, 367–370 (1981)

Model Transformation Languages and Their Implementation by Bootstrapping Method

Janis Barzdins, Audris Kalnins, Edgars Rencis, and Sergejs Rikacovs

University of Latvia, IMCS, 29 Raina boulevard, Riga, Latvia
janis.barzdins@mii.lu.lv, audris.kalnins@mii.lu.lv, edgars.rencis@lumii.lv,
sergejs.rikacovs@lumii.lv

Dear Boris, You are the father of Computer Science in Latvia.
Thank you for this.

Abstract. In this paper a sequence of model transformation languages L0, L1, L2 is defined. The first language L0 is very simple, and for this language it is easy to build an efficient compiler to C++. The next language L1 is an extension of L0, and it contains powerful pattern definition facilities. The last language L2 is of sufficiently high level and can be used for implementation of traditional pattern-based high level model transformation languages, as well as for the development of model transformations directly. For languages L1 and L2 efficient compilers have been built using the bootstrapping method: L1 to L0 in L0, and L2 to L1 in L1. The results confirm the efficiency of model transformation approach for domain specific compiler building.

1 Introduction

A well known fact is that the heart of the most advanced software engineering technology MDA [1] is model transformation languages. In recent years the main emphasis has been on the development of industrial transformation languages [2,3,4,5,6,7]. For most of the transformation languages there is an implementation. However, there has been no thorough research on transformation language implementation, especially on the efficiency aspects. On the other hand, there have been only a few attempts to use transformation languages for defining their compilers (to use bootstrapping) [5,7,8]. It is a little bit strange taking into the account that the main idea of MDA is to use transformation languages for transforming formal design models also a sort of language. Most of the MDA success stories are related to Domain Specific Languages there the corresponding transformations are in fact compilers. One of the goals of this paper is demonstrate the usability and efficiency of transformation languages namely for defining compilers for transformation languages.

The other goal is to propose a very simple, but at the same time sufficiently high level, transformation language L2 which can be used in practice for direct development of model transformations.

A. Avron et al. (Eds.): Trakhtenbrot/Festschrift, LNCS 4800, pp. 130–145, 2008.

The main results of this paper are the following:

- a sequence of transformation languages L0, L1, L2 is offered and each language is obtained from the previous one by adding some features. The final language L2 is of pretty high level (it contains a kind of patterns, loops, etc.)
- the first of languages L0 is very simple. It contains only the basic transformation facilities and its complete description can be given in less than two pages (see Section 2). For this language it is easy to build an efficient compiler to C++
- a compiler from L_{i+1} to L_i (i = 0,1) can be easily specified in L_i (this can be done also in L0). This acknowledges the efficiency of using transformation languages for building their compilers as long as an appropriate for bootstrapping language sequence has been found
- the last language in the sequence L2 is of sufficiently high level for traditional pattern-based high level model transformation languages (such as MOLA [6]) to be compiled to it in a natural way, with the compiler also being easily definable in L2.

The language L0 and henceforth also L_i include also the basic facilities for defining metamodels, in order to make these languages self-contained.

2 The Base Language L0

The purpose of this section is to give a brief overview of the transformation language **L0**. This language is a rather low level procedural textual language, with control structures mostly taken from assembler-like languages (and syntax influenced by C++). The basic setting of L0 is as for any transformation language - we process a model, which is an instance of metamodel (MOF style). But the language constructs which are specific to model transformations have been chosen to be as simple as possible.

Basically these constructs give the programmer the ability to:

- iterate through instances (both links and objects),
- create/delete objects and links,
- read/write (change) object attribute values.

An elementary unit of L0 transformation program is a **command** (an imperative statement). L0 transformation program itself is a **transformation**, which contains several parts:

- global variable definition part
- native subprogram (function or procedure) declaration part (used C++ library function headers)
- L0 subprogram definition part. It is expected that exactly one subprogram in this part is labeled with the reserved word **main**. The subprogram labeled with **main** defines the entry point of the transformation. An L0 subprogram definition also consists of several parts:

- Subprogram header
- Local variable definitions
- Keyword **begin**;
- Subprogram body definition
- Keyword **end**;

L0 contains the following kinds of commands:

1. **transformation** <transformationName>; This command starts a transformation definition.
2. **endTransformation**; The command ends a transformation definition.
3. **pointer** <pointerName> : <className>; Defines a pointer to objects of class <className>.
4. **var** <varName> : <ElementaryTypeName>; <ElementaryTypeName> is one of Boolean, Integer, Real, String. Defines a variable of elementary type.
5. **procedure** <procName>(<paramList>); Subprogram header, the (formal) parameter list can be empty. Parameter list consists of formal parameter definitions separated by ",". A parameter definition consists of its name, the parameter type (the type can be an elementary type or a class from the metamodel), and the passing method (parameters can be passed by reference or by value). If the parameter is passed by reference, its type name is preceded by the **&** character.
6. **function** <funcName>(<paramList>): <returnType>; Return type name can be an elementary type name or class name.
7. **begin**; Starts subprogram body definition.
8. **end**; Ends subprogram body definition.
9. **return**; Returns execution control to caller.
10. **return** <identifier>; Return the value of <identifier> to the caller, the type must coincide with the function return type. <identifier> is an elementary variable name or pointer name.
11. **call** <subProgName>(<actPrmList>); The actual parameter list, which can be empty, consists of identifiers separated by ",". An identifier can be a variable name, a pointer name, or a subprogram parameter name.
12. **setVarF** <variable>=<funcName>(<actPrmList>); This command can be used to obtain the value of the function result. The result is of an elementary type and is assigned to a variable. The variable type must coincide with the function return type.
13. **first** <pointer> : <className> **else** <label>; Positions <pointer> to an arbitrary (the first one in an implementation dependent ordering) object of <className>. Typically, this command in combination with the **next** command is used to traverse all objects of the given class (including subclass objects). If <className> does not have objects, <pointer> becomes **null**, and execution control is transferred to the <label>. The <className> in this command must be the same as (or a subclass of) the class used in pointer definition; if it is a subclass, then the pointer value set is narrowed (for the subsequent executions of **next**).

14. **first** <pointer1> : <className> **from** <pointer2> **by** <roleName> **else** <label>; Similar to the previous command, the difference is that it positions <pointer1> to an arbitrary class object, which is reachable from <pointer2> by the link <roleName>. Similarly, this command in combination with the **next** command is used to traverse all objects linked to an object by the given link type.

15. **next** <pointer> **else** <label>; Gets the next object, which satisfies conditions, formulated during the execution of the corresponding **first** and which has not been visited (iterated) with this variable yet. If there is no such object, the <pointer> becomes **null**, and execution control is transferred to <label>.

16. **goto** <label>; Unconditionally transfers control to <label>, <label> should be located in the current subprogram.

17. **label** <labelName>; Defines a label with the given name.

18. **addObj** <pointer>:<className>; Creates a new object of the class <className>.

19. **addLink** <pointer1>.<roleName>.<pointer2>; Creates a new link (of type specified by <roleName>) between the objects pointed to by the <pointer1> and <pointer2> , respectively.

20. **deleteObj** <pointer>; Deletes the object, which is pointed to by <pointer>.

21. **deleteLink** <pointer1>.<roleName>.<pointer2>; Deletes link, whose type is specified by <roleName>, between objects pointed to by <pointer1> and <pointer2>, respectively.

22. **setPointer** <pointer1>=<pointer2>; Sets <pointer1> to the object, which is pointed to by <pointer2>; in place of <pointer2> the **null** constant can be used.

23. **setPointerF** <pointer>=<funcName>(<actPrmList>); Sets <pointer> to the object, which is returned by <funcName>.

24. **setVar** <variable> = <binExpr>; Sets <variable> to <binExpr> value. <binExpr> is a binary expression consisting of the following elements: elementary variables, subprogram parameters (of elementary types), literals, object attributes and standard operators (+, -, *, /, &&, ||, !).

25. **setAttr** <pointer>.<attrName>=<binExpr>; Sets the value of attribute <attrName> (of the object, pointed to by <pointer>) to the <binExpr> value.

26. **type** <pointer> == <className> **else** <label>; If the type of the pointed object is identical to the <className>, then control is transferred to the next command, else control is transferred to <label>. In place of the equality symbol == an inequality symbol != can be used. This command is used for determining the exact subclass of an object.

27. **var** <variable>==<binExpr> **else** <label>; If the condition is not true then control is transferred to <label>. In place of equality symbol other (<, <=, >, >=, !=) relational operators compatible with argument types can be used.

28. **attr** <pointer>.<attrName> == <binExpr> **else** <label>; If condition is not true then control is transferred to <label>. Other relational operators (<, <=, >, >=, !=) can be used too.

29. **link** <pointer1>.<roleName>.<pointer2> **else** <label>; Checks whether there is a link (with the type specified by <roleName>) between the objects pointed to by <pointer1> and <pointer2>, respectively.

30. **pointer** <pointer1>==<pointer2> **else** <label>; Checks whether the objects pointed to by <pointer1> and <pointer2>, respectively, are identical. Instead of <pointer2> **null** can be used, and the inequality symbol can be used too.

Actually L0 contains also commands for building the relevant metamodel; for details see `http://Lx.mii.lu.lv/`.

It is easy to see that the language L0 contains only the very basic facilities for defining transformations. At the same time, it obviously is **complete** in the sense of its functional capabilities. This is confirmed by the fact that high level transformation languages such as MOLA can be successfully compiled to it. We omit this result in the form of a theorem, but all informal justifications of this thesis are in place. Namely this is why we call L0 the basic transformation language. We start our bootstrap approach with this language.

We conclude this section with a very simple example of L0 - a transformation which builds a representation B of a directed graph (where edge connection points are also objects) from the simplest one A (where only nodes and edges are present). Figure 1 presents the metamodel for both representations.

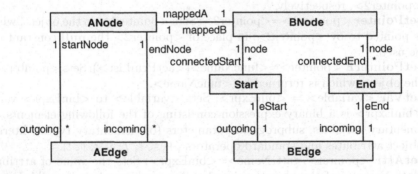

Fig. 1. Metamodel for the example

The L0 program performing the transformation:

```
transformation Graphs;
main procedure Graph2Graph();
pointer a : ANode;
pointer b : BNode;
pointer aEd : AEdge;
pointer bEd : BEdge;
pointer edgeStart : Start;
pointer edgeEnd : End;
pointer aEdgeStNode : ANode;
```

```
pointer aEdgeEnNode : ANode;
pointer mapBNode : BNode;
begin;
//copy nodes;
first a : ANode else aNodeProcessed;
label loopANode;
addObj b : BNode;
addLink a . mappedB . b;
next a else aNodeProcessed;
goto loopANode;
label aNodeProcessed;
//copy edges;
first aEd : AEdge else aEdgesProc;
label loopAEdge;
addObj bEd : BEdge;
addObj edgeStart : Start;
addObj edgeEnd : End;
addLink bEd.eStart.edgeStart;
addLink bEd.eEnd.edgeEnd;
//quit if not found;
first aEdgeStNode:ANode from aEd by startNode else aEdgesProc;
first mapBNode:BNode from aEdgeStNode by mappedB else
aEdgesProc;
addLink edgeStart.node.mapBNode;
first aEdgeEnNode:ANode from aEd by endNode else aEdgesProc;
first mapBNode:BNode from aEdgeEnNode by mappedB else
aEdgesProc;
addLink edgeEnd . node. mapBNode;
next aEd else aEdgesProc;
goto loopAEdge;
label aEdgesProc;
end;
endTransformation;
```

3   Implementation of L0

The language L0 can be implemented in several ways. The first problem is how to
store and access the persistent data the metamodel and its instances. Obviously,
a kind of data store is required for this. A traditional relational database could
be used, but they typically have no adequate low level API. Another alternative
could be an in-memory data store, such as RDF-oriented Sesame [9] or an MOF-
oriented one (EMF [10], MDR [11]). However, for this approach we have selected
our own metamodel-based in-memory repository [12], which has an appropriate
low level API. Being developed over many years for other goals - generic metamodel based tool building [13], this repository occurred to be efficient enough
for implementing L0.

The API of this repository is implemented as a C++ function library. This library offers: a) a system of low-level data retrieval functions that is complete for low-level data query programming; b) a selected set of more complicated widely usable data searching functions. By means of a sophisticated indexing mechanism, these more complicated functions are also efficiently implemented.

The API of this repository includes two groups of functions:

1. Meta-model management functions for creating, modifying, and deleting classes, attributes and associations, querying about their properties, class inheritance, etc. However, meta-model management functions are used relatively seldom, the most heavily used functions belong to the next group.
2. Instance management. This group of functions, in its turn, also can be subdivided in two groups:
   (a) functions for creating instances, assigning attribute values, creating links between instances, modifying and deleting instances and links, querying about instance attributes and links. For example:
   ```
   long CreateObject(long ObjTypeId); // returns objId
   int DeleteObjectHard(long ObjId);
   int CreateLink(long LinkTypeId, long ObjId1, long ObjId2);
   int DeleteLink(long LinkTypeId, long ObjId1, long ObjId2);
   ```
   (b) efficient searching functions (internally these functions use sophisticated indexing mechanisms):
   ```
   int GetObjectNum(long ObjTypeId);
   long GetObjectIdByIndex(long ObjTypeId, int Index);
   int GetLinkedObjectNum(long ObjId, long LinkTypeId);
   long GetLinkedObjectIdByIndex(long ObjId, long LinkTypeId,
                                 int Index);
   ```

If a repository with such API is available, then building an L0 compiler (to C++) is quite a straightforward job. Such a compiler has been built by one of the authors of this paper (S. Rikacovs) in two months (not including L0 debugging facilities). The main advantage of using this repository is that the instance management functions in L0 (**first** and **next**, including the **by** link options) have close counterparts in the repository API.

The implementation efficiency is also sufficiently high. First, some experiments show that efficiency loss with respect to the same transformation manually coded in C++ is no more than 1.5 times. Another aspect is efficiency of the selected repository for typical transformations, where another group of experiments [12] show that the selected repository is at least as efficient as Ocsame [9] data store for typical instance retrieval operations.

4   The Language L1

The crucial component of any advanced transformation language is some sort of pattern definition facilities. This way, the transformation language L1 is obtained from L0 by adding pattern definition facilities of a specific new form. In selecting the pattern definition method we were guided by two conflicting requirements. On the one hand, the pattern concept must be practically usable. On the other hand, it must have a simple and efficient implementation by compiler (traditional patterns, e.g. in [4,5,6] not always have this property). One of the main results of this paper is the proposed pattern definition facility, which satisfies both requirements. The main component of pattern specification is a facility for defining expressions over environments of model objects. Our approach is based on a new kind of expressions - **begin-end expressions**, which are defined as **command blocks** of the kind:

begin <commandSequence> **end**.

Namely, if we execute the block on the given object environment, and reach the **end** command, then the expression value is defined to be true, otherwise it is false.

For example, the expression (block):

```
begin
  attr p.age==23;
  attr p.occupation=="Student";
end
```

has the value true if and only if the pointer p (of type Person) points to an instance, whose attribute age has the value 23 and the attribute occupation has the value "Student".

Some more comments on begin-end expressions must be given - what is meant by not reaching the **end**. If during the block execution we reach an undefined **else** branch of a command (there is no **else** keyword or it is not followed by a label, this is permitted for all **else**-containing commands of L0) then the expression is defined to have the value false. A similar way is to use a **goto** command without label (but it is forbidden to use a label not defined in the block).

Now, when the begin-end expressions are described, it is possible to define the language L1 precisely.

The language **L1** differs from L0 in commands **first** and **next** extended by **suchthat** part containing a begin-end expression:

```
first <pointer> : <className>  suchthat <BeginEndExpression>
else <label>;

next <pointer>  suchthat <BeginEndExpression>
else <label>;
```

Now we will explain in some details the role of begin-end expressions for pattern definition and compare them to other facilities for pattern definition. Let us assume that we have the class diagram ("metamodel") in Figure 2. Such a class diagram can be treated also as a signature for formula definition in many-sorted first order logic (MS FOL) - an association corresponds to a binary predicate and an attribute to a function. We want to define certain patterns

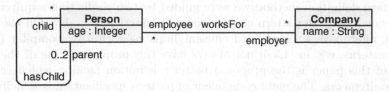

Fig. 2. Metamodel for pattern examples

for p:Person, i.e., constraints which should be satisfied by appropriate Person instances.

To get a deeper insight into the situation, we will define these patterns in several languages, starting from the natural language. Let us consider an example:

p is a Person, whose age is 50 and who works for (i.e., its employer is) the Company "UniBank".

The same pattern can be specified graphically in the MOLA transformation language:

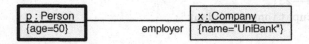

(in other transformation languages this can be done in a similar way).

In MS FOL the same pattern can be represented by the following formula F(p) (the free variable p has the type Person):

$$p.age = 50 \quad \&$$
$$\exists x : Company(x.name = \text{``}UniBank\text{''} \quad \& \quad worksFor(p,x)) . \quad (1)$$

The same pattern can be specified also by a begin-end expression, where p is a pointer variable with the type Person:

```
begin
  attr p.age==50;
  first x:Company suchthat
    begin
    attr x.name=="UniBank";
    link p.employer.x;
  end;
end;
```

Let us note that in this context "**first** x: **suchthat** " is equivalent to "**exists** x: **suchthat** ".

Now let us consider a more complicated example:

p is a Person, who has a child working for the Company "UniBank".

This corresponds to the following MOLA pattern:

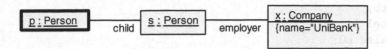

The corresponding MS FOL formula F(p) is:

$$\exists s : Person(hasChild(p, s) \quad \& $$
$$\exists x : Company(x.name = ``UniBank" \quad \& \quad worksFor(s, x))) \ . \qquad (2)$$

The corresponding begin-end expression is:

```
begin
  first s:Person suchthat
  begin
    link p.child.s;
    first x:Company suchthat
    begin
      attr x.name=="UniBank";
      link s.employer.x;
    end;
  end;
end;
```

Now let us consider a significantly more complicated example:

p is a Person, whose all adult (not younger than 18) children work for the Company "UniBank".

It is difficult to specify such a pattern in a graphical pattern definition language. At the same time it can be specified quite easily as a MS FOL formula and also as a begin-end expression.

The corresponding MS FOL formula F(p) is:

$$\forall s : Person(s.age >= 18 \quad \& \quad hasChild(p, s) \quad \supset$$
$$\exists x : Company(x.name = ``UniBank") \quad \& \quad worksFor(s, x)) \ . \qquad (3)$$

The corresponding begin-end expression is:

```
begin
  first s:Person suchthat
  begin
    link p.child.s;
    attr s.age>=18;
    first x:Company suchthat
    begin
      attr x.name=="UniBank";
      link s.employer.x;
    end
    else L1;
```

```
    goto;
    label L1;
  end else L0;
  goto;
  label L0;
end;
```

It is easy to check that we can reach the final **end** iff p points to a Person, which satisfies the abovementioned constraint. This begin-end expression actually corresponds to the following MS FOL formula (which is equivalent to the formula above):

$$\neg \exists s : Person(hasChild(p,s) \quad \& \quad s.age >= 18 \quad \&$$
$$\neg \exists x : Company(x.name = \text{``}UniBank\text{''} \quad \& \quad worksFor(s,x))) \; . \quad (4)$$

MS FOL apparently is one of the most universal languages for defining patterns. However, existing transformation languages avoid the use of MS FOL formulas for pattern definition. The reason is that for such a universal pattern specification no satisfactory (non-exponential) pattern matching algorithm is known (most probably, such an algorithm does not exist). Therefore existing transformation languages limit in a natural way their pattern definition mechanisms in accordance with their graphical capabilities.

A natural question arises about the relation between our begin-end expressions and MS FOL formulas in the context of pattern definition. The answer is that for pattern definition the power of begin-end expressions is not less than that of MS FOL formulas. We will not go into details of this problem. Let us note only that the proof of this assertion (after the corresponding concepts are made precise enough) is not complicated - it is sufficient to trace the inductive definition of MS FOL formulas.

However, in order to give a deeper insight into begin-end expressions, we explain a small fragment of this proof. Let $F(p)$ and $G(p)$ be MS FOL formulas with p as the free variable. We assume that we have already built begin-end expressions $E_{F(p)}$ and $E_{G(p)}$ which define the same patterns. Namely,

$$E_{F(p)} \equiv \quad \textbf{begin} <\text{commandSequence for } F> \textbf{end}$$

and

$$E_{G(p)} \equiv \quad \textbf{begin} <\text{commandSequence for } G> \textbf{end}.$$

Let us consider the formula $F(p)\&G(p)$. It is easy to see that the following begin-end expression defines an equivalent pattern:

begin <commandSequence for F> <commandSequence for G> **end**

Now let us consider the formula $\neg F(p)$. The corresponding begin-end expression can be obtained in the following way. Those else-branches inside $E_{F(p)}$ which have no label are completed by a certain fixed label, let's say L. The same action is applied to goto's without label (such commands are permitted in L1). This action is not applied to begin-end expressions which are inside nested suchthat parts. Let us denote the transformed begin-end sequence by <commandSequence for ¬F>. The sought for begin-end expression has the following form:

begin <commandSequence for ¬*F*> **goto; label L; end.**

It is easy to see that we can reach the label L (which is the last one in this block and therefore reaching it means that the whole expression assumes the value true) iff the original expression for F had the value false.

The other inductive steps for MS FOL formula definition can be treated in a similar way.

In reality begin-end expressions are even more powerful than pure MS FOL, since begin-end expressions can contain also operations on elementary variables.

A question arises why our begin-end expressions are superior to MS FOL for specifying patterns. There are three essential reasons for this:

1. A begin-end expression specifies the command execution order during the pattern matching (i.e., the order in which the instances are traversed).
2. When a pattern is matched all its elements are assigned an identity which can be used further for referencing these elements (a similar approach is used in all graphical pattern languages).
3. Begin-end expressions can be easily compiled to L0 (the obtained L0 fragment directly implements the pattern matching for the expression).

5   The Final Language L2 and Its Usage

The language **L2** is obtained from L1 by extending it with a **foreach** command (loop) and the **if-then-else** command:

foreach <loopVariable> : <className> **suchthat**
<BeginEndExpression> **do** <L2commSequence> **end;**

if <BeginEndExpression> **then do** <L2commSequence> **end else do**
<L2commSequence> **end;**

The loop semantics is quite natural: the loop variable traverses all instances of the class, which satisfy the suchthat condition, for each such instance the do-end block is executed (explicit jumping out of the loop body is prohibited). The foreach command may be used also inside a suchthat block.

The metamodel of L languages is given in Figure 3 (dashed association corresponds to element of L1, bold classes/associations to L2).

The language L2 has at least two important usage areas. On the one hand, it can be used as a practical model transformation language. On the other hand, practical high level model transformation languages can be adequately compiled to it, and the compiler itself can be written in L2 (we consider this kind of usage the main one). Currently such a schema has been successfully applied for building an efficient implementation of MOLA [6], but the same approach could be applied also for implementing MOF QVT [2] and other transformation languages. The main issue for such compilations is how to map "completely declarative" traditional patterns to patterns with the specified search order in L languages. In some sense the basic idea for such a mapping is given in [14].

6   Implementation of L1 and L2

The languages L1 and L2 have been implemented according to the bootstrapping principles described in the introduction.

A compiler from L1 to L0 has been implemented in L0 (as a set of recursive procedures). It contains about 200 lines of L0 and has been written in one month (by E. Rencis). Though L1 includes a pattern definition mechanism even more powerful than that of MS FOL, implementation of L1 patterns is relatively simple since the search order of pattern elements is precisely specified in the language. Actually the command sequence defining a begin-end expression can quite easily be transformed into an equivalent sequence of L0 commands, using recursion for nested expressions.

To illustrate the idea, we will show briefly the schema how the L1 command

```
first <pointer> : <className>  suchthat <BeginEndExpression>
else <label>;
```

can be compiled to L0 commands. By means of **first**, **next** and **goto** commands a simple loop is organized which scans all instances of the given class. The "body" of this loop contains slightly modified commands form the begin-end expression commands with missing (or empty) else-branch are "redirected" to a new label in the else-case. Then reaching this new label would mean that this **suchthat** fails on the given instance and the next instance must be tried. If, on the contrary, the end of the loop body is reached, the given instance satisfies the whole **suchthat** and the job is done. If a command within the expression body is not an L0 command, but a true L1 command, the same procedure is applied recursively. This compilation schema is illustrated in Table 1.

The compiler from L2 to L1 is also relatively simple (about 560 lines of L0).

Both L1 and L2 compilers rely on the metamodel of L languages (Figure 3). Compilation of L_{i+1} to L_i actually converts into a transformation of models (i.e., L_{i+1} programs) corresponding to the given metamodel. As it was already mentioned, this transformation occurrs to be relatively simple.

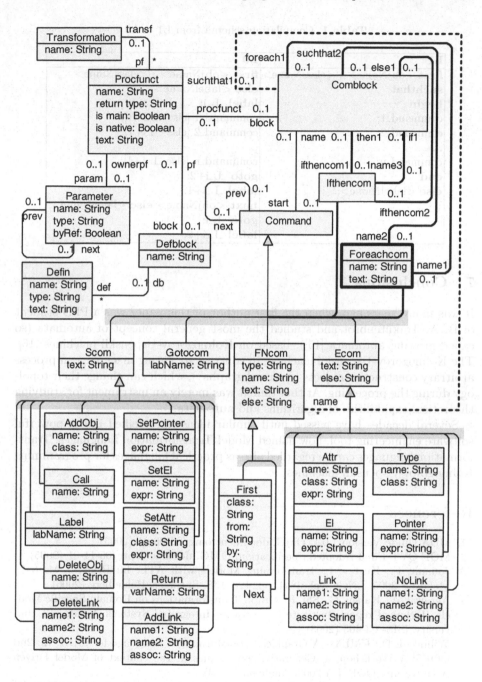

Fig. 3. Metamodel of L languages

Table 1. Compilation schema from L1 to L0

L1	L0
first \<objName\> : \<className\>	**first** \<objName\> : \<className\>
suchthat	**else** \<labelName\>;
begin	**label** _L_i;
command_1;	command_1 [**else** _L_i+1];
command_2;	command_2 [**else** _L_i+1];
...	...
command_n;	command_n [**else** _L_i+1];
end	**goto** _L_i+2;
else \<labelName\>;	**label** _L_i+1;
	next \<objName\> **else** \<labelName\>;
	goto _L_i;
	label _L_i+2;

7   Conclusions

It was many years ago, when the first author of this paper was a PhD student
of B. A. Trakhtenbrot and studied the most general concept of automata (so
called growing automata [15]), based on Kolmogorov-Uspenskii machines [16].
The Kolmogorov-Uspenskii machine, in contrast to Turing machine, can process
arbitrary constructive objects ("colored graphs"), which can change their topology during the processing. At that time it was merely an instrument for studying
theoretical capabilities of algorithms and automata.

Several decades have passed until similar ideas have reified into a powerful
software engineering tool, now named Model Transformation Languages. Transformation languages can be regarded also as practical languages for programming
Kolmogorov-Uspenskii machines.

References

1. MDA Guide Version 1.0.1. OMG, document omg/03-06-01 (2003)
2. MOF QVT Final Adopted Specification, OMG, document ptc/05-11-01 (2005)
3. Jouault, F., Kurtev, I.: Transforming Models with ATL. In: Bruel, J.-M. (ed.)
   MoDELS 2005. LNCS, vol. 3844, pp. 128–138. Springer, Heidelberg (2006)
4. Agrawal, A., Karsai, G., Shi, F.: Graph Transformations on Domain-Specific Models. Technical report, Institute for Software Integrated Systems, Vanderbilt University, ISIS- 03-403 (2003)
5. Willink, E.D.: UMLX - A Graphical Transformation Language for MDA. In: 2nd
   OOPSLA Workshop on Generative Techniques in the context of Model Driven
   Architecture, OOPSLA 2003, Anaheim (2003)
6. Kalnins, A., Barzdins, J., Celms, E.: Model Transformation Language MOLA. In:
   Aßmann, U., Aksit, M., Rensink, A. (eds.) MDAFA 2003. LNCS, vol. 3599, pp.
   62–76. Springer, Heidelberg (2005)
7. Clark, T., et al.: Language Driven Development and MDA, BPTrends. MDA Journal (October 2004)

8. Bezivin, J., et al.: The ATL Transformation-based Model Management Framework. Research Report No 03.08, IRIN, Universite de Nantes (2003)
9. Broekstra, J., Kampman, A., Harmelan, F.V.: Sesame: A Generic Architecture for Storing and Querying RDF and RDF Schema. In: Proc. International Semantic Web Conference, Sardinia, Italy (2002)
10. Budinsky, F., et al.: Eclipse Modeling Framework. Addison-Wesley, Reading (2003)
11. Metadata Repository (MDR), http://mdr.netbeans.org/
12. Barzdins, J., et al.: Towards Semantic Latvia. Communications of the 7th International Baltic Conference on Databases and Information Systems (Baltic DB&IS 2006), Vilnius, pp. 203–218 (2006)
13. Kalnins, A., Barzdins, J., Celms, E., et al.: The first step towards generic modeling tool. In: Proceedings of the 5th International Baltic Conference on Databases and Information Systems, Tallin, vol. 2, pp. 167–180 (2002)
14. Kalnins, A., Barzdins, J., Celms, E.: Efficiency Problems in MOLA Implementation. In: 19th International Conference OOPSLA 2004, Workshop "Best Practices for MDSD", Vancouver, Canada (October 2004)
15. Barzdin, J.M.: Universality problems in the theory of growing automata. Dokl. Akad. Nauk SSSR (in Russian) 157(3) (1964) (English translation in: Soviet Math. Dokl. 9, 535–537 (1964)
16. Kolmogorov, A.N., Uspensky, V.A.: To the Definition of an Algorithm. Uspekhi. Mat. Nauk (in Russian) 13(4), 3–28 (1958) (English translation in: AMS Translations, ser. 2, 21, 217–245 (1963))

Modal Fixed-Point Logic and Changing Models

Johan van Benthem[1] and Daisuke Ikegami[2]

[1] Universiteit van Amsterdam Plantage Muidergracht 24,
1018 TV, Amsterdam, The Netherlands
johan@science.uva.nl
http://staff.science.uva.nl/~johan/
[2] Universiteit van Amsterdam Plantage Muidergracht 24,
1018 TV, Amsterdam, The Netherlands
ikegami@science.uva.nl

This paper is dedicated to Professor Boris Trakhtenbrot, whose work and
spirit have inspired us and so many others in our field.

Abstract. We show that propositional dynamic logic and the modal
μ-calculus are closed under product modalities, as defined in current
dynamic-epistemic logics. Our analysis clarifies the latter systems, while
also raising some new questions about fixed-point logics.

1 Basic Closure Properties of Logics

Standard first-order logic has some simple but important closure properties.
First, it is closed under *relativization*: for every formula ϕ and unary predicate
letter P, there is a formula $(\phi)^P$ which says that ϕ holds in the sub-model
consisting of all objects satisfying P. One usually thinks of relativization as
a syntactic operation which transforms the given formula by relativizing each
quantifier $\exists x$ to $\exists x(Px \wedge$ and each quantifier $\forall x$ to $\forall x(Px \rightarrow$. But one can also
think of evaluating the original formula itself, but then in a changed semantic
model. The connection between the two viewpoints is stated in

Fact 1 (Relativization Lemma)

$$\mathbf{M}, s \vDash (\phi)^P \iff \mathbf{M} \mid P, s \vDash \phi.$$

where $\mathbf{M} \mid P$ is the restriction of the model $\mathbf{M}$ to its sub-model defined by
the predicate (or formula with one free variable) P. Relativization is a useful
property of abstract logics, and it is used extensively in proofs of Lindström
theorems. Also useful is closure under *predicate substitutions* $[\psi/P]\phi$, which may
again be read as either a syntactic operation, or as a shift to evaluation in a
suitably changed model, via the following well-known

Fact 2 (Substitution Lemma)

$$\mathbf{M}, s \vDash [\psi/P]\phi \iff \mathbf{M}[P := \psi^{\mathbf{M}}], s \vDash \phi.$$

A. Avron et al. (Eds.): Trakhtenbrot/Festschrift, LNCS 4800, pp. 146–165, 2008.

where $\mathbf{M}[P:=\psi^{\mathbf{M}}]$ is the model $\mathbf{M}$ with the denotation of the predicate letter P changed as indicated. Substitutions may be viewed as *translations* of basic predicates into newly defined ones.

Even more ambitious operations on models occur in the theory of *relative interpretation* between theories. E.g., embedding the first-order ordering theory of the rational numbers into that of the integers requires taking rationals as ordered pairs of relatively prime integers (a definable subset of the full Cartesian product $\mathbb{Z} \times \mathbb{Z}$), and redefining their order $<$ accordingly. Thus, we now also have a *product construction* where certain definable tuples become the new objects. As is easy to see, the first-order language is also closed under such product constructions - in a sense which we will not spell out. For our purpose here, we will define a precise sense of 'product closure' in terms of modal logic below, returning to the general situation at the end.

The three mentioned properties also hold of many languages extending first-order logic, such as LFP(FO), first-order logic with added fixed-point operators. But as we just said, our focus in this note will be on *modal languages*, which are rather fragments of a full first-order logic over directed graphs with unary predicates, although we also add fixed-point operators later on. For such modal languages, and especially vividly, in their epistemic interpretation as logics of knowledge and information flow, the above properties acquire special meanings of independent interest.

2 Closure Properties of Modal Languages

2.1 Epistemic Logic

Take a modal language with proposition letters, Boolean operators, and universal modalities $[i]$ which we read as stating what agent i knows, or maybe better: what is true to the best of i's information. More precisely, in epistemic pointed graph models $\mathbf{M}$ with actual world s, representing the information of a group of agents:

$$\mathbf{M}, s \vDash [i]\phi \iff \text{for all } t, \text{if } sR_i t, \text{then } \mathbf{M}, t \vDash \phi.$$

2.2 Public Announcement and Definable Submodels

In this epistemic setting, taking the relativization of the current model $\mathbf{M}, s$ to its sub-model $\mathbf{M} \mid P, s$ consisting of all points satisfying the formula P is the natural rendering of an informational event $!P$ of *public announcement* that P is currently true. Thus, model change reflects information update. The language of *public announcement logic* PAL extends epistemic logic, making these updates explicit by adding modal operators $[!P]$ for truthful announcement actions:

$$\mathbf{M}, s \vDash [!P]\phi \iff \text{if } \mathbf{M}, s \vDash P, \text{then } \mathbf{M} \mid P, s \vDash \phi.$$

Here is the relevant completeness result.

Theorem 1. *PAL is axiomatized by the minimal modal logic for the new operators [i] plus four reduction axioms*:

$$[!P]q \quad \leftrightarrow \quad P \rightarrow q \quad \text{for atomic facts } q,$$
$$[!P]\neg\phi \quad \leftrightarrow \quad P \rightarrow \neg[!P]\phi,$$
$$[!P]\phi \wedge \psi \quad \leftrightarrow \quad [!P]\phi \wedge [!P]\psi,$$
$$[!P][i]\phi \quad \leftrightarrow \quad P \rightarrow [i](P \rightarrow [!P]\phi).$$

We can read these principles as a complete recursive analysis of what agents know after they have received new information. But as was pointed out in van Benthem 2000 [4], this completeness theorem due to Plaza and Gerbrandy really just states the standard recursive clauses for performing syntactic relativization of modal formulas. Thus the technical question becomes which modal languages are closed under relativization.

This is not always the case. E.g., consider an epistemic language with an operator of common knowledge (everyone knows that everyone knows that, and so on $\cdots$.), or semantically:

$$\mathbf{M}, s \vDash C_G\phi \iff \text{for all worlds } t \text{ reachable from } s \text{ by some finite}$$
$$\text{sequence of } \sim_i \text{ steps } (i \in G), \mathbf{M}, t \vDash \phi.$$

This amounts to adding an operator of *reflexive-transitive closure* over the union of all individual accessibility relations. This infinitary operation takes us from the basic modal language into a fragment of so-called *propositional dynamic logic* (PDL). It can be shown that this fragment does not have the relativization property: indeed, the formula $[!p]C_Gq$ is not definable without modalities $[!p]$. Van Benthem, van Eijck & Kooi 2006 [5] proved this and go on to propose richer epistemic languages, using richer fragments of PDL which do have relativization closure, using so-called 'conditional common knowledge' $C_G(\phi, \psi)$ which says that ϕ is true in every world reachable with steps staying inside the ψ-worlds.

Remark 1. These observations are reminiscent of the fact that languages with generalized quantifiers may lack relativization closure. An example is first-order logic with the added quantifier "for most objects". To get the closure, one needs to add a truly binary quantifier "Most ϕ are ψ".

2.3   General Observation and Product Update

Public announcement is just one mechanism of information flow. In real-life scenarios, different agents often have different powers of observation. To model this, *dynamic-epistemic logic* (DEL) works with *event models*

$$\mathbf{A} = (E, \{R_i\}_i, \text{PRE}).$$

Here the precondition function maps events e to precondition formulas PRE_e which must hold in order for the event to occur. Just as worlds in epistemic models, events can be related by accessibility relations $\{R_i\}$ for agents. Now 'product update' turns a current model $\mathbf{M}, s$ into a model $\mathbf{M} \times \mathbf{A}, (s, e)$ recording the information of different agents after some event e has taken place in the epistemic setting represented by $\mathbf{A}$. Product update redefines the universe of relevant possible worlds, and the epistemic accessibility relations between them:

$\mathbf{M} \times \mathbf{A}$ has domain $\{(s, a) \mid s$ a world in $\mathbf{M}, a$ an event in $\mathbf{A}, (\mathbf{M}, s) \vDash \mathrm{PRE}_a\}$.

The new uncertainties satisfy $(s, a)R_i(t, b)$ if both $sR_i t$ and $aR_i b$.

The valuation for proposition letters on (s, e) is just as that for s in $\mathbf{M}$.

Here uncertainty among new worlds $(s, a), (t, b)$ can only come from old uncertainty among s, t via indistinguishable events a, b. In general, this product construction can blow up the size of the input model $\mathbf{M}$ - it does not just go to a definable sub-model. In what follows, we will assume that the event models are finite, though infinitary versions are possible.

Despite the apparent complexity of this product construction, there is a natural matching dynamic epistemic language DEL with a new modality $[\mathbf{A}, e]$:

$$\mathbf{M}, s \vDash [\mathbf{A}, e]\phi \iff \text{ if } \mathbf{M}, s \vDash \mathrm{PRE}_e, \text{ then } \mathbf{M} \times \mathbf{A}, (s, e) \vDash \phi.$$

Theorem 2. *DEL is completely axiomatizable.*

Proof. The argument, due to Baltag, Moss & Solecki 1998 [2], is as follows. The atomic and Boolean reduction axioms involved are like the earlier ones for public announcement, but here is the essential clause for the knowledge modality:

$$[\mathbf{A}, e][i]\phi \iff \mathrm{PRE}_e \to \bigwedge_{eR_i f \text{ in } \mathbf{A}} [i][\mathbf{A}, f]\phi.$$

By successive application of such principles, all dynamic modalities can be eliminated to obtain a standard epistemic formula.   □

We sum this up, somewhat loosely, by stating the following:

Fact 3. *Basic epistemic logic is product-closed.*

But again, the situation gets more complicated when we add common knowledge. In this case, no reduction to the language without $[\mathbf{A}, e]$ modalities is possible. Van Benthem, van Eijck & Kooi 2006 [5] solve this problem by moving to the language E-PDL which is just the propositional dynamic logic version of epistemic logic, but now allowing the formation of arbitrary 'complex agents' using the standard PDL program vocabulary:

basic agents i, tests $?\phi$ on arbitrary formulas ϕ of the language, unions, compositions, and Kleene iteration.

They provide an explicit axiomatization for the dynamic-epistemic version of this with added modalities $[\mathbf{A}, e]\phi$. Thus E-PDL has a completeness theorem like the earlier ones; but cf. Section 4 for remaining desiderata.

For present purposes, however, we summarize the gist of this result as follows: 'E-PDL is closed under the product construction'. In what follows, for convenience, we use obvious existential counterparts to the earlier universal modalities. Here is the central observation of the above paper:

Theorem 3. *For all $\phi \in$ E-PDL, and all action models* **A** *with event a, the formula $\langle \mathbf{A}, a \rangle \phi$ has an equivalent formula in E-PDL.*

Public announcements $!P$ are special action models with just one event with precondition P, equally visible to all agents. Thus, the theorem also says that E-PDL, or PDL, is closed under relativization - as observed earlier in van Benthem 2000 [4]. In addition, E-PDL has been shown to be closed under predicate substitutions in Kooi 2007 [10].

The point of the current paper is to analyze this situation more formally, in terms of general closure properties of modal languages, and their fixed-point extensions. In particular, we provide a new proof of Theorem 3 clarifying its background in modal fixed-point logic.

3   Closure Under Relativization for Modal Standard Languages

It is easy to see that the basic modal language is closed under relativization. The procedure relativizes modalities, just as one does with quantifiers in first-order logic. Likewise, we already mentioned that propositional dynamic logic is closed under relativization. This requires an operation which also transforms program expressions, as follows:

$$([\pi]\phi)^P = [\pi \,|\, P](\phi)^P.$$

Here one must also *relativize programs* π to programs $\pi \,|\, P$, as follows:

$$i \,|\, P = ?P; i; ?P$$
$$?\phi \,|\, P = ?(\phi \wedge P)$$
$$(\pi \cup \theta) \,|\, P = \pi \,|\, P \cup \theta \,|\, P$$
$$(\pi; \theta) \,|\, P = \pi \,|\, P; \theta \,|\, P$$
$$(\pi^*) \,|\, P = (\pi \,|\, P)^*.$$

Finally, consider the most elaborate modal fixed-point language, the so-called μ-*calculus*. Formulas $\phi(q)$ with only positive occurrences of the proposition letter q define a monotonic set transformation in any model **M**:

$$F_\phi^{\mathbf{M}}(X) = \{s \in \mathbf{M} \mid (\mathbf{M}[q := X], s) \vDash \phi.\}$$

The formula $\mu q \bullet \phi(q)$ defines the smallest fixed point of this transformation, which can be computed in ordinal stages starting from the empty set as a first approximation. Likewise, $\nu q \bullet \phi(q)$ defines the greatest fixed point of $F_\phi^{\mathbf{M}}$, with ordinal stages starting from the whole domain of **M** as a first approximation. Both exist for monotone maps, by the Tarski-Knaster theorem (Bradfield and Stirling 2006 [8]). For convenience, we assume that each occurrence of a fixed-point operator binds a unique proposition letter. Here is our first observation.

Fact 4. *The modal μ-calculus is closed under relativization.*

Proof. We show the universal validity of the following interchange law

$$\langle !P \rangle \mu q \bullet \phi(q) \leftrightarrow P \wedge \mu q \bullet \langle !P \rangle \phi(q). \tag{1}$$

Here the occurrences of q are still syntactically positive in $\langle !P \rangle \phi(q)$ - in an obvious sense. Now to prove (1), compare the following identities, for all sets $X \subseteq P^{\mathbf{M}}$:

$$\begin{aligned}
F^{\mathbf{M}}_{\langle !P \rangle \phi}(X) &= \{s \in \mathbf{M} \mid \mathbf{M}[q := X], s \vDash \langle !P \rangle \phi(q)\} \\
&= \{s \in \mathbf{M} \mid P \mid (\mathbf{M} \mid P)[q := X], s \vDash \phi(q)\} \\
&= F^{\mathbf{M} \mid P}_{\phi}(X).
\end{aligned}$$

It should be clear that the approximation maps on both sides now work in exactly the same way. $\square$

Still, there is a difference with standard fixed-point logic. One usually thinks of, e.g., a smallest fixed-point formula $\mu q \bullet \phi(q)$ as defining the limit of a sequence of ordinal approximations starting from the empty set, whose successor stages are computed by substitution of earlier ones:

$$\phi^0 = \bot, \quad \phi^{\alpha+1} = \phi(\phi^{\alpha}/q).$$

But this analogy breaks down between the two sides of the above equation (2). The approximation sequences defined in a direct manner will diverge. Consider the modal formulas

$$\phi(q) = \Box q, \quad P = \Diamond \top$$

in a model consisting of the numbers $1, 2, 3$ in their natural order. Both sequences in equation (2) start with the empty set, defined by $\bot$, but then they diverge:

for $\langle !\Diamond\top \rangle \mu q \bullet \Box q$:	for $\Diamond\top \wedge \mu q \bullet \langle !\Diamond\top \rangle \Box q$:
$\langle !\Diamond\top \rangle \Box\bot$, only true at 2	$\Diamond\top \wedge \langle !\Diamond\top \rangle \Box\bot$, only true at 2
$\langle !\Diamond\top \rangle \Box\Box\bot$, true at $1,2$	$\Diamond\top \wedge \langle !\Diamond\top \rangle \Box \langle !\Diamond\top \rangle \Box\bot$, only true at 2.

The reason for the divergence is that the formula on the right-hand side keeps prefixing formulas with dynamic model-changing modalities, so that we are now evaluating in models of the form $(\mathbf{M} \mid P) \mid P$, etc.

The general observation explaining this divergence involves another basic closure property of logical languages that we mentioned in Section 1, viz. closure under *substitutions*:

Fact 5. *The Substitution Lemma fails even for the basic modal language when announcement modalities* $\langle !P \rangle$ *are added.*

E.g., consider again our three-point model $\mathbf{M}$, with a proposition letter p true at 2 only, and let ϕ be the formula $\langle !\Diamond\top\rangle\Diamond p$. Now consider the substitution $[(\langle !\Diamond\top\rangle\top)/p]$. First consider the model after performing this substitution: it will assign p to $\{1,2\}$. Hence $[p := (\langle !\Diamond\top\rangle\top)^{\mathbf{M}}]\langle !\Diamond\top\rangle\Diamond p$ will be true in 1. Next perform the substitution syntactically to obtain the formula $\langle !\Diamond\top\rangle\Diamond(\langle !\Diamond\top\rangle\top)$: this is true nowhere in the model $\mathbf{M}$.

Since the modal language is simply translatable into first-order logic, a similar observation holds for first-order logic with relativization operators $(\phi)^P$ added as part of its syntax. The resulting language does not satisfy the usual Substitution Lemma, since the model-changing operators $()^P$ create new contexts where formulas can change their truth values. So, model-changing operators are nice devices, but they exact a price.

Remark 2 (Alternative dynamic definitions of substitution)

Fact 5 holds for the straightforward operational definition of substitutions $[\phi/p]\psi$ as syntactically replacing each occurrence of p in ψ by an occurrence of ϕ. However, there is an alternative. In line with earlier approaches in 'dynamic semantics' of first-order logic (cf. van Benthem 1996 [3]), Kooi 2007 [10] treats substitutions $[\phi/p]$ as modalities changing the current model in its denotation for p. These new modalities satisfy obvious recursive axioms pushing them through Booleans and standard modal operators. To push them also through public announcement modalities, one can first rewrite the latter via their PAL recursion axioms, and only then apply the substitution to the components. Van Eijck 2007 [9] shows how this provides an alternative syntactic operational definition of substitution, working inside out. One first reduces innermost PAL or DEL formulas to their basic modal equivalents, and then performs standard syntactic substitution in these. Though not compositional, this procedure is effective. When applied to the two approximation sequences in our earlier problematic example, these would now come out being the same after all.

Thus, dynamic modal languages are closed under semantic substitutions, but finding the precise corresponding syntactic operation in their static base language requires some care.

4 Closure of Dynamic Logic Under Products

Theorem 3 said that the language E-PDL is closed under the product operation $\langle \mathbf{A}, e \rangle \phi$. The proof in van Benthem, van Eijck & Kooi 2006 [5] uses special arguments involving Kleene's Theorem for finite automata and program transformations. We provide a new proof which provides further insight by restating the situation within modal fixed-point logic.

First, consider the obvious inductive proof of Theorem 3, the 'Main Reduction'. Its steps follow the construction of the formula ϕ. The atomic case,

Booleans $\neg$, $\vee$, and basic epistemic modalities $\langle i \rangle$ are taken care of by the standard DEL reduction axioms. The remaining case is that of formulas $\langle \mathbf{A}, a \rangle \langle \pi \rangle \psi$ with an E-PDL modality involving a complex epistemic program π. To proceed, we need a deeper analysis of program structure. The following result can be proved together with Theorem 3 by a simultaneous induction:

Theorem 4. *For all* $\mathbf{A}, a$, *and programs* $\pi' \in$ *E-PDL, there exist E-PDL programs* $T_{a,b}^{\pi'}$ *(for each* $b \in A$) *such that, for all E-PDL formulas* ψ,

$$\mathbf{M}, s \vDash \langle \mathbf{A}, a \rangle \langle \pi' \rangle \psi \iff \mathbf{M}, s \vDash \bigvee_{b \in \mathbf{A}} \langle T_{a,b}^{\pi'} \rangle \langle \mathbf{A}, b \rangle \psi.$$

Proof. We use induction on the construction of the program π'.

Case 1: $\pi' = i$.

$$\langle \mathbf{A}, a \rangle \langle i \rangle \psi \iff \mathrm{PRE}_a \text{ and } \bigvee_{aR_i b \text{ in } A} \langle i \rangle \langle \mathbf{A}, b \rangle \psi.$$

This can be brought into our special form by setting

$$T_{a,b}^i = ?\mathrm{PRE}_a ; i \text{ if } aR_i b \text{ in } A, \text{ and } T_{a,b}^i = \bot, \text{ otherwise.}$$

Case 2: $\pi' = ?\alpha$ for some formula α.

$$\langle \mathbf{A}, a \rangle \langle ?\alpha \rangle \psi \iff \langle \mathbf{A}, a \rangle (\alpha \wedge \psi) \iff$$
$$\langle \mathbf{A}, a \rangle \alpha \wedge \langle \mathbf{A}, a \rangle \psi \iff \langle ?(\langle \mathbf{A}, a \rangle \alpha) \rangle \langle \mathbf{A}, a \rangle \psi.$$

Here the less complex formula $\langle \mathbf{A}, a \rangle \alpha$ can be taken to be in the language of E-PDL already, by the simultaneous induction proving Theorem 3. It is easy to then define the correct transition predicates $T_{a,b}^{?\alpha}$ for all events $b \in \mathbf{A}$.

Case 3: $\pi' = \alpha \cup \beta$ for some formulas α, β.

$$\langle \mathbf{A}, a \rangle \langle \alpha \cup \beta \rangle \psi \iff \langle \mathbf{A}, a \rangle (\langle \alpha \rangle \psi \vee \langle \beta \rangle \psi) \iff$$
$$\langle \mathbf{A}, a \rangle \langle \alpha \rangle \psi \vee \langle \mathbf{A}, a \rangle \langle \beta \rangle \psi \overset{\text{ind.hyp.}}{\iff} \bigvee_{b \in \mathbf{A}} \langle T_{a,b}^\alpha \rangle \langle \mathbf{A}, b \rangle \psi \vee \bigvee_{b \in \mathbf{A}} \langle T_{a,b}^\beta \rangle \langle \mathbf{A}, b \rangle \psi$$

and, by recombining parts of this disjunction, using the valid PDL-equivalence $\langle \alpha \rangle \psi \vee \langle \beta \rangle \psi \leftrightarrow \langle \alpha \cup \beta \rangle \psi$, we get the required normal form.

Case 4: $\pi' = \alpha ; \beta$ for some formulas α, β.

$$\langle \mathbf{A}, a \rangle \langle \alpha ; \beta \rangle \psi \iff \langle \mathbf{A}, a \rangle \langle \alpha \rangle \langle \beta \rangle \psi \overset{\text{ind. hyp.1}}{\iff}$$
$$\bigvee_{b \in \mathbf{A}} \langle T_{a,b}^\alpha \rangle \langle \mathbf{A}, b \rangle \langle \beta \rangle \psi \overset{\text{ind. hyp.2}}{\iff} \bigvee_{b \in \mathbf{A}} (\langle T_{a,b}^\alpha \rangle \bigvee_{c \in \mathbf{A}} \langle T_{b,c}^\beta \rangle \langle \mathbf{A}, c \rangle \psi)$$

and here, using the minimal logic of PDL again, substituting one special form in another once more yields a special form. E.g., we have the equivalence
$$\langle \alpha \rangle (\langle \beta \rangle p \vee \langle \gamma \rangle q) \iff \langle \alpha ; \beta \rangle p \vee \langle \alpha ; \gamma \rangle q.$$

Case 5: $\pi' = \pi^*$ for some program π.

The crux lies in this final case: combinations with Kleene iterations $\langle \mathbf{A}, a \rangle \langle \pi^* \rangle \psi$ do not reduce as before. But even so, we can analyze them in the same style, using a *simultaneous fixed-point operator* $\mu \mathbf{q}_b \bullet$ defining the propositions $\langle \mathbf{A}, b \rangle \langle \pi^* \rangle \psi$ for all events $b \in \mathbf{A}$ in one fell swoop. The need for this simultaneous recursion explains earlier difficulties in the literature with reduction axioms for common knowledge with product update. To find the right schema, first recall the PDL fixed-point equation for Kleene iteration:

$$\langle \mathbf{A}, a \rangle \langle \pi^* \rangle \psi \iff \langle \mathbf{A}, a \rangle (\psi \vee \langle \pi \rangle \langle \pi^* \rangle \psi) \iff$$

$$\langle \mathbf{A}, a \rangle \psi \vee \langle \mathbf{A}, a \rangle \langle \pi \rangle \langle \pi^* \rangle \psi \overset{\text{ind. hyp.}}{\iff} \langle \mathbf{A}, a \rangle \psi \vee \bigvee_{b \in \mathbf{A}} \langle T^\pi_{a,b} \rangle \langle \mathbf{A}, b \rangle \langle \pi^* \rangle \psi.$$

Here, again because of the simultaneous inductive proof with Theorem 3, we can think of the first disjunct as being some formula α_a of E-PDL. The result of this unpacking are simultaneous equivalences of the form (with propositional variables q_a for each $a \in \mathbf{A}$):

$$q_a \leftrightarrow \alpha_a \vee \bigvee_{b \in \mathbf{A}} \langle T^\pi_{a,b} \rangle q_b. \tag{$*$}$$

Lemma 1. *The denotations of the modal formulas $\langle \mathbf{A}, a \rangle \langle \pi^* \rangle \psi$ in a model $\mathbf{M}$ are precisely the a-projections of the smallest fixed-point solution to the simultaneous equations $(*)$.*

Proof (Lemma 1). Here, smallest fixed-points for simultaneous equations in the μ-calculus are computed just as those for single fixed-point equations. lemma 1 follows by a simple induction, showing that the standard meanings of the modal formulas $\langle \mathbf{A}, a \rangle \langle \pi^* \rangle \psi$ in a model $\mathbf{M}$ are contained in any solution for the simultaneous fixed-point equation.

We calculate the meaning of the least fixed-point of $(*)$ through the approximation procedure and show it is equal to that of $\langle \mathbf{A}, a \rangle \langle \pi^* \rangle \psi$ ($a \in \mathbf{A}$).

From now on, we identify formulas by their truth sets in $\mathbf{M}$, reading ϕ as $\{m \in \mathbf{M} \mid (\mathbf{M}, m) \vDash \phi\}$. For simplicity, we rewrite $(*)$ as follows:

$$q_i = \alpha_i \vee \bigvee_{1 \leq j \leq n} \langle T^\pi_{i,j} \rangle q_j \quad (1 \leq i \leq n)$$

Let F be the monotone operator from $\mathcal{P}(\mathbf{M})^n$ to itself induced by the right hand side of $(*)$, where n is the number of elements in $\mathbf{A}$. More precisely, for $\mathbf{X} = (X_1, \cdots, X_n) \in \mathcal{P}(\mathbf{M})^n$, $F(\mathbf{X}) = (Y_1, \cdots, Y_n)$ where for each $1 \leq i \leq n$,

$$Y_i = \{ m \in \mathbf{M} \mid (\mathbf{M}[\{q_j := X_j\}_{j=1,\cdots,n}], m) \vDash \alpha_i \vee \bigvee_{1 \leq j \leq n} \langle T^\pi_{i,j} \rangle q_j \}.$$

Next, for $\boldsymbol{X} \in \mathcal{P}(\mathbf{M})^n$, define $\langle F^\xi(\boldsymbol{X}) \mid \xi \in \mathrm{On}\rangle$ as follows:

$$F^0(\boldsymbol{X}) = \boldsymbol{X}$$
$$F^{\xi+1}(\boldsymbol{X}) = F(F^\xi(\boldsymbol{X}))$$
$$F^\xi(\boldsymbol{X}) = \bigcup_{\eta < \xi} F^\eta(\boldsymbol{X}) \text{ if } \xi \text{ is a limit ordinal.}$$

For any $m \in \omega$, we can prove the following equation by induction on m.

$$F^m(\bot) = \Big\{ \bigvee_{1 \le j_1, j_2, \cdots, j_{m-1} \le n} [\alpha_i \vee \langle T^\pi_{i,j_1}\rangle \alpha_{j_1} \vee \langle T^\pi_{i,j_1}\rangle\langle T^\pi_{j_1,j_2}\rangle \alpha_{j_2}$$
$$\vee \cdots \vee \langle T^\pi_{i,j_1}\rangle\langle T^\pi_{j_1,j_2}\rangle \cdots \langle T^\pi_{j_{m-2},j_{m-1}}\rangle \alpha_{j_{m-1}}]\Big\}_{1\le i \le n}.$$

Hence

$$F^\omega(\bot) = \Big\{(\exists m < \omega)(\exists j_1, \cdots, j_{m-1})\ \langle T^\pi_{i,j_1}\rangle \cdots \langle T^\pi_{j_{m-2},j_{m-1}}\rangle \alpha_{j_{m-1}}\Big\}_{1\le i \le n},$$

which implies $F^\omega(\bot) = F^{\omega+1}(\bot)$: the least fixed-point is reached in ω steps.

Therefore we only have to show that $\{\langle \mathbf{A}, a_i\rangle\langle\pi^*\rangle\psi\}_{1\le i \le n} = F^\omega(\bot)$.

Recall that, by the defining property of $T^\pi_{i,j}$, for any E-PDL formula ψ', any $1 \le i \le n$ and any state s in $\mathbf{M}$,

$$\mathbf{M}, s \vDash \langle \mathbf{A}, a_i\rangle\langle\pi\rangle\psi' \iff \mathbf{M}, s \vDash \bigvee_{1\le j \le n} \langle T^\pi_{i,j}\rangle\langle\mathbf{A}, a_j\rangle\psi'.$$

By using this condition repeatedly, we get the following equivalence: for any n-tuple s of elements in M and any i with $1 \le i \le n$,

$$\mathbf{M}, s_i \vDash \langle \mathbf{A}, a_i\rangle\langle\pi^*\rangle\psi$$
$$\iff (\exists m \in \omega)\ \mathbf{M}, s_i \vDash \langle \mathbf{A}, a_i\rangle\langle\pi\rangle^m\psi$$
$$\iff (\exists m \in \omega)(\exists j_1)\ \mathbf{M}, s_i \vDash \langle T^\pi_{i,j_1}\rangle\langle\mathbf{A}, a_{j_1}\rangle\langle\pi\rangle^{m-1}\psi$$
$$\iff (\exists m \in \omega)(\exists j_1, j_2)\ \mathbf{M}, s_i \vDash \langle T^\pi_{i,j_1}\rangle\langle T^\pi_{j_1,j_2}\rangle\langle\mathbf{A}, a_{j_2}\rangle\langle\pi\rangle^{m-2}\psi$$
$$\iff \cdots$$
$$\iff (\exists m \in \omega)(\exists j_1, \cdots, j_m)\ \mathbf{M}, s_i \vDash \langle T^\pi_{i,j_1}\rangle\langle T^\pi_{j_1,j_2}\rangle \cdots \langle T^\pi_{j_{m-1},j_m}\rangle\langle\mathbf{A}, a_{j_m}\rangle\psi$$
$$\iff (\exists m \in \omega)(\exists j_1, \cdots, j_m)\ \mathbf{M}, s_i \vDash \langle T^\pi_{i,j_1}\rangle\langle T^\pi_{j_1,j_2}\rangle \cdots \langle T^\pi_{j_{m-1},j_m}\rangle\alpha_{j_m}$$
$$\iff s_i \in \big(F^\omega(\bot)\big)_i$$

where $\big(F^\omega(\bot)\big)_i$ is the i-th coordinate of $F^\omega(\bot)$. Hence

$$(\forall i)\big(\mathbf{M}, s_i \vDash \langle\mathbf{A}, a_i\rangle\langle\pi^*\rangle\psi\big) \iff s \in F^\omega(\bot),$$

which is what we desired. □

What really happens here is this. Computing the explicit solutions for the predicates q_i after ω steps, one gets the countable disjunction over all finite 'path formulas' of the form $\langle T_{i,j_i}^{\pi}; T_{j_1,j_2}^{\pi}; \cdots; T_{j_n,k}^{\pi}\rangle \alpha_k$. And the latter are exactly the meanings of the original propositions $\langle \mathbf{A}, a \rangle \langle \pi^* \rangle \psi$.

But we are not done yet. What we need to show next is that the solutions obtained in this way are actually in the language E-PDL! The following lemma tells us the relevant fact about the μ-calculus. Simultaneous fixed-point equations of the above special disjunctive shape $(*)$ can be solved one by one, and the solutions lie inside dynamic logic.

Lemma 2. *Any system of simultaneous fixed-point equations of $(*)$ has an explicit minimal solution for each q_a in E-PDL. Moreover, the solutions retain the special disjunctive form described in Theorem 4.*

Proof (Lemma 2). The inductive procedure producing explicit E-PDL solutions works line by line - like Gaussian Elimination in a system of linear equations.

- Case 1. There is only one q-variable, as with public announcements.
  The line reads $q_1 \leftrightarrow \alpha_1 \vee \langle \beta_{1,1} \rangle q_1$. The explicit solution works just as in standard dynamic logic, in the

$$q_1 = \langle \beta_{1,1}^* \rangle \alpha_1.$$

- Case 2. There are n lines in the recursion schema, with $n > 1$.
  We first solve for the variable q_1 as in Case 1 - obtaining an explicit E-PDL formula $\sigma_1(q_2, \cdots, q_n)$ in the other recursion variables. We then substitute this solution in the remaining $n - 1$ equations, and solve these inductively. Finally, the solutions thus obtained for the $q_2, \cdots, q_n$ are substituted in $\sigma_1(q_2, \cdots, q_n)$ to also solve for q_1.

Some syntactic checking will show that these solutions remain in the syntactic format described in Theorem 4. But of course, we also need to show that this is really a solution for the above fixed-point equations $(*)$, and indeed the smallest one. To prove that, we formulate the algorithm more formally in the following way (cf. Arnold & Niwinski [1] for a more extensive treatment).

For any monotone operator G, let G_* denote the least fixed point of G. Let $F \colon \mathcal{P}(\mathbf{M})^n \to \mathcal{P}(\mathbf{M})^n$ be the monotone operator induced by the n equations in $(*)$. Now take any $X_2, \cdots, X_n \in \mathcal{P}(\mathbf{M})$ and fix them. Next, define $F_{X_2, \cdots, X_n} \colon \mathcal{P}(\mathbf{M}) \to \mathcal{P}(\mathbf{M})$ as follows:

$$F_{X_2, \cdots, X_n}(X_1) = \big(F(X_1, \cdots, X_n) \big)_1,$$

where $(\boldsymbol{X})_i$ is the i-th coordinate of $\boldsymbol{X}$. Since F is monotone, $F_{X_2, \cdots, X_n}$ is also monotone. Then define $F_{X_3, \cdots, X_n} \colon \mathcal{P}(\mathbf{M}) \to \mathcal{P}(\mathbf{M})$:

$$F_{X_3, \cdots, X_n}(X_2) = \big(F((F_{X_2, \cdots, X_n})_*, X_2, \cdots, X_n) \big)_2.$$

This is also monotone because F and the function $(X_2, \cdots, X_n) \mapsto (F_{X_2, \cdots, X_n})_*$ are both monotone. Continue this process until we define $F_{\emptyset}$. Then the solution of the earlier 'Gaussian' algorithm is the unique F_*' such that

$$(F_*')_i = \big(F_{(F_*')_{i+1}, \cdots, (F_*')_n} \big)_* \quad (1 \leq i \leq n).$$

Note how we compute the rightmost fixed-point first here, and then substitute leftward. Hence all we have to show is the following:

Claim 1. $F_* = F'_*$.

The proof is in Arnold & Niwinski [1] (see Section 1.4. in this book). To make our paper self-contained, we will put a proof in an Appendix below.   □

This concludes the proofs of Theorems 3 and 4.   □

Illustration 1. *We compute the solutions for the update model* $\mathbf{A} =$

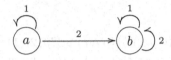

$$PRE_a = p, \qquad PRE_b = \top$$

This describes a security scenario where agent 1 correctly observes that event a is taking place, while agent 2 mistakenly believes that b occurs. Here is a description of the non-trivial common knowledge for $1, 2$ arising from this scenario, by writing out the fixed point equation for $\langle \mathbf{A}, a \rangle \langle (1 \cup 2)^* \rangle r$.

By step 1 in the proof,

$$T_{a,a}^1 = ?PRE_a; 1 = ?p; 1, \quad T_{a,b}^1 = \bot$$
$$T_{b,a}^1 = \bot, \qquad\qquad T_{b,b}^1 = ?PRE_b; 1 = 1$$
$$T_{a,a}^2 = \bot, \qquad\qquad T_{a,b}^2 = ?PRE_a; 2 = ?p; 2$$
$$T_{b,a}^2 = \bot, \qquad\qquad T_{b,b}^2 = ?PRE_b; 2 = 2.$$

Then by step 3,

$$T_{a,a}^{1 \cup 2} = T_{a,a}^1 \cup T_{a,a}^2 = ?p; 1, \quad T_{a,b}^{1 \cup 2} = T_{a,b}^1 \cup T_{a,b}^2 = ?p; 2$$
$$T_{b,a}^{1 \cup 2} = T_{b,a}^1 \cup T_{b,a}^2 = \bot, \quad T_{b,b}^{1 \cup 2} = T_{b,b}^1 \cup T_{b,b}^2 = 1 \cup 2.$$

Now put

$$q_a = \langle \mathbf{A}, a \rangle \langle (1 \cup 2)^* \rangle r, \quad q_b = \langle \mathbf{A}, b \rangle \langle (1 \cup 2)^* \rangle r.$$

Then by $(*)$,

$$q_a = \langle \mathbf{A}, a \rangle r \vee \langle T_{a,a}^{1 \cup 2} \rangle q_a \vee \langle T_{a,b}^{1 \cup 2} \rangle q_b$$
$$= (PRE_a \wedge r) \vee \langle ?p; 2 \rangle q_b \vee \langle ?p; 1 \rangle q_a$$
$$= ((p \wedge r) \vee \langle ?p; 2 \rangle q_b) \vee \langle ?p; 1 \rangle q_a$$

and

$$q_b = \langle \mathbf{A}, b \rangle r \vee \langle T_{b,a}^{1 \cup 2} \rangle q_a \vee \langle T_{b,b}^{1 \cup 2} \rangle q_b$$
$$= r \vee \langle 1 \cup 2 \rangle q_b$$

Since the order of eliminating variables does not influence the solutions, we first solve q_b as follows:

$$q_b = \langle (1 \cup 2)^* \rangle r.$$

By substituting this solution in the above equation for q_a,

$$q_a = \big((p \wedge r) \vee \langle ?p; 2 \rangle \langle (1 \cup 2)^* \rangle r\big) \vee \langle ?p; 1 \rangle q_a$$
$$= \big((p \wedge r) \vee \langle ?p; 2; (1 \cup 2)^* \rangle r\big) \vee \langle ?p; 1 \rangle q_a.$$

Hence

$$q_a = \langle (?p; 1)^* \rangle \big((p \wedge r) \vee \langle ?p; 2; (1 \cup 2)^* \rangle r\big)$$
$$= \langle (?p; 1)^* \rangle (p \wedge r) \vee \langle (?p; 1)^*; ?p; 2; (1 \cup 2)^* \rangle r$$

We can easily check that these q_a, q_b satisfy the equations we gave by an independent semantic argument.

Remark 3. The calculation in this example is really just the following well-known fact about the modal μ-calculus:

Let $\phi(q_1, q_2)$, $\psi(q_1, q_2)$ be positive formulas in the modal μ-calculus. Then the simultaneous least fixed points of these formulas is

$$\big(\mu q_1.\phi(q_1, \mu q_2.\psi(q_1, q_2)), \mu q_2.\psi(\mu q_1.\phi(q_1, q_2), q_2)\big).$$

In the proof of Claim 1 (cf. the Appendix), we only use the condition that F is monotone. This means we can generalize the result as follows:

Corollary 1. *The modal μ-calculus is closed under the formation of simultaneous fixed-point operators.*

5 Closure of the μ-Calculus Under Products

Finally, we show how the preceding analysis also extends to the μ-calculus itself, where it even becomes simpler.

Theorem 5. *The μ-calculus is closed under product operators.*

Proof. We prove the statement by induction on the complexity of formulas. We only consider the fixed point case, as the others go like before.

Our main task is to analyze fixed-point computations in product models $\mathbf{M} \times \mathbf{A}$ in terms of similar computations in the original model $\mathbf{M}$. The following idea turns out to work here. Let X be a subset of $\mathbf{M} \times \mathbf{A}$. Modulo the event preconditions possibly ruling out some pairs, we can describe X, without loss of information, in terms of the sequence of its projections to the events in $\mathbf{A}$, viewed as a finite set of indices. Thus, we can describe the computation in $\mathbf{M} \times \mathbf{A}$

by means of a finite set of computations in M. The following set of definitions and observations makes this precise.

Take any Kripke model $\mathbf{M}$ and any event model $\mathbf{A}$. Let n be the number of elements of $\mathbf{A}$ and let $\mathbf{A} = \{a_j\}_{1 \leq j \leq n}$. There are canonical mappings $\pi \colon \mathcal{P}(\mathbf{M})^n \to \mathcal{P}(\mathbf{M} \times \mathbf{A})$ and $\iota \colon \mathcal{P}(\mathbf{M} \times \mathbf{A}) \to \mathcal{P}(\mathbf{M})^n$ with $\pi \circ \iota = \mathrm{id}$:

$$\pi(X) = \bigcup_{1 \leq j \leq n} (X_j \times \{a_j\}) \cap (\mathbf{M} \times \mathbf{A}),$$

$$\iota(Y) = \{Y_j\}_{1 \leq j \leq n},$$

where $Y_j = \{x \in \mathbf{M} \mid (x, a_j) \in Y\}$.

Given a positive formula $\phi(q)$ in the modal μ-calculus, let $F_\phi^{\mathbf{M} \times \mathbf{A}} \colon \mathcal{P}(\mathbf{M} \times \mathbf{A}) \to \mathcal{P}(\mathbf{M} \times \mathbf{A})$ be the monotone function induced by $\phi(q)$. Define $F^{\phi(q)} \colon \mathcal{P}(\mathbf{M})^n \to \mathcal{P}(\mathbf{M})^n$ as follows:

$$F^{\phi(q)} = \iota \circ F_\phi^{\mathbf{M} \times \mathbf{A}} \circ \pi.$$

We claim that $F_\phi^{\mathbf{M} \times \mathbf{A}}$ is monotone if and only if $F^{\phi(q)}$ is monotone. Suppose $F_\phi^{\mathbf{M} \times \mathbf{A}}$ is monotone. Since π, ι are monotone and compositions of monotone functions are monotone, $F^{\phi(q)}$ is also monotone. To prove the converse, suppose $F^{\phi(q)}$ is monotone. Pick any $X, Y \in \mathcal{P}(\mathbf{M} \times \mathbf{A})$ with $X \subseteq Y$. First note that $F_\phi^{\mathbf{M} \times \mathbf{A}}(X) \subseteq F_\phi^{\mathbf{M} \times \mathbf{A}}(Y)$ holds if and only if $\iota \circ F_\phi^{\mathbf{M} \times \mathbf{A}}(X) \subseteq \iota \circ F_\phi^{\mathbf{M} \times \mathbf{A}}(Y)$ holds. Hence all we have to check is $\iota \circ F_\phi^{\mathbf{M} \times \mathbf{A}}(X) \subseteq \iota \circ F_\phi^{\mathbf{M} \times \mathbf{A}}(Y)$. But

$$\begin{aligned}
\iota \circ F_\phi^{\mathbf{M} \times \mathbf{A}}(X) &= \iota \circ F_\phi^{\mathbf{M} \times \mathbf{A}}\big(\pi \circ \iota(X)\big) = \iota \circ F_\phi^{\mathbf{M} \times \mathbf{A}} \circ \pi\big(\iota(X)\big) \\
&= F^{\phi(q)}\big(\iota(X)\big) \subseteq F^{\phi(q)}\big(\iota(Y)\big) = \iota \circ F_\phi^{\mathbf{M} \times \mathbf{A}} \circ \pi\big(\iota(Y)\big) \\
&= \iota \circ F_\phi^{\mathbf{M} \times \mathbf{A}}\big(\pi \circ \iota(Y)\big) = \iota \circ F_\phi^{\mathbf{M} \times \mathbf{A}}(Y),
\end{aligned}$$

where the above inclusion follows from the monotonicity of $F^{\phi(q)}$ and ι.

Moreover, there is a further canonical correspondence: if X is an $F^{\phi(q)}$-fixed point, then $\pi(X)$ is an $F_\phi^{\mathbf{M} \times \mathbf{A}}$-fixed-point, and if Y is an $F_\phi^{\mathbf{M} \times \mathbf{A}}$-fixed-point, then $\iota(Y)$ is an $F^{\phi(q)}$-fixed-point. Hence the least $F^{\phi(q)}$-fixed-point corresponds to the least $F_\phi^{\mathbf{M} \times \mathbf{A}}$-fixed-point.

Remark 4 (Relating fixed-point computations in different models). The argument above may be seen as a special case of the "Transfer Lemma" (Theorem 1.2.15) in Arnold & Niwinski [1]. This lemma only uses our ι function, while we added the function π for clarity, to restrict an input to the inverse image of ι – which is why the equation $\pi \circ \iota = \mathrm{id}$ holds. For further background to this kind of argument, cf. Bloom and Ésik [7].

So far, we have seen that the least $F_\phi^{\mathbf{M} \times \mathbf{A}}$-fixed-point can be correlated with the least $F^{\phi(q)}$-fixed-point in a natural way. Our next task is to show that $\langle \mathbf{A}, a \rangle \mu q . \phi(q)$ is actually definable in the modal μ-calculus. For that purpose, first note that $\langle \mathbf{A}, a_j \rangle \, \mu q \bullet \phi(q)$ defines the j-th coordinate of the least $F_\phi^{\mathbf{M} \times \mathbf{A}}$-fixed-point. By the definition of ι, it is also the j-th coordinate of the least $F^{\phi(q)}$-fixed-point. Now, since the modal μ-calculus is closed under simultaneous fixed-point operators by Corollary 1, if we can express $F^{\phi(q)}$ by a formula of the modal μ-calculus with positive variables, we are done.

To prove this, we generalize the syntactic analysis employed in Section 4 to formulas with many variables $q = q_1, \cdots, q_m$. For any formula $\phi(q)$ in the modal μ-calculus, define $F_{\phi(q)}^{\mathbf{M} \times \mathbf{A}} : \mathcal{P}(\mathbf{M} \times \mathbf{A})^m \to \mathcal{P}(\mathbf{M} \times \mathbf{A})$ as follows:

$$F_{\phi(q)}^{\mathbf{M} \times \mathbf{A}}(\mathbf{Y}) = \{(s,a) \mid ((\mathbf{M} \times \mathbf{A})[q_k := Y_k], (s,a)) \vDash \phi(q)\},$$

where $\mathbf{Y} \in (\mathbf{M} \times \mathbf{A})^m$.

Claim 2. *For any formula $\phi(q)$ in the modal μ-calculus, there are formulas ψ_ϕ such that $F^{\phi(q)} = F_{\psi_\phi}^{\mathbf{M}}$ where $F^{\phi(q)} : \mathcal{P}(\mathbf{M})^{m \cdot n} \to \mathcal{P}(\mathbf{M})^n$ and*

(*) *For any $1 \le k \le m$, if all the occurrences of q_k in ϕ are positive (negative resp.), then for each $1 \le j, j' \le n$, all the occurrences of $p_{k,j}$ in $(\psi_\phi)_{j'}$ are positive (negative resp.),*

Proof (Claim 2). In the following definitions, we only display the essential argument variables needed to understand the function values. We prove the statement by induction on the complexity of ϕ. As in the proof of Lemma 1, we identify formulas with their truth sets. Also, if ψ is a sequence of formulas, ψ_j is the j-th coordinate of ψ.

- Case 1: $\phi = p$ (p is not in q).

$$F^{\phi(q)} = \left(p \wedge \mathrm{PRE}_{a_1}, \cdots, p \wedge \mathrm{PRE}_{a_n} \right).$$

Hence $(\psi_{\phi(q)})_j = p \wedge \mathrm{PRE}_{a_j}$. It is easy to check (*).
- Case 2: $\phi = q_k$ (q_k is the k-th coordinate of q).

$$F^{\phi(q)}(\mathbf{X}) = \{X_{k,j} \wedge \mathrm{PRE}_{a_j}\}_{1 \le j \le n}.$$

Hence $(\psi_{\phi(q)})_j = p_{k,j} \wedge \mathrm{PRE}_{a_j}$, where $p_{k,j}$ is the j-th variable in the k-th block corresponding to q_k. It is also easy to check (*).
- Case 3: $\phi = \phi_1 \wedge \phi_2$.

$$F^{\phi(q)} = \psi_{\phi_1} \wedge \psi_{\phi_2}.$$

Hence $\psi_{\phi(q)} = \psi_{\phi_1} \wedge \psi_{\phi_2}$. It is easy to check (*).
- Case 4: $\phi = \neg \phi'$.

$$F^{\phi(q)} = \{\neg(\psi_{\phi'})_j \wedge \mathrm{PRE}_{a_j}\}_{1 \le j \le n}.$$

Hence $(\psi_{\phi(q)})_j = \neg(\psi_{\phi'})_j \wedge \mathrm{PRE}_{a_j}$. It is easy to check (*) by our inductive hypothesis, and the simultaneous definition for positive and negative occurrences.

– Case 5: $\phi = \langle i \rangle \phi'$.
For any $1 \leq j \leq n$ and $x \in M$,

$$x \in \left(F^{\phi(q)}(\boldsymbol{X}) \right)_j \iff$$

$$(1 \leq \exists j' \leq n) \, (\exists y \in \mathbf{M}) \, \left(x R_i y \wedge a_j R_i a_{j'} \wedge y \in \left(F^{\phi'(q)}(\boldsymbol{X}) \right)_{j'} \right).$$

To see that this is true, observe that the condition $y \in \left(F^{\phi'(q)}(\boldsymbol{X}) \right)_{j'}$ implies $(y, a_{j'}) \in \mathbf{M} \times \mathbf{A}$. Therefore, we can put

$$\left(\psi_{\phi(q)} \right)_j = \bigvee_{a_j R_i a_{j'}} \langle i \rangle \left(\psi_{\phi'(q)} \right)_{j'}.$$

– Case 6: $\phi = \mu q' \bullet \phi'$, where all the occurrences of q' are positive in ϕ'.

$$F^{\phi(q)}(\boldsymbol{X}) = \left\{ \left(F^{\mathbf{M} \times \mathbf{A}}_{\mu q' \bullet \phi'(q', q)}(\pi(\boldsymbol{X})) \right)_j \right\}_{1 \leq j \leq n}$$

$$= \left\{ \left((F^{\mathbf{M} \times \mathbf{A}}_{\phi'(q', q)}(\pi(\boldsymbol{X})))_* \right)_j \right\}_{1 \leq j \leq n}$$

$$= \left(\boldsymbol{X}' \mapsto F^{\mathbf{M}}_{\psi_{\phi'}}(\boldsymbol{X}', \boldsymbol{X}) \right)_*,$$

where $(F(\cdot))_*$ is the least F-fixed-point. By induction hypothesis, all the occurrences of p'_j are positive in $(\psi_{\phi'})_{j'}$ for any $1 \leq j, j' \leq n$, where $\boldsymbol{p}'$ corresponds to q'. Since the modal μ-calculus is closed under simultaneous fixed-point operators, we can put $\psi_{\phi(q)} = \mu \boldsymbol{p}' \bullet \psi_{\phi'}(\boldsymbol{q})$, that are also in the modal μ-calculus. Since μ-operators do not change the positivity (negativity) of variables not bounded by them, (∗) also holds in this case.   □

The proof of the last case explains why we needed to 'blow-up' in the number of variables in Claim 2. Also, we proved the claim for arbitrary formulas (not only for positive ones) because otherwise we cannot use the induction hypothesis in Case 4 (if ϕ is positive, then ϕ' must be negative).   □

Remark 5 (Effective reduction axioms)
As in Fact 4, we could also an explicit reduction axiom for $\langle \mathbf{A}, a_j \rangle \mu q. \phi(q)$ by taking the j-th coordinate of the simultaneous fixed-point expression $\mu \boldsymbol{q}. \psi_{\phi(q)}$. Since our proof is effective, we can effectively compute the shape of the axiom.

The common point of the proofs of Theorems 3,4 and Theorem 5 is that both E-PDL and the modal μ-calculus are closed under simultaneous fixed point operators (in the case of E-PDL, such operators have the special form of (∗)). The proof of that fact is essentially the same (it is that of Claim 1) but the case of the full μ-calculus is easier because we have arbitrary μ-operators, while in E-PDL, we have to check if the solution is also in E-PDL.

6   Conclusions and Further Directions

T he preceding results place current modal logics of information update in a more general light, relating their 'reduction axiom' approach for obtaining conservative dynamic extensions of existing static logics to abstract closure properties of fixed-point logics. Our observations also suggest a number of more general issues, of which we mention a few.

Fine-structure of the μ-calculus. Our results show that product closure holds for basic modal logic, propositional dynamic logic PDL, and the μ-calculus itself. We think that there are further natural fragments with this property, including the μ-ω-calculus, which only allows fixed-points whose computations stop uniformly by stage ω. Another case to look for product closure is the hierarchy of nested fixed-point alternation. Our proof removes modal product operators by means of simultaneous fixed-points, which can then be removed by nested single ones, but we have not yet analyzed its precise syntactic details.

On another matter, our proof method in Section 4 suggests that PDL is distinguished inside the μ-calculus as the smallest fragment closed under some very simple 'additive' fixed-point equations. This seems related to the fact that the semantics of dynamic logic only describes linear computation traces, and no more complex constructs, such as arbitrary finite trees. Can this equational observation be turned into a characterization of PDL?

Connections with automata theory. The first proof of product closure for PDL in van Benthem & Kooi 2004 [6] used finite automata to serve as 'controllers' restricting state sequences in product models $M \times A$. The second, different proof in Van Benthem, van Eijck & Kooi 2006 [5] involved a non-trivial use of Kleene's Theorem for regular languages, and hence again a connection with finite automata. What is the exact connection of this proof with our special unwinding of simultaneous 'disjunctive' fixed-point equations inside PDL? Can Kleene's Theorem be interpreted as a normal-form result in fixed-point logic?

There may also be a more general automata-based take on our arguments, given the strong connection between automata theory and μ-calculus.

Martin Otto (p.c.) has proposed using the product closure of MSOL and the bisimulation invariance of the mu-calculus with added product modalities for an alternative proof of our Theorem 5, by an appeal to the Janin - Walukiewicz Theorem.

Logical languages and general product closure. Finally, we know now that many modal languages are product-closed. What about logical systems in general? We would like to have an abstract formulation which applies to a wider class of logical systems, such as first-order logic and its extensions in abstract model theory. We feel that product closure is a natural requirement on expressive power, especially given its earlier motivation in terms of relative interpretability. But the correct formulation may have to be stronger than our notion in this paper. Even in the modal case, our proofs would also go through if we allowed, say, definable substitutions for atomic proposition letters in product models. Also, one might also try to split our modal notion into full product closure plus predicate substitutions, treating our use of preconditions as a case of definable domain relativization.

There may also be a connection here with the Feferman-Vaught Theorem, and product constructions reducing truth in the product to truth of related statements in the component models. After all, our proof of uniform definability of dynamic modal operators $\langle \mathbf{A}, a \rangle \phi$ induces an obvious translation relating truth of ϕ in a product model $\mathbf{M} \times \mathbf{A}$ to that of some effective translation of ϕ in the component model $\mathbf{M}$.

Finally, one way of seeing how strong product closure really is would be to ask a converse question. For instance, assume that a fragment of the μ-calculus is product-closed. Does it follow that it is closed under simultaneous fixed-points?

In all, our results, though somewhat technical and limited in scope, seem to provide a vantage point for raising many interesting new questions.

Acknowledgements

We thank Balder ten Cate, Jan van Eijck, Martin Otto, Olivier Roy, Luigi Santocanale, and Yde Venema for their remarks and suggestions on this draft.

References

1. Arnold, A., Niwinski, D.: Rudiments of μ-calculus. Studies in Logic and the Foundations of Mathematics, vol. 146. North-Holland, Amsterdam (2001)
2. Baltag, A., Moss, L.S., Solecki, S.: The logic of public announcements, common knowledge, and private suspicions, vol. 30 (1999)
3. van Benthem, J.: Exploring Logical Dynamics. CSLI Publications, Stanford University (1996)
4. van Benthem, J.: Information update as relativization. Technical report, Institute for Logic, Language and Computation, Universiteit van Amsterdam (2000)
5. van Benthem, J., van Eijck, J., Kooi, B.: Logics of communication and change. Inform. and Comput. 204(11), 1620–1662 (2006)
6. van Benthem, J., Kooi, B.: Reduction axioms for epistemic actions. In: Proceedings Advances in Modal Logic, Department of Computer Science, pp. 197–211 (2004) (University of Manchester, Report UMCS-04 9-1, R. Schmidt, I. Pratt-Hartmann, M. Reynolds, H. Wansing (eds.)
7. Bloom, S.L., Ésik, Z.: Iteration theories: the equational logic of iterative prcesses. Springer, Heidelberg (1993)
8. Bradfield, J., Stirling, C.: Modal mu-calculi. In: Handbook of Modal Logic. Studies in Logic and Practical Reasoning, vol. 3, Elsevier Science, Amsterdam (2006)
9. van Eijck, J.: The proper definition of substitution in dynamic logics. Working note. CWI, Amsterdam (2007)
10. Kooi, B.: Expressivity and completeness for public update logics via reduction axioms. J. Appl. Non-Classical Logic 17(2), 231–253 (2007)

Appendix

Proof (Claim 1) By the property of F'_*, it suffices to show the following:

$$(F_*)_i = \big(F_{(F_*)_{i+1}, \cdots, (F_*)_n}\big)_* (1 \leq i \leq n).$$

We prove that by induction on i.

– Case 1: $i = 1$.
Since

$$F_{(F_*)_2,\cdots,(F_*)_n}((F_*)_1) = (F(F_*))_1 = (F_*)_1,$$

$(F_*)_1$ is a fixed point of $F_{(F_*)_2,\cdots,(F_*)_n}$. Since $\big(F_{(F_*)_2,\cdots,(F_*)_n}\big)_*$ is the least fixed point of $F_{(F_*)_2,\cdots,(F_*)_n}$, $\big(F_{(F_*)_2,\cdots,(F_*)_n}\big)_* \subseteq (F_*)_1$.
Since $\big(F_{(F_*)_2,\cdots,(F_*)_n}\big)_* \subseteq (F_*)_1$ and F is monotone,

$$F\big((F_{(F_*)_2,\cdots,(F_*)_n})_*, (F_*)_2, \cdots, (F_*)_n\big) \subseteq F((F_*)_1, (F_*)_2, \cdots, (F_*)_n)$$
$$= F_*.$$

Hence

$$\Big(F\big((F_{(F_*)_2,\cdots,(F_*)_n})_*, (F_*)_2, \cdots, (F_*)_n\big)\Big)_j \subseteq (F_*)_j \quad (2 \leq j \leq n).$$

Combining this with

$$\Big(F\big((F_{(F_*)_2,\cdots,(F_*)_n})_*, (F_*)_2, \cdots, (F_*)_n\big)\Big)_1$$
$$= F_{(F_*)_2,\cdots,(F_*)_n}\big((F_{(F_*)_2,\cdots,(F_*)_n})_*\big) = (F_{(F_*)_2,\cdots,(F_*)_n})_*,$$

we get

$$\Big(F\big((F_{(F_*)_2,\cdots,(F_*)_n})_*, (F_*)_2, \cdots, (F_*)_n\big)\Big)_j$$
$$\subseteq \Big(\big((F_{(F_*)_2,\cdots,(F_*)_n})_*, (F_*)_2, \cdots, (F_*)_n\big)\Big)_j$$

for any $1 \leq j \leq n$, which means that $\big((F_{(F_*)_2,\cdots,(F_*)_n})_*, (F_*)_2, \cdots, (F_*)_n\big)$ is an F-prefixed point.
Since F_* is the least F-prefixed point, F_* is a subset of $\big((F_{(F_*)_2,\cdots,(F_*)_n})_*, (F_*)_2, \cdots, (F_*)_n\big)$, which implies $(F_*)_1 \subseteq (F_{(F_*)_2,\cdots,(F_*)_n})_*$.
– Case 2: $i > 1$.
By the induction hypothesis,

$$F_{(F_*)_{i+1},\cdots,(F_*)_n}((F_*)_i) = (F(F_*))_i = (F_*)_i.$$

Therefore, $(F_*)_i$ is a fixed point of $F_{(F_*)_{i+1},\cdots,(F_*)_n}$. Since $\big(F_{(F_*)_{i+1},\cdots,(F_*)_n}\big)_*$ is the least fixed point of $F_{(F_*)_{i+1},\cdots,(F_*)_n}$, $\big(F_{(F_*)_{i+1},\cdots,(F_*)_n}\big)_* \subseteq (F_*)_i$.
Let F_j $(1 \leq j \leq i)$ be the ones uniquely determined by the following equations:

$$F_j = \big(F_{F_{j+1},\cdots,F_i,(F_*)_{i+1},\cdots,(F_*)_n}\big)_*, \quad (1 \leq j \leq i-1)$$
$$F_i = \big(F_{(F_*)_{i+1},\cdots,(F_*)_n}\big)_*.$$

By the same argument as before, we can prove

$$\big(F(F_1, \cdots, F_i, (F_*)_{i+1}, \cdots, (F_*)_n)\big)_j \subseteq F_j \quad (1 \leq j \leq i),$$
$$F_j \subseteq (F_*)_j \quad (1 \leq j \leq i).$$

Hence

$$F\big(F_1, \cdots, F_{i-1}, (F_{(F_*)_{i+1}, \cdots, (F_*)_n})*, (F_*)_{i+1}, \cdots, (F_*)_n\big)$$
$$\subseteq \big(F_1, \cdots, F_{i-1}, (F_{(F_*)_{i+1}, \cdots, (F_*)_n})*, (F_*)_{i+1}, \cdots, (F_*)_n\big),$$

which means $\big(F_1, \cdots, F_{i-1}, (F_{(F_*)_{i+1}, \cdots, (F_*)_n})*, (F_*)_{i+1}, \cdots, (F_*)_n\big)$ is an F-prefixed point. Since F_* is the least F-prefixed point,
$F_* \subseteq \big(F_1, \cdots, F_{i-1}, (F_{(F_*)_{i+1}, \cdots, (F_*)_n})*, (F_*)_{i+1}, \cdots, (F_*)_n\big)$, which implies
$(F_*)_i \subseteq \big(F_{(F_*)_{i+1}, \cdots, (F_*)_n}\big)_*$.   □

Fields, Meadows and Abstract Data Types

Jan Bergstra[1], Yoram Hirshfeld[2], and John Tucker[3]

[1] University of Amsterdam
[2] Tel Aviv University
[3] University of Wales, Swansea

To Boaz, until 120!

Abstract. Fields and division rings are not algebras in the sense of "Universal Algebra", as inverse is not a total function. Mending the inverse by any definition of 0^{-1} will not suffice to axiomatize the axiom of inverse $x^{-1} \cdot x = 1$, by an equation. In particular the theory of fields cannot be used for specifying the abstract data type of the rational numbers.

We define equational theories of *Meadows* and of *Skew Meadows*, and we prove that these theories axiomatize the equational properties of fields and of division rings, respectively, with $0^{-1} = 0$. Meadows are then used in the theory of Von Neumann regular ring rings to characterize strongly regular rings as those that support an inverse operation that turns it into a skew meadow. To conclude, we present in this framework the specification of the abstract type of the rational numbers, as developed by the first and third authors in [2]

1 Universal Algebra

Model theory is the study of general structures and their logical properties. The theory of Algebra studies the behavior of operations on a set. Abstract data types are syntactical objects that were created for representing mathematical objects to the computer in a language that it understands. The theory of Universal Algebras is the playground where these seemingly unrelated disciplines meet and interact.

Wikipedia defines Universal Algebra by: "Universal algebra studies common properties of all algebraic structures,... the axioms in universal algebra often take the form of equational laws" [15]. More traditional sources on the subjects are [1,3,4,7,10]. In Universal Algebra, as in general model theory, we specify a signature that declares the kind of constructs that are investigated. For universal algebras there are only constant names and function names with prescribed arity (and no relation names). We are interested in the class of all the structures where these names are interpreted as individuals and functions of the appropriate arity. Next we specify a collection T of (usually finitely many) equations among terms with variables. The class of algebras (models) for these equational theory, is the class $\mathcal{M}(T)$ of structures for which the equations hold universally.

There are two major features to an equational theory and its models:

A. Avron et al. (Eds.): Trakhtenbrot/Festschrift, LNCS 4800, pp. 166–178, 2008.
© Springer-Verlag Berlin Heidelberg 2008

– The class $\mathcal{M}(T)$ is closed under substructures, under homomorphism, and under Cartesian products. In particular every algebra has a minimal subalgebra, in which every element interprets a fixed term (without variables). An algebra in which every element interprets a term is called *a prime algebra*.
– Among the prime algebras there is one that is called the *initial algebra*, and which is maximal among the prime algebras with respect to homomorphisms: every prime algebra is a homomorphic image of the initial algebra, via a unique homomorphism (in particular, the initial algebra is unique up to isomorphism). A canonical way to represent the initial algebra is as the term algebra modulo the congruence generated by the equations in the theory.

Here are some examples:

1. The signature has one constant "e" called "the unit element" one binary operation "$\cdot$" called "product", and one unary operation called "inverse". T_1 is the theory of groups, with the three equations: $(x\cdot y)\cdot z = x\cdot(y\cdot z)$, $e\cdot x = x$ and $x^{-1}\cdot x = e$. The theory T_2 includes in addition the equation $x\cdot y = y\cdot x$. $\mathcal{M}(T_1)$ is the class of groups. $\mathcal{M}(T_2)$ is the class of commutative groups. The initial algebra in both cases is the trivial group, since the equations force all the terms to be equal to e.
2. Add to the signature n new constants, and don't change the axioms. The two new classes $\mathcal{M}(T_3)$ and $\mathcal{M}(T_4)$ are the same classes as $\mathcal{M}(T_1)$ and $\mathcal{M}(T_2)$, with arbitrary (not necessarily distinct) elements interpreting the new constants. The initial algebras are the free group with n generators, and the free commutative group with n generators, which are just the term algebras modulo the congruence generated by the group equations.
3. The signature has two constants 0 and 1, two binary operations, addition $+$ and multiplication $\cdot$, and one unary operation, "minus", $-$. T_5 is the set of ring equations, and T_6 is the set of equations for commutative rings. The classes of algebras are the class of rings and the class of commutative rings, respectively.

The initial algebra in both of the classes T_5 and T_6 is the ring Z of integers: It is easy to prove by structural induction on the constant terms (without variables) that the equational theory implies that each is equal to 0, some $\underline{n}$ or some $-\underline{n}$, where $\underline{n}$ is a sum of n copies of 1.

2 Algebraically Specified Abstract Data Types

The last example is suggestive: The initial structure for the ring equations is the structure Z of the integers. Every integer has a name in the language (actually many names), and the operations of addition, multiplication, and subtraction have names in the language. The syntactical rules for generating constant terms correspond to applying the operations on the integers. Thus we specified a notation system for the integers and their operations, in the setting of a universal algebra with its initial algebra. This is nice, as the equational theory on the one hand, and all the models of the theory, whether prime or general, can be used to study and better understand the structure that is behind the formal definition of the data type. Algebraic specification of abstract data types started with [5]; recent surveys are [8,9]; [16] is a comprehensive introduction to the subject.

Here are two more equational theories whose initial algebras are interesting data types:

Example 1 (Equational number theory). The signature has one constant "0", one unary function symbol, the successor symbol "S", and two binary function symbols for addition and multiplication. The equational theory says:

1. $x + 0 = x$
2. $x + S(y) = S(x + y)$
3. $x \cdot 0 = 0$
4. $x \cdot S(y) = x \cdot y + x$

(Following the formal rules of term construction would involve including more parentheses in each of the terms, but we do not bother).

For every n we denote by $\underline{n}$ the term

$$\underbrace{SS \cdots S}_{n \text{ times}}(0).$$

The proof that the initial algebra is the set of natural numbers, with the operations that we named, is done in two steps. First we show by induction on natural numbers that $\underline{n} + \underline{m} = \underline{n + m}$ and $\underline{n} \cdot \underline{m} = \underline{n \cdot m}$. Then we show by structural induction over closed terms that every closed term is equal to some $\underline{n}$ in every algebra in the class.

We can extend the signature by additional operations, and add equations that will ensure that the initial algebra is the set of natural numbers with the familiar operations. For example we can add predecessor P and the two equations:

$$P(0) = 0 \quad \text{and} \quad P(S(x)) = x$$

Or we can add the exponential function $E(x, y)$ and the equations:

$$E(x, 0) = S(0) \quad \text{and} \quad E(x, S(y)) = E(x, y) \cdot x$$

We note that this approach raises interesting questions about the types that are specified, and the class of models of the theory that specifies them. In particular, for the theory that specifies the integers (commutative ring theory) there are different prime algebras, the different rings Z_n of integers modulo n. The initial algebra is Z, because all the other prime algebras are homomorphic images of Z. In contrast, for the equational number theory, N is the unique prime structure. What is the significance of this fact? Are the other algebras in the class of models of equational number theory of any interest? Is the equational number theory complete, in that it entails every equation that holds in the natural numbers (or in all the first order models of Peano's axioms)? For the particular case of number theory the two last questions are items in an extensive body of research into the nature of the natural numbers and the limitations of formal logic as means to treat them.

Example 2 (String Equations (Lists)). Let $a_1, \cdots, a_n$ be n letters. The signature has one constant: Λ, and at least $n + 1$ unary functions, Pop and a_i. Possibly also a binary function $Cat(,)$, and the equational theory says:

1. $Pop(\Lambda) = \Lambda$
2. $Pop(a_i(x)) = x$
3. $Cat(\Lambda, x) = x$
4. $Cat(a_i(x), y) = a_i(Cat(x, y))$

This is just a minor variation of the previous example, starting with Λ instead of 0, and with n different successor functions a_i instead of S, Pop instead of predecessor and Cat instead of $+$.

3   Fields, Meadows, and Skew Meadows

The most familiar abstract algebraic structure is a field. The most common abstract data type, alongside strings and the natural numbers, is the type of the rational numbers. Why didn't we say anything about them until now?

The sad truth is that the class of fields is not, and cannot be made into a class of models of an equational theory, simply because it is not closed under cartesian product, as the product of two fields has zero divisors.

We have no satisfactory account of who noticed this problem first and how it was dealt with in the past. Moore in 1920 [11] already discussed a weaker notion of inverse, and Penrose [12] made it into the "Moore Penrose Pseodoinverse". Independently Von Neuman rings were discussed [13] and a similar weak inverse was introduced in a different context. However the weak inverse was not introduced to address the problem that in a field 0 has no inverse, but the problem that in non fields elements that have no inverse may still poses something close to inverse. The fact that also 0 happens to have a pseodoinverse was not considered to be of interest. In the theory of algebraic type specifications the fact that a field was not a universal algebra was probably a major motive for the attempts to extend the notion of Universal Algebra to many sorted universal algebra, and try and replace equational theories by conditional equational theories, where simple implications are permitted. Both notions are very sensible, both as general concepts and with regards to fields: It is natural to treat 0 as a special kind, and on the other hand the axioms of fields involve only a very simple conditional statement. Unfortunately no modification of the notion of Universal Algebra can retain the main features of Universal Algebras stated in section 1. We prefer therefore to stay within the framework of universal algebra, weakening the field axioms to an equational theory. This leads to the following questions:

- What is the equational theory of fields, what is the class of its models, and what is the initial algebra in the class?
- Is this initial algebra the field of rational numbers? If not can we specify the rational numbers as the initial algebra for some equational theory?

Chronologically the second question was answered first, in [2], defining the notion of a meadow and using Lagrange four squares theorem. This is discussed in section 6. The first issue produced a new interesting algebraic theory, the theory of *meadows* (and of *skew meadows*), which fills in a gap between field (and skew field) theory, and advanced ring theory. Meadows (and skew meadows) are the main subject of the paper.

The commutative case of fields and meadows is simpler and it is the only part that is relevant to equational axioms for fields and to the specification of the type of rational numbers (or any other particular algebraic field). The non commutative case, that of skew meadows, is harder, interesting, relevant to division rings and to Von Neuman rings, and it includes the commutative case as a special case. We will investigate algebras that are not necessarily commutative.

3.1   Meadows and Skew Meadows

Strictly speaking a field is not an algebra at all, since the inverse function is not a total function, as 0 has no inverse. We intend to make the inverse function total by defining 0^{-1}. Since ring theory implies that $0 \cdot x = 0$ any definition of 0^{-1} will fails to satisfy the equation $x^{-1} \cdot x = 1$, and we must look for an alternative equational theory. The key observation is that we can weaken the inverse equation to one of "local inverse", which will hold for 0 because of the fact that $x \cdot 0 = 0$.

The signature for fields and of meadows has two constants 0 and 1, two unary functions , minus and inverse (written as x^{-1}), and two binary functions for addition and multiplication. We denote by $\mathcal{R}ing$ the ring equations. The addition of the axiom of commutativity $x \cdot y = y \cdot x$ makes it a commutative ring.

What can be said about the inverse function? The key observation is the fact that no matter how 0^{-1} is defined, the following equations will hold in every division ring:

$$(x^{-1} \cdot x) \cdot x = x \text{ and } x \cdot (x^{-1} \cdot x) = x$$

Should we specify both of the equations, or does one of them imply the other? If so, does it make a difference which one of the two we choose? With hindsight we choose the first equation, and we will return to this question later.

From the rest of the properties of the inverse function in fields we choose to specify reflexivity: $(x^{-1})^{-1} = x$. We note that we are now forced to define $0^{-1} = 0$, because $0^{-1} = (0^{-1})^{-1} \cdot 0^{-1} \cdot 0^{-1} = 0 \cdot 0^{-1} \cdot 0^{-1}$ and $0 \cdot x = 0$ is implied by the ring equations.

This leads us to the following definition:

Definition 1. *The theory of* Skew meadows *has the following equations:*

1. $\mathcal{R}ing$, the ring equations;

2. $\mathcal{R}ef$, *reflexivity*, $(x^{-1})^{-1} = x$;

3. $\mathcal{R}il$, *a restricted inverse equation*, $(x^{-1} \cdot x) \cdot x = x$.

We denote by $\mathcal{R}il'$ the dual axiom $x \cdot (x^{-1} \cdot x) = x$.

The theory of meadows is the theory of skew meadows with the addition of the axiom of commutativity, $x \cdot y = y \cdot x$. A model of the axioms will be called *a skew meadow*, and a commutative skew meadow is called *a meadow*. Every division ring, and every product of division rings is a skew meadow. To make this statement precise, from now on, "a field" or " a division ring" means a *totalized algebra*, where the inverse function is is extended by the definition $0^{-1} = 0$. We will prove that an algebra in the signature of fields is a skew meadow if and only if it is isomorphic to a substructure of division rings. From this we will deduce that the equational theory of (skew) meadows equals to the equational theory of (skew) fields, and that finite skew fields are commutative (and in fact products of finite fields).

3.2 Some Properties of Skew Meadows

We list some properties that follow from the equations of skew meadows. Firstly, since $x^{-1} \cdot x \cdot x = x$, we know that if $x \cdot x = 0$ then $x = 0$, so that:

(1) There are no non trivial nilpotent elements in a skew meadow.

If e is idempotent ($e \cdot e = e$), and x is arbitrary then simple computation shows that $e \cdot x \cdot (1 - e)$ and $(1 - e) \cdot x \cdot e$ are idempotent. therefore $e \cdot x \cdot (1 - e) = 0$ and $(1 - e) \cdot x \cdot e = 0$, which shows that $e \cdot x = e \cdot x \cdot e$ and $x \cdot e = e \cdot x \cdot e$. We conclude that $e \cdot x = x \cdot e$, so that:

(2) Idempotent elements are central, they commute with every element.

In the following computation we underline the part that is modified according to the axiom $\mathcal{R}il$:

$$x \cdot x^{-1} \cdot \underline{x} \cdot x^{-1} = \underline{x \cdot x^{-1} \cdot x^{-1}} \cdot x \cdot x \cdot x^{-1} = \underline{x^{-1} \cdot x \cdot x} \cdot x^{-1} = x \cdot x^{-1}$$

and we conclude that

(3) $x \cdot x^{-1}$ (and substituting x^{-1} for x also $x^{-1} \cdot x$) is idempotent.

By (3) we have $x \cdot x^{-1} = x \cdot \underline{x^{-1} \cdot x} \cdot x^{-1}$, and also $x^{-1} \cdot x = x^{-1} \cdot \underline{x \cdot x^{-1}} \cdot x$.

By (2) the first underlined expression commutes with its right neighbor and the second underlined expression commutes with its left neighbor, and in both cases we obtain the product $x \cdot x^{-1} \cdot x^{-1} \cdot x$. Therefore

(4) $x \cdot x^{-1} = x^{-1} \cdot x$

In particular $\underline{x \cdot x^{-1}} \cdot x = x^{-1} \cdot x \cdot x = x$, so that

(5) The equation $\mathcal{R}il'$, $x \cdot x^{-1} \cdot x = x$ holds in every skew meadow.

The following theorem will connect skew meadows (and in particular meadows) with products of division rings (respectively, with products of fields).

Theorem 1. *Let S be a skew meadow and $x \neq 0$ an element. There is a homomorphism from S onto a division ring that respects addition, multiplication and inverse, and that does not map x to 0.*

Proof (outline). The proof has three ingredients:

1. Write $e = x \cdot x^{-1}$. By the axioms and by what was already established we know that e is idempotent, $e \cdot x = x \cdot e = x$ and $e \cdot x^{-1} = x^{-1} \cdot e = x^{-1}$. It follows that:
   - $e \cdot R$ is a skew meadow in which x is invertible.
   - The map $H(z) = e \cdot z$ is a ring homomorphism from R onto $e \cdot R$, with $H(x) = x$.
2. If J is a maximal two sided ideal in a meadow then the quotient ring is a ring with division, and invertible elements are not mapped to 0. The significant fact here is that if a is not in J then there is some b such that $a \cdot b$ is equivalent to 1 modulo J, so that every element in the quotient is invertible. We denote by f the idempotent $a \cdot a^{-1}$. If $f \in J$ then also $a \notin J$. Hence $J + f \cdot R$ is a two sided ideal properly extending J, and therefore it is R. Therefore there are $i \in J$ and $r \in R$ such that $1 = i + f \cdot r = i + a \cdot a^{-1} \cdot r$, which shows that $a^{-1} \cdot r$ satisfies the requirement.
3. Composing H with the quotient mapping we obtain a ring homomorphism onto a division ring that does not map x to 0. We then note that by cancellation in division rings every ring homomorphism from a meadow into a division ring preserves also inverses.   □

3.3   Skew Meadows and Division Rings

From the previous theorem we conclude:

Theorem 2

1. *A structure in the signature of meadows is a skew meadow if and only if it is isomorphic to a substructure of a product of division rings, and it is a meadow if and only if it is isomorphic to a substructure of a product of fields.*
2. *The equational theory of skew meadows entails all the equations that are true in division rings. The equational theory of meadows entails all the equations that are true in fields.*

Proof (outline). Every product of division rings, or substructure of one, is a skew meadow, since division rings are skew meadows and products and substructures preserve equations. On the other hand if R is a skew meadow, then for every $x \in R$ there is some division ring D_x and a homomorphism $h_x : R \longrightarrow D_x$, such that $h_x(x) \neq 0$. We combine these homomorphisms to a homomorphism into the Cartesian product:

$$H: R \longrightarrow \prod_{x \in R} D_x$$

Such that $H(z)$ has $h_x(z)$ at the xth entry. Since $h_z(z) \neq 0$ the kernel of the homomorphism is trivial. This proves the first claim. Every equation that holds universally in division rings holds also in products of division rings, and in sub-structures of products of division rings. Therefore also in all the skew fields.   □

We can do a little better:

Theorem 3. *Every finite skew meadow is commutative, and is in fact isomorphic to a finite product of finite fields.*

Proof (outline). Assume that the skew meadow $\mathcal{M}$ is finite, and check the division rings whose Cartesian product was used for the embedding of $\mathcal{M}$. Each was a homomorphic image of $\mathcal{M}$, and hence a finite division ring. By Wedderburn's theorem [?] it is necessarily commutative. There was one component corresponding to any element of $\mathcal{M}$, so that it is a finite product of fields. It probably can be shown that if the minimal number of components is taken then the isomorphism is onto the product. We took a different path, interesting for its own sake, showing that a finite meadow is the Cartesian product of its minimal ideals, and these ideals are fields (each with its private unit, which is an idempotent that generates the ideal in the meadow).   □

4   Strongly Von Neumann Regular Rings

Long before Meadows were introduced, Von Neumann investigated function rings [13], and he found that some of them, in particular every ring of matrices over a field, satisfy the *(Von Neumann) Regularity property*:

$$\forall x \exists y (x \cdot y \cdot x = x)$$

Every skew meadow is necessarily a regular ring. The converse is not true: A ring of matrices over a finite field is a finite regular ring which is not commutative, and therefore it cannot support an inverse function that makes it a skew meadow.

Von Neumann regular rings were, and are, the subject of intensive investigation. Goodearl's book [6], is a good source on the subject. The property which is dual to regularity and which interests us is called *strong regularity* in [6]:

Definition 2. *A ring is* strongly regular *if it satisfies the axiom:*

$$\forall x \exists y (y \cdot x \cdot x = x)$$

Using ingredients from the proof of Theorem 3.5 of [6], we prove the analogue to Theorem 32

Theorem 4. *Let R be a strongly regular ring, and let $a \neq 0$ be an element. There is a ring homomorphism from R onto a division ring which does not map a to 0.*

Proof (outline). The proof is quite different from the proof of theorem 32. We note that in strongly regular rings there are no nilpotent elements; if $x^2 = 0$ then $x = 0$ by $\forall x \exists y (y \cdot x \cdot x = x)$. It follows that $a^n \neq 0$ for every n . Using Zorn's lemma we find a maximal two sided ideal J that does not contain any power of a, and we show that it is a *prime ideal*, i.e, if $x \cdot R \cdot y \subseteq J$ then $x \in J$ or $y \in J$. Indeed if $xRy \subseteq J$ and $x, y \notin J$ then a has a power in either of the two ideals generated by J and either one of x, y. Therefore there are j, j' in J, and $x_1, \cdots, x_n, x_1', \cdots, x_n'$ and $y_1, \cdots, y_{n'}, y_1', \cdots, y_{n'}'$ such that

$$a^m = j + x_1 \cdot x \cdot x_1' + \cdots + x_n \cdot x \cdot x_n'$$

and

$$a^{m'} = j' + y_1 \cdot y \cdot y_1' + \cdots + y_{n'} \cdot y \cdot y_{n'}'$$

Multiplying the two expressions on the right hand side we see that all the summands that do not include j or j' are of the form $(x_i \cdot x \cdot x_i')(y_{i'} \cdot y \cdot y_{i'}')$. Since the inner subterm $x \cdot x_i' \cdot y_{i'} \cdot y$ is in J by $x \cdot R \cdot y \subseteq J$, these summands are also in J, contradicting the fact that $a^{m+m'}$ is not in J.

The image of a in R/J is not 0 and it remains to show that R/J is a division ring. We show first that it has no zero divisors. If $[x] \neq 0$ and $[y] \neq 0$ in R/J then $x \notin J$ and $y \notin J$. By primeness there is some z such that $x \cdot z \cdot y \notin J$, so that $[x \cdot z \cdot y] \neq 0$. Strong regularity is preserved under homomorphisms, so that there are no nilpotent elements, and the square is also not 0. I.e, $[x \cdot z \cdot y] \cdot [x \cdot z \cdot y] = [x] \cdot [z] \cdot [y] \cdot [x] \cdot [z] \cdot [y] \neq 0$. In particular $[y] \cdot [x] \neq 0$.

We conclude by observing that a strongly regular ring with no zero divisors is a division ring. For every $x \neq 0$ there is some y such that $y \cdot x \cdot x = x$ and therefore $(y \cdot x - 1) \cdot x = 0$. Since $x \neq 0$ we conclude that $y \cdot x - 1 = 0$. Thus every non zero element has an inverse. This concludes the proof.   □

We note for further use that the last step in the proof shows that if in a ring R we have $y \cdot x \cdot x = x$ or $x \cdot y \cdot x = x$, and $h(x) \neq 0$ for a homomorphism onto a division ring, $h : R \to D$ then $h(y) = (h(x))^{-1}$.

We have now the following characterization for strongly regular rings:

Theorem 5

1. *A ring R is strongly regular if and only if it supports an inverse function that makes the ring into a skew meadow.*
2. *Such an inverse function is unique.*

Proof (outline). One direction is clear: A skew meadow is clearly a strongly regular ring.

As before We choose for every element x of the ring a division ring D_x, such that R is mapped onto D_x by a homomorphism h_x with $h_x(x) \neq 0$. We denote by H the monomorphism from R into the Cartesian product of the division rings, that maps every element z to the sequence that has $h(x)(z)$ as its xth entry. Let R' be the image of R:

$$H : R \longrightarrow R' \subseteq \prod_{x \in R} D_x$$

R is isomorphic to a substructure of a product of division rings. But this does not conclude our quest! The isomorphism is not with respect to the inverse function, and it is not clear at all that R' is closed under the inverse function. However the particulars of the monomorphism are enough to prove the following claim:

Assume that $y \cdot x \cdot x = x$ in R. Then

- $x \cdot y \cdot x = x$.
- If we put $y' = y \cdot x \cdot y$ then $x \cdot y \cdot x = x$ and $y' \cdot x \cdot y' = y'$
- $H(y') = (H(x))^{-1}$

All these facts hold easily in R'. For every z, $h_z(y \cdot x)$ is 1 if $h_z(x) \neq 0$, and it is 0 if $h_z(x) = 0$. In either case $H(y \cdot x)$ is a sequence of zeros and ones that commutes with every element in the Cartesian product. The second item follows easily from the first one and we prove the third: For every z, if $h_z(x) \neq 0$ then $h_z(y') = (h_z(x))^{-1}$ by $x \cdot y \cdot x = x$. And if $h_z(x) = 0$ then $h_z(y') = 0$ by $y' = y' \cdot x \cdot y'$. Therefore in every entry $h_z(y') = (h_z(x))^{-1}$.

Thus R is ring isomorphic to a substructure of the product $\prod_{x \in R} D_x$ that is closed under the inverse function, and is therefore a skew meadow. The isomorphism induces an inverse operation on R and makes R into a skew meadow.

The uniqueness of the inverse function follows from the fact that a homomorphism of a skew meadow onto a division ring preserves also inverses. It follows that if there are two inverse function on R then the homomorphism on each D_x reduces both inverse functions to the inverse in D_x. Therefore under the ring monomorphism into $\prod_{x \in R} D_x$ both inverses are identified with the inverse function of $\prod_{x \in R} D_x$. □

5   What about Regular Rings and the Weaker Axiom, $\mathcal{R}il'$?

With hindsight, it would be natural to declare that $\mathcal{R}il$ is "the right axiom" and discard $\mathcal{R}il'$ as a corollary which curiously is quite weaker than $\mathcal{R}il$. Because of Von Neumann's theory emphasis on the condition $\forall x \exists y [x \cdot (x^{-1} \cdot x) = x]$, and not on strong regularity, we cannot ignore the corresponding property $x \cdot (x^{-1} \cdot x) = x$. We know a little about it, and we mainly list open questions. We will call a ring with an inverse function a *"weak skew meadow"* if it satisfies the axioms of skew meadows with the axiom $\mathcal{R}il'$ replacing $\mathcal{R}il$.

1. Not every weak skew meadow is a skew meadow. In fact there are finite non commutative weak skew meadows: By a careful choice of inverses we can define a reflexive inverse function in the ring of two by two matrices, over a field of characteristics different from 2, which satisfies $x \cdot x^{-1} \cdot x = x$.
2. We do not know if a ring has at most one inverse function that makes it a weak skew meadow (we know that a skew meadow can not support a second unary operation that makes it into a weak skew meadow because the proof of Theorem 43 applies also in this case).

3. We do not know if every regular ring supports an inverse function that makes it into a weak skew meadow.

There are some natural properties that follow from the axioms of skew meadows, and their status with respect to weak skew meadow axioms is unclear. For each of these properties and for any combination of them, there are three possibilities: The property may be implied by the weak skew meadow axioms, or it may complete the weak axioms and imply the strong skew meadow axioms, or its addition may describe a new class that lies between the class of weak meadows and that of strong meadow. In the last case the new axiom may or may not imply the non existence of a finite non commutative algebra in the class. Here are some natural properties that are taken for granted in division rings and skew meadows and their status is unclear for weak skew meadows:

1. $x \cdot x^{-1} = x^{-1} \cdot x$.
2. $(x \cdot y)^{-1} = y^{-1} x^{-1}$.
3. $e^{-1} = e$ if e is idempotent (this property is not an equational property).
4. The inverse function is unique (not an equational property).

In weak skew meadows, item (1) implies, trivially, strong regularity. Items (2,3) together imply a little less trivially strong regularity. It is not clear if one of the items (2,3,4) by itself or if any combination other than (2,3) implies regularity, or at least prevents the theory from having finite non commutative models. We conclude that we are far from understanding the difference between the seemingly similar axioms $(x^{-1} \cdot x) \cdot x = x$ and $x \cdot x^{-1} \cdot x = x$.

6   Algebraic Specification of the Rational Field

We return to the question that started it all: Which equational theory has the rational field as its initial structure? The answer was given by the first and the third authors in [2]. We suggest to call it *"the theory of Lagrange meadows"*. It is the equational theory of (commutative) skew meadows together with the following equation

$$(1 + x^2 + y^2 + z^2 + v^2) \cdot (1 + x^2 + y^2 + z^2 + v^2)^{-1} = 1 .$$

Theorem 6. *The field Q of rational numbers is the initial algebra in the class of Lagrange meadows, and moreover any prime algebra in the class is isomorphic to Q.*

Proof (outline). By Lagrange four squares theorem every natural number n can be written as the sum of four squares so that if $n \neq 0$ we can write $\underbrace{1 + \cdots + 1}_{n}$ in the form $(1 + x^2 + y^2 + z^2 + v^2)$. This assures that for every natural n the term $\underline{n} = \underbrace{1 + \cdots + 1}_{n}$ is not 0. It is not hard to prove by structural induction that in Lagrange meadows every fixed term equals to 0 or to a term of the form $\underline{n} \cdot (\underline{m})^{-1}$, which is different from 0. Therefore the prime algebra in every Lagrange meadow is the field of rational numbers, and we conclude that the field of rational numbers is the initial algebra for the equational theory of Lagrange meadows.   □

Thus the abstract type of rational number is algebraically specified as an initial algebra.

7 Conclusion

The definition of meadows and skew meadow enabled us to contribute in three different areas:

- In the theory of *Universal Algebras* we identified the equational theory of fields and rings with division, and proved that it was just the equational theory for meadows and skew meadows.
- In *Ring Theory* we characterized the strongly regular Von Neuman rings as those that support a (necessarily unique) inverse function that makes it into a skew meadow. The inverse function is easily defined in term of any choice function that associates with every element x a partner y for which $y \cdot x \cdot x = x$.
- In the theory of *Abstract Types* we specified the type of the rational numbers as the algebra specified by the axioms of Lagrange meadows.

This seems like an invitation to an interesting domain of research that can shed light on the theories of universal algebras, of regular rings and of algebraic specification of data types. In particular, what is the significance of the fact that an algebraic theory has a single prime algebra, as in the case of Lagrange meadows that specify the rational field, but in contrast to the theory of meadows in general? Is there any additional importance to the equational theory of meadows, and of Lagrange meadows?

References

1. Bergman, G.M.: An Invitation to General Algebra and Universal Constructions. Henry Helson (1998)
2. Bergstra, J.A., Tucker, J.V.: The rational numbers as an abstract data type. J. ACM 54(2), Article 7 (April 2007)
3. Burris, S.N., Sankappanavar, H.P.: A Course in Universal Algebra, free online edition. Springer, Heidelberg (1981)
4. Cohn, P.M.: Universal Algebra. D. Reidel Publishing, Dordrecht (1981)
5. Gougen, J.A., Thacher, J.W., Wagner, E.G.: An initial algebra approach to the specification, correctness and implementation of abstract data types. In: Yeh, R.T. (ed.) Current Trends in Programing Methodology, VI, Data Structuring, pp. 80–149. Prentice Hall, Englewood Cliffs (1978)
6. Goodearl, K.R.: Von Neumann regular rings. Pitman, London, San Francisco, Melburne (1979)

7. Graetzer, G.: Universal Algebra. Hobby, David, and Ralph McKenzie (1988)
8. Loeckx, J., Ehrich, H.D., Wolf, M.: Specification of Abstract Data Types. and Teubner. Wiley and Teubner, Chichester (1996)
9. Loeckx, J., Ehrich, H.-D., Wolf, M.: Algebraic specification of abstract data types. In: Abramsky, S., Gabbay, D.M., Maibaum, T.S.E. (eds.) Handbook of Logic in Computer Science, Oxford University Press, Oxford (2000)
10. Meinke, K., Tucker, J.V.: Universal Algebra. In: Abramsky, S., Gabbay, D., Maibaum, T. (eds.) Handbook of logic in computer science I, Mathematical structures, pp. 189–411 (1992)
11. Moore, E.H.: On the reciprocal of the general algebraic matrix. Bulletin of the American Mathematical Society 26, 394–395 (1920)
12. Penrose, R.: A generalized inverse for matrices. Proceedings of the Cambridge Philosophical Society 51, 406–413 (1955)
13. von Neumann, J.: Continuous geometries. Princeton University Press, Princeton (1960)
14. Maclagan-Wedderburn, J.H.: A theorem on finite algebras. Transactions of the American Mathematical Society 6, 349–352 (1905)
15. Wikipedia, Universal Algebra,
   http://en.wikipedia.org/wiki/Universal_algebra
16. Wirsing, M.: Algebraic Specification. In: Leeuwen, J.v. (ed.) Handbook of Theoretical Computer Science. Formal Models and Sematics, vol. B, Elsevier and MIT Press (1990)

Why Sets?*

Andreas Blass[1],[**] and Yuri Gurevich[2]

[1] Mathematics Department, University of Michigan, Ann Arbor, MI 48109, USA
[2] Microsoft Research, One Microsoft Way, Redmond, WA 98052, USA

Dedicated to Boaz Trakhtenbrot on the occasion of his 85th birthday.

Abstract. Sets play a key role in foundations of mathematics. Why? To what extent is it an accident of history? Imagine that you have a chance to talk to mathematicians from a far-away planet. Would their mathematics be set-based? What are the alternatives to the set-theoretic foundation of mathematics? Besides, set theory seems to play a significant role in computer science; is there a good justification for that? We discuss these and some related issues.

1 Sets in Computer Science

Quisani: I wonder why sets play such a prominent role in foundations of mathematics. To what extent is it an accident of history? And I have questions about the role of sets in computer science.

Author[1]: Have you studied set theory?

Q: Not really but I came across set theory when I studied discrete mathematics and logic, and I looked into Enderton's book [21] a while ago. I remember that ZFC, first-order Zermelo-Fraenkel set theory with the axiom of choice, became for all practical purposes the foundation of mathematics. I can probably reconstruct the ZFC axioms.

A: Do you remember the intuitive model for ZFC.

Q: Let me see. You consider the so-called cumulative hierarchy of sets. It is a transfinite hierarchy, so that you have levels $0, 1, \ldots, \omega, \omega + 1, \ldots$ On the level zero, you have the empty set and possibly some atoms. On any other level α you have the sets of objects that occur on levels $< \alpha$. Intuitively the process never ends. To model ZFC, you just go far enough in this hierarchy so that all axioms are satisfied. Is that correct, more or less?

---

* This is a revised version of an article originally published in the Bulletin of the European Association for Theoretical Computer Science, Number 84, October 2004, and republished here with permission of the Association.

** Blass was partially supported by NSF grant DMS–0070723 and by a grant from Microsoft Research. This paper was written during a visit to Microsoft Research.

[1] As in our previous conversations with Quisani, we simplified the record of the conversation by blending the two authors into one who prefers "we" to "I".

A. Avron et al. (Eds.): Trakhtenbrot/Festschrift, LNCS 4800, pp. 179–198, 2008.

A: More or less. ZFC is intended to describe the whole, never-ending universe of sets obtained in the cumulative hierarchy, but technically this universe is not a model because it's not a set. That's the reason for stopping at a stage where all the axioms are satisfied. (By Gödel's second incompleteness theorem, the existence of such a stage is an assumption that goes beyond ZFC, but it is a rather mild additional assumption.)

You should be careful about the phrase "far enough ... so that all axioms are satisfied" because "far enough" seems to suggest that any sufficiently large number of steps will do. But in fact, once you've got the axioms satisfied, you can't just go on for another step or two; you need to add many more levels to get the axioms satisfied again. You should stop at some level where your model, consisting of the sets created so far, has the closure properties required by the axioms.

Q: OK. Turning to computer science, I read at the Z users website [58] the following: "The formal specification notation Z (pronounced "zed"), useful for describing computer-based systems, is based on Zermelo-Fraenkel set theory and first order predicate logic." And I was somewhat surprised.

A: Were you surprised that they use the ZF system rather than ZFC, the Zermelo-Fraenkel system with the axiom of choice? As long as we consider only finite families of sets, the axiom of choice is unnecessary. That is, one can prove in ZF that, if X is a finite family of nonempty sets, then there is a function assigning to each set $S \in X$ one of its members. Furthermore, there is a wide class of statements, which may involve infinite sets, but for which one can prove a metatheorem saying that any sentence in this class, if provable in ZFC, is already provable in ZF; see [49, Sect. 1] for details. This class seems wide enough to cover anything likely to arise in computer science, even in its more abstract parts.

Q: That is an interesting issue in its own right but I was surprised by something else. Set theory wasn't developed to compute with. It was developed to be a foundation of mathematics.

A: There are many things that were developed for one purpose and are used for another.

Q: Sure. But, because set theory was so successful in foundations of mathematics, there may be an exaggerated expectation of the role that set theory can play in foundations of computer science. Let me try to develop my thought. What makes set theory so useful in foundations of mathematics? I see two key aspects. One aspect is that the notion of set is intuitively simple.

A: Well, it took time and effort to clarify the intuition about sets and to deal with set-theoretic paradoxes; see for example [27] and [36]. But we agree that the notion of set is intuitively simple.

Q: The other aspect is that set theory is very expressive and succinct: mathematics can be faithfully and naturally translated into set theory. This is extremely important. Imagine that somebody claims a theorem but you don't understand some notions involved. You can ask the claimer to define the notions more and more precisely. In the final account, the whole proof can be reduced to ZFC, and then the verification becomes mechanical.

Can sets play a similar role in computing? I see a big difference between the reduction to set theory in mathematics and in computing. The mathematicians do not actually translate their stuff into set theory. They just convince themselves that their subject is translatable.

A: Bourbaki [10] made a serious attempt to actually translate a nontrivial portion of mathematics into set theory, but it is an exception.

Q: Right. In computing, such translations have to be taken seriously. If you want to use a high-level language that is compiled to some set-theoretic engine, then a compiler should exist in real life, not only in principle. I guess all this boils down to the question whether the datatype of sets can be the basic datatype in computing. Can sets and set operations be implemented efficiently? Can other data be succinctly interpreted in set theory.

A: There has been an attempt made in this direction [55].

Q: Yes, and most people remained unconvinced that this was the way to go. Sequences, or lists, are appropriate as the basic datastructure.

A: We know one example where sets turned out to be more succinct than sequences as the basic datastructure.

Q: Tell me about it.

A: OK, but bear with us as we explain the background. We consider computations where inputs are finite structures, for example graphs, rather than strings.

Q: Every such structure can be presented as a string.

A: That is true. But we restrict attention to computing properties that depend only on the isomorphism type of the input structure. For example, given a bipartite graph, decide whether it has a matching. Call such properties *invariant queries*.

Q: Why the restriction?

A: Because we are interested in queries that are independent of the way the input structure is presented or implemented. Consider, for example, a database query. You want that the result depends on the database only and not on exactly how it is stored.

Q: Fine; what is the problem?

A: The original problem was this: Does there exist a query language L such that

(**Restrained**) every query that can be formulated in L is an invariant query computable in polynomial time, and

(**Maximally expressive**) every polynomial-time computable invariant query can be formulated in L.

Q: How can one ensure that all L-queries are invariant?

A: Think about first-order logic as a query language. Every first-order sentence is a query. First-order queries are *pure* in the sense that they give you no means to express a property of the input structure that is not preserved by isomorphisms. Most restrained languages in the literature are pure in that same sense.

Q: But, in principle, can a restrained language allow you to have non-invariant intermediate results? For example, can you compute a particular matching, throw away the matching and return "Yes, there is a matching"?

A: Yes, a restrained language may have non-invariant intermediate results. In fact, Ashok Chandra and David Harel, who raised the original problem in [11], considered Turing machines M that are invariant in the following sense: If M accepts one string representation of the given finite structure then it accepts them all. They asked whether there is a decidable set L of invariant polynomial time Turing machines such that, for every invariant polynomial time Turing machine T_1, there is a machine $T_2 \in L$ that computes the same query as T_1 does. In the case of a positive answer, such an L would be restrained and maximally expressive.

Q: Hmm, a decidable set of Turing machines does not look like a language.

A: One of us conjectured [31] that there is no query language, even as ugly as a decidable set of Turing machines, that is restrained and maximally expressive.

Q: But one can introduce, I guess, more and more expressive restrained languages.

A: Indeed. In particular, the necessity to deal with invariant database queries led to the introduction of a number of restrained query languages [1] including the polynomial-time version of the language while$_{new}$. In [8], Saharon Shelah and the two of us proposed a query language, let us call it BGS, that is based on set theory. BGS is pure in the sense discussed above. A polynomial time bounded version of BGS, let us call it Ptime BGS, is a restrained query language.

Q: In what sense is BGS set-theoretic?

A: It is convenient to think of BGS as a programming language. A state of a BGS program includes the input structure A, which is finite, but the state itself is an infinite structure. It contains, as elements, all hereditarily finite sets built from the elements of A. These are sets composed from the elements of A by repeated use of the pairing operation $\{x, y\}$ and the union operation $\bigcup(x) = \{y \ : \ \exists z \, (y \in z \in x)\}$. BGS uses standard set theoretic operations and employs comprehension terms $\{t(x) \ : \ x \in r \ \land \ \varphi(x)\}$. In any case, to make a long story short, it turned out that Ptime BGS was more expressive than the Ptime version of the language while $_{\text{new}}$ that works with sequences; see [9] for details. For the purpose at hand, sets happened to be more efficient than sequences.

Q: I don't understand this. A set s can be easily represented by a sequence of its elements.

A: Which sequence?

Q: Oh, I see. You may have no means to define a particular sequence of the elements of s and you cannot pick an arbitrary sequence because this would violate the purity of BGS.

A: Right. You may want to consider all $|s|!$ different sequences of the elements of s. This does not violate the purity of BGS. But, because of the polynomial time restriction, you may not have the time to deal with $|s|!$ sequences.

On the other hand, a sequence $[a_1, a_2, \ldots, a_k]$ can be succinctly represented by a set $\{[i, a_i] \ : \ 1 \le i \le k\}$. Ordered pairs have a simple set-theoretic representation due to Kuratowski: $[a, b] = \{\{a, b\}, \{a\}\}$.

Q: I agree that, in your context, sets are more appropriate than sequences.

A: It is also convenient to have the datatype of sets available in software specification languages.

Q: But closer to the hardware level, under the hood so to speak, we cannot deal with sets directly. They have to be represented e.g. by means of sequences.

A: You know hardware better than we do. Can one build computers that deal with sets directly?

Q: A good question. The current technology would not support a set oriented architecture.

A: What about quantum or DNA-based computing?

Q: I doubt that these new paradigms will allow us to deal with sets directly but your guess is as good as mine.

2 Sets in Mathematics

Q: Let me return to the question why sets play such a prominent role in the foundation of mathematics. But first, let me ask a more basic question: Why do we need foundations at all? Is mathematics in danger of collapsing? Most mathematicians that I know aren't concerned with foundations, and they seem to do OK.

A: Well, you already mentioned the fact that an alleged proof can be made more and more detailed until it becomes mechanically verifiable.

Q: Yes, but I'd hope that this could be done with axioms that talk about all the different sorts of objects mathematicians use – real numbers, functions, sequences, Hilbert spaces, etc. – and that directly reflect the facts that mathematicians routinely use. What's the advantage of reducing everything to sets?

A: We see three advantages. First, people have already explicitly written down adequate axiomatizations of set theory. The same could probably be done for the sort of rich theory that you described, but it would take a nontrivial effort. Besides, new sorts of objects keep entering the mathematical world.

Second, when proving that a statement is consistent with ordinary mathematics, one only has to produce a model of set theory in which the statement is true. Without the set theoretic foundation, one would have to construct a model of a much richer theory.

Third, the reduction of mathematics to set theory means that the philosopher who wants to understand the nature of mathematical concepts needs only to understand one concept, namely sets.

By the way, if someone developed mathematics on the basis of a simple concept other than sets, then these advantages would apply to that alternative foundation also.

Q: These advantages make sense but they also show why a typical mathematician never has to use the reduction to set theory. Actually, the third advantage is not entirely clear to me; it seems that by reducing mathematics to set theory the philosopher can lose some of its semantic or intuitive content. Consider a proof that complex polynomials have roots, and imagine a set-theoretic formalization of it.

A: It's not a matter of the philosopher's understanding particular mathematical results or the intuition behind them, but rather understanding the general nature of abstract, mathematical concepts.

Q: Anyway, granting the value of a reduction of mathematics to a simple foundation, why should it be set theory? For example, since sequences are so important in computing, it's natural to ask whether they could replace sets in the foundations of mathematics. Similarly, Dijkstra has suggested that multisets, also known as bags, are a natural, basic notion and should play a more prominent role in mathematics [20].

A: Both transfinite sequences and multisets have recently been proposed as foundations for mathematics in [19], where axiomatizations are given and the basic theories developed. It is too soon to say how useful or how widely accepted these foundations will be. The axiom systems proposed in [19] can be interpreted in ZFC and vice versa, so they could be regarded as just providing an alternative view of the usual universe of sets, but such alternatives may turn out to be useful aids to the intuition and they may lead to technical simplifications in some topics (and complications in others).

2.1   Adequacy of Sets

Q: This notion of interpretations seems to be crucial for foundations. The way we use set theory as a foundation is by interpreting into it the richer theories – of real numbers, functions, sequences, Hilbert spaces, etc. – that mathematicians really work with. So if another theory, say of transfinite lists, and set theory are each interpretable in the other, then they can serve as foundations for exactly the same body of mathematics, just by composing interpretations.

A: That's right, but of course composing may lead to more complicated interpretations.

Q: The crucial point, though, is that set theory can serve, via suitable, possibly complicated interpretations, as a foundation for all of ordinary mathematics. So it seems the problem of foundations for mathematics is completely solved; in other words, the study of foundations of mathematics is dead.

A: Not so fast! There are things in ordinary mathematics that "stick out" of the set-theoretic foundation.

Q: Like what?

A: If we take "stick out" in the strong sense of not even being expressible in the usual set-theoretic framework, then category theory provides an example. One wants the categories of all groups, all topological spaces, etc., and these aren't sets.

Q: So you would need proper classes, right.

A: Actually, you'd need more, since you also want things like the category of all functors from topological spaces to groups. There have been various proposals for reformulating category theory to fit into a set-like framework, ZFC with some additional axioms, but they end up talking about the category of small groups, small topological spaces, etc., where "small" amounts to considering only things below a certain stage of the cumulative hierarchy. In one of these proposals, that of Feferman [22], results proved about, say, small groups automatically imply the same results about all groups, but there is still no category of all groups.

Q: You referred to the strong sense of "stick out", so I suppose there's a weak sense.

A: That would refer to questions that can be formulated in the ZFC context but cannot be settled on the basis of the ZFC axioms. There are a great many such questions, not only in set theory itself but in topology, algebra, and analysis; see [52] for a brief description of some examples. And sometimes even the inability of ZFC to prove certain facts depends on assumptions beyond ZFC,

Q: That last statement is confusing. Give me an example.

A: Solovay [57] proved that the following theory is consistent: ZF (without the axiom of choice) plus "all sets of real numbers are Lebesgue measurable" plus the axiom of dependent choice (a weak form of the axiom of choice that is sufficient for all "nice" results in analysis, like the countable additiviity of Lebesgue measure, but not for "pathology" like non-measurable sets). For people who won't give up the axiom of choice, he also showed the consistency of ZFC plus all *definable* sets of real numbers are Lebesgue measurable (where "definable" refers to definitions by formulas of set theory in which real numbers and ordinal numbers can appear as parameters). But his proof for these results used the assumption that ZFC is consistent with the existence of an inaccessible cardinal (a certain sort of large cardinal, whose existence cannot be proved in ZFC). And Shelah [56] proved that there is no way to eliminate the assumption about an inaccessible cardinal from Solovay's proof. So the inability of ZFC to explicitly define (even with real and ordinal parameters) a specific set and prove that it isn't measurable is established subject to an assumption about cardinal numbers that themselves go beyond what ZFC can provide.

Q: I find the stronger sort of sticking out to be more interesting, because it seems to require new foundational concepts, not just new axioms.

A: New axioms are an interesting topic too. Are they really needed? And if so, then what axioms are appropriate? And why? There's a wide-ranging discussion of such issues in the collection [26].

Q: Why is it that almost all mathematical concepts can be represented set-theoretically, and even the exception you cited, category theory, seems to stick out in a way that doesn't suggest fundamental new concepts?

A: We don't know. It might be a historical accident. That is, maybe it is just the mathematics developed by human beings until now that is (almost) covered by set theory, but not necessarily the mathematics of the future or of the inhabitants of far-away planets. Or the set-theoretic interpretability of our mathematics might be due to the structure of human brains; so the human race's future mathematics would admit a set-theoretic foundation but that of alien races might not. Or set-theoretic interpretability might be a really intrinsic property of all rigorous, mathematical thought. Or there might be other explanations; feel free to dream some up.

2.2   Non-ZF Sets

Q: Returning to the intuitive idea of sets, is ZFC still the only game in town?

A: It's the biggest game, but there are others. For example, there are theories of sets and proper classes which extend ZFC. The most prominent ones are the von Neumann-Bernays-Gödel theory (NBG) and the Morse-Kelley theory (MK). In both cases the idea is to continue the cumulative hierarchy for one more step. The collections created at that last step are called proper classes.

Q: Wait a minute! You said that ZFC is intended to describe the whole cumulative hierarchy of sets. So how can there be another step? And if there is one more step, why not two or many?

A: We admit that this extra step doesn't quite make sense philosophically, in the light of the intended meaning of "set" in the ZFC axioms, but it is convenient technically. Consider some property of sets, for example the property of having exactly three members. It is convenient to refer to the multitude of the sets with this property as a single object. If this object isn't a set then it is a proper class.

There is also a less known but rather elegant extension of ZFC due to Ackermann [2]. It uses a distinction between sets and classes, but not the same distinction as in NBG or MK. For Ackermann, what makes a class a set is not that it is small but rather that it is defined without reference to the totality of all sets. It turns out [43,51] that, despite the difference in points of view, Ackermann's set theory plus an axiom of foundation is equivalent to ZF in the sense that they prove the same theorems about sets. Lévy [43] showed how to interpret Ackermann's axioms by taking an initial segment of the cumulative hierarchy as the domain of sets and a much longer initial segment as the domain of classes.

Q: Are there set theories that contradict ZFC?

A: Yes. One is Quine's "New Foundations" (NF), named after the article [50] in which it was proposed. Another is Aczel's set theory with the anti-foundation axiom [3,6].

Quine's NF is axiomatically very simple. It has the axiom of extensionality (just as in ZF) and an axiom schema of comprehension, asserting the existence of $\{x : \varphi(x)\}$ whenever $\varphi(x)$ is a stratified formula. "Stratified" means that one can attach integer "types" to all the variables so that, if $v \in w$ occurs in $\varphi(x)$, then $\text{type}(v) + 1 = \text{type}(w)$, and if $v = w$ occurs then $\text{type}(v) = \text{type}(w)$.

Q: This looks just like simple type theory.

A: Yes, but the types aren't part of the formula; stratification means only that there exist appropriate types. The point is that this restriction of comprehension seems sufficient to avoid the paradoxes.

Q: I see that it avoids Russell's paradox, since $\neg(x \in x)$ isn't stratified, but how do you know that it avoids all paradoxes?

A: We only said it seems to avoid paradoxes. Nobody has yet deduced a contradiction in NF, but nobody has a consistency proof (relative to, say, ZFC or even ZFC with large cardinals). But Jensen [34] has shown that NF becomes consistent if one weakens the extensionality axiom to allow atoms. Rosser [53] has shown how to develop many basic mathematical concepts and results in NF. For lots of information about NF and (especially) the variant NFU with atoms, see Randall Holmes's web site [32].

Q: How does NF contradict the idea of the cumulative hierarchy?

A: The formula $x = x$ is stratified, so it is an axiom of NF that there is a universal set, the set of all sets. No such thing can exist in the cumulative hierarchy, which is never completed.

Q: And what about anti-foundation?

A: This theory is similar to ZFC, but it allows sets that violate the axiom of foundation. For example, you can have a set x such that $x \in x$; you can even have $x = \{x\}$.

Q: And you could have $x \in y \in x$ and even $x = \{y\} \wedge y = \{x\}$, right?

A: Yes, but the anti-foundation axiom imposes tight controls on these things. There is only one x such that $x = \{x\}$. Using that x as the value of both x and y you get $x = \{y\} \wedge y = \{x\}$, and this pair of equations has no other solutions. The axiom says, very roughly, that if you propose some binary relation to serve (up to isomorphism) as the membership relation in a transitive set, then, as long as it's consistent with the axiom of extensionality, it will be realized exactly once. It turns out that this axiomatic system and ZFC, though they prove quite different things, are mutually interpretable. That is, one can define, within either of the two theories, strange notions of "set" and "membership" that satisfy the axioms of the other theory.

2.3   Categories

Q: What about possible replacements for sets as the fundamental concept for mathematics? You mentioned that category theory sticks out of the standard set-theoretic framework, and I've heard people say that category theory itself could replace set theory as a foundation for mathematics. But I don't understand them. A category consists of a set (or class) of objects, plus morphisms and additional structure. So category theory presupposes the notion of set. How can it serve as a foundation by itself?

A: The idea that the objects (and morphisms) of a category must be viewed as forming a set seems to be an artifact of the standard, set-theoretic way of presenting general structures, namely as sets with additional structure. One can write down the axioms of category theory as first-order sentences and then do proofs from these axioms without ever mentioning sets (or classes).

Q: Sure, but unless you're a pure formalist, you have to wonder what these first-order sentences mean. How can you explain their semantics without invoking the traditional notion of structures for first-order logic, a notion that begins with "a non-empty *set* called the universe of discourse (or base set) ..."?

A: This seems like another artifact of the set-theoretic mind-set, insisting that the semantics of first-order sentences must be expressed in terms of sets. People understood first-order sentences long before Tarski introduced the set-theoretic definition of semantics. Think of that set-theoretic definition as representing, within set theory, a pre-existing concept of meaning, just as Dedekind cuts or Cauchy sequences represent in set theory a pre-existing concept of real number.

Q: Hmmm. I'll have to think about that. It still seems hard to imagine the meaning of a first-order sentence without a set for the variables to range over. But let's suppose, at least for the sake of the discussion, that the axioms of category theory make sense without presupposing sets. Those axioms seem much too weak to serve as a foundation; after all, they have a model with one object and one morphism.

A: That's right. For foundational purposes, one needs axioms that describe not just an arbitrary category but a category with additional structure, so that its objects can represent the entities that mathematicians study.

Q: That sounds reasonable but vague. What sort of axioms are we talking about here?

A: There have been two approaches. One is to axiomatize the category of categories and the other is to axiomatize a version of the category of sets.

Q: The first of these sounds more like a genuinely category-theoretic foundation; the second mixes categories and sets.

A: Yes, but the first has had relatively little success.

Q: Why? What's its history?

A: The idea was introduced by Lawvere in [40]. He proposed axioms, in the first-order language of categories, to describe the category of categories, and to provide tools adequate for the formalization of mathematics. But three problems arose. First, as pointed out by Isbell in his review [33], the axioms didn't quite accomplish what was claimed for them. That could presumably be fixed by modifying the axioms. But there was a second problem: Although some of the axioms were quite nice and natural, others were rather unwieldy, and there were a lot of them. As a result, it looked as if the axioms had just been rigged to simulate what can be done in set theory. That's related to the third problem: The representation of some mathematical concepts in terms of categories was done by, in effect, representing them in terms of sets and then treating sets as discrete categories (categories in which the only morphisms are the identity morphisms, so the category is essentially just its set of objects). This third point should not be over-emphasized; some concepts were given very nice category-theoretic definitions. For example, the natural number system is the so-called coequalizer of a pair of morphisms between explicitly described finite categories. But the use of discrete categories for some purposes made the whole project look weak.

Q: So what about the other approach, axiomatizing the category of sets?

A: That approach, also pioneered by Lawvere [39], had considerably more success, for several reasons. First, many of the basic concepts and constructions of set theory (and even of logic, which underlies set theory) have elegant descriptions in the language of categories; specifically, they can be described as so-called adjoint functors. In the category of sets, adjoint functors provide definitions of disjoint union, cartesian product, power set, function set (i.e., the set of all functions from X to Y), and the set of natural numbers, as well as the logical connectives and quantifiers.

Q: That covers quite a lot. What other advantages does the category of sets have – or provide?

A: There is a technical advantage, namely that the axioms admit a natural weakening that describes far more categories than just the category of sets. These categories, called topoi or toposes, resemble the category of sets in many ways (including the availability of all of the constructions listed above, except that the existence of the set of natural numbers is usually not included in the definition of topos) but also differ in interesting ways (for example, the connectives and quantifiers may obey intuitionistic rather than classical logic), and there are many topoi that look quite different from the category of sets (not only non-standard models of set theory but also categories of sheaves, categories of sets with a group acting on them, and many others). As a result, set-theoretic arguments can often be applied in topoi in order to obtain results about, for example, sheaves. These ideas were introduced by Lawvere and Tierney in [42]; see [35] and [44] for further information.

Q: I don't know what sheaves are. In any case, I care mostly about foundations, so this technical advantage doesn't do much for me. What more can the category of sets do for the foundations of mathematics?

A: One can argue that the notion of abstract set described in this category-theoretic approach is closer to ordinary mathematical practice than the cumulative hierarchy described by the Zermelo-Fraenkel axioms.

Q: What is this notion of abstract set? The ZF sets look pretty abstract to me.

A: The phrase "abstract set" refers (in this context) to abstracting from any internal structure that the elements of a set may have. A typical set in the cumulative hierarchy has, as elements, other sets, and there may well be membership relations (or more complicated set-theoretic relations) between these elements. Abstract set theory gets rid of all this. As described in [41], an abstract set "is supposed to have elements, each of which has no structure, and is itself to have no internal structure, except that the elements can be distinguished as equal or unequal, and to have no external structure except for the number of elements."

Q: How is this closer to ordinary mathematical practice than the cumulative hierarchy view of sets?

A: One way to describe the difference is that the abstract view gets rid of unnecessary structure.

Q: What unnecessary structure? Give me some examples.

A: The Kuratowski representation of ordered pairs $[a, b]$ as $\{\{a\}, \{a, b\}\}$ has the side effect that a is an element of an element of the pair. This double-element relationship is an artifact of the particular coding of pairs and is not intrinsic to the notion of ordered pair.

For another example, in any of the usual set-theoretic representations of the real numbers, the basic facts about $\mathbb{R}$ depend on information about, say, members of members of real numbers – information that mathematicians would never refer to except when giving a lecture on the set-theoretic representation of the real numbers. The abstract view discards this sort of information. Of course, some structural information is needed – unlike abstract sets, the real number system has internal structure. But the relevant structure is postulated directly, say by the axioms for a complete ordered field, not obtained indirectly as a by-product of irrelevant structure.

Q: So if an abstract-set theorist wanted to talk about a set from the cumulative hierarchy, with all the structure imposed by that hierarchy, he would include that structure explicitly, rather than relying on the hierarchy to provide it.

A: Exactly. If x is a set in the cumulative hierarchy, then one can form its transitive closure t, the smallest set containing x and containing all members of its members. Then t with the membership relation $\in$ (restricted to t) is an abstract representation of t. It no longer matters what the elements of t were, because any isomorphic copy of the structure $(t, \in)$ contains the same information and lets you recover x.

Q: I have a couple of additional questions abbout the abstract approach to the real number system. First, it seems that we're getting farther from category theory and closer to set theory, especially with the completeness axiom, which talks about arbitrary subsets of $\mathbb{R}$.

A: Sets are certainly an essential ingredient of the completeness axiom for the reals, but they can still be abstract sets; we don't need the cumulative hierarchy here. As we already mentioned, it is possible to describe in category-theoretic terms the notion of power set. So category theory, in particular the notion of adjoint functor, allows one to formulate the notion of "real number system in a topos" without importing any notion of cumulative hierarchy.

Q: My second question concerns the notion of simply *postulating* the desired properties of $\mathbb{R}$, rather than *proving* them as the traditional set-theoretic approach does. The postulational approach looks like cheating.

A: It's not a matter of postulating *rather than* proving but rather postulating *separately from* proving. What is involved here is a separation of two concerns. The first concern is saying what the real number system is; here the abstract approach says $\mathbb{R}$ is a complete ordered field (not, for example, that it is the set of Dedekind cuts, or the set of equivalence classes of Cauchy sequences, or anything of that sort). The second concern is proving that such a thing exists in suitable categories. The suitable categories here are topoi, and the existence proof for $\mathbb{R}$ would use a construction like Dedekind cuts. The work of constructing $\mathbb{R}$ doesn't disappear in the category-theoretic approach, but it is separated from the definition of what $\mathbb{R}$ is.

Category-theorists often emphasize (not just for $\mathbb{R}$ but for all sorts of other things) the distinction between "what is it?" and "how do you construct it?"

Q: Well if this category-theoretic view of abstract sets is so wonderful, why isn't everybody using it?

A: There are (at least) four answers to your question. One is a matter of history. The cumulative hierarchy view of sets has been around explicitly at least since 1930 [60], and Zermelo's part of ZFC (all but the replacement and foundation axioms) goes back to 1908 [59]. ZFC has had time to demonstrate its sufficiency as a basis for ordinary mathematics. People have become accustomed to it as the foundation of mathematics, and that includes people who don't actually know what the ZFC axioms are. There is, however, a chance that the abstract view of sets will gain ground if students learn basic mathematics from books like [41].

A second reason is the simplicity of the primitive notion of set theory, the membership predicate. Perhaps, we should say "apparent simplicity," in view of the complexity of what can be coded in the cumulative hierarchy. But still, the idea of starting with just $\in$ and defining everything else is philosophically appealing. Another way to say this is that, in developing mathematics, one certainly needs the concepts of "set" and "membership"; if everything else can be developed from just an iteration of these (admittedly a transfinite iteration), why not take advantage of it?

Third, there is a technical reason. Although topos theory provides an elegant view of the set-theoretic constructions commonly used in mathematics, serious uses of the replacement axiom don't look so nice in category-theoretic terms. (By serious uses of replacement, we mean something like the proof of Borel determinacy [45], which provably [28] needs uncountably many iterations of the power set operation.) But such serious uses are still quite rare.

Q: OK, what's the fourth answer to why people aren't using the category-theoretic view of abstract sets?

A: The fourth answer is that they *are* using this point of view but just don't realize it. Mathematicians talk about ZFC as the foundation of what they do, but in fact they rarely make explicit use of the cumulative hierarchy. That hierarchy enters into their work only as an invisible support for the structures they really use – like the complete ordered field $\mathbb{R}$. When you look at what these people actually say and write, it is entirely consistent with the category-theoretic viewpoint of abstract sets equipped with just the actually needed structure.

2.4   Functions

Q: The discussion of categories, with their emphasis on morphisms alongside objects, reminds me of a way in which functions could be considered more basic than sets.

A: More basic? "As basic" seems reasonable, if one doesn't insist on representing functions set-theoretically (using ordered pairs), but in what sense do you mean "more basic"?

Q: This came up when I was a teaching assistant for a discrete mathematics class. Sets were one of the topics, and several students had trouble grasping the idea that, for example, a thing a and the set $\{a\}$ are different, or that the empty set is one thing, not nothing. They thought of a set as a physical collection, obtained by bringing the elements together, not as a separate, abstract entity.

A: Undergraduate students aren't the only people who had such difficulties; see [36] for some relevant history. But what does this have to do with functions?

Q: Well, I found that I could clarify the problem for these students by telling them to think of a set S as a black box, where you can put in any potential element x and it will tell you "yes" if $x \in S$ and "no" otherwise. So I was explaining the notion of set in terms of functions, essentially identifying a set with its characteristic function. The black-box idea, i.e., functions, seemed to be something the students could understand directly, whereas sets were best understood via functions.

A: It seems that functions are obviously abstract, so the students aren't tempted to identify them with some concrete entity, whereas they are tempted to do that with sets.

Q: That may well explain what happened with my students.

If one takes seriously the idea of functions being more basic than sets, then it seems natural to develop a theory of functions as a foundation for mathematics. Has that been tried?

A: Yes, although sometimes the distinction between using sets and using functions as the basic notion is rather blurred.

Q: Blurred how?

A: Well, the set theory now known as von Neumann-Bernays-Gödel (NBG) was first introduced by von Neumann [47,48] in terms of functions. But he minimizes the significance of using functions rather than sets. Not only do the titles of both papers say "Mengenlehre" (i.e., "set theory") with no mention of functions, but von Neumann explicitly writes that the concepts of set and function are each easily reducible to the other and that he chose functions as primitive solely for technical simplicity.[2] And when Bernays [7] recast the theory in terms of sets and classes (the form in which NBG is known today), he described his work as "a modification of a system due to von Neumann," the purpose of the modification being "to remain nearer to the structure of the original Zermelo system and to utilize at the same time some of the set-theoretic concepts of the Schröder logic and of *Principia Mathematica*." Bernays doesn't mention that the primitive concept has been changed from function to set (and class). The tone of Bernays's introduction gives the impression that the change is not regarded as a significant change in content but rather as a matter of connecting with earlier work (Zermelo, Schröder, Russell, and Whitehead) and of technical convenience (Bernays mentions a "considerable simplification" vis à vis von Neumann's system).

---

[2] Wir haben statt dem Begriffe der Menge hier den Begriff der Funktion zum Grundbegriffe gemacht: die beiden Begriffe sind ja leicht aufeinander zurückzuführen. Die technische Durchführung gestaltet sich jedoch beim Zugrundelegen des Funktionsbegriffes wesentlich einfacher, allein aus diesem Grunde haben wir uns für denselben entschieden. [48, page 676]

Q: Von Neumann claimed that functions were technically simpler than sets, and Bernays claimed the opposite?

A: Yes. Of course, the set-based system that von Neumann had in mind for his comparison may have been more complex than Bernays's system. Presumably part of Bernays's work was to make the set-based approach simpler.

By the way, Gödel [30] modified Bernays's formulation slightly; in particular, he used a single membership relation, whereas Bernays had distinguished between membership in sets and membership in classes. Gödel describes his system as "essentially due to P. Bernays and ... equivalent to von Neumann's system ...." In the announcement [29], Gödel stated his consistency result in terms of von Neumann's system.

Q: So it seems we can think of von Neumann's function-based axiom system as being in some sense the same as the set-based system now known as NBG. But are there function-based foundations that aren't just variants of more familiar set-based systems?

A: The lambda calculus [4,5] and its variations fit that description. The idea here is that one works in a world of functions, with application of a function to an argument as a primitive concept. There is also the primitive notion of lambda-abstraction; given a description of a function using a free variable v, say some meaningful expression A involving v, one can produce a term $\lambda v A$ (which most mathematicians would write as $v \mapsto A$), denoting the function whose value at any v is given by A. In the untyped lambda calculus, one takes the functions to be defined at all arguments. That way, one doesn't need to specify sets as the domains of the functions; every function has universal domain. The typed lambda calculus is less antagonistic to sets; its functions have certain types as their domains and codomains.

Q: I've seen that the lambda calculus is used in computer science. In particular, Church's original statement [13] of his famous thesis identified the intuitive concept of computability with definability in the lambda calculus.[3] Also, lambda calculus plays a major role in denotational semantics. But how does it relate to foundations of mathematics?

A: Church [12] originally intended the lambda calculus as an essential part (the other part being pure logic) of a foundational system for mathematics. The other pioneers of lambda calculus, albeit in the equivalent formulation using combinators, were Schönfinkel [54] and Curry [16,17,18], and they also had foundational objectives. Unfortunately, Church's system turned out to be inconsistent [37], and the system proposed by Curry was not strong enough to serve as a general

---

[3] Church's official formulation in [13, Sect. 7] is in terms of recursiveness rather than lambda-definability, but these were proven equivalent earlier in the paper. Much earlier in the paper, Church writes in footnote 3 that the definition can be given in two ways, and he then lists lambda-definability before recursiveness. So whether Quisani is right here depends on whether the footnote counts as the original statement of the thesis or whether one must wait until Sect. 7.

foundation for mathematics. (Schönfinkel's system was also weak, being intended just as a formulation of first-order logic.)

Q: So this approach to foundations was a dead end.

A: Not really; the task is neither dead nor ended. The original plans didn't succeed, but there has been much subsequent work, which has succeeded to a considerable extent, and which may have more successes ahead of it. Church himself developed not only the pure lambda calculus [15] (essentially the lambda part of his earlier inconsistent system, but without the logical apparatus that led to the inconsistency) but also a typed lambda calculus [14] that is essentially equivalent to the simple theory of types but expressed in terms of functions and lambda abstraction instead of sets and membership. The typed lambda calculus also provides a good way to express the internal logic of topoi (and certain other categories) [38]. It forms the underlying framework of the system developed by Martin-Löf [46] as a foundation for intuitionistic mathematics. There is also a considerable body of work by Feferman (for example [23,24,25]) on foundational systems that incorporate versions of the lambda calculus and that have both constructive and classical aspects.

Q: So if you meet mathematicians from a far-away planet, would you expect their mathematics to be set-based?

A: Not necessarily but we wouldn't be surprised if their mathematics is set-based. We would certainly expect them to have a set theory, but it might be quite different from the ones we know, and it might not be their foundation of mathematics.

Acknowledgment

We thank Akihiro Kanamori and Jan Van den Bussche for promptly reading a draft of the original version of this paper and making helpful remarks. We also thank Nachum Dershowitz for some last-minute improvements of this revised version.

References

1. Abiteboul, S., Hull, R., Vianu, V.: Foundations of Databases. Addison-Wesley, Reading (1995)
2. Ackermann, W.: Zur Axiomatik der Mengenlehre. Math. Ann. 131, 336–345 (1956)
3. Aczel, P.: Non-Well-Founded Sets. CSLI Lecture Notes 14, Center for the Study of Language and Information. Stanford Univ. (1988)
4. Barendregt, H.: The Lambda Calculus. Its Syntax and Semantics. Studies in Logic and the Foundations of Mathematics, vol. 103. North-Holland, Amsterdam (1984)
5. Barendregt, H.: The impact of the lambda calculus in logic and computer science. Bull. Symbolic Logic 3, 181–215 (1997)
6. Barwise, J., Moss, L.: Vicious Circles. On the mathematics of non-wellfounded phenomena. CSLI Lecture Notes 60, Center for the Study of Language and Information. Stanford Univ. (1996)

7. Bernays, P.: A system of axiomatic set theory – Part I. J. Symbolic Logic 2, 65–77 (1937)

8. Blass, A., Gurevich, Y., Shelah, S.: Choiceless polynomial time. Annals of Pure and Applied Logic 100, 141–187 (1999)

9. Blass, A., Gurevich, Y., Van den Bussche, J.: Abstract state machines and computationally complete query languages. Information and Computation 174, 20–36 (2002)

10. Bourbaki, N.: Elements of Mathematics: Theory of Sets. Translation from French. Addison-Wesley, Reading (1968)

11. Chandra, A., Harel, D.: Structure and complexity of relational queries. J. Comput. and System Sciences 25, 99–128 (1982)

12. Church, A.: A set of postulates for the foundation of logic. Ann. Math (2) 33, 346–366 (1933) and 34, 839–864 (1932)

13. Church, A.: An unsolvable problem of elementary number theory. Amer. J. Math. 58, 345–363 (1936)

14. Church, A.: A formulation of the simple theory of types. J. Symbolic Logic 5, 56–68 (1940)

15. Church, A.: The Calculi of Lambda-Conversion. Annals of Mathematics Studies, vol. 6. Princeton Univ. Press, Princeton (1941)

16. Curry, H.: Grundlagen der kombinatorischen Logik. Amer. J. Math. 52, 509–536, and 789–834 (1930)

17. Curry, H.: The combinatory foundations of mathematical logic. J. Symbolic Logic 7, 49–64 (1942)

18. Curry, H., Feys, R.: Combinatory Logic, vol. 1. North-Holland, Amsterdam (1958)

19. Deiser, O.: Orte, Listen, Aggregate. Habilitationsschrift, Freie Universität Berlin (2006)

20. Dijkstra, E.W.: Sets are unibags. Handwritten note EWD786 (April 1981), http://www.cs.utexas.edu/users/EWD/ewd07xx/EWD786a.PDF

21. Enderton, H.: Elements of Set Theory. Academic Press, London (1977)

22. Feferman, S.: Set-theoretical foundations of category theory. In: Mac Lane, S. (ed.) Reports of the Midwest Category Seminar, III. Lecture Notes in Mathematics, vol. 106, pp. 201–247. Springer, Heidelberg (1969)

23. Feferman, S.: A language and axioms for explicit mathematics. In: Crossley, J. (ed.) Algebra and Logic. Lecture Notes in Mathematics, vol. 450, pp. 87–139. Springer, Heidelberg (1975)

24. Feferman, S.: Constructive theories of functions and classes. In: Boffa, M., van Dalen, D., McAloon, K. (eds.) Logic Colloquium 1978. Studies in Logic and the Foundations of Mathematics, vol. 97, pp. 159–224. North-Holland, Amsterdam (1980)

25. Feferman, S.: Toward useful type-free theories, I. J. Symbolic Logic 49, 75–111 (1984)

26. Feferman, S., et al.: Does mathematics need new axioms. Bull. Symbolic Logic 6, 401–446 (2000)

27. Fraenkel, A., Bar-Hillel, Y., Lévy, A.: Foundations of Set Theory. Studies in Logic and the Foundations of Mathematics, vol. 67. North-Holland, Amsterdam (1973)

28. Friedman, H.: Higher set theory and mathematical practice. Ann. Math. Logic 2, 325–357 (1970)

29. Gödel, K.: The consistency of the axiom of choice and of the generalized continuum hypothesis. Proc. Nat. Acad. Sci. U.S.A. 24, 556–557 (1938)

30. Gödel, K.: The Consistency of the Axiom of Choice and the Generalized Continuum Hypothesis with the Axioms of Set Theory. Annals of Mathematics Studies, vol. 3. Princeton Univ. Press, Princeton (1940)
31. Gurevich, Y.: Logic and the challenge of computer science. In: Börger, E. (ed.) Current Trends in Theoretical Computer Science, pp. 1–57. Computer Science Press (1988)
32. Holmes, R.: New Foundations Home Page, http://math.boisestate.edu/~holmes/holmes/nf.html
33. Isbell, J.: Review of [40]. Mathematical Reviews 34, 7332 (1967)
34. Jensen, R.: On the consistency of a slight(?) modification of Quine's NF. Synthèse 19, 250–263 (1969)
35. Johnstone, P.T.: Topos Theory. London Math. Soc. Monographs, vol. 10. Academic Press, London (1977)
36. Kanamori, A.: The empty set, the singleton, and the ordered pair. Bull. Symbolic Logic 9, 273–298 (2003)
37. Kleene, S., Rosser, J.B.: The inconsistency of certain formal logics. Ann. Math. 36(2), 630–636 (1935)
38. Lambek, J., Scott, P.: Introduction to Higher Order Categorical Logic. Cambridge Studies in Advanced Mathematics, vol. 7. Cambridge Univ. Press, Cambridge (1986)
39. Lawvere, F.W.: An elementary theory of the category of sets. Proc. Nat. Acad. Sci. U.S.A. 52, 1506–1511 (1964)
40. Lawvere, F.W.: The category of categories as a foundation for mathematics. In: Proc. Conf. Categorical Algebra (La Jolla, CA, 1995), pp. 1–20. Springer, Heidelberg (1966)
41. Lawvere, F.W., Rosebrugh, R.: Sets for Mathematicians. Cambridge University Press, Cambridge (2003)
42. Lawvere, F.W., Tierney, M.: Quantifiers and sheaves. In: Actes du Congrès International des Mathématiciens (Nice, 1970), Tome 1, pp. 329–334. Gauthier-Villars (1971)
43. Lévy, A.: On Ackermann's set theory. J. Symbolic Logic 24, 154–166 (1959)
44. Mac Lane, S., Moerdijk, I.: Sheaves in geometry and logic. A first introduction to topos theory. Universitext. Springer, Heidelberg (1994)
45. Martin, D.A.: Borel determinacy. Ann. Math (2) 102(2), 363–371 (1975)
46. Martin-Löf, P.: An intuitionistic theory of types: predicative part. In: Rose, H.E., Shepherdson, J.C. (eds.) Proceedings of the Logic Colloquium (Bristol, July, 1973). Studies in Logic and the Foundations of Mathematics, vol. 80, pp. 73–118. North-Holland, Amsterdam (1975)
47. von Neumann, J.: Eine Axiomatisierung der Mengenlehre. J. Reine Angew. Math. 154, 219–240 (1925) (English translation in From Frege To Gödel. In: van Heijenoort, J. (ed.) A Source Book in Mathematical Logic, 1879–1931, 393–413, Harvard University Press (1967))
48. von Neumann, J.: Die Axiomatisierung der Mengenlehre. Math. Z. 27, 669–752 (1928)
49. Platek, R.: Eliminating the continuum hypothesis. J. Symbolic Logic 34, 219–225 (1969)
50. Van Orman Quine, W.: New foundations for mathematical logic. Amer. Math. Monthly 44, 70–80 (1937)
51. Reinhardt, W.: Ackermann's set theory equals ZF. Ann. Math. Logic 2, 189–249 (1970)

52. Roitman, J.: The uses of set theory. Math. Intelligencer 14, 63–69 (1992)
53. Rosser, J.B.: Logic for Mathematicians. McGraw-Hill, New York (1953)
54. Schönfinkel, M.: Über die Bausteine der mathematischen Logik. Math. Ann. 92, 305–316 (1924)
55. Schwartz, J.T., Dewar, R.B.K., Dubinsky, E., Schonberg, E.: Programming with Sets: An Introduction to SETL. Springer, Heidelberg (1986)
56. Shelah, S.: Can you take Solovay's inaccessible away. Israel J. Math. 48, 1–47 (1984)
57. Solovay, R.: A model of set theory in which every set of reals is Lebesgue measurable. Ann. Math. 92(2), 1–56 (1970)
58. Z users website, http://vl.zuser.org/
59. Zermelo, E.: Untersuchungen über die Grundlagen der Mengenlehre I. Mathematische Annalen 65, 261–281 (1908)
60. Zermelo, E.: Über Grenzzahlen und Mengenbereiche, Neue Untersuchungen über die Grundlagen der Mengenlehre. Fundamenta Mathematicae 16, 29–47 (1930)

The Church-Turing Thesis
over Arbitrary Domains*

Udi Boker and Nachum Dershowitz

School of Computer Science, Tel Aviv University, Ramat Aviv 69978, Israel
udiboker@tau.ac.il, nachum.dershowitz@cs.tau.ac.il

For Boaz, pillar of a new discipline.

Abstract. The Church-Turing Thesis has been the subject of many variations and interpretations over the years. Specifically, there are versions that refer only to functions over the natural numbers (as Church and Kleene did), while others refer to functions over arbitrary domains (as Turing intended). Our purpose is to formalize and analyze the thesis when referring to functions over arbitrary domains.

First, we must handle the issue of domain representation. We show that, prima facie, the thesis is not well defined for arbitrary domains, since the choice of representation of the domain might have a non-trivial influence. We overcome this problem in two steps: (1) phrasing the thesis for entire computational models, rather than for a single function; and (2) proving a "completeness" property of the recursive functions and Turing machines with respect to domain representations.

In the second part, we propose an axiomatization of an "effective model of computation" over an arbitrary countable domain. This axiomatization is based on Gurevich's postulates for sequential algorithms. A proof is provided showing that all models satisfying these axioms, regardless of underlying data structure, are of equivalent computational power to, or weaker than, Turing machines.

1 Introduction

Background. In 1936, Alonzo Church and Alan Turing each formulated a claim that a particular model of computation completely captures the conceptual notion of "effective" computability. Church [5, p. 356] proposed that effective computability of numeric functions be identified with Gödel and Herbrand's general recursive functions, or – equivalently, as it turned out [5] – with Church and Kleene's lambda-definable functions of positive integers. Similarly, Turing [31] suggested that his computational model, namely, Turing machines, could compute anything that might be mechanically computable, but his interests extended beyond numeric functions.

---

* This research was supported by the Israel Science Foundation (grant no. 250/05) and was carried out in partial fulfillment of the requirements for the Ph.D. degree of the first author.

A. Avron et al. (Eds.): Trakhtenbrot/Festschrift, LNCS 4800, pp. 199–229, 2008.

Church's original thesis concerned functions over the natural numbers with their standard interpretation [5, p. 346, including fn. 3] (emphasis ours):

> The purpose of the present paper is to propose a definition of effective calculability. As will appear, this definition of effective calculability can be stated in either of two equivalent forms, (1) that a *function of positive integers* will be called effectively calculable if it is λ-definable..., (2) that a *function of positive integers* shall be called effectively calculable if it is recursive....

Kleene, when speaking about Church's Thesis, also refers to functions over the natural numbers [13, pp. 58, 60] (emphasis ours):

> We entertain various proposition *about natural numbers* ... This heuristic fact [all recognized effective functions turned out to be general recursive], as well as certain reflections on the nature of symbolic algorithmic processes, led Church to state the following thesis. The same thesis is implicit in Turing's description of computing machines.
>
> THESIS I. Every effectively calculable function (effectively decidable predicate) is general recursive.

Turing, on the other hand, explicitly extends the notion of "effective" beyond the natural numbers [32, fn. p. 166] (emphasis added):

> We shall use the expression "computable function" to mean a function calculable by a machine, and we let "effectively calculable" refer to the intuitive idea without particular identification with one of these definitions. *We do not restrict the values taken by a computable function to be natural numbers*; we may for instance have computable propositional functions.

But for Turing, even numerical calculations operate on their string representation.

Turing's model of computability was instrumental in the wide acceptance of Church's Thesis. As Trakhtenbrot explained [30]:

> This is the way the miracle occurred: the essence of a process that can be carried out by purely mechanical means was understood and incarnated in precise mathematical definitions.

The Problem. Let f be some decision function (a Boolean-valued function) over an arbitrary countable domain D. What does one mean by saying that "f is computable"? One most likely means that there is a Turing machine M, such that M computes f, *using some string representation of the domain D*. But what are the allowed string representations? Obviously, allowing an arbitrary representation (any injection from D to Σ^*) is problematic – it will make any decision function "computable". For example, by permuting the domain of machine codes, the halting function can morph into the simple parity function, which returns true when the input number is even, representing a halting machine, and false otherwise). Thus, under a "strange" representation the function becomes eminently "computable" (see Sect. 2.1). Another approach is to allow only "natural" or "effective" representations. However, in the context of defining computability, one is obliged to resort to a vague and undefined notion of "naturalness" or of "effectiveness", thereby defeating the very purpose of characterizing computability.

Our Solution. Our approach to overcoming the representation problem is to ask about effectiveness of a set of functions over the domain of interest, rather than of a single function. As Myhill observed [19], undecidability is a property of *classes* of problems, not of individual problems. In this sense, the halting function is undecidable in conjunction with an interpreter (universal machine) for Turing machine programs that uses the same representation. The Church-Turing Thesis, interpreted accordingly, asserts that there is no effective computational model that is more inclusive than Turing machines.

Nonetheless, there remains a potentially serious problem. Let M be a computational model (computing a set of functions) over some countable domain D. Might it be the case that the set of functions that M computes is equal to the Turing-computable functions under one string representation, but strictly contains it under a different representation? Generally speaking, this could indeed be the case when comparing arbitrary computational models. For example, the standard two-counter machine model (2CM) is strictly contained in some models, while it also strictly contains them – all depending on the choice of domain representation.

Fortunately, this cannot be the case with Turing machines (nor with the recursive functions), as we have demonstrated in [4], where we proved that Turing machines are "complete" in the sense that if some model is equivalent to, or weaker than, Turing machines under one representation, then no other representation (no matter how "strange") can make it stronger than Turing machines. Hence, the Church-Turing Thesis is well-defined for arbitrary computational models.

Due to this completeness of Turing machines, we can also sensibly define what it means for a string representation of an arbitrary domain to be "effective".

Axiomatization. Equipped with a plausible interpretation of the Church-Turing Thesis over arbitrary domains, we investigate the general class of "effective computational models". We proffer an axiomatization of this class, based on Yuri Gurevich's postulates for a sequential algorithm [11]. The thesis is then proved, in the sense that a proof is provided that all models satisfying these axioms are of equivalent power to, or weaker than, Turing machines.

Gurevich's postulates are a natural starting point for computing over arbitrary domains. They are applicable for computations over any mathematical structure and aim to capture any sequential algorithm. Nevertheless, while the computation steps are guaranteed to be algorithmic, that is, effective, the initial states are not. In addition, the postulates refer to a single algorithm, while effectiveness should consider, as explained above, the whole computational model. We address the effectiveness of the initial state by adding a fourth axiom to the three of Gurevich. The effectiveness of an entire computational model is addressed by providing a minimal criterion for two sequential algorithms to be in the same model.

This direction of research follows Shoenfield's suggestion [25, p. 26]:

> [I]t may seem that it is impossible to give a proof of Church's Thesis. However, this is not necessarily the case.... In other words, we can write down some axioms about computable functions which most people would agree are evidently true. It might be possible to prove Church's Thesis from such axioms.

In fact, Gödel has also been reported (by Church in a letter to Kleene cited by Davis in [7]) to have thought "that it might be possible ... to state a set of axioms which would embody the generally accepted properties of [effective calculability], and to do something on that basis".

Thanks to Gurevich's Abstract State Machine Theorem, showing that sequential abstract state machines (ASMs) capture all (ordinary, sequential) algorithms (those algorithms that satisfy the three Abstract State Machine postulates), we get a third definition of an effective computational model: A model that consists of ASMs that share initial states satisfying the initial-state axiom.

The specifics of our effectiveness axiom may perhaps be arguable. Nevertheless, it demonstrates the possibility of such an axiomatization of effectiveness for *arbitrary domains*, and provides evidence for the validity of the Church-Turing Thesis, regardless of underlying data structure and internal mechanism of the particular computational model.

The relationship of the three approaches to characterizing effectiveness over arbitrary domains is summarized in Sect. 3.4 and depicted in Fig. 1.

Axioms of Effectiveness. We understand an "effective computational model" to be some set of "effective procedures". Since all procedures of a specific computational model should have some common mechanism, a minimal requirement is that they share the same domain representation ("base structure"). Any "effective procedure" should satisfy four postulates (formally defined as Axioms 1–4 in Sects. 3.2–3.3):

1. **Sequential Time.** *The procedure can be viewed as a set of states, specified initial states, and a transition function from state to state.*

   This postulate reflects the view of a computation as some transition system, as suggested by Knuth [14, p. 7] and others. Time is discrete; transitions are deterministic; transfinite sequences are not relevant.

2. **Abstract State.** *Its states are (first-order) structures sharing the same finite vocabulary. States are closed under isomorphism, and the transition function preserves isomorphism.*

   Formalizing the states of the transition system as logical structures follows the proposal of Gurevich [11, p. 78]. This is meant to be fully general, allowing states to contain all salient features.

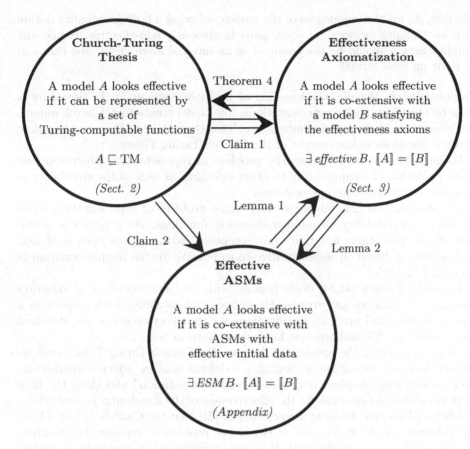

Fig. 1. Equivalent characterizations of an *extensional effectiveness* of a computational model over an arbitrary domain

3. **Bounded Exploration.** *There is a finite bound on the number of vocabulary-terms that affect the transition function.*

   This postulate ensures that the transition system has effective behavior. Informally, this means that it can be described by a finite text that explains the algorithm without presupposing any special knowledge.

4. **Initial Data.** *The initial state comprises only finite data in addition to the domain representation. The latter is isomorphic to a Herbrand universe.*

   The fourth postulate restricts procedures to be wholly effective by insisting on the effectiveness of the initial data, in addition to the effectiveness of the algorithm.

The freedom to add any finite data is obvious, but why do we limit the domain representation to be isomorphic to a Herbrand universe? There are two limitations here: (a) every domain element has a name (a closed term); and (b) the name of each element is unique. Were we to allow unnamed domain elements, then a computation could not be referred to, nor repeated, hence would not be effective. As for the uniqueness of the names, allowing a built-in equality notion with an "infinite memory" of equal pairs is obviously non-effective. Hence, the equality notion should be the result of some internal mechanism, one that can be built up from scratch.

Previous Work. Usually, the handling of multiple domains in the literature is done by choosing specific representations, like Gödel numbering, Church numerals, unary representation of numbers, etc. This is also true of the usual handling of representations in the context of the Church-Turing Thesis.

Richard Montague [18] raises the problem of representation when applying Turing's notion of computability to other domains, as well as the circularity in choosing a "computable representation".

Stewart Shapiro [24] raises the very same problem of representation when applying computability to number-theoretic functions. He suggests a definition of an "acceptable notation" (an acceptable string representation of natural numbers), based on some intuitive concepts. We discuss Shapiro's notion in Sects. 2.1 and 2.5.

Klaus Weihrauch [33,34] deals heavily with the representation of arbitrary domains by numbers and strings. He defines computability with respect to a representation, and provides justifications for the effectiveness of the standard representations. We elaborate on his justifications in Sect. 2.5.

After overcoming the problem of defining the Church-Turing Thesis over arbitrary domains, we suggest, in Sect. 2.5, a definition of an "effective representation", resembling Shapiro's notion of "acceptable notation" and along the lines of Weihrauch's justifications for the effectiveness of the standard representations.

Michael Rescorla claims in a recent paper [21] that the Church-Turing Thesis has inherent circularity because of the above problem of representing numbers by strings. He is not satisfied with Shapiro's definition of an acceptable notation, finding it insufficiently general.

A more general approach for comparing the power of different computational models would be to allow any representation based on an injective mapping between their domains. This is done, for example, by Rogers [22, p. 27], Sommerhalder [29, p. 30], and Cutland [6, p. 24]. A similar approach is used for defining the effectiveness of an algebraic structure by Froehlich and Shepherdson [9], Rabin [20], and Mal'cev [16]. Our notion of comparing computational power is very similar to this.

To the best of our knowledge, our work in [2,4] was the first to point out and handle the possible influence of the representation on the extensionality of computational models.

As for the axiomatization of effectiveness, several different approaches have been taken over the years. Turing [31] already formulated some principles for effective sequential deterministic symbol manipulation: finite internal states; finite symbol space; external memory that can be represented linearly; finite observability; and local action.

Robin Gandy [10], and later Sieg and Byrnes [28], define a model whose states are described by hereditarily finite sets. Effectiveness of Gandy machines is achieved by bounding the rank (depth) of states, insisting that they be unambiguously assemblable from individual "parts" of bounded size, and requiring that transitions have local causes.[1]

In [3], we proposed effectiveness axioms, but gave no proof that the axioms yield the same definition of effectiveness as does the Church-Turing Thesis. Whereas Turing machine states involve a linear sequence of symbols, and Gandy machine states are hereditarily finite sets, our axioms are meant to apply to arbitrary (countable) domains.

In [8], Gurevich and the second author provide an axiomatization of Church's Thesis based on the Abstract State Machine Thesis. They handle only numeric functions, ignoring the issue of effective computation over arbitrary domains, but allowing the use of domains richer than just the numbers.

Overview. The first part of this paper, Sect. 2, deals with the issue of domain representation. In Sect. 2.1, we show that checking for the computability of a single function over an arbitrary domain is problematic due to the influence of the domain representation. As a result, we interpret the Church-Turing Thesis for entire computational models. In Sect. 2.2 we define the notion of power comparison between computational models, required for the above interpretation of the thesis. In Sect. 2.3 we show that the representation might generally have an influence even on entire computational models. In Sect. 2.4 we solve the above problem with relation to the Church-Turing Thesis, by taking advantage of a "completeness" property enjoyed by the recursive functions and Turing machines with respect to domain representations. We conclude this part by discussing, in Sect. 2.5, what are in fact "effective representations".

The second part, Sect. 3, proposes an axiomatization of an "effective model of computation" over an arbitrary countable domain. In Sect. 3.2, we axiomatize "sequential procedures" along the lines of Gurevich's postulates for a sequential algorithm. In Sect. 3.3, we axiomatize "effective models" on top of sequential procedures by adding a fourth axiom, requiring the effectiveness of the initial state. We then show, in Sect. 3.4, that Turing machines, which constitute an effective model, are at least as powerful as any effective model. We conclude with a brief discussion. The proofs of this part are given in the Appendix.

We employ Gurevich's most general "Abstract State Machines" (ASMs) [11] as our programming paradigm.[2] Gurevich's ASM Theorem [11] shows that (sequential) ASMs capture (sequential) algorithms, the latter defined axiomatically. As a result, we get a third definition of an effective computational model over an arbitrary domain, namely programmable as an ASM satisfying the extra initial state axiom. See Sect. 3.4 and Fig. 1.

Terminology. When we speak of the recursive functions, denoted $\mathbb{REC}$, we mean the partial recursive functions. Similarly, the set of Turing machines, denoted TM, includes both halting and non-halting machines; we use $\mathbb{TM}$ for the set of

---

[1] The explicit bound on rank is removed in Sieg's more recent work [26,27].
[2] Some of the problems of incorporating the Gandy model under the abstract state machine rubric are dealt with in [1].

string functions computed by TMs. We use the term "domain" of a computational model and of a (partial) function to denote the set of elements over which it operates, not only those for which it is defined. By "image", we mean the values that a function actually takes: $\text{Im } f := \{f(x) \mid x \in \text{Dom } f\}$.

2   Arbitrary Domains

Simply interpreted, the Church-Turing Thesis is not well defined for arbitrary domains: the choice of domain representation might have a significant influence on the outcome. We explore below the importance of the domain representation and suggest how to overcome this problem.

2.1   Computational Model Versus Single Function

A single function over an arbitrary domain cannot be classified as computable or not. Its computability depends on the representation of the domain.[3] For example, as mentioned above, the (uncomputable) halting function over the natural numbers (sans the standard order) is isomorphic to the simple parity function, under a permutation of the natural numbers that maps the usual codes of halting Turing machines to strings ending in "0", and the rest of the numbers to strings ending with "1". The result is a computable standalone "halting" function.

An analysis of the classes of number-theoretic functions that are computable relative to different notations (representations) is provided by Shapiro [24, p. 15]:

> It is shown, in particular, that the class of number-theoretic functions which are computable relative to every notation is too narrow, containing only rather trivial functions, and that the class of number-theoretic functions which are computable relative to some notation is too broad containing, for example, every characteristic function.

An intuitive approach is to restrict the representation only to "natural" mappings between the domains. However, when doing so in the scope of defining "effectiveness" one must use a vague and undefined notion.

This problem was already pointed out by Richard Montague on 1960 [18, pp. 430–431]:

> Now Turing's notion of computability applies directly only to functions on and to the set of natural numbers. Even its extension to functions defined on (and with values in) another denumerable set S cannot be accomplished in a completely unobjectionable way. One would be inclined to choose a one-to-one correspondence between S and the set of natural

---

[3] There are functions that are inherently uncomputable, regardless of the domain representation. For example, a permutation of some countable domain, in which the lengths of the orbits are exactly the standard encodings of the non-halting Turing machines.

numbers, and to call a function f on S computable if the function of natural numbers induced by f under this correspondence is computable in Turing's sense. But the notion so obtained depends on what correspondence between S and the set of natural numbers is chosen; the sets of computable functions on S correlated with two such correspondences will in general differ. The natural procedure is to restrict consideration to those correspondences which are in some sense 'effective', and hence to characterize a computable function on S as a function f such that, for some effective correspondence between S and the set of natural numbers, the function induced by f under this correspondence is computable in Turing's sense. But the notion of effectiveness remains to be analyzed, and would indeed seem to coincide with computability.

Stewart Shapiro suggests a definition of "acceptable notation", based on several intuitive concepts [24, p. 18]:

> This suggests two informal criteria on notations employed by algorithms:
> (1) The computist should be able to *write* numbers in the notation. If he has a particular number in mind, he should (in principle) be able to write and identify tokens for the corresponding numeral.
> (2) The computist should be able to *read* the notation. If he is given a token for a numeral, he should (in principle) be able to determine what number it denotes.
> It is admitted that these conditions are, at best, vague and perhaps obscure.

Michael Rescorla argues that the circularity is inherent in the Church-Turing Thesis [21]:

> My argument turns largely upon the following constraint: a successful conceptual analysis should be *non-circular....* I will suggest that purported conceptual analyses involving Church's thesis generate a subtle yet ineliminable circularity: they characterize the intuitive notion of computability by invoking the intuitive notion of computability.... So that syntactic analysis can illuminate the computable number-theoretic functions, we correlate syntactic entities with non-syntactic entities like numbers. We endow the syntax with a primitive semantics. I submit that, in providing this semantics, we must deploy the intuitive notion of computability. Specifically, we must demand that the semantic correlation between syntactic entities and non-syntactic entities itself be computable. But then the proposed analysis does not illuminate computability non-circularly.

A possible solution is to allow any representation (injection between domains), while checking for the effectiveness of an entire computational model. That is, to check for the computability of a function together with the other functions that are computable by that computational model. The purpose lying behind this idea is to view the domain elements as arbitrary objects, deriving all their meaning from the model's functions. For example, it is obvious that the halting function has a meaning only if one knows the order of the elements of its domain. In that case, the successor function provides the meaning for the domain elements.

A variation of this solution is to allow any representation (injection between domains), provided that the image of the injection is computable. We consider both variations.

Adopting the above approach of checking for computability of an entire computational model, we interpret the Church-Turing Thesis as follows:

Thesis A. *All "effective" computational models are of equivalent power to, or weaker than, Turing machines.*

By "effective", in quotes, we mean effective in its intuitive sense.

To understand this thesis, it remains for us to define what it means to be "equivalent to, or weaker than". That is, we must define a method by which to compare computational power of computational models.

For maximum generality, we do not want to limit computational models to any specific mechanism; hence, we allow a model to be any object, as long as it is associated with the set of functions that it implements. We consider only deterministic computations, as originally envisioned in Hilbert's program (see [8]). As models may have non-terminating computations, we deal with sets of partial functions. For convenience, we assume that the domain and range (codomain) of functions are identical.

Definition 1 (Computational Model)

- *A computational model B over domain D is any object associated with a set of partial functions $f : D \to D$. This set of functions is called the* extensionality *of the computational model, denoted $[\![B]\!]$.*
- *We write* Dom B *for the domain over which model B operates.*

2.2 Comparing Computational Power

Since we are dealing with models that operate over different domains, we adopt the quasi-ordering on extensional power developed in [2,4]. Basically, we say that model A is at least as powerful as model B if there is some representation via which A contains all the functions of B. A representation may be any injection between the domains (a generalization to mappings other than injections can be found in [2]). A variation of the above requires that the "stronger" model would also be able to compute the image of the representation. We formalize the comparison notion below.

A computational model is associated with a set of functions (see Definition 1), and its representation over a different domain is just the result of some renaming of the underlying domain elements.

Definition 2 (Representation)

Domain. *Let D_A and D_B be two domains (arbitrary sets of atomic elements). A representation of D_B over D_A is an injection $\rho : D_B \to D_A$ (i.e. ρ is total and one-one). We write Im ρ for the image of the representation (the values in D_B that ρ takes).*

Function and Relation. *Representations naturally extend to functions and relations, which are sets of tuples of domain elements: $\rho(f) := \{\langle \rho(x_1), \ldots, \rho(x_n) \rangle \mid \langle x_1, \ldots, x_n \rangle \in f\}$.*

Model. *Representations also naturally extend to (the extensionalities of) computational models, which are sets of functions: $\rho(B) := \{\rho(f) \mid f \in [\![B]\!]\}$.*

Since representations are allowed to be arbitrary injections, they might not cover the target domain. Hence, when we compare a model A with a representation of some model B, we should restrict A to the image of the representation.

Definition 3 (Restriction)

1. *A restriction of a function f over domain D to a subdomain $C \subseteq D$, denoted $f\!\restriction_C$, is the subset of tuples of f in which all elements are in C. That is, $f\!\restriction_C := f \cap C^{n+1}$, for f of arity n.*
2. *We write $\rho(f) \in [\![A]\!]$ as shorthand for $\exists g \in [\![A]\!].\, \rho(f) = g\!\restriction_{\mathrm{Im}\,\rho}$, meaning that the function f belongs to the (restriction of) the computational model A via representation ρ.*

We can now provide the appropriate comparison notion.

Definition 4 (Computational Power)

- *Model A is (computationally) at least as powerful as model B, denoted $A \succsim B$, if there is a representation ρ such that $\rho(B) \subseteq \{f\!\restriction_{\mathrm{Im}\,\rho} \mid f \in [\![A]\!]\}$. In such a case, we also say that model A simulates model B (via representation ρ).*
- *Models A and B are (computationally) equivalent if $A \succsim B \succsim A$.*

This is the notion of "implemented" used in [12, p. 52] and of "incorporated" used in [29, p. 29].

Proposition 1. *The computational power relation $\succsim$ between models is a quasi-order. Computational equivalence is an equivalence relation.*

Turing-computable functions simulate the recursive functions via a unary representation of the natural numbers. The (untyped) λ-calculus (Λ) is computationally equivalent to the recursive functions ($\mathbb{REC}$), via Church numerals, on the one hand, and via Gödelization, on the other.

One may reasonably require that for a model A to be at least as powerful as a model B it should also be able to compute the image of the representation (see [2]). In such a case we get the following variation of the power comparison notion:

Definition 5 (Representational Power)

- *Model A is* (representationally) *at least as powerful as model B, denoted $A \sqsupseteq B$, if there is a representation ρ such that $\rho(B) \subseteq \{f \restriction_{\text{Im } \rho} \mid f \in [\![A]\!]\}$ and there is a total function $f \in [\![A]\!]$, such that* $\text{Im } f = \text{Im } \rho$.
- *Models A and B are* (representationally) *equivalent if $A \sqsupseteq B \sqsupseteq A$.*

In what follows, we use both computational comparisons ($\succsim$) and representational comparisons ($\sqsupseteq$), preferring the more general one whenever possible.[4]

Our interpretation of the Church-Turing Thesis (Thesis A) agrees with Rabin's definition of a computable group [20, p. 343]:

> DEFINITION 3. An *indexing* of a set S is a one to one mapping $i : S \rightarrow I$ such that $i(S)$ is a recursive subset of I. ...
> DEFINITION 4. An indexing i of a group G is *admissible* if the function m from $i(G) \times i(G)$ into $i(G)$ ... is a computable function. ...
> DEFINITION 5. A group is *computable* if it possesses at least one *admissible* indexing.

Rabin defines computability for groups, fields, and rings; however, the idea naturally generalizes to any algebraic structure, as done by Lambert [15, p. 594]:

> Following Rabin ... we let an (*admissible*) *indexing* for structure $\mathfrak{U}$ be a 1-1 function $\kappa : A \rightarrow \omega$ such that
> (i) $K = \text{range } \kappa$ is recursive;
> (ii) each $\kappa^*(F_a)$ and $\kappa^*(R_a)$ are recursive relative to K ..., where κ^* applied to an (unmixed) operation or relation in A is the operation or relation in ω naturally induced by κ. ...
> $\mathfrak{U}$ is *computable* iff there is an indexing for $\mathfrak{U}$.

Similar notions were also presented by Froehlich and Shepherdson [9] and Mal'cev [16].

2.3   Influence of Representations

It turns out that even when dealing with entire computational models, we are not yet on terra firma. The representation of the domain still allows for the possibility that a model be equivalent to one of its strict supermodels. That is, a representation might allow to "enlarge" a model, adding some "new" functions to it.

---

[4] Specifically, the "completeness" property (Definition 6) is defined using $\succsim$, which makes it stronger. Accordingly, the theorem that Turing machines and recursive functions are complete (Theorem 1) applies also to the analogous case with $\sqsupseteq$ instead. Likewise, when we show that Turing machines are at least as powerful as any effective model (Theorem 4, per Definition 17), we use the $\sqsupseteq$ notion, which provides a stronger result. Thus, the theorem applies also to the analogue $\succsim$. On the other hand, Claims 1 and 2 are stated with respect to $\sqsupseteq$, and do not necessarily hold for $\succsim$.

Consider two-counter machines. It is known that two-counter machines cannot compute the function $\lambda x.2^x$.[5] On the other hand, since two-counter machines can simulate all the recursive functions via some proper injective representation (viz. $n \mapsto 2^n$; see, for example, [17]), it follows that two-counter machines can "enlarge" their computational power via some representations.

A reasonable direction might have been to restrict the representation to bijections between domains. However, while it works for this example, it turns out that there are models equivalent to some of their supermodels even with bijective representations [4]. Hence, there are models isomorphic to some of their strict supermodels.

This places a question mark on the definition of Turing-computability and on the meaning of the Church-Turing Thesis. Can it be that the recursive functions are isomorphic to a larger set of functions?! Can we find a string representation of the natural numbers via which we have Turing machines to compute all the recursive functions *plus* some additional functions?

In the next section, we put firmer ground beneath the definition of the Church-Turing Thesis, by showing that Turing machines, as well as the recursive functions, enjoy a special "completeness" property.

2.4 Completeness

As seen above, a model can be of equivalent power to its strict supermodel. There are, however, models that are not susceptible to such an anomaly; these are referred to as "complete" models, among which are Turing machines and the recursive functions.

Definition 6 (Completeness). *A model is* complete *if it is not of equivalent power to any of its strict supermodels. That is, A is complete if $A \succsim B$ and $[\![B]\!] \supseteq [\![A]\!]$ imply that $[\![A]\!] = [\![B]\!]$ for any B.*

A supermodel of the recursive functions (or Turing machines) is a "hypercomputational" model.

Definition 7 (Hypercomputational Model). *A model H is* hypercomputational *if it simulates a model that strictly contains the recursive functions.*

Theorem 1 ([4]). *The recursive functions and Turing machines are complete. They cannot simulate any hypercomputational model.*

(The completeness of the recursive functions proved in [4] refers only to unary functions, but it is quite straightforward to extend it to any arity.)

Note that the completeness property is defined with computational comparison $\succsim$, which makes it a stronger property. Accordingly, Turing machines and the recursive functions are also complete with respect to representational comparison $\sqsupseteq$.

---

[5] This was shown by Rich Schroeppel in [23], and independently by Frances Yao and others.

The Church-Turing Thesis, as interpreted in Sect. 2.1, matches the intuitive understanding only due to this completeness of the recursive functions and Turing machines. Were the thesis defined in terms of two-counter machines (2CM), for example, it would make no sense: a computational model is not necessarily stronger than 2CM even if it computes strictly more functions.[6]

2.5 Effective Representations

What is an effective representation? We argued above that a "natural representation" must be a vague notion when used in the context of defining effectiveness. We avoided the need of restricting the representation by checking the effectiveness of entire computational models. But what if we adopt the Church-Turing Thesis; can we then define what is an effective string representation?

Simply put, there is a problem here. Turing machines operate only over strings. Thus a string representation, which is an injection from some domain D to Σ^*, is not itself computable by a Turing machine. All the same, when we consider, for example, string representations of natural numbers, we can obviously say regarding some of them that they are effective. How is that possible? The point is that we look at a domain as having some structure. For the natural numbers, we usually assume their standard order. A function over the natural numbers without their order is not really well-defined. As we saw, the halting function and the simple parity function are exactly the same (isomorphic) function when numbers are unordered.

Hence, even when adopting the Church-Turing Thesis, a domain without any structure cannot have an effective representation. It is just a set of arbitrary elements. However, if the domain comes with a generating mechanism (as the natural numbers come with the successor) we can consider effective representations.

Due the completeness of the recursive functions and Turing machines, we can define what is an effective string representation of the natural numbers (with their standard structure). A similar definition can be given for other domains, besides the natural numbers, provided that they come with some finite means of generating them all, akin to successor for the naturals.

Definition 8. *An* effective representation *of the natural numbers by strings is an injection* $\rho : \mathbf{N} \rightarrow \Sigma^*$*, such that* $\rho(s)$ *is Turing-computable* $(\rho(s) \in \mathbb{TM})$*, where s is the successor function over* $\mathbf{N}$*.*

That is, a representation of the natural numbers is effective if the successor function is Turing-computable via this representation.

Remark 1. One may also require that the image of the representation ρ is totally Turing computable, meaning, that the question whether some string is in Im ρ is decidable.

---

[6] In fact, the lambda calculus also suffers from incompleteness in this sense, and would, therefore, not be a suitable candidate in terms of which to characterize generic effectivity.

We justify the above definition of an effective representation by showing that:
(a) every recursive function is Turing-computable via any effective representation; (b) every non-recursive function is not Turing-computable via any effective representation; and (c) for every non-effective representation there is a recursive function that is not Turing-computable via it.

Theorem 2

(a) *Let f be a recursive function and $\rho : \mathbf{N} \to \Sigma^*$ an effective representation. Then $\rho(f) \in \mathrm{TM}$.*
(b) *Let g be a non-recursive function and $\rho : \mathbf{N} \to \Sigma^*$ an effective representation. Then $\rho(g) \notin \mathrm{TM}$.*
(c) *Let $\eta : \mathbf{N} \to \Sigma^*$ be a non-effective representation. Then there is a recursive function f, such that $\eta(f) \notin \mathrm{TM}$.*

Proof Let $\xi : \mathbf{N} \to \Sigma^*$ be some standard bijective representation via which $\xi(\mathrm{REC}) = \mathrm{TM}$ (see, for example, [12, p. 131]). The point is that, once $\rho(s) \in \mathrm{TM}$, there are Turing-computable functions for switching between the ρ and the ξ representations. That is, $\rho \circ \xi^{-1}, \xi \circ \rho^{-1} \in \mathrm{TM}$. It can be done by a Turing machine that enumerates in parallel over both representations until reaching the required string.
(a) Since $f \in \mathrm{REC}$, it follows that there is a function $f' \in \mathrm{TM}$, such that $f = \xi^{-1}(f') = \xi^{-1} \circ f' \circ \xi$. Thus, $\rho(f) = \rho \circ f \circ \rho^{-1} = \rho \circ \xi^{-1} \circ f' \circ \xi \circ \rho^{-1}$. Hence, $\rho(f) \in \mathrm{TM}$ by the closure of TM under functional composition.
(b) Assume by contradiction that $g \notin \mathrm{REC}$ but $\rho(g) \in \mathrm{TM}$. Let g' be the corresponding function under the ξ representation. That is, $g' = \xi \circ \rho^{-1} \rho(g) \circ \rho \circ \xi^{-1}$. We have by the closure of TM under functional composition that $g' \in \mathrm{TM}$. Since $\xi^{-1}(g') \in \mathrm{REC}$, it is left to show that $\xi^{-1}(g') = g$ for getting a contradiction: $\xi^{-1}(g') = \xi^{-1} \circ g' \circ \xi = \xi^{-1} \circ \xi \circ \rho^{-1} \rho(g) \circ \rho \circ \xi^{-1} \circ \xi = \rho^{-1} \rho(g) \circ \rho = \rho^{-1} \rho \circ g \circ \rho^{-1} \circ \rho = g$.
(c) By the definition of recursive representation, the successor is such a function. □

To see the importance of the completeness property for the definition of an effective representation, one can check that an analogous definition cannot be provided with two-counter machines as the yardstick.

Our definition of an effective representation resembles Shapiro's notion of an "acceptable notation". He proposes three necessary "semi-formal" criteria for an acceptable notation [24, p. 19]:

(1a) If the computist is given a finite collection of distinct objects, then he can (in principle) write and identify tokens for the numeral which denotes the cardinality of the collection.
(1b) The computist can *count* in the notation. He is able (in principle) to write, in order, tokens for the numerals denoting any finite initial segment of the natural numbers.
(2a) If the computist is given a token for a numeral p and a collection of distinct objects, then he can (in principle) determine whether the

denotation of p is smaller than the cardinality of the collection and, if it is, produce a subcollection whose cardinality is the denotation of p.

Our notion coincides with Shapiro's second criterion (1b). It also goes along with Weihrauch's justifications for the effectiveness of the standard "numberings" (representation by natural numbers). He defines a standard numbering of a word set (the words over $\{a, b\}$, for example, are enumerated in the following order: $\varepsilon, a, b, aa, ab, ba, bb, aaa, aab, \ldots$), and then proves three claims for justifying the effectiveness of the numbering [33, p. 80–81]:

A numbering $\nu : \mathbf{N} \rightarrow W(\Sigma)$ is neither a word function nor a number function, hence neither of our two definitions of computability is applicable to ν. Nevertheless standard numberings $\nu : \mathbf{N} \rightarrow W(\Sigma)$ are intuitively effective. The following lemma expresses several effectivity properties of standard numberings of word sets.

LEMMA (*effectivity of standard numberings of word sets*)
Let Σ, Γ and Δ be alphabets with $\Delta = \Sigma \cup \Gamma$. Let ν_Σ (ν_Γ) be a standard numbering of $W(\Sigma)$ ($W(\Gamma)$).

(1) Define $S, V : \mathbf{N} \rightarrow \mathbf{N}$ by $S(x) := x + 1$, $V(x) := x \dot{-} 1$.
Define $S_\Sigma, V_\Sigma : W(\Sigma) \rightarrow W(\Sigma)$ by
$$S_\Sigma := \nu_\Sigma S \nu_\Sigma^{-1}, \quad V_\Sigma := \nu_\Sigma V \nu_\Sigma^{-1}.$$
Then S_Σ and V_Σ are computable.

(2) Let $b \in \Sigma$. Define $h^b : W(\Sigma) \rightarrow W(\Sigma)$, $S^b : W(\Sigma) \rightarrow \{1, 2\}$ and $pop : W(\Sigma) \rightarrow W(\Sigma)$ by
$h^b(w) = wb$, $S^b(w) := (1$ if $w = xb$ for some $x \in W(\Sigma)$, 2 otherwise),
and $pop(\varepsilon) := \varepsilon$, $pop(wc) := w$.
Define $h^b_\Sigma : \mathbf{N} \rightarrow \mathbf{N}$, $S^b_\Sigma : \mathbf{N} \rightarrow \{1, 2\}$ and $pop_\Sigma : \mathbf{N} \rightarrow \mathbf{N}$ by
$h^b_\Sigma := \nu_\Sigma^{-1} h^b \nu_\Sigma$, $pop_\Sigma := \nu_\Sigma^{-1} pop \, \nu_\Sigma$, $S^b_\Sigma := S^b \nu_\Sigma$.
Then h^b_Σ, pop_Σ, and S^b_Σ are computable.

(3) The following functions $p : W(\Delta) \rightarrow W(\Delta)$ and $q : \mathbf{N} \rightarrow \mathbf{N}$ are computable:
$p(w) := (\nu_\Sigma \nu_\Gamma^{-1}(w)$ if $w \in W(\Gamma), \varepsilon$ otherwise),
$q(j) := (\nu_\Gamma^{-1} \nu_\Sigma(j)$ if $\nu_\Sigma(j) \in W(\Gamma), 0$ otherwise).

Our notion resembles Weihrauch's first and second claims (the second claim concerns the construction of strings, and plays the rôle of the successor when reversing the "numbering" for representing numbers by strings).

3   An Axiomatization of Effective Models

Section 2 formalized the Church-Turing Thesis over arbitrary domains. We now provide additional evidence for the thesis by validating it against a class of "effective computational models", axiomatized on top of Gurevich's postulates for a sequential algorithm [11].

Gurevich's postulates are applicable for computations over any mathematical structure (of first order) and aim to capture any sequential algorithm. This makes them a natural candidate for axiomatizing effectiveness over arbitrary domains. Yet, there are several problems:

1. The postulates concern algorithms and not computations with input and output.
2. Initial states are not limited; thus, they might not be effective.
3. The postulates consider a single algorithm and not an entire computational model.

We address the first issue, in Sect. 3.2, by adding special input and output constants, and allowing a single initial state, up to differences in input. The second issue is addressed by adding Axiom 4, which limits the initial data. The third issue is addressed, in Sect. 3.3, by requiring all functions of the same model to share the same domain representation.

A proof is provided in the Appendix, showing that this axiomatization yields the same definition of effectiveness as the Church-Turing Thesis does. It is based on Gurevich's Abstract State Machine Theorem [11], showing that sequential abstract state machines (ASMs) capture sequential algorithm. As a result, we get three equivalent definitions of an effective computational model over an arbitrary domain. See Fig. 1.

We start in Sect. 3.2, with an axiomatization of "sequential procedures", along the lines of Gurevich's sequential algorithms [11]. Next, we axiomatize, in Sect. 3.3, "effective procedures" as a subclass, satifying an "effectivity axiom". We then show, in Sect. 3.4, the equivalence of Turing machines to the class of effective models. We conclude this part with a brief discussion.

3.1 Structures

The states of a procedure should be a full instantaneous description of all its relevant features. We represent them by (first order) *structures*, using the standard notion of structure from mathematical logic. For convenience, these structures will be *algebras*; that is, having purely functional vocabulary (without relations).

Definition 9 (Structures)

- A domain D is a (nonempty) set of elements.
- A vocabulary $\mathcal{F}$ is a collection of function names, each with a fixed finite arity.
- A term of vocabulary $\mathcal{F}$ is either a nullary function name (constant) in $\mathcal{F}$ or takes the form $f(t_1, \ldots, t_k)$, where f is a function name in $\mathcal{F}$ of positive arity k and $t_1, \ldots, t_k$ are terms.
- A structure S of vocabulary $\mathcal{F}$ is a domain D together with interpretations $[\![f]\!]_S$ over D of the function names $f \in \mathcal{F}$.
- A location of vocabulary $\mathcal{F}$ over a domain D is a pair, denoted $f(\bar{a})$, where f is a k-ary function name in $\mathcal{F}$ and $\bar{a}$ is a k-tuple of elements of D. (If f is a constant, then $\bar{a}$ is the empty tuple.)

- *The* value *of a location* $f(\bar{a})$ *in a structure* S, *denoted* $[\![f(\bar{a})]\!]_S$, *is the domain element* $[\![f]\!]_S(\bar{a})$.
- *It is often useful to indicate a location by a (ground) term* $f(t_1, \ldots, t_k)$, *standing for* $f([\![t_1]\!]_S, \ldots, [\![t_k]\!]_S)$.
- *Structures* S *and* S' *with vocabulary* $\mathcal{F}$ *sharing the same domain* coincide *over a set* T *of* $\mathcal{F}$-*terms if* $[\![t]\!]_S = [\![t]\!]_{S'}$ *for all terms* $t \in T$.

It is convenient to think of a structure S as a memory, or data-storage, of a kind. For example, for storing an (infinite) two dimensional table of integers, we need a structure S over the domain of integers, having a single binary function name f in its vocabulary. Each entry of the table is a location. The location has two indices, i and j, for its row and column in the table, marked $f(i, j)$. The content of an entry (location) in the table is its value $[\![f(i,j)]\!]_S$.

Definition 10 (Update). *An* update *of location* l *over domain* D *is a pair, denoted* $l := v$, *where* v *is an element of* D.

Definition 11 (Structure Mapping). *Let* S *be structure of vocabulary* $\mathcal{F}$ *over domain* D *and* $\rho : D \to D'$ *an injection from* D *to domain* D'. *A mapping of* S *by* ρ, *denoted* $\rho(S)$, *is a structure* S' *of vocabulary* $\mathcal{F}$ *over* D', *such that* $\rho([\![f(\bar{a})]\!]_S) = [\![f(\rho(\bar{a}))]\!]_{S'}$ *for every location* $f(\bar{a})$ *in* S.

Structures S and S' of the same vocabulary over domains D and D', respectively, are *isomorphic*, denoted $S \simeq S'$, if there is a bijection $\pi : D \leftrightarrow D'$, such that $S' = \pi(S)$.

3.2   Sequential Procedures

Our axiomatization of a "sequential procedure" is very similar to that of Gurevich's sequential algorithm [11], with the following two main differences, allowing for the computation of a specific function, rather than expressing an abstract algorithm:

- The vocabulary includes special constants "*In*" and "*Out*".
- Initial states are identical, except for changes in *In*.

Axiom 1 (Sequential Time). *The procedure can be viewed as a collection* $\mathcal{S}$ *of states, a sub-collection* $\mathcal{S}_0 \subseteq \mathcal{S}$ *of initial states, and a transition function* $\tau : \mathcal{S} \to \mathcal{S}$ *from state to state.*

Axiom 2 (Abstract State)

- States. *All states are first-order structures of the same finite vocabulary* $\mathcal{F}$.
- Input. *There are nullary function names In and Out in* $\mathcal{F}$. *All initial states* ($\mathcal{S}_0 \subseteq \mathcal{S}$) *share a domain* D, *and are equal up to changes in the value of In. (For convenience, the initial states can be referred to, collectively, as* $\mathcal{S}_0$.)
- Isomorphism Closure. *The procedure states are closed under isomorphism. That is, if there is a state* $S \in \mathcal{S}$, *and an isomorphism* π *via which* S *is isomorphic to a* $\mathcal{F}$-*structure* S', *then* S' *is also a state in* $\mathcal{S}$.

- Isomorphism Preservation. *The transition function preserves isomorphism. That is, if states S and S' are isomorphic via π, then $\tau(S)$ and $\tau(S')$ are also isomorphic via π.*
- Domain Preservation. *The transition function preserves the domain. That is, the domain of S and $\tau(S)$ is the same for every state $S \in \mathcal{S}$.*

Axiom 3 (Bounded Exploration). *There exists a finite set T of "critical" terms, such that $\Delta(S, \tau(S)) = \Delta(S', \tau(S'))$ if S and S' coincide over T, for all states $S, S' \in \mathcal{S}$, where $\Delta(S, S') = \{l := v' \mid [\![l]\!]_S \neq [\![l]\!]_{S'} = v'\}$ is a set of updates turning S into S'.*

The isomorphism constraints reflects the fact that we are working at a fixed level of abstraction. See [11, p. 89]:

> A structure should be seen as a mere representation of its isomorphism type; only the isomorphism type matters. Hence the first of the two statements: distinct isomorphic structures are just different representations of the same isomorphic type, and if one of them is a state of the given algorithm A, then the other should be a state of A as well.

Domain preservation simply ensures that a specific "run" of the procedure is over a specific domain. (Should it be necessary, one could always combine many domains into one.) The bounded-exploration axiom ensures that the behavior of the procedure is effective. This reflects the informal assumption that the program of an algorithm can be given by a finite text [11, p. 90].

Definition 12 (Runs)

1. *A run of procedure with transition function τ is a finite or infinite sequence $S_0 \leadsto_\tau S_1 \leadsto_\tau S_2 \leadsto_\tau \cdots$, where S_0 is an initial state and every $S_{i+1} = \tau(S_i)$.*
2. *A run $S_0 \leadsto_\tau S_1 \leadsto_\tau S_2 \leadsto_\tau \cdots$ terminates if it is finite or if $S_i = S_{i+1}$ from some point on.*
3. *The terminating state of a terminating run $S_0 \leadsto_\tau S_1 \leadsto_\tau S_2 \leadsto_\tau \cdots$ is its last state if it is finite, or its stable state if it is infinite.*
4. *If there is a terminating run beginning with state S and terminating in state S', we write $S \leadsto_\tau^! S'$.*

Definition 13 (Procedure Extensionality). *Let P be sequential procedure over domain D. The extensionality of P, denoted $[\![P]\!]$, is the partial function $f : D \to D$, such that $f(x) = [\![Out]\!]_{S'}$ whenever there's a run $S \leadsto_\tau^! S'$ with $[\![In]\!]_S = x$, and is undefined otherwise.*

Equality, Booleans and Undefined. In contradistinction with Gurevich's ASM's, we do not have built in equality, Booleans, or undefined in the definition of procedures. That is, a procedure need not have Boolean values (True, False) or connectives ($\neg, \wedge, \vee$) pre-defined in its vocabulary; rather, they may be defined like any other function. It also should not have a special term for undefined values, though the value of the function implemented by the procedure is not defined when its run doesn't terminate. The equality notion is also not presumed in the procedure's initial state, as it compises infinite data. Nevertheless, since every domain element has a unique construction, it follows that an effective procedure may implement the equality notion with only finite initial data. A detailed description of this implementation is given in Sect. A.3.

3.3   Effective Models

A sequential procedure may be equipped with any oracle, given as an operation of the initial state. Hence, the extensionality of such a procedure might not be effective. As a result, we are interested only in sequential procedures that use effective oracles. Since we are defining effectiveness, we get an inductive definition, allowing initial states to include functions that are the extensionality of effective procedures. The starting point must be operations that are very simple and inherently effective. These basic operations must then be finite. We begin, then, with sequential procedures, in which the initial state has finite data in addition to the domain representation ("base structure"). This constraint is formalized in Axiom 4, below.

Different procedures of the same computational model have some common mechanism. The level of shared configuration between the model's procedures may vary, but they must obviously share the same domain representation. Hence, we define an "effective model" to be some set of "effective procedures" that share the same "base structure".

We formalize the finiteness of the initial data by allowing the initial state to contain an "almost-constant structure".

Definition 14 (Almost-Constant Structure). *A structure F is almost constant if all but a finite number of locations have the same value.*

Since we are heading for a characterization of effectiveness, the domain over which the procedure actually operates should have countably many elements, which have to be nameable. Hence, without loss of generality, one may assume that naming is via terms.

Definition 15 (Base Structure). *A structure S of finite vocabulary $\mathcal{F}$ over a domain D is a base structure if every domain element is the value of a unique $\mathcal{F}$-term. That is, for every element $e \in D$ there exists a unique $\mathcal{F}$-term t such that $[\![t]\!]_S = e$.*

A base structure is isomorphic to the standard tree term algebra (Herbrand universe) of its vocabulary.

Proposition 2. *Let S be a base structure over vocabulary G and domain D, then:*

- *The vocabulary G has at least one nullary function.*
- *The domain D is countable.*
- *Every domain element is the value of a unique location of S.*

Example 1. A structure over the natural numbers with constant *zero* and unary function *successor*, interpreted as the regular successor, is a base structure.

Example 2. A structure over binary trees with constant *nil* and binary function *cons*, interpreted as in Lisp, is a base structure.

We are now in position to formalize the fourth axiom, requiring the effectiveness of the initial state. It is an inductive definition, allowing any function that can be implemented by a (simpler) effective procedure.

Definition 16 (Structure Union). *Let S' and S'' be two structures with domain D and with vocabularies $\mathcal{F}'$ and $\mathcal{F}''$, respectively. A structure S over D is the union of S' and S'', denoted $S = S' \uplus S''$, if its vocabulary is the disjoint union $\mathcal{F} = \mathcal{F}' \uplus \mathcal{F}''$, and if $[\![l]\!]_S = [\![l]\!]_{S'}$ for locations l in S' and $[\![l]\!]_S = [\![l]\!]_{S''}$ for locations in S''.*

Axiom 4 (Initial Data). *The initial state consists of:*

- *a fixed base structure BS (the domain representation);*
- *a fixed almost-constant AS structure (finite initial data); and*
- *a fixed effective structure ES over the base structure BS (effective oracles);*

in addition to an input value In over BS that varies from initial state to initial state. That is, the initial state S_0 is the union $BS \uplus AS \uplus ES \uplus \{In\}$, for some base structure BS, almost-constant structure AS, and effective structure ES.

The effective structure contains finite many functions that are the extensionality of effective procedures over the same domain representation. This allows the procedure to use an algorithm at any abstraction level, as long as we can assure that the underlying oracles are effective.

As already mentioned, there are two aspects to the requirement that the domain representation be isomorphic to a Herbrand universe: every domain element has a name, and names are unique. Were one to allow unnamed domain elements, then a computation cannot be referred to, nor repeated, hence would not be effective. As for the uniqueness of the names, allowing a built-in equality notion with an "infinite memory" of equal pairs is obviously not effective. Hence, the equality notion should be the product of some internal effective mechanism, and thus needs to be a part of the computational model.

An *effective procedure* must satisfy Axioms 1–4.

Definition 17 (Effective Model). *An* effective model *is a set of effective procedures (objects satisfying Axioms 1–4) that share the same base structure.*

To sum up:

Thesis B. *All "effective" computational models are effective models (per Definition 17).*

3.4   Effective Equals Computable

In the sense of our above definition of effectiveness (Definition 17) we have that:

Theorem 3. *Turing machines are an effective model.*

Furthermore,

Theorem 4. *Turing machines are representationally at least as powerful as any effective model.*

That is, TM $\sqsupseteq E$ for every model E satisfying the effectiveness axioms.

Note that we use representational comparison $\sqsupseteq$, which provides a stronger result. Accordingly, Turing machines are also computationally at least as powerful ($\gtrsim$) as any effective model.

The proofs of Theorems 3 and 4 are quite straightforward but somewhat lengthy, so are relegated to the appendix. They make usage of Abstract State Machines, which operate over arbitrary domains, and are based on Gurevich's Abstract State Machine Theorem [11], showing that sequential abstract state machines (ASMs) capture sequential algorithms, defined axiomatically.

Definition 18 (Effective State Model). *An ASM model satisfying the initial data restrictions is called an* Effective State Model *(or ESM).*

This suggests the following variant thesis:

> **Thesis C.** *Every "effective" computational model is behaviorally equivalent to an ESM.*

If we adopt the variation of the comparison notion that requires the "stronger" model to be able to compute the image of the representation (Definition 5), we get a closer relationship between the three definitions of effectiveness (Theses A–C): When considering only the extensionality of computational models (that is, the set of functions that they compute) we have that the three effectiveness criteria (Theses A–C) are equivalent.

Definition 19 (Effective Looks). *A model A looks effective if the set of functions that it computes may be represented by Turing-computable functions. That is, if $A \sqsubseteq$ TM.*

Claim 1. *A model A looks effective if and only if there exists an effective model B, such that $[\![A]\!] = [\![B]\!]$.*

Thanks to Gurevich's Abstract State Machines Theorem [11], we have the analogous claim with respect to ASMs:

Claim 2. *A model A looks effective if and only if there is an ESM B, such that $[\![A]\!] = [\![B]\!]$.*

Claims 1 and 2 are not proved herein. (Their proofs are based on the proofs of Theorems 3 and 4, as well as aspects of the proof of Theorem 2.)

The resulting relationship between the different characterizations of effectiveness is depicted in Fig. 1.

3.5 Discussion

Necessity. An effective procedure should satisfy, by our definitions, Axioms 1–4. In the introduction, we argued for the necessity of the postulates from the intuitive point of view of effectiveness. Moreover, omitting any of them allows for models that compute more than Turing machines:

1. The Sequential Time Axiom is necessary if we wish to analyze computation, which is a step-by-step process. Allowing for transfinite computations, for example, would allow a model to precompute all members of a recursively-enumerable set.
2. In the context of effective computation, there is no room for infinitary functions, for example. Without closure under isomorphism there would be no value to the Bounded-Exploration Axiom, allowing the assigning of any desired value to the *Out* location.
3. By omitting the Bounded-Exploration Axiom, a procedure need not have any systematical behavior, hence may "compute" any function by simply assigning the desired value at the *Out* location. That is, for each initial state S there is a state S', such that $\tau(S) = S'$ and $[\![Out]\!]_{S'}$ is the "desired" value.
4. Omitting the Initial-Data Axiom, one may "compute" any function (e.g. a halting oracle), by simply having all its values in the initial state. Such functions could also be encoded in equalities between locations, were the initial data not (isomorphic to) a free term algebra.

Algorithm versus Model. In [11], Gurevich proved that any algorithm satisfying his postulates can be represented by an Abstract State Machine. But an ASM is designed to be "abstract", so is defined on top of an arbitrary structure that may contain *non-effective* functions. Hence, it itself may compute non-effective functions. We have adopted Gurevich's postulates, but added an additional postulate (Axiom 4) for effectiveness: an algorithm's initial state may contain only finite data and known effective operations in addition to the domain representation. Different runs of the same procedure share the same initial data, except for the input; different procedures of the same model share a base structure. We proved that – under these assumptions – the class of all effective procedures is of equivalent computational power to Turing machines.

Acknowledgement

The second author thanks Félix Costa for his gracious hospitality and for substantive comments on a draft of this work.

References

1. Blass, A., Gurevich, Y.: Background, Reserve, and Gandy Machines. In: Clote, P.G., Schwichtenberg, H. (eds.) CSL 2000. LNCS, vol. 1862, Springer, Heidelberg (2000)

2. Boker, U., Dershowitz, N.: How to Compare the Power of Computational Models. In: Cooper, S.B., Löwe, B., Torenvliet, L. (eds.) CiE 2005. LNCS, vol. 3526, pp. 54–64. Springer, Heidelberg (2005)

3. Boker, U., Dershowitz, N.: Abstract effective models. In: Fernández, M., Mackie, I. (eds.) New Developments in Computational Models: Proceedings of the First International Workshop on Developments in Computational Models (DCM 2005), Lisbon, Portugal (July 2005), Electronic Notes in Theoretical Computer Science, 135(3), 15–23 (2006)

4. Boker, U., Dershowitz, N.: Comparing computational power. Logic Journal of the IGPL 14(5), 633–648 (2006)

5. Church, A.: An unsolvable problem of elementary number theory. American Journal of Mathematics 58, 345–363 (1936)

6. Cutland, N.: Computability: An Introduction to Recursive Function Theory. Cambridge University Press, Cambridge (1980)

7. Davis, M.: Why Gödel didn't have Church's Thesis. Information and Control 54(1/2), 3–24 (1982)

8. Dershowitz, N., Gurevich, Y.: A natural axiomatization of Church's Thesis. Bulletin of the ASL (to appear), available as Technical report MSR-TR-2007-85, Microsoft Research, Redmond, WA (July 2007)

9. Froehlich, A., Shepherdson, J.: Effective procedures in field theory. Philosophical Transactions of the Royal Society of London 248, 407–432 (1956)

10. Gandy, R.: Church's thesis and principles for mechanisms. In: Barwise, J., et al. (eds.) The Kleene Symposium. Studies in Logic and The Foundations of Mathematics, vol. 101, pp. 123–148. North-Holland, Amsterdam (1980)

11. Gurevich, Y.: Sequential abstract state machines capture sequential algorithms. ACM Transactions on Computational Logic 1, 77–111 (2000)

12. Jones, N.D.: Computability and Complexity from a Programming Perspective. The MIT Press, Cambridge, MA (1997)

13. Kleene, S.C.: Recursive predicates and quantifiers. Transactions of the American Mathematical Society 53(1), 41–73 (1943)

14. Knuth, D.E.: The Art of Computer Programming. Fundamental Algorithms, vol. 1. Addison-Wesley, Reading, MA (1968)

15. Lambert Jr., W.M.: A notion of effectiveness in arbitrary structures. The Journal of Symbolic Logic 33(4), 577–602 (1968)

16. Mal'cev, A.: Constructive algebras I. Russian Mathematical Surveys 16, 77–129 (1961)

17. Minsky, M.L.: Computation: Finite and Infinite Machines. Prentice-Hall, Englewood Cliffs, NJ (1967)

18. Montague, R.: Towards a general theory of computability. Synthese 12(4), 429–438 (1960)

19. Myhill, J.: Some philosophical implications of mathematical logic. Three classes of ideas 6(2), 165–198 (1952)

20. Rabin, M.O.: Computable algebra, general theory and theory of computable fields. Transactions of the American Mathematical Society 95(2), 341–360 (1960)

21. Rescorla, M.: Church's thesis and the conceptual analysis of computability. Notre Dame Journal of Formal Logic 48(2), 253–280 (2007)

22. Rogers Jr., H.: Theory of Recursive Functions and Effective Computability. McGraw-Hill, New York (1966)

23. Schroeppel, R.: A two counter machine cannot calculate 2^N. Technical report, Massachusetts Institute of Technology, Artificial Intelligence Laboratory (1972) (viewed November 28, 2007), ftp://publications.ai.mit.edu/ai-publications/pdf/AIM-257.pdf
24. Shapiro, S.: Acceptable notation. Notre Dame Journal of Formal Logic 23(1), 14–20 (1982)
25. Shoenfield, J.R.: Recursion Theory. Lecture Notes in Logic, vol. 1. Springer, Heidelberg (1991)
26. Sieg, W.: Church without dogma—Axioms for computability. In: Löwe, B., Sorbi, A., Cooper, S.B. (eds.) New Computational Paradigms: Changing Conceptions of What is Computable, pp. 18–44. Springer, Heidelberg (2007)
27. Sieg, W.: Computability: Emergence and analysis of a mathematical notion. In: Irvine, A. (ed.) Handbook of the Philosophy of Mathematics (to appear)
28. Sieg, W., Byrnes, J.: An abstract model for parallel computations: Gandy's thesis. The Monist 82(1), 150–164 (1999)
29. Sommerhalder, R., van Westrhenen, S.C.: The Theory of Computability: Programs, Machines, Effectiveness and Feasibility. Addison-Wesley, Workingham, England (1988)
30. Trakhtenbrot, B.A.: Comparing the Church and Turing approaches: Two prophetical messages. In: Herken, R. (ed.) The Universal Turing Machine: A half-century survey, pp. 603–630. Oxford University Press, Oxford (1988)
31. Turing, A.M.: On computable numbers, with an application to the Entscheidungsproblem. Proceedings of the London Mathematical Society 42, 230–265 (1936), Corrections in vol. 43, pp. 544–546 (1937), Reprinted in Davis, M. (ed.), The Undecidable, Raven Press, Hewlett, NY (1965)
32. Turing, A.M.: Systems of logic based on ordinals. Proceedings of the London Mathematical Society 45, 161–228 (1939)
33. Weihrauch, K.: Computability. EATCS Monographs on Theoretical Computer Science, vol. 9. Springer, Berlin (1987)
34. Weihrauch, K. (ed.): Computable Analysis – An introduction. Springer, Berlin (2000)

A   Proofs of Two Theorems

We provide here proofs of Theorems 3 and 4. First, we require some additional definitions and lemmata.

A.1   Programmable Machines

In Sect. 3.2, we axiomatized sequential procedures. To link these procedures with Turing machines, we define some mediators, named "programmable procedures," along the lines of Gurevich's Abstract State Machines (ASMs) [11]. We then show that sequential procedures and programmable procedures are equivalent (Lemma 1).

A "programmable procedure" is like a sequential procedure, with the main difference that its transition function should be given by a finite "flat program" rather than satisfy some constraints.

Definition 20 (Flat Program). *A flat program P of vocabulary $\mathcal{F}$ has the following syntax:*

if $x_{11} \doteq y_{11}$ and $x_{12} \doteq y_{12}$ and $\ldots$ $x_{1k_1} \doteq y_{1k_1}$
    then $l_1 := v_1$

if $x_{21} \doteq y_{21}$ and $x_{22} \doteq y_{22}$ and $\ldots$ $x_{2k_2} \doteq y_{2k_2}$
    then $l_2 := v_2$

$\vdots$

if $x_{n1} \doteq y_{n1}$ and $x_{n2} \doteq y_{n2}$ and $\ldots$ $x_{nk_n} \doteq y_{nk_n}$
    then $l_n := v_n$

where each $\doteq$ is either '$=$' or '$\neq$', $n, k_1, \ldots, k_n \in \mathbf{N}$, and all the x_{ij}, y_{ij}, l_i, and v_i are $\mathcal{F}$-terms.

Each line of the program is called a rule. *The part of a rule between the* if *and the* then *is the* condition, *l_i is its* location, *and v_i is its* value.

The activation *of a flat program P on an $\mathcal{F}$-structure S, denoted $P(S)$, is a set of updates $\{l := v \mid$ there is a rule in P, whose condition holds (under the standard interpretation), with location l and value $v\}$, or the empty set if the above set includes two values for the same location.*

Coding Style. To make flat programs more readable, let

```
% comment
if cond-1
    stat-1
    stat-2
else
    stat-3
```

stand for

```
if cond-1 then stat-1
if cond-1 then stat-2
if not cond-1 then stat-3
```

and, similarly, for other such abbreviations.

Definition 21 (Programmable Procedure). *A programmable procedure is composed of: $\mathcal{F}, In, Out, D, \mathcal{S}, \mathcal{S}_0$, and P, where all but the last component is as in a sequential procedure (see Sect. 3.2), and P is a flat program of $\mathcal{F}$.*

The run *of a programmable procedure and its* extensionality *are defined as for sequential procedures (Definitions 12 and 13), where the transition function τ is given by $\tau(S) = S' \in \mathcal{S}$ such that $\Delta(S, S') = P(S)$.*

A.2   Sequential Equals Programmable

We show that every programmable procedure is sequential (satisfying the three axioms), and every sequential procedure is programmable. This result is derived directly from the main lemma of [11].

Lemma 1. *Every programmable procedure is sequential. That is, let A be a programmable procedure with states S and a flat program P, then there exists a sequential procedure B with the same elements of A, except for having a transition function τ instead of the program P, such that $\Delta(S, \tau(S)) = P(S)$ for every $S \in \mathcal{S}$.*

Proof. Let $A = \langle \mathcal{F}, In, Out, D, \mathcal{S}, \mathcal{S}_0, P \rangle$ be an arbitrary programmable procedure. Define the finite set of critical $\mathcal{F}$-terms T to include all terms and subterms of P. Define a transition function $\tau : \mathcal{S} \rightarrow \mathcal{S}$ by $\tau(S) = S'$ such that $\Delta(S, S') = P(S)$. To show that $B = \langle \mathcal{F}, In, Out, D, \mathcal{S}, \mathcal{S}_0, \tau \rangle$ is a sequential procedure such that $\Delta(S, \tau(S)) = P(S)$ for every $S \in \mathcal{S}$ it remains to show that B satisfies the constraints defined for τ in a sequential procedure. Since the flat program P includes only terms in T (and doesn't refer directly to domain elements), it obviously follows that τ satisfies the isomorphism constraint. Since T includes all the terms of P, as well as the subterms of the location-terms of P, it obviously follows that states that coincide over T have the same set of updates by τ. Thus, τ satisfies the bounded-exploration constraint. □

Lemma 2. *Every sequential procedure is programmable. That is, let B be a sequential procedure with states S and a transition function τ, then there exists a programmable procedure A with the same elements of B, except for having a flat program P instead of τ, such that $\Delta(S, \tau(S)) = P(S)$ for every $S \in \mathcal{S}$.*

This follows directly from Gurevich's proof that for every sequential algorithm there exists an equivalent sequential abstract state machine [11, Lemma 6.11].

A.3   Effective Equals Computable

We prove now that Turing machines are of equivalent computational power to all effective models.

Turing Machines are Effective. First, we show that the class of effective procedures is at least as powerful as Turing machines, as the latter is an effective model.

Proof (of Theorem 3). We consider Turing machines with two-way infinite tapes. By way of example, let the tape alphabet be $\{0, 1\}$. So domain elements are comprised of an internal machine state and an infinite tape, containing finitely many 0's and 1's, and the rest blank, and a read/write head somewhere along the tape.

A Turing machine state (instantaneous description) contains three things: *Left*, a finite string containing the tape section left of the reading head; *Right*, a finite string with the tape section to the right to the read head; and q, the internal state of the machine. The read head points to the first character of *Right*.

Turing machines can be viewed as an effective model with the following components:

Domain: The domain consists of all finite strings over $0, 1$. That is the domain $D = \{0, 1\}^*$.

Base structure: Constructors for the finite strings: the constant symbol @ and unary function symbols $Cons\_0$ and $Cons\_1$. Thus, @ has the empty string, ε, as its permanent value.

Almost-constant structure:

- Input and Output (nullary functions): *In*, *Out*. The value of *In* at the initial state is the content of the tape, as a string over $\{0, 1\}^*$.
- Constants for the alphabet characters and TM-states (nullary): 0, 1, q_0, q_1, $\ldots$, q_k. Their actual values are of no significance, as long as they are all different.
- Variables to keep the current status of the Turing machine (nullary): *Left*, *Right*, and q. Their initial values are: $Left = \varepsilon$, $Right = \varepsilon$, and $q = q\_0$.

Effective structure:

- Functions to examine the tape (unary functions): *Head* and *Tail*. Their initial values are as in the standard implementation of *Head* and *Tail*. Their effective implementation is given below, after the description of the Turing machine model.
- The Boolean equality notion $=$. Note that the standard equality notion contains infinite data, thus cannot be contained in the almost-constant structure, nor in the base structure. Nevertheless, since every domain element has a unique construction, the equality notion can be effectively implemented with only finite initial data. This implementation is explained after the implementation of *Head* and *Tail*.

Transition function: By Lemma 1, every programmable procedure is a sequential procedure. Thus, a programmable procedure that satisfies the initial-data postulate is an effective procedure. Every Turing machine is an effective procedure with a flat program looking like this:

```
if q = q_0  % TM's state q_0
   if Head(Right) = 0
      % write 1, move right, switch to q_3
      Left  := Cons_1(Left)
      Right := Tail(Right)
      q := q_3
   if Head(Right) = 1
      % write 0, move left, switch to q_1
      Left  := Tail(Left)
      Right := Cons_0(Right)
      q := q_1
   if Right = @
      % write 0, move left, switch to q_2
```

```
      Left := Tail(Left)
      Right := Cons_0(Right)
      q := q_2
if q = q_1   % TM's state q_1
      ...
if q = q_k   % the halting state
   Out := Right
```

In the above description of Turing machines as an effective model we've used the functions *Head* and *Tail*. We show now their effectiveness.

The implementation sequentially enumerates all strings, assigning their *Head* and *Tail* values, until encountering the input string. Note that it uses the equality notion, which is shown to be effective afterwards.

It uses the same base structure and almost-constant structure described above, with the addition of the following nullary functions (Name = initial value): $New = \varepsilon$, $Backward = 0$, $Forward = 1$, $AddDigit = 0$, and $Direction = \varepsilon$.

```
% Sequentially constructing the Left variable
% until it equals to the input In, for filling
% the values of Head and Tail.
% The enumeration is: empty string, 0, 1, 00, 01, ...
if Left = In % Finished
   Right := Left
   Left := @
else % Keep enumerating
   if Direction = New % default val
      if Left = @ % @ -> 0
         Left := Cons_0(Left)
         Head(Cons_0(Left)) := 0
         Tail(Cons_0(Left)) := Left
      if Head(Left) = 0 % e.g. 110 -> 111
         Left := Cons_1(Tail(Left))
         Head(Cons_1(Tail(Left))) := 1
         Tail(Cons_1(Tail(Left))) := Tail(Left)
      if Head(Left) = 1 % 01->10; 11->000
         Direction := Backward
         Left := Tail(Left)
         Right := Cons_0(Right)
   if Direction = Backward
      if Left = @ % add rightmost digit
         Direction := Forward
         AddDigit := True
      if Head(Left) = 0 % change to 1
         Left := Cons_1(Tail(Left))
         Direction := Forward
      if Head(Left) = 1 % keep backwards
         Left := Tail(Left)
         Right := Cons_0(Right)
   if Direction = Forward % Gather right 0s
      if Right = @ % finished gathering
```

```
      Direction := New
      if AddDigit = 1
         Left := Cons_0(Left)
         Head(Cons_0(Left)) := 0
         Tail(Cons_0(Left)) := Left
         AddDigit = 0
   else
         Left := Cons_0(Left)
         Right := Tail(Right)
         Head(Cons_0(Left)) := 0
         Tail(Cons_0(Left)) := Left
```

The equality notion. The standard equality notion has infinite data, thus cannot be given in the initial state. However, since the domain elements are uniquely constructed, it follows that it can be effectively implemented using only finite initial data. The implementation scheme is quite similar to the above implementation of *Head* and *Tail*. Initially, the value of the equality function is $\perp$ at all locations. The implementation sequentially enumerates the strings, assigning *True* as the value of equality of each string with itself and *False* for comparisons with all preceeding strings. This continues until the process gives one of the defined values to the equality operation applied to the inputs. $\square$

Effective Procedures are Computable. Next, we show that all effective models are equal to or weaker than Turing machines by mapping every effective model to a **while**-like computer program (CP). The computer program may be of any programming language known to be of equivalent power to Turing machines, as long as it operates over the natural numbers and includes the syntax and semantics of flat programs.

Lemma 3. *Every infinite base structure S of vocabulary $\mathcal{F}$ over a domain D is isomorphic to a computable structure S' of the same vocabulary over $\mathbf{N}$. That is, there is a bijection $\pi : D \leftrightarrow \mathbf{N}$ such that for every location $f(\overline{a})$ of S we have that $[\![f(\overline{a})]\!]_S = \pi^{-1}([\![f(\pi(\overline{a}))]\!]_{S'})$.*

Proof. Let S be a base structure of vocabulary $\mathcal{F}$ over a domain D. Let $\mathcal{T}$ be the domain of all $\mathcal{F}$-terms, and $\tilde{S}$ the standard free term algebra (structure) of $\mathcal{F}$. Since all structure functions are total, it follows that every $\mathcal{F}$-term has a value in D, and by Proposition 2, every element $e \in D$ is the value of a unique $\mathcal{F}$-term. Therefore, there is bijection $\varphi : D \leftrightarrow \mathcal{T}$, such that $\varphi^{-1}(t) = [\![t]\!]_S$ for every $t \in \mathcal{T}$. Hence, S and $\tilde{S}$ are isomorphic via φ. Since $\mathcal{F}$ is finite, it follows that its set of terms $\mathcal{T}$ is recursive. Define a computable enumeration $\eta : \mathcal{T} \leftrightarrow \mathbf{N}$. Define a structure S' of vocabulary $\mathcal{F}$ over $\mathbf{N}$ by the following computable recursion: $[\![f(n_1, \ldots, n_k)]\!]_{S'} = \eta(f(\eta^{-1}(n_1), \ldots, \eta^{-1}(n_k)))$. That is, for computing the value of a function f on a tuple $\overline{n}$ the program should recursively find the terms of $\overline{n}$, and then compute the enumeration of the combined term. By the construction of S' we have that S' and $\tilde{S}$ are isomorphic via η. Hence, S' and S are isomorphic via $\varphi \circ \eta$. $\square$

Lemma 4. *Computer programs* (CP) *are at least as powerful, representationally, as any effective model.*

Proof. We show that for every effective model E over domain D there is a bijection $\pi : D \to \mathbf{N}$ such that CP $\sqsupseteq_\pi E$.

When the effective model E has a finite base structure, then the computability is obvious due to the finite number of possible procedures. We consider then the infinite case; let E be an effective model over a domain D with an infinite base structure BS. By Lemma 3 there is a bijection $\pi : D \leftrightarrow \mathbf{N}$, such that the structure $BS' := \pi(BS)$ is computable. Let P_{BS} be a computer program implementing BS'. For each effective procedure $e \in E$, let AS_e be its almost-constant structure. Since AS_e is almost constant, it follows that $AS'_e := \pi(AS_e)$ is computable; let P_{AS_e} be a computer program implementing AS'_e. Analogously, we have by induction a computer program P_{ES_e} implementing the effective structure of e. By Lemma 2, the transition function of every effective procedure $e \in E$ can be defined by a flat program P_e. For every effective procedure $e \in E$, define a computer program $P'_e = P_{BS} \cup P_e \cup P_{AS_e} \cup P_{ES_e}$. Since $BS' = \pi(BS)$, $AS'_e = \pi(AS_e)$ and $ES'_e = \pi(ES_e)$, it follows that $[\![P'_e]\!] = \pi([\![e]\!])$. Therefore, there is a bijection $\pi : D \leftrightarrow \mathbf{N}$, such that for every effective procedure $e \in E$ there is a computer program $P'_e \in$ CP such that $[\![P'_e]\!] = \pi([\![e]\!])$. Hence, CP $\sqsupseteq E$. $\qquad\square$

We are now in position to prove that Turing machines are at least as powerful as any effective model.

Proof (of Theorem 4). By Lemma 4, computer programs (CP) are representationally at least as powerful as any effective model, while Turing machines (TM) are of equivalent power to computer programs. (There are standard bijections between Σ^* and $\mathbf{N}$.) $\qquad\square$

Generalized Categorial Dependency Grammars

Michael Dekhtyar[1] and Alexander Dikovsky[2,*]

[1] Dept. of Computer Science, Tver State University, Tver, Russia, 170000
Michael.Dekhtyar@tversu.ru
[2] LINA CNRS 2729, Université de Nantes, 2, rue de la Houssinière BP 92208 F 44322
Nantes cedex 3 France
Alexandre.Dikovsky@univ-nantes.fr

To our dear Teacher Boris Trakhtenbrot.

Abstract. Generalized Categorial Dependency Grammars (gCDG) studied in this paper are genuine categorial grammars expressing projective and discontinuous dependencies, stronger than CF-grammars and non-equivalent to mild context-sensitive grammars. We show that gCDG are parsed in polynomial time and enjoy good mathematical properties.

1 Introduction

Dependency grammars (DGs) are formal grammars assigning *dependency structures* to the sentences of the language they define. A dependency structure (DS) of a sentence is an oriented graph whose nodes are the words of the sentence and whose arcs are labelled with dependency names. In other words, they are structures on sentences in terms of various binary relations on words. If two words v_1 and v_2 are related by dependency d (denoted $v_1 \xrightarrow{d} v_2$), then v_1 is the *governor* and v_2 is the *subordinate*. Intuitively, the dependency d encodes constraints on lexical and grammatical features of v_1 and v_2, on their precedence, pronominalization, context, etc. which together mean that "v_1 licenses v_2" (see [24] for a detailed presentation). For instance, in the DS of the sentence *In the beginning was the Word* in Fig. 1, *was* $\xrightarrow{pred}$ *Word* stands for the predicative dependency between the copula *was* and the subject *Word*. From such basic dependency relations some derived relations are defined, e.g. the dependency

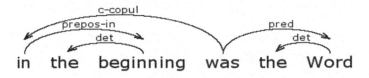

Fig. 1.

* This work was sponsored by the Russian Fundamental Studies Foundation (Grant 05-01-01006-a).

A. Avron et al. (Eds.): Trakhtenbrot/Festschrift, LNCS 4800, pp. 230–255, 2008.

relation $v_1 \longrightarrow v_2 \equiv \exists d \ (v_1 \xrightarrow{d} v_2)$ and its reflexive-transitive closure (*dominance*) $v_1 \xrightarrow{*} v_2$. For instance, in the DS in Fig. 1, both occurrences of the article *the* are dominated by *was*: $was \xrightarrow{*} the_2$, $was \xrightarrow{*} the_5$, but only the first one is dominated by *in*: $in \xrightarrow{*} the_2$.

The idea of such explicit representation of syntactic relations in sentences is by far more ancient than that of the constituent structure and goes back at least to the early grammars of the Arabic language, which used the notions of governor and subordinate (Kitab al-Usul of Ibn al-Sarrang, (d. 928)). Modern theories of syntax use various DSs. Prevailing is the tradition, going back to L. Tesnire [29], to use only the tree-like DS: *dependency trees* (DTs). There are also approaches where general dependency structures are used (cf. [17,28]). Sometimes (this is the case of [17]), it is due to combining in the same structure several relations of different nature, for instance, the surface syntactic relations and the co-reference relations. Another difference point is the word order (WO) included or not into the DSs. Some important properties of DSs cannot be expressed without the WO, first of all, *projectivity*. This property is defined in terms of the *projection* $D(v)$ of a word v in DS D of a sentence w: $D(v) = \{v' \in D \mid v \xrightarrow{*} v'\}$. D is *projective* if the projections of all words in w are **continuous intervals** of w. So the DS in Fig. 1 is projective. Meanwhile, non projective DSs are frequent in natural languages. E.g., both DTs in Fig. 2 are non projective. The non projectivity is

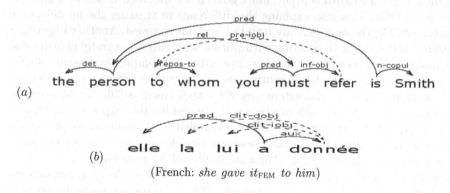

(French: *she gave it*FEM *to him*)

Fig. 2.

always due to *discontinuous* dependencies, i.e. the dependencies in which the governor v_g is separated from the subordinate v_s by a word not dominated by v_g (see [13] for more details). In Fig. 2, the discontinuous dependencies are represented by dotted arrows. When the dependencies are emancipated from the WO (cf. [6]), it is only done to define more exactly the WO constraints. Contrary to this, we suppose that the DSs are linearly ordered by the WO.

There is a great many definitions of DGs: from generating to constraint based (see [20,21] for references and discussion). Our definition goes back to the early valency/precedence style definitions [16,14] having much in common with those of the classical categorial grammars [1,2]. Both are lexicalized, use syntactic types in the place of rewriting rules, naturally fit compositional semantic structures and are equivalent to CF-grammars if only the weak expressive power is concerned and the core syntax is considered. But as far as it concerns the strong expressive power, many fundamental differences appear between these formalisms. It is true that there is a simple translation from phrase structures with head selection to projective DTs and back (see [15,27] or [13] for more details), which conforms with the direct simulation of core dependency grammars by the classical CGs [14]. Unfortunately, this technical resemblance does not preserve the intended syntactic types. The reason is that the syntactic functions corresponding to the dependencies are different from those of the heads in the syntagmatic structures originating from the X-bar theory [18]. Basically, the difference is that the type of a constituent head determines its syntagmatic (phrase) valencies, whereas a dependency represents a valency of the governor word in one subordinate word. It reflects its lexical and syntactic class, its position with respect to the governor, its semantic role, pronominalization, etc. (see [24] for more details). In particular, this means that the dependency types should be more numerous and specific than the syntagmatic ones and not prone to type raising. Essential distinctions are also in treating verb and noun modifiers, which in dependency surface syntax are *subordinate* and *iterated*. The canonical CGs' elimination rules imply dependencies from the functional type words to the argument type words. So, in the absence of type raising, the adjectives, whose canonical type in English is $[n/n]$, must govern the modified nouns and not vice versa as in DGs. This also explains the difference in treating the modifiers. In DGs (cf. [28,23]) the modifiers are iterated and not recursed. Another important difference is that DTs, in contrast with phrase structures, naturally capture discontinuous surface word order. Rather expressive and complex extensions of CGs are needed to cope with the discontinuous and naturally oriented dependencies simulation (e.g. multi-modal extensions of Lambek calculus [26,25]). Meanwhile, as it was shown in [9,10], both can be naturally and feasibly expressed in DGs in terms of *polarized dependency valencies* controlled by a simple principle, which enables a discontinuous dependency between two closest words having the same valency with the opposite signs ("first available" (**FA**) principle).

Below, we study a class of *generalized categorial dependency grammars* established on the base of the **FA** principle. These grammars prove to be very expressive. At the same time, they are parsed in practical polynomial time and can be naturally linked with the underspecified semantics defined in [11] (this subject will be treated elsewhere).

2   Syntactic Types

Dependency type of a word (to be called *category*) represents its governor-subordinate valencies. There are two basic ideas of how to transform dependencies into categories. The first idea, proposed in [12], consists in decomposing each dependency $Gov \xrightarrow{d} Sub$ into two parts: Gov and Sub. The first

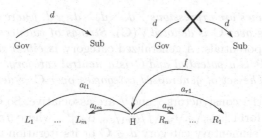

Fig. 3.

becomes the *argument-type d*, whereas the second, the *value-type d* (see Fig. 3). Grouping together, for a word H, the value type f corresponding to the incoming dependency f and the argument types corresponding to the outgoing left dependencies $a_{l1}, \ldots, a_{lm}$ and right dependencies $a_{r1}, \ldots, a_{rm}$ (in this order) we obtain the category $[a_{lm}\backslash\ldots\backslash a_{l1}\backslash f/a_{r1}/\ldots/a_{rn}]$ of H (denoted $H \mapsto [a_{lm}\backslash\ldots\backslash a_{l1}\backslash f /a_{r1}/\ldots/a_{rn}]$). For instance, the DT in Fig. 1 determines the types:

$$in \mapsto [c-copul/prepos-in], \quad the \mapsto [det],$$
$$beginning \mapsto [det\backslash prepos-in], \quad was \mapsto [c-copul\backslash S/pred],$$
$$Word \mapsto [det\backslash pred].$$

The second idea put forward in [9,10], consists in interpreting discontinuous dependencies as polarized valencies using four *polarities*: left and right positive $\nwarrow, \nearrow$ and left and right negative $\swarrow, \searrow$. For each polarity v, there is the unique "dual" polarity $\breve{v}$: $\breve{\nwarrow} = \swarrow, \breve{\swarrow} = \nwarrow, \breve{\nearrow} = \searrow, \breve{\searrow} = \nearrow$. Intuitively, the argument type $\nwarrow d$ can be seen as the valency of a word whose subordinate through dependency d is situated *somewhere* on the left. The dual value type $\swarrow d$ can be seen as the valency of a word whose governor through the same dependency d is situated *somewhere* on the right. Together, the *paired* dual valencies $\swarrow d, \nwarrow d$ (respectively, $\nearrow d, \searrow d$) define the discontinuous dependency d. For instance, the DT in Fig. 2b determines the types:

$$elle \mapsto [pred], \quad la \mapsto [\swarrow clit-dobj],$$
$$lui \mapsto [\swarrow clit-iobj], \quad a \mapsto [pred\backslash S/aux],$$
$$donne \mapsto [\nwarrow clit-iobj\backslash \nwarrow clit-dobj\backslash aux]$$

Speaking about *generalized categories*, we will factor out from them the polarized subtypes. For instance, $[\nwarrow clit\!-\!iobj\backslash \nwarrow clit\!-\!dobj\backslash aux]$ and $[\swarrow clit\!-\!iobj]$ will become respectively $[aux]^{\nwarrow clit\!-\!iobj \nwarrow clit\!-\!dobj}$ and $[\varepsilon]^{\swarrow clit\!-\!iobj}$. Here is a definition of the generalized categories.

Definition 1. *Let* $\mathbf{C}$ *be a set of* elementary (dependency) categories. $S \in \mathbf{C}$ *is the selected category of sentences. For each* $d \in \mathbf{C}$, *the category* $d*$ *is* iterated.

All elementary categories and ε *are* neutral. *If a category* C *is neutral and a category* α *is elementary or iterated, then the categories* $[\alpha\backslash C]$ *and* $[C/\alpha]$ *are also neutral. There are no other neutral categories. The set of neutral categories over* $\mathbf{C}$ *is denoted* $nCat(\mathbf{C})$.

Polarized valencies are expressions $\nearrow d$, $\searrow d$, $\nwarrow d$, $\nearrow d$, *where* $d \in \mathbf{C}$. *The set of polarized valencies over* $\mathbf{C}$ *is denoted* $V(\mathbf{C})$. *Strings of valences* $P \in Pot(\mathbf{C})=_{df}$ $V(\mathbf{C})^*$ *are called* potentials. *A generalized category is either neutral or has the form* C^P, *where* P *is a potential and* C *is a neutral category. We will omit the empty potential. The set of generalized categories over* $\mathbf{C}$ *is denoted* $gCat(\mathbf{C})$.

We suppose that the constructors $\backslash$ and $/$ are associative. So every generalized category has the form $[\alpha_{lm}\backslash...\backslash\alpha_{l1}\backslash f/\alpha_{r1}/.../\alpha_{rn}]^P$, where $f \in \mathbf{C} \cup \{\varepsilon\}$, each α_{li} and α_{rj} is an elementary category $d \in \mathbf{C}$ or its iteration $d*$, $m, n \geq 0$ and $P \in Pot(\mathbf{C})$.

In [9] a simple and natural principle of pairing dual polarized valencies was proposed called **First Available (FA)**-principle: *the closest dual valences with the same name are paired.*

Definition 2. *An occurrence of dual polarized valencies* v *and* $\breve{v}$ *in a potential* $P_1 v P \breve{v} P_2$ *satisfies the* **FA***-principle if* P *has no occurrences of* v *and* $\breve{v}$.

3   Generalized Categorial Dependency Grammar

Categorial dependency grammars are *lexicalized* in the same sense as the conventional categorial grammars: they have a few language non-specific rules constituting a dependency calculus and a language specific lexicon defining the words using dependency types.

Definition 3. *A generalized categorial dependency grammar (gCDG) is a system* $G = (W, \mathbf{C}, S, \delta)$, *where* W *is a finite set of words,* $\mathbf{C}$ *is a finite set of elementary categories containing the selected category* S, *and* δ *- called* lexicon *- is a finite substitution on* W *such that* $\delta(a) \subset gCat(\mathbf{C})$ *for each word* $a \in W$.

The generalized dependency calculus consists of the following rules.[1]

L[1]. $C^{P_1}[C\backslash\beta]^{P_2} \vdash [\beta]^{P_1 P_2}$
I[1]. $C^{P_1}[C^*\backslash\beta]^{P_2} \vdash [C^*\backslash\beta]^{P_1 P_2}$
Ω[1]. $[C^*\backslash\beta]^P \vdash [\beta]^P$
D[1]. $\alpha^{P_1(\nearrow C)P(\searrow C)P_2} \vdash \alpha^{P_1 P P_2}$, if $(\nearrow C)P(\searrow C)$ satisfies **FA**

Intuitively, the rule **L**[1] corresponds to the classical elimination rule of categorial grammars. Eliminating the argument subtype C it constructs the (projective) dependency C in which the governor is the word with the functional type and the subordinate is the word with the argument type. At the same time, it concatenates the potentials of these types (if any). The rules **I**[1], **Ω**[1] derive the iterated (projective) dependencies. **I**[1], analogous to the rule **L**[1], may derive $k > 0$ dependencies C and **Ω**[1] corresponds to the case $k = 0$. It is the rule **D**[1] which derives discontinuous dependencies. It pairs and eliminates dual valencies $\nearrow C, \searrow C$ (or $\nearrow C, \searrow C$) and creates the discontinuous dependency C between the words whose types have these polarized valencies. This calculus naturally induces the immediate provability relation $\vdash$ on the strings of generalized dependency types $\Gamma_1 \vdash \Gamma_2$ underlying the following definition of languages.

---
[1] We show only left argument rules. The right argument rules are symmetric.

Definition 4. *For a gCDG $G = (W, \mathbf{C}, S, \delta)$, let $G(D, w)$ denote the relation: D is the DS of a sentence w constructed in the course of a proof $\Gamma \vdash S$ for some $\Gamma \in \delta(w)$. In particular, we will use notation $w = w(D)$ for the DS D of w. The DS-language generated by G is the set of dependency structures*

$$\Delta(G) =_{df} \{D \mid \exists w \; G(D, w)\}$$

and the language *generated by G is the set of sentences*

$$L(G) =_{df} \{w \mid \exists D \; G(D, w)\}.$$

$\mathcal{D}(gCDG)$ and $\mathcal{L}(gCDG)$ will denote the families of DS-languages and languages generated by these grammars.

Example 1. For instance, in the gCDG G_{abc} :

$$a \mapsto A^{\swarrow A}, [A\backslash A]^{\swarrow A}, \quad b \mapsto [B/C]^{\searrow A}, [A\backslash S/C]^{\searrow A}, \quad c \mapsto C, [B\backslash C],$$

$G_{abc}(D^{(3)}, a^3b^3c^3)$ holds for the DS $D^{(3)}$ in Fig. 4 and the string $a^3b^3c^3$ due to the types assignment

$$a^3b^3c^3 \mapsto A^{\swarrow A}[A\backslash A]^{\swarrow A}[A\backslash A]^{\swarrow A}[A\backslash S/C]^{\searrow A}[B/C]^{\searrow A}[B/C]^{\searrow A}C[B\backslash C][B\backslash C]$$

and the proof in Fig. 5.

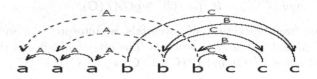

Fig. 4.

It is not difficult to prove:

$$
\frac{\dfrac{[A]^{\swarrow A}[A\backslash A]^{\swarrow A}}{\dfrac{[A]^{\swarrow A \swarrow A}}{[A]^{\swarrow A \swarrow A \swarrow A}}\,(\mathbf{L}^l)}\;\;[A\backslash A]^{\swarrow A}\;(\mathbf{L}^l)\quad [A\backslash S/C]^{\searrow A}\quad \dfrac{\dfrac{[B/C]^{\searrow A}C}{B^{\searrow A}}\,(\mathbf{L}^r)\;\;\dfrac{[B\backslash C]}{C^{\searrow A}}\,(\mathbf{L}^l)}{\dfrac{B^{\searrow A \searrow A}}{\dfrac{C^{\searrow A \searrow A}}{\dfrac{[A\backslash S]^{\searrow A \searrow A \searrow A}}{\dfrac{[S]^{\swarrow A \swarrow A \swarrow A \searrow A \searrow A \searrow A}}{S}\,(\mathbf{D}^l \times 3)}}\,(\mathbf{L}^l)}}
$$

Fig. 5.

Proposition 1. $L(G_{abc}) = \{a^n b^n c^n \mid n > 0\}$.

Remark 1. It should be noted that a type assignment to a string may have multiple correctness proofs. As a consequence, even a *rigid gCDG*, i.e. a gCDG assigning one type per word, may generate various DSs for the same string, as it is shown in the next example.

Example 2. The rigid gCDG G_r : $x \mapsto [S/S]$, $y \mapsto S$, $z \mapsto [S \backslash S]$ generating the regular language $x^* y z^*$, has two different proofs for $[S/S]S[S \backslash S] \vdash S$. As a result, the string xyz has the two DTs shown in Fig. 6.

Fig. 6.

An important particularity of gCDG is the property of independence of neutral and polarized valencies in the proofs, expressed using two *projections* of generalized categories.

Definition 5. *Local and* valency *projections* $\|\gamma\|_l$, $\|\gamma\|_v$ *are defined as follows:*
1. $\|\varepsilon\|_l = \|\varepsilon\|_v = \varepsilon$; $\|\alpha\gamma\|_l = \|\alpha\|_l \|\gamma\|_l$ *and* $\|\alpha\gamma\|_v = \|\alpha\|_v \|\gamma\|_v$ *for* $\alpha \in gCat(\mathbf{C})$ *and* $\gamma \in gCat(\mathbf{C})^*$.
2. $\|C^P\|_l = C$ *and* $\|C^P\|_v = P$ *for* $C^P \in gCAT(\mathbf{C})$.

To speak about "well-bracketing" of potentials, we interpret $\swarrow d$ and $\nearrow d$ as *left brackets* and $\nwarrow d$ and $\searrow d$ as *right brackets*. The sets of all left and right bracket valencies are denoted $V^l(\mathbf{C})$ and $V^r(\mathbf{C})$. $V(\mathbf{C}) =_{df} V^l(\mathbf{C}) \cup V^r(\mathbf{C})$.

Definition 6. *Pairs* $(\alpha, \breve{\alpha})$ *are called* correct. *For a dependency d and a potential P, let* $P \upharpoonright d$ *be the result of deleting the occurrences of all valencies but* $\swarrow d$, $\nearrow d$, $\nwarrow d$ *and* $\searrow d$. *Then P is* balanced *if* $P \upharpoonright d$ *is well bracketed in the usual sense for every d.*

This property can be incrementally checked using the following values.

Definition 7. *For a (neutral or polarized) valency v and a category projection* γ, $|\gamma|_v$ *will denote the number of occurrences of v in* γ. *For a potential P, a left-bracket valency* $v \in V^l(\mathbf{C})$, *and the dual right-bracket valency* $\breve{v} \in V^r(\mathbf{C})$,
$$\Delta_v(P) = max\{|P'|_v - |P'|_{\breve{v}} \mid P' \text{ is a suffix of } P\}$$
$$\Delta_{\breve{v}}(P) = max\{|P'|_{\breve{v}} - |P'|_v \mid P' \text{ is a prefix of } P\},$$
express respectively the deficit *of right and left v−brackets in P (i.e. the maximal number of right and left bracket v-valencies which need to be added to P on the right (left) so that it became balanced.[2]*

The following facts are easy to prove:

---
[2] Having in mind that there is $P' = \varepsilon$, the values $\Delta_{\breve{v}}(P)$ and $\Delta_v(P)$ are non-negative.

Lemma 1. *1. A potential P is balanced iff* $\sum\limits_{v \in V(\mathbf{C})} \Delta_v(P) = 0$.

2. For all potentials P_1, P_2, and every $v \in V^l(\mathbf{C}), \breve{v} \in V^r(\mathbf{C})$,

$$\Delta_v(P_1 P_2) = \Delta_v(P_2) + max\{\Delta_v(P_1) - \Delta_{\breve{v}}(P_2), 0\},$$
$$\Delta_{\breve{v}}(P_1 P_2) = \Delta_{\breve{v}}(P_1) + max\{\Delta_{\breve{v}}(P_2) - \Delta_v(P_1), 0\}$$

3. A potential P is balanced iff for every category α^P there is a proof $\alpha^P \vdash \alpha$ using only the rules $\mathbf{D}^l$ and $\mathbf{D}^r$.

Finally, we will denote by **c** the projective core of the generalized dependency calculus, consisting of the rules **L**, **I** and $\mathbf{\Omega}$. $\vdash_{\mathbf{c}}$ will denote the provability relation in this sub-calculus. Now we can state the *property of projections independence.*

Theorem 1. *Let $G = (W, \mathbf{C}, S, \delta)$ be a gCDG. $x \in L(G)$ iff there is a string of categories $\gamma \in \delta(x)$ such that:*
1. $\|\gamma\|_l \vdash_{\mathbf{c}} S$,
2. $\|\gamma\|_v$ is balanced.

Proof. The theorem is proved by induction on the proof length. We will prove $(\Rightarrow)$, the inverse being similar. Let $x \in L(G)$ due to an assignment $\delta : x \mapsto \gamma \in gCat(\mathbf{C})^*$ and a proof $\gamma \vdash S$. Let n be the length of this proof.
I. $n = 0$. Then $\gamma = \|\gamma\|_l = S$, $\|\gamma\|_v = \varepsilon$ and the statement is trivially true.
II. $n > 0$. Then the proof has the form $\gamma \vdash^R \gamma' \vdash S$, where R is the first rule applied in the proof.
Case 1. $R = \mathbf{D}$. In this case, R does not affect the local projection. So $\|\gamma\|_l = \|\gamma'\|_l$ and, therefore, $\|\gamma\|_l \vdash_{\mathbf{c}} S$ by induction hypothesis. On the other hand, $\|\gamma'\|_v$ results from $\|\gamma\|_v$ by elimination of a correct pair of polarized valencies satisfying the **FA**-principle. This means that $\|\gamma'\|_v$ is balanced iff $\|\gamma\|_v$ is so.
Case 2. $R \neq \mathbf{D}$. In this case, R does not affect the valency projection. So $\|\gamma\|_v = \|\gamma'\|_v$ and, therefore, $\|\gamma\|_v$ is balanced. On the other hand, $\gamma' \vdash S$ implies $\|\gamma'\|_l \vdash_{\mathbf{c}} S$ by induction hypothesis. So $\|\gamma\|_l \vdash^R \|\gamma'\|_l \vdash_{\mathbf{c}} S$. $\qquad\square$

4 Expressive Power of GCDG

gCDG are very expressive. The Example 1 shows that they can generate non-CF languages. In fact, they have the same weak expressive power as the *Dependency Structure Grammars* (DSG), a class of generating rule based dependency grammars introduced in [9,10] and simplified and studied in [3]. Below we cite the key definitions from [3].

The DSG use generalized DS over a mixed vocabulary of *terminals* W and *nonterminals* N. In these DS, one connected component[3] is selected as *head component* and some node in this component is selected as *DS head*. We will call *headed* the DS with such selection (hDS). In the two-component hDS in

---

[3] Slightly abusing the standard graph-theoretic terminology, we call *connected component* of a DS D any its maximal subgraph corresponding to connected components of the *non-oriented* graph resulting from D after cancellation of its arcs' orientation.

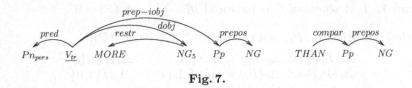

Fig. 7.

Fig. 7, the underlined node is head. The following composition $D[v \backslash D_1]$ (and simultaneous composition $D[v_1, \ldots, v_n \backslash D_1, \ldots, D_n]$) is defined on hDS.

Definition 8. *Let* $\delta_1 = \{D_0, D_1, \ldots, D_k\}$ *be a hDS. Let a nonterminal A have an occurrence in* δ_1: $w(\delta_1) = xAy$ *and* δ_2 *be a hDS with the head* n_0. *Then the composition of* δ_2 *into* δ_1 *in the selected occurrence of A, denoted* $\delta_1[A \backslash \delta_2]$, *is the hDS* δ *resulting from the union of* δ_1 *and* δ_2 *by unifying A and* n_0 *and by defining the order and labeling by the string substitution of* $w(\delta_2)$ *in the place of A in* $w(\delta_1)$. *Formally:*

1. $nodes(\delta) =_{df} (nodes(\delta_1) - \{A\}) \cup nodes(\delta_2)$.
2. $arcs(\delta) =_{df} arcs(\delta_2) \cup (arcs(\delta_1) - \{d \in arcs(\delta_1) | \exists n(d = (A, n) \lor d = (n, A))\}) \cup \{(n_0, n) | \exists n((A, n) \in arcs(\delta_1))\} \cup \{(n, n_0) | \exists n((n, A) \in arcs(\delta_1))\}$.
3. *The order of nodes(δ) is uniquely defined by equation* $w(\delta) = xw(\delta_2)y$.

$$D_1 = A \;\; \underline{B} \qquad D_2 = a \;\;\;\; A \qquad D_3 = b \;\;\;\; B \;\;\; c$$

$$D_1[A, B \backslash D_2[A \backslash D_2], D_3[B \backslash D_3]] = a \;\;\; a \;\;\; A \;\;\; b \;\;\; b \;\;\; \underline{B} \;\;\; c \;\;\; c$$

Fig. 8.

In Fig. 8 is shown an example of such composition.[4] The FA-principle is used in DSG in the form of valency neutralization:

Definition 9. *For potentials* $\Gamma = \Gamma_1 v \Gamma_2 \breve{v} \Gamma_3$ *and* $\Gamma' = \Gamma_1 \Gamma_2 \Gamma_3$ *such that* $v = (\nearrow A)$, $\breve{v} = (\searrow A)$ *or* $v = (\swarrow A), \breve{v} = (\nwarrow A)$, *v is* neutralized *by* $\breve{v}$ *in* Γ *(denoted* $\Gamma \twoheadrightarrow_{FA} \Gamma'$*) if* Γ_2 *has no occurrences of v and* $\breve{v}$. *This reduction of potentials* $\twoheadrightarrow_{FA}$ *is terminal and confluent. So each potential* Γ *has a unique FA-normal form denoted* $[\Gamma]_{FA}$. *The product* $\odot$ *of potentials defined by:* $\Gamma_1 \odot \Gamma_2 =_{df} [\Gamma_1 \Gamma_2]_{FA}$ *is clearly associative. So we obtain the* monoid of potentials $\mathbf{P} = (Pot(\mathbf{C}), \odot)$ *with the unit* ε.

Definition 10. *A* **Dependency Structure Grammar** (DSG) *G has the rules* $r = (A \to D)$ *with* $A \in N$ *and hDS D with assigned potentials:* $[\Gamma_X^L]X[\Gamma_X^R]$

---

[4] We use nonterminals $label(v)$ in the place of v when no conflicts.

(the left and right potentials Γ_X^L and Γ_X^R may be assigned to each nonterminal X in D [5]).

Derivation trees of G result from the derivation trees T of the cf-grammar $\{A \to w(D) \mid A \to D \in G\}$ by defining potentials $\pi(T,n)$ of nodes n:
1. *$\pi(T,n) = \varepsilon$ for every terminal node n;*
2. *$\pi(T,n) = \Gamma_1 \odot \ldots \odot \Gamma_k$, for every node n with sons $n_1, \ldots, n_k$ derived by rule $r = (A \to D)$, in which $w(D) = X_1 \ldots X_k$ and $\Gamma_i = \Gamma_i^L \odot \pi(T,n_i) \odot \Gamma_i^R$, where $[\Gamma_i^L]X_i[\Gamma_i^R]$ are the rule potential assignments. A hDS is generated in the node n by the composition: $hDS(T,n) = D[X_1 \ldots X_k \backslash hDS(T,n_1), \ldots, hDS(T,n_k)]$. Every pair of dual valencies neutralized at this step corresponds to a discontinuous dependency added to this hDS.*

A derivation tree T is complete if the potential of its root S is neutral: $\pi(T,S) = \varepsilon$. We set $G(D,w)$ if there is a complete derivation tree T of G from the axiom S such that $D = hDS(T,S)$ and $w = w(D)$.

$\Delta(G) = \{D \mid \exists w \in W^+ \; G(D,w)\}$ *is the DS-language generated by G.*

$L(G) = \{w \in W^+ \mid \exists D \; G(D,w)\}$ *is the language generated by G.*

For instance, the following four-rule DSG:

$G_1: \quad S \to a[\swarrow a] \; \underline{S} \quad \mid \quad A \quad c \qquad A \to [\nwarrow a]b \; A \quad c \quad \mid \quad [\nwarrow a]b$

generates the language $L(G_1) = \{a^n b^n c^n \mid n > 0\}$. Its complete derivation tree of the string $a^3 b^3 c^3$ is shown in Fig. 9.

Clearly, $\mathcal{L}(CF) \subseteq \mathcal{L}(DSG)$. So this example shows that $\mathcal{L}(CF) \subsetneq \mathcal{L}(DSG)$. In [3] it is shown that DSG have Greibach normal form. Using this fact, it is shown that $\mathcal{L}(DSG) \subseteq \mathcal{L}(gCDG)$.[6] On the other hand, it is also proved that $\mathcal{D}(gCDG) \subseteq \mathcal{D}(DSG)$. In particular, this means that the gCDG and the DSG have the same weak generative power:

Theorem 2. *[3] $\mathcal{L}(CF) \subsetneq \mathcal{L}(gCDG) = \mathcal{L}(DSG)$.*

In [9] a measure of discontinuity σ was defined which will be called *valency deficit*. Intuitively, its value is the maximal potential size in a derivation. For instance, for the gCDG it is defined as follows.

Definition 11. *Let $G = (W, \mathbf{C}, S, \delta)$ be a gCDG. For a proof $p = (\Gamma \vdash S)$, where $\Gamma \in \delta(w)$ and $w \in W^+$, its valency deficit $\sigma(\Gamma, p)$ is the maximal size of a potential used in p. $\sigma_G(w)$ is the minimal value of $\sigma(\Gamma, p)$ among all $\Gamma \in \delta(w)$. Finally, $\sigma_G(n) = max\{\sigma_G(w) \mid |w| \leq n\}$.*

The examples of gCDG G_{abc} and DSG G_1 show that the valency deficit of these grammars cannot be bounded by a constant. As it is shown in [9], the dependency grammars with bounded valency deficit generate CF-languages. This theorem can be easily extended to gCDG and DSG. Let $\mathcal{L}^{\sigma < const}(gCDG)$ and $\mathcal{L}^{\sigma < const}(DSG)$ denote the classes of languages generated respectively by gCDG and DSG with bounded valency deficit.

---

[5] For instance, $A \to [\nwarrow d_1]\underline{B}[\nearrow d_2] \; C$ denotes the rule $A \to \underline{B} \; C$ with assignment $[\nwarrow d_1]B[\nearrow d_2]$. We omit empty potentials.

[6] In [3] is used an equivalent notational variant of gCDGs.

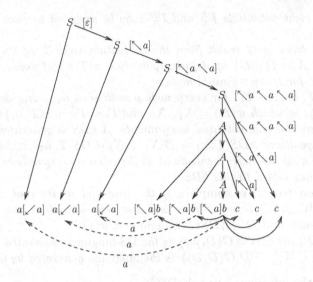

Fig. 9.

Theorem 3. $\mathcal{L}^{\sigma<const}(gCDG) = \mathcal{L}^{\sigma<const}(DSG) = \mathcal{L}(CF)$.

Let us consider some more examples.

Example 3. Let $W = \{a_1, \ldots, a_m\}$ and $L^{(m)} = \{a_1^n a_2^n \ldots a_m^n \mid n \geq 1\}$. Let us consider the gCDG

$$
G^{(m)} : \begin{cases}
a_1 \mapsto [S/A_1]^{\nearrow A_2}, [A_1/A_1]^{\nearrow A_2}, [A_1/A_2]^{\nearrow A_2}, \\
\ldots \quad \ldots \quad \ldots \\
a_i \mapsto [A_i/A_i]^{\searrow A_i \nearrow A_{i+1}}, [A_i/A_{i+1}]^{\searrow A_i \nearrow A_{i+1}}, 2 \leq i < m, \\
\ldots \quad \ldots \quad \ldots \\
a_m \mapsto [A_m/A_m]^{\searrow A_m}, [A_m]^{\searrow A_m}
\end{cases}
$$

It is not difficult to prove the proposition

Proposition 2. $L(G^{(m)}) = L^{(m)}$ for all $m \geq 2$.

Meanwhile, as it is well known, the languages $L^{(m)}$ are mild context sensitive and cannot be generated by basic TAGs starting from $m > 4$ (see [19]).

Example 4. Let us consider the language MIX consisting of all permutations of the strings $a^n b^n c^n, n > 0 : MIX = \{w \in \{a,b,c\}^+ \mid |w|_a = |w|_b = |w|_c\}$. Emmon Bach conjectures that MIX is not a mild CS language. At the same time, this language is generated by the following gCDG:

gCDG G_{MIX}		
left	right	middle
$a \mapsto [S]^{\searrow B \searrow C}$	$a \mapsto [S]^{\nearrow C \nearrow B}$	$a \mapsto [S]^{\searrow B \nearrow C}, [S]^{\searrow C \nearrow B}$
$a \mapsto [S \setminus S]^{\searrow B \searrow C}$	$a \mapsto [S \setminus S]^{\nearrow C \nearrow B}$	$a \mapsto [S \setminus S]^{\searrow B \nearrow C}, [S \setminus S]^{\searrow C \nearrow B}$
$b \mapsto [\varepsilon]^{\nearrow B}$	$b \mapsto [\varepsilon]^{\searrow B}$	
$c \mapsto [\varepsilon]^{\nearrow C}$	$c \mapsto [\varepsilon]^{\searrow C}$	

Proposition 3. *[3]. $L(G_{MIX}) = MIX$.*

As it is well known, the copy language $L_{copy} = \{ww \mid w \in \{a,b\}^+\}$ is generated by a basic TAG. On the other hand, it is conjectured in [12,3] that L_{copy} cannot be generated by gCDG and DSG. As we will see below, the gCDG-language are parsed in polynomial time. This means that this family of grammars represents an interesting alternative for the mild CS grammars (see the diagram in Fig. 10 presenting a comparison of the two families in weak generative power). The gCDG-languages have good operation closure properties. In particular, they form an AFL. To show this fact, we need some preliminary propositions.

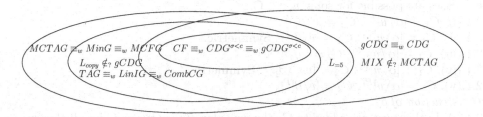

Fig. 10.

Lemma 2. *For each gCDG G there is a weakly equivalent gCDG G' in which the axiom type S is not an argument subtype of a category.*

Proof. Otherwise, just add a new axiom S' and double the categories $[\alpha \backslash S / \beta]$ with new categories $[\alpha \backslash S' / \beta]$. □

Lemma 3. *For each gCDG G there is a weakly equivalent gCDG G' in which there are no categories with empty value type: $[\alpha \backslash \varepsilon / \beta]^P$.*

Proof. If there is one: $t = [\alpha \backslash \varepsilon / \beta]^P$, then add a new elementary type d_t, replace t in all type assignments with the new category $t' = [\alpha \backslash d_t / \beta]^P$ and then, in the resulting grammar, substitute the new category $[d_t * \backslash A_l \backslash \dots A_1 \backslash d_t * \backslash V / d_t * / B_k \dots / B_1 / d_t *]^{P_1}$ for each category $[A_l \backslash \dots A_1 \backslash V / B_k \dots / B_1]^{P_1}$. Regardless of the fact that the resulting gCDG is greater than G, it is weakly equivalent to G and has one empty value type less. □

Theorem 4. *The family $\mathcal{L}(gCDG)$ is an AFL.*

Proof. We suppose that gCDGs satisfy the conditions of Lemmas 2,3.
1. $\mathcal{L}(gCDG)$ is closed under ε-free homomorphisms. Let $G = (W, \mathbf{C}, S, \delta)$ be a gCDG with $W = \{a_1, \dots, a_n\}$ and $h : W \to X^+$ be a homomorphism such that $h(a_i) = x_{i0} \dots x_{im_i}, m_i \geq 0, x_{ij} \in X$, for all $0 \leq j \leq m_i, 1 \leq i \leq n$. The new gCDG G_h keeps all elementary types of G including S which is also its axiom. Besides them, it has a new elementary type d_{ij} for all $0 \leq j \leq m_i, 1 \leq i \leq n$. Its lexicon δ_h is defined as follows: if $h(a_i) = x_{i0} \dots x_{im_i}$, then for every category $\alpha \in \delta(a_i)$, it has the assignment $\delta_h : x_{i0} \mapsto [\alpha / d_{im_i} / \dots / d_{i1}]$. In particular, $\delta_h : x_{i0} \mapsto \alpha$,

if $m_i = 0$. Besides that, there are also the assignments $\delta_h : x_{ij} \mapsto d_{ij}$ for all $1 \le j \le m_i, 1 \le i \le n$. Clearly, $L(G_h) = h(L(G))$. $\qquad\qquad\square$

2. $\mathcal{L}(gCDG)$ is closed under the inverses of homomorphisms. First of all, let us remark that to prove this proposition it suffices to prove it for the homomorphisms $h : X \to W^*$, $X \cap W = \emptyset$, differing from a bijection $h : X \leftrightarrow W$ by no more than one assignment which is either of the form $h(x) = ab$, $a, b \in W$, or of the form $h(x) = \varepsilon$. Let $G = (W, \mathbf{C}, S, \delta)$ be the original gCDG and gCDG $G_{h^{-1}} = (X, \mathbf{C}_1, S, \delta_{h^{-1}})$ the gCDG to construct.

2.1. Let $h(x) = ab$, $a, b \in W$, $C_1 \in \delta(a)$ and $C_2 \in \delta(b)$. Then $\delta_{h^{-1}} = \delta \cup \delta_x$, where δ_x is defined below depending on the form of categories C_1, C_2. The following five cases are possible for some $u, v \in \mathbf{C}$:

2.1.(i). $C_1 = [\alpha/u]^{P_1}$, $C_2 = [v/\beta]^{P_2}$.

2.1.(ii). $C_1 = [\alpha\backslash u]^{P_1}$, $C_2 = [v\backslash\beta]^{P_2}$. Symmetric.

2.1.(iii). $C_1 = [\alpha/u*]^{P_1}$, $C_2 = [v/\beta]^{P_2}$.

2.1.(iv). $C_1 = [\alpha\backslash u]^{P_1}$, $C_2 = [v * \backslash\beta]^{P_2}$. Symmetric.

2.1.(v). $C_1 = [\alpha\backslash u]^{P_1}$, $C_2 = [v/\beta]^{P_2}$.

Construction of δ_x:

2.1.(i). In this case, are added to $\mathbf{C}_1$ the new elementary types d_v for all elementary types $d \in \mathbf{C}$ and is added to δ_x the assignment:

$x \mapsto [\alpha/\beta]^{P_1 P_2}$ if $u = v$, and whatever are u, v ($u = v$ included), are also added the following assignments:

$x \mapsto [\alpha/u_v/\beta]^{P_1 P_2}$,

$y \mapsto [\alpha'\backslash d_v/\beta']^P$ for each assignment $y \mapsto [\alpha'\backslash v/d/\beta']^P \in \delta$,

$y \mapsto [\alpha'\backslash e_v\backslash d_v/\beta']^P$ for each assignment $y \mapsto [\alpha'\backslash e\backslash d/\beta']^P \in \delta$.

2.1.(iii). The only difference with the preceding case is in the form of the second assignment:

$x \mapsto [\alpha/u_v * /\beta]^{P_1 P_2}$.

2.1.(v). In this case, three subsets are added to δ : $\delta_{x,fork}$, $\delta_{x,left}$ and $\delta_{x,right}$. We will construct $\delta_{x,fork}$ and $\delta_{x,right}$ ($\delta_{x,left}$ is symmetric to $\delta_{x,right}$).

Construction of $\delta_{x,fork}$:

Are added to $\mathbf{C}_1$ new elementary types: f_{lr}, $d_{a(b)}$ and $d_{(a)b}$ for all elementary types $l, r, d \in \mathbf{C}$, and are added the following assignments:

$x \mapsto [\alpha\backslash l_{a(b)}\backslash f_{lr}/r_{(a)b}/\beta]^{P_1 P_2}$ for all $l, r \in \mathbf{C}$,

$y \mapsto [\alpha'\backslash d_{a(b)}/\beta']^P$ for each assignment $y \mapsto [\alpha'\backslash d/u/\beta']^P \in \delta$,

$y \mapsto [\alpha'\backslash d_{(a)b}/\beta']^P$ for each assignment $y \mapsto [\alpha'\backslash v\backslash d/\beta']^P \in \delta$,

$y \mapsto [\alpha'\backslash d_{a(b)}/e_{a(b)}/\beta']^P$ for each assignment $y \mapsto [\alpha'\backslash d/e/\beta']^P \in \delta$,

$y \mapsto [\alpha'\backslash e_{(a)b}\backslash d_{(a)b}/\beta']^P$ for each assignment $y \mapsto [\alpha'\backslash e\backslash d/\beta']^P \in \delta$,

$y \mapsto [\alpha'/f_{lr}/\beta']^P$ for each assignment $y \mapsto [\alpha'/r/l/\beta']^P \in \delta$, where $\alpha' \ne \varepsilon$,

$y \mapsto [\alpha'\backslash f_{lr}\backslash\beta']^P$ for each assignment $y \mapsto [\alpha'\backslash r\backslash l\backslash\beta']^P \in \delta$, where $\beta' \ne \varepsilon$,

$y \mapsto [\alpha'/f_{dd} * /\beta']^P$ for each assignment $y \mapsto [\alpha'/d * /\beta']^P \in \delta$, where $\alpha' \ne \varepsilon$ and $d = l = r$,

$y \mapsto [\alpha'\backslash f_{dd} * \backslash\beta']^P$ for each assignment $y \mapsto [\alpha'\backslash d * \backslash\beta']^P \in \delta$, where $\beta' \ne \varepsilon$ and $d = l = r$.

Construction of $\delta_{x,right}$:

For all $d \in \mathbf{C}$, are added to $\mathbf{C}_1$ new elementary types: $d_{(\varepsilon)ab}$ and $d_{(\varepsilon)a|b}$ (symmetric types $d_{ab(\varepsilon)}$ and $d_{a|b(\varepsilon)}$ for $\delta_{x,left}$), and are added the following assignments:

$x \mapsto [\alpha \backslash d_{(\varepsilon)ab}/\beta]^{P_1 P_2}$ for all $d \in \mathbf{C}$ ($x \mapsto [\alpha \backslash d_{ab(\varepsilon)}/\beta]^{P_1 P_2}$ for $\delta_{x,left}$),

$y \mapsto [\alpha' \backslash d_{(\varepsilon)ab}/\beta']^P$ for each assignment $y \mapsto [\alpha' \backslash v \backslash u \backslash d/\beta']^P \in \delta$,

$y \mapsto [\alpha' \backslash e_{(\varepsilon)ab} \backslash d_{(\varepsilon)ab}/\beta']^P$ for each assignment $y \mapsto [\alpha' \backslash e \backslash d/\beta']^P \in \delta$,

$y \mapsto [\alpha' \backslash d_{(\varepsilon)a|b}/\beta']^P$ for each assignment $y \mapsto [\alpha' \backslash v \backslash d/\beta']^P \in \delta$,

$y \mapsto [\alpha' \backslash e_{(\varepsilon)a|b} \backslash d_{(\varepsilon)a|b}/\beta']^P$ for each assignment $y \mapsto [\alpha' \backslash e \backslash d/\beta']^P \in \delta$,

$y \mapsto [\alpha' \backslash e_{(\varepsilon)a|b} \backslash d_{(\varepsilon)ab}/\beta']^P$ for each assignment $y \mapsto [\alpha' \backslash u \backslash d/\beta']^P \in \delta$,

$y \mapsto [\alpha' \backslash d_{(\varepsilon)ab} \backslash \beta']^P$ for each assignment $y \mapsto [\alpha' \backslash d \backslash \beta']^P \in \delta$,

$y \mapsto [\alpha'/d_{(\varepsilon)ab}/\beta']^P$ for each assignment $y \mapsto [\alpha'/d/\beta']^P \in \delta$,

$y \mapsto [\alpha' \backslash d_{(\varepsilon)ab} * \backslash \beta']^P$ for each assignment $y \mapsto [\alpha' \backslash d * \backslash \beta']^P \in \delta$,

$y \mapsto [\alpha'/d_{(\varepsilon)ab} * /\beta']^P$ for each assignment $y \mapsto [\alpha'/d * /\beta']^P \in \delta$.

2.2. Let $h(x) = \varepsilon$. Then $\mathbf{C}_1 = \mathbf{C} \cup \{d_x\}$, $\delta_{h^{-1}}(x) = d_x$ and each assignment $\delta : c \mapsto [l_m \backslash \ldots \backslash l_1 \backslash v/r_1/ \ldots /r_n]$, $c \in W$, is replaced by the assignment $\delta_{h^{-1}} : c \mapsto [d_x * \backslash l_m \backslash \ldots \backslash l_1 \backslash d_x * \backslash v/d_x * /r_1/ \ldots /r_n/d_x *]$.   □

3. $\mathcal{L}(gCDG)$ is closed under intersection with regular languages. Let $G = (W, \mathbf{C}, S, \delta)$ be a gCDG and A be a FA with states $Q = \{q_{in}, q_1, ..., q_k, q_{fin}\}$, where q_{in} is initial, $q_{fin} \neq q_{in}$ is final and in every transition $a\ q \rightarrow q'$, $a \in W, q \neq q_{fin}$ and $q' \neq q_{in}$. Then in the new gCDG $G_A = (W, \mathbf{C}_A, S, \delta_A)$ $\mathbf{C}_A = \mathbf{C} \cup Q$ and for every assignment $\delta : a \mapsto C^P$ and every transition $a\ q \rightarrow q'$, the new assignments δ_A are defined as $\delta_A : a \mapsto C^{P P_1 P_2}$, where:

$$P_1 = \begin{cases} (\nearrow q_{in})(\nearrow q'), & if\ q = q_{in} \\ (\searrow q)(\searrow q_{fin}), & if\ q = q_{fin} \\ (\searrow q)(\nearrow q'), & otherwise \end{cases} \qquad P_2 = \begin{cases} (\searrow q_{in})(\nearrow q_{fin}), & if\ C = [\alpha \backslash S/\beta] \\ \varepsilon, & otherwise \end{cases} \quad □$$

4-6. The proofs of closure under union, concatenation and Kleene + are standard and obvious.   □

5   Categorial Dependency Grammars

Generalized CDG are useful for formal study of the grammars and languages. However, they are not flexible enough for designing real application grammars. Their main drawback is that in order to fix the exact position of a distant subordinate, one needs to violate the tree-likeness of the DS (cf. the DS in Fig. 4). In order to eliminate this defect, we will use two more valency types: that of the *anchored* distant subordinate $\#$ and that of the host word $\flat$. When a distant right subordinate s through a dependency d should be positioned immediately on the left of its host word h, the latter must have in its type the argument $\flat^l(\searcher d)$: $h \mapsto [\flat^l(\searcher d) \backslash \beta]$ whereas the former must have the value type $\#^l(\searcher d)$: $s \mapsto [\beta_1 \backslash \#^l(\searcher d)/\beta_2]$. After the argument valencies β_1 and β_2 of s will be saturated, the value type $\#^l(\searcher d)$ becomes adjacent to the category $[\flat^l(\searcher d) \backslash \beta]$, the host argument $\flat^l(\searcher d)$ of this category is eliminated and the polarized valency $\searcher d$ loses its anchor marker and falls under the **FA**-principle. These new types need a change in the dependency calculus. Below we present an extended calculus we call *sub-commutative*. The new DGs using this calculus will be called *Categorial Dependency Grammars* (CDG). We constrain the dependency types in order that the CDG generate only the DSs *with the single governor per word*.

Type constraints: In the categories $[L_1 \setminus \cdots \setminus L_i \setminus C / R_j / \cdots / R_1]$:
(i)   the value type C can be neutral, or negative ($\swarrow C, \searrow C$) or anchored
($\#^l(\swarrow C), \#^r(\swarrow C), \#^l(\searrow C), \#^r(\searrow C)$),
(ii) the argument types L_i, (R_j) can be neutral or positive ($\searrow C, \nearrow C$) or host
($\flat^l(\swarrow C), \flat^l(\searrow C)$, respectively $\flat^r(\swarrow C), \flat^r(\searrow C)$).

Definition 12. Sub-commutative dependency calculus.[7] *[8]*
$\mathbf{L^1}$. $C[C \setminus \beta] \vdash [\beta]$
$\mathbf{I^1}$. $C[C^* \setminus \beta] \vdash [C^* \setminus \beta]$
$\mathbf{\Omega^1}$. $[C^* \setminus \beta] \vdash [\beta]$
$\mathbf{V^1}$. $[\alpha \setminus \beta] \vdash \alpha[\beta], \alpha \in \{(\searrow C), \flat^l(\swarrow C), \flat^l(\searrow C)\}$
$\mathbf{A^1}$. $\#^l(\alpha)\flat^l(\alpha) \vdash \alpha, \alpha \in \{(\swarrow C), (\searrow C)\}$
$\mathbf{C^1}$. $\alpha\beta \vdash \beta\alpha, \alpha \in \{(\swarrow C), (\searrow C), (\nearrow C), (\searrow C)\}$, *where*
     $\beta = \flat(v)$ or β *has no occurrences of* $\alpha, \breve{\alpha}, \#(\alpha), \flat(\alpha)$
$\mathbf{D^1}$. $(\swarrow C)(\searrow C) \vdash \varepsilon$

In this calculus, the new rule $\mathbf{D^1}$ creates a discontinuous dependency for *adjacent* dual dependencies. At the same time, the rule $\mathbf{C^1}$ permutes the polarized valencies with other types when the permutation does not violate the **FA**-principle. The rule $\mathbf{V^1}$ decomposes complex types and the rule $\mathbf{A^1}$ eliminates the anchor markers if the corresponding anchor and host types are adjacent.

Remark 2. In contrast with the CDGs of [12], which generate only DTs, the DSs generated by the sub-commutative CDGs may have cycles, as shows the following example:

$[(\swarrow A)/(\nearrow B)]_1[(\searrow A)\setminus(\searrow B)]_2 S_3 \vdash (\swarrow A)_1(\nearrow B)_1[(\searrow A)\setminus(\searrow B)]_2 S_3 \vdash (\swarrow A)_1(\nearrow B)_1(\searrow A)_2(\searrow B)_2 S_3 \vdash (\swarrow A)_1(\searrow A)_2(\nearrow B)_1(\searrow B)_2 S_3 \vdash S_3$.

A simple sufficient condition of acyclicity of CDGs can be formulated in terms of a well-founded order on dependency types. But more important is that these CDGs can naturally express adjacency of distant subordinates without violation of the single governor condition.

Example 5

$$G_2 = \begin{cases} a \mapsto \#^l(\swarrow A), \ [\flat^l(\swarrow A)\setminus\#^l(\swarrow A)] \\ b \mapsto [(\searrow A)\setminus B \ / \ C], \ [\flat^l(\swarrow A)\setminus(\searrow A)\setminus S/C] \\ c \mapsto C, \ [B \setminus C] \end{cases}$$

It is not difficult to prove that $L(G_2) = \{a^n b^n c^n \mid n > 0\}$. For instance, $a^3 b^3 c^3 \in L(G_2)$ due to the types assignment:

$a^3 b^3 c^3 \mapsto \#^l(\swarrow A)[\flat^l(\swarrow A)\setminus\#^l(\swarrow A)][\flat^l(\swarrow A)\setminus\#^l(\swarrow A)][\flat^l(\swarrow A)\setminus(\searrow A)\setminus S/C][(\searrow A)\setminus B/C][(\searrow A)\setminus B/C]C[B\setminus C][B\setminus C]$

and the proof shown in Fig. 11. This proof determines the DT shown in Fig. 12.

---

[7] We show only left rules. The right rules are symmetric.

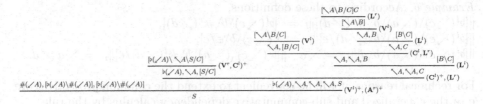

Fig. 11.

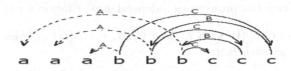

Fig. 12.

CDGs also enjoy the property of projections independence. The *local* projection of CDGs, preserves elementary and host argument subtypes. Intuitively, it reflects not only projective dependencies of words, but also their adjacency through anchor-host relations. The *valency projection* of CDGs, similar to that of gCDGs, preserves only polarized valency argument and value subtypes. Intuitively, it reflects only discontinuous dependencies.

Let $V^-(\mathbf{C})$ and $V^+(\mathbf{C})$ denote the sets of all negative polarized types: $\swarrow d$, $\searrow d$ (respectively, all positive polarized types: $\searrow d$, $\nearrow d$), where $d \in \mathbf{C}$. We set:

$Host^l(\mathbf{C}) =_{df} \{\flat^l(\alpha) \mid \alpha \in V^-(\mathbf{C})\}$, $Anc^l(\mathbf{C}) =_{df} \{\#^l(\alpha) \mid \alpha \in V^-(\mathbf{C})\}$,

$Host^r(\mathbf{C}) =_{df} \{\flat^r(\alpha) \mid \alpha \in V^-(\mathbf{C})\}$, $Anc^r(\mathbf{C}) =_{df} \{\#^r(\alpha) \mid \alpha \in V^-(\mathbf{C})\}$,

$Host(\mathbf{C}) =_{df} Host^l(\mathbf{C}) \cup Host^r(\mathbf{C})$, $Anc(\mathbf{C}) =_{df} Anc^l(\mathbf{C}) \cup Anc^r(\mathbf{C})$.

Definition 13. Local projection $\|\gamma\|_l$ of $\gamma \in Cat(\mathbf{C})^*$ is defined as follows:

l1. $\|\varepsilon\|_l = \varepsilon$; $\|C\gamma\|_l = \|C\|_l \|\gamma\|_l$ for $C \in Cat(\mathbf{C})$ and $\gamma \in Cat(\mathbf{C})^*$.

l2. $\|C\|_l = C$ for $C \in \mathbf{C} \cup \mathbf{C}^* \cup Anc(\mathbf{C})$.

l3. $\|C\|_l = \varepsilon$ for $C \in V^+(\mathbf{C}) \cup V^-(\mathbf{C})$.

l4. $\|[\alpha]\|_l = \|\alpha\|_l$ for all $\alpha \in Cat(\mathbf{C})$.

l5. $\|[a\backslash\alpha]\|_l = [a\backslash \|\alpha\|_l]$ and $\|[\alpha/a]\|_l = [\|\alpha\|_l/a]$ for $a \in \mathbf{C} \cup \mathbf{C}^* \cup Host(\mathbf{C})$ and $\alpha \in Cat(\mathbf{C})$.

l6. $\|[(\searrow a)\backslash\alpha]\|_l = \|[\alpha/(\nearrow a)]\|_l = \|\alpha\|_l$ for all $a \in \mathbf{C}$ and $\alpha \in Cat(\mathbf{C})$.

Valency projection $\|\gamma\|_v$ of $\gamma \in Cat(\mathbf{C})^*$ is defined as follows:

v1. $\|\varepsilon\|_v = \varepsilon$; $\|C\gamma\|_v = \|C\|_v \|\gamma\|_v$ for $C \in Cat(\mathbf{C})$ and $\gamma \in Cat(\mathbf{C})^*$.

v2. $\|C\|_v = \varepsilon$ for $C \in \mathbf{C} \cup \mathbf{C}^*$.

v3. $\|C\|_v = C$ for $C \in V^+(\mathbf{C}) \cup V^-(\mathbf{C})$.

v4. $\|\#(C)\|_v = C$ for $C \in V^-(\mathbf{C})$.

v5. $\|[\alpha]\|_v = \|\alpha\|_v$ for all $[\alpha] \in Cat(\mathbf{C})$.

v6. $\|[a\backslash\alpha]\|_v = \|[\alpha/a]\|_v = \|\alpha\|_v$ for $a \in \mathbf{C} \cup \mathbf{C}^* \cup Host(\mathbf{C})$.

v7. $\|[a\backslash\alpha]\|_v = a \|\alpha\|_v$, if $a \in V^+(\mathbf{C})$.

v8. $\|[\alpha/a]\|_v = \|\alpha\|_v \, a$, if $a \in V^+(\mathbf{C})$.

Example 6. According to these definitions,

$$\|[b^l(\diagdown c)\backslash(\diagdown a)\backslash b\backslash \#^r(\diagup d)]\|_l = [b^l(\diagdown c)\backslash b\backslash \#^r(\diagup d)],$$
$$\|[b^l(\diagdown c)\backslash(\diagdown a)\backslash b\backslash(\diagup d)/e]\|_l = [b^l(\diagdown c)\backslash b\backslash \varepsilon/e],$$
$$\|[b^l(\diagdown c)\backslash(\diagdown a)\backslash b\backslash d]\|_v = \diagdown a, \quad \|[b^l(\diagdown c)\backslash(\diagdown a)\backslash b\backslash \#^l(\diagup d)/e]\|_v = \diagdown a \diagup d.$$

For technical reasons, it will be convenient to extend the common projective core $\mathbf{c}$ of the generalized and sub-commutative dependency calculus by the rule
$\mathbf{E}^l.$ $\#^l(\alpha)[b^l(\alpha)\backslash\beta] \vdash \beta$ for $\alpha \in V^-(\mathbf{C})$.
The resulting extension will be denoted by $\mathbf{p}$ and the corresponding provability relation will be denoted by $\vdash_{\mathbf{p}}$.
Now we can state the projections independence criterion for the CDGs.

Theorem 5. *Let $G = (W, \mathbf{C}, S, \delta)$ be a CDG. $x \in L(G)$ iff there is a string of categories $\gamma \in \delta(x)$ such that:*
1. $\|\gamma\|_l \vdash_{\mathbf{p}} S$,
2. $\|\gamma\|_v$ is balanced.

Proof. Evidently, we can ignore the dependencies.
($\Rightarrow$) Let $x \in L(G)$ and $\delta : x \mapsto \gamma$ be an assignment for which there exists a proof $\gamma \vdash \ldots \vdash \gamma^k \vdash \gamma^n = S$ for some $n \geq 0$. We will prove by induction on k that for each $0 \leq k \leq n$ the following two assertions hold:
(i) $\|\gamma\|_l \vdash_{\mathbf{p}} \|\gamma^k\|_l$,
(ii) each correct pair $(\alpha, \check{\alpha})$ is eliminated in $\|\gamma^k\|_v$ iff it is eliminated in $\|\gamma\|_v$.

Let us suppose that the conditions (i) and (ii) are satisfied for some $k < n$ and prove that they will be satisfied for $k + 1$ as well.
Let $\gamma^k \vdash^R \gamma^{k+1}$ (immediately derived by rule R).
If $R = \mathbf{L}^l$, then $\gamma^k = \Gamma_1 C[C\backslash\beta]\Gamma_2$ and $\gamma^{k+1} = \Gamma_1\beta\Gamma_2$ for some Γ_1, Γ_2 and β. Passing to their local projections, we obtain: $\|\gamma^k\|_l = \|\Gamma_1\|_l C[C\backslash\|\beta\|_l]\|\Gamma_2\|_l \vdash_{\mathbf{p}}$ $\|\Gamma_1\|_l\|\beta\|_l\|\Gamma_2\|_l = \|\gamma^{k+1}\|_l$ and $\|\gamma^k\|_v = \|\gamma^{k+1}\|_v$. Then $\|\gamma\|_l \vdash_{\mathbf{p}} \|\gamma^k\|_l \vdash^{\mathbf{L}^l} \|\gamma^{k+1}\|_l$ and both conditions (i) and (ii) are satisfied for $k + 1$.
If $R = \mathbf{A}^l$, then $\gamma^k = \Gamma_1\#^l(\alpha)[b^l(\alpha)\backslash\beta]\Gamma_2$ and $\gamma^{k+1} = \Gamma_1(\alpha)\beta\Gamma_2$ for some Γ_1, Γ_2 and $\#^l(\alpha) \in Anc(\mathbf{C})$, $b^l(\alpha) \in Host(\mathbf{C})$. Then by definition of projections, we get: $\|\gamma^k\|_l = \|\Gamma_1\|_l\#^l(\alpha)[b^l(\alpha)\backslash\|\beta\|_l]\|\Gamma_2\|_l \vdash_{\mathbf{p}}^{\mathbf{E}^l} \|\Gamma_1\|_l\|\beta\|_l\|\Gamma_2\|_l = \|\gamma^{k+1}\|_l$ and $\|\gamma^k\|_v = \|\Gamma_1\|_v (\alpha)\|\beta\|_v\|\Gamma_2\|_v = \|\gamma^{k+1}\|_v$. So (i) and (ii) are satisfied for $k + 1$.
If $R = \mathbf{C}^l$, then $\gamma^k = \Gamma_1 C\alpha \Gamma_2$ and $\gamma^{k+1} = \Gamma_1 \alpha C \Gamma_2$ for some $\alpha \in (\diagdown \mathbf{C} \cup \diagdown \mathbf{C})$ and $C \in Cat(\mathbf{C})$. Clearly, in this case $\|\gamma^k\|_l = \|\gamma^{k+1}\|_l$. Now, since C has no occurrences of $\alpha, \#(\alpha)$ or $\check{\alpha}$, then the correct pair $(\alpha, \check{\alpha})$ is eliminated in $\|\gamma^{k+1}\|_v$ by rule $\mathbf{D}^l$ iff it is eliminated in $\|\gamma^k\|_v$ by this rule. The projections of $\|\gamma^{k+1}\|_v$ and $\|\gamma^i\|_v$ on any other pair $(\beta, \check{\beta})$ of polarized valencies are not affected by this step, so they do not change. Therefore, both conditions (i) and (ii) are satisfied for $k + 1$.
If $R = \mathbf{D}^l$, then $\gamma^k = \Gamma_1(\diagup C)(\diagdown C)\Gamma_2$ and $\gamma^{k+1} = \Gamma_1\Gamma_2$. Clearly, $\|\gamma^k\|_l = \|\gamma^{k+1}\|_l$. As to the valency projection $\|\gamma^{k+1}\|_v = \|\Gamma_1\|_v\|\Gamma_2\|_v$, it is obtained from the projection $\|\gamma^k\|_v = \|\Gamma_1\|_v(\diagup C)(\diagdown C)\|\Gamma_2\|_v$ by eliminating the correct valency pair $(\diagup C)(\diagdown C)$. Therefore, this pair is eliminated in $\|\gamma^{k+1}\|_v$ iff it is eliminated in $\|\gamma^k\|_v$. The projections of $\|\gamma^{k+1}\|_v$ and $\|\gamma^k\|_v$ on any other pair

$(\beta, \breve{\beta})$ of polarized valencies rest intact. Therefore, both conditions (i) and (ii) are satisfied for $k + 1$.

The proof steps via "right" rules $\mathbf{L^r}$, $\mathbf{A^r}$, $\mathbf{P^r}$, $\mathbf{C^r}$, and $\mathbf{D^r}$ are proved similarly. Iterative dependency rule $\mathbf{I^l}$ is treated as the rule $\mathbf{L^l}$. The case of the rule $\mathbf{\Omega^l}$ is trivial and the rules $\mathbf{V^l}, \mathbf{V^r}$ just do not affect the projections.

So we prove that (i) and (ii) are satisfied for each $k = 0, \ldots, n$. Since $\|\gamma^n\|_l = S$ and $\|\gamma^n\|_v = \varepsilon$, the assertions 1 and 2 of the theorem are true.

($\Leftarrow$) Now let us suppose that

1. $\|\gamma\|_l \vdash_{\mathbf{P}} S$ and
2. each correct pair $(\alpha, \breve{\alpha})$ is eliminated in $\|\gamma\|_v$ for an assignment $\delta : x \mapsto \gamma$.

We will show that $\gamma \vdash S$, which implies that $x \in L(G)$. To do this, we will suppose that $\|\gamma\|_l \vdash_{\mathbf{P}}^n S$ for some $n \geq 0$ and show the existence of a proof $\gamma \vdash S$ by induction on n.

If $n = 0$, then $\gamma = \Gamma_1 S \Gamma_2$ for some balanced potentials Γ_1 and Γ_2. Then, the needed proof has the form $\gamma = \Gamma_1 S \Gamma_2 \vdash \Gamma_1 \Gamma_2 S \vdash S$. In the first part of this proof only the commutativity rules are used. The second part exists by Lemma 1.3. Suppose that the assertion is valid for $n \leq k$. Let us prove it for $n = k + 1$. Let $\|\gamma\|_l \vdash_{\mathbf{P}}^{k+1} S = \|\gamma\|_l \vdash_{\mathbf{P}}^R \gamma_l' \vdash_{\mathbf{P}}^k S$, where R is the first applied rule.

If $R = \mathbf{L^l}$, then $\|\gamma\|_l = \Gamma_1 C [C \backslash \beta] \Gamma_2$ and $\gamma_l' = \Gamma_1 \beta \Gamma_2$ for some $\Gamma_1, \Gamma_2, \beta$. Clearly, $\gamma = \tilde{\Gamma}_1 C \Delta [C \backslash \tilde{\beta}] \tilde{\Gamma}_2$, where $\|\tilde{\Gamma}_1\|_l = \Gamma_1, \|\tilde{\beta}\|_l = \beta, \|\tilde{\Gamma}_2\|_l = \Gamma_2$ and Δ is a potential. Now we can construct the proof $\gamma = \tilde{\Gamma}_1 C \Delta [C \backslash \tilde{\beta}] \tilde{\Gamma}_2 \vdash \tilde{\Gamma}_1 \Delta C [C \backslash \tilde{\beta}] \tilde{\Gamma}_2$ $\vdash^{\mathbf{L^l}} \tilde{\Gamma}_1 \Delta \tilde{\beta} \tilde{\Gamma}_2 = \gamma'$, where in the first part of the proof only commutativity rules are used in order to permute C and Δ. Regarding the projections of γ', we can see that $\|\gamma'\|_l = \gamma_l'$ and $\|\gamma'\|_v = \|\gamma\|_v$. Therefore, $\|\gamma'\|_v$ is balanced and the assertion follows by induction.

If $R = \mathbf{E^l}$, then $\|\gamma\|_l = \Gamma_1 \#^l(\alpha) [\flat^l(\alpha) \backslash \beta] \Gamma_2$ and $\gamma_l' = \Gamma_1 \beta \Gamma_2$ for some Γ_1, Γ_2, β and $\#^l(\alpha) \in Anc(\mathbf{C})$, $\flat^l(\alpha) \in Host(\mathbf{C})$. By definition of projections, $\gamma = \tilde{\Gamma}_1 \#^l(\alpha) \Delta [\flat^l(\alpha) \backslash \tilde{\beta}] \tilde{\Gamma}_2$, $\|\gamma\|_v = \|\tilde{\Gamma}_1\|_v \alpha \|\Delta\|_v \|\tilde{\beta}\|_v \|\tilde{\Gamma}_2\|_v$ is balanced, $\|\tilde{\Gamma}_1\|_l = \Gamma_1, \|\tilde{\Gamma}_2\|_l = \Gamma_2, \|\tilde{\beta}\|_l = \beta$, and $\Delta \in (V^+(\mathbf{C}) \cup V^-(\mathbf{C}))^*$.

Let us consider the proof:

$$(1) \quad \gamma = \tilde{\Gamma}_1 \#^l(\alpha) \Delta [\flat^l(\alpha) \backslash \tilde{\beta}] \tilde{\Gamma}_2 \vdash^{\mathbf{V^l}} \tilde{\Gamma}_1 \#^l(\alpha) \Delta \flat^l(\alpha) \tilde{\beta} \tilde{\Gamma}_2 \vdash \tilde{\Gamma}_1 \#^l(\alpha) \flat^l(\alpha) \Delta \tilde{\beta} \tilde{\Gamma}_2$$
$$\vdash^{\mathbf{A^l}} \tilde{\Gamma}_1 \alpha \Delta \tilde{\beta} \tilde{\Gamma}_2,$$

in which in the part $\vdash$ only the commutativity rules are used in order to permute $\flat^l(\alpha)$ and Δ. This means that $\|\tilde{\Gamma}_1 \alpha \Delta \tilde{\beta} \tilde{\Gamma}_2\|_l = \Gamma_1 \beta \Gamma_2 = \gamma_l'$ and $\|\tilde{\Gamma}_1 \alpha \Delta \tilde{\beta} \tilde{\Gamma}_2\|_v = \|\tilde{\Gamma}_1\|_v \alpha \|\Delta\|_v \|\tilde{\beta}\|_v \|\tilde{\Gamma}_2\|_v = \|\gamma\|_v$ is balanced. Therefore, by induction the proof (1) can be completed with a proof $\tilde{\Gamma}_1 \alpha \Delta \tilde{\beta} \tilde{\Gamma}_2 \vdash S$.   □

Corollary 1. $\mathcal{L}(CDG) \subseteq \mathcal{L}(gCDG)$.

Proof. This corollary follows from Theorems 1,5 using the type simulation, in which the types $\#^l(\alpha)$, $\flat^l(\alpha)$ are replaced by the new *primitive* type $<\#^l(\alpha)>$, in the place of each assignment $\delta : x \mapsto [\beta_1 \backslash \#^l(\alpha) / \beta_2]^P$ there is the assignment $\delta : x \mapsto [\beta_1 \backslash <\#^l(\alpha)> / \beta_2]^{P\alpha}$ and in the place of $\delta : x \mapsto [\flat^l(\alpha) \backslash \beta]^P$ there is the assignment $\delta : x \mapsto [<\#^l(\alpha)> \backslash \beta]^{P\alpha}$ (similar for other orientations).   □

6 Parsing Complexity

The general parsing problem $pars(G, s, w) \equiv$ "$w \in L(G)$ and s is a syntactic structure assigned to w by G" is not necessarily polynomial time when all problems $pars_{G_0}(s, w) \equiv pars(G_0, s, w)$ for particular G_0 are so (as this is the case of CF-grammars). If there is no uniform bound on the number of polarized valencies, then the general parsing problem for gCDGs is hard.

Theorem 6. *The general parsing problem $G(D, w)$ for gCDGs is NP-complete.*

Proof. The NP-hardness can be proved by the following polynomial reduction of $3 - CNF$. Let $\Phi = C_1 \wedge \ldots \wedge C_m$ be a CNF with clauses C_j including three literals l_1^j, l_2^j, l_3^j and $l_k^j \in \{x_1, \neg x_1, \ldots, x_n, \neg x_n\}$. We define from Φ the CDG $G(\Phi) = (W, \mathbf{C}, S, \delta)$, in which $W = \{\Phi, C_1, \ldots, C_m, x_1, \ldots, x_n, y_1, \ldots y_n\}$, $\mathbf{C} = \{S, A, 1_0, 1_1, 2_0, 2_1, \ldots, n_0, n_1\}$ and $\delta(\Phi) = [(A\backslash)^n\backslash S]$, $\delta(x_i) = \{[A/(\nearrow i_0)], [A/(\nearrow i_1)]\}$, $\delta(y_i) = \{(\searrow i_0), (\searrow i_1)\}$, $\delta(C_j) = \{cat(l_1^j), cat(l_2^j), cat(l_3^j)\}$, where $cat(x_i) = [(\searrow i_1)/(\nearrow i_1)]$, $cat(\neg x_i) = [(\searrow i_0)/(\nearrow i_0)]$. Let also $w(\Phi) = x_1 x_2 \ldots x_n \Phi C_1 C_2 \ldots C_m y_1 y_2 \ldots y_n$.

Assertion. Φ is satisfiable iff $(\exists D : DT)\ G(\Phi)(D, w(\Phi))$.

This assertion follows from the fact that $G(\Phi)(D, w(\Phi))$ does not hold iff at least for one $i, 1 \leq i \leq n$, the category $[A/(\nearrow i_0)]$ is chosen in some $\delta(C_j)$ and $[A/(\nearrow i_1)]$ is chosen in some other $\delta(C_k)$. On the other hand, this conflict cannot be avoided iff Φ is not satisfiable. □

In practice, the inventory of polarized valencies is finite and fixed for DGs of particular languages. Due to the projections independence property, each particular gCDG parsing problem turns out to be polynomial time. In [8] we have described a polynomial time parsing algorithm for CDGs. It was implemented in Lisp by Darin and Hristian Todorov.[8] Below we will present a parsing algorithm for gCDG. It was implemented in $C\#$ and optimized by Ilya Zaytsev.

Preliminaries. Let us fix for the rest a gCDG $G = (W, \mathbf{C}, S, \delta)$. We will first define two failure functions used for the algorithm optimization.

Let $w = w_1 w_2 \ldots w_n \in W^+$, $\alpha \in V^l(\mathbf{C})$ and $1 \leq i \leq n$. Then

$$\pi^L(\alpha, i) = max\{\Delta_\alpha(\|\Gamma\|_v) \mid \Gamma \in \delta(w_1 \ldots w_i)\}$$

is the *left failure function* and for $\alpha \in V^r(\mathbf{C})$,

$$\pi^R(\alpha, i) = max\{\Delta_\alpha(\|\Gamma\|_v) \mid \Gamma \in \delta(w_{n-i+1} \ldots w_n)\}$$

is the *right failure function*. We set $\pi^L(\alpha, 0) = \pi^R(\alpha, 0) = 0$. It is not difficult to prove the following properties of these functions.

Lemma 4. *(i) Let $1 \leq i \leq n - 1$. Then*
$\pi^L(\alpha, i + 1) =$

---

[8] The analyses shown in the figures are carried out by this algorithm.

$$max\{\Delta_\alpha(P) + max\{\pi^L(\alpha, i) - \Delta_{\check\alpha}(P), 0\} \mid P = \|\gamma\|_v, \gamma \in \delta(w_{i+1})\},$$
$$\pi^R(\alpha, i+1) =$$
$$max\{\Delta_{\check\alpha}(P) + max\{\pi^R(\alpha, i) - \Delta_\alpha(P), 0\} \mid P = \|\gamma\|_v, \gamma \in \delta(w_{n-i+1})\}.$$

(ii) *If $\Gamma \vdash S$ for some $\Gamma = \gamma_1...\gamma_n \in \delta(w)$, then*
$$\Delta_\alpha(\|\gamma_i...\gamma_j\|_v) \leq \pi^R(\check\alpha, n-j), \quad \Delta_{\check\alpha}(\|\gamma_i...\gamma_j\|_v) \leq \pi^L(\alpha, i-1)$$
for all $1 \leq i \leq j \leq n$, all $\alpha \in V^l(\mathbf{C})$ and $\check\alpha \in V^r(\mathbf{C})$.

Algorithm description. **gCdgAnalyst** is a standard dynamic programming parsing algorithm. It applies to a gCDG $G = (W, \mathbf{C}, S, \delta)$ with left polarized valencies $V^l(\mathbf{C}) = \{v_1, ..., v_p\}$ and dual right valencies $V^r(\mathbf{C}) = \{\check v_1, ..., \check v_p\}$ and to a string $w = w_1 w_2 ... w_n \in W^+$ and fills up a $n \times n$ triangle matrix M with *items*. Each cell $M[i, j]$, $i \leq j$, corresponds to the string interval $w_i...w_j$ and contains a finite set of items. Each item codes a generalized type C^P and has the form $\langle C, \Delta^L, \Delta^R, I^l, I^r \rangle$, where:

- C is a neutral category $C \in nCat(\mathbf{C})$,
- $\Delta^L = (\Delta_{v_1}, ..., \Delta_{v_p})$ and $\Delta^R = (\Delta_{\check v_1}, ..., \Delta_{\check v_p})$ are integer vectors whose component i contains the corresponding deficits of right (left) non-paired v-brackets in the potential P (see Definition 7),
- I^l, I^r are left and right angle items from which I is calculated (for I in diagonal $M[i, i]$, $I^l = I^r = \emptyset$).

Algorithm gCdgAnalyst
```
// Input: gCDG G, string w = w₁...wₙ
// Output: ⟨"yes", DSD⟩ iff w ∈ L(G)
{
    CalcFailFuncL();
    CalcFailFuncR();
    for (k = 1, ..., n)
    {
        Propose( k )
    }
    for (l = 2, ..., n)
    {
        for (i = 1, ..., n − l)
        {
            j := i + l − 1;
            for (k = i, ..., j − l)
            {
                SubordinateL(i, k, j);
                SubordinateR(i, k, j);
            }
        }
    }
    if (I = ⟨S, (0, 0, ..., 0), (0, 0, ..., 0), Iˡ, Iʳ⟩ ∈ M[1, n])
        return ⟨"yes", Expand(I)⟩;
```

```
//procedure Expand( I ) calculates the output DS
else
        return ⟨"no", ∅⟩;
}

CalcFailFuncL()
{
  foreach (v ∈ V^l(C))
  {
      π^L[v, 0] := 0;
      for (i = 1, . . . , n)
      {
          π_max := 0;
          foreach (C^P ∈ δ(w_i))
          {
              π_max := max{π_max, Δ_v(P) + max{π^L[v, i − 1] − Δ_v̆(P), 0}};
          }
          π^L[v, i] := π_max;
      }
  }
}

CalcFailFuncR()
{
  foreach (v̆ ∈ V^r(C))
  {
      π^R[v, 0] := 0;
      for (i = 1, . . . , n)
      {
          π_max := 0;
          foreach (C^P ∈ δ(w_{n−i+1}))
          {
              π_max := max{π_max, Δ_v̆(P) + max{π^R[v, i − 1] − Δ_v(P), 0}};
          }
          π^R[v, i] := π_max;
      }
  }
}

AddItem( M[i, j], ⟨C, Δ^L, Δ^R, I^l, I^r⟩ )
{
  M[i, j] := M[i, j] ∪ {⟨C, Δ^L, Δ^R, I^l, I^r⟩};
  if (C = [C' * \β])
  {
      AddItem( M[i, j], ⟨[β], Δ^L, Δ^R, I^l, I^r⟩ );
  }
  if (C = [β/C'*])
```

```
    {
        AddItem( M[i, j], ⟨[β], Δ^L, Δ^R, I^l, I^r⟩ );
    }
}

//For 1 ≤ i ≤ n
Propose( i )
{
    (loop) foreach (C^P ∈ δ(w_i))
    {
        foreach (v ∈ V^l(C))
        {
            Δ^L[v] := Δ_v(P);
            if (Δ^L[v] > π^R[v̆, n − j]) next (loop);
            Δ^R[v̆] := Δ_v̆(P);
            if (Δ^R[v̆] > π^L[v, i − 1]) next (loop);
        }
        AddItem( M[i, i], ⟨C, Δ^L, Δ^R, ∅, ∅⟩ );
    }
}


//For 1 ≤ i ≤ k ≤ j ≤ n
SubordinateL( i, k, j )
{
    (loop) foreach (I_1 = ⟨α_1, Δ_1^L, Δ_1^R, I_1^l, I_1^r⟩ ∈ M[i, k],
                    I_2 = ⟨α_2, Δ_2^L, Δ_2^R, I_2^l, I_2^r⟩ ∈ M[k + 1, j])
    {
        foreach (v ∈ V^l(C))
        {
            Δ^L[v] := Δ_2^L(v) + max{Δ_1^L(v) − Δ_2^R(v), 0};
            if (Δ^L[v] > π^R[v̆, n − j]) next (loop);
            Δ^R[v̆] := Δ_1^R(v̆) + max{Δ_2^R(v̆) − Δ_1^L(v̆), 0};
            if (Δ^R[v̆] > π^L[v, i − 1]) next (loop);
        }
        if ( α_1 = C and α_2 = [C\β] )
        {
            AddItem( M[i, j], ⟨[β], Δ^L, Δ^R, I_1, I_2⟩ );
        }
        elseif ( (α_1 = C and α_2 = [C * \β]) or α_1 = [ε] )
        {
            AddItem( M[i, j], ⟨α_2, Δ^L, Δ^R, I_1, I_2⟩ );
        }
    }
}
SubordinateR( i, k, j ) is similar.
```

Correctness. Correctness of CalcFailFuncL() and CalcFailFuncR() follows from Lemma 4.

Example 7. Let $W = \{a, b\}$, $\mathbf{C} = \{A, B\}$, $\delta(a) = \{[A/A]^{\nearrow B \swarrow B}, [B * \backslash A]^{\nwarrow A \nearrow A}\}$, $\delta(b) = \{[B]^{\nearrow A \nwarrow B}, [\varepsilon]^{\nearrow A \swarrow B}\}$. Then $V^l(\mathbf{C}) = \{\nearrow A, \swarrow B\}$, $V^r(\mathbf{C}) = \{\searrow A, \nwarrow B\}$. For the string $w = abba$, the category potentials are presented in the table:

a	b	b	a
$\swarrow B \;\swarrow B$	$\nearrow A \;\nwarrow B$	$\nearrow A \;\nwarrow B$	$\swarrow B \;\swarrow B$
$\searrow A \;\searrow A$	$\nearrow A \;\swarrow B$	$\nearrow A \;\swarrow B$	$\nwarrow A \;\nwarrow A$

CalcFailFuncL() and CalcFailFuncR() will calculate the following values:

i	0	1	2	3	4
$\pi^L[\nearrow A, i]$	0	0	1	2	2
$\pi^L[\swarrow B, i]$	0	2	3	4	6

i	4	3	2	1	0
$\pi^R[\searrow A, i]$	2	0	1	2	0
$\pi^R[\nwarrow B, i]$	2	2	1	0	0

Theorem 7. *Let $G = (W, \mathbf{C}, S, \delta)$ be a gCDG and $w = w_1 w_2 ... w_n \in W^+$. Then for any $1 \le i \le k \le j \le n$, an item $I = \langle \theta, \Delta^L, \Delta^R, I^l, I^r \rangle$ falls to $M[i,j]$ iff there is $\Gamma = \Gamma_1 \gamma_i ... \gamma_j \Gamma_2 \in \delta(w)$ such that $\gamma_i ... \gamma_j \in \delta(w_i ... w_j)$ and*
(i) $\gamma_i ... \gamma_j \vdash \theta$,
(ii) $\Delta^L[\alpha] = \Delta_\alpha(\|\gamma_i ... \gamma_j\|_v)$, $\Delta^R[\alpha] = \Delta_{\check\alpha}(\|\gamma_i ... \gamma_j\|_v)$ for all $\alpha \in V^l(\mathbf{C})$, $\check\alpha \in V^r(\mathbf{C})$,
(iii) $\gamma_i ... \gamma_j$ satisfies the condition (ii) of Lemma 4.

Proof. Let $l = j - i + 1$.
($\Rightarrow$) Let $I \in M[i,j]$. We will show that there is Γ satisfying the conditions of the theorem by induction on l.
1. If $l = 1$, then I is put to $M[i,i]$ by Propose(i). In this case, the conditions $(i) - (iii)$ are trivially satisfied.
2. Let us suppose that the theorem is true for all $l' < l$. Then $i < j$. In this case, $I \in M[i,j]$ implies that there is $i \le k < j$ such that I was put in $M[i,j]$ by SubordinateL(i, k, j) or by SubordinateR(i, k, j). Let it be by SubordinateL(i, k, j). Then there must be $I_1 = \langle \theta_1, \Delta_1^L, \Delta_1^R, I_1^l, I_1^r \rangle \in M[i, k]$ $I_2 = \langle \theta_2, \Delta_2^L, \Delta_2^R, I_2^l, I_2^r \rangle \in M[k+1, j]$ satisfying $(i) - (iii)$. Therefore, $\gamma_i ... \gamma_k \vdash \theta_1$, $\gamma_{k+1} ... \gamma_j \vdash \theta_2$ and by definition of SubordinateL(i, k, j), $\theta_1 = C, \theta_2 = [C \backslash \beta], \theta = \beta$ or $\theta_1 = C, \theta_2 = [C * \backslash \beta], \theta = \theta_2$, or $\theta_1 = \varepsilon, \theta = \theta_2$. In all of these cases, $\gamma_i ... \gamma_k \gamma_{k+1} ... \gamma_j \vdash \theta$. Given that $\Gamma_1' \gamma_i ... \gamma_k \Gamma_2' \in \delta(w)$, $\Gamma_1'' \gamma_{k+1} ... \gamma_j \Gamma_2'' \in \delta(w)$, $\gamma_i ... \gamma_k \in \delta(w_i ... w_k)$, $\gamma_{k+1} ... \gamma_j \in \delta(w_{k+1} ... w_j)$, we see that $\Gamma_1' \gamma_i ... \gamma_j \Gamma_2'' \in \delta(w)$, $\gamma_i ... \gamma_j \in \delta(w_i ... w_j)$ and $\gamma_i ... \gamma_j \vdash \theta$. Point (ii) directly follows from Lemma 4(i). Finally, $I \in M[i,j]$ means that I does not violate the necessary condition in Lemma 4(ii).
($\Leftarrow$) By induction on l immediately following the definition of **gCdgAnalyst**. $\square$

Complexity. For a gCDG $G = (W, \mathbf{C}, S, \delta)$, let $l_G = |\delta|$ be the number of category assignments in the lexicon, $a_G = max\{k \mid \exists x \in W([\alpha_k \backslash ... \backslash \alpha_1 \backslash C/\beta]^P \in \delta(x) \vee [\beta \backslash C/\alpha_1/.../\alpha_k]^P \in \delta(x))\}$ be the maximal number of argument subtypes in assigned categories, $p_G = |V^l(\mathbf{C})| = |V^r(\mathbf{C})|$ be the number of polarized valencies and $\Delta_G = max\{\Delta_\alpha(P) \mid \exists x \in W(C^P \in \delta(x) \vee \alpha \in V(\mathbf{C}))\}$ be the maximal valency deficit in assigned categories. In the complexity bound below n will denote the length of the input string $n = |w|$.

Theorem 8. *Algorithm* **gCdgAnalyst** *has time complexity*
$$O(l_G \cdot a_G^2 \cdot (\Delta_G \cdot n)^{2p_G} \cdot n^3).$$

Proof. A category $\gamma \in \delta(x)$ may be cancelled to no more than a_G^2 different categories. So the maximal number of matrix cell elements is $l_G \cdot a_G^2$. The valency deficits are bounded by the maximal value of the failure functions. So the maximal deficit of a polarized valency is $\Delta_G \cdot n$. Therefore, the number of different valency deficit vectors is bounded by $(\Delta_G \cdot n)^{2p_G}$. Filling one matrix cell needs visiting n cells. There are $\frac{n^2}{2}$ cells in M. This proves the time bound. $\square$

Remark 3. 1. When G has no polarized valencies, the parsing time is evidently $O(n^3)$. Due to Theorem 3, every gCDG G with bounded valency deficit $\sigma < c$ can be translated into an equivalent gCDG G_c without polarized valencies (so with parsing time $O(n^3)$). Of course, the size of G_c is exponential: $|G_c| = O(|G| \cdot c^{p_G})$.
2. In practice, the failure functions significantly lower the time complexity.

7 Concluding Remarks

The main advantage of gCDG as compared to other formal models of surface syntax is that they allow to define the dependencies of all kinds, local and long, projective and discontinuous, in the same elegant and completely local manner. On the one hand, they are genuine categorial grammars and, as such, they are completely lexicalized and use types in the place of rules. On the other hand, they keep the traditional valency / polarity style peculiar to all dependency grammars. The CDG which, in fact, constitute a subclass of gCDG, can be used in real applications. As we have shown, gCDG have a practical polynomial time parsing algorithm and enjoy good mathematical properties. They are learnable from positive data (see [4]) and equivalent to rule based DSG [3]. At the same time, a more detailed study of their expressiveness is needed, in particular, a comparison with the mild CS grammars [19] and the pregroup grammars [22].[9].

Very important is the question, whether the **FA**-principle is universal. There are evidences that it is adequate for many languages with, so to say, rigid WO, e.g. English, French, Spanish, Italian, German, Japan and many others. It seems adequate even for the languages with elaborated morphology and flexible WO, such as Russian, Turkish and some others. However, this principle does not apply to the constructions with serial infinitive phrase subjects (so called *cross-serial dependencies* [5]) in Dutch. The gCDGs with the **FA**-principle can be seen as uni-modal DGs. To cover these complex constructions, one should use other polynomially implementable modes and pass to *multimodal* gCDGS.

---

[9] The pregroup grammars are weakly equivalent to CF-grammars [7] At the same time, the types they assign to words are often close to the projective dependency types of the CDG.

References

1. Bar-Hillel, Y.: A quasi-arithmetical notation for syntactic description. Language 29(1), 47–58 (1953)
2. Bar-Hillel, Y., Gaifman, H., Shamir, E.: On categorial and phrase structure grammars. Bull. Res. Council Israel 9F, 1–16 (1960)
3. Béchet, D., Dikovsky, A., Foret, A.: Dependency structure grammars. In: Blache, P., et al. (eds.) LACL 2005. LNCS (LNAI), vol. 3492, pp. 18–34. Springer, Heidelberg (2005)
4. Béchet, D., et al.: On learning discontinuous dependencies from positive data. In: Proc. of the 9th Intern. Conf. Formal Grammar 2004 (FG 2004), pp. 1–16 (2004), http://cs.haifa.ac.il/~shuly/fg04/
5. Bresnan, J., et al.: Cross-serial dependencies in Dutch. Linguistic Inquiry 13(4), 613–635 (1982)
6. Bröker, N.: Separating surface order and syntactic relations in a dependency grammar. In: Proc. COLING-ACL, Montreal, pp. 174–180 (1998)
7. Buszkowski, W.: Lambek grammars based on pregroups. In: de Groote, P., Morrill, G., Retoré, C. (eds.) LACL 2001. LNCS (LNAI), vol. 2099, Springer, Heidelberg (2001)
8. Dekhtyar, M., Dikovsky, A.: Categorial dependency grammars. In: Moortgat, M., Prince, V. (eds.) Proc. of Intern. Conf. on Categorial Grammars, Montpellier, pp. 76–91 (2004)
9. Dikovsky, A.: Grammars for local and long dependencies. In: Proc. of the Intern. Conf. ACL 2001, Toulouse, France, pp. 156–163. ACL & Morgan Kaufman (2001)
10. Dikovsky, A.: Polarized non-projective dependency grammars. In: de Groote, P., Morrill, G., Retoré, C. (eds.) LACL 2001. LNCS (LNAI), vol. 2099, pp. 139–157. Springer, Heidelberg (2001)
11. Dikovsky, A.: Linguistic meaning from the language acquisition perspective. In: Jäger, G., et al. (eds.) Proc. of the 8th Intern. Conf. Formal Grammar 2003, FG 2003, Vienna, Austria, Vienna Techn. Univ. (2003)
12. Dikovsky, A.: Dependencies as categories. In: Duchier, D., Kruijff, G.J.M. (eds.) Recent Advances in Dependency Grammars. COLING 2004 Workshop, pp. 90–97 (2004)
13. Dikovsky, A., Modina, L.: Dependencies on the other side of the curtain. Traitement Automatique des Langues (TAL) 41(1), 79–111 (2000)
14. Gaifman, H.: Dependency systems and phrase structure systems. Report p-2315, RAND Corp. Santa Monica (CA) (1961) (Published in: Information and Control 8(3), 304–337 (1965))
15. Gladkij, A.V.: Lekcii po Matematičeskoj Lingvistike dlja Studentov NGU [Course of Mathematical Linguistics. Novosibirsk State University (Russ.)]. (French transl. Leçons de linguistique mathématique. facs. 1, 1970, Dunod). Novosibirsk State University (1966)
16. Hays, D.: Grouping and dependency theories. Research memorandum RM-2646, The RAND Corporation (1960), Published in Proc. of the National Symp. on Machine Translation, Englewood Cliffs (NY), pp. 258–266 (1961)
17. Hudson, R.A.: Word Grammar. Basil Blackwell, Oxford-New York (1984)
18. Jackendoff, R.: X' Syntax: A Study of Phrase Structure. MIT Press, Cambridge, MA (1977)
19. Joshi, A.K., Shanker, V.K., Weir, D.J.: The convergence of mildly context-sensitive grammar formalisms. In: Sells, P., Shieber, S., Wasow, T. (eds.) Foundational issues in natural language processing, pp. 31–81. MIT Press, Cambridge, MA (1991)

20. Kahane, S. (ed.): Les grammaires de dépendance. In: Kahane, S. (ed.) Traitement automatique des langues, Paris, Hermes, vol. 41, n. 1/2000 (2000)
21. Kruijff, J.J.M., Duchier, D. (eds.): Recent Advances in Dependency Grammars. In: Kruijff, J.J.M., Duchier, D. (eds.) Proceedings of COLING Workshop, Geneva (August 2004)
22. Lambek, J.: Type grammar revisited. In: Lecomte, A., Perrier, G., Lamarche, F. (eds.) LACL 1997. LNCS (LNAI), vol. 1582, pp. 1–27. Springer, Heidelberg (1999)
23. Lombardo, V., Lesmo, L.: An Earley-type recognizer for dependency grammar. In: Proc. 16th COLING, pp. 723–728 (1996)
24. Mel'čuk, I.: Dependency Syntax. SUNY Press, Albany, NY (1988)
25. Moortgat, M., Morrill, G.V.: Heads and phrases. Type calculus for dependency and constituent structure. Ms OTS, Utrecht (1991)
26. Morrill, G.V.: Type Logical Grammar. Categorial Logic of Signs. Kluwer Academic Publishers, Dordrecht (1994)
27. Robinson, J.J.: Dependency structures and transformational rules. Language 46(2), 259–285 (1970)
28. Sleator, D., Temperly, D.: Parsing English with a Link Grammar. In: Proc. IWPT 1993, pp. 277–291 (1993)
29. Tesnière, L.: Éléments de syntaxe structurale. Librairie C. Klincksieck, Paris (1959)

Temporal Verification of Probabilistic Multi-Agent Systems*

Michael I. Dekhtyar[1], Alexander Ja. Dikovsky[2], and Mars K. Valiev[3]

[1] Dept. of CS, Tver St. Univ., Tver, Russia, 170000
Michael.Dekhtyar@tversu.ru
[2] LINA, Université de Nantes, France
Alexandre.Dikovsky@univ-nantes.fr
[3] Keldysh Inst. for Appl. Math., Moscow, Russia, 125047
valiev@keldysh.ru

To Boris Avraamovich, our teacher in life and research, on the occasion of his 85th Anniversary, with the deep gratitude.

Abstract. Probabilistic systems of interacting intelligent agents are considered. They have two sources of uncertainty: uncertainty of communication channels and uncertainty of actions. We show how such systems can be polynomially transformed to finite state Markov chains. This allows one to transfer known results on verifying temporal properties of the finite state Markov chains to the probabilistic multi-agent systems of the considered type.

1 Introduction

Lately, there has been increasing interest in the area of software multi-agent systems (MAS). The range of applications of MAS is very broad and extends from operating system interfaces, processing of satellite imaging data and WEB navigation to air traffic control, business process management and electronic commerce. The states and interaction rules of agents in MAS may be very complicate. This makes the behavior of MAS (as well as of other concurrent software systems) badly predictable and leads to necessity of developing formal means to analyze this behavior.

There is a number of papers on this matter in the literature which deal with different models of agents, multi-agent systems and specification languages describing their behavior. In particular, in [17,19] a behavior is considered for abstract agents with no internal structure, in [3,11] agents are specified by formulas of some temporal logics. Another popular approach to describing agents is based on "Believe-Desire-Intention" model initiated in [14] (see also [4,5,20]). In our previous papers [8,9] we considered verification complexity for MAS constructed on the base of IMPACT-architecture introduced in [16].

---

* This work was sponsored by the Russian Fundamental Studies Foundation (Grants 07-01-00637-a and 08-01-00241-a).

A. Avron et al. (Eds.): Trakhtenbrot/Festschrift, LNCS 4800, pp. 256–265, 2008.

In all these papers it is assumed that all agents operate with a complete and certain view of the world, and information transfer from one agent to another is lossless and takes some determined time. However, in many real-world applications, these assumptions are not satisfied, and agents have only a partial, uncertain view of what is true in the world.

In [10] a model of probabilistic agents is proposed in which the main cause of uncertainty in an agent is due to its state being uncertain. There may be also other sources of uncertainty in MAS. Here we consider two of them: uncertainty of communication channels between agents of the system and uncertainty of actions. Namely, we assume that times of delivering messages through channels can be probabilistic, and some messages can be lost. Moreover, the actions can have alternatives which are executed with some probabilities. However, we assume that the choice of actions to execute at each step is deterministic, i.e. the MAS considered here are not concurrent in the sense used by M. Vardi [18].

The main result of this paper is that any such probabilistic MAS can be effectively transformed into a finite state Markov chain with polynomially computable probabilities of transitions. There is a number of papers devoted to research of complexity of verifying dynamic properties of finite state Markov chains. Our transformation of MAS to Markov chains permits to apply results of these papers to the problem of verifying behavior of MAS with probabilistic channels and actions.

Let us mention some of these papers. The research of complexity of verification problem for finite state Markov chains was initiated in the abovementioned paper by Vardi. His results on the complexity of verification of linear temporal logic (LTL) formulas on Markov chains and decision processes were improved in [7]. Analogous results for probabilistic logics of branching time (PCTL and PCTL*) were obtained in [12,2].

The paper is organized as follows. Section 2 contains a syntactic definition of our variant of probabilistic MAS. In Sect. 3 we describe operational semantics of these MAS. In Sect. 4 we present an algorithm of computing transition probabilities for Markov chains corresponding to MAS. Section 5 contains the results on the complexity of verification of probabilistic MAS obtained by applying the results of [7,12].

2 Probabilistic MAS

There are a lot of readings and definitions of intelligent agents and multi-agent systems (see e.g. [15,16,21]). Here we consider the verification of behavior properties for MAS which basically conform to the so called IMPACT architecture introduced and described in detail in the book [16].

A multi-agent system A contains a finite set $\{A_1, ..., A_n\}$ of interacting intelligent agents. Any agent A has an internal database (DB) I_A consisting of a finite set of ground atoms (i.e. expressions of the form $p(c_1, \ldots, c_k)$, where p is a predicate symbol, $c_1, \ldots, c_k$ are constants; we suppose that the set of constants used by any MAS is bounded) and a message box $MsgBox_A$. Current contents of the internal DB and the message box of the agent A constitute its local state $IM_A = < I_A, MsgBox_A >$.

The agents of $\mathbf{A}$ interact by sending messages of the form $msg(Sender, Receiver, Msg)$ to other agents where $Sender$ and $Receiver$ are agents (the source and the destination of the message), and Msg is a ground atom transferred.

For any pair of agents A and B in $\mathbf{A}$ there is a communication channel CH_{AB}, which receives messages sent to B by A. After some time these messages are transferred to the message box of B. We consider the length of the transfer time of the messages as a random variable identified by a discrete finite probability distribution. $p_{AB}(t)$ denotes the probability that B receives a message sent to B by A in exactly $t \geq 1$ steps after its sending (so, a constant t_0 is connected with $\mathbf{A}$ such that $p_{AB}(t) = 0$ for all A, B and $t > t_0$).

We assume that random variables for different messages are independent, and $\sum_{t=1}^{\infty} p_{AB}(t) \leq 1$. The difference $1 - \sum_{t=1}^{\infty} p_{AB}(t)$ defines the probability that the message will be lost in the channel. If $p_{AB}(1) = 1$ then any message sent to B by A will be received by the destination in the next time instant. If $p_{AB}(1) = 1$ for all agents of MAS we have synchronous variant of multi-agent systems. Such systems were considered in [8,9]. If $p_{AB}(1) = 0.5$, $p_{AB}(2) = 0.4$ and $p_{AB}(t) = 0$ when $t > 2$, then the half of messages sent to B by A will be received in the next time, $4/10$ of them will be on the path 2 steps, and average $1/10$ of them will be lost in the channel.

The current state of CH_{AB} contains all the messages sent to B by A which are not received by B; they are marked by time they are in the channel. For the current state of the channel we use the same notation as for the channel, i.e. $CH_{AB} = \{(Msg, t)|$ the message Msg is in this channel during t steps of execution$\}$. For brevity we use also notations CH_{ij} and p_{ij} for $CH_{A_i A_j}$ and $p_{A_i A_j}$, respectively.

Each agent A is capable of performing a number of parameterized actions constituting its action base ACT_A. Any (parameterized) action has a name of the form $a(X_1, \ldots, X_m)$ and a set of alternatives: $a^1 = < ADD_a^1(X_1, \ldots, X_m)$, $DEL_a^1(X_1, \ldots, X_m)$, $SEND_a^1(X_1, \ldots, X_m) >, \ldots, a^k = < ADD_a^k(X_1, \ldots, X_m)$, $DEL_a^k(X_1, \ldots, X_m)$, $SEND_a^k(X_1, \ldots, X_m) >$. A probabilistic distribution $p_a(j)$, $1 \leq j \leq k$, is defined on these alternatives for a such that $\sum_{j=1}^{k} p_a(j) = 1$. The sets $ADD_a^j(X_1, \ldots, X_m)$ and $DEL_a^j(X_1, \ldots, X_m)$ consist of atoms of the form $p(t_1, \ldots, t_r)$, where p is an r-ary predicate (for some r) in the signature of the internal DB, $t_1, \ldots, t_r$ are variables $X_1, \ldots, X_m$ or constants. These sets determine updates of the internal DB (adding and deleting facts) when the corresponding action is executed. The set $SEND_a^j(X_1, \ldots, X_m)$ consists similarly of atoms of the form $msg(A, B, p(t_1, \ldots, t_r))$, determining messages which will be sent by A to other agents. Let $c_1, \ldots, c_m$ be constants. Let us denote by $ADD_a^j(c_1, \ldots, c_m)$ the set of facts obtained by substitution of $c_1, \ldots, c_m$ instead of $X_1, \ldots, X_m$ into atoms of $ADD_a^j(X_1, \ldots, X_m)$. The sets $DEL_a^j(c_1, \ldots, c_m)$ and $SEND_a^j(c_1, \ldots, c_m)$ are defined similarly. The ground atoms $a(c_1, \ldots, c_m)$ are called $ground\ action\ names$ (or simply, $ground\ actions$).

For example, let an agent *Accountant* works with a DB *Salary,* and 0.1 is a probability that she misses a directive by *Boss* to change the salary to workers. Then the probabilistic parameterized action

sc(Name, Position, OldSum, NewSum)

has two alternatives sc^1 and sc^2, the first alternative being executed with probability 0.9, and second with probability 0.1. The sets ADD^2, DEL^2, $SEND^2$ are empty, and

ADD^1={salary(Name, Position, NewSum)},

DEL^1 ={salary(Name, Position, OldSum)},

$SEND^1$ = {(Boss, salary_changed(Name, Position, NewSum))}

The policy of the agent A for choosing actions to execute depends on the current local state of A and is determined by a pair $< LP_A, Sel_A >$. Here LP_A is a logical program which determines a set *Perm* $(= Perm_{A,t})$ of ground action names permitted for execution at current time t. The obligation operator Sel_A selects from *Perm* a ground action $a(c_1, \ldots, c_q)$. We assume that Sel_A is a polynomially computable function. Then one of alternatives for the action $a(c_1, \ldots, c_q)$ (say, a^j) should be chosen with probability $p_a(j)$ to be currently executed.

This execution goes in the following way:

1) the next state of the internal base of A is obtained from the current state by deleting all the facts belonging to $DEL_a^j(c_1, \ldots, c_q)$, and then adding all the facts belonging to $ADD_a^j(c_1, \ldots, c_q)$;

2) simultaneously with changing internal DB the executing of the alternative a^j leads to changes of states of the communication channels. Namely, to any channel CH_{AB}, $B \neq A$, pairs of the form *(Ms, 0)* are added such that $msg(A, B, Ms) \in SEND_a^j(c_1, \ldots, c_q)$.

For example, let the *Accountant* agent has to execute the action

sc(smith, engineer, 3500, 5000).

Then 0.1 is the probability that *Accountant* does not nothing in this step, and with probability 0.9 the fact

salary(smith, engineer, 3500)

will be deleted from the internal DB, and the fact

salary(smith, engineer, 5000)

will be added to it. Moreover, the entry

(salary_changed(smith, engineer, 5000), 0)

will be placed into the channel $CH_{accountant\ boss}$.

To complete the definition of A and one-step semantics for it we should define LP_A, and how it does determine the current value of the set *Perm*.

As LP_A we consider logic programs with the clauses of the form

H :- $L_1, \ldots, L_n$,

where $n \geq 0$, the head H is an action atom, the literals L_i are either action literals, or (extensional) internal DB literals, or atoms of the form *msg(Sender, A, Msg)* or their negations **not** *msg(Sender, A, Msg)*, or calls of some built-in polynomially computable predicates.

We suppose that the program clauses are *safe* in the sense that all variables in the head H occur *positively* in the body $L_1,...,L_n$, and, moreover, the program LP_A is stratified [1]. Then for any local state $state= < I_A, MsgBox_A >$ the program

$$LP_{A,state} = LP_A \cup I_A \cup MsgBox_A,$$

determining the set of actions which can be currently executed, is also stratified.

It is well known (see [1]) that stratified logic programs have a unique minimal model. Let $M_{A,state}$ denote such a model for $LP_{A,state}$. The standard fixpoint computation procedure constructs this model in polynomial time with respect to the size of groundization $gr(LP_{A,state})$ of $LP_{A,state}$ (remember that we suppose polynomial computability of all built-in predicates). Note that the size of $gr(LP_{A,state})$ can be exponential with respect to the size of $LP_{A,state}$.

Then the set $Perm$ of actions permitted for current execution is defined as the set of ground action names contained in $M_{A,state}$. Let Sem denote the function defining $Perm$ from $LP_{A,state}$.

3   The Probabilistic MAS Behavior

The global state S of the system $\mathbf{A}$ includes local states of its agents and states of all channels:

$$S =< I_1, \ldots, I_n; CH_{1,2}, CH_{2,1}, \ldots, CH_{n-1,n}, CH_{n,n-1} >.$$

Let $\mathbf{S}_A$ denote the set of all the global states of $\mathbf{A}$. Then the one-step semantics of $\mathbf{A}$ defines a transition relation $S \Rightarrow_A S'$, and probabilities $p_{i,j}(t)$ induce probabilities $p(S, S')$ of these transitions.

The transition $S \Rightarrow_A S'$ starts with changes in channels and message boxes. Namely, the time counters of all the messages in channels are increased by 1, then into message box $MsgBox_j$ of any agent A_j the facts $msg(A_i, A_j, Msg)$ are placed with probability $p_{i,j}(t)$, for any Msg and i such that $(Msg, t) \in CH_{i,j}$. The pairs (Msg, t_0) can be considered as lost and are deleted from $CH_{i,j}$. After this any agent $A_i \in \mathbf{A}$ determines the set $Perm_{A_i} = Sem(LP_{A_i,state})$ of actions permitted to be currently executed, and a ground action $a_i(c_1,\ldots,c_q)$ to be executed is selected from $Perm_i$ by using the selection function Sel_{A_i}. After this an alternative a_i^j for a_i is chosen with probability $p_{a_i}(j)$, all the facts in $DEL_{a_i}^j(c_1,\ldots,c_q)$ are deleted from I_i, and all the facts in $ADD_{a_i}^j(c_1,\ldots,c_q))$ are added to it. Moreover, the communication channels $CH_{i,m}$ are complemented by entries $(ms, 0)$ such that messages $msg(A_i, A_m, ms)$ are in $SEND_{a_i}^j(c_1,\ldots,c_q))$. The message boxes of all the agents are emptied (in fact this does not restrict generality since all needed data can be transferred earlier from message boxes into internal DBs).

So, the transition $S \Rightarrow_A S'$ is computed by the following probabilistic algorithm:

A-step (Input: S **; Output:** S' **)**
(1) **FOR EACH** $A_i, A_j \in \mathbf{A}$ $(i \neq j)$ **DO**
(2)   **FOR EACH** $(Msg, t) \in CH_{i,j}$ **DO**
(3)   **BEGIN** $CH_{i,j} := (CH_{i,j} \setminus \{(Msg, t)\});$

(4)   IF $t \leq t_0$ THEN $CH_{i,j} := (CH_{i,j} \cup \{(Msg, t+1)\}$ END;
(5) FOR EACH $A_i, A_j \in \mathbf{A}(i \neq j)$ DO
(6)   FOR EACH $(Msg, t) \in CH_{i,j}$ DO with probability $p_{i,j}(t)$
(7)   BEGIN $CH_{i,j} := (CH_{i,j} \setminus \{(Msg, t)\})$;
(8) $MsgBox_j := MsgBox_j \cup \{msg(A_i, A_j, Msg)\}$
(9)   END;
(10) FOR EACH $A_i \in \mathbf{A}$ DO
(11)   BEGIN $Perm_i := Sem(LP_{A_i, state})$;
(12)   Let $Sel_{A_i}(Perm_i)$ be $a_i(c_1, \ldots, c_q)$;
(13)   Let $a_i^1, \ldots, a_i^k$ be all the alternatives of a_i;
(14)   Let us choose an alternative $a_i^j, 1 \leq j \leq k$, for a_i
             with probability $p_{a_i}(j)$;
(15) $I_i' := ((I_i \setminus DEL_{a_i}^j(c_1, \ldots, c_q))$
                $\cup ADD_{a_i}^j(c_1, \ldots, c_q))$;
(16)   FOR EACH $(m \neq i)$ DO
(17) $CH_{i,m}' := (CH_{i,m}$
                $\cup \{(ms, 0)|msg(A_i, A_m, ms) \in SEND_{a_i}^j(c_1, \ldots, c_q)\})$;
(18) $MsgBox_i := \emptyset$;
(19)   END;
(20) RETURN S'.

This definition of semantics for MAS permits to connect a finite Markov chain $\mathbf{MC(A)}$ with any MAS $\mathbf{A}$. The states of $\mathbf{MC(A)}$ are global states from S_A, and probabilities $p_A(S, S')$ of transitions from S to S' can be computed by the algorithm described in the next section. The behavior of $\mathbf{A}$ for an initial global state S^0 is described by a tree $t_A(S^0)$ of possible trajectories of this chain with root labelled by S^0. Nodes of this tree are labelled by global states of $\mathbf{A}$, and from any node on the level t labelled by S goes an edge labelled by $p_A(S, S')$ to a node labelled by S' if $p_A(S, S') > 0$.

Note that the cardinality of the set of states of the Markov chain $\mathbf{MC(A)}$ is exponential with respect to the size of $\mathbf{A}$ in the worse case, if $\mathbf{A}$ is ground, and even double exponential if $\mathbf{A}$ is non-ground.

4   Probabilistic MAS as Finite Markov Chains

We note that all the stochastics in the program $\mathbf{A}$-step is concentrated in lines 6-9 and 14, which determine as messages transfer to message boxes with accord to probabilities $p_{i,j}(t)$ and which alternative of the action a_i is chosen. We assume that all the probabilistic choices in these lines are independent.

The following effective procedure permits to compute the probability $p_{\mathbf{A}}(S, S')$ of transition $S \Rightarrow_A S'$:

Algorithm *Prob(S, S')*
(1) FOR EACH $A_i, A_j \in \mathbf{A}$ $(i \neq j)$ DO
(2)   BEGIN $M[i, j] := \{(m, t)|((m, t) \in CH_{i,j})$ &
                $((m, t+1) \notin CH_{i,j}')\}$;

(3) $p_{i,j} := \prod\{p_{i,j}(t)|((m,t) \in M[i,j]\}$;
(4)   **END**;
(5) **FOR EACH** $A_j \in \mathbf{A}$ **DO**
(6)   **BEGIN** $MsgBox_j := \emptyset$;
(7)   **FOR EACH** $A_i \in \mathbf{A}(i \neq j)$ **DO**
(8) $MsgBox_j := MsgBox_j$
           $\cup \{msg(A_i, A_j, m)|\exists t((m,t) \in M[i,j])\}$
(9)   **END**;
(10)**FOR EACH** $A_i \in \mathbf{A}$ **DO**
(11)   **BEGIN** $Perm_i := Sem(LP_{A_i, state})$;
(12)   Let $Sel_{A_i}(Perm_i)$ be $a_i(c_1, \ldots, c_q)$;
(13)   Let $a_i^1, \ldots, a_i^k$ be all the alternatives of a_i;
(14) $p_i := \sum_j \{p_{a_i}(j)|I_i' = ((I_i \setminus DEL_{a_i}^j(c_1, \ldots, c_q))$
           $\cup ADD_{a_i}^j(c_1, \ldots, c_q))$ **and** $(\bigwedge_{m \neq i}\{ms|(ms,0) \in CH_{i,m}'\}$
           $= \{ms|msg(A_i, A_m, ms) \in SEND_{a_i}^j(c_1, \ldots, c_q))\}\}$;
(15)   **END**;
(16) $p_\mathbf{A}(S, S') := \prod\{p_{i,j}|1 \leq i, j \leq n; j \neq i\} * \prod_{i=1}^n p_i$;
(17) **RETURN** $p_\mathbf{A}(S, S')$.

Note that $M[i,j]$ in the line 2 is the set of entries of the channel $CH_{i,j}$ which are put into $MsgBox_{A_j}$. Then $p_{i,j}$ in the line 3 is the probability of the event: the set of messages from A_i to A_j included into $MsgBox_{A_j}$ is equal to $M[i,j]$. Moreover, p_i in the line 14 is the probability to obtain a new internal state I_i' from I_i after applying the action $a_i(c_1, \ldots, c_q)$.

Theorem 1. *The algorithm* **Prob(S, S')** *computes probability $p(S, S')$ of transition $S \Rightarrow_A S'$ in time polynomial on sum of sizes of MAS* **A** *and states S and S' , i.e. on $|\mathbf{A}| + |S| + |S'|$ (we include into the size $|\mathbf{A}|$ of MAS the sizes of all signatures, of the set of constants, of the agent descriptions with their action bases and groundizations of agents' programs and of probability distributions for action alternatives and communication channels).*

5   Complexity of Verifying Dynamic Properties of MAS

Traditionally behavior properties of discrete dynamic systems are specified in some variants of temporal logics, see e.g [6]. There are two basic types of such logics: of linear time and of branching time. Normally, states of Markov chains are considered as non-structured. So, dynamic properties of such Markov chains can be adequately represented by formulas of propositional versions of these logics. States of MAS have a structure of finite models. Hence it is natural to extend logics for specifying their dynamic properties by introducing first-order features (as in [8,9]). Namely, the extension is that ordinary closed first-order formulas in signature of internal databases of agents (called *basic state formulas*) can be used in formulas instead of propositional variables. Then the possibility of transferring results on complexity of verifying finite Markov chains to probabilistic MAS stems from the well-known fact that the basic state formulas can be verified on finite models of states in polynomial space (or even in polynomial time for formulas of bounded quantifier depth)..

The problem of verification of dynamic properties for logics of linear and branching time are formulated in a somewhat different way.

- *Linear time*: for a given probabilistic MAS A, its initial state S^0 and formula F of FLTL describing a property of trajectories to find the measure (probability) $p_A(S^0, F)$ of the set of trajectories of the tree $t_A(S^0)$ which satisfy F. If this probability is equal to 1 we say that the pair (A, S^0) satisfies F.

- *Branching time*: A main role in branching time logics play formulas expressing properties of states (not trajectories). The measure of satisfiability of such formulas does not express stochastic properties of behavior of the system. Because of this it was proposed in [12]: to replace in formulas quantifiers on trajectories by probability bounds. E.g. formula $[Gf]_{>p}$ means that the measure of trajectories starting in the current state with all their states satisfying f is greater than p. The logic obtained is called PCTL.

Now we can state some of numerous results on complexity of verifying dynamic properties of MAS which can be obtained by transferring corresponding results from Markov chains.

- *Linear time* In this case we can apply Theorem 3.1.2.1 of the paper [7]. This theorem states existence of two algorithms: 1) testing if a given finite Markov chain M satisfies a formula F of PLTL in time $O(|M|2^{|F|})$, or in space polynomial in $|F|$ and polylogarithmic in $|M|$, and 2) computing the probability $p_M(F)$ of satisfaction F on M in time exponential in $|F\}$ and polynomial in $|M|$.

To apply this theorem we need only to use Theorem 1 and the remark from the end of the Sect. 3 on estimates of the size of $MC(A)$ with respect to the size of A. We give here only few of corollaries.

Theorem 2. *(1) There exists an algorithm which checks satisfiability of a formula F from FLTL in a state S of a ground probabilistic MAS A in polynomial space on $|A|$ and $|F|$.*

(2)There exists an algorithm which computes probability $p_A(S^0, F)$ for any ground probabilistic MAS A and formula F in time exponential both in $|A|$ and $|F|$.

(3) There exists an algorithm which computes probability $p_A(S^0, F)$ for any (non-ground) probabilistic MAS A and formula F in time exponential in $|F|$ and double exponential in $|A|$.

- *Branching time*: In [12] an algorithm is constructed which decides whether a formula F of PCTL is satisfied in a Markov chain M. The time complexity of this algorithm is $O(|M|^3 * |F|$. From this we obtain (using the first-order extension FPCTL instead of PCTL)

Theorem 3. *(1) There exists an algorithm which checks satisfiability of a formula F from FPCTL in a state S of a ground probabilistic MAS A in exponential time on $|A|$ and linear time on $|F|$.*

*(2)There exists an algorithm which checks satisfiability of a formula F from FPCTL in a state S of a (non-ground) probabilistic MAS **A** in time double exponential on $|A|$ and linear on $|F|$.*

We note that the estimates for $|MC(A)|$ above were given for worse case. However, in many cases these estimates can be drastically decreased (from exponential to polynomial or from double exponential to exponential). E.g., if arities of predicates in internal DBs, action bases and messages are bounded, then the cardinality of set of global states for nonground MAS is bounded by some exponential of a polynomial. So, in the assertion (3) of Theorem 2 words "double exponential in $|A|$" can be replaced by "exponential of a polynomial of $|A|$". Moreover, it may happen under constructing $MC(A)$ that many global states of **A** are not reachable or not acceptable. This can also lead to a serious decreasing of complexity of problem of verification.

6   Conclusion

In this paper we showed how probabilistic multi-agent systems can be transformed to finite state Markov chains. This permitted to obtain some results on complexity of verifying dynamic properties of MAS by applying corresponding results for finite Markov chains known from the literature. Note that we considered here only MAS with deterministic selection of actions. Of course, it is also interesting to consider verification problem for MAS with non-deterministic selection of actions. It seems that in this case results on verifying concurrent Markov chains (Markov decision processes) [18,7,13] can be applied.

References

1. Apt, K.R.: Logic programming. In: van Leeuwen, J. (ed.) Handbook of Theoretical Computer Science, Formal Models and Semantics, ch. 10, vol. B, pp. 493–574. Elsevier Science Publishers B.V, Amsterdam (1990)
2. Aziz, A., et al.: It usually works: The temporal logic of stochastic systems. In: Wolper, P. (ed.) CAV 1995. LNCS, vol. 939, pp. 155–165. Springer, Heidelberg (1995)
3. Barringer, H., et al.: METATEM: An Introduction. Formal Aspects of Computing 7, 533–549 (1995)
4. Bordini, R., et al.: Model checking AgentSpeak. In: AAMAS 2003, pp. 409–416 (2003)
5. Benerecetti, M., Guinchiglia, F., Serafini, L.: Model checking multiagent systems. Technical Report # 9708-07. Instituto Trentino di Cultura (1998)
6. Clarke, E.M., Grumberg, O., Peled, D.: Model checking. MIT Press, Cambridge (2000)
7. Courcoubetis, C., Yannakakis, M.: The complexity of probabilistic verification. J. ACM 42(4), 857–907 (1995)
8. Dekhtyar, M., Dikovsky, A., Valiev, M.: On feasible cases of checking multi-agent Systems Behavior. Theoretical Computer Science 303(1), 63–81 (2003)

9. Dekhtyar, M.I., Dikovsky, A.Ja., Valiev, M.K.: On complexity of verification of interacting agents' behavior. Annals of Pure and Applied Logic 141, 336–362 (2006)
10. Dix, J., Nanni, M., Subrahmanian, V.S.: Probabilistic agent reasoning. ACM Transactions of Computational Logic 1(2), 201–245 (2000)
11. Giordano, L., Martelli, A., Schwind, C.: Verifying communication agents by model checking in a temporal action Logic. In: Alferes, J.J., Leite, J.A. (eds.) JELIA 2004. LNCS (LNAI), vol. 3229, pp. 57–69. Springer, Heidelberg (2004)
12. Hansson, H., Jonsson, B.: A logic for reasning about time and reliability. Formal Aspects of Computing 6(5), 512–535 (1994)
13. Marta, K.: Model Checking for probability and time: from theory to practice. In: Proc. 18th IEEE Symposium on Logic in Computer Science, pp. 351–360 (2003)
14. Rao, A.S., Georgeff, M.P.: Modeling rational agents within a BDI architecture. In: Proc. 2nd Intern. Conf. on Principles of Knowledge Representation and Reasoning, Morgan Kaufman Publishers, San Francisco (1991)
15. Shoham, Y.: Agent oriented programming. Artificial Intelligence 60, 51–92 (1993)
16. Subrahmanian, V.S., Bonatti, P., Dix, J., et al.: Heterogeneous agent systems. MIT Press, Cambridge (2000)
17. van der Hoek, W., Wooldridge, M.: Tractable multiagent planning for epistemic goals. In: AAMAS 2002, Bologna, Italy (2002)
18. Vardi, M.Y.: Automatic verification of probabilistic concurrent finite state programs. In: Proceedings of 26th IEEE Symposium on Foundations of Computer Science, pp. 327–338. IEEE, New York
19. Wooldridge, M., Dunne, P.E.: The Computational complexity of Agent Verification. In: Meyer, J.-J.C., Tambe, M. (eds.) ATAL 2001. LNCS (LNAI), vol. 2333, Springer, Heidelberg (2002)
20. Wooldridge, M., et al.: Model Checking Multiagent systems with MABLE. In: Proc. of the First Intern. Conf. on Autonomous Agents and Multiagent Systems (AAMAS 2002), Bologna, Italy (July 2002)
21. Wooldridge, M., Jennings, N.: Intelligent agents: Theory and practice. The Knowledge Engineering Review 10(2) (1995)

Linear Recurrence Relations
for Graph Polynomials

Eldar Fischer* and Johann A. Makowsky**

Department of Computer Science,
Technion–Israel Institute of Technology, Haifa, Israel
{eldar,janos}@cs.technion.ac.il

For Boaz (Boris) Abramovich Trakhtenbrot
on the occasion of his 85th birthday.

Abstract. A sequence of graphs G_n is *iteratively constructible* if it can be built from an initial labeled graph by means of a repeated fixed succession of elementary operations involving addition of vertices and edges, deletion of edges, and relabelings. Let G_n be a iteratively constructible sequence of graphs. In a recent paper, [27], M. Noy and A. Ribò have proven linear recurrences with polynomial coefficients for the Tutte polynomials $T(G_i, x, y) = T(G_i)$, i.e.

$$T(G_{n+r}) = p_1(x, y)T(G_{n+r-1}) + \ldots + p_r(x, y)T(G_n).$$

We show that such linear recurrences hold much more generally for a wide class of graph polynomials (also of labeled or signed graphs), namely they hold for all the extended MSOL-definable graph polynomials. These include most graph and knot polynomials studied in the literature.

1   Introduction

Among Boaz' celebrated papers we find two papers dealing with Monadic Predicate Calculus and finite automata [30,31,21], and therein the theorem known today as the Büchi-Elgot-Trakhtenbrot Theorem. It states that the regular languages are exactly those sets of words which are definable in Monadic Predicate Calculus, also known as Monadic Second Order Logic MSOL.

There are innumerous papers dealing the importance of Monadic Predicate Calculus for algorithmic questions. One of the crucial properties of MSOL is the fact that the MSOL-theories of two structures determine uniquely the MSOL-theory of the disjoint union of the two structures, and also of many other sum-like

* Partially supported by a Grant of the Fund for Promotion of Research of the Technion–Israel Institute of Technology.
** Partially supported by a Grant of the Fund for Promotion of Research of the Technion–Israel Institute of Technology and an ISF Grant 2007–2009.

A. Avron et al. (Eds.): Trakhtenbrot/Festschrift, LNCS 4800, pp. 266–279, 2008.
© Springer-Verlag Berlin Heidelberg 2008

compositions of the two structures. This is generally known as the Feferman-Vaught theorem for Monadic Second Order Logic MSOL.[1] In [23] many algorithmic applications of this property of MSOL are discussed. We present here yet another application of Monadic Predicate Calculus to algorithmic questions, namely to graph polynomials. A sequence of graphs G_n is *iteratively constructible*[2] if it can be built from an initial graph by means of a repeated fixed succession of elementary operations involving addition of vertices and edges, and deletion of edges. Let G_n be a iteratively constructible sequence of graphs. In a recent paper, [27], M. Noy and A. Ribò have proven linear recurrences with polynomial coefficients for the Tutte polynomials $T(G_i, x, y)$, i.e. recurrences of the form

$$T(G_{n+r}, x, y) = p_1(x, y) \cdot T(G_{n+r-1}, x, y) + \ldots + p_r(x, y) \cdot T(G_n, x, y).$$

Particular cases were studied previously in [4]. We show in Theorem 1 that such linear recurrences hold much more generally for a wide class of graph polynomials (also of labeled or signed graphs), namely the MSOL-definable graph polynomials introduced in [10] and further studied[3] in [25,23,26,24]. These include the classical chromatic polynomial and the Tutte polynomial, the matching polynomials, the interlace polynomials, the cover polynomial, certain Farrell polynomials, and the various colored Tutte polynomials studied by Bollobás and Riordan, [6]. Because of the latter, our result can also be applied to the computation of the Jones polynomials and Kauffman brackets for iteratively constructible knots and links. Actually, a close inspection of the literature reveals that virtually all graph polynomials studied in the literature fall into this class, [24]. Only the interlace polynomials seem to be an exception. For those one has to use an extended logic CMSOL obtained from MSOL by adding modular counting quantifiers, cf. [1,2,8].

Our proof is based on a further refinement of Makowsky's Splitting Theorem for MSOL-definable graph polynomials from [23]. All graphs and logical structures in this paper are *finite*.

2   Guiding Examples

We consider six iteratively constructed graph sequences of undirected simple graphs. These sequences are constructed from an initial graph by the repeated application of a deterministic graph operation.

2.1   Six Graph Sequences and Their Iterative Constructions

For a graph $G = (V, E)$ we denote by $\bar{G}$ the complement graph $\bar{G} = (V, V^2 - E - Diag(V))$, with $Diag(V) = \{(v, v) : v \in V\}$ We look at the following six graph sequences:

---

[1] Strictly speaking the extension of the Feferman-Vaught theorem for First Order Logic FOL to MSOL emerged explicitly only in later papers of H. Läuchli, Y. Gurevich and S. Shelah [22,29,19].

[2] In [27] they are called *recursively constructible*.

[3] The papers [23,26,24] contain extensive bibliographies on graph polynomials.

(i) The sequence E_n of empty graphs with vertex set $V(E_n) = \{0, \ldots, n-1\}$ and edge set $E(E_n) = \emptyset$. $E_1 = P_1$ is an isolated vertex $\{0\}$. E_{n+1} is obtained from E_n by the disjoint union of $E_n \sqcup E_1$.

(ii) The sequence K_n of cliques on n vertices. $K_1 = E_1$ and $K_{n+1} = \overline{E_{n+1}}$. Iteratively we have $K_{n+1} = K_n \bowtie K_1$ where $\bowtie$ denotes the join operation.

(iii) The sequence P_n of paths on n vertices, i.e. the graphs with vertex set $V(P_n) = \{0, \ldots, n-1\}$ and edge set $E(P_n) = \{(i, i+1) : 0 \leq i \leq n-2\}$. $E_1 = P_1$ is an isolated vertex. P_{n+1} is obtained from P_n by setting $V(P_{n+1}) = V(P_n) \sqcup \{n\}$ and $E(P_{n+1}) = E(P_n) \cup \{(n-1, n)\}$. To specify this sequence with an iterated graph operation we look at $\bar{P}_n$ obtained from P_n by distinguishing a vertex of degree one. Then we have $\bar{P}_{n+1} = \eta(\bar{P}_n \sqcup \bar{P}_1)$ where η put an edge between the two distinguished elements, and leaves only the vertex coming from $\bar{P}_1$ as the distinguished element, and P_{n+1} is obtained from $\bar{P}_{n+1}$ by ignoring the distinguished element.

(iv) The sequence C_n of circles on n vertices, i.e. the graphs with vertex set $V(C_n) = \{0, \ldots, n-1\}$ and edge set $E(C_n) = \{(i, i+1) : 0 \leq i \leq n-2\} \cup \{(n-1, 0)\}$. C_1 is a single vertex with a loop. C_{n+1} is obtained from C_n by setting $V(C_{n+1}) = V(C_n) \sqcup \{n\}$ and $E(C_{n+1}) = E(P_{n+1}) \cup \{(0, n)\}$. To specify this sequence with an iterated graph operation we can use a vertex replacement operation where a distinguished vertex of C_n is replace by $\bar{P}_2$.

(v) The sequence L_n of ladders on $2n$ vertices, i.e. the graphs with vertex set $V(L_n) = \{0, \ldots, 2n-1\}$ and edge set

$$E(L_n) = \{(2i, 2i+2) : 0 \leq i \leq n-2\} \cup$$
$$\{(2i+1, 2i+3) : 0 \leq i \leq n-2\} \cup$$
$$\{(2i, 2i+1) : 0 \leq i \leq n-1\}$$

The reader can easily describe how L_{n+1} is obtained from L_n by a suitable choice of distinguished elements and appropriate vertex replacements. A formal definition is presented in Sect. 3, cf. Proposition 2.

(vi) Similarly, the sequence W_n of wheels on $n+1$ vertices can be obtained. Here W_n is the graph with vertex set $V(W_n) = \{0, \ldots, n\}$ and edge set

$$E(W_n) = \{(i, i+1) : 0 \leq i \leq n-2\} \cup \{(n-1, 0)\}$$
$$\cup \{(i, n) : 0 \leq i \leq n-1\}$$

For $G = (V, E)$ let $I(G)$ be the graph with $V(I(G)) = V \sqcup E$ and $E(I(G)) = \{(v, e) \in V \times E : \text{there is an } u \text{ with } (v, u) = e\}$. The sequence $I(G_n)$ is often much more complicated to describe iteratively than the sequence G_n. In particular we shall see in the next section, Corollary 1, that the sequence $I(K_n)$ is not iteratively constructible in the sense we have in mind. A general definition of *iteratively constructed* and *iteratively constructible* classes is given in Sect. 3.4.

2.2   The Matching Polynomial

For a graph G, the *matching polynomial* $\mu(G, x) \in \mathbb{Z}[x]$ is defined by

$$\mu(G, x) = \sum_k m_k(G) \cdot x^k$$

where $m_k(G)$ is the number of k-matchings of G.

To compute $\mu(P_n, x)$ we use auxiliary polynomials

$$\mu^+(P_n, x) = \sum_k m_k^+(P_n) \cdot x^k$$

and

$$\mu^-(P_n, x) = \sum_k m_k^-(P_n) \cdot x^k$$

where $m_k^+(P_n)$ and $m_k^-(P_n)$ is the number of k-matchings of P_n which includes, respectively excludes the last vertex.

Clearly we have

$$m_k(P_n) = m_k^+(P_n) + m_k^-(P_n)$$

hence

$$\mu(P_n, x) = \mu^+(P_n, x) + \mu^-(P_n, x).$$

It is easy to see that

$$\mu^-(P_{n+1}) = \mu^-(P_n) + \mu^+(P_n)$$
$$\mu^+(P_{n+1}) = x \cdot \mu^-(P_n)$$

Let[4] $\bar{\mu}_n = (\mu^-(P_n), \mu^+(P_n))^t$. We get

$$A\bar{\mu}_n = \bar{\mu}_{n+1}$$

with

$$a_{1,1} = 1, a_{1,2} = 1, a_{2,1} = x, a_{2,2} = 0$$

The characteristic polynomial of A is

$$det(\lambda \mathbf{1} - A) = \lambda^2 - \lambda - x$$

so we get the linear recurrence relation (independent of n)

$$\mu(P_{n+2}) = \mu(P_{n+1}) + x \cdot \mu(P_n)$$

2.3 The Vertex-Cover Polynomial

For a graph G, the *vertex-cover polynomial* $vc(G, x) \in \mathbb{Z}[x]$ is defined by

$$vc(G, x) = \sum_k vc_k(G) \cdot x^k$$

where $vc_k(G)$ is the number of k-vertex-covers of G. In [12] the following recurrence relations are derived:

(i) $vc(P_{n+1}, x) = x \cdot vc(P_n, x) + x \cdot vc(P_{n-1}, x)$

---

[4] $\bar{a}^t$ denotes the transposed vector of the vector $\bar{a}$.

(ii) $vc(C_{n+1}, x) = x \cdot vc(C_n, x) + x^2 \cdot vc(C_{n-2}, x)$

(iii) Let $Loop_n$ be the graph which consists of n isolated loops.
$$vc(Loop_{n+1}, x) = x \cdot vc(Loop_n, x) = x^n$$

(iv) For the wheel graph W_n we have
$$vc(W_{n+1}, x) = x \cdot vc(W_n, x) + x^n =$$
$$x \cdot vc(W_n, x) + x \cdot vc(Loop_n, x)$$

Using the characteristic polynomial of the matrix, $A = (a_{i,j})$ with
$$a_{1,1} = a_{1,2} = a_{2,2} = x \text{ and } a_{2,1} = 0$$
we get
$$vc(W_{n+1}, x) = 2x \cdot vc(W_n, x) - x^2 \cdot vc(W_{n-1}, x)$$

2.4   The Tutte Polynomial

We deal now with multi-graphs (multiple edges and loops are allowed). For a graph $G = (V(G), E(G))$ we denote by $k(G)$ the number of connected components of G. We define the *rank* $r(G)$ of G by,
$$r(G) = |V(G)| - k(G)$$
and the *nullity* $n(G)$ of G by
$$n(G) = |E(G)| - |V(G)| + k(G).$$

For $F \subseteq E(G)$ we put $\langle F \rangle = (V(G), F)$, the spanning subgraph of G with edges in F. We write $k\langle F \rangle_G, r\langle F \rangle_G, n\langle F \rangle_G$ for the number of connected components, the rank and the nullity of $\langle F \rangle_G$. We omit the G in $\langle F \rangle_G$, when the context is clear.

The Tutte polynomial is now defined as
$$T_G(x, y) = \sum_{F \subseteq E} (x-1)^{r\langle E \rangle - r\langle F \rangle} (y-1)^{n\langle F \rangle}$$

There is a rich literature on the Tutte polynomial, cf. [5]. In [4], the question was studied, for which iteratively constructed sequences the Tutte polynomial can be computed with linear recurrence relations. Positive answers and explicit formulas were given for, among others, the paths P_n, the circles C_n, the ladders L_n, and the wheels W_n. To describe this phenomenon the authors called these sequences T recursive, indicating that the Tutte polynomial T could be computed by a linear recurrence relation. In [27], a fairly general method is described by which one can obtain many iteratively constructed sequences of graphs, which are T-recursive. This method is reminiscent of graph grammars. We shall see that is no coincidence.

Instead of using the existing formal setting of graph grammars as described in [28], Noy and Ribó give an ad hoc definition of *repeated fixed succession of elementary operations*, which can be applied to a graph with a *context*, i.e. a labeled graph.

Definition 1. *Let F denote such an operation. Given a graph (with context) G, we put*

$$G_0 = G, G_{n+1} = F(G_n)$$

Then the sequence

$$\mathcal{G} = \{G_n : n \in \mathbb{N}\}$$

is called iteratively constructible *using F, or an F-iteration sequence.*

A precise version of a generalization of this definition is given in the next section, Definition 3.

2.5 The General Strategy

Given a graph polynomial $\mathfrak{P}$, such as the matching polynomial, the vertex-cover polynomial or the Tutte polynomial, and a sequence of iteratively constructible graphs G_n using an operation F, we want to compute $\mathfrak{P}(G_n)$ for all n.

To compute $\mathfrak{P}(G_{n+1})$, we try to find, depending on $\mathfrak{P}$ and, possibly, on G_0 and F, but *independently of* n, an $m \in \mathbb{N}$, auxiliary polynomials $\mathfrak{P}_i(G_{n+1}), i \leq m$, and a matrix $Q = (q_{i,j}) \in \mathbb{Z}[\bar{x}]^{m \times m}$, such that

$$\mathfrak{P}_j(G_{n+1})(\bar{x}) = \sum_i q_{i,j}(\bar{x}) \cdot \mathfrak{P}_i(G_n)(\bar{x})$$

Then we use the *characteristic polynomial of Q* to convert this into a *linear recurrence* relation.

We shall give very general sufficient conditions on the definability of $\mathfrak{P}$ and F, which will allow us to carry through such an argument.

3 Enter Logic

3.1 The Logic MSOL

Let us define some basics for the reader less familiar with Monadic Second Order Logic. A *vocabulary* τ is a set of constant, function and relation symbols. A *one-sorted τ-structure* is an interpretation of a vocabulary over *one* fixed set, the *universe*. Interpretations of constant symbols are elements of the universe, interpretations of function symbols are functions, and interpretations of relation symbols are relations of the prescribed arity. τ-terms are formed using individual variables, constant symbols and function symbols from τ. Interpretations of terms are elements of the universe. In first order logic FOL we have atomic formulas which express equality between terms and assert basic relations between terms. We are allowed to form boolean combination of formulas and to quantify existentially and universally over elements of the universe. In second order logic SOL we are allowed, additionally, to quantify over relations and functions of some fixed arity (number of arguments). In monadic second order logic MSOL, quantification over relations is restricted to unary relations, and quantification over functions is not allowed. The quantifier rank r of a formula in MSOL is defined like for FOL and without distinguishing between first order and second order quantification. An excellent reference for our logical background is [14].

3.2 MSOL-Polynomials

To understand better what many of the graph polynomials have in common we have to look closer at the way they are defined. Besides their recursive definition, like in the case of the Tutte polynomial and its close relatives, cf. [5,6,7], they usually also have an equivalent (up to some transformation) *static definition* as some kind of generating function. The matching polynomial e.g. can be written as

$$\sum_{M \subseteq E} x^{|M|} = \sum_{M \subseteq E} \prod_{e \in M} x$$

where M ranges over all subsets of edges which have no vertex in common i.e. subsets of edges which are matchings. The property of being a matching can be expressed in first order logic FOL with M a new relation variable, or in monadic second order logic MSOL, where M is a unary set variable ranging over subsets of edges.

Without going into the more delicate details, the MSOL-definable polynomials are in a polynomial ring $\mathcal{R}[\bar{x}]$ and are typically of the form

$$g(G, \bar{x}) = \sum_{A:\phi(A)} \prod_{v:v \in A} t(v) \tag{1}$$

where A is a unary relation variable, $\phi(A)$ is an MSOL-formula with A as a parameter,[5] and $t(v)$ is a term in $\mathcal{R}[\bar{x}]$ which may depend uniformly on v.

Alternatively, and more precisely, one can give an inductive definition of MSOL-polynomials as follows: First one introduces MSOL-monomials as being of the form $\prod_{v:v \in A} t(v)$, and then one closes under addition and multiplication, and under summations of the form $\sum_{A:\phi(A)} t(A)$ and multiplications of the form $\prod_{v:A(v)} t(\bar{v})$. Note that this gives more polynomials than just those of the form given in 1, due to nesting of summations and multiplications. To get a normal form of the type 1 one has to allow full second order logic, rather than MSOL.

In [26] the class of *extended* MSOL-*polynomials* is introduced. In the extended case the basic combinatorial polynomials are also included. More precisely, for every $\phi(\bar{v}) \in$ SOL(τ) and τ-structure $\mathcal{M}$ we define the cardinality of the set defined by ϕ:

$$card_{\mathcal{M}, \bar{v}}(\phi(\bar{v})) = |\{\bar{a} \in M^m : \langle \mathcal{M}, \bar{a} \rangle \models \phi(\bar{a})\}|.$$

The *extended* MSOL(τ)-*polynomials* are defined inductively by allowing as extended MSOL-monomials additionally:
For every $\phi(\bar{v}) \in$ MSOL(τ) and for every $x \in \mathbf{x}$, the polynomials

$$x^{card_{\mathcal{M}, \bar{v}}(\phi(\bar{v}))}, \quad x_{(card_{\mathcal{M}, \bar{v}}(\phi(\bar{v})))}, \quad \binom{x}{card_{\mathcal{M}, \bar{v}}(\phi(\bar{v}))}$$

---

[5] It may be a subgraph or induced subgraph generated by A, or a spanning subgraph generated by A, if A is a subset of edges. But the main point is that it be definable and the definition is part of ϕ.

are MSOL-definable $\mathcal{M}$-monomials. The first two are the exponentiation and the falling factorial respectively. The last is the real continuation of the number of subsets of a fixed size, see [18].[6]

Example 1. For a graph G and a non-negative integer n, let $P(G,n)$ denote the number of proper vertex colorings of G. It is well known that $P(G,n)$ is a polynomial in n, which is called the *chromatic polynomial* of G. To see this one uses a recursive definition. The static definition, given in [3,13] is not an MSOL-definable polynomial, but it is an extended MSOL-definable polynomial.

The extended MSOL-polynomials play an important role in the study initiated in [26]. The choice of extended MSOL-monomials was dictated by the characterization theorems proved in [26].

It is straight forward to see that all the results of [23], stated for MSOL-polynomials, are also valid for extended for MSOL-polynomials. In particular this applies to Theorem 2 in the sequel.

3.3   MSOL-Smooth Operations

Let $\mathfrak{A}$ and $\mathfrak{B}$ be two τ-structures. We write $\mathfrak{A} \equiv_r^{MSOL} \mathfrak{B}$, if $\mathfrak{A}$ and $\mathfrak{B}$ cannot be distinguished by $MSOL(\tau)$-formulas of quantifier rank r.

A unary operation F on τ-structures is MSOL-*smooth* if whenever $\mathfrak{A} \equiv_r^{MSOL} \mathfrak{B}$, then also $F(\mathfrak{A}) \equiv_r^{MSOL} F(\mathfrak{B})$.

The operation F should be MSOL-smooth for the presentation of the graphs, for which the polynomial is MSOL-definable. The presentation matters. For forming the cliques K_n we need the operation of adding a vertex connected to all previous vertices. This is MSOL-smooth for graphs $G = (V, E)$ with an edge relation E, but not for two sorted graphs $I(G) = (V \cup E, R)$. with vertices and edges as disjoint universes, and an incidence relation R.

3.4   Iteration Operations

We shall define inductively a large class of unary iteration operations which are MSOL-smooth on τ-structures enhanced with a fixed number of labels or colors.

For $k \in \mathbb{N}$, a k-τ-structure is a τ-structure with k additional unary relations $C_1^A, \ldots C_k^A$, called *colors*. We denote by τ_k the vocabulary $\tau \cup \{C_1, \ldots, C_k\}$.

Definition 2. *The following are the* basic operations *on k-τ-structures:*

$Add_i(\mathfrak{A})$: *For $i \leq k$, add a new element to A of color C_i.*

$\rho_{i,j}(\mathfrak{A})$: *For $i, j \leq k$, recolor all elements of A of color i with color j.*

$\eta_{R,i_1,\ldots,i_m}(\mathfrak{A})$: *For an m-ary relation symbol $R \in \tau$ and for each $a_1 \in C_{i_1}^A, \ldots, a_m \in C_{i_m}^A$ add the tuple $(a_1, \ldots, a_m)$ to R^A.*

$\delta_{R,i_1,\ldots,i_m}(\mathfrak{A})$: *For an m-ary relation symbol $R \in \tau$ and for each $a_1 \in C_{i_1}^A, \ldots, a_m \in C_{i_m}^A$ delete the tuple $(a_1, \ldots, a_m)$ from R^A.*

---

[6] The choice of these three combinatorial functions as the basic functions in the definition of extended MSOL-polynomials seems natural. However, we have not addressed the question whether this choice is *complete* in a sense yet to be defined.

Quantifier-free transductions: *For each $R \in \tau_k$ of arity $\alpha(R)$ let $\phi_R(x_1, \ldots, x_{\alpha(R)})$ be a quantifier-free τ_k formula with free variables as indicated. A quantifier free transduction redefines all the predicates R^A in $\mathfrak{A}$ by ϕ_R^A.*

Duplication: *The unary operation which associates with a graph $\mathfrak{A}$ the disjoint union of two copies of $\mathfrak{A}$.*

Remark 1. Note that the binary operation of the disjoint union of two τ-structures is not a basic operation for the purpose of iteration operations. It is, however, one of the basic operations in the induction definition of graphs of tree-width at most k or clique-width at most k, cf. [23].

Proposition 1. *All the basic operations are MSOL-smooth.*

We now state the key definitions for our main result.

Definition 3

(i) *An operation F on τ_k-structures is MSOL-elementary if F is a finite composition of any of the basic operations on τ_k-structures.*

(ii) *Let F be MSOL-elementary. Given a graph G, we put*

$$G_0 = G, G_{n+1} = F(G_n)$$

Then the sequence

$$\mathcal{G} = \{G_n : n \in \mathbb{N}\}$$

is called iteratively constructed *using F, or an F-iteration sequence.*

(iii) *A sequence of graphs G_n is* iteratively constructible *if it is an F-iteration sequence for some MSOL-elementary operation F.*

Proposition 2. *All the sequences E_n, K_n, P_n, C_n, L_n, W_n are F-iteration sequences for some MSOL-elementary operation.*

Proof. We sketch the proof for the ladders L_n, and leave the remaining cases to the reader. H_1 is L_1 with the vertices colored by the colors C_1 and C_2 respectively. H_n will be the ladder L_n with the vertices $2n - 1$ and $2n$ colored with colors C_1 and C_2 respectively, and all the other vertices colored by C_0. To construct H_{n+1} we add two isolated vertices colored with the colors C_3 and C_4 respectively. Then we connect the vertices colored by C_1 and C_3, C_2 and C_4, and C_3 and C_4. Finally we recolor C_1 and C_2 by C_0, and then C_3 by C_1 and C_4 by C_2.   □

The following is from [23, Sect. 2]:

Proposition 3. *Let F be MSOL-elementary and $\mathfrak{A}$ and $\mathfrak{B}$ two τ_k structures with $\mathfrak{A} \equiv_r^{MSOL} \mathfrak{B}$. Then $F(\mathfrak{A}) \equiv_r^{MSOL} F(\mathfrak{B})$, hence, F is a MSOL-smooth.*

Proof. The proof is straightforward from our definitions.   □

The basic operations Add_i, $\rho_{i,j}$ and $\eta_{E,i,j}$ are the basic operations used to inductively define the class of graphs of clique-width at most k. The other operations are generalizations thereof. For the vocabulary of graphs, it was shown in [9] that any class of graphs, defined inductively using these operations and starting with a finite set of graphs, is of bounded clique-width. Hence we have the following:

Proposition 4. *Let F be an MSOL-elementary operation for k-graphs, and $\mathcal{G}$ be an F-iteration sequence. Then $\mathcal{G}$ has bounded clique-width.*

Remark 2. If we exclude the use of duplication in Proposition 4, we get that $\mathcal{G}$ has bounded linear clique-width. The notion of linear clique-width is introduced in [20].

It was shown in [11,17] that the class of square grids and the class $I(K_n)$ are of unbounded clique-width. Therefore we conclude:

Corollary 1. *The sequences $I(K_n)$, $Grid_{n,n}$ are not F-iteration sequences for any F which is MSOL-elementary.*

Remark 3. The notion of a *iteratively constructible* sequence of graphs, as defined in [27], cf. 1, is a special case of our F-iteration sequences for an MSOL-elementary operation F.

Remark 4. We have not attempted here to classify all the MSOL-smooth unary operations on τ-structures. Although we think that there are MSOL-smooth unary operations which are provably not MSOL-elementary, we have no example at hand. Related questions were studied in [9].

3.5   Main Result

Our main result can now be stated.

Theorem 1. *Let*

(i) F be an MSOL-smooth operation on τ_k-structures;
(ii) $\mathfrak{P}$ be an extended MSOL(τ)-definable τ-polynomial;
(iii) $\mathcal{A} = \{A_n : n \in \mathbb{N}\}$ be an F-iteration sequence of τ-structures.

Then $\mathcal{A}$ is $\mathfrak{P}$-iterative, i.e. there exists $\beta \in \mathbb{N}$, and polynomials $p_1, \ldots, p_\beta \in \mathbb{Z}[\bar{x}]$ such that for sufficiently large n

$$\mathfrak{P}(G_{n+\beta+1}) = \sum_{i=1}^{\beta} p_i \cdot \mathfrak{P}(G_{n+i})$$

3.6   Proof of Theorem 1

The proof of Theorem 1 uses first the splitting theorem for graph polynomials from [23]. Its scenario is as follows.

A binary operation on $k - \tau$-structures $\bowtie_X$ is MSOL-*smooth*, if whenever $\mathfrak{A} \equiv_r^{\mathrm{MSOL}} \mathfrak{B}$, and $\mathfrak{A}' \equiv_r^{\mathrm{MSOL}} \mathfrak{B}'$, then also

$$\mathfrak{A} \bowtie_X \mathfrak{A}' \equiv_r^{\mathrm{MSOL}} \mathfrak{B} \bowtie_X \mathfrak{B}'.$$

Here X is used to indicate the dependence on the particular choice of the MSOL-smooth binary operation.

Let $\mathfrak{P}_i^r$, $i \in I_r$, the set of all extended MSOL-definable graph polynomials with defining formulas of quantifier rank at most r. I_r is finite of size α_r, as there are, up to logical equivalence, only finitely many formulas of fixed quantifier rank r.

A sharpened form of the splitting theorem [23, Theorem 6.4] now states the following:

Theorem 2 (Bilinear Splitting Theorem). *Let* $\bowtie_X$ *be an* MSOL-*smooth binary operation. There exists* $A(X) = (a_{i,k,\ell}(X)) \in \{0,1\}^{\alpha_r \times \alpha_r \times \alpha_r}$ *such that*

$$\mathfrak{P}_i^r(\mathfrak{A} \bowtie_X \mathfrak{B}) = \sum_{k,\ell \leq \alpha_r} a_{i,k,\ell}(X) \cdot \mathfrak{P}_k^r(\mathfrak{A}) \cdot \mathfrak{P}_\ell^r(\mathfrak{B})$$

Proof (Sketch). The Bilinear Splitting Theorem is a refinement of the Feferman-Vaught Theorem for MSOL. Its proof is exactly as the proof of [23, Theorem 6.4]. The generalization to extended MSOL-definable polynomials is straight forward.   □

The next step in the proof consists of a characterization of the MSOL-elementary operations F.

Proposition 5. *Let* F *be an* MSOL-*elementary operation on* τ_k-*structures where exactly* m *many new elements are added. Then there exists a* τ_k-*structure* $\mathfrak{C}_F$ *of size* m *and a binary* MSOL-*smooth operation* $\bowtie_F$ *such that for all* τ_k-*structures* $\mathfrak{A}$ *we have*

$$F(\mathfrak{A}) = \mathfrak{C}_{\mathfrak{F}} \bowtie_F \mathfrak{A}$$

Proof (Sketch). The proof is by induction on the sequence of basic operations used in the definition of F.   □

Now we define

$$\mathfrak{q}_{i,\ell} = \sum_k a_{i,k,\ell} \cdot \mathfrak{P}_k^r(\mathfrak{C})$$

and use the Bilinear Splitting Theorem. We obtain:

$$\mathfrak{P}_i^r(F(\mathfrak{A})) = \sum_{\ell \leq \alpha_r} \mathfrak{q}_{i,\ell} \cdot \mathfrak{P}_\ell^r(\mathfrak{B})$$

The matrix $Q = (\mathfrak{q}_{i,\ell})$ is a matrix of polynomials. To obtain Theorem 1 we compute the characteristic polynomial

$$\chi(Q) = \sum_{i=0}^{\alpha_r} q_i \lambda^i$$

of Q and obtain the required linear recurrence relation with

$$\beta = \alpha_q \text{ and } p_i = -q_i \text{ for } i = 0, \ldots \alpha_r - 1$$

where q_i are the coefficients of $\chi(Q)$. Note that $q_{\alpha_r} = 1$.
This completes the proof of Theorem 1.

4 Conclusions and Further Research

We have introduced the class of MSOL-elementary operations F of τ-structures
and their associated F-iteration sequences. We have shown that a very wide class
of graph polynomials, and even of polynomial invariants of general τ-structures,
can be computed on F-iteration sequences by linear recurrence relations. This
explains a widely observed, but not systematically studied phenomenon.

As a consequence of our method we get immediately the following.

Corollary 2. *Let F be an MSOL-smooth operation, G_n be an F-iteration sequence and $\mathfrak{P}$ an extended MSOL-definable graph polynomial. Then $\mathfrak{P}(G_n)$ can
be computed in polynomial time in n.*

Proof (Sketch). We first observe that the size of G_n is linear in n. besides that,
proof is the same as the one given in [10,23].   □

Although our method is theoretically computable, it is not effective for several
reasons:

– The number α_r is too large for practical use.
– The boolean array $a_{i,k,\ell}$ cannot be efficiently computed. In fact, its computation is non-elementary, cf. [16].

It is likely that in practice, for explicitly given F, a much smaller recurrence
relation can be found explicitly. This remains a challenging topic for further
research.

Our results can be extended in several ways:

(i) We can add one more MSOL-smooth basic operation, provided the vocabu-
     lary τ contains only unary and binary relation symbols.
     Fuse: The graph $fuse_i(G)$ is obtained from G by identifying all vertices of
         color C_i and leaving all the resulting edges with the exception of the
         resulting loops.
     A detailed discussion of this and further operations may be found in [23, Sect.
     3] and in [9]. We do not go into further details, due to space limitations.
(ii) The logic MSOL can be extended to the logic CMSOL where we have addi-
     tionally modular counting quantifiers $C_{m,n}x\phi(x)$ for each $m, n \in \mathbb{N}$, which
     say that there are, modulo m, exactly n elements satisfying $\phi(x)$. The Split-
     ting Theorem in [23] is proven for CMSOL. Clearly CMSOL is a sublogic
     of SOL. The interlace polynomials, [1,2] are CMSOL-definable, but it is not
     known whether they are MSOL-definable.

(iii) In [15] the notion of clique-width was generalized further and the notion of patch-width was introduced. For F an MSOL-elementary operation the F-iteration sequences of arbitrary τ-structures are of bounded patch-width. It remains to be investigated whether there are interesting polynomial invariants of τ-structures, where our method leads to useful results.

Our method does not apply to square grids $G_{n,n}$. But they are obviously regularly constructed in some way, and it is to be expected that for most graph polynomials $\mathfrak{P}$, some recurrence relation does exist to compute the values of $\mathfrak{P}(G_{n,n})$. Can one find a general theorem which captures this intuition?

Acknowledgments

We would like to thank the referee for has very careful reading of the first version of this paper. We would also like to thank the editors of this volume, Arnon Avron, Nachum Dershowitz, and Alexander Rabinovich, for their encouragement and patience.

References

1. Arratia, R., Bollobás, B., Sorkin, G.B.: The interlace polynomial: a new graph polynomial. Journal of Combinatorial Theory, Series B 92, 199–233 (2004)
2. Arratia, R., Bollobás, B., Sorkin, G.B.: A two-variable interlace polynomial. Combinatorica 24(4), 567–584 (2004)
3. Biggs, N.: Algebraic Graph Theory, 2nd edn. Cambridge University Press, Cambridge (1993)
4. Biggs, N.L., Damerell, R.M., Sand, D.A.: Recursive families of graphs. J. Combin. Theory Ser. B 12, 123–131 (1972)
5. Bollobás, B.: Modern Graph Theory. Springer, Heidelberg (1999)
6. Bollobás, B., Riordan, O.: A Tutte polynomial for coloured graphs. Combinatorics, Probability and Computing 8, 45–94 (1999)
7. Chung, F.R.K., Graham, R.L.: On the cover polynomial of a digraph. Journal of Combinatorial Theory, Ser. B 65(2), 273–290 (1995)
8. Courcelle, B.: A multivariate interlace polynomial (December 2006) (preprint)
9. Courcelle, B., Makowsky, J.A.: Fusion on relational structures and the verification of monadic second order properties. Mathematical Structures in Computer Science 12(2), 203–235 (2002)
10. Courcelle, B., Makowsky, J.A., Rotics, U.: On the fixed parameter complexity of graph enumeration problems definable in monadic second order logic. Discrete Applied Mathematics 108(1–2), 23–52 (2001)
11. Courcelle, B., Olariu, S.: Upper bounds to the clique–width of graphs. Discrete Applied Mathematics 101, 77–114 (2000)
12. Dong, F.M., et al.: The vertex-cover polynomial of a graph. Discrete Mathematics 250, 71–78 (2002)
13. Dong, F.M., Koh, K.M., Teo, K.L.: Chromatic Polynomials and Chromaticity of Graphs. World Scientific, Singapore (2005)
14. Ebbinghaus, H.D., Flum, J.: Finite Model Theory. In: Perspectives in Mathematical Logic, Springer, Heidelberg (1995)

15. Fischer, E., Makowsky, J.A.: On spectra of sentences of monadic second order logic with counting. Journal of Symbolic Logic 69(3), 617–640 (2004)
16. Frick, M., Grohe, M.: The complexity of first-order and monadic second-order logic revisited. Annals of Pure and Applied Logic 130(1), 3–31 (2004)
17. Golumbic, M.C., Rotics, U.: On the clique-width of some perfect graph classes. Internation Journal of Foundations of Computer Science 11, 423–443 (2000)
18. Graham, R., Knuth, D., Patashnik, O.: Concrete Mathematics, 2nd edn. Addison-Wesley, Reading (1994)
19. Gurevich, Y.: Modest theory of short chains, I. Journal of Symbolic Logic 44, 481–490 (1979)
20. Gurski, F., Wanke, E.: On the relationship between NLC-width and linear NLC-width. Theoretical Computer Science 347(1–2), 76–89 (2005)
21. Kobrinski, N.E., Trakhtenbrot, B.A.: Introduction to the Theory of Finite Automata. In: Studies in Logic and the Foundations of Mathematics, North-Holland, Amsterdam (1965)
22. Läuchli, H.: A decision procedure for the weak second order theory of linear order. In: Logic Colloquium 1966, pp. 189–197. North-Holland, Amsterdam (1968)
23. Makowsky, J.A.: Algorithmic uses of the Feferman-Vaught theorem. Annals of Pure and Applied Logic 126, 1–3 (2004)
24. Makowsky, J.A.: From a zoo to a zoology: Towards a general theory of graph polynomials. Theory of Computing Systems, published online first (July 2007)
25. Makowsky, J.A., Mariño, J.P.: Farrell polynomials on graphs of bounded treewidth. Advances in Applied Mathematics 30, 160–176 (2003)
26. Makowsky, J.A., Zilber, B.: Polynomial invariants of graphs and totally categorical theories. MODNET Preprint No. 21 server (2006),
    http://www.logique.jussieu.fr/modnet/Publications/Preprint%20
27. Noy, M., Ribó, A.: Recursively constructible families of graphs. Advances in Applied Mathematics 32, 350–363 (2004)
28. Rozenberg, G. (ed.): Handbook of graph grammars and computing by graph transformations, Foundations, vol. 1. World Scientific, Singapore (1997)
29. Shelah, S.: The monadic theory of order. Annals of Mathematics 102, 379–419 (1975)
30. Trakhtenbrot, B.: Finite automata and the logic of monadic predicates. Doklady Akademy Nauk SSSR 140, 326–329 (1961)
31. Trakhtenbrot, B.: Some constructions in the monadic predicate calculus. Doklady Akademy Nauk SSSR 138, 320–321 (1961)

Artin's Conjecture and Size of Finite Probabilistic Automata*

Rūsiņš Freivalds

Institute of Mathematics and Computer Science, University of Latvia,
Raiņa bulvāris 29, Rīga, LV-1459, Latvia

I was Prof. Boris Trakhtenbrot's doctoral student in Novosibirsk, 1968-1970. I learned from him much, including the notion of probabilistic automata. I thank him for getting me interested in Theoretical Computer Science.

Abstract. Size (the number of states) of finite probabilistic automata with an isolated cut-point can be exponentially smaller than the size of any equivalent finite deterministic automaton. The result is presented in two versions. The first version depends on Artin's Conjecture (1927) in Number Theory. The second version does not depend on conjectures but the numerical estimates are worse. In both versions the method of the proof does not allow an explicit description of the languages used. Since our finite probabilistic automata are reversible, these results imply a similar result for quantum finite automata.

1 Introduction

M. O. Rabin proved in [14] that if a language is recognized by a finite probabilistic automaton with n states, r accepting states and isolation radius δ then there exists a finite deterministic automaton which recognizes the same language and the deterministic automaton may have no more than $(1 + \frac{r}{\delta})^n$ states. However, how tight is this bound? Rabin gave an example of languages in [14] where probabilistic automata indeed had size advantages but these advantages were very far from the exponential gap predicted by the formula $(1 + \frac{r}{\delta})^n$. Unfortunately, the advantage proved by Rabin's example was only linear, not exponential. Is it possible to diminish the gap? Is the upper bound $(1 + \frac{r}{\delta})^n$ tight or is Rabin's example the best possible?

R. Freivalds in [5] constructed an infinite sequence of finite probabilistic automata such that every automaton recognizes the corresponding language with the probability $\frac{3}{4}$, and if the probabilistic automaton has n states then the language cannot be recognized by a finite deterministic automaton with less than $\Omega(2^{\sqrt{n}})$ states. This did not close the gap between the lower bound $\Omega(2^{\sqrt{n}})$ and the purely exponential upper bound $(1 + \frac{r}{\delta})^n$ but now it was clear that the size advantage of probabilistic versus deterministic automata may be super-polynomial.

* Research supported by Grant No.05.1528 from the Latvian Council of Science and European Commission, contract IST-1999-11234.

A. Avron et al. (Eds.): Trakhtenbrot/Festschrift, LNCS 4800, pp. 280–291, 2008.

A. Ambainis [1] constructed a new sequence of languages and corresponding sequence of finite probabilistic automata such that every automaton recognizes the corresponding language with the probability $\frac{3}{4}$ and if the probabilistic automaton has n states then the language cannot be recognized by a finite deterministic automaton with less than $\Omega(2^{\frac{n \log \log n}{\log n}})$ states. On the other hand, the languages in [5] were in a single-letter alphabet but for the languages in [1] the alphabet grew with n unlimitedly.

This paper gives the first ever purely exponential distiction between the sizes of probabilistic and deterministic finite automata. Existence of an infinite sequence of finite probabilistic automata is proved such that all of them recognize some language with a fixed probability $p > \frac{1}{2}$ and if the probabilistic automaton has n states then the language cannot be recognized by a finite deterministic automaton with less than $\Omega(a^n)$ states for a certain $a > 1$. This does not end the search for the advantages of probabilistic finite automata over deterministic ones. We still do not know the best possible value of a. Moreover, the best estimate proved in this paper is proved under assumption of the well-known Artin's conjecture in Number Theory. Our final Theorem 3 does not depend on any open conjectures but the estimate is worse, and the description of the languages used is even less constructive. These seem to be the first results in Finite Automata depending on open conjectures in Number Theory.

The essential proofs are non-constructive. Such an approach is not new. A good survey of many impressive examples of non-constructive methods is by J. Spencer [16]. Technically, the crucial improvement over existing results and methods comes from our usage of mirage codes to construct finite probabilistic automata. Along this path of proof, it turned out that the best existing result on mirage codes (Theorem A below) is not strong enough for our needs. The improvement of Theorem A is based on the notion of Kolmogorov complexity. It is well known that Kolmogorov complexity is not effectively computable. It turned out that non-computability of Kolmogorov complexity allows to prove the existence of the needed mirage codes and it is enough for us to prove an exponential gap between the size of probabilistic and deterministic finite automata recognizing the same language. On the other hand, some results of abstract algebra (namely, elementary properties of group homomorphisms) are also used in these proofs.

2   Number-Theoretical Conjectures

By p we denote an odd prime number, i.e. a prime greater than 2. To prove the main theorems we consider several lemmas. Most of them are valid for arbitrary p but we are going to use them only for odd primes of a special type.

Consider the sequence

$$2^0, 2^1, 2^2, \ldots, 2^{p-2}, 2^{p-1}, 2^p, \ldots$$

and the corresponding sequence of the remainders of these numbers modulo p

$$r^0, r^1, r^2, \ldots, r^{p-2}, r^{p-1}, r^p, \ldots \tag{1}$$

$(r_k \equiv 2^k \pmod{p})$. For arbitrary p, the sequence (1) is periodic. Since $r_0 = 1$ and, by the Fermat Little Theorem, $r_{p-1} \equiv 2^{p-1} \equiv 1 \pmod{p}$, one may think that $p - 1$ is the least period of the sequence (1).

This is not the case. For instance, $2^{7-1} \equiv 1 \pmod 7$. but also $2^3 \equiv 1 \pmod 7$. However, sometimes $p - 1$ can be the least period of the sequence (1). In this case, 2 is called a primitive root modulo p. More generally, a number a is called a primitive root modulo p if and only if a is a relatively prime to p and $p - 1$ is the least period in the sequence of remainders modulo p of the numbers

$$a^0 = 1, a^1, a^2, \ldots, a^{p-2}, a^{p-1}, a^p, \ldots$$

Emil Artin made in 1927 a famous conjecture the validity of which is still an open problem.

Artin's Conjecture. [3] *If a is neither -1 nor a square, then a is a primitive root for infinitely many primes.*

Moreover, it is conjectured that density of primes for which a is a primitive root equals $A = 0.373956\ldots$. In 1967, C. Hooley [9] proved that Artin's conjecture follows from the Generalized Riemann hypothesis. D. R. Heath-Brown [10] proved that Artin's conjecture can be wrong no more than for 2 distinct primes a.

3 Linear Codes

Linear codes is the simplest class of codes. The alphabet used is a fixed choice of a finite field $GF(q) = F_q$ with q elements. For most of this paper we consider a special case of $GF(2) = F_2$. These codes are binary codes.

A generating matrix G for a linear $[n, k]$ code over F_q is a k-by-n matrix with entries in the finite field F_q, whose rows are linearly independent. The linear code corresponding to the matrix G consists of all the q^k possible linear combinations of rows of G. The requirement of linear independence is equivalent to saying that all the q^k linear combinations are distinct. The linear combinations of the rows in G are called codewords. However we are interested in something more. We need to have the codewords not merely distinct but also as far as possible in terms of Hamming distance. Hamming distance between two vectors $v = (v_1, \ldots, v_n)$ and $w = (w_1, \ldots, w_n)$ in F_{q^k} is the number of indices i such that $v_i \neq w_i$.

The textbook [7] contains

Theorem A. *For any integer $n \geq 4$ there is a $[2n, n]$ binary code with a minimum distance between the codewords at least $n/10$.*

However the proof of the theorem in [7] has a serious defect. It is non-constructive. It means that we cannot find these codes or describe them in a useful manner. This is why P. Garret calls them mirage codes.

If q is a prime number, the set of the codewords with the operation "component-wise addition" is a group. Finite groups have useful properties. We single out Lagrange's Theorem. The order of a finite group is the number of elements in it.

Lagrange's Theorem (see e.g. [7]). *Let GR be a finite group. Let H be a subgroup of GR. Then the order of H divides the order of G.*

Definition 1. *A generating matrix G of a linear code is called* **cyclic** *if along with an arbitrary row $(v_1, v_2, v_3, \ldots, v_n)$ the matrix G contains also a row $(v_2, v_3, \ldots, v_n, v_1)$.*

We would have liked to prove a reasonable counterpart of Theorem A for cyclic mirage codes, but this attempt fails. Instead we consider binary generating matrices of a bit different kind. Let p be an odd prime number, and x be a binary word of length p. The generating matrix $G(p, x)$ has p rows and $2p$ columns. Let $x = x_1 x_2 x_3 \ldots x_p$. The first p columns (and all p rows) make a unit matrix with elements 1 on the main diagonal and 0 in all the other positions. The last p columns (and all p rows) make a cyclic matrix with $x = x_1 x_2 x_3 \ldots x_p$ as the first row, $x = x_p x_1 x_2 x_3 \ldots x_{p-1}$ as the second row, and so on.

Lemma 1. *For arbitrary x, if $h_1 h_2 h_3 \ldots h_p h_{p+1} h_{p+2} h_{p+3} \ldots h_{2p}$ is a codeword in the linear code corresponding to $G(p, x)$, then $h_p h_1 h_2 \ldots h_{p-1} h_{2p} h_{p+1} h_{p+2} \ldots h_{2p-1}$ is also a codeword.*

There are 2^p codewords of the length $2p$. If the codeword is obtained as a linear combination with the coefficients $c_1, c_2, \ldots, c_p$ then the first p components of the codeword equal $c_1 c_2 \ldots c_p$. We denote by $R(x, c_1 c_2 \ldots c_p)$ the subword containing the last p components of this codeword.

Lemma 2. *If $c_1 c_2 \ldots c_p = 000 \ldots 0$, then $R(x, c_1 c_2 \ldots c_p) = 000 \ldots 0$, for arbitrary x.*

Definition 2. *We will call a word* **trivial** *if all its symbols are equal. Otherwise we call the word* **nontrivial**.

Lemma 3. *If $c_1 c_2 \ldots c_p$ is trivial, then $R(x, c_1 c_2 \ldots c_p)$ is trivial for arbitrary x.*

Proof. Every symbol of $R(x, c_1 c_2 \ldots c_p)$ equals $x_1 + x_2 + \ldots + x_p \pmod 2$.

Lemma 4. *If x is trivial, then $R(x, c_1 c_2 \ldots c_p)$ is trivial for arbitrary $c_1 c_2 \ldots c_p$.*

Definition 3. *A word $x = x_1 x_2 \ldots x_p$ is called a* **cyclic shift** *of the word $y = y_1 y_2 \ldots y_p$ if there exists i such that $x_1 = y_i, x_2 = y_{i+1}, \ldots, x_p = y_{i+p}$ where the addition is modulo p. If $(i, p) = 1$, then we say that this cyclic shift is* **nontrivial**.

Lemma 5. *If x is a cyclic shift of y, then $R(x, c_1 c_2 \ldots c_p)$ is a cyclic shift of $R(y, c_1 c_2 \ldots c_p)$.*

Lemma 6. *If p is an odd prime, x is a nontrivial word and y is a nontrivial cyclic shift of x, then $x \neq y$.*

Lemma 7. *If p is an odd prime and $c_1 c_2 \ldots c_p$ is nontrivial, then the set $T_{c_1 c_2 \ldots c_p} = \{ R(x, c_1 c_2 \ldots c_p) | x \in \{0, 1\}^p$ and $R(x, c_1 c_2 \ldots c_p)$ nontrivial $\}$ has a cardinality which is a multiple of p.*

Proof. Immediately from Lemmas 5 and 6.

For arbitrary fixed $c_1c_2 \ldots c_p$, the set $\{R(x, c_1c_2 \ldots c_p) | x \in \{0,1\}^p\}$ with algebraic operation "component-wise addition modulo z" is a group. We denote this group by B. By D we denote the group of all 2^p binary words of the length p with the same operation.

Lemma 8. *For arbitrary $c_1c_2 \ldots c_p$, x and y,*
$R(x, c_1c_2 \ldots c_p) + R(y, c_1c_2 \ldots c_p) = R(x + y, c_1c_2 \ldots c_p)$.

In other words, for arbitrary $c_1c_2 \ldots c_p$, the map $D \to B$ defined by $x \to R(x, c_1c_2 \ldots c_p)$ is a group homomorphism. (Definition and properties of group homomorphisms can be found in every textbook on group theory. See e.g. [4].) The **kernel** of the group homomorphism is the set $ker_0 = \{x | R(x, c_1c_2 \ldots c_p) = 000 \ldots 0\}$.

The image of the group homomorphism is the set B. For arbitrary $z \in B$, by ker_z we denote the set $ker_z = \{x | R(x, c_1c_2 \ldots c_p) = z\}$.

From Lemma 8 we easily get

Lemma 9. *For arbitrary $z \in B$, $card(ker_z) = card(ker_0)$.*

Lemma 10. *For arbitrary $z \in B$, $card(ker_z) = \frac{card(D)}{card(B)}$.*

Lemma 11. *If x contains $(p-1)$ zeroes and 1 one, and $c_1c_2 \ldots c_p$ is nontrivial, then $R(x, c_1c_2 \ldots c_p)$ is nontrivial.*

Proof. For such an x, the number of ones in $R(x, c_1c_2 \ldots c_p)$ is the same as the number of ones in $c_1c_2 \ldots c_p$.

Lemma 12. *If p is an odd prime such that 2 is a primitive root modulo p and $c_1c_2 \ldots c_p$ is nontrivial, then the set $S_{c_1c_2\ldots c_p} = \{R(x, c_1c_2 \ldots c_p) | x \in \{0,1\}^p\}$ is either of cardinality 1 or of cardinality 2.*

Proof. By Lagrange's Theorem the order 2^p of the group B divides the order of the group D. Hence the order of B is 2^b for some integer b. The neutral element of these groups is the word $000 \ldots 0$. It belongs to every subgroup. There are two possible cases:

1. $111 \ldots 1$ is in B,
2. $111 \ldots 1$ is not in B.

In the case 1 $card(T_{c_1c_2\ldots c_p}) = card(B) - 2$, and by Lemmas 7 and 10 $card(T_{c_1c_2\ldots c_p})$ is a multiple of p. Hence $2^b = card(B) \equiv 2 \pmod{p}$ and $2^{b-1} \equiv 1 \pmod{p}$. Since 2 is a primitive root modulo p, either $2^{b-1} = 2^{p-1}$ or $2^{b-1} = 2^0$. If $2^{b-1} = 2^{p-1}$, then $2^b = 2^p$ and for this fixed $c_1c_2 \ldots c_p$ the map $x \to R(x, c_1c_2 \ldots c_p)$ takes distinct x'es into distinct $R(x, c_1c_2 \ldots c_p)$'s. If $2^{b-1} = 2^0$, then $2^b = 2$ and $B = \{000 \ldots 0, 111 \ldots 1\}$, but this is impossible by Lemma 11.

In the case 2 $card(T_{c_1c_2...c_p}) = card(B) - 1$ and by Lemma 7 $card(T_{c_1c_2...c_p})$ is a multiple of p. Hence $2^b \equiv 1 \pmod{p}$. Since 2 is a primitive root modulo p, either $2^b = 2^{p-1}$ or $2^b = 2^0$. If $2^b = 2^{p-1}$, then $card(B) = 2^{p-1}$ and, by Lemma 10, for arbitrary $z \in T_{c_1c_2...c_p}$, $card(\ker_z) = 2$. If $2^b = 2^0$, then $B = \{000...0\}$ but this is impossible by Lemma 11.

4 Kolmogorov Complexity

The theorems in this section are well-known results in spite of the fact that it is not easy to find exact references for all of them.

Definition 4. *We say that the numbering $\Psi = \{\Psi_0(x), \Psi_1(x), \Psi_2(x), ...\}$ of 1-argument partial recursive functions is* **computable** *if the 2-argument function $U(n, x) = \Psi_n(x)$ is partial recursive.*

Definition 5. *We say that a numbering Ψ is reducible to the numbering η if there exists a total recursive function $f(n)$ such that, for all n and x, $\Psi_n(x) = \eta_{f(n)}(x)$.*

Definition 6. *We say that a computable numbering φ of all 1-argument partial recursive functions is a* **Gödel numbering** *if every computable numbering (of any class of 1-argument partial recursive functions) is reducible to φ.*

Theorem ([15]). *There exists a Gödel numbering.*

Definition 7. *We say that a Gödel numbering ϑ is a* **Kolmogorov numbering** *if for arbitrary computable numbering Ψ (of any class of 1-argument partial recursive functions) there exist constants $c > 0, d > 0$, and a total recursive function $f(n)$ such that:*

1. *for all n and x, $\Psi_n(x) = \vartheta_{f(n)}(x)$,*
2. *for all n, $f(n) \leq c \cdot n + d$.*

Kolmogorov Theorem ([11]). *There exists a Kolmogorov numbering.*

5 New Mirage Codes

In the beginning of Section 3 we introduced a special type generating matrices $G(p, x)$ where p is an odd prime and x is a binary word of length p. Now we introduce two technical auxiliary functions. If z is a binary word of length $2p$, then $d(z)$ is the subword of z containing the first p symbols, and $e(z)$ is subword of z containing the last p symbols. Then $z = d(z)e(z)$.

There exist many distinct Kolmogorov numberings. We now fix one of them and denote it by η. Since Kolmogorov numberings give indices for all partial recursive functions, for arbitrary x and p, there is an i such that $\eta_i(p) = x$. Let

$i(x, p)$ be the minimal i such that $\eta_i(p) = x$. It is easy to see that if $x_1 \neq x_2$, then $i(x_1, p) \neq i(x_2, p)$. We consider all binary words x of the length p and denote by $x(p)$ the word x such $i(x, p)$ exceed $i(y, p)$ for all binary words y of the length p different from x. It is obvious that $i \geq 2^p - 1$.

Until now we considered generating matrices $G(p, x)$ for independently chosen p and x. From now on we consider only odd primes p such that 2 is a primitive root modulo p and the matrices $G(p, x(p))$. We wish to prove that if p is sufficiently large, then Hamming distances between two arbitrary codewords in this linear code is at least $\frac{4p}{19}$.

We introduce a partial recursive function $\mu(z, \epsilon, p)$ defined as follows. Above when defining $G(p, x)$ we considered auxiliary function $R(x, c_1 c_2 \ldots c_p)$. To define $\mu(z, \epsilon, p)$ we consider all 2^p binary words x of the length p. If z is not a binary word of length $2p$, then $\mu(z, \epsilon, p)$ is not defined. If ϵ is not in $\{0, 1\}$, then $\mu(z, \epsilon, p)$ is not defined. If z is a binary word of length $2p$ and $\epsilon \in \{0, 1\}$, then we consider all $x \in \{0, 1\}^p$ such that $R(x, d(z)) = e(z)$. If there are no such x, then $\mu(z, \epsilon, p)$ is not defined. If there is only one such x, then $\mu(z, \epsilon, p) = x$. If there are two such x, then

$$\mu(z, \epsilon, p) = \begin{cases} \text{the first such } x \text{ in the lexicographical order, for } \epsilon = 1 \\ \text{the second such } x \text{ in the lexicographical order, for } \epsilon = 0 \end{cases}$$

If there are more than two such x, then $\mu(z, \epsilon, p)$ is not defined.

Now we introduce a computable numbering of some partial recursive functions. This numbering is independent of p.

For each p (independently from other values of p) we order the set of all the 2^{2p} binary words z of the length $2p$: $z_0, z_1, z_2, \ldots, z_{2^{2p}-1}$. We define z_0 as the word $000 \ldots 0$. The words $z_1, z_2, \ldots, z_{2^{2p}-1}$ are words with exactly one symbol 1. We strictly follow a rule "if the word z_i contains less symbols 1 than the word z_j, then $i < j$". Words with equal number of the symbol 1 are ordered lexicographically. Hence $z_{2^{2p}-1} = 111 \ldots 1$.

For each p, we define

$$\Psi_0(p) = \mu(z_0, 0, p)$$
$$\Psi_1(p) = \mu(z_0, 1, p)$$
$$\Psi_2(p) = \mu(z_1, 0, p)$$
$$\Psi_3(p) = \mu(z_1, 1, p)$$
$$\Psi_4(p) = \mu(z_2, 0, p)$$
$$\Psi_5(p) = \mu(z_2, 1, p)$$
$$\ldots$$
$$\Psi_{2^{2p+1}-2}(p) = \mu(z_{2^{2p}-1}, 0, p)$$
$$\Psi_{2^{2p+1}-1}(p) = \mu(z_{2^{2p}-1}, 1, p)$$

For $j \geq 2^{2p+1}$, $\Psi_j(p)$ is undefined.

We have fixed a Kolmogorov numbering η and we have just constructed a computable numbering Ψ of some partial recursive functions.

Lemma 13. *There exist constants $c > 0$ and $d > 0$ (independent of p) such that for arbitrary i there is a j such that*

1. *$\Psi_i(t) = \eta_j(t)$ for all t, and*
2. *$j \leq ci + d$.*

Proof. Immediately from Kolmogorov Theorem.

We consider generating matrices $G(p, x(p))$ for linear codes where p is an odd prime such that 2 is a primitive root modulo p, and, as defined above, $x(p)$ is a binary word of length p such that $\eta_i(p) = x(p)$ implies $i \geq 2^p - 1$. We denote the corresponding linear code by $LC_2(p)$.

Now we prove several lemmas showing that, if p is sufficiently large, then Hamming distances between arbitrary two codewords are no less than $\frac{4p}{19}$.

Lemma 14. *For every linear code, there is a codeword $000\ldots0$.*

Proof. The codeword $000\ldots0$ is obtained by using coefficients $c_1 c_2 \ldots c_p = 000\ldots0$.

Lemma 15. *For every linear code, if there exists a pair of distinct codewords with Hamming distance less than d, then there is a codeword with less than d symbols 1 in it.*

Proof. If x_1 and x_2 are codewords, then $x_1 \oplus x_2$ also is a codeword.

Lemma 16. *If p is sufficiently large, and a codeword in $LC_2(p)$ contains less than $\frac{4p}{19}$ symbols 1, then the codeword is $000\ldots0$.*

Proof. Assume from the contrary that there is a codeword $z \neq 000\ldots0$ containing less than $\frac{4p}{19}$ symbols 1. Above we introduced an ordering $z_0, z_1, z_2, \ldots, z_{2^{2p}-1}$ of all binary words of the length $2p$. Then $z = z_i$ where

$$i \leq \binom{2p}{0} + \binom{2p}{1} + \binom{2p}{2} + \cdots + \binom{2p}{\lfloor \frac{4p}{19} \rfloor}.$$

Hence $i = o(2^p)$. On the other hand, the choice of $x(p)$ implies that $i \geq 2^p - 1$. Contradiction.

Lemma 17. *If p is sufficiently large, then the Hamming distance between any two distinct codewords in $LC_2(p)$ is no less than $\frac{4p}{19}$.*

Proof. By Lemmas 16 and 15.

6   Probabilistic Reversible Automata

M. Golovkins and M. Kravtsev [8] introduced probabilistic reversible automata (PRA) to describe the intersection of two classes of automata, namely, the classes of the 1-way probabilistic and quantum automata. The paper [8] describes several versions of these automata. We concentrate here on the simplest and the least powerful class of PRA.

$\Sigma = \{a_1, a_2, \ldots, a_m\}$ is the input alphabet of the automaton. Every input word is enclosed into end-marker symbols $\#$ and $\$$. Therefore the working alphabet is defined as $\Gamma = \Sigma \cup \{\#, \$\}$. $Q = \{q_1, q_2, \ldots, q_n\}$ is a finite set of states. Q is presented as a union of two disjoint sets: Q_A (accepting states) and Q_R (rejecting states). At every step, the PRA is in some probability distribution $(p_1, p_2, \ldots, p_n)$ where $p_1 + p_2 + \ldots + p_n = 1$. As the result of reading the input $\#$, the automaton enters the initial probability distribution $(p_1(0), p_2(0), \ldots, p_n(0))$. $M_1, M_2, \ldots, M_m$ are doubly-stochastic matrices characterising the evolution of probability distributions.

If at some moment t the probability distribution is

$$(p_1(t), p_2(t), \ldots, p_n(t))$$

and the input symbol is a_u, then the probability distribution

$$(p_1(t+1), p_2(t+1), \ldots, p_n(t+1))$$

equals $(p_1(t), p_2(t), \ldots, p_n(t)) \cdot M_u$. If after having read the last symbol of the input word x the automaton has reached a probability distribution $(p_1, p_2, \ldots, p_n)$ then the probability to accept the word equals

$$prob_x = \Sigma_{i \in Q_A} p_i$$

and the probability to reject the word equals

$$1 - prob_x = \Sigma_{i \in Q_R} p_i.$$

We say that a language L is recognized with bounded error with an interval (p_1, p_2) if $p_1 < p_2$ where $p_1 = sup\{prob_x | x \notin L\}$ and $p_2 = inf\{prob_x | x \in L\}$.

We say that a language L is recognized with a probability $p > \frac{1}{2}$ if the language is recognized with interval $(1 - p, p)$.

In the previous section we constructed a binary generating matrix $G(p, p(x))$ for a linear code. Now we use this matrix to construct a probabilistic reversible automaton $R(p)$.

The matrix $G(p, x(p))$ has $2p$ columns and p rows. The automaton $R(p)$ has $4p + 1$ states, $2p$ of them being accepting and $2p + 1$ being rejecting. The input alphabet consists of 2 letters.

The (rejecting) state q_0 is special in the sense that the probability to enter this state and the probability to exit from this state during the work equals 0. This state always has the probability $\frac{17}{36}$. The states $q_1, q_2, \ldots, q_{4p}$ are related to the columns of $G(p, x(p))$ and should be considered as $2p$ pairs $(q_1, q_2), (q_3, q_4), \ldots, \ldots (q_{4p-1}, q_{4p})$ corresponding to the $2p$ columns of $G(p, x(p))$. The states $q_1, q_3, q_5, q_7, \ldots, q_{4p-1}$ are accepting and the states $q_2, q_4, q_6, q_8, \ldots, q_{4p}$ are rejecting. The initial probability distribution is as follows:

$$\begin{cases} \frac{17}{36}, \text{ for } q_0, \\ \frac{19}{72p}, \text{ for each of } q_1, q_3, \ldots, q_{4p-1} \\ 0, \text{ for each of } q_2, q_4, \ldots, q_{4p}. \end{cases}$$

The processing of the input symbols a, b is deterministic. Under the input symbol a the states are permuted as follows:

$$q_1 \rightarrow q_3 \qquad q_2 \rightarrow q_4 \qquad q_{2p+1} \rightarrow q_{2p+3} \qquad q_{2p+2} \rightarrow q_{2p+4}$$
$$q_3 \rightarrow q_5 \qquad q_4 \rightarrow q_6 \qquad q_{2p+3} \rightarrow q_{2p+5} \qquad q_{2p+4} \rightarrow q_{2p+6}$$
$$q_5 \rightarrow q_7 \qquad q_6 \rightarrow q_8 \qquad q_{2p+5} \rightarrow q_{2p+7} \qquad q_{2p+6} \rightarrow q_{2p+8}$$
$$\ldots \qquad\qquad \ldots \qquad\qquad \ldots \qquad\qquad \ldots$$
$$q_{2p-3} \rightarrow q_{2p-1} \quad q_{2p-2} \rightarrow q_{2p} \quad q_{4p-3} \rightarrow q_{4p-1} \quad q_{4p-2} \rightarrow q_{4p}$$
$$q_{2p-1} \rightarrow q_1 \qquad q_{2p} \rightarrow q_2 \qquad q_{4p-1} \rightarrow q_{2p+1} \qquad q_{4p} \rightarrow q_{2p+2}$$

The permutation of the states under the input symbol b depends on $G(p, x(p))$. Let

$$G(p, x(p)) = \begin{pmatrix} g_{11} & g_{12} & \cdots & g_{1\ 2p} \\ g_{21} & g_{22} & \cdots & g_{2\ 2p} \\ \cdots & \cdots & \cdots & \cdots \\ g_{p1} & g_{p2} & \cdots & g_{p\ 2p} \end{pmatrix}$$

For arbitrary $i \in \{1, 2, \ldots, p\}$,

$$\begin{cases} q_{2i-1} \rightarrow q_{2i-1} \ , & \text{if } g_{1i} = 0 \\ q_{2i} \rightarrow q_{2i} \quad\ , & \text{if } g_{1i} = 0 \\ q_{2i-1} \rightarrow q_{2i} \quad , & \text{if } g_{1i} = 1 \\ q_{2i} \rightarrow q_{2i-1} \ , & \text{if } g_{1i} = 1. \end{cases}$$

In order to understand the language recognized by the automaton $R(p)$ we consider the following auxiliary mapping W from the words in $\{a, b\}^*$ into the set of binary $2p$-vectors defined recursively:

1. $CW(\Lambda) = g_{11}g_{12} \ldots g_{1\ 2p}$
2. if $CW(w) = h_1 h_2 h_3 \ldots h_p h_{p+1} h_{p+2} h_{p+3} \ldots h_{2p}$ then

$$\begin{cases} CW(wa) = h_p h_1 h_2 \ldots h_{p-1} h_{2p} h_{p+1} h_{p+2} \ldots h_{2p-1} \text{ and} \\ CW(wb) = (h_1 \oplus g_{11})(h_2 \oplus g_{12})(h_3 \oplus g_{13}) \ldots (h_{2p} \oplus g_{1\ 2p}). \end{cases}$$

The next two lemmas can be proved by induction over the length of w.

Lemma 18. *For arbitrary word $w \in \{a, b\}^*$, $CW(w)$ is a codeword in the linear code corresponding to the generating matrix $G(p, x(p))$.*

Lemma 19. *Let w be an arbitrary word in $\{a, b\}^*$, and $CW(w) = h_1 h_2 \ldots h_{2p}$. Then the probability distribution of the states in $R(p)$ is*

$$\begin{cases} \frac{17}{36} & , \text{ for } g_0, \\ \frac{19}{72p} & , \text{ for } g_{2i-1} \text{ if } h_i = 0, \\ 0 & , \text{ for } g_{2i} \text{ if } h_i = 0, \\ 0 & , \text{ for } g_{2i-1} \text{ if } h_i = 1, \\ \frac{19}{72p} & , \text{ for } g_{2i} \text{ if } h_i = 1. \end{cases}$$

We introduce a language

$$L_{G(p, x(p))} = \{w | w \in \{a, b\}^* \& CW(w) = 000 \ldots 0\}.$$

Lemma 20. *If 2 is a primitive root modulo p and p is sufficiently large, then the automaton $R(p)$ recognizes the language $L_{G(p,x(p))}$ with the probability $\frac{19}{36}$.*

Lemma 21. *For arbitrary p and arbitrary deterministic finite automaton A recognizing $L_{G(p,x(p))}$ the number of states of A is no less than 2^p.*

Lemmas 20 and 21 imply

Theorem 1. *If 2 is a primitive root for infinitely many distinct primes then there exists an infinite sequence of regular languages $L_1, L_2, L_3, \ldots$ in a 2-letter alphabet and a sequence of positive integers $p(1), p(2), p(3), \ldots$ such that for arbitrary j:*

1. *any deterministic finite automaton recognizing L_j has at least $2^{p(j)}$ states,*
2. *there is a probabilistic reversible automaton with $(4p(j)+1)$ states recognizing L_j with the probability $\frac{19}{36}$.*

7   Without Conjectures

In 1989 D. R. Heath-Brown [10] proved Artin's conjecture for "nearly all integers". We use the following corollary from Heath-Brown's Theorem:

Corollary from Heath-Brown Theorem ([10]). *At least one integer a in the set $\{3, 5, 7\}$ is a primitive root for infinitely many primes p.*

Above we constructed a binary linear code, the binary generating matrix $G(p, x(p))$ of which incorporated a binary word $x(p)$ with maximum complicity in the Kolmogorov numbering η. Now we are going to modify the construction to get generating matrices $G_3(p, x_3(p))$, $G_5(p, x_5(p))$, $G_7(p, x_7(p))$ for ternary, pentary and septary linear codes $LC_3(p)$, $LC_5(p)$ and $LC_7(p)$, respectively. The constructions remain essentially the same only the words x and $c_1 c_2 \ldots c_p$ now are in $\{0, 1, 2\}^p$, $\{0, 1, 2, 3, 4\}^p$ or $\{0, 1, \ldots, 6\}^p$, resp., and the summation is modulo 3, 5, 7, resp. Recall that by Heath-Brown's Theorem [10] there exists $u \in \{3, 5, 7\}$ such that u is a primitive root for infinetely many distinct primes.

Theorem 1 can be re-formulated as follows.

Theorem 2. *Assume Artin's Conjecture. There exists an infinite sequence of regular languages $L_1, L_2, L_3, \ldots$ in a 2-letter alphabet and an infinite sequence of positive integers $z(1), z(2), z(3), \ldots$ such that for arbitrary j.*

1. *there is a probabilistic reversible automaton with $(z(j)$ states recognizing L_j with the probability $\frac{19}{36}$,*
2. *any deterministic finite automaton recognizing L_j has at least $(2^{1/4})^{z(j)} = (1.1892071115\ldots)^{z(j)}$ states.*

Corollary from Heath-Brown's Theorem allows us to prove the following counterpart of Theorem 2.

Theorem 3. *There exists an infinite sequence of regular languages L_1, L_2, L_3, ... in a 2-letter alphabet and an infinite sequence of positive integers $z(1), z(2)$, $z(3), \ldots$ such that for arbitrary j:*

1. *there is a probabilistic reversible automaton with $z(j)$ states recognizing L_j with the probability $\frac{68}{135}$,*
2. *any deterministic finite automaton recognizing L_j has at least $(7^{\frac{1}{14}})^{z(j)} = (1.1149116725\ldots)^{z(j)}$ states.*

References

1. Ambainis, A.: The complexity of probabilistic versus deterministic finite automata. In: Nagamochi, H., et al. (eds.) ISAAC 1996. LNCS, vol. 1178, pp. 233–237. Springer, Heidelberg (1996)
2. Ambainis, A., Freivalds, R.: 1-way quantum finite automata: strengths, weaknesses and generalizations. In: Proc. IEEE FOCS 1998, pp. 332–341 (1998)
3. Artin, E.: Beweis des allgemeinen Reziprozitätsgesetzes. Mat. Sem. Univ. Hamburg B.5, 353–363 (1927)
4. Aschbacher, M.: Finite Group Theory (Cambridge Studies in Advanced Mathematics), 2nd edn. Cambridge University Press, Cambridge (2000)
5. Freivalds, R.: On the growth of the number of states in result of the determinization of probabilistic finite automata. Avtomatika i Vichislitel'naya Tekhnika (Russian) (3), 39–42 (1982)
6. Gabbasov, N.Z., Murtazina, T.A.: Improving the estimate of Rabin's reduction theorem. Algorithms and Automata, Kazan University, (Russian) pp. 7–10 (1979)
7. Garret, P.: The Mathematics of Coding Theory. Pearson Prentice Hall, Upper Saddle River (2004)
8. Golovkins, M., Kravtsev, M.: Probabilistic Reversible Automata and Quantum Automata. In: H. Ibarra, O., Zhang, L. (eds.) COCOON 2002. LNCS, vol. 2387, pp. 574–583. Springer, Heidelberg (2002)
9. Hooley, C.: On Artin's conjecture. J.ReineAngew.Math 225, 229–220 (1967)
10. Heath-Brown, D.R.: Artin's conjecture for primitive roots. Quart. J. Math. Oxford 37, 27–38 (1986)
11. Kolmogorov, A.N.: Three approaches to the quantitative definition of information. Problems in Information Transmission 1, 1–7 (1965)
12. Kondacs, A., Watrous, J.: On the power of quantum finite state automata. In: Proc. IEEE FOCS 1997, pp. 66–75 (1997)
13. Paz, A.: Some aspects of probabilistic automata. Information and Control 9(1), 26–60 (1966)
14. Rabin, M.O.: Probabilistic Automata. Information and Control 6(3), 230–245 (1963)
15. Rogers Jr., H. (ed.): Theory of Recursive Functions and Effective Computability. McGraw Hill Book Company, New York (1967)
16. Spencer, J.: Nonconstructive methods in discrete mathematics. In: Rota, G.-C. (ed.) Studies in Combinatorics (MAA Studies in Mathematics), vol. 17, pp. 142–178 (1978)

Introducing Reactive Kripke Semantics
and Arc Accessibility

Dov M. Gabbay

King's College London, London WC2R 2LS, U.K.
dov.gabbay@kcl.ac.uk

To Boaz!

Abstract. Ordinary Kripke models are not reactive. When we evaluate (test/measure) a formula A at a model $\mathbf{m}$, the model does not react, respond or change while we evaluate. The model is static and unchanged. This paper studies Kripke models which react to the evaluation process and change themselves during the process. The additional device we add to Kripke semantics to make it reactive is to allow the accessibility relation to access itself. Thus the accessibility relation $\mathcal{R}$ of a reactive Kripke model contains not only pairs $(a, b) \in \mathcal{R}$ of possible worlds (b is accessible to a, i.e. there is an accessibility *arc* from a to b) but also pairs of the form $(t, (a, b)) \in \mathcal{R}$, meaning that the arc (a, b) is accessible to t, or even connections of the form $((a, b), (c, d)) \in \mathcal{R}$.

This new kind of Kripke semantics allows us to characterise more axiomatic modal logics (with one modality $\square$) by a class of reactive frames. There are logics which cannot be characterised by ordinary frames but which can be characterised by reactive frames.

We also discuss the manifestation of the 'reactive' idea in the context of automata theory, where we allow the automaton to react and change it's own definition as it responds to input, and in graph theory, where the graph can change under us as we manipulate it.

1 Motivation and Background

1.1 The Reactive Idea[1]

Traditional modal logic uses possible world semantics with accessibility relation R. When we evaluate a formula such as $B = \square p \wedge \square^2 q$ in a Kripke model $\mathbf{m} = (S, R, a, h)$ (S is the set of possible worlds, $a \in S, R \subseteq S^2$ and h is the assignment) the model $\mathbf{m}$ does not change in the course of evaluation of B. We say the model $\mathbf{m}$ is not *reactive*. It stays the same during the process of evaluation.

To make this point absolutely clear, consider the situation in Fig. 1 below[2]

---

[1] An earlier version of this paper was presented in CombLog 04, July 28–30, 2004, and published in the proceedings [11]. See www.cs.math.ist.ut.pt/comblog04/talks.html

[2] Single arrows indicate point-to-point accessibility, double arrows indicate point-to-arc accessibility.

A. Avron et al. (Eds.): Trakhtenbrot/Festschrift, LNCS 4800, pp. 292–341, 2008.
© Springer-Verlag Berlin Heidelberg 2008

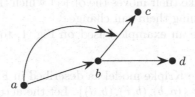

Fig. 1.

To evaluate $a \vDash \Box^2 q$, we have to check $b \vDash \Box q$. We can also check another formula at b, say, $b \vDash \Box p$. In either case the world accessible to b are c and d.

We *do not* say that since $b \vDash \Box q$ started its evaluation at world a as $a \vDash \Box^2 q$ and continued to $b \vDash \Box q$, then the accessible worlds to b are now different. In other words the model does not react to our starting the evaluation of $a \vDash \Box^2 q$ by changing the accessible worlds at b (for example, see the double arrow in Fig. 1, it indicates that we can disconnect the accessibility of node c to b) and therefore allowing us to see a different set of accessible worlds when we continue the evaluation of $b \vDash \Box q$.

The evaluation of $\Box$ at b *does not* depend, in traditional Kripke semantics, on how we "got" to b.

This paper addresses the case where the semantics does change (or react) under us as we evaluate a formula. This idea makes the evaluation of a wff at a world t dependent on the route leading to t. Thus we get a new kind of semantics, the reactive semantics. This semantics is stronger than ordinary Kripke semantics as there are axiomatic propositional logics with one modality $\Box$ which cannot be characterised by a class of traditional Kripke models but is complete for a class of reactive Kripke models.

The idea of dynamic interaction in logic, classical or non-classical, is not new. Interactive logic has been done in many guises. The most well known example is the game theoretic semantics for the classical quantifiers, and hence for modalities, which has extensive applications in natural language analysis and in theoretical computer science. A related interactive system is a dialogue logic with its vast applications to logic and to argumentation theory. The paradigm of these systems is different from what we are offering here. The interactive paradigm involves two players over a fixed object (a model or a database), who have their respective goals and rules which allow them to manipulate the object and each other in order to win the interaction. Our paradigm is different. As we discuss later in Sect. 2, we consider systems which react a nd change because of built-in faults and remedies which get activated while they are used. Applied in this context we can roughly say that we allow the object as well as the rules controlling the players to react to the moves of the players and try to influence them. As the players make their moves the object which they manipulate as well as the game rules governing them can change.[3]

Meanwhile let us give an example based on Fig. 1, to give the reader some immediate comparison.

Example 1. Consider the Kripke model as described in Fig. 1. It has the nodes $S = \{a, b, c, d\}$ with $R = \{(a, b), (b, c), (b, d)\}$. Let the actual world be a.

Let the assignment h for the atom q be $h(q) = \{a, b, d\}$. Thus q is false at world c but holds at worlds a, b and d.

Let us evaluate $a \vDash_h^1 \Box^2 q$ in a traditional manner (use superscript 1 for traditional)

$$a \vDash_h^1 \Box^2 q \text{ iff } b \vDash_h \Box q \text{ iff both } c \vDash_h^1 q \text{ and } d \vDash_h^1 q \text{ iff false.}$$

If we allow the model to be reactive (use superscript 2) we get

$$a \vDash_h^2 \Box^2 q \text{ iff } b \vDash_h^2 \Box q \text{ iff } d \vDash_h^2 q \text{ iff true.}$$

under $\vDash^2$, once we passed through node a, c was no longer accessible from b when we came to evaluate $b \vDash^2 \Box q$.[4]

The reader might wish to say that we have two models here, one with c accessible to b and one with c not accessible to b.

This view does not work in general. Suppose the reactive model switches bRc on and off each time we make a move. Then whether bRc holds or not depends on how many moves we make to get to b. To illustrate this point, let us modify the model and assume that aRa holds and evaluate $a \vDash \Box^3 q$.

$$a \vDash_h^2 \Box^3 q \text{ iff both } (*) \ a \vDash_h^2 \Box^2 q \text{ and } (**) \ b \vDash_h^2 \Box^2 q \text{ hold.}$$

Since we made only one move, the node c is not accessible to b.

Let us continue:

$$(*) \ a \vDash_h^2 \Box^2 q \text{ iff both } (*1) \ a \vDash_h^2 \Box q \text{ and } (*2) \ b \vDash_h \Box q \text{ hold.}$$

Since we now made two moves, we have that c is again accessible to b and hence $(*2)$ is evaluated at b with bRc active.

On the other hand, $(**)$ is evaluated at b with bRc not active.

We shall see later that this reactive semantics is not reducible to the traditional semantics.

---

[3] If the reader insists on the interactive game theoretical point of view then we can regard the reactive paradigm as adding an additional player to the game which is trying to modify and obstruct the two existing players. Whether this point of view is technically correct and/or useful is a matter for investigation towards the end of the paper, after we define our reactive semantics. This additional player may not have any strategy beyond interfering with the game of the other two players. It may be regarded as a faulty environment which changes unpredictably. Michael Gabbay suggested that perhaps we can study a third player whose winning strategy is to deprive the other two players from having a winning strategy.

[4] Think of node a transmitting a signal along the double arrow and disconnecting the arrow from b to c.

Let us now check how a game theoretical semantics would evaluate $\Box^2 q$ in this model. We have two players **A** and **B**. **A** claims that $\Box^2 q$ holds and **B** claims it does not hold. **B** moves by choosing an accessible point (e.g. b) and challenges **A** to claim that $\Box q$ holds at b. The game goes on and **A** must have a strategy for winning. It is clear that the model is fixed in this game.

We can add a third player **C** who makes changes to the model (and the game rules, if you want) in reaction to the moves made by **A** and **B** and we expect a winning strategy from **A** against both **B** and **C**.

1.2   Examples Motivating the Reactive Idea

Before we continue with more technical material, let us motivate our idea of reactive semantics and consider some case studies.

Airline Example. We begin with a very simple and familiar example. Consider Fig. 2

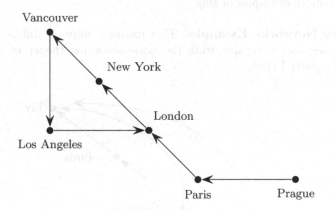

Fig. 2.

Figure 2 gives the possible flight routes for the aeroplanes of TUA (Trans Universal Airlines). It is well known that many features of a flight depend on the route. These include the cost of tickets, as well as the right to take passengers at an airport. The right to take passengers at an airport depends on the flight route to that airport and on bilateral agreements between the airlines and governments. Thus, for example, flights to New York originating in London, may take on passengers in London to disembark in New York. However, a flight starting at Paris going to New York through London may not be allowed to pick up passengers in London to go onto New York. It is all a matter of agreements and landing rights. It is quite possible, however, that on the route Prague–Paris–London–New York, the airline is allowed to take passengers in London to disembark in New York. We can describe the above situation in Fig. 3.

The double-headed arrow from Paris to the arc London→New York indicates a cancellation of the 'passenger' connection from London to New York. The

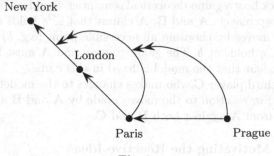

Fig. 3.

double-headed arrow from Prague to the double arrow arc emanating from Paris indicates a cancellation of the cancellation.

Figure 3 looks like a typical reactive Kripke model, where we have arcs leading into arcs.

Let us see more examples of this.

Inheritance Networks Example. This example offers a different point of view of arc semantics, coming from the non-monotonic theory of inheritance networks. Consider Fig. 4.

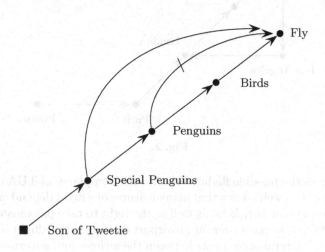

Fig. 4.

In Fig. 4, the circular nodes are predicates, such as Fly, Birds, etc. The arrows indicate inheritance, so for example, we have $\forall x(\text{Bird}(x) \rightarrow \text{Fly}(x))$. The arrows with a bar indicate blockage, for example $\forall x(\text{Penguin}(x) \rightarrow \neg\text{Fly}(x))$. The square nodes indicate instantiation, so son of Tweetie is a special penguin.

Figure 4 is the kind of figure one finds in papers on inheritance networks. The figure indicates that Penguins are Birds, that Birds Fly but that Penguins do not Fly. However, special Penguins do Fly and the son of Tweetie, a rare

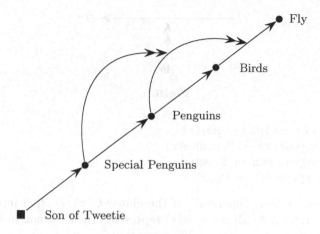

Fig. 5.

bird, is a special Penguin, and therefore does Fly. The arrow with the bar on it
blocks the information from flowing from the Penguin node to the Fly node. The
theory of inheritance networks spends a lot of effort on algorithms that allow us
to choose between paths in the network so that we can come up with the desired
intuitively correct answers. In the case of Fig. 4 we want to get that the son
of Tweetie does fly, since we have the most specific information about him. It
is not important to us in this paper to take account of how inheritance theory
deals with this example. We want to look at the example from our point of view,
using our notation, as in Fig. 5.

In Fig. 5 the double headed arrow $\twoheadrightarrow$ emanating from Penguins attacks the
arrow from Birds to Fly, and the double arrow emanating from Special penguins
attacks the double arrow emanating from Penguins and attacking the arrow
from Birds to Fly. This is not how inheritance theory would deal with this
situation but we are not doing inheritance theory here. Our aim is to motivate
our approach and what we need from the inheritance example is just the idea of
the algorithmic flow of information during the dynamic evaluation process.[5]

We have already put forward the reactive and dynamic idea of evaluation in
earlier papers and lectures (see [18]). A typical example we give is to consider
$t \vDash \Diamond A$. In modal logic this means that there is a possible world s such that we
have $s \vDash A$ we take a more dynamic view of it.

We ask: where is s? How long does it take to get to it? and how much does it
cost to get there?

The reader should recall the way circumscription theory deals with the
Tweetie example, see [20, Sect. 4.1, especially p. 324]. We write

- Birds $(x) \wedge \neg Ab_1(x) \rightarrow$ Fly(x)
- Penguins$(x) \rightarrow$ Birds(x)

---

[5] It is our intention to explore whether our idea of double headed arrows cancelling
other arrows can simplify inheritance theory algorithms.

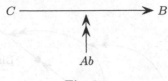

Fig. 6.

- Penguins$(x) \wedge \neg Ab_2(x) \rightarrow Ab_1(x)$
- Special Penguins$(x) \rightarrow$ Penguins(x)
- Special Penguins (son of Tweetie)
- Special Penguins $(x) \rightarrow Ab_2(x)$.

"$Ab(x)$" stands for "x is abnormal". If the clause $C(x) \rightarrow B(x)$ represents the arc $C \rightarrow B$ then $C(x) \wedge \neg Ab(x) \rightarrow B(x)$ represents the situation in Fig. 6

A Technical Example. It is now time to give a technical example. Consider Fig. 7 below. This figure displays a past flow of time. The node t is the present moment and a single headed arrow from one node to another, say from s to t, means that t is in the immediate future of s. We use the modality $\Box$ to mean 'always in the immediate past'. Thus the accessibility relation R of Fig. 7 is as follows:

- $tRs, tRb, bRs, sRa, tRt, bRb, sRs$ and aRa.

The double-headed arrows cancel the accessibility relation.

Let us calculate $t \vDash \Box^3 q$ in Fig. 7.

Initial Position: Starting point is t and all arrows are active.
Step 1: Send double arrow signal from t to all destinations inverting the active/inactive status of all destination arrows. Then go to all accessible worlds (in this case s and b) and evaluate $\Box^2 q$ there. If the result is positive 1 and at all nodes, then send 'success' back to node t.
Step 2: Evaluate $\Box^2 q$ at nodes b and s.
Subcase 2.b. Evaluation at b: First we send a double arrow signal from b to all destinations reversing the activation status of these destinations. Thus the

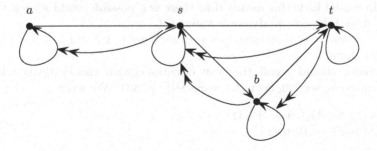

Fig. 7.

single arrow from s to s will be re-activated and we will evaluate $\Box q$ at s at the next step 3 (with s accessible to itself). b is not accessible to itself because its arrow has been deactivated by t at Step 1.

Subcase 2.s. Evaluation at s: First we send a double arrow signal to reverse the status arrow from a to a. Then we evaluate $\Box^2 q$ at s with s not accessible to itself, since the arrow from s to s was deactivated by t at step 1.

This can go on, but we shall not continue as we trust that the reader has got the idea by now.

Note that if we start at t and evaluate $B = \Box^3 q \wedge \Box^2 q$, we will get that $\Box q$ must be evaluated at s in two ways. One with s accessible to itself (coming from b via $t \vDash \Box^3 q$) and once with s not accessible to itself (coming from t via $t \vDash \Box^2 q$).

Let us now calculate $t \vDash \Diamond^2 q$ in Fig. 7.

Initial Position: Starting point is t and all arrows are active.

Step 1: Send double arrow signal from t to all destinations inverting the active/inactive status of all destination arrows. Then go to one of the accessible worlds (in this case s or b) and evaluate $\Diamond^2 q$ there. If the result is positive 1 at this node, then send 'success' back to node t.

Step 2: Evaluate $\Diamond^2 q$ at nodes b or s.

Subcase 2.b. Evaluation at b: First we send a double arrow signal from b to all destinations reversing the activation status of these destinations. Thus the single arrow from s to s will be re-activated and we will evaluate $\Diamond q$ at s at the next step 3 (with s accessible to itself). b is not accessible to itself because its arrow has been deactivated by t at Step 1.

Subcase 2.s. Evaluation at s: First we send a double arrow signal to reverse the status arrow from a to a. Then we evaluate $\Diamond^2 q$ at s with s not accessible to itself, since the arrow from s to s was deactivated by t at step 1.

This can go on, but we shall not continue as we trust that the reader has got the idea by now. For the case of $\Diamond$ we make a non-deterministic choice. The model is not sensitive to whether we come to a point because we are evaluating $\Box$ or a $\Diamond$. If we want this kind of sensitivity we can have arrows of the form $\twoheadrightarrow_\Box$ and $\twoheadrightarrow_\Diamond$.

Such distinctions may be desirable in dealing with quantifier games, where changes may be different for the cases of $\forall$ and $\exists$.

Tax Example. Having explained the technical side of our reactive (changing) semantics, let us give some real examples.

House prices in London have gone up a great deal. An average upper middle class family is liable to pay inheritance tax on part of the value of their house (if the house is valued over £500,000, for example, then there is tax liability on £250,000). Some parents solved the problem by giving the house as a gift to their children. If at least one of the parents remains alive for seven years after the transaction, then current rules say that there is no tax. Consider therefore the following scenario:

1. current date is April 2004
2. parents gave house as a gift to children in 1996
3. parents continued to live in house as guests of the children

(1)–(3) above imply that (4):

4. if parents both die in March 2004, then no tax is liable.

To continue the story, there were rumours that the tax people were going to change the rules in April 2004, declaring that if parents remain living in the house after it was given as a gift, then the gift does not count as such an there is tax liability. The rumours also said that this law is going to apply *retrospectively*.[6]

Thus we have that (5) holds:

5. If parents both die on March 2005, then tax is liable.

We assume that (4) still holds even after the new law as we cannot imagine that the UK tax inspector would be opening closed old files and demanding more tax.

The way to represent (4) and (5) is to use two dimensional logic. We write $t \vDash_s A$ to mean at time t A is true given the point of view proposed or held at time s.

Thus $2005 \vDash_{2003} \neg(5)$ holds, because from the 2003 laws point of view (before legislation) no tax is liable ((5) says tax is liable).But $2005 \vDash_{2004} (5)$ also holds, because according to 2004 legislation tax is liable.

So far we have no formal problem and no need for our new semantics, because we can write

- $t \vDash_t \Box A$ iff for all future $s, s \vDash_s A$.

In other words we evaluate sentences at time t according to the point of view held at the very same time t.

The problem arises when we want to formalise the following scenario. The parents die in 2003. The lawyer is dealing with the estate. We do not know when he is going to finish. When he submits the paperwork then the tax liability at 2003 is judged according to the time of submission. Now the second index s in $t \vDash_s A$ behaves like a reactive model as we are evaluating

$$2003 \vDash_{\text{time lawyer submits}} (4).$$

Salesman Example. Consider the simple graph of Fig. 8

A salesman wants to traverse this graph in such a way that he doesn't pass through the same edge twice. Such problems are very common in graph theory. The simplest way of implementing this restriction is to cancel an edge once it has been used. Figure 9 will do the job.

---

[6] Some countries, like Austria, for example, would *never* legislate retrospectively. They regard this as a cultural taboo.

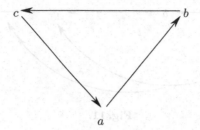

Fig. 8.

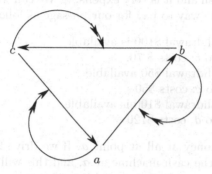

Fig. 9.

One cannot always implement the salesman problem in this way. It depends on the graph we deal with. However, it is one more reason for considering arc accessibility.[7]

Resource Example. Consider a road system as in Fig. 10.

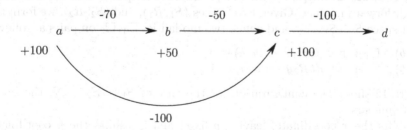

Fig. 10.

---

[7] For the salesman example we need accessibility from arcs to arcs. So in Fig. 9, rather than have the double arrow $(b, (a, b))$, we need the double arrow $((a, b), (a, b))$. The arc cancels itself as we go through it.

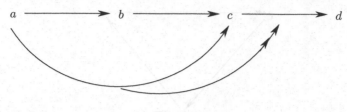

Fig. 11.

Assume that whenever we drive through a road we need to pay a fee to pass. Payment must be in cash and it is very expensive. We can withdraw money from cash machines along the way to pay for our passage as follows:

Point a: Cash withdrawal $100 is available.
            Road a to b: costs $ 70.
Point b: Cash withdrawal $50 available.
            Road b to c: costs $50
Point c: Cash withdrawal $100 is available.
            Road c to d: costs $120.

We start with no money at all at point a. If we drive from a to c directly, we can get $100 from the cash machine at a, and this will get us to c with no money left and so we cannot get at c enough cash to pay passage from c to d. However, if we pass through b, we can withdraw money at b and later at c and we will have enough to pay for the passage from c to d.

Figure 10 is essentially an ordinary annotated graph describing the resource situation but the qualititative situation (where money considerations are hidden) can be described in Fig. 11. The upshot of Fig. 11 is that the arc from a to c sends a signal to cancel the arc from c to d. This is also the first case where we have double arrows going from arc to arc.

Flow Products Example. Ordinary products of Kripke frames are defined in a straightforward manner. Given two frames (S_1, R_1), and (S_2, R_2), we form the product space $S = S_1 \times S_2$, and define two modalities $\Box_1, \Box_2$ on pairs as follows:

- $(a, b) \vDash \Box_1 A$ iff $\forall x(aR_1x \to (x, b) \vDash A)$
- $(a, b) \vDash \Box_2 A$ iff $\forall y(bR_2y \to (a, y) \vDash A)$.

Figure 12 shows the configuration for the case of $S_1 = S_2 = N$, the set of natural numbers

$\Box_1$ shifts the x coordinate, leaving b fixed and $\Box_2$ shifts the y coordinate, leaving a fixed.

We have, for example, among other things that $\Box_1\Box_2 = \Box_2\Box_1$. Products spaces are used whenever we deal with two independent modal or temporal aspects. This is a very active area of many dimensional modal logics, also related to classical predicate logic with a fixed number of variables. See our book [16].

The authors introduce in [18] the concept of flow products. Imagine the x axis is space (measured in kilometers) and the y axis is time measured in hours. Any

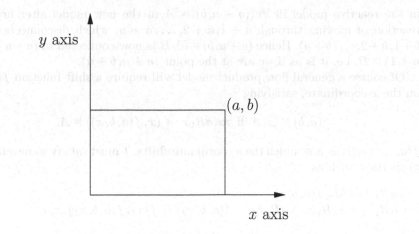

Fig. 12.

shift in space will necessarily cause a shift in time. Assuming speed of 1 mile per hour, we get

- $(a, b) \vDash \square_1 A$ iff $\forall u((a + u, b + u) \vDash A)$
- $(a, b) \vDash \square_2 A$ iff $\forall u((a, b + u) \vDash A)$.

When we move in space time shifts.

We can view this as a reactive model. Imagine for any point (a, b) the following double arrows exist

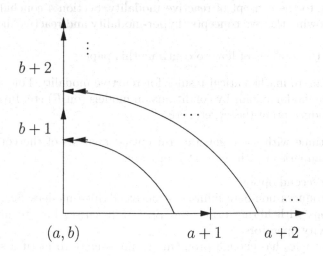

as we move from a to $a + 1$, the connection from b to $b + 1$ is switched off. Therefore we get

$$(a, b) \vDash \square_1 A$$

in the reactive model iff $\forall u((a+u,b) \vDash A$ in the new model after firing the reaction of moving through $a+1, a+2, \ldots, a+u$, which disconnects b from $b+1, b+2, \ldots, b+u)$. Hence $(a+u,b) \vDash \Box_2 B$ is now connected to $(a+u+1, b+u+1) \vDash B$, i.e. it is as if we are at the point $(a+u, b+u)$.

Of course a general flow product model will require a shift function $f(a,b,x)$ on the x-coordinate, satisfying

$$(a,b) \vDash \Box_1 A \text{ iff } \forall x(aR_1 x \to (x, f(a,b,x)) \vDash A.$$

$f(a,b,x)$ tells us how much the y coordinate shifts. f must satisfy some additivity properties, such as

- $aR_1 x \to bR_2 f(a,b,x)$
- $aR_1 x_1 \wedge x_1 R_1 x_2 \wedge aR_1 x_2 \to f(a,b,x_2) = f(x_1, f(a,b,x_1), x_2)$.

1.3   Plan of This Paper

The next section, Sect. 2, describes the reactive paradigm in general and proceeds to give some specific examples from different application areas. These areas are studied in separate research papers. We also discuss several options for defining reactivity. The rest of the paper should give some technical results in the area of modal logic. There is a lot to do here and the main bulk of the results is postponed to a sequel paper. In order to give some formal results in this paper, we want to show that the class of switch reactive Kripke models can characterise more syntactic modal axiomatic systems than the class of ordinary Kripke models. This is done in Sect. 4. To prepare the ground for this result we need to devote Sect. 3 first to hyper-modalities, being a related but not an identical concept to the concept of reactive modality. Section 4 concludes with two examples showing the two concepts (hyper-modality and reactive modality) are different.

We now have two options of how to continue this paper:

Option 1. Get more mathematical results for reactive modality. For example, characterise some modal axioms by conditions on models (on R) etc. Investigate complexity, introduce proof theory, etc, etc.

Option 2 Continue with some general conceptual analysis of the concept of reactivity and comparison with relevant literature.

We follow the second option.

Section 5 introduces non-deterministic reactive Kripke models, Sect. 6 discusses connections with fibring logics and Sect. 7 discusses how to introduce dedicated reactivity operators.

Each of these topics has enough problems to investigate to merit a spearate research paper.

We continue with four appendices which give further results and comparison with the literature.

2   The Reactive Paradigm in General

Consider a system S containing components and some internal connections and procedures governing the interactive behaviour of these components in response to external inputs. We can think for example of S as a graph (to be traversed or manipulated) or as an automaton which responds to inputs and changes states, or as a Kripke model in which a formula is evaluated, or as a bureaucratic system responding to an application form or even as more familiar daily objects such as a washing machine, a television set or a car.

We can represent such a system as a network containing possibly labelled vertices represeting the components and arrows between the vertices represeting possible "control flows" within the system of whatever the system internal processes do.

Figure 13 is a typical situation we want to consider:

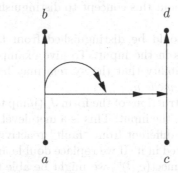

Fig. 13.

The arrows from a to b and from c to d indicate that internally whatever the system does in response to input or command, there is a possible path where component a passes the "flow" to component b and similarly maybe in another part of the system, there is a connection between c and d.

So far we have nothing more than a possible network graphic representation of some system.

Now comes the reactive idea. Consider the possibility that the system develops faults due to overuse or stress. The double arrow from the arc (a, b) to the arc (c, d) (which we denote by $((a, b), (c, d))$) indicates how the fault may develop. When we pass from a to b we put internal pressure on the connection from c to d. If significant pressure builds up on the connectiuon from c to d it may fail. For example everyone who drove a car long distance during a very hot day knows that the engine may overheat. This is an example of a fault. In general these faults are predictable and are predetermined by the construction of the system. They are represented in Fig. 13 by the double arrows.

In fact, Fig. 13 also indicates a remedy to this pressure, this is a double arrow from the arc (a, b) to the double arrow $((a, b), (c, d))$, namely $((a, b), ((a, b), (c, d)))$.

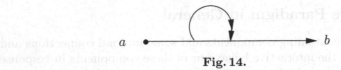

$a \bullet \longrightarrow b$

Fig. 14.

The double arrows represent known weaknesses or remedies in the system. We can look at the double arrows as indicating the ways in which the system reacts and adjusts itself under pressure. The simplest model is a once only bridge in Fig. 14

When we cross the bridge from a to b, a signal (representing stress) is sent to the bridge to collapse and so after we cross the bridge the arc from a to b is cancelled.

The above is the basic intuition behind the idea of reactivity. Think of it as "fault" reactivity, or better it is "fault-remedy" reactivity.

We now elaborate more about this concept to distinguish it from other somewhat similar concepts.

The "fault" reactivity should be distinguished from the idea of incuding dynamic metalevel operators in the input. To give example of such metalevel operators, consdier the possibility that the system may have several modes of operoations say $mode\ 1, \ldots, mode\ n$.

The input may contain instructions of the form J_i (jump to mode i) which tells the system how to operate on the input. This is a metalevel instruction inducing a change in the system. It is different from "fault" reactivity, though the latter can be case-by-case represented in it. If we replace double arrows to (c, d) by the metalevel instruction "disconnect(c, d)", we might be able to simulate the fault induced by an input by adding (interleaving) some "disconnect" instructions to the input to form a new input.

This option will be discussed later in the paper on a case-by-case analysis. It is not reactivity but "fibring" of metalevel instructions giving the illusion of reactivity.

The notion of fault reactivity should also be distinguished from the notion of "sabotage", independently introduced by J. van Benthem [4] and further developed by P. Rohde [21]. The notion is similar; we envisage a saboteur disconnecting components within the system. There are both conceptual and tecnnical differences between the two notions ("fault" reactivity and "sabotage" reactivity) and this will be analysed in detail in Appendix D. For the moment we note that faults and remedies are built into the system while sabotage is not and also sabotage is presented as a metalevel connective and fault is not. For more detail, see Appendix D.[8]

It turns out that as we apply this idea in various application areas where different non-reactive systems are used we can get new kinds of systems, the reactive systems, which can have better applicability in each respective area.

---

[8] Some sabotage may be non-deterministic like water damage running through the system. This can be modelled as a non-deterministic reactive system. The possible options for water damage can be forseen!

We give five immediate examples:

1. Ordinary Kripke models become reactive Kripke models which can change during the process of evaluation, affording a wider class of models for modal logic.
2. Ordinary non-deterministic automata become reactive automata whose transition table changes every time they make a transition, see [6]. Although it is shown in [6] that every reactive automaton is equivalent to another ordinary non-reactive automaton, the reactive automata can perform tasks (recognise inputs) with much less states and represent substantial savings in computational costs!
3. An ordinary graph becomes a reactive graph where there are also arrows from edges to edges, see [17]. This area is very promising and is now under intensive investigation.
4. Reactive grammars and rewrite systems [2]. For reactive rewrite systems we get new hierarchies of languages. Again this area is actively investigated.
5. A proof system becomes a reactive proof system which changes as we advance in the proof process, see [9,2].

In each case a new class of models/graphs/automata/proof theory is introduced and some natural questions can be asked.

The following are some sample questions:

1. What is the expressive power/complexity of the new class and what can it do or not do?
2. What do traditional investigations yield when applied to this new class? (E.g. completeness theorems, cut elimination, correspondence theory, axioms, hierarchies of automata, theorems about cycles or classes of graphs, etc, etc.)
3. How does this concept relate to other dynamic/change concepts already existing in the literature?
4. Can the new models be reduced to known models via traditional interpretations? What is the edge/advantage we have by using the new reactive models?
5. Comparison and evaluation of the options for representing reactivity in the object level vs. metalevel.

Let us give some concrete examples, which will help us get a feel for the notion of reactivity.

Example 2 (Modal logic). Here we take the reactive Kripke models of the previous sections. The system is the evaluation process (for a given formula). The reactive parameter is a Kripke model with a point of evaluation t. The environment (or the faults in the system) is realised by the double arrows $\twoheadrightarrow$, which change the accessibility relation every time we make a move (i.e. every time we continue with the evaluation).

This example generalises slightly, giving it a familiar everyday meaning. Think of a network of nodes representing various components of a system. The arrows represent connections between components and a path through this graph following the arrows represents a way in which the system can be used. Imagine that the system is prone to faults and failures. So the use of one component may send a (double arrow) signal to other components and weaken or influence them.

We know this to be true for many real life systems. So the reactivity of a network is just a measure of the ways it can fail and disappoint you!

Example 3 (Quantifier games). Recall the basic game semantics for the quantifiers. Given a classical model $\mathbf{m}$ and a formula, say $\Psi = \forall x \forall y \exists z (\varphi(x, y, z)$ with $\varphi(x, y, z)$ quantifier free, we play a game over the model between two players, $\mathbf{A}$ (claiming that Ψ holds) and player $\mathbf{B}$ claiming that Ψ does not hold. At step 0, $\mathbf{A}$ puts forward Ψ. At step 1 $\mathbf{B}$ challenges by choosing a in the domain of $\mathbf{m}$, and continues to choose an element b in the domain of $\mathbf{m}$. It is the task of player $\mathbf{A}$ to supply a c such that $\varphi(a, b, c)$ holds. Player $\mathbf{A}$ has a winning strategy iff he has a function f such that for any a, b the element $c = f(a, b)$ is such that $\mathbf{m} \vDash \varphi(a, b, f(a, b))$.

In this case the moves are the choices of elements. The reactive environment player $\mathbf{E}$ has a tinkering function $\tau : (a, \mathbf{m}) \mapsto \mathbf{m}_a^\tau$ which changes any model $\mathbf{m}$ and an element a in the domain of $\mathbf{m}$ into a new model $\mathbf{m}_a$ with the same domain.

Thus in order to win, $\mathbf{A}$ has to have a winning function $\lambda \tau \lambda x y f_\tau(x, y)$ such that

$$\forall \tau \forall a \forall b \; \mathbf{m}_{a,b}^\tau \vDash \varphi(a, b, f_\tau(a, b)).$$

So as we can see, this definition covers the modal logic case, if viewed through its translation to classical logic.

For consider a reactive Kripke model $\mathbf{m}$ with actual world a. Consider the evaluation $a \vDash \Box^2 \Diamond q$. This is translated into classical logic as

$$\Psi = \forall x \forall y \exists z (aRx \wedge xRy \rightarrow yRz \wedge Q(z)).$$

The function τ for changing the model is governed by the double arrows $\twoheadrightarrow$ and we clearly have a special case of the quantifier games tinkering.

We get in this case the formula Ψ':

$$\Psi' = \forall x \forall y \exists z [aRx \wedge xR_a y \rightarrow yR_{a,x} z \wedge Q(z)]$$

where R_t denotes the new accessibility relation obtained from R after implementing the double arrows emanating from it.

Example 4 (Automata). A non-deterministic automaton $\mathcal{A} = (S, M, a, F, \Sigma)$ is characterised by a set of states S, an initial state $a \in S$, a set $F \subseteq S$ of final states and an alphabet Σ. M is a function giving for each state $t \in S$ and a letter $\sigma \in \Sigma$ a new set of states $M(t, \sigma) \subseteq S$, which are the states that the automaton $\mathcal{A}$ can non-deterministically move to.

Another way to view the automaton is as a multi-modal Kripke model of the form (S, R_σ, a, F), where R_σ for $\sigma \in \Sigma$ is defined by $xR_\sigma y$ iff $y \in M(x, \sigma)$.

A word of the form $(\sigma_1, \ldots, \sigma_n)$ is said to be recognisable by the automaton $\mathcal{A}$ iff there exists a sequence of states $x_1, \ldots, x_n$ such that $a R_{\sigma} x_1 \wedge \ldots \wedge x_{n-1} R_{\sigma_n} x_n$ holds and $x_n \in F$. If the atom q is assigned the set F, then $(\sigma_1, \ldots, \sigma_n)$ is recognisable by $\mathcal{A}$ iff $\mathcal{A} \vDash \Diamond_{\sigma_1}, \ldots, \Diamond_{\sigma_n} q$, where $\Diamond_{\sigma}$ corresponds to $R_{\sigma}, \sigma \in \Sigma$.

Thus naturally we can define a reactive automaton by adding double arrows $\twoheadrightarrow_{\sigma}$ for every $\sigma \in \Sigma$ in the places we want. Figure 15 is an example of such an automaton. Let $\Sigma = \{\sigma_1, \sigma_2\}$, $S = \{a, b, c\}$ and $F = \{c\}$.

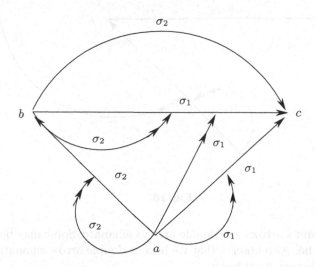

Fig. 15.

We have

$$R_{\sigma_1} = \{(a, c), (b, c)\}$$
$$R_{\sigma_2} = \{(a, b), (b, c)\}$$

Figure 15 indicates the available double arrows. For example, the figure shows that as we move through node a in response to the letter σ_1 we disconnect (a, c) and (b, c) from the R_{σ_1} relation. If we move through node a in response to the letter σ_2 then we disconnect (a, b) from R_{σ_2}.

It is interesting to note that according to this figure, if we move out of state b in response to the letter σ_2 then we disconnect the connection (b, c) in the R_{σ_1} relation but not in the R_{σ_2} relation.

So there can be an interplay between the modalities here.

The reader should note that we can get a new hierarchy of automata here by taking the usual hierarchy and making it reactive. We shall pursue this idea in a subsequent papers, [12,6].

The next remark, however, lists some possible options for creating reactivity.

Remark 1 (Options for the reactive property). We saw that the basic idea of a reactive action is to change the model every time a move is made. To explain our options on how to change the model consider Fig. 16 below.

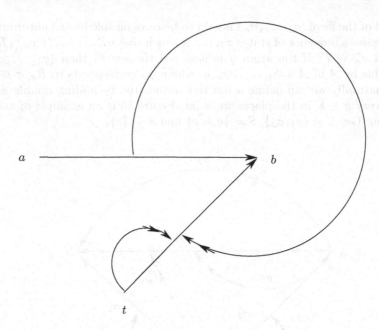

Fig. 16.

From the point t arrows and double arrows emanate. Some may be active and some may not be. Also observe that we have a double arrow emanating from an arrow. Let us list what we have:

1. $t \to b$
2. $t \twoheadrightarrow (t \to b)$
3. $a \to b$
4. $(a \to b) \twoheadrightarrow (t \to b)$.

Our first list of options relates to what kind of arrows we allow. Do we want only arrows emanating from a point or do we also allow arrows emanating from arcs?

Items 1–3 above emanate from points and item 4 emanates from arcs.

The second list of options has to do with how we use the arrows to change the model as we pass through a point or an arc.

The first possibility is the switch-like use of the arrows. At any given moment some of the arrows are active (on) and some are dormant (off). When we pass through a point or an arc a signal is sent along the arrows emanating from the said point or arc and reverses the status of the target arrows at the destination. This is a simple switch action. In general, the reactivity can be more intelligent. The second possibility is to allow each node and arc from which arrows emanate to decide, depending on the state (on off) of the target arrows, which of the target arrows to switch on and which to switch off. Care must be taken to ensure that this decision process is of the same complexity as the original non-reactive model. So for example in the case of reactive finite state automaton of Example 4, the

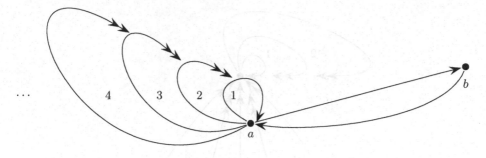

Fig. 17.

decision of each node which double arrows to activate should be done by another automaton.

Consider Fig. 17 below

Suppose we start with all arcs being switched on at the starting point a. As we move out of a to b, a switch behaviour will switch off all arcs $1, 2, 3, \ldots$. If we move through node a again, these arcs will be switched on again.

A more intelligent option might switch them on selectively one at a time, which each passage through the node a. This would allow us to use them as markers emulating aspects of a stack. Note that the idea is intuitively sound, and is not just a technicality. It makes sense to give nodes some intelligence to decide how to react, based on the situation it 'sees'.

The above options were deterministic. The most general option was to attach an automaton at each exit point of arrows (of any kind) to decide what to switch on and what to switch off. It is also possible to make these decisions non-deterministic or probabilistic. Consider Fig. 16. We can make all double arrows in this figure non-deterministic. So as we pass, for example, from node a to node b the double arrow to the arc $t \to b$ (namely $(a \to b) \twoheadrightarrow (t \to b)$), if active, may or may not (non-deterministically) send a signal.

Similarly we may attach probabilities to such connection, say 0.7, and so with 0.7 probability a signal will be sent. We shall elaborate more about this option in a subsequent paper [12].

Example 5 (Intelligent switches). Previous examples used on and off switches. The present example uses an intelligent switching system. In fact, we build on Example 4 (automata) and show in the present example how to simulate a stack automaton. We make sure the intelligent switching process is also done by a finite automaton. Consider the model described in Fig. 18

This model has two relations R_{σ_1} and R_{σ_2}. It corresponds to an automaton with alphabet $\{q_1^+, q_1^-, q_2^+, q_2^-, t, a\}$ and stack letter α. The initial state is a and the terminal state is t. It is designed to recognise the words of the form $\sigma_i^m \sigma_j^m, i \neq j, m \geq 1$.

The starting state is a. Upon seeing σ_i the automaton moves to state q_i^+ and writes α in the stack. It continues to write α as long as it sees σ_i. When it sees $\sigma_j, j \neq i$ it moves to q_i^- and starts deleting from the stack.

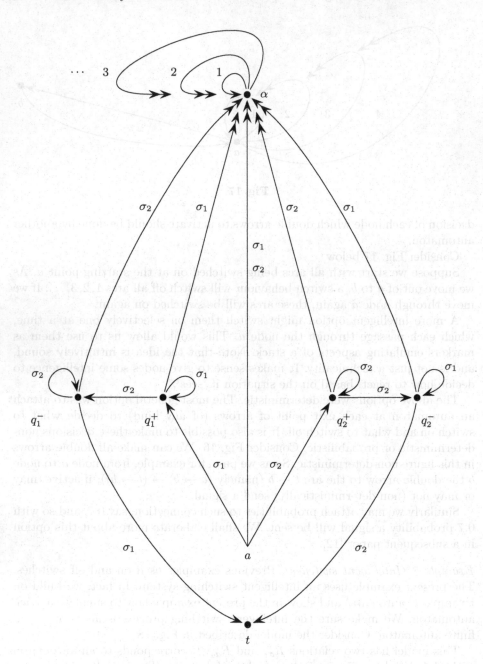

Fig. 18.

The double arrows are intelligent. The ones emanating from node a or node q^+ activate the first highest non-active arrow or double arrow at α. The ones emanating from a q^- deactivate the highest active double arrow from α. If

Table 1.

state	stack at α	Input letter	Reaction
a	all arrows not active	σ_i	move to state q_i^+ and activate arrow 1 at α
a	some arrows active	any	do not care. Case will not arise
q_i^+	arrows $1,\ldots,m$ are active $m \geq 1$	σ_i	stay at q_i^+ and activate arrow $m+1$
q_i^+	arrows $1,\ldots,m$ are active	$\sigma_j, j \neq i$	move to q_i^-
q_i^+	not the above	any	don't care. Case will not arise
q_i^-	arrows $1,\ldots,m$ are active $m \geq 1$	$\sigma_j, j \neq i$	stay at q_i^- and deactivate arrow m.
q_i^-	arrows $1,\ldots,m$ are active, $m \geq 1$	σ_i or no input	move to t and deactivate arrow m
q_i^-	all arrows at α are not active or arrows $2,\ldots,m$ are active $m \geq 2$ and arrow 1 not active	no input or input σ_i	move to t, activate arrow $m+1$. If no arrow at α is active then activate arrow 2.
q_i^-	same as previous	$\sigma_j, j \neq i$	remain at q_i^-, activate arrow $m+1$ or if no arrows at α are active then activate arrow 2.
q_i^-	Different from previous	any	don't care. Case will not arise.
t	any	no input. terminal position	no reaction. terminal position.

q^- sees that all arrows at α are not active it starts to activate arrow 2 and higher. So if arrow 1 is active then q^- deactivates the top arrow and if arrow 1 is not active then q^- activates the first non active arrow above arrow 1. We need to assume that only a finite number of arrows and double arrows from α are active at any given time.

The following table describes the moves of the automaton of Fig. 18. Note that only arrows at α may switch. The initial position is state a with all arrows at α not active. The terminal state is t.

So let us simulate an input computation.

Starting state

All double arrows and arrows are active except those at α which are not active.

Input step 1

We can assume without loss of generality that we get σ_1. So we move along R_{σ_1} from node a to node q_1^+. This move activates arrow 1 at α. The more σ_1 we see the more double arrows at α are activated in sequence. So if we see a total of m σ_1s, i.e. σ_1^m we get that arrows $1, \ldots, m$ are active. The first time we see σ_2 we move from q_1^+ to q_1^- along R_{σ_2}. There is no R_{σ_2} double arrow from q_1^+ to α.

If we continue to get σ_2 at q_1^- we cancel an arrow at α. The minute we get σ_1 again or the input finishes we move to a terminal state t. If the number of σ_2 is equal to σ_1, we end up stopping at t with no α arrow active. If the number of σ_2 is less than σ_1, we end up stopping with some α arrows active. If the number of σ_2 is larger than σ_1 we are faced with a situation where the q_1^- automaton sees no arrow connections at α at all and needs to decide what to do. We can tell it to activate arrow 2, leaving arrow 1 not active and to continue to activate arrows $3, 4, \ldots$ as long as the input is σ_2. If the input is empty or q_1^- sees σ_1 again then it stops.

Example 6 (Options for interpreting necessity when double arrows emanate from arcs). Consider Fig. 19 below:

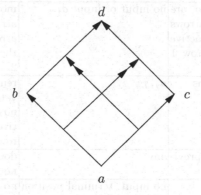

Fig. 19.

In this figure we have two double arrows emanating from arcs. We ask: What semantic meaning can we give to $\Box$?

We want to evaluate

$$a \vDash \Box\Box\bot$$

How do we go about it? Do we go to every accessible point one-by-one and check whether $\Box\bot$ holds? Or do we go to all points *simultaneously*?

When double arrows emanate from points, there is no difference in the mode of operation, but when they emanate also from arcs, then there is a difference

Case 1.   Separate evalaution of $\Box$
When we move from a to c, the arc (c, d) is still connected and so we have that $c \vDash \neg\Box\bot$. Similarly when we move from a to b, the arc (b, d) is still connected and so we have that $b \vDash \neg\Box\bot$.

Therefore, according to the separate evaluation of $\Box$, we get that $a \vDash \Box\neg\Box\bot$.

Case 2.   Simultaneous evaluation of $\Box$

If we move to b and to c simultaneously and evaluate at b and at c both arcs (b, d) and (c, d) are disconnected and so we have $b \vDash \Box\bot$ and $c \vDash \Box\bot$ and hence $a \vDash \Box\Box\bot$.

It makes more sense to adopt the separate evaluation of $\Box$ because of the traditional connection of $\Box$ with $\Diamond$, namely $\Box = \neg\Diamond\neg$ we have:

$a \vDash \Diamond A$ iff for some accessible point s, when we move to s we ahve $s \vDash A$.

We also have $a \vDash \Box A$ iff for all accessible points s when we move to s we have $s \vDash A$.

To preserve the duality $\Diamond = \neg\Box\neg$ we must adopt the separate evaluation of $\Box$.

3   Connection with Hyper-modalities

In [10], we introduced hyper-modal logics. We showed that such modalities cannot always be characterised by a class of Kripke frames. However, there is hope that our new reactive semantics might provide frames for some of these modalities. This section and the next study the connection.

A hyper-modality $\Box$ is a modality which changes its nature depending on where in a formula it appears. So for example, in the formula $B = \Box^3 q \wedge \Box^2 q$, the inner modalities may not have the same meaning as the outer ones. To illustrate this point consider the arrangement of Fig. 20

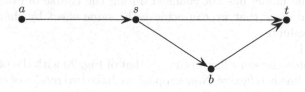

Fig. 20.

t is *now*, s is in the past of t and so are a and b. We consider two past operators

$-\ H_{\mathbf{K}} A$ saying A was true at all the immediately past moments of time

and

$-\ H_{\mathbf{T}} A$ saying $A \wedge H_{\mathbf{K}} A$

We let $\Box$ alternate between $H_{\mathbf{K}}$ and $H_{\mathbf{T}}$, starting with $H_{\mathbf{T}}$.

Thus $B = \Box^3 q \wedge \Box^2 q$ reads

$$H_{\mathbf{T}} H_{\mathbf{K}} H_{\mathbf{T}} q \wedge H_{\mathbf{T}} H_{\mathbf{K}} q.$$

Let us evaluate B at t. We write $t \vDash_{\mathbf{K}} A$ when we are evaluating A at a $\mathbf{K}$ mode and $t \vDash_{\mathbf{T}} A$ when we are evaluating A at the $\mathbf{T}$ mode.

Writing the above in full we have:

(*1) $t \vDash_{\mathbf{K}} \Box A$ iff for all immediately past points s we have $s \vDash_{\mathbf{T}} A$.

(*2) $t \vDash_{\mathbf{T}} \Box A$ iff $t \vDash_{\mathbf{K}} A$ and for all immediately past points s we have $s \vDash_{\mathbf{K}} A$.

Let us now evaluate $t \vDash (\square^3 q \wedge \square^2 q)$ in the flow of Fig. 20. We have (remember we start with $\vDash_{\mathbf{T}}$):

- $t \vDash_{\mathbf{T}} \square^3 q$ iff first $s \vDash_{\mathbf{K}} \square^2 q$ and second $b \vDash_{\mathbf{K}} \square^2 q$ and third $t \vDash_{\mathbf{K}} \square^2 q$ iff first $a \vDash_{\mathbf{T}} \square q$ and second $s \vDash_{\mathbf{T}} \square q$ and third $b \vDash_{\mathbf{T}} \square q$ and $s \vDash_{\mathbf{T}} \square q$.
- $t \vDash_{\mathbf{T}} \square^2 q$ iff first $s \vdash_{\mathbf{K}} \square q$, and second $b \vDash_{\mathbf{K}} \square q$ and third $t \vDash_{\mathbf{K}} \square q$.

Since both $t \vDash_{\mathbf{T}} \square^3 q$ and $t \vDash_{\mathbf{T}} \square^2 q$ must hold, we see that we need to evaluate both $s \vDash_{\mathbf{T}} \square q$ and $s \vDash_{\mathbf{K}} \square q$.

This means that we cannot make the evaluation of $\square q$ at s be dependent solely on the properties of the set $\{y \mid yRs\}$. We do need the dependency on the $\mathbf{T}$ and $\mathbf{K}$ modes.

Indeed, we axiomatise in [10] a modal logic with only the connective $\square$ with the property that this logic can be characterised by the two $\mathbf{K}$ and $\mathbf{T}$ modes but it cannot be characterised by any class of frames. This shows that mode shifting is a genuinely stronger instrument of defining modal logics than imposing conditions on the accessibility relation R. We will to show in this paper that this logic can be characterised by a class of reactive models.

So much for a short survey of the ideas of [10]. See Appendix A for formal definitions of hyper-modal logics. Let us now proceed to show the connection with the reactive semantics of this paper.

First observe that the mode described above change the meaning of $\square$. The modes do not change the semantics. In other words, the geometry of Fig. 20 remained fixed. The model has not changed during the course of evaluation of $t \vDash B$. We want to show that we can achieve the same effect by changing the semantics as we evaluate.

Consider Fig. 21.

Figure 21 describes the same flow of time as that of Fig. 20 with the addition of the property that time is reflexive. Now suppose we have two modes of evaluation $\vDash_{\mathbf{T}}$, where we evaluate $\square$ in the reflexive mode (i.e. Fig. 21), and $\vDash_{\mathbf{K}}$, where we evaluate in the irreflexive model (i.e. in Fig. 20).

Let us spell it out clearly:

(*3) $t \vDash_{\mathbf{K}} \square A$ iff first deactivate all reflexive arrows in the accessibility relation R and then ask for $s \vDash_{\mathbf{T}} A$ to hold at every s which is accessible to t.

(*4) $t \vDash_{\mathbf{T}} \square A$ iff first reactivate all cancelled reflexive arrows in R and then ask for $s \vDash_{\mathbf{K}} A$ to hold at every s accessible to t.

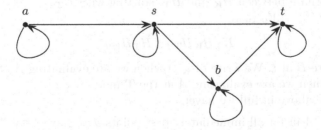

Fig. 21.

Clearly the evaluation of $t \vDash B$ will end up the same whether we view it as shifting the meaning of $\Box$ or shifting the underlying accessibility relation in the model.

Let us view the changing of the semantics as disconnecting or reconnecting arrows (accessibility) in the model.

This we have already done in the technical example of Subsection 1.3. See Fig. 7.

It is not difficult to see how a general arc model can be constructed in which $\Box$ alternates between a **T** and a **K** modality.

To do this properly, we need first a formal definition of the reactive semantics for modal logic.

4 Switch Reactive Kripke Models

We now give a definition of what we call switch reactive Kripke models, giving rise to what we call switch arc modal logics. In these models the reactive double arrows are just on/off switches, and the emanate from points.

Definition 1. *Let S be a set of possible worlds and $a \in S$ is the actual world. An arc-accessibility relation on S is defined as follows:*

1. *The set of all arcs $\mathcal{A}$ is defined by*
   1.1. *$S \subseteq \mathcal{A}$, these are 0 level arcs.*
   1.2. *If $\alpha \in \mathcal{A}$ is an n level arc and $s \in S$ then $\beta = (s \to \alpha)$ is an $n+1$ level arc.*
2. *A subset $\mathcal{R} \subseteq \mathcal{A}$ is an arc relation.*
3. *An arc-Kripke model has the form $(S, \mathcal{R}, \mathcal{R}^*, a, h)$, where $\mathcal{R}$ and $\mathcal{R}^*$ are an arc-accessibility relations and $\mathcal{R} \subseteq \mathcal{R}^*$ and h is an assignment giving to each $t \in S$ and each atom q a value $h(t, q) \in \{0, 1\}$. The part of the model $(S, \mathcal{R}, \mathcal{R}^*, a)$ is called the model frame and $a \in S$ is the actual world of the frame.*
4. *Note that in this definition all arcs emanate from points.*

Definition 2. *Let $\mathbf{m} = (S, \mathcal{R}, \mathcal{R}^*, a, h)$ be a model.*
   Define $a \vDash A$, by structural induction

1. *$a \vDash q$ if $h(a, q) = 1$, for q atomic.*
2. *$a \vDash A \wedge B$ iff $a \vDash A$ and $a \vDash B$.*
3. *$a \vDash \neg A$ iff $a \nvDash A$.*
4. *$a \vDash \Box A$ iff for all s in S such that $(a, s) \in \mathcal{R}$ we have that $s \vDash A$ in the model $\mathbf{m}_s = (S, \mathcal{R}_a, \mathcal{R}^*, s, h)$, where $\mathcal{R}_a$ is obtained from $\mathcal{R}$ as follows:*

$$\mathcal{R}_a = \mathcal{R} - \{\alpha \mid (a \to \alpha) \in \mathcal{R} \wedge \alpha \in \mathcal{R}\} \cup \{\alpha \mid (a \to \alpha) \in \mathcal{R} \wedge \alpha \notin \mathcal{R} \wedge \alpha \in \mathcal{R}^*\}.$$

   This is a switch satisfaction clause for $\Box$.
5. *A model $\mathbf{m} = (S, \mathcal{R}, \mathcal{R}^*, a, h)$ is said to be of level $\leq n$, $n \geq 1$, if all its arcs in $\mathcal{R}^*$ are of level $\leq n$.*
   For example, a model of level ≤ 2 can contain either arcs of the form $t \to s$ or of the form $r \to (t \to s)$, where $r, t, s, \in S$.

6. Let $\mathbf{K}^n_A, n \geq 1$ be the set of wffs valid in the class of all reactive Kripke models of level $\leq n$. Note that $\mathbf{K}^1_A$ is ordinary modal $\mathbf{K}$.
   Let $\mathbf{K}_A$ Be the set of wffs valid in the class of all reactive Kripke models. This is the basic reactive analogue of modal $\mathbf{K}$.

7. A modal logic $\mathbf{L}$ is said to be a reactive modal logic if for some class $\mathcal{K}$ of reactive models $\mathbf{L} = \{A \mid A$ is valid in all models of $\mathcal{K}\}$.

8. Note that to be more precise these models use only arcs emanating from points and the reactivity acts like a switch.

9. Note that the evaluation of $a \vDash \Box A$ and $a \vDash \Box B$ for any two wffs A and B is done independently of one another and in parallel, as indicated in item 4 above. We have $\Box(A \wedge B) \leftrightarrow \Box A \wedge \Box B$ and both A and B in this case get evaluated at the model with $\mathcal{R}_a$.
   This gives us scope to introduce the connective "$\wedge$", where $X \wedge Y$ reads A and then y. So $a \vDash \Box A \wedge \Box B$ will evaluate $a \vDash \Box A$ first. In the course of the evaluation of $\Box A$ the relation $\mathcal{R}$ will become $\mathcal{R}_a$ and then we evaluate $a \vDash \Box B$ in the model $(S, \mathcal{R}_a, \mathcal{R}^*, a, h)$.

Example 7. Consider the two point model of Fig. 22 with $S = \{a, b\}$:
   Let $\mathcal{R} = \mathcal{R}^* = \{a \to b, b \to a, a \to (a \to b), a \to (a \to (a \to b)), b \to (b \to a), b \to (b \to (b \to a))\}$.
   Consider a model
$$\mathbf{m} = (S, \mathcal{R}, \mathcal{R}^*, a, h).$$
Here $\mathcal{R}_a = \{b \to a, b \to (b \to a), a \to (a \to (a \to b))\}$.

$$\mathcal{R}_{a,b} = \{b \to (b \to a), a \to (a \to (a \to b))\}$$
$$\mathcal{R}_{a,b,a} = \{b \to (b \to a), a \to (a \to (a \to b)), a \to (a \to b)\}.$$

Note that $a \to b$ is not restored until $\mathcal{R}_{a,b,a,b,a}$.

Definition 3. Let (S, R) be a set S with a binary relation $R \subseteq S^2$. Let τ be a Horn clause theory in the language with R containing universal clauses of the form

(universal closure) $(\bigwedge_{i=1}^n x_i R y_i \to x R y)$. Let $R^\tau \supseteq R$ be defined as the smallest extension of R such that $(S, R^\tau) \vDash \tau$. R^τ can be constructed by induction as the closure of R under all instances of τ as follows:

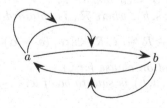

Fig. 22.

1. Let $R_0 = R$

2. Let $R_{n+1} = \{(a,b) \mid$ for some clause L in τ of the form (universal closure) $(\bigwedge x_i R y_i \to xRy)$ and a substitution θ to the variables of L such that $\theta(y_i), \theta(x_i) \in S, \theta(x) = a, \theta(y) = b$, we have that $\bigwedge \theta(x_i) R_n \theta(y_i)$ holds.

3. Let $R_\tau = \bigcup_n R_n$.

Definition 4

1. Let $\mathbb{N}$ be the set of natural numbers $\{0, 1, 2, \ldots\}$. Let $\mathbb{N}^*$ be the set of all finite sequences of natural numbers including the empty sequence $\varnothing$. Define $\alpha < \beta$, for $\alpha, \beta \in \mathbb{N}^*$ by

$$\alpha < \beta = (\text{ definition}) \text{ for some } m \in \mathbb{N}, \beta = \alpha * (m),$$

   where $*$ is concatenation.
   Let $<^*$ be the transitive closure of $<$.

2. A tree T is a nonempty subset of $\mathbb{N}^*$ such that if $\beta \in T$ and $\alpha < \beta$ then $\alpha \in T$.

3. Let τ be a Horn theory on $<$.
   Let $(T, <)$ be a tree and let $(T, <^\tau)$ be its τ-closure. Define $\mathcal{R}_\tau$ as follows:
   - $(\alpha \to \beta) \in \mathcal{R}_\tau$ if $(\alpha, \beta) \in <^\tau$.
   - $(\gamma \to (\alpha \to \beta)) \in \mathcal{R}_\tau$ whenever $\gamma <^\tau_* \alpha$ and $\gamma <^\tau_* \beta$ hold where $<^\tau_*$ is the transitive closure of $<^\tau$.

4. Let $\mathcal{K}_\tau$ be the class of all models of the form $(T, <, \mathcal{R}_\tau, \delta, h)$, where T is a tree and $\delta \in T$.

5. Let H_τ be a set of all wffs A such that A holds in any model of $\mathcal{K}_\tau$.

Example 8. Let $\tau = \{\forall x(xRx)\}$. Then the models of $\mathcal{K}_\tau$ have the form $(T, <, \mathcal{R}_\tau, \delta, h)$, where $\mathcal{R}_\tau = < \cup \{\gamma \to (\alpha \to \alpha) \mid \gamma \leq^* \alpha\}$. This is so since $<^\tau$ is $< \cup \{(\alpha, \alpha) \mid \alpha \in T\}$.

It is easy to see that the meaning of $\square$ alternates between $\mathbf{K}$ and $\mathbf{T}$ modalities because $\mathcal{R}$ switches the reflexive arcs on and off. Thus the logic H_τ of our example is the same as the logic H_1^S of Sect. 3 of [10].

We know from [10] that the following is a Hilbert axiomatisation of H_τ. We make use of our irreflexivity rule, see [7].

Axioms: (E, F are wffs without $\square$).

1. $A \wedge \square A$, where A is a substitution instance of a truth functional tautology.
2. $\square(A \to B) \to (\square A \to \square B)$
3. $\square(\square(A \to B) \to (\square A \to \square B))$
4. $\Diamond \top$
5. $\neg E \wedge \square^2 E \wedge Y \to \Diamond(\neg E \wedge Y)$, where $Y = A$ or $Y = \square A$, for A without $\square$.
6. $\neg E \wedge \square^2 E \wedge \Diamond(\neg E \wedge A) \wedge \Diamond(\neg E \wedge B) \to \Diamond(\neg E \wedge A \wedge B)$.
7. $\neg E \wedge \square^2 E \to \Diamond(\neg E \wedge A) \vee \Diamond(\neg E \wedge \neg A)$
8. $\square A \wedge \neg E \wedge \square^2 E \to \Diamond(\neg E \wedge A)$.
9. $\square X \wedge \neg E \wedge \square^2 E \to \Diamond(\neg E \wedge \square(\neg \wedge \square^2 F \to \Diamond \neg F \wedge X))$

10. $\neg E \wedge \Box^2 E \wedge \Diamond A \wedge \Diamond(\neg E \wedge \Box Y \wedge \Box \Box Y' \wedge \neg A \wedge \Box \Box X) \to \Diamond(A \wedge X \wedge Y \wedge \Box Y')$
    where Y, Y' are without $\Box$
11. $\neg E \wedge \Box^2 E \wedge \Diamond(C \wedge E \wedge Y) \to \Diamond(\neg E \wedge \Diamond(Y \wedge E \wedge \Diamond(C \wedge E)))$

Rules

MP: $$\frac{\vdash A; \vdash A \to B}{\vdash B}$$

IRR: $$\frac{\vdash \neg q \wedge \Box^2 q \to A}{\vdash A}$$
where q is an atom not in A.

2-necessitation: $$\frac{\vdash A}{\vdash \underset{n}{\Box^2} A}$$

IRR^n: $$\frac{\vdash \bigwedge_{m=1} \beta_m^m \to A}{\vdash A}$$
where β_m^m are as defined below and q_j^i are all not in A.

The following defines β_j^i:
Let q_j^i be a double indexed sequence of atoms. Let

1. $\beta_1^i(q_1^i) = \neg q_1^i \wedge \Box^2(q_1^i)$.
2. $\beta_2^i(q_1^i, q_2^i) = \neg q_2^i \wedge \Box^2 q_2^i \wedge \Diamond(\neg q_2^i \wedge \Box \beta_1^i)$.
3. $\beta_{n+1}^i(q_1^i, \ldots, q_{n+1}^i) = \neg q_{n+1}^i \wedge \Box^2 q_{n+1}^i \wedge \Diamond(\neg q_{n+1}^i \wedge \Box \beta_n^i)$

Example 9. We now give an example of a class of reactive Kripke models characterising a logic which cannot be presented as a hyper-modal logic. Consider one point models of the following form, see Fig. 23

Let
$$\text{arc}_0 = a \to a$$
$$\vdots$$
$$\text{arc}_{n+1} = (a \to \text{arc}_n)$$

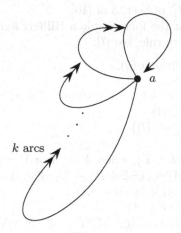

k arcs

Fig. 23.

Consider the models $\mathbf{m}_k$ of the form

$$\mathbf{m}_k = (\{a\}, \{\mathrm{arc}_0, \mathrm{arc}_k\}, \{\mathrm{arc}_n \mid n \le k\}, a)$$

The model $\mathbf{m}_k$ is a one node model with arcs as in Fig. 23, where only the arcs $a \to a$ and $a \to (a \to \ldots (a \to a) \ldots)$ (k arrows) are active.

The following points are clear for the class of models $\{\mathbf{m}_k\}$.

1. $\Diamond^n \Box \bot$ is consistent for all n.
2. The following holds in the logic $\mathbf{L}$ of this class

$$\vDash \Diamond^n \Box \bot \to \bigwedge_{m \le n} (\Box^m A \Leftrightarrow A)$$

for all wffs A without $\Box$.

We claim the logic $\mathbf{L}$ cannot be presented as a hyper-modal logic.

For suppose there is a class of traditional Kripke models of the form (S, R, a) and a sequence $(\Psi_1, \ldots, \Psi_k)$ of conditions characterising modalities $\Box_1, \ldots, \Box_k$ such that the evaluation of $\Diamond$ alternates according to this sequence (see Definition 9).

Then for a high enough n, the meaning of $\Diamond$ in $\Diamond^n \Box \bot$ starts repeating itself.

Let (S, R, a) be a model of $\Diamond^n \Box \bot$, for n large enough. By property 2, all $\Diamond^m, m \le n$ satisfy $\vDash_a \Diamond^m A \Leftrightarrow A$, for A without $\Box$. Thus we must have $\Psi_i(a, R) = \{(a, a)\}$. Therefore, how can we also have $\vDash_a \Diamond^n \Box \bot$?

This example shows that there is a class of reactive models of finite level defining a modal logic $\mathbf{L}$ which is not a hyper-modal logic.

Example 10. We now exhibit a hyper-modal logic which cannot be characterised by a class of finite level reactive models. Consider the situation in Fig. 24

Let $\Psi_1 = \{(a, a)\}$ and let $\Psi_2 = \{(a, a), (a, b)\}$. Consider the sequence (Ψ_1, Ψ_1, Ψ_2). This means that $\Box$ is interpreted as seeing only the a node twice and then it can also see b once before repeating. Let $\mathbf{L}$ be the hyper-modal logic defined by this set up.

To implement this logic by reactive models we need to switch the arc $a \to b$ on and off by other arcs in the repeating sequence $(-, -, +)$.

Figure 25 can help visualise the situation:

Fig. 24.

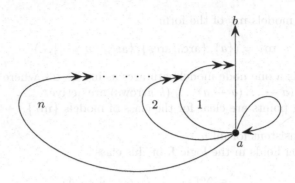

Fig. 25.

We note that if the connection $a \to b$ must alternate as $(-, -, +)$ then connection $a \to (a \to b)$ (i.e. arc_1 in Fig. 25) must alternate $(-, +, +)$ and arc_2 must alternate $(+, -, +)$ and so on.

However in any finite level model, the highest level arcs cannot alternate. Hence modalities of the form $\square^n$ for high enough n cannot be implemented in any given model. We conjecture that if we allow models of unbounded level then all reasonable hyper-modalities can be implemented. In our case the sequence $(-, -, +)$ can be implemented by a single model of infinite level.

Remark 2. Let us give some thought to the problem of how to axiomatise the basic reactive modal logic $\mathbf{K}_{\mathcal{A}}$ (see Definition 1, item 6). Let Δ be a consistent complete theory of this logic. Then for some model $\mathbf{m} = (S, \mathcal{R}, \mathcal{R}^*, a, h)$ we have $\Delta = \{A \mid a \vDash A\}$. So Δ must contain within it enough information to reconstruct a canonical model with a suitable arc accessibility. How this can be is not immediately clear. In ordinary modal logic we can construct accessibility $\Theta R \Theta'$ by letting it be

$$\forall \alpha(\square \alpha \in \Theta \to \alpha \in \Theta)$$

We can do the same here and this will give us the arrows but how do we define the double arrows?

Coming to think of it, we still have to check whether $\mathcal{K}_{\mathcal{A}}$ is axiomatisable or is RE?

5   Non-deterministic Reactive Kripke Models

Let us begin by considering again the basic situation of Fig. 1. This is a simple reactive Kripke model. Let us make it non-deterministic. This would mean that as we move from node a to node b, the double arrow $a \twoheadrightarrow (b \to c)$, may or may not (non-deterministically) fire. So to evaluate $a \vDash \square^2 q$, we need to evaluate $b \vDash \square q$. If the double arrow does fire, then the connection $b \to c$ is cancelled and only $b \to d$ is active. If the double arrow does not fire, then $b \to c$ is also active when we evaluate $b \vDash \square q$.

So at this model (the non-deterministic model of Fig. 1), there are two possible non-deterministic valuations of $a \vDash \Box^2 q$. One with $b \rightarrow c$ active at the time of the evaluation of $b \vDash \Box q$ and one where it is not active. We say that a non-deterministic model $\mathbf{m}$ can satisfy a wff A if there is at least one non-deterministic valuation of A in the model in which A gets the value $\top$. We say that A holds in the model if it holds (gets $\top$) under all non-deterministic valuations in the model.

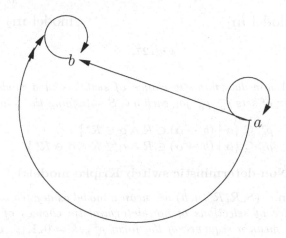

Fig. 26.

Example 11. Consider the model of Fig. 26.

Then as a non-deterministic model we have that $\Diamond^n \Box \bot$ can always be made false, by never activating the double arrow in the evaluation. But it can also be satisfied by activating the double arrow at the $(n-1)$st step and then the model becomes irreflexive at point b and $\Box \bot$ then holds at b. Thus we see that such models can satisfy both a wff A and its negation $\neg A$, because different non-deterministic options can come to bear. However, only one of A and $\neg A$ can hold in the model. We can see already that we need to be careful with this concept. Consider the axiom $\Box(A \wedge B) \rightarrow \Box A \wedge \Box B$. This is valid in any reactive Kripke model (provided the evaluations of $\Box A$ and of $\Box B$ are done in parallel, see Definition 2). Not so in a non-deterministic model for consider $\Diamond \Box (p \wedge \neg p) \wedge \Diamond \Box (q \wedge \neg q)$ where p and q are different atoms. When we evaluate $a \vDash \Diamond \Box (p \wedge \neg p)$ in the model of Fig. 26, we may take the non-deterministic choice of not firing the double arrow and thus get a value false. When we do it again for the second copy $\Diamond \Box (q \wedge \neg q)$, we may choose to fire and thus get value true. We are going to have to formulate our definitions carefully here.

The single non-deterministic model of Fig. 26 is equivalent to the two ordinary Kripke models of Fig. 27, $\mathbf{m}_1$ and $\mathbf{m}_2$

During the evaluation at node a we can change our view of whether node b is reflexive or not.

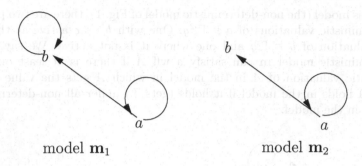

model $\mathbf{m}_1$ model $\mathbf{m}_2$

Fig. 27.

Definition 5. *A non-deterministic choice of switches in a model* $\mathbf{m} = (S, \mathcal{R}, \mathcal{R}^*, a, h)$ *is a pair of sets* ρ_a^+, ρ_a^- *for each* $b \in S$ *satisfying the following:*

$$\rho_b^+ \subseteq \{\alpha \mid (b \to \alpha) \in \mathcal{R} \wedge \alpha \in \mathcal{R}^*\}$$
$$\rho_b^- \subseteq \{\alpha \mid (b \to \alpha) \in \mathcal{R} \wedge \alpha \notin \mathcal{R} \wedge \alpha \in \mathcal{R}^*\}$$

Definition 6 (Non-determinstic switch Kripke models)

1. *By a model* $\mathbf{m} = (S, \mathcal{R}, \mathcal{R}^*, a, h)$ *we mean a model as defined in Definition 1.*
2. *By a sequence of selections of non-deterministic choices of switches in a model* $\mathbf{m}$ *we mean a sequence of the form* $\rho_{a,k}^\pm, k = 0, 1, \ldots$ *where for each* k, $\rho_{a,k}^\pm$ *is a non-deterministic choice as defined in the previous definition.*
3. *Given a wff* A *and a node* a *in a model* $\mathbf{m}$, *we define the universal non-deterministic evaluation* $a|\equiv_k A$ *in parallel to Definition 2 as follows (where* k *is a natural number representing a universal ticking clock meaning that this is the kth time that a* $\Box$ *is evaluated at the node* a).
    (a) $a \models_k q$ *if* $h(a, q) = 1$, *for* q *atomic and any* $k \geq 0$.
    (b) $a \models_k A \wedge B$ *iff* $a \models_k A$ *and* $a \models_k B$
    (c) $a \models_k \neg A$ *iff* $A \not\models_k A$
    (d) $a \models_k \Box A$ *iff for all* $s \in S$ *such that* $(a, s) \in \mathcal{R}$, *we have that* $s \models_{k+1} A$ *in the model* $\mathbf{m}_s^k$, *where*

$$\mathbf{m}_s^k = (S, \mathcal{R}_a^k, \mathcal{R}^*, s, h)$$

   and where

$$\mathcal{R}_a^k = (\mathcal{R} - \rho_{a,k}^+) \cup \rho_{a,k}^-.$$

    (e) *A formula* A *is said to be satisfiable in a model* $\mathbf{m}$ *if for some non-deterministic choice sequence* $\rho_{t,k}^\pm, t \in S, k = 0, 1, 2, \ldots$. *We have that* $a \models_0 A$. *Note that it is possible to have that both* $a \models_0 A$ *and* $a \models_0 \neg A$.
    (f) *A formula is said to hold in a model* $\mathbf{m}$ *if its negation is not satisfiable in* $\mathbf{m}$.
    (g) *A modal logic* $\mathbf{L}$ *is said to be a non-deterministically reactive if for some class of reactive Kripke models we have*

$$\mathbf{L} = \{A \mid A \text{ is non-deterministically valid in all models in the class}\}$$

Example 12 This example explains the 'ticking clock' of the previous definition.

Consider the frame of Fig. 26, and consider $a \vDash \Diamond\Box(p \wedge \neg p)$ and again $a \vDash \Diamond\Box(q \wedge \neg q)$.

In the first evaluation we activate the double arrow immediately and get true and in the second evaluation we do not activate it and get false. To avoid such a situation, we introduce a universal counting clock, counting the number of nested modalities. We now write

$$a\vDash_0 \Diamond\Box(P \wedge \neg p) \text{ iff } a\vDash_1 \Box(p \wedge \neg p) \text{ or } b\vDash_1 \Box(p \wedge \neg p)$$

We require always the same non-deterministic choice for any evaluation of the form $t\vDash_k \Box$, with the same t and k.

If we have this we get that $\Box(A \wedge B) \to \Box A \wedge \Box B$ holds.

6   Connection with Fibring Logics

This section gives some methodological remarks and points out a connection with fibring logics; see [9,12].

6.1   Methodological Considerations

Consider two modal logics $\mathbf{L}_1$ and $\mathbf{L}_2$. Assume $\mathbf{L}_1$ has a $\mathbf{K}$ modality, $\Box_1$ and let $\mathbf{L}_2$ have a $\mathbf{K4}$ modality, $\Box_2$. We know that $\mathbf{L}_1$ is complete for all Kripke frames of the form (S_1, R_1, a_1), with $a_1 \in S_1, R_1 \subseteq S_1^2$, and $\mathbf{L}_2$ is complete for all Kripke frames of the form (S_2, R_2, a_2) with $a_2 \in S_2, R_2 \subseteq S_2^2$ and R_2 is a transitive (irreflexive) relation. When we combine these two logics in a way which is generally called *fusion* (or *dovetailing* in [9]), we get completeness for the class of all Kripke models of the form (S, R_1, R_2, a) with R_2 transitive irreflexive and where the truth table for $\Box_i$ is defined as follows:

(∗1) $t \vDash \Box_i q$ iff $\forall s(tR_i s \to s \vDash q)$ for $i = 1, 2$.

The above combination of $\mathbf{L}_1$ and $\mathbf{L}_2$ assumes that the languages $\mathbf{L}_1$ and $\mathbf{L}_2$ are disjoint, and to the extent that they share the classical connectives, they agree on these connectives.

Assume now that $\mathbf{L}_1$ and $\mathbf{L}_2$ contain an additional modality $\Box$, i.e. $\mathbf{L}_1 = \mathbf{L}_2(\Box, \Box_1)$ and $\mathbf{L}_2 = \mathbf{L}_2(\Box, \Box_2)$, where $\Box$ is a $\mathbf{T}$ modality. Now the models for $\mathbf{L}_i$ have the form (S_i, R, R_i, a_i) where $R \subseteq S_i^2$ and R is reflexive. Obviously $\mathbf{L}_1$ and $\mathbf{L}_2$ which contain the *same* symbol for modality $\Box$, must agree on the nature of this modality (namely that it is a $\mathbf{T}$ modality). We can therefore safely combine them (in the same way as before) and get a logic $\mathbf{L}_{1+2}$ with $\Box, \Box_2, \Box_2$. This logic is complete for models of the form (S, R, R_1, R_2, a) where $R \subseteq S^2$ is reflexive and $R_2 \subseteq S^2$ is irreflexive and transitive, and $R_1 \subseteq S^2$.

The question arises what can we do if $\mathbf{L}_1$ and $\mathbf{L}_2$ do not agree on the modality $\Box$? What if $\mathbf{L}_1$ has $\Box$ as $\mathbf{K}$ modality while $\mathbf{L}_2$ has $\Box$ as a $\mathbf{T}$ modality? How do we do this? To put it bluntly, can we combine two systems for the same modality $\Box$, where one says $\Box$ is a **K** modality and the other one says that $\Box$ is a **T** modality?

To make the problem more acute we can have modalities with contradictory conditions on R. For example, $\mathbf{L_1}$ can ask $\forall x \exists y x R y$ (the axiom $\Box \Diamond \top$) and $\mathbf{L_2}$ can ask $\exists x \forall y \neg x R y$ (the axiom $\Diamond \Box \bot$).

One way to combine two different systems for the same symbols is to timeshare the symbol, i.e. $\Box$ can sometimes be an $\mathbf{L_1}$ modality and sometimes an $\mathbf{L_2}$ modality. This is the way processes share the same resource; they timeshare according to a certain protocol μ.

So let us examine modal logics with a single modality $\Box$, where the meaning of $\Box$ is timeshared between, say a **K** and a **T** modality.

The questions we ask are the following:

(Q1)   How do we define and execute the timeshare?
(Q2)   What is the semantics corresponding to (Q1)?

Let us start by choosing something simple. Let the meaning of $\Box$ alternate between a **T** and a **K** modality. Thus a wff of the form

$$\alpha = \Box(\Box q \wedge \Box\Box\Box p)$$

will read as

$$\beta = \Box_{\mathbf{T}}(\Box_{\mathbf{K}} q \wedge \Box_{\mathbf{K}}\Box_{\mathbf{T}}\Box_{\mathbf{K}} p).$$

Thus the meaning of $\Box$ alternates according to its nesting depth in the formula.

So now we know how to read α, namely we read it as β, but what is the semantics for β? Being a mono-modal logic, we expect a Kripke semantics with a single accessibility relation R, i.e. models of the form (S, R, a). So how can R be and not be reflexive alternatively? This is impossible!

To this end we need a new point of view on how we see our semantics. Let us consider the notion of a *mode of search* for a possible world. Namely let $\Psi(x, y)$ be (a possibly higher order) binary wff on the classical model (S, R, a). Let

(∗2) $t \vDash_\Psi \Box q$ iff for all s such that $\Psi(t, s)$ holds we have $s \vDash q$.

Ψ is the mode of search for worlds s. The following are examples:

- For **K** modality $\Psi_{\mathbf{K}}(x, y) \equiv x R y$
- For **T** modality $\Psi_{\mathbf{T}}(x, y) \equiv x = y \vee x R y$
- For **K4** modality $\Psi_{\mathbf{K4}}(x, y) \equiv$ for some $n \geq 0$ we have $x R^n y$.
  Note that $\Psi_{\mathbf{K4}}$ is not first-order.

The truth condition for a system with alternating **K** and **T** modality can be written as follows:

(∗3)
- $t \vDash_{\mathbf{K}} \Box q$ iff for all $s, \Psi_{\mathbf{K}}(t, s) \to s \vDash_{\mathbf{T}} q$.
- $t \vDash_{\mathbf{T}} \Box q$ iff for all $s, \Psi_{\mathbf{T}}(t, s) \to s \vDash_{\mathbf{K}} q$.

The logic with (S, R, a) where $\square$ alternates in meaning as above can be seen as one way of fibring disagreeing modalities via a timesharing protocol.

Let us use the following terminology:

- use the term *C-fibring* for *Compromise Fibring* when we combine two logics which do not agree on the common part.
- Use the words *μ-Protocol* for any procedures μ for timesharing the common symbols.
- A modality which changes its meaning in the sentence is called a *hypermodality*, see [10]

We now continue and discuss another 'twist' in our way of thinking. Obviously when $\square$ is interpreted as $\square_{\mathbf{T}}$, we use $\Psi_{\mathbf{T}}$. How about modalities complete for classes of models whose relations cannot be nicely defined via some Ψ (even second order)? We certainly can form models of the form (S, R_1, R_2, a) and let the meaning of $\square$ alternate between R_1 and R_2. However, this is not neat, because with one symbol $\square$ we expect semantics with one symbol R or some generalisation of it.

We need some inspiration at this junction. So here it is:

We talk about *reactive models*. These are models whose accessibility relation changes as we go along. Figure 28 illustrates the basic mechanism:

We evaluate $t \vDash \square q$ at moment t. We assume that at each moment, either all the R_1 or all the R_2 (but not both) connections are on. When we move from $t \vDash \square q$ to s, i.e. to $s \vDash q$, we send a signal from t ($\twoheadrightarrow$) to all other R_1 and R_2 connections. If a connection receives the signal and it is on, then it turns off. If it is off then it turns on. So when we come to s, we switch from R_1 to R_2 or from R_2 to R_1 depending which one is active. In general, we we deal with reactive models as in Sect. 2.

Now in order to be able to compromise-fibre two logics with the same modality but different reactive Kripke semantics, we need to allow arrow $\twoheadrightarrow$ to cancel

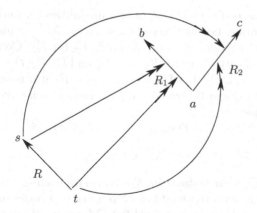

Fig. 28.

other arrows up to arbitrary finite iterations. This yields Definitions 1 and 1 of Sect. 3.

We have now completed our methodological motivation for reactive Kripke models. For more on compromise fibring, see [12].

6.2 Formal Compromise Fibring

It is now time to give a formal definition of C-fibring.

Definition 7

1. Let (S, R, a) be a Kripke frame. By a finite path we mean a sequence of elements of the form $St = (a, a_1, \ldots, a_n)$ such that $aRa_1 \wedge a_1 Ra_2 \wedge \ldots a_{n-1} Ra_n$ holds. We also include the empty sequence as a path.
2. Let $t \in S$, we define the model $(S_{[t]}, R_{[t]}, t)$ as follows.
   (a) Let $S_{[t]}$ be the smallest subset $S' \subseteq S$ such that $t \in S'$ and if $x \in S'$ and xRy then $y \in S'$.
   (b) Let $R_{[t]}$ be $R \upharpoonright S_{[t]}$.

Definition 8. Let $\mathcal{K}_1, \ldots, \mathcal{K}_n$ be n classes of Kripke frames. Let $\mathbf{L}_1, \ldots, \mathbf{L}_n$ be the logics they define as frames in a language with the syntactical modality $\square$. We define a C-fibred general model for $\square$ as follows:

1. $(S, R, a, \mathbf{f})$ is a C-fibred model for $\mathbf{L}_1 + \mathbf{L}_2 + \ldots + \mathbf{L}_n$ iff $\mathbf{f}$ is a function giving for each finite path $t = (a, a_1, \ldots, a_n)$ a value $\mathbf{f}(t) = R_t^{\mathbf{f}} \subseteq R_{[a_n]}$ such that for some $m \leq n$, the model $(S_{[a_n]}, R_t^{\mathbf{f}}, a_n)$ is in $\mathcal{K}_m$.
2. We define satisfaction in $(S, R, a, \mathbf{f})$ as follows (under an assignment h to the atoms) for $t = (a, a_1, \ldots, a_n)$:
   − $a_n \vDash_t \square q$ iff for all $s \in S_{[t]}$ such that $tR_t^{\mathbf{f}}s$ we have that $s \vDash_{t*(s)} q$, where $*$ is concatenation of sequences.
3. Note that the condition $R_t^{\mathbf{f}} \subseteq R_{[a_n]}$ is not restrictive because we can always start with $R = S^2$.
4. Let $\mathbf{L}_1 + \ldots + \mathbf{L}_n$ be the logic defined by all models $(S, R, a, \mathbf{f})$.

Remark 3. Note that ordinary fibring of two modalities $\square_1$ and $\square_2$ can be captured by this semantics. Let us pretend that $\square_1 = \square_2 = \square$ and that $\mathbf{L}_1$ reads $\square$ as $\square_1$ for a class of models $\mathcal{K}_1$ and similarly $\mathbf{L}_2$ for $\mathcal{K}_2$. Consider all models with frames (S, R_1, R_2, a) where $(S, R_1, t) \in \mathcal{K}_1$ and $(S, R_2, t) \in \mathcal{K}_2$ for all $t \in S$. Let R be defined as $R_1 \cup R_2$. Let $\mathbf{f}(a, a_1, \ldots, a_n) = R_1$ if n is odd and R_2 if n is even. Then $\square_1$ and $\square_2$ can be translated in a *context sensitive way* as seen by the following example.

Let $\alpha = \square_1(\square_1 q \wedge \square_2 \square_1 p)$. Then α is translated into β.

$$\beta = \square(\square^2 q \wedge \square\square p).$$

We translate each $\square_i$ from outside in. We translate each $\square_i$ as either $\square$ or $\square\square$ depending whether it gives us an odd or even nesting of boxes as required by the index i. **This translation is not faithful. More work needs to be done on this idea.**

7   Dedicated Reactivity Connectives

Previous sections dealt with a modality $\Box$ that both sent the evaluation to accessible worlds and at the same time activated all reactivity double arrows. Let us check and see how we can separate these two functionalities using two separate connectives. Consider the simple model in Fig. 29

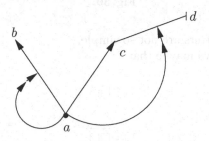

Fig. 29.

When we evaluate $a \vDash \Box q$, we move to the evaluation of $b \vDash q$ and $c \vDash q$. The node a also sends a signal along the double arrows at a and as a result the arcs (a, b) and (c, d) are disconnected.

Let us separate these two operations. Let us look at the fault disconnection signal separately from the ordinary modality operation. Let us have an ordinary modality $\overrightarrow{\Box} A$ and a reactive modality $\overset{\rightarrow}{\Box} A$ operating as follows:

1. $a \vDash \overrightarrow{\Box} A$ iff for all y such that aRy we have $y \vDash A$. This is an ordinary modality moving along the single arrows.
2. $a \vDash \overrightarrow{\Box} A$ iff $a \vDash A$ in the model obtained by letting all reactive double arrows do their job!
   So this is a modality which changes the model (set of possible worlds and its accessibility relation). It is like a **Jump** operator of the form

$$a \vDash_{\text{model } 1} \textbf{Jump}\ A \text{ iff } a \vDash_{\text{model } 2} A.$$

The move from one model to the other is done by activating the double arrows emanating from a.

So, for the situation in Fig. 29 we have that

$$a \vDash \overrightarrow{\Box} A \text{ in the model of Fig. 29}$$
$$\text{iff}$$
$$a \vDash A \text{ in the model of Fig. 30}$$

Having isolated the reactive operation, we would like to check the interdefinability properties of $\Box, \overrightarrow{\Box}$ and $\overrightarrow{\Box}$.

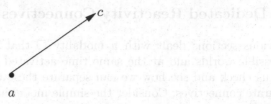

a

Fig. 30.

We will see that matters are not so simple.
We would like to have maybe that

$$\Box = \overrightarrow{\Box}\,\overrightarrow{\Box}$$

or

$$\Box = \overrightarrow{\Box}\,\overrightarrow{\Box}$$

or some combination

$$\Box = combination\ (\overrightarrow{\Box}, \overrightarrow{\Box})$$

However this is not so simple because $\Box$ fires both $\overrightarrow{\Box}$ and $\overrightarrow{\Box}$ simultaneously. If we do $\overrightarrow{\Box}$ first, we no longer can disconnect the (c, d) arc. If we do $\overrightarrow{\Box}$ first, we no longer can get to b.

If we use $a \vDash \Box\top$, we do not get the effect of $\overrightarrow{\Box}$ because we move to the accessible worlds b and c.

Thus, no matter what we do, we have a problem interdefining these connecteives.

In fact $\overrightarrow{\Box}$ is a metalevel connective, operating on the model (disconnecting accessibility arrows) and is different in nature from the evaluation of the ordinary connective $\overrightarrow{\Box}$. Being action metalevel connective we can talk naturally about the simultanety of several actions. Let "∥" indicate simultaneous operation of actions, being another metalevel connective.

Then we can write

$$\Box = \overrightarrow{\Box} \parallel \overrightarrow{\Box}$$

The difference between the connectives is even more serious. Consider Fig. 31.

Figure 31 is like Fig. 29 except the "fault" double arrows emanate from the arcs and not from the point. Do we want $\overrightarrow{\Box}$ to fire all double arrows from arcs or only one of them? Note that when we write $a \vDash \Diamond A$, we either go to $c \vDash A$ in which case only (c, d) is disconnected or we go to $b \vDash A$ in which case only (a, b) is disconnected.

It looks like we need, in addition to the modalities $\overrightarrow{\Box}$ and $\overrightarrow{\Diamond} = \neg\,\overrightarrow{\Box}\,\neg$ six additional fault connectives to be able to express all our options:

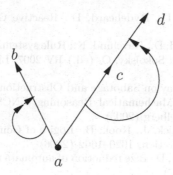

Fig. 31.

1. $a \vDash \vec{\Box}_{\text{point}} A$ activates all double arrows emanating from the point a.
2. $a \vDash \vec{\Box}_{\text{arc}} A$ activates all double arrows emanating from all arcs beginning at a.
3. $a \vDash \Diamond_{\substack{\text{point} \\ \to \text{one}}} A$ activates at least one double arrow emanating from a
4. $a \vDash \Diamond_{\substack{\text{arc} \\ \to \text{all}}} A$ activates at least one double arrow emanting from one arc coming out of a
5. $a \vDash \Diamond_{\text{arc}} A$ activates *all* double arrows emanting from at least one arc coming out of a.
6. $a \vDash (X\|Y)A$. Execute XA and YA in parallel, simultaneously.
   Thus we have
   $$- \; \Box = \vec{\Box} \; \| \; \vec{\Box}_{\text{point}} \; \| \; \vec{\Box}_{\substack{\text{arc} \\ \to \text{all}}}$$
   $$- \; \Diamond = \vec{\Diamond} \; \| \; \vec{\Box}_{\text{point}} \; \| \; \Diamond_{\text{arc}}$$
   Note that we can use $\|$ also as a connective and write $(\Box A)\|(\Box B)$, which means evaluate $\Box A$ and $\Box B$ in parallel at the same time so they will interfere with each other. Compare with $\Box A \wedge \Box B$ which do not interefere.

These are metalevel deletion operators brought into the object level by the use of connectives. Their detailed logical study is a big story on its own and needs to be done within a wider context. Deletion appears in abduction in logic and in handling databases as well as deletion of elements from models. See our anti-formula and anti-element papers [3,6].

Acknowledgements

I am grateful to David Makinson and to the referee for valuable comments.

References

1. Abramsky, S.: Semantics of interaction: an introduction to game semantics. In: Dybjer, P., Pitts, A. (eds.) Proceedings of the 1996 CLiCS Summer School, Isaac Newton Institute, pp. 1–31. Cambridge University Press, Cambridge (1997)

2. Barringer, H., Gabbay, D., Rydeheard, D.: Reactive Grammars, paper 303 (in preparation, 2007)
3. Barringer, H., Rydeheard, D., Havelund, K.: Rule systems for run-time monitoring: from Eagle to RuleR. In: Sokolsky, O. (ed.) RV 2007. LNCS, vol. 3839, Springer, Heidelberg (2007)
4. van Benthem, J.: An Essay on Sabotage and Obstruction. In: Hutter, D., Stephan, W. (eds.) Mechanizing Mathematical Reasoning. LNCS (LNAI), vol. 2605, pp. 268–276. Springer, Heidelberg (2005)
5. van Benthem, J., van Eijck, J., Kook, B.: Logics of Communciation and Change. Informationa nd Computation, 1620–1662 (2006)
6. Crochemore, M., Gabbay, D.: Size reduction of automata using reactive links, paper 304 (in preparation, 2007)
7. Gabbay, D.M.: An irreflexivity lemma with applications to axiomatizations of conditions on tense frames. In: Monnich, U. (ed.) Aspects of Philosophical Logic, pp. 67–89. D. Reidel (1981)
8. Gabbay, D.M.: Labelled Deductive Systems. Oxford University Press, Oxford (1996)
9. Gabbay, D.M.: Fibring Logics. Oxford University Press, Oxford (1998)
10. Gabbay, D.M.: A theory of hypermodal logics: mode shifting in modal logic. Journal of Philosophical Logic 31, 211–243 (2002)
11. Gabbay, D.M.: Reactive Kripke semantics and arc accessibility. In: Carnielli, W., Dionesio, F.M., Mateus, P. (eds.) Proceedings of CombLog 2004, Centre for Logic and Computation, University of Lisbon, pp. 7–20 (2004)
12. Gabbay, D.M.: More on reactive Kripke semantics and proof theory (in preparation)
13. Gabbay, D.M., Rodrigues, O., Woods, J.: Belief Contraction, Anti-formulas and Resource Overdraf: Part I. Logic Journal of the IGPL 10, 601–652 (2002)
14. Gabbay, D.M., Rodrigues, O., Woods, J.: Belief Contraction, Anti-formulas and Resource Overdraf: Part II. In: Gabbay, D.M., et al. (eds.) Logic, Epistemology and the Unityof Science, pp. 291–326. Kluwer, Dordrecht (2004)
15. Gabbay, D.M., Hogger, C.J., Robinson, J.A. (eds.): Handbook of Logic in Artificial Intelligence and Logic Programming, Nonmonotonic Reasoning and Uncertain Reasoning, vol. 3. Oxford University Press, Oxford (1994)
16. Gabbay, D.M., et al.: Many Dimensional Modal Logics. Elsevier, Amsterdam (2004)
17. Gabbay, D., Marcelino, S.: Theory of reactive graphs, paper 306 (in preparation 2007)
18. Gabbay, D.M., Shehtman, V.: Flow products of modal logics. Draft incorporated in [16] (2000)
19. Horty, J.F.: Some direct theories of nonmonotonic inheritance. In: [15], pp. 111–188
20. Lifschitz, V.: Circumscription. In [15], pp. 297–352.
21. Rohde, P.: Moving in a Crumbling Network: The Balanced Case. In: Marcinkowski, J., Tarlecki, A. (eds.) CSL 2004. LNCS, vol. 3210, pp. 1–25. Springer, Heidelberg (2004)

Appendix: Further Topics

A   Hypermodalities

Since we are comparing in this paper the notions of hyper-modal logics and reactive modal logics, we need to give here the exact definition of a hyper-modal logic.

Our starting point is a general Kripke model of the form $\mathbf{m} = (S, R, a, h)$. R is an arbitrary binary relation on S.

We are going to introduce evaluation modes into such semantics. It is convenient to regard the relation xRy as a classical formula $\Psi_{\mathbf{K}}(x, R, a, y)$ in the language of the relation R, the individual variables x, y and actual world constant a as follows

- $\Psi_{\mathbf{K}}(x, R, a, y) =_{\text{def}} xRy$

we have

- $t \vDash \Box A$ iff $\forall s(\Psi_{\mathbf{K}}(t, s) \to s \vDash A)$.

We can refer to $\Psi_{\mathbf{K}}$ as the mode of evaluation for $\Box$. It is fixed in the semantics and does not change. Intuitively it tells us, for a world x, how to evaluate $\Box A$ at x, namely where to look for worlds y where $y \vDash A$ must hold. The subscript $\mathbf{K}$ indicates that this formula is used in the case of $\mathbf{K}$ modality.

We can think of different formulas Ψ for the mode. Consider for example:

- $\Psi_{\mathbf{T}}(x, R, a, y) =_{\text{def}} xRy \vee x = y$
- $\Psi_{\mathbf{K4}}(x, R, a, y) =_{\text{def}} (\exists n \geq 1)xR^n y$.
  Where $xR^n y$ is defined by induction as:
  - $xR^0 y$ iff $x = y$
  - $xR^{n+1}y$ iff $\exists z(xRz \wedge zR^n y)$.
- $\Psi_{\mathbf{KB}}(x, R, a, y) =_{\text{def}} (xRy \vee yRx)$.

Clearly $\Psi_{\mathbf{K4}}(x, R, a, y)$ is not a first-order formula. It defines the transitive closure of R.

One can think of Ψ as changing the accessibility relation from R to $\lambda x \lambda y \Psi$ (x, y). Another way of looking at Ψ is that it gives us a new mode of how to use R in evaluating the truth value of $\Box A$. The latter view is more convenient to use because we will be shifting modes during the evaluation.

Let us write $\vDash_i$, to mean that the mode Ψ_i is used in the evaluation. Then $\vDash_{\mathbf{K}}$ for arbitrary frames (S, R, a) yields the logic $\mathbf{K}$, $\vDash_{\mathbf{T}}$ yields the logic $\mathbf{T}$, $\vDash_{\mathbf{KB}}$ yields the logic $\mathbf{KB}$ and $\vDash_{\mathbf{K4}}$ yields the logic $\mathbf{K4}$.

Note that our starting point is a frame (S, R, a) with an arbitrary R. We define $\Psi_i(x, R, a, y)$ as a binary relation and use it to evaluate $\Box$. Thus in traditional terms the frame we are using is (S, Ψ_i, a) not (S, R, a). When we shift modalities, i.e. change from $t \vDash_i \Box A$ to $s \vDash_j \Box A$ it is like shifting from (S, Ψ_i, a) to (S, Ψ_j, a). We now give a formal definition of hypermodality.

We treat the simple case is where the number of modes is a finite set μ and there is a function ε for shifting modes. This case is given in the next definition.

Definition 9 (Mode shifting). *Let $\mu = \{\Psi_0, \ldots, \Psi_k\}$ be a set of modes and let ε be a function assigning to each $0 \leq i \leq k$ a value $0 \leq \varepsilon(i) \leq k$.*

Let (S, R, a, h) be a Kripke model. We define the following (μ, ε) satisfaction in the model

- $t \vDash_i \Box A$ iff $\forall s(\Psi_i(t, s) \to s \vDash_{\varepsilon(i)} A)$.
- We say A is true in the model if $a \vDash_0 A$.

Definition 10

1. Let $\mathcal{K}$ be a class of models of the form (S, R, a, h). Let (μ, ε) be a mode system. We write $\mathcal{K} \vDash_{(\mu, \varepsilon)} A$ iff for every model (S, R, a, h) in $\mathcal{K}$ we have $a \vDash_0 A$.

2. Let $\mathbf{L}$ be a logic complete for a class $\mathcal{K}$ of Kripke models of the form (S, R, a, h). Let $\mathcal{K}[\mu, \varepsilon]$ be $\{A \mid \mathcal{K} \vDash_{(\mu, \varepsilon)} A\}$. We sometimes write $\mathbf{L}[\mu, \varepsilon]$ for $\mathcal{K}[\mu, \varepsilon]$, when the implicit dependence on $\mathcal{K}$ is clear.

Obviously the nature of hypermodal logic depends on (μ, ε) and its abstract properties and also on the class $\mathcal{K}$ of models chosen.

B   Traversing a Graph

Definition 11. By a graph G we mean a set S with a binary relation $R \subseteq S^2$. Let $a \in S$ be the starting point. We write $G = (S, R, a)$.

Definition 12.   1. By a Horn closure condition in the language of R we mean a clauses of the form

$$C : \bigwedge_i x_i R y_i \wedge \bigwedge_j u_j \neq v_j \to x R y.$$

2. A substitution θ from $Z = \{x_i, y_i, u_j, v_j, x, y\}$ into S is a function θ assigning values $\theta(z) \in S$ for each $z \in Z$.
3. We say a relation $R^* \subseteq S^2$ satisfies the clause C if for all θ, if $\theta(x_i) R^* \theta(y_i)$ holds and $\theta(u_j) \neq \theta(v_j)$ holds then $\theta(x) R^* \theta(y)$ also holds. We write $G^* = (S, R^*, a) \vDash C$.
   Let τ be a set of clauses. We say $G \vDash \tau$ iff $G \vDash C$ for all $C \in \tau$.

Lemma 1. Let $G = (S, R, a)$ be a graph and τ a set of clauses. Then there exists the smaller $R^* \supseteq R$ such that $G^* = (S, R^*, a) \vDash \tau$.

Definition 13. Let $G = (S, R, a)$ be a graph and let $\mu = (\tau_1, \ldots, \tau_k)$ be a sequence of sets of clauses. Let $R_1^*, \ldots, R_k^*$ be the closures of R under τ_i resp. Define a μ-path H through S as follows.
   The first element of H is $a_0 = a$. The next element of H is a_1 such that $a_0 R_1^* a_1$ holds. We say a_1 is a τ_1 choice.
   Assume a_n is a τ_n choice then a_{n+1} is such that $a_n R_{n+1}^* a_{n+1}$ where $R_{n+1}^* = R_{i+1}^*$ if a_n is a τ_i choice, $i < k$ and $R_{n+1}^* = R_1^*$ if a_n is a τ_k choice.

Definition 14 (A reactive graph). *Let $G = (S, R, a)$ be a graph. Write $R = \{(x, y) \mid xRy \text{ holds}\}$. Define W as follows:*

$$
\begin{aligned}
W_0 &= S \times S && \text{connections of level } 0 \\
W_{n+1} &= S \times W_n && \text{connections of level } n + 1 \\
W &= \bigcup_n W_n
\end{aligned}
$$

We consider any set of connections of level ≥ 1 as a switch.
A subset $\bar{R}$ of W can be used to describe S paths H as follows.

1. *initial element of H is $a_0 = a$ initial set of $\bar{R}_1 = \bar{R}$.*
2. *Assume R_n is defined and $\bar{R}_n \subseteq \bar{R}$.*
3. *We define $\bar{R}_{n+1}$.*
   Let $R_{n+1} = \bar{R}_n = \{\alpha \mid (a_n, \alpha) \in R_n\} \cup \{\beta \mid (a_n, \beta) \in (\bar{R}) - R_n\}$.
   We assume (a_n, a_{n+1}) is in $\bar{R}_n$.

Conjecture: Let H_μ be the all possible paths defined on $G = (S, R, a)$ using $\mu = (\tau_1, \ldots, \tau_k)$. Then there exists an $\bar{R} \subseteq W$ that yields the same paths.

Definition 15. *1. A multigraph has the form $G = (V, E, s, r)$ where V is a set of vertices, E a set of edges, and $s, r : E \to V$ are maps describing the source and range of edges.*
2. A family of sets of edges, $R_H = R_h = (E_h, s_h, r_h)$, is called compatible if: for

$$e \in E_h \cap E_{h'}$$

then we have

$$s_h(e) = s_{h'}(e) \text{ and } r_h(e) = r_{h'}(e) .$$

So it make sense considering a set of edges that is the union of the elements R_H as

$$\bigcup R_H = \left(\bigcup_{h \in H} E_h, s, r \right)$$

where $s(e) = s_h(e)$ and $s(e) = s_h(e)$ for any $h \in H$. In a similar way we define intersection of sets of edges and the predicate $\in$ to edges and sets of edges of this form.

Definition 16. *A reactive multigraph has the form $G = (V, R_H)$ where $R_H = \{R_h = (E_h, s_h, r_h) \mid h \in H\}$ is a compatible family of set of edges based on V indexed on H and (V, R_h, a) is a multigraph for all h. The elements of H are sequences of edges and it satisfies:*

$$\varepsilon \in H \text{ the empty path, corresponding to the empty sequence of edges}$$

$$p = e_1 \ldots e_n \in H \text{ iff } e_1 \ldots e_{n-1} \in H \text{ and } r_{e_1 \ldots e_{n-1}}(e_{n-1}) = s_{e_1 \ldots e_{n-1}}(e_n)$$

H is called the set of reactive paths of G.

Definition 17 (Switch reactivity)

1. Let $G = (V, R)$ be a multigraph and $R = (E, s, r)$. Define the total set of possible switches W_R as follows:

$$W_0 = E$$
$$\vdots$$
$$W_{n+1} = \{(e, \alpha) \mid e \in E, \alpha \in W_n\}$$
$$\vdots$$
$$W_R = \bigcup_n W_n.$$

2. A switch graph $\mathbf{G}$ has the form $(V, R, \mathbf{R}_\varepsilon, \mathbf{R})$ where (V, R) is a multigraph, which set of edges "possible in the graph", $\mathbf{R}_\varepsilon \subseteq \mathbf{R} \subseteq W_R$, $\mathbf{R}_\varepsilon$ is the initial configuration of the switch graph.

3. Let $\mathbf{R}' \subseteq \mathbf{R}$ and let $e \in E$. Define $\mathbf{R}'_e$ as follows:

$$\mathbf{R}'_e = \mathbf{R}' - \{\alpha \mid (e, \alpha) \in \mathbf{R}'\} \cup \{\beta \mid (e, \beta) \in \mathbf{R} \wedge (e, \beta) \notin \mathbf{R}'\}.$$

We say α is deactivated and β is activated by the passing through the node x.

4. Let $H = (e_1 \ldots e_n)$. We let $\mathbf{R}_H$ be defined as follows:

$$\mathbf{R}_{e_1} = (\mathbf{R}_\varepsilon)_{e_1}$$
$$\vdots$$
$$\mathbf{R}_{e_1 \ldots e_{m+1}} = (\mathbf{R}_{e_1 \ldots e_m})_{e_{m+1}}$$

Here we see $e \in E$ as edge $(e, s_{|e}, r_{|e})$.

5. A switch graph has level n of reactivity if $\mathbf{R} \subseteq \bigcup_{i \leq n} W_i$.

A switch graph is a particular case of a reactive graph, with:

$$R_h = \mathbf{R}_h \cap R$$

Problem 1 Can every reactive multigraph be represented by a switch graph?

C   Reactive Proof Theory

We give an example of how to do reactive proof theory. A full analysis will be done in [12]. The proper environment for developing reactive proof theory is the methodology of Labelled Deductive Systems, see [8]. See also [3] where reactive rules were independently introduced.

The usual natural deduction propositional system has elimination rules and introduction rules. In principle an elimination rule has the form

$$\boxed{\quad \text{ER} \quad \dfrac{A_1, \ldots, A_n}{B} \quad \text{where } A_1, \ldots, A_n, B, \text{ are well formed formulas.} \quad}$$

and an introduction rule has the form of a subproof:

$\mathbb{IR}$: To show $\varphi(A_1, \ldots, A_n, B)$, start a subproof with $A_1, \ldots, A_n$ as additional assumptions and conclude the subproof successfully with obtaining B, where $\varphi, A_1, \ldots, A_n, B$ are well formed formulas. φ is built up from $A_1, \ldots, A_n, B$.

Well known examples of such rules are the implicational rules:

$$\Rightarrow \mathbb{E} \qquad \frac{A, A \Rightarrow B}{B}$$

$$\Rightarrow \mathbb{I} \qquad \text{To show } A \Rightarrow B, \text{ assume } A \text{ and prove } B$$

Definition 18 (Proofs of level $\leq n$). *Let $\mathbf{S}$ be a set of proof rules.*

1. *A line (in a proof) is a sequence of the form*

$$\ell : A, J, \rho, \alpha$$

   where ℓ is a line reference (line number), A is the formula of the line, J is the justification of the line, and ρ is the set of proof rules active at this line and α is a set of line references accessible at the line. ρ changes only in a reactive proof system. In an ordinary proof system ρ is always the full set of rules of the logic, and α contains all previous line references.
   The justification in $\ell : B, J, \rho, \alpha$ can be either the word "assumption" or the phrase "A is obtained using the elimination rule $\mathbb{ER} : \dfrac{A_1, \ldots, A_n}{A}$, where A_i are formulas obtained in lines $\ell_i, i = 1, \ldots, n$ and ℓ_i are accessible", or the phrase "$A = \varphi(A_1, \ldots, A_n, B)$ where A is a formula of an introduction rule $\mathbb{IR}$ and the justification is a subproof π whose assumptions are $A_1, \ldots, A_n$ together with whatever is accessible at ℓ".

2. *A (non-reactive) proof π of level 0 is a sequence of numbered lines containing formulas and justifications of the form*

   line number: wff, Justification

   such that the beginning of the sequence contains formulas justified as "assumptions" and any subsequent line has the form $\ell : B, \mathbb{R}$ where $\mathbb{R}$ is an elimination rule of the form $\dfrac{A_1, \ldots, A_n}{B}$ here $\ell_i : A_i$ Justification are previous lines in the sequence. We say that the sequence is a proof of the formula of the last line from the initial sequence of assumptions.

3. *A proof of level $\leq n+1$ is defined as a sequence π of lines containing formulas and justifications where the initial elements of the sequence are all justified as "assumptions" and each subsequent formula in line ℓ is either justified from previous lines using an elimination rule, as described in (2) above, or is a formula of the form $\varphi(A_1, \ldots, A_n, B)$ appearing in an introduction rule, justified by a proof π_φ of level $\leq n$, of B from the assumptions $A_1, \ldots, A_n$ and whatever is accessible to ℓ.*

If the last line of π contains the formula E and the initial sequence of the proof are the assumptions $D_1, \ldots, D_k$, then we say that π is a level $\leq n + 1$ proof of E from $D_1, \ldots, D_k$.

Definition 19 (Reactive proof system). *A reactive system has reactive rules of the form*

$$\boxed{\begin{array}{l} \mathbf{R}_n : (\mathbb{R}, r_i^+, r_j^-) \\ i = 1, \ldots, k^+, j = 1, \ldots, k^-, n = 1, \ldots, m, r_i^{\pm} \leq m. \end{array}}$$

where $\mathbb{R}$ is a rule (elimination or introduction). The reading of $\mathbf{R}_n$ is that if we use rule $\mathbb{R}$ then the rules $\mathbf{R}_{r_i^+}, i = 1, \ldots, k^+$ should be activated and rules $\mathbf{R}_{r_j^-}, j = 1, \ldots, k^-$ should be deactivated.

The notion of a proof is modified as follows.

At the start of the proof (line 1) we add a third component ρ indicating which rules are active and a fourth component α indicating which previous lines are accessible for the purpose of justification. The general nature of the logic will have procedures for telling us given a line ℓ in the proof and its α, what will be the α of the next line? For example if in line ℓ, the justification is "assumption", then line ℓ is available to be used in justification of the next line or for example if line ℓ is justified by an elimination rule of the form $\dfrac{A_1, \ldots, A_n}{B}$, where A_i are from accessible lines ℓ_i, resp. then lines ℓ_i are no longer accessible at the next line (line $\ell + 1$). Lines which are assumptions do not change which rules are active and which are not. A line which uses a rule $\mathbf{R}_n$ as justification (either elimination or introduction) activates or deactivates the other rules as indicated in the rule itself and the result of the change are the rules ρ which are available for the next line of the proof. The changes in accessibility α can depend on the logic at hand. For example as we have mentioned, a resource logic may make assumptions not accessible once they are used in a justification.

Example 13. Take the ordinary natural deduction rules for classical logic including

$$\boxed{\dfrac{A, A \Rightarrow B}{B}}$$

and

$$\boxed{\dfrac{\perp}{B}}$$

and

$$\boxed{\text{To show } A \Rightarrow B \text{ assume } A \text{ and prove } B}$$

Assume that using modus ponens deactivates the negation rule.

Thus $A, A \Rightarrow \perp \not\vdash C$ by modus ponens alone using level 0 proofs because once we use modus ponens to get $\perp$ we do not have the negation rule anymore. We could work our way around this in this case, using level 1 proofs as follows:

1. A, assumption
2. $A \Rightarrow \perp$, assumption
3. $\perp \Rightarrow C$
   using $\Rightarrow$ introduction rule,
   3.1. $\perp$, assumption
   3.2. C from 3.1 and the negation rule
   3.3. Exit with $\perp \Rightarrow C$ proved
4. $\perp$, from 1 and 2 using modus ponens. Note that the negation rule which was used in the subproof at 3.2 does not cancel the modus ponens in the main proof.
5. C from 4 and 5 using modus ponens again.

D J. van Benthem's Sabotage Modal Logic

J. van Benthem introduced sabotage modal logic in [4]. The idea was put forward by van Benthem (2002) and is based on games played on graphs where one player is trying to traverse the graph while another tries to delete edges to make the first player fail.

This is another instance of reactive behaviour, conceputally conceived as sabotage. In the context of Kripke models (S, R, a, h), van Benthem introduces two modalities, the usual evaluation modality (denoted in our notation by $\overrightarrow{\Box}$) and a sabotage modality denoted by $\diamondsuit$.

There are several versions for $\diamondsuit$ [9]

$a \vDash \diamondsuit_{\text{point}} A$ iff $a \vDash A$ in a new model in which some point $s \in S, s \neq a$ is deleted.

We can use the convenient notation $a \vDash_s A$.

$a \vDash \diamondsuit_{\text{arc}} A$ iff $a \vDash A$ in a new model in which some arc $(t, s) \in R$ is deleted. Again we can use the notation $a \vDash_{(t,s)} A$.

Philip Rohde studied properties of these logics, with regard to complexity.

Another recent variation recommended by Benedikt Löwe, is to have the sabotage modality localised by turning the system two dimensional.

$a, t \vDash \diamondsuit_{\text{arc}} A$ iff for some s such that tRs we have $a, s \vDash_{(t,s)} A$. [10]

---

[9] van Benthem has modalities indexed also by actions, but this is irrelevant to the system's conceptual aspects.

[10] To clarify the notation, let

- $\sharp(z_1, \ldots, z_n) \equiv \bigwedge_{i \neq j} z_i \neq z_j$
- $\sharp((x_1, y_1), \ldots, (x_m, y_m)) = \bigwedge_{i \neq j} (x_i \neq x_j \vee y_i \neq y_j)$

Let $t \vDash_{z_1, \ldots, z_n} A$ mean that A holds at t, in the model where all the different points $z_1, \ldots, z_n$ have been deleted.

We now compare sabotage modality with our "fault-remedy" modality to give the reader a view of what is going on.

It is obvious that Professor van Benthem and I share the idea of reactivity, of working against a system which changes under you, whether by built-in weaknesses and remedies or by sabotage. The actual modal logics produced emerge from the original point of view taken[11]

I also adopt the point of view that deletion is a metalevel notion and that there should be a logical discipline and stylised machinery for bringing it into the object level. This should be seen in a wider context of dynamic evaluations and dynamic operators on which there is a lot of literature. See [5] for a recent application, and the references there.

The rest of this comparison is simply technical, comparing formal options between teh fault-remedy modality and the sabotage modality.

1. Fault modality can have arrow to arrow deletions of many levels. This comes from the original point of view.
2. Switch procedures also come naturally in this context.

Similarly, $t \vDash_{(x_1,y_1),...,(x_m,y_m)} A$ means that A holds at t in the model where all the different arcs $(x_1, y_1), \ldots, (x_n, y_n)$ have been deleted.

Thus for point sabotage we have

- $t \vDash_{z_1,...,z_n} \Diamond A$ iff $\exists s \exists y (tRs \wedge \sharp(t, s, z_1, \ldots, z_n, y)$ and $s \vDash_{z_1,...,z_n,y} A)$.

For arc sabotage we have:

- $t \vDash_{(x_1,y_1),...,(x_n,y_n)} \Diamond A$ iff $\exists s \exists u \exists v (tRs \wedge uRv \wedge \sharp((x_1, y_1), \ldots, (x_n, y_n), (t, s)) \wedge \sharp((x_1, y_1), \ldots, (x_n, y_n), (u, v))$ and $s \vDash_{(x_1,y_1),...,(x_n,y_n),(u,v)} A)$.

---

[11] van Benthem says in [4] (this paper was already written in 2002) as follows:

"In particular, the logical model-checking angle suggests a study of evaluation of first-order logic on structures *which change under evaluation*. E.g. an object might become unavailable once drawn from a domain, or a fact might change when inspected (think of measurement in quantum mechanics)".

In comparison, I say in [7]

"This paper addresses the case where the semantics does change (or react) under us as we evaluate a formula. This idea makes the evaluation of a wff at a world t dependent on the route leading to t. Thus we get a new kind of semantics, the reactive semantics."

and later in the paper, I continue to say

"we have put forward the reactive and dynamic idea of evaluation in earlier papers and lectures. A typical example we give is to consider $t \vDash \Diamond A$. In modal logic this means that there is a possible world s such that $s \vDash A$. We take a more dynamic view of it. We ask: where is s? How long does it take to get to it? and how much does it cost to get there?"

3. Fault modality is a simultaneous parallel combination of metalevel changes and evaluation. This makes it possibly technically different from the logic where the functionalities are separated.

   We can simulate sabotage by double arrows and non-determinism. The operator $\bar{\Diamond}_{arc}$, deleting an arc, can be simulated, for example, by looking at double arrows going from each node to all arcs and non-deterministically activating one double arrow every time we make a move.

4. Conceptually the fault-remedy concept is universal, while the notion of sabotage is agent specific. Consider a case of one agent sabotaging the work of several other agents. Additional special effort is required by the saboteur to make it look like a fault (accident).[12]

---

[12] Think of the numerous detective novels where the murderer tries to make the murder look like an accident, and how the detective can show it was not.

On Partially Wellfounded Generic Ultrapowers

Moti Gitik[1] and Menachem Magidor[2]

[1] School of Mathematical Sciences, Tel Aviv University, Tel Aviv 69978, Israel
gitik@post.tau.ac.il
[2] Institute of Mathematics, The Hebrew University of Jerusalem,
Jerusalem 91904, Israel
menachem@math.huji.ac.il

Dedicated to Boaz Trakhtenbrot on the occasion of his 85-th birthday.

Abstract. We construct a model without precipitous ideals but so that for each $\tau < \aleph_3$ there is a normal ideal over $\aleph_1$ with generic ultrapower wellfounded up to the image of τ.

1 Introduction

Let κ be a regular uncountable cardinal. For $f, g \in {}^{\kappa}On$ set

$$f <^* g \text{ iff } \{\alpha < \kappa \mid f(\alpha) < g(\alpha)\} \text{ contains a closed unbounded subset.}$$

The Galvin-Hajnal rank $\|g\|$ of a function $g \in {}^{\kappa}On$ is defined as follows

$$\|g\| = \sup\{\|f\| + 1 \mid f <^* g\}.$$

By induction on α, the αth canonical function h_α is defined (if it exists) as the $<^*$-least function greater than each $h_\beta, \beta < \alpha$. If h_α exists then it is unique modulo the nonstationary ideal over κ. First κ^+ canonical functions always exist. Hajnal (see [4], 27.11) showed that already in L the ω_2nd canonical function for $\kappa = \omega_1$ does not exist. By Jech and Shelah [6], the existence of ω_2nd canonical function is not a large cardinal property. Note that the existence of $f \in {}^{\kappa}\kappa$ with $\|f\| = \kappa^+$ does not necessary imply the existence of κ^+ canonical function over κ. Just, for example, in L there are many functions of the rank ω_2 without the least such function. On the other hand non existence of such f implies large cardinals. Thus, Donder and Koepke [1] showed that then $\kappa \geq \aleph_2$ implies $0^\dagger$ exists and $\kappa = \aleph_1$ implies $\aleph_2$ is almost $< \aleph_1$-Erdős cardinal in the core model $\mathcal{K}$.

An ideal I over κ is called precipitous if every its generic ultrapower is well founded. It is not hard to see that if every generic ultrapower of I is well founded up to the image of $(2^\kappa)^+$ then I is precipitous.

Suppose now that for each $\tau < (2^\kappa)^+$ there is an ideal over κ with generic ultrapowers well founded up to the image τ. Does this imply the existence of a precipitous ideal?

Our aim is to provide a negative answer. We will show the following:

A. Avron et al. (Eds.): Trakhtenbrot/Festschrift, LNCS 4800, pp. 342–350, 2008.
© Springer-Verlag Berlin Heidelberg 2008

Theorem 1. *Suppose that*

1. $2^{\aleph_1} = \aleph_2$
2. *there is an $\aleph_1$-Erdős cardinal*
3. *there is a function $f : \omega_1 \to \omega_1$ with $\|f\| \geq \omega_2$.*

Then *for every $\tau < \omega_3$ there exists a normal ideal over $\aleph_1$ with a generic ultrapower wellfounded up to the image of τ.*

Remark 2. 1. Note that in general it is impossible to allow $\tau = \omega_3$. Thus, the cardinality of the forcing is only ω_2. Hence, if a generic ultrapower is wellfounded up to the image of $\tau = (\omega_3)^V$, then it is fully wellfounded (just taking a big enough elementary submodel (in V) of cardinality ω_2 arbitrary functions to those with the ranges being subsets of ω_3). But this implies an inner model in which ω_1 is a measurable cardinal, see [4]. The original V does not need to have even an inner model with a Ramsey cardinal.
2. The assumption 3 is not very restrictive. Thus by [1], if there is no such a function, then $\aleph_2$ is almost $< \aleph_1$-Erdős cardinal in the core model $\mathcal{K}$. In the last case we can assume that $V = \mathcal{K}$ or just collapse first a non $< \aleph_1$-Erdős cardinal in $\mathcal{K}$ to be new $\aleph_2$.
3. Note that up to $(\aleph_2)^V$ (not its image!) a generic ultrapower by the nonstationary ideal is always wellfounded , just due to the existence of canonical functions. It is possible (consistently) to get to the image of $\aleph_1$ using the canonical functions, if the nonstationary ideal on $\aleph_1$ is $\aleph_2$-saturated or consistently using a weaker assumptions as was shown in [7].
4. It is an open question whether any large cardinal hypothesis implies (directly, not consistently) the existence of a precipitous ideal on $\aleph_1$. In view of 1, a kind of "almost" precipitousness follows from $\aleph_1$-Erdős cardinal.
5. We do not know whether $\aleph_1$-Erdős cardinal is needed for the conclusion of 1. Note only that it is easy to show that $\aleph_1$ must be a weakly compact limit of weakly compact cardinals in L (just the tree property and a generic elementary embedding). Also, if $\aleph_1 = \aleph_1^{\mathcal{K}}$ then at least $0^{\sharp}$ exists.
6. We do not know if the analog of the theorem holds once $\aleph_1$ is replaced by a bigger cardinal.

2   The Game

Let λ be an $\aleph_1$-Erdős cardinal. Fix some $\tau < \lambda$.

Consider the following game $\mathcal{G}_\tau$:

Player I starts by picking a stationary subset A_0 of $\aleph_1$. Player II chooses a function $f_1 : A_0 \to \tau$ and either a partition $\langle B_n | n < \omega \rangle$ of A_0 into at most countably many pieces or a sequence $\langle B_\alpha | \alpha < \aleph_1 \rangle$ of disjoint subsets of $\aleph_1$ so that

$$\nabla_{\alpha < \omega_1} B_\alpha \supseteq A_0.$$

The first player then supposed to respond by picking an ordinal $\alpha_2 < \lambda$ and a stationary set A_2 which is a subset of A_0 and of one of B_n's or B_α's.

At the next stage the second player supplies again a function $f_3 : A_2 \to \tau$ and either a partition $\langle B_n | n < \omega \rangle$ of A_2 into at most countably many pieces or a sequence $\langle B_\alpha | \alpha < \aleph_1 \rangle$ of disjoint subsets of $\aleph_1$ so that

$$\nabla_{\alpha < \omega_1} B_\alpha \supseteq A_2.$$

The first player then supposed to respond by picking a stationary set A_4 which is a subset of A_2 and of one of B_n's or B_α's on which everywhere f_1 is either above f_3 or equal f_3 or below f_3. In addition he picks an ordinal $\alpha_4 < \lambda$ such that

$$\alpha_2 < \alpha_4 \text{ iff } f_1 \restriction A_4 < f_3 \restriction A_4.$$

Intuitively, α_{2n} pretends to represent f_{2n-1} in a generic ultrapower.

Continue further in the same fashion.

Player I wins if the game continues infinitely many moves. Otherwise Player II wins. Clearly it is a determined game.

Let us argue that the second player cannot have a winning strategy.

Lemma 3. *For each $\tau < \lambda$ Player II does not have a winning strategy in the game $\mathcal{G}_\tau$.*

Proof. Suppose otherwise. Let σ be a strategy of two. Find a set $X \subset \lambda$ of cardinality $\aleph_1$ such that σ does not depend on ordinals picked from X. In order to get such X let us consider a structure

$$\mathfrak{A} = \langle H(\lambda), \in, \lambda, \tau, \mathcal{P}(\aleph_1), \mathcal{G}, \sigma \rangle.$$

Let X be a set of $\aleph_1$ indiscernibles for $\mathfrak{A}$.

Pick now a countable elementary submodel M of $H(\chi)$ for $\chi > \lambda$ big enough with $\sigma, X \in M$. Let $\alpha = M \cap \omega_1$. Let us produce an infinite play in which the second player uses σ. This will give us the desired contradiction.

Consider the set $S = \{ f(\alpha) | f \in M, f \text{ is a partial function from } \omega_1 \text{ to } \tau \}$. Obviously, S is countable. Hence we can fix an order preserving function $\pi : S \to X$.

Let one start with $A_0 = \omega_1$. Consider $\sigma(A_0)$. Clearly, $\sigma(A_0) \in M$. It consists of a function $f_1 : A_0 \to \tau$ and, say a sequence $\langle B_\xi | \xi < \aleph_1 \rangle$ of disjoint subsets of $\aleph_1$ so that

$$\nabla_{\xi < \omega_1} B_\xi \supseteq A_0.$$

Now, $\alpha \in A_0$, hence there is $\xi^* < \alpha$ such that $\alpha \in B_{\xi^*}$. Then $B_{\alpha^*} \in M$, as $M \supseteq \alpha$. Hence, $A_0 \cap B_{\xi^*} \in M$ and $\alpha \in A_0 \cap B_{\xi^*}$. Let $A_2 = A_0 \cap B_{\xi^*}$. Pick $\alpha_2 = \pi(f_1(\alpha))$.

Consider now the answer of two which plays according to σ. It does not depend on α_2, hence it is in M. Let it be a function $f_3 : A_2 \to \tau$ and, say a sequence $\langle B_\xi | \xi < \aleph_1 \rangle$ of disjoint subsets of $\aleph_1$ so that

$$\nabla_{\xi < \omega_1} B_\xi \supseteq A_2.$$

As above find $\xi^* < \alpha$ such that $\alpha \in B_{\xi^*}$. Then $B_{\alpha^*} \in M$, as $M \supseteq \alpha$. Hence, $A_2 \cap B_{\xi^*} \in M$ and $\alpha \in A_2 \cap B_{\xi^*}$. Let $A'_2 = A_2 \cap B_{\xi^*}$. Split it into three sets $C_<, C_=, C_>$ such that

$$C_< = \{\nu \in A_2'|f_3(\nu) < f_1(\nu)\},$$
$$C_= = \{\nu \in A_2'|f_3(\nu) = f_1(\nu)\},$$
$$C_> = \{\nu \in A_2'|f_3(\nu) > f_1(\nu)\}.$$

Clearly, α belongs to only one of them, say to $C_<$. Set then $A_4 = C_<$. Then, clearly, $A_4 \in M$, it is stationary and $f_3(\alpha) < f_1(\alpha)$. Set $\alpha_4 = \pi(f_3(\alpha))$.

Continue further in the same fashion.   □

It follows that the first player has a winning strategy.

3   The Construction of an Ideal

Let $\tau < \aleph_3$. We like to construct an ideal on $\aleph_1$ with a generic ultrapower wellfounded up to the image of τ.

Fix a winning strategy σ for Player I in the game $\mathcal{G}_\tau$.

Set $I = \{X \subseteq \omega_1 \mid \sigma \text{ never picks } X\}$.

Lemma 4. *I is a normal proper ideal over ω_1.*

Proof. Let us show for example the ω_1-completeness. Thus let that $\langle B_n|n < \omega\rangle$ be a partition of a set $A \in I^+$. Consider a game according to σ in which A appears as a move of the player one. Let two to answer by $\langle B_n|n < \omega\rangle$ (and arbitrary function). Then the answer of one according to σ will be a subset of one of B_n's. But this means that this B_n is I-positive.   □

Fix a sequence $\langle h_\alpha|\alpha < \aleph_2\rangle$ of the first $\aleph_2$ canonical functions from ω_1 to ω_1.

We would like to have a function that represents $(\aleph_2)^V$ in a generic ultrapower. If there exists the $\aleph_2$nd canonical function then it will be as desired. Here we do not assume its existence, but rather a weaker property that there is $f : \omega_1 \to \omega_1$ with $\|f\| = \omega_2$. Clearly, such f is above each $h_\alpha, \alpha < \omega_2$ (modulo the nonstationary ideal). The problem is that there may be many such f's without the least one. The way to overcome this will be to find an ideal $J \supseteq I$ which has have the J-least function above all canonical functions.

Proceed as follows. Set

$$S = \{f \in {}^{\omega_1}\omega_1 \mid \|f\| \geq \omega_2\}.$$

Basically we let Player II to play functions in S and Player I to respond using the strategy σ. Find a function $h \in S$, a finite play $t = \langle t_1, ..., t_n\rangle$ and an ordinal η such that

1. t was played according σ
2. h was picked by Player II at his last move t_{n-1}
3. Player I responded with η
4. there is no continuation of t, with Player I using σ, in which a response to a function from S less than η.

Note that such $\eta \geq \omega_2$, since otherwise Player II can easily win by playing h_η at the very next move. Then Player I should respond respond by some $\eta_1 < \eta$ on which II respond by h_{η_1} etc.

Also note that such h is not necessary unique, but any other function attached to η which appears further in the game will be equal to h on the corresponding set.

Set now

$$J = \{X \subseteq \omega_1 \mid X \text{ is never picked by } \sigma \text{ in the continuation of } t\}.$$

The proof of the next lemma repeats those of Lemma 4.

Lemma 5. *J is a normal proper ideal over ω_1 extending I.*

Lemma 6. *Generic ultrapowers by J are wellfounded at least up to $(\omega_2)^V + 1$. Moreover $(\omega_2)^V$ is represented by h.*

Proof. Just note, that by the choice of h and the definition of J, the only functions that are below h on a J-positive set are the canonical functions $h_\alpha, \alpha < \omega_2$. □

Assume without loss of generality that for each $\alpha < \aleph_2$ we have $h_\alpha(\nu) < h(\nu)$, for each $\nu < \omega_1$. Also fix for each $\nu < \omega_1$ a function $H_\nu : \omega \to_{onto} h(\nu)$.

Let

$$A_{n\alpha} = \{\nu < \omega_1 \mid H_\nu(n) = h_\alpha(\nu)\}.$$

Lemma 7. *Let $X \in J^+$. Then for each $n < \omega$ there is $\alpha < \omega_2$ such that $X \cap A_{n\alpha} \in J^+$.*

Proof. By 6 a generic ultrapower with J is wellfounded up to $\omega_2^V + 1$ and ω_2^V is represented by h.

Let $G \subseteq J^+$ be a generic ultrafilter with $X \in G$ and $j : V \to M_G = V \cap {}^{\omega_1 >}V/G$ be the corresponding elementary embedding. We may assume that the ordinals of M up to $[h]_G$ are just ω_2^V. Consider $H = [\langle H_\nu | \nu < \omega_1^V \rangle]_G$. Then, $H : \omega \to_{onto} \omega_2^V$ in M_G. So, for some $\alpha < \omega_2^V$ we have $H(n) = \alpha$. But then $X \cap A_{n\alpha} \in G$ and be are done. □

The following lemma is similar.

Lemma 8. *Let $X \in J^+$. Then for each $m < \omega$ there is $n > m$ so that $|\{\alpha < \omega_2 \mid X \cap A_{n\alpha} \in J^+\}| = \aleph_2$.*

Proof. Just otherwise X or its extension will force that the range of H (as in 7) will be bounded in ω_2^V. □

Now we will use an argument similar to those of [3] in order to extend J to an ideal with the desired property.

Let $\langle f_\alpha \mid \alpha < \aleph_2 \rangle$ be an enumeration of the set of all functions from ω_1 to τ (recall that τ is a fixed ordinal less than $\aleph_3$ and $2^{\aleph_1} = \aleph_2$). Fix an enumeration $\langle X_\alpha \mid \alpha < \aleph_2 \rangle$ of J-positive sets.

By 8 there is $n < \omega$ such that

$$|\{\alpha < \omega_2 \mid A_{n\alpha} \in J^+\}| = \aleph_2.$$

Suppose for simplicity that $n = 0$. Let

$$\langle A_{0\tau(\xi)} \mid \xi < \omega_2\rangle$$

be a one to one enumeration of this set.

We construct by induction a sequence of ordinals $\langle \xi_{0\alpha} | \alpha < \omega_2\rangle$ and a sequence of J positive sets $\langle C_{0\alpha} | \alpha < \omega_2\rangle$. Let $\alpha < \omega_2$. If there is $\xi < \omega_2$ such that $\xi \neq \xi_{0\beta}$ for each $\beta < \alpha$ and $X_\alpha \cap A_{0\tau(\xi)} \in J^+$, then let $\xi_{0\alpha}$ be the least such ξ. We would like now to attach an ordinal to the function f_α. So let us play the game $\mathcal{G}$ (which continues t)where the player one uses the strategy σ until the stage at which the player one plays $X_\alpha \cap A_{0\tau(\xi)}$. All the previous move do not matter much here, but we fix some such play. Let the player two respond by $X_\alpha \cap A_{0\tau(\xi)}$ and f_α. The strategy σ provides then the answer of the player one. It consists of a subset $C_{0\alpha}$ of $X_\alpha \cap A_{0\tau(\xi)}$ and an ordinal $\eta_{0\alpha}$.

Let

$I_{0\alpha} = \{X \subseteq \omega_1 \mid \sigma$ never picks X in all possible continuations of the play started above.$\}$

If there is no such ξ then

$$X_\alpha \subseteq \nabla_{\varepsilon < \omega_1} A_{0\tau(\xi_{\beta_\varepsilon})},$$

where $\langle \beta_\varepsilon | \varepsilon < \omega_1\rangle$ is an enumeration of α. Let then $\xi_{0\alpha}$ be the least ordinal above all $\xi_{0\beta}$ with $\beta < \alpha$. Replace X_α be $\aleph_1$ and then proceed with it as above.

Set $I_0 = \bigcap\{I_{0\alpha} | \alpha < \aleph_2\}$. Then I_0 is a normal ideal over $\aleph_1$, since each of $I_{0\alpha}$ is such.

The next lemma follows from the construction above.

Lemma 9. *For each $X \in J^+$ we have $X \in I_{0\alpha}$, for some $\alpha < \aleph_2$ or $X \subseteq \{\nu < \omega_1 | \; \exists \beta < \nu \;\; \nu \in A_{0\zeta_\beta}\}$ mod J, for some sequence $\langle \zeta_\beta | \beta < \omega_1\rangle$ of ordinals below ω_2.*

As in [3] we can now deduce the following:

Lemma 10. *Let $X \subseteq \omega_1$. Then $X \in I_0$ iff $X \subseteq \{\nu < \omega_1 | \; \exists \beta < \nu) \;\; \nu \in Y_\beta\}$ mod J, for some sequence $\langle Y_\beta | \beta < \omega_1\rangle$ such that for some sequence $\langle \alpha_\beta | \beta < \omega_1\rangle$ of ordinals below ω_2 we have $Y_\beta \subseteq A_{0\tau(\xi_{\alpha_\beta})}$ and $Y_\beta \in I_{0\alpha_\beta}$.*

Let now $n = 1$. Fix some $\gamma < \omega_2$. We apply 8 to find the least $n_\gamma \geq 1$ such that the set

$$|\{\alpha < \omega_2 | A_{n_\gamma \alpha} \in I_{0\gamma}^+\}| = \aleph_2.$$

Let

$$\langle A_{n_\gamma \tau(\xi)} | \xi < \omega_2\rangle$$

be a one to one enumeration of this set. For each $\xi < \omega_2$ we would like to attach an ordinal to a restriction of f_ξ to an $I_{0\gamma}$ positive subset of $A_{n_\gamma \tau(\xi)}$.

Proceed as above. Define recursively sequences $\langle \xi_{\langle 0\gamma,1\alpha\rangle} | \alpha < \omega_2 \rangle$ and $\langle C_{\langle 0\gamma,1\alpha\rangle} | \alpha < \omega_2 \rangle$.

At stage α consider the α-th set X_α in $I_{0\gamma}$. If there is $\xi < \omega_2$ such that $\xi \neq \xi_{\langle 0\gamma,1\beta\rangle}$, for each $\beta < \alpha$ and $X_\alpha \cap A_{n_\gamma\tau(\xi)} \in I_{0\gamma}^+$, then let $\xi_{\langle 0\gamma,1\alpha\rangle}$ be the least such ξ. We would like to shrink $I_{0\gamma}$ below $X_\alpha \cap A_{n_\gamma\tau(\xi_{\langle 0\gamma,1\alpha\rangle})}$ in order to decide an ordinal which will correspond to f_α. As above we fix a play according to σ which is a continuation of the previous play (the one from the definition of $I_{0\gamma}$ reaching $X_\alpha \cap A_{n_\gamma\tau(\xi_{\langle 0\gamma,1\alpha\rangle})}$. Let the second player plays at his next move $X_\alpha \cap A_{n_\gamma\tau(\xi_{\langle 0\gamma,1\alpha\rangle})}$ and f_α. Apply the strategy σ. It supplies an $I_{0\gamma}$ positive subset $C_{\langle 0\gamma,1\alpha\rangle}$ of $X_\alpha \cap A_{n_\gamma\tau(\xi_{\langle 0\gamma,1\alpha\rangle})}$ and an ordinal $\eta_{0\gamma,1\alpha}$. This will be the ordinal corresponding to $f_\alpha \restriction C_{\langle 0\gamma,1\alpha\rangle}$.

Let $I_{\langle 0\gamma,1\alpha\rangle} = \{X \subseteq \omega_1 \mid \sigma \text{ never picks } X \text{ in all possible continuations of the play started above.}\}$

If there is no such ξ then let $\xi_{\langle 0\gamma,1\alpha\rangle}$ be the least ordinal above all $\xi_{\langle 0\gamma,1\beta\rangle}$ for $\beta < \alpha$. Take ω_1 instead of X_α and run the construction above.

Set $I_1 = \bigcap\{I_{0\gamma,1\alpha\rangle} \mid \gamma, \alpha < \aleph_2\}$. Then I_1 is a normal ideal over $\aleph_1$, since each of $I_{\langle 0\gamma,1\alpha\rangle}$ is such.

Continue similar and define I_s and I_n for each $n < \omega$ and $s \in [\omega \times \omega_2]^{<\omega}$. Let F_s and F_n be the corresponding dual filters. Finally set

$$I_\omega = \text{ the closure under } \omega \text{ unions of } \bigcup_{n<\omega} I_n.$$

Let F_ω be the corresponding dual filter.

The following lemmas of [3] transfer directly to the preset context.

Lemma 11. $F \subseteq F_0 \subseteq ... \subseteq F_n \subseteq ... \subseteq F_\omega$ and $I \subseteq J \subseteq I_0 \subseteq ... \subseteq I_n \subseteq ... \subseteq I_\omega$.

Lemma 12

$$F_\omega = \left\{ X \subseteq \omega_1 \,\middle|\, \exists\langle X_n | n < \omega\rangle \forall n < \omega\, X_n \in F_n \quad X = \bigcap_{n<\omega} X_n \right\}$$

and

$$I_\omega = \left\{ X \subseteq \omega_1 \,\middle|\, \exists\langle X_n | n < \omega\rangle \forall n < \omega\, X_n \in I_n \quad X = \bigcup_{n<\omega} X_n \right\}$$

Lemma 13. I_ω is a proper ω_1-complete filter over ω_1.

Lemma 14. If $\langle Y_\beta | \beta < \omega_1 \rangle$ is a sequence of sets in I_ω then the set

$$Y = \{\nu < \kappa \mid \exists\beta < \nu \quad \nu \in Y_\beta\}$$

is in I_ω as well and hence I_ω is normal.

Lemma 15. A set X is in I_ω^+ iff $X \in F_s$, for some $s \in [\omega \times \omega_2]^{<\omega}$.

Now we are ready to show the desired result.

Theorem 16. *Let G be a generic subset of I_ω^+ and $j_G : V \to M_G = V \cap {}^{\omega_1}V/G$ be the corresponding elementary embedding. Then M_G is wellfounded at least up to $j_G(\tau)$.*

Proof. Suppose that $\langle \check{g}_n | n < \omega \rangle$ is a sequence of I_ω^+-names of old (in V) functions from $\omega_1 \to \tau$.

Let $G \subseteq I_\omega^+$ be a generic ultrafilter. Pick a set $X_0 \in G$ and a function

$$g_0 : \omega_1 \to \tau$$

in V such that

$$X_0 \Vdash_{I_\omega^+} \check{g}_0 = \check{g}_0.$$

Let $\alpha_0 < (\omega_2)^V$ be so that $f_{\alpha_0} = g_0$.

Apply Lemma 15 to X_0. There is a sequence s_0 with F_{s_0} defined and so that $X_0 \in F_{s_0}$. Recall now the definition of the filters $F_{s_0 \frown \langle |s_0| \alpha \rangle}$ which extend F_{s_0} at the very next stage of the construction. There will be $\beta_0 < \kappa^+$ and $n_0 > |s_0|$ such that $A_{n_0 \tau(\alpha_0)} \in F_{s_0 \frown \langle |s_0| \beta_0 \rangle}$. Denote by η_0 the the ordinal attached to f_{α_0} at the level of s_0 in the construction of $F_{s_0 \frown \langle |s_0| \beta_0 \rangle}^+$. By shrinking if necessary we can assume that $A_{n_0 \tau(\alpha_0)} \cap X_0 \in F_t$ implies that the sequence $s_0 \frown \langle |s_0| \beta_0 \rangle$ is an extension of the sequence t or vice verse. Without loss of generality we can assume that $A_{n_0 \tau(\alpha_0)} \cap X_0 \in G$, just otherwise replace X_0 by arbitrary positive subset and use density.

Continue now below $A_{n_0 \tau(\alpha_0)} \cap X_0$ and pick $X_1 \in G$ such that for some function

$$g_1 : \kappa \to \tau$$

in V we have

$$X_1 \Vdash_{F_\omega^+} \check{g}_1 = \check{g}_1.$$

Let $g_1 = f_{\alpha_1}$. Again, by 15, there is a sequence s_1 extending s_0 with F_{s_1} defined and so that $X_1 \in F_{s_1}$. Recall now the definition of the filters $F_{s_1 \frown \langle |s_1| \alpha \rangle}$ which extend F_{s_1} at the very next stage of the construction. There will be $\beta_1 < \kappa^+$ and $n_1 > |s_1|$ such that $A_{n_1 \tau(\alpha_1)} \in F_{s_1 \frown \langle |s_1| \beta_1 \rangle}$. Denote by η_1 the the ordinal attached to f_{α_1} at the level of s_1 in the construction of $F_{s_1 \frown \langle |s_1| \beta_1 \rangle}^+$. By shrinking if necessary we can assume that $A_{n_1 \tau(\alpha_1)} \cap X_1 \in F_t$ implies that the sequence $s_1 \frown \langle |s_1| \beta_1 \rangle$ is an extension of the sequence t or vice verse. Without loss of generality we can assume that $A_{n_1 \tau(\alpha_1)} \cap X_1 \in G$, just otherwise replace X_1 by arbitrary positive subset and use density.

Continue the process for each $n < \omega$. There will be $k < m < \omega$ with $\rho_k \leq \rho_m$. Then the set

$$\{\nu \in X_m \cap A_{n_m \alpha_m} | f_{\alpha_k}(\nu) \leq f_{\alpha_m}(\nu)\} \in F_{s_m \frown \langle |s_m| \beta_m \rangle}.$$

But $X_m \cap A_{n_m \alpha_m} \in G$ as well. Then,

$$\{\nu \in X_m \cap A_{n_m \alpha_m} | f_{\alpha_k}(\nu) \leq f_{\alpha_m}(\nu)\} \in G,$$

just no elements of G can be outside of $X_m \cap A_{n_m \alpha_m}$ (mod $F \subseteq F_\omega$) since all of them are in F_t's for sequences t which are subsequences of s_n, for some $n < \omega$. $\qquad \square$

Actually the argument provides a bit more information. Thus the following holds:

Theorem 17. *Assume that $2^{\aleph_1} = \aleph_2$ and $\|f\| = \omega_2$, for some $f : \omega_1 \to \omega_1$. Suppose that Player I has a winning strategy in the game $\mathcal{G}_\tau$, for some $\tau < \aleph_3$, then there is a normal ideal on $\aleph_1$ with a generic ultrapower wellfounded up to the image of τ.*

Proof. Note that the construction of I_ω above relays only on the strategy for the player one in the game $\mathcal{G}_\tau$. □

The opposite direction is true as well:

Theorem 18. *Suppose that J is a normal ideal on $\aleph_1$ with a generic ultrapower well founded up to the image of τ (for some ordinal τ), then Player I has a winning strategy in the game $\mathcal{G}_\tau$.*

Proof. Just start with ω_1 or any J-positive set. At a stage $2n - 1 (n > 0)$ the second player responds with a function $f : A_{2n-2} \to \tau$ and, say, a sequence $\langle B_\alpha | \alpha < \aleph_1 \rangle$ such that

$$\nabla_{\alpha < \omega_1} B_\alpha \supseteq A_{2n-2}.$$

Then one of B_α's should have the intersection with A_{2n-2} in J^+ (J is normal and we assume that $A_{2n-2} \in J^+$). Pick the least α such that $A_{2n-2} \cap B_\alpha \in J^+$. Shrink then $A_{2n-2} \cap B_\alpha$ to a set deciding the value of $[f]_{\dot{G}}$ in the generic ultrapower. Let A_{2n} be such a set.

The above defines a winning strategy for the player one in the game $\mathcal{G}_\tau$. □

Acknowledgement

We would like to thank A. Ferber and A. Rinot for their comments and remarks.

References

1. Donder, H.-D., Koepke, P.: On the consistency strength of 'Accessible' Jonsson Cardinals and of the Chang Conjecture, APAL 25, pp. 233–261 (1983)
2. Foreman, M.: Ideals and Generic Elementary Embeddings, in Handbook of Set Theory (to appear)
3. Gitik, M.: On normal precipitous ideals, www.math.tau.ac.il/~gitik
4. Jech, T.: Set Theory, 3rd ed.
5. Jech, T., Prikry, K.: On ideals of sets and the power set operation. Bull. Amer. Math. Soc. 82(4), 593–595 (1976)
6. Jech, T., Shelah, S.: A note on canonical functions. Israel J. Math. 68, 376–380 (1989)
7. Larson, P., Shelah, S.: Bounding by canonical functions, with CH. J. Math. Logic 3(2), 193–215 (2003)
8. Mitchell, W.: The covering lemma, in Handbook of Set Theory (to appear)

Some Results on the Expressive Power and Complexity of LSCs*

David Harel**, Shahar Maoz, and Itai Segall

The Weizmann Institute of Science, Rehovot, Israel
{dharel,shahar.maoz,itai.segall}@weizmann.ac.il

This paper is dedicated to Prof. Boaz Trakhtenbrot,
with deep admiration and respect.

Abstract. We survey some of the main results regarding the complexity and expressive power of Live Sequence Charts (LSCs). We first describe the two main semantics given to LSCs: a trace-based semantics and an operational semantics. The expressive power of the language is then examined by describing translations into various temporal logics. Some limitations of the language are also discussed. Finally, we survey complexity results, mainly due to Bontemps and Schobbens, regarding the use of LSCs for model checking, execution, and synthesis.

1 Introduction

Live Sequence Charts (LSCs, or LSC for the language) [9] constitute a visual formalism for inter-object scenario-based specification and programming. The language extends classical Message Sequence Charts (MSC) [21], mainly by adding universal and existential modalities. LSC distinguishes between behaviors that may happen in the system (existential, cold) and those that must happen (universal, hot). A universal chart contains a *prechart*, which specifies the scenario which, if successfully executed, forces the system to satisfy the scenario given in the actual chart body.

An executable (operational) semantics for LSC was defined in [18]. Thus, LSC can be viewed not only as a specification language but also as a high-level programming language for reactive systems.

Since its original definition, the language has been the subject of much work, e.g., in the contexts of scenario-based testing [25,26], synthesis [3,13,15], execution (play-out) [18], formal verification [22,33], specification and verification of hardware [6], telecommunication systems [8], biological systems [11], specification mining [27], and compilation into aspects [12,29]. Also, recently, in [16], a

* The research was supported in part by The John von Neumann Minerva Center for the Development of Reactive Systems at the Weizmann Institute of Science and by a Grant from the G.I.F., the German-Israeli Foundation for Scientific Research and Development.

** Part of this author's work carried out during a visit to the School of Informatics at the University of Edinburgh, which was supported by a grant from the EPSRC.

A. Avron et al. (Eds.): Trakhtenbrot/Festschrift, LNCS 4800, pp. 351–366, 2008.
© Springer-Verlag Berlin Heidelberg 2008

UML2 compliant and slightly generalized variant of LSC was defined, allowing
the embedding of LSC into the UML standard [35].

In this paper we survey some results regarding the expressive power and succinctness of the language, as well as complexity results for various problems
related to using LSC for specification and programming.

2   Language Overview

The LSC language was originally defined by Damm and Harel in [9]. The language has two types of charts: *universal* (annotated by a solid borderline) and
existential (annotated by a dashed borderline). Universal charts are used to specify restrictions over all possible system runs. A universal chart typically contains
a *prechart*, that specifies the scenario which, if successfully executed, forces the
system to satisfy the scenario given in the actual chart body. Existential charts
specify sample interactions between the system and its environment, and must
be satisfied by at least one system run. They thus do not force the application to
behave a certain way in all cases, but rather state that there is at least one set
of circumstances under which a certain behavior occurs. Existential charts can
be used to specify system tests, or simply to illustrate longer (non-restricting)
scenarios that provide a broader picture of the behavioral possibilities to which
the system gives rise.

Most constructs in the language, e.g., messages and conditions, also have a
hot/cold modality. Hot behaviors are mandatory and must be satisfied by any
system run. Cold behaviors, on the other hand, are provisional, and may be
satisfied. For example, a hot message must eventually be sent, while a cold
message may or may not be sent.

An example of a universal LSC is given in Fig. 1. The chart in the example
is adopted from [24], and is part of a specification for a cellular phone. The
chart requires that whenever the user closes the Cover, the Chip will send the
message StartRing(Silent) to the Speaker and later the speaker will turn
silent as designated by the self message Sound(Silent). The Display will set
its state to Time and later set its background to Green. An LSC induces a
partial order that is determined by the order along an instance line, by the fact
that a message can be received only after it is sent, and by taking into account
that a synchronous message blocks the sender until receipt. Thus in Fig. 1, the
message ChangeBackground(Green) must occur after message SetState(Time),
but both are unordered with respect to the messages StartRing(Silent) and
Sound(Silent).

An example of an existential LSC is given in Fig. 2. The chart states that there
is a possible run of the system where the user presses Click on the Send Key
and eventually the Chip receives an ACK from the environment ENV. The SYNC
condition restricts the order between the two messages, which are otherwise
unordered.

We give here a restricted and simplified trace-based semantics for a kernel subset of LSC. The original LSC semantics was given in [9]. In subsequent work the

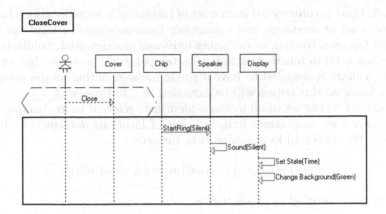

Fig. 1. An example of a universal chart

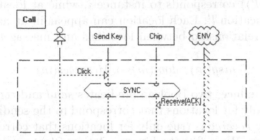

Fig. 2. An example of an existential chart

semantics of (restricted subsets or extensions of) the language was given using temporal logics (see, e.g., [24]) or various types of automata (see, e.g., [16,23]). An operational semantics, explicated in the play-out algorithm, was given in [18].

2.1 Basic Definitions

The following definitions are adopted from [24]. We assume the LSC specification relates to an *object system* composed of a set of objects $\mathcal{O} = \{O_1 \ldots O_n\}$. An object system corresponds to an implementation, and our goal in providing semantics for LSCs is to define when a given object system satisfies an LSC specification. The instance identifiers in the LSC charts refer to objects from $\mathcal{O}$, and possibly also the environment, denoted *env*. The LSC specifies the behavior of the system in terms of the message communication between the objects in the system. We want to define the notion of satisfiability of an LSC specification. In other words, we want to capture the languages $\mathcal{L} \subseteq A^* \cup A^\omega$ generated by the object systems that satisfy the LSC specification. The alphabet A used defines message communication between objects, $A = \mathcal{O} \times (\mathcal{O}.\Sigma)$, where Σ is the alphabet of messages.

An LSC chart is constructed from a set of instances, a set of locations in those instances, a set of messages, and a mapping from messages to locations. Each chart also has an activation mode, either universal or existential. Similarly, each message has a *temp* function, defining its temperature, as either hot or cold. For now, a chart is assumed to have a single message acting as the activation message. Later on this notion will be extended to a full prechart.

Let $inst(m)$ be the set of all instance-identifiers referred to in chart m. With each instance i we associate a finite number of locations $dom(m, i) \subseteq \{0, \ldots, l\_max(i)\}$. We collect all locations of m in the set

$$dom\,(m) = \{\langle i, l\rangle \mid i \in inst(m) \wedge l \in dom(m, i)\}.$$

The messages appearing in m are triples

$$Messages(m) = dom(m) \times \Sigma \times dom(m),$$

where $(\langle i, l\rangle, \sigma, \langle i', l'\rangle)$ corresponds to instance i, while at location l, sending σ to instance i' at location l'. Each location can appear in at most one message in the chart. The relationship between locations and messages is given by the mapping

$$msg(m) : dom(m) \rightarrow Messages(m)$$

The msg function induces two Boolean predicates *send* and *receive*. The predicate *send* is true only for locations that correspond to the sending of a message, while the predicate *receive* is true only for locations that correspond to the receiving of a message. We define the binary relation $R(m)$ on $dom(m)$ to be the smallest relation satisfying the following axioms and closed under transitivity and reflexivity:

- order along an instance line:

$$\forall \langle i, l\rangle \in dom(m), l < l\_max(i) \Rightarrow \langle i, l\rangle R(m)\langle i, l + 1\rangle$$

- order induced from message sending:

$$\forall msg \in Messages(m), msg = (\langle i, l\rangle, \sigma, \langle i', l'\rangle) \Rightarrow$$

$$\langle i, l\rangle R(m)\langle i', l'\rangle$$

- messages are synchronous; they block the sender until receipt:

$$\forall msg \in Messages(m), msg = (\langle i, l\rangle, \sigma, \langle i', l'\rangle) \Rightarrow$$

$$\langle i', l'\rangle R(m)\langle i, l + 1\rangle$$

We say that the chart m is *well-formed* if the relation $R(m)$ is acyclic. We assume all charts to be well-formed, and use $\leq_m$ to denote the partial order $R(m)$.

We denote the *preset* of a location $\langle i, l \rangle$ containing all elements in the domain of a chart smaller than $\langle i, l \rangle$ by

$$^\bullet\langle i, l \rangle = \{\langle i', l' \rangle \in dom(m) | \langle i', l' \rangle \leq_m \langle i, l \rangle\}.$$

We denote the partial order induced by the order along an instance line by $\prec_m$; thus $\langle i, l \rangle \prec_m \langle i', l' \rangle$ iff $i = i'$ and $l < l'$.

A *cut* through m is a set c of locations, one for each instance, such that for every location $\langle i, l \rangle$ in c, the preset $^\bullet\langle i, l \rangle$ does not contain a location $\langle i', l' \rangle$ such that $\langle j, l_j \rangle \prec_m \langle i', l' \rangle$ for some location $\langle j, l_j \rangle$ in c. A cut c is specified by the locations in all of the instances in the chart:

$$c = (\langle i_1, l_1 \rangle, \langle i_2, l_2 \rangle, ..., \langle i_n, l_n \rangle)$$

For a chart m with instances $i_1, ..., i_n$ the *initial cut* c_0 has location 0 in all the instances. Thus, $c_0 = (\langle i_1, 0 \rangle, \langle i_2, 0 \rangle, ..., \langle i_n, 0 \rangle)$. We denote $cuts(m)$ the set of all cuts through the chart m.

For chart m, some $1 \leq j \leq n$ and cuts c, c', with

$$c = (\langle i_1, l_1 \rangle, \langle i_2, l_2 \rangle, ..., \langle i_n, l_n \rangle), c' = (\langle i_1, l_1' \rangle, \langle i_2, l_2' \rangle, ..., \langle i_n, l_n' \rangle)$$

we say that c' is a $\langle j, l_j \rangle$-*successor* of c, and write $succ_m(c, \langle j, l_j \rangle, c')$, if c and c' are both cuts and

$$l_j' = l_j + 1 \ \wedge \ \forall i \neq j, l_i' = l_i$$

Notice that the successor definition requires that both c and c' are cuts, so that advancing the location of one of the instances in c is allowed only if the obtained set of locations remains unordered.

A *run* of m is a sequence of cuts, $c_0, c_1, ..., c_k$, satisfying the following:

- c_0 is an initial cut.
- for all $0 \leq i < k$, there is $1 \leq j_i \leq n$, such that $succ_m(c_i, \langle j_i, l_{j_i} \rangle, c_{i+1})$.
- in the final cut c_k all locations are maximal.

$Runs(m)$ is the set of all runs of m.

Assume the natural mapping f between $(dom(m) \cup env) \times \Sigma \times dom(m)$ to the alphabet A, defined by

$$f(\langle i, l \rangle, \sigma, \langle j, l' \rangle) = (O_i, O_j.\sigma)$$

Intuitively, the function f maps a location to the sending object and to the message of the receiving object. With this notation in mind, $f(Messages(m))$ will be used to denote the letters in A corresponding to messages that are restricted by chart m:

$$f(Messages(m)) = \{f(v) \mid v \in Messages(m)\}$$

Definition 1. *Let* $c = c_0, c_1, ..., c_k$ *be a run. The execution trace, or simply the trace of* c, *written* $w = trace(c)$, *is the word* $w = w_1 \cdot w_2 \cdots w_k$ *over the alphabet* A, *defined by:*

$$w_i = \begin{cases} f(msg(m)(\langle j, l_j \rangle)) & if \ succ_m(c_{i-1}, \langle j, l_j \rangle, c_i) \ \wedge \ send(\langle j, l_j \rangle) \\ \epsilon & otherwise \end{cases}$$

We define the trace language *generated by chart* m, $\mathcal{L}_m^{trc} \subseteq A^*$, *to be*

$$\mathcal{L}_m^{trc} = \{w \mid \exists (c_0, c_1, ..., c_k) \in \text{Runs}(m) \text{ s.t. } w = trace(c_0, c_1, ..., c_k)\}$$

There are two additional notions that we associate with an LSC, its *mode* and its *activation message*. These are defined as follows:

$$mod : m \to \{existential, universal\}$$

$$amsg : m \to dom(m) \times \Sigma \times dom(m)$$

The activation message of a chart designates when a scenario described by the chart should start, as we describe below. The charts and the two additional notions are now put together to form a specification. An *LSC specification* is a triple

$$LS\langle M, amsg, mod \rangle,$$

where M is a set of charts, and $amsg$ and mod are the activation messages and modes of the charts, respectively.

The *language* of the chart m, denoted by $\mathcal{L}_m \subseteq A^* \cup A^\omega$, is defined as follows:

For an existential chart, $mod(m) = existential$, we require that the activation message is relevant (i.e., sent) at least once, and that the trace will then satisfy the chart:

$$\mathcal{L}_m = \big\{w = w_1 \cdot w_2 \cdots \mid \exists i_0, i_1, ..., i_k \text{ and } \exists v = v_1 \cdot v_2 \cdots v_k \in \mathcal{L}_m^{trc}, \text{ s.t.}$$
$$(i_0 < i_1 < ... < i_k) \wedge (w_{i_0} = f(amsg(m))) \wedge$$
$$(\forall j, 1 \leq j \leq k, w_{i_j} = v_j) \wedge$$
$$(\forall j', i_0 \leq j' \leq i_k, j' \notin \{i_0, i_1, ..., i_k\} \Rightarrow w_{j'} \notin f(Messages(m)))\big\}$$

The formula requires that the activation message is sent once $(w_{i_0} = f(amsg(m)))$, and then the trace satisfies the chart; i.e., there is a subsequence belonging to the trace language of chart m ($v = v_1 \cdot v_2 \cdots v_k = w_{i_1} \cdot w_{i_2} \cdots w_{i_k} \in \mathcal{L}_m^{trc}$), and all the messages between the activation message until the end of the satisfying subsequence ($\forall j', i_0 \leq j' \leq i_k$) that do not belong to the subsequence ($j' \notin \{i_0, i_1, ..., i_k\}$) are not restricted by the chart m ($w_{j'} \notin f(Messages(m))$).

For a universal chart, $mod(m) = universal$, we require that each time the activation message is sent the trace will satisfy the chart:

$$\mathcal{L}_m = \big\{w = w_1 \cdot w_2 \cdots \mid \forall i, w_i = f(amsg(m)) \Rightarrow \exists i_1, i_2, ..., i_k \text{ and}$$
$$\exists v = v_1 \cdot v_2 \cdots v_k \in \mathcal{L}_m^{trc}, \text{ s.t. } (i < i_1 < i_2 < ... < i_k) \wedge$$
$$(\forall j, 1 \leq j \leq k, w_{i_j} = v_j) \wedge$$
$$(\forall j', i \leq j' \leq i_k, j' \notin \{i_1, ..., i_k\} \Rightarrow w_{j'} \notin f(Messages(m)))\big\}$$

The formula requires that after each time the activation message is sent $(\forall i, w_i = f(amsg(m)))$, the trace will satisfy the chart m (this is expressed in the formula in a similar way to the case for an existential chart).

Now come the main definitions, which finalize the semantics of the language by connecting it with an object system:

Definition 2. *A system S satisfies the LSC specification* $LS = \langle M, amsg, mod \rangle$, *written* $S \models LS$, *if*:

1. $\forall m \in M, \quad mod(m) = universal \Rightarrow \mathcal{L}_S \subseteq \mathcal{L}_m$
2. $\forall m \in M, \quad mod(m) = existential \Rightarrow \mathcal{L}_S \cap \mathcal{L}_m \neq \emptyset$

In this short introduction, we assumed that a chart has an activation message. The extension of this notion to a prechart is omitted here. Informally, a chart containing a prechart must be satisfied whenever its prechart is satisfied. We also assumed all messages are hot, therefore all cuts must progress. However, when introducing cold messages, a cut containing only cold messages may progress, but need not.

The kernel language of LSC, introduced in [9], contains several constructs, in addition to the messages formally introduced above. These include:

- *Conditions*, which act as requirements on the state of the system at a given point in time. Like messages, conditions too can have a hot/cold modality, defining the effect of a false condition. A false hot condition is a violation of the requirements, whereas a false cold condition merely induces an immediate normal exit from the chart (or enclosing subchart).
- *Subcharts* are the main structuring mechanism in the LSC language. A subchart is a well-formed fragment of a chart. Along with conditions, it can also be used to define branching constructs like if-then-else.
- *Variables*, whose scope is local to an LSC. One can use assignments to assign values to variables. Expressions within conditions may include variables.

2.2 Different Variants and Additional Constructs

The above definitions constitute a kernel subset of the LSC language. A number of variants and extensions have been suggested and used in different kinds of work and in different contexts. We list some of these variants and extensions below.

The first variation to be discussed refers to the question of how often a universal LSC should be activated. The most general case is that of an *invariant* LSC, which calls for the LSC to be activated whenever the prechart is completed, regardless of the state of the system. This means that multiple copies of the same chart may be active simultaneously, if the prechart is completed several times. Two restrictions to this mode are *initial* and *iterative* (see, for example, in [5]). The initial mode indicates that the LSC is activated at system start only; i.e., it is intended to describe a start-up or initialization sequence. The iterative mode allows only one incarnation of the chart at a time, i.e., as long as a chart is active its prechart is not monitored for further satisfactions.

Another variant, suggested in [18], is that of *strict* vs. *tolerant* (or *weak*) semantics. A strict LSC restricts the occurrence of the messages used in the LSC to exactly those points in time where they are supposed to occur according to the scenario. Any message appearing out-of-order in a strict LSC is considered a violation. In the weak interpretation, the specification is satisfied if each necessary message occurs at least once where it is supposed to, and additional occurrences of it are ignored.

A variety of extensions have also been suggested to the kernel subset of the language. We now list some of them.

- *Symbolic messages* were introduced in [30]. In a symbolic message, the arguments passed by the message are symbolic, thus a single message in an LSC can stand for several different instantiations of it in the system. The actual arguments used in a specific run can be stored in LSC-local variables, so that they can be used again in the same chart. See also Chapter 7 in [18].
- In a real system, multiple objects can be instances of the same class. A *symbolic instance* in an LSC represents an entire class, or rather, any instance of the class, instead of a single concrete object. Symbolic instances were first suggested in [30], and are also covered in Chapter 15 of [18].
- A *co-region* is a sequence of locations belonging to the same instance, in which the partial order requirement is relaxed, i.e., locations within a co-region may appear in any order.
- *Forbidden messages and conditions* were introduced in [18], allowing one to state behaviors that are forbidden while an LSC (or a part of it) is active. Similarly, one may add restrictions on message sending, besides the ones derived from the LSC's partial order, using a *restricts* clause.
- *Timing constraints* on LSCs are considered in [23] and [17]. In [23], LSCs can be annotated by timers and by delay intervals, thus allowing one to express timing constraints on pairs of events that are either on the same instance line, or are connected by a message. In [17], on the other hand, a single clock object with one property, `Time`, and a single method, `Tick`, are introduced. This, together with the rich LSC language, suffices for specifying a wide variety of timing constraints (see Chapter 16 of [18]).

2.3 Scenario-Based Execution

The semantics described so far is a trace-based semantics, defining when a trace of events is in the language of the LSC specification. However, in [18,19], the play-out approach is presented. In this approach, the LSC specification can be directly executed, without any intermediate steps. Play-out is implemented in the *Play-Engine* tool. The play-out process calls for the Play-Engine to continuously monitor the applicable precharts of all universal charts, and whenever successfully completed, to execute their bodies. A full operational semantics is supplied in Appendix A of [18], defining how an LSC specification can be executed. We quote some of the main definitions from there.

The operational semantics is given as a transition system

$$Sem(S) = \langle V, V_0, S_D, S_M, \Delta[S_O, S_C] \rangle$$

where V is the set of possible configurations (states) of $Sem(S)$, V_0 is the initial configuration, $S_D \subseteq S_U$ is the set of driving LSCs, $S_M \subseteq S_U \cup S_E$ is the set of monitored LSCs, and $\Delta \subseteq \mathcal{V} \times (\mathcal{E} \cup \bigcup_{L \in \mathcal{S}} E_L) \times \mathcal{V}$ is the set of allowed transitions. We require that $\mathcal{S}_D \cap \mathcal{S}_M = \emptyset$.

A state $V \in \mathcal{V}$ is defined as

$$V = \langle \mathcal{RL}, \mathcal{ML}, Violating \rangle$$

where $\mathcal{RL}$ is a set of live copies of 'driving' LSCs, $\mathcal{ML}$ is a set of live copies of monitored LSCs, and $Violating$ indicates by $True$ or $False$ whether the state is a violating one.

The initial configuration contains no copies of driving LSCs and no copies of monitored LSCs, and is defined as:

$$V_0 = \langle \emptyset, \emptyset, False \rangle$$

The transition relation Δ is parameterized by two sets. The first, $\mathcal{S}_O$, is the set of original LSCs to which Δ should be applied. The second, $\mathcal{S}_C$, is the set of live copies that currently exist. This set contains only copies of LSCs from $\mathcal{S}_O$. The two sets are instantiated with either $(\mathcal{S}_D, \mathcal{RL})$ or $(\mathcal{S}_M, \mathcal{ML})$.

Δ is described as a set of rules to its set parameters and to $Violating$ with respect to a given event e. Since the set parameters are instantiated also by $\mathcal{RL}$ and $\mathcal{ML}$, which are taken from a state V, the result of applying Δ is a new state V' consisting of the modified components. In other words, Δ defines the result of executing an event e in a given system state. We skip the formal definition of Δ. The idea behind its rules is to advance any cut that needs to be advanced by executing e, to open new live copies of charts for which the prechart has become relevant, and to update $Violating$ to state whether there has been a violation.

The same definitions are used in [18] both for describing how a specification can be used for testing (quite similarly to the trace-based semantics described above), and for actual execution. The execution mechanism works in phases of *step* and *super-step*. The input to a step is a system event e. The procedure for a step phase consists of applying the transition relation onto the event e and, if the event represents a property change, changing the state of the object model according to the new value in the message.

In the super-step phase, the Play-Engine continuously executes the steps associated with internal events — i.e., those that do not originate with the user, the environment, the $Clock$ or external objects — until it reaches a 'stable' state where no further such events can be carried out.

The execution algorithm proposed in [18] is naïve, in the sense that when facing multiple choices for a step, none of them causing an immediate violation, it chooses one arbitrarily. Its choice might lead to a contradiction in the future, while perhaps there could have been a different choice that would have avoided it. This problem is addressed by the smart play-out algorithm proposed in [14], in which a legal super-step is found using a model checker. The specification is translated into a model, and the model checker is fed with this model along with the claim that no legal super-step exists. If one does exist, it will be given as a counter-example to the claim. In [20] the problem is translated into an AI planning problem, and an extended planner is used in order to find all legal supersteps from a given system state, up to a predefined length.

The operational semantics given above expresses the same ideas as the trace-based semantics of section 2.1, but in a manner more suitable for execution. The operational semantics somewhat restricts the trace-based semantics to those cases that are interesting in the context of execution. In a sense, all "interesting" traces can be generated by the operational semantics. When equipped with the smart play-out approach, it is also sound, in the sense that every supserstep generated by it is also a legal trace in the trace-based semantics.

The Play-Engine [18] is an interpreter based execution engine for an LSC specification. The specification is executed directly, with no intermediate code being generated. An implementation of play-out by compilation into aspects was suggested in [29] and is implemented in a compiler called S2A [12]. This work is defined for the slightly generalized and UML2-compliant variant of LSC given in [16], in which, unlike the version supported by the Play-Engine where precharts are monitored and main-charts are executed, the hot/cold modality is orthogonal to a new monitor/execute modality.

3 Expressive Power

The expressive power of LSC was studied in [3,10,13,24] by suggesting translations from fragments of the language into various Temporal Logics.

A first embedding of a kernel subset of the language (which omits variables, for example) into CTL* was given in [13]. For this kernel subset the embedding is a strict inclusion, since given the single level quantification mechanism of LSCs, the language cannot express general formulas with alternating path quantifiers.[1]

This embedding was improved in [24] to support a wider subset of the language and in a more efficient way. Specifically, it was shown that existential charts can be expressed using the branching temporal logic CTL, while universal charts are in the intersection of linear temporal logic and branching temporal logic LTL ∩ CTL. Below we give the basic and then the improved explicit translations from [24].

Definition 3 ([24]). *Let* $w = m_1 m_2 m_3 \ldots m_k$ *be a finite trace. Let* $R = \{e_1, e_2, e_3 \cdots e_l\}$ *be a set of events. The temporal logic formula* ϕ_w^R *is defined as:*

$$\phi_w^R = NU\left(m_1 \wedge \left(X\left(NU\left(m_2 \wedge \left(X(NU(m_3 \ldots))\right)\right)\right)\right)\right),$$

where the formula N *is given by* $N = \neg e_1 \wedge \neg e_2 \ldots \wedge \neg e_l$.

Definition 4 ([24]). *Let* $LS = \langle M, amsg, mod \rangle$ *be an LSC specification. For a chart* $m \in M$, *we define the formula* ψ_m *as follows.*

- *If* $mod(m) = universal$, *then* $\psi_m = AG\left(amsg(m) \to X\left(\bigvee_{w \in \mathcal{L}_m^{trc}} \phi_w^R\right)\right)$.
- *If* $mod(m) = existential$, *then* $\psi_m = EF\left(\bigvee_{w \in \mathcal{L}_m^{trc}} \phi_w^R\right)$.

---

[1] It shouldn't be too difficult to extend LSCs to allow certain kinds of quantifier alternation, as noted in [9]. However, as in [9], this was not done there either, since it was judged to have been too complex and unnecessary for real world usage of sequence charts.

(for a universal chart m, R includes the events appearing in the prechart and in the main chart.)

In the above, the formula for a universal chart is in LTL. However, it can be large, due to the possibility of having many different traces for the chart, which affects the number of clauses in the disjunction, and also due to the similarity of clauses at the different sides of the implication operator. In the improved translation given below, the resulting temporal logic formulas are much more succinct, i.e., polynomial vs. exponential in the number of locations.

We consider the case where both the prechart and the main chart consist only of message communication, and denote by $p_1, \cdots p_k$ the events appearing in the prechart, and by $m_1, \cdots m_l$ the events appearing in the main chart. Denote by e_i any of these events, either in the prechart or in the main chart. We write $e_i \prec e_j$ if e_i precedes e_j in the partial order induced by the chart, and $e_i \not\prec e_j$ if e_i and e_j are unordered.

Definition 5 ([24])

$$\psi_m = G \left((\bigwedge_{p_i \prec p_j} \phi_{p_i,p_j} \wedge \bigwedge_{\forall p_i, m_j} \phi_{p_i,m_j} \wedge \bigwedge_{p_i \not\prec p_j} \neg\chi_{p_j,p_i}) \rightarrow \right.$$

$$\left. (\bigwedge_{m_i \prec m_j} \phi_{m_i,m_j} \wedge \bigwedge_{m_j \text{ is maximal}} Fm_j \wedge \bigwedge_{\forall e_i, m_j} \neg\chi_{e_i,m_j}) \right)$$

$$\phi_{x_i,x_j} = \neg x_j U x_i$$
$$\chi_{x_i,x_j} = (\neg x_i \wedge \neg x_j)U(x_i \wedge X((\neg x_i \wedge \neg x_j)U x_i))$$

Here the formula ϕ_{x_i,x_j} specifies that x_j must not happen before x_i, which eventually occurs. The formula $\neg\chi_{x_i,x_j}$ specifies that x_i must not occur twice before x_j occurs.

Note that this translation is polynomial in the number of messages appearing in the chart, while the translation in Definition 4 may be exponential in that number. However, the above translation assumes that a message does not appear more than once in the same chart. Whether an efficient translation exists for the most general case is left open in [24]. A construction given in [3] provides a polynomial translation for the more general case of deterministic LSCs, i.e., where a message may occur more than once in a chart but all appearances of the same message are ordered.

Using a characterization by Maidl for the common fragment of LTL and CTL [28] and a theorem by Clarke and Draghicesku [7], it is shown in [24] that the formulas given in Definition 5 have equivalent CTL formulas. Finally, [24] considers also the extension of the above to support conditions and bounded iterations. An explicit translation that supports these, however, is left in [24] for future work.

A different translation of LSC into TL, which supports variables but considers activation only by *activation condition* and not the general case of precharts, was given by Damm, Toben, and Westphal in [10]. To support variables, the work defines a translation of LSC into a fragment of first-order CTL*. Specifically, a translation is defined from bounded LSCs (i.e., where conditions and local invariants only appear in simultaneous regions with messages) into *(deterministic) communication sequence first-order prenex CTL** (DCSCTL), a syntactically characterized fragment of CTL*. The translation is shown to be tight, i.e., a translation back from DCSCTL into LSC is constructively defined, thus establishing an equivalence.

Restricted to messages, the two pieces of work surveyed above [10,24] coincide. They consider different subsets of the LSC language. Neither of them handles explicit time.

3.1 Limitations

As mentioned above, given the single level quantification mechanism of LSCs, the language cannot express general formulas with alternating path quantifiers. However, as shown in [10], the embedding of LSC into CTL* is strict even without resorting to the nesting of path quantifiers. The question of whether adding constructs not included in the above work (e.g., bounded iterations, specifically within precharts) will make LSCs equivalent in expressive power to LTL remains open in [10]. A similar result is provided in [1] where it is shown that the language $\Sigma^* aa\Sigma^\omega$ that is expressible using a deterministic Büchi automaton (DBA) and by an LTL formula ($F(a \wedge Xa)$) is not expressible in LSC.

4   Complexity Results

In this section we survey the complexity results for the three main applications of LSCs, i.e., model checking, execution (play-out), and synthesis. Essentially, all results mentioned are due to Bontemps and Schobbens in [2] and [3].

4.1   Model Checking

Our first problem is that of model-checking. In this setting, one is given a system implementation (either centralized or distributed) in some formal language, e.g., I/O automata, and an LSC specification, and we want decide whether the system satisfies the specification. The complexity of this problem grows along two axes: centralized vs. distributed systems, and closed vs. open environments (i.e., whether the system is a stand-alone one or interacts with an environment).

Theorem 1 ([2]). *Closed Centralized Model Checking (CCMC) is complete for co-NP.*

Proof. Membership in co-NP is proved by guessing a counter-example, which is a path in the system automaton that violates an LSC.

Hardness is proved by reducing the complement of the traveling salesman problem (CoTSP) (see [31]) to CCMC. Given a weighted graph, an automaton is built such that a tour in the graph corresponds to a set of automata transitions. The automaton is equipped with a counter that sums the weights of the edges in the tour. The fact that all tours have length $\geq k$ is encoded in an LSC. Its prechart is matched when all vertices have occurred exactly once, and the main chart makes sure the value of the counter is $\geq k$. A tour of length $< k$ exists iff the automaton violates the LSC. $\qquad\square$

Theorem 2 ([2]). *Open centralized model checking (OCMC), closed distributed model checking (CDMC) and open distributed model checking (ODMC) are all complete for PSPACE.*

Proof. (For CDMC) Membership is proved by building a nondeterministic PSPACE Turing machine deciding on the complement of the distributed model checking problem, and relying on coPSPACE=PSPACE, according to Savitch's theorem [32].

The hardness proof takes a DPSPACE Turing machine and builds a set of automata, A_i, one for each cell tape. Each automaton records the letter in its cell, and whether the tape head is located on it or not. Each transition of the Turing machine is encoded by transitions in the relevant automaton. The LSC states that whenever a run starts it must halt. This causes the system to satisfy the LSC iff the Turing machine halts. $\qquad\square$

4.2 Reachability and Smart Play-Out

When considering the complexity of play-out, there are two main problems to be considered, reachability, and smart play-out. In the reachability problem, an LSC specification and a single existential LSC are given, and one wants to decide whether, under the constraints of the former, the latter can be satisfied. In smart play-out, the environment has executed several steps, and the system should find a superstep, i.e., a series of steps that satisfies the specification.

Theorem 3 ([2]). *Reachability is PSPACE-complete.*

Proof. Membership is proved by transforming the LSC specification into an LTL formula, Φ_u, and the claim that the existential formula can not be satisfied into another LTL formula, ϕ_e, and checking whether $\Phi_u \rightarrow \phi_e$ is valid. This solves the complement of the reachability problem. The solution for LTL is in PSPACE according to [34]. Note that membership can also be proved by considering [14], in which the problem is reduced to model-checking, which is known to be in PSPACE.

Hardness is proved by encoding the execution of a DPSPACE Turing machine on the blank input as an LSC specification. The existential LSC calls for the execution to start and to halt. A halting run of the Turing machine exists iff the existential LSC can be satisfied. $\qquad\square$

Theorem 4. *Smart play-out is PSPACE-complete.*

Proof. The theorem can be proved by adapting the reachability proof above. In other work, not yet published, the same claim is proved by a reduction from QBF. $\qquad\square$

4.3 Synthesis and Consistency

The most complex class of problems considered here is that of synthesis. In this class of problems, we would like to know whether the objects participating in the LSC specification can actually be implemented consistently. This problem is also termed "agent design".

A related problem is that of consistency; i.e., deciding whether the specification has no internal contradictions. A formal definition of a consistent system is given in [13]. Informally, a system is consistent if there exists a non-empty regular language, $\mathcal{L}$, s.t. (1) all universal charts are satisfied by all traces in $\mathcal{L}$; (2) every trace in $\mathcal{L}$ is extendible if a new message is sent from the environment; and (3) each existential chart is satisfied by some trace in $\mathcal{L}$. In [13] it is shown that a system is consistent if and only if it is satisfiable (i.e., can be synthesized).

As in previous sections, two versions of the synthesis problem are considered; a centralized one, in which a single automaton is built, and a distributed one, in which each object has its own automaton.

Theorem 5 ([1]). *Centralized synthesis is EXPTIME-complete.*

Proof. Membership follows from the exponential time algorithms proposed in [4] and [15].

Hardness is proved by encoding an alternating PSPACE Turing machine as an LSC, similar to the construction in Theorem 3, in which existential and universal moves are distinguished.   □

An interesting question regarding the centralized synthesis problem deals with the size of the synthesized automaton. This is also answered in [3], where it is shown that there exists a family $(\phi_n)_{n>0}$ of LSC specifications, such that any implementation of ϕ_n requires memory of size $2^{\Omega(n \log n)}$. It is worth noting that this proof uses *co-region* constructs, which relax the ordering of events. A co-region succinctly encodes an exponential number of orderings.

Finally, [3] considers the problem of distributed synthesis, in which each object has to be synthesized separately. The question of whether such a synthesis exists is undecidable. This is proved by reducing Post's correspondence problem to the problem of deciding whether the specification is not distributively implementable.

5   Conclusion

The language of LSC has been a subject of much work. We surveyed here some theoretical results regarding the expressive power of the language and the complexity of some of its main applications.

References

1. Bontemps, Y.: Relating Inter-Agent and Intra-Agent Specifications (The Case of Live Sequence Charts). PhD thesis, Facultés Universitaires Notre-Dame de la Paix, Institut d'Informatique (University of Namur, Computer Science Dept) (April 2005)

2. Bontemps, Y., Schobbens, P.-Y.: The Complexity of Live Sequence Charts. In: Sassone, V. (ed.) FOSSACS 2005. LNCS, vol. 3441, pp. 364–378. Springer, Heidelberg (2005)

3. Bontemps, Y., Schobbens, P.-Y.: The Computational Complexity of Scenario-Based Agent Verification and Design. J. Applied Logic 5(2), 252–276 (2007)

4. Bontemps, Y., Schobbens, P.-Y., Löding, C.: Synthesis of Open Reactive Systems from Scenario-Based Specifications. Fundam. Inform. 62(2), 139–169 (2004)

5. Brill, M., et al.: Live Sequence Charts: An Introduction to Lines, Arrows, and Strange Boxes in the Context of Formal Verification. In: Ehrig, H., et al. (eds.) INT 2004. LNCS, vol. 3147, pp. 374–399. Springer, Heidelberg (2004)

6. Bunker, A., Gopalakrishnan, G., Slind, K.: Live Sequence Charts Applied to Hardware Requirements Specification and Verification: A VCI Bus Interface Model. Software Tools for Technology Transfer 7(4), 341–350 (2005)

7. Clarke, E.M., Draghicescu, I.A.: Expressibility Results for Linear-Time and Branching-Time Logics. In: de Bakker, J.W., de Roever, W.-P., Rozenberg, G. (eds.) Linear Time, Branching Time and Partial Order in Logics and Models for Concurrency. LNCS, vol. 354, pp. 428–437. Springer, Heidelberg (1989)

8. Combes, P., Harel, D., Kugler, H.: Modeling and Verification of a Telecommunication Application Using Live Sequence Charts and the Play-Engine Tool. In: Peled, D.A., Tsay, Y.-K. (eds.) ATVA 2005. LNCS, vol. 3707, pp. 414–428. Springer, Heidelberg (2005)

9. Damm, W., Harel, D.: LSCs: Breathing Life into Message Sequence Charts. J. on Formal Methods in System Design 19(1), 45–80 (2001); Preliminary version In: Ciancarini, P., Fantechi, A., Gorrieri, R. (eds.) Proc. 3rd IFIP Int. Conf. on Formal Methods for Open Object-Based Distributed Systems (FMOODS 1999), pp. 293–312. Kluwer Academic Publishers, Dordrecht (1999)

10. Damm, W., Toben, T., Westphal, B.: On the Expressive Power of Live Sequence Charts. In: Reps, T., Sagiv, M., Bauer, J. (eds.) Wilhelm Festschrift. LNCS, vol. 4444, pp. 225–246. Springer, Heidelberg (2007)

11. Fisher, J., et al.: Combining State-Based and Scenario-Based Approaches in Modeling Biological Systems. In: Danos, V., Schachter, V. (eds.) CMSB 2004. LNCS (LNBI), vol. 3082, pp. 236–241. Springer, Heidelberg (2005)

12. Harel, D., Kleinbort, A., Maoz, S.: S2A: A Compiler for Multi-modal UML Sequence Diagrams. In: Dwyer, M.B., Lopes, A. (eds.) FASE 2007. LNCS, vol. 4422, pp. 121–124. Springer, Heidelberg (2007)

13. Harel, D., Kugler, H.: Synthesizing State-Based Object Systems from LSC Specifications. Int. J. of Foundations of Computer Science 13(1), 5–51 (2002) Also: Harel, D., Kugler, H.: Synthesizing State-Based Object Systems from LSC Specifications. In: Yu, S., Păun, A. (eds.) CIAA 2000. LNCS, vol. 2088, pp. 1–33. Springer, Heidelberg (2001) Preliminary version appeared as technical report MCS99-20, Weizmann Institute of Science (1999)

14. Harel, D., et al.: Smart Play-out of Behavioral Requirements. In: Aagaard, M.D., O'Leary, J.W. (eds.) FMCAD 2002. LNCS, vol. 2517, pp. 378–398. Springer, Heidelberg (2002)

15. Harel, D., Kugler, H., Pnueli, A.: Synthesis Revisited: Generating Statechart Models from Scenario-Based Requirements. In: Kreowski, H.-J., et al. (eds.) Formal Methods in Software and Systems Modeling. LNCS, vol. 3393, pp. 309–324. Springer, Heidelberg (2005)

16. Harel, D., Maoz, S.: Assert and Negate Revisited: Modal Semantics for UML Sequence Diagrams. Software and Systems Modeling (SoSyM) (to appear, 2007)

17. Harel, D., Marelly, R.: Playing with Time: On the Specification and Execution of Time-Enriched LSCs. In: MASCOTS, pp. 193–202. IEEE Computer Society, Los Alamitos (2002)
18. Harel, D., Marelly, R.: Come, Let's Play: Scenario-Based Programming Using LSCs and the Play-Engine. Springer, Heidelberg (2003)
19. Harel, D., Marelly, R.: Specifying and Executing Behavioral Requirements: The Play-In/Play-Out Approach. Software and Systems Modeling (SoSyM) 2(2), 82–107 (2003)
20. Harel, D., Segall, I.: Planned and Traversable Play-Out: A Flexible Method for Executing Scenario-Based Programs. In: Grumberg, O., Huth, M. (eds.) TACAS 2007. LNCS, vol. 4424, pp. 485–499. Springer, Heidelberg (2007)
21. ITU. International Telecommunication Union Recommendation Z.120: Message Sequence Charts. Technical report (1996)
22. Klose, J., et al.: Check It Out: On the Efficient Formal Verification of Live Sequence Charts. In: Ball, T., Jones, R.B. (eds.) CAV 2006. LNCS, vol. 4144, pp. 219–233. Springer, Heidelberg (2006)
23. Klose, J., Wittke, H.: An Automata Based Interpretation of Live Sequence Chart. In: Margaria, T., Yi, W. (eds.) ETAPS 2001 and TACAS 2001. LNCS, vol. 2031, pp. 512–527. Springer, Heidelberg (2001)
24. Kugler, H., et al.: Temporal Logic for Scenario-Based Specifications. In: Halbwachs, N., Zuck, L.D. (eds.) TACAS 2005. LNCS, vol. 3440, pp. 445–460. Springer, Heidelberg (2005)
25. Kugler, H., Stern, M.J., Hubbard, E.J.A.: Testing Scenario-Based Models. In: Dwyer, M.B., Lopes, A. (eds.) FASE 2007. LNCS, vol. 4422, pp. 306–320. Springer, Heidelberg (2007)
26. Lettrari, M., Klose, J.: Scenario-Based Monitoring and Testing of Real-Time UML Models. In: Gogolla, M., Kobryn, C. (eds.) UML 2001. LNCS, vol. 2185, pp. 317–328. Springer, Heidelberg (2001)
27. Lo, D., Maoz, S., Khoo, S.-C.: Mining Modal Scenario-Based Specification from Execution Traces of Reactive Systems. In: Proc. 22nd IEEE/ACM Int. Conf. on Automated Software Engineering (ASE 2007), pp. 465–468 (2007)
28. Maidl, M.: The Common Fragment of CTL and LTL. In: FOCS, pp. 643–652 (2000)
29. Maoz, S., Harel, D.: From Multi-Modal Scenarios to Code: Compiling LSCs into AspectJ. In: Proc. 14th Int. ACM/SIGSOFT Symp. Foundations of Software Engineering (FSE-14), Portland, Oregon, pp. 219–230 (November 2006)
30. Marelly, R., Harel, D., Kugler, H.: Multiple Instances and Symbolic Variables in Executable Sequence Charts. In: Proc. 17th ACM Conf. on Object-Oriented Prog., Systems, Lang. and App. (OOPSLA 2002), Seattle, WA, pp. 83–100 (2002)
31. Papadimitriou, C.H.: Computational Complexity. Addison-Wesley, Reading (1994)
32. Savitch, W.J.: Relationships Between Nondeterministic and Deterministic Tape Complexities. J. Comput. Syst. Sci. 4(2), 177–192 (1970)
33. Schinz, I., et al.: The Rhapsody UML Verification Environment. In: Proc. of the 2nd Int. Conf. on Software Engineering and Formal Methods (SEFM 2004), pp. 174–183. IEEE Computer Society Press, Washington, DC, USA (2004)
34. Sistla, A.P., Clarke, E.M.: The Complexity of Propositional Linear Temporal Logics. J. ACM 32(3), 733–749 (1985)
35. UML. Unified Modeling Language Superstructure Specification, v2.0. OMG spec., OMG (August 2005), http://www.omg.org

Finite Dimensional Vector Spaces Are Complete for Traced Symmetric Monoidal Categories

Masahito Hasegawa[1], Martin Hofmann[2], and Gordon Plotkin[3]

[1] RIMS, Kyoto University
hassei@kurims.kyoto-u.ac.jp
[2] LMU München, Institut für Informatik
hofmann@ifi.lmu.de
[3] LFCS, University of Edinburgh
gdp@inf.ed.ac.uk

Abstract. We show that the category **FinVect**$_k$ of finite dimensional vector spaces and linear maps over any field k is *(collectively) complete* for the traced symmetric monoidal category freely generated from a signature, provided that the field has characteristic 0; this means that for any two different arrows in the free traced category there always exists a strong traced functor into **FinVect**$_k$ which distinguishes them. Therefore two arrows in the free traced category are the same if and only if they agree for all interpretations in **FinVect**$_k$.

1 Introduction

This paper is affectionately dedicated to Professor B. Trakhtenbrot on the occasion of his 85th birthday. Cyclic networks of various kinds occur in computer science, and other fields, and have long been of interest to Professor Trakhtenbrot: see, e.g., [15,9,16,8]. In this paper they arise in connection with Joyal, Street and Verity's *traced monoidal categories* [6]. These categories were introduced to provide a categorical structure for cyclic phenomena arising in various areas of mathematics, in particular knot theory [17]; they are (balanced) monoidal categories [5] enriched with a *trace*, a natural generalization of the traditional notion of trace in linear algebra that can be thought of as a 'loop' operator.

In computer science, specialized versions of traced monoidal categories naturally arise as recursion/feedback operators as well as cyclic data structures. In particular, Hyland and Hasegawa independently observed a bijective correspondence between Conway (Bekič, or dinatural diagonal) fixpoint operators [1,11] and traces on categories with finite products [2,3]. Thus, the notion of trace very neatly characterises the well-behaved fixpoint operators commonly used in computer science. More generally, traced symmetric monoidal categories equipped with the additional structure of a cartesian center can be used for modelling recursive computation created from cyclic data structures, see *ibid*. In this context, freely generated traced symmetric monoidal categories can be characterised as categories of cyclic networks, and so are of particular interest (see [14] for a related treatment).

A. Avron et al. (Eds.): Trakhtenbrot/Festschrift, LNCS 4800, pp. 367–385, 2008.

We characterise the equivalence of arrows in free traced symmetric monoidal categories via interpretations in the very familiar setting of linear algebra: the category $\mathbf{FinVect}_k$ of finite dimensional vector spaces and linear maps over a field k. Specifically, we show (Theorem 4) that if k has characteristic 0 then $\mathbf{FinVect}_k$ is *(collectively) complete* for the traced symmetric monoidal category freely generated from a signature; this means that for any two different arrows in the free traced category there always exists a structure-preserving functor into $\mathbf{FinVect}_k$ which distinguishes them. Therefore two arrows in the free traced category are the same if and only if they agree for all interpretations in $\mathbf{FinVect}_k$.

In order to show this, we present the freely generated traced symmetric monoidal category in terms of networks modulo suitable isomorphisms, and reduce the problem to that of finding suitable interpretations of these networks in $\mathbf{FinVect}_k$. This problem is then further reduced to considering a certain class of networks: those over a one-sorted signature and with no inputs or outputs. Finally, given any two such networks X and Y, we construct interpretations $[\![-]\!]^{\mu_X}$ and $[\![-]\!]^{\mu_Y}$ such that, ignoring some trivial cases, $[\![X]\!]^{\mu_X} = [\![Y]\!]^{\mu_X}$ and $[\![X]\!]^{\mu_Y} = [\![Y]\!]^{\mu_Y}$ jointly imply that X and Y are isomorphic.

One motivation for our work was previous completeness results for the cartesian case, where the monoidal product is the categorical one. As remarked above, in that case trace operators correspond to Conway fixpoint operators. However, the mathematically natural model categories, such as that of pointed directed complete posets and continuous functions, obey further equations, and the relevant notion is that of an iteration operator [1,11]. It is shown in [11] that any category with an iteration operator satisfying a mild non-triviality condition is collectively complete for the theory of iteration operators. It would be interesting to investigate conditions for the collective completeness of a symmetric monoidal category for trace operators. Another direction which may be of interest would be to look for completeness results for various classes of symmetric monoidal categories equipped with some natural combinations of (co)units and (co)diagonals; see [4] for a discussion of possible such combinations.

A closely related research thread is that of higher-order structures. Concerning coherence problems in category theory, Mac Lane conjectured that the category of vector spaces over a field is complete for the symmetric monoidal closed category freely generated by a set of atoms. This was proved in a more general form by Soloviev [12]; his proof-theoretic approach differs substantially from our model-theoretic one. In the cartesian case one considers the typed λ-calculus, where there is a good deal of work, starting with Friedman's completeness theorem: see [10] and the references given there for further developments. The combination of higher-order structure and traces could be an interesting subject for investigation; specifically one might consider the case of traced symmetric monoidal closed categories.

Organisation of this paper. The rest of this paper is organised as follows. In Sect. 2 we recall the notion of traced symmetric monoidal category, and describe the trace on $\mathbf{FinVect}_k$. Section 3 is devoted to a theory of cyclic networks, which provide a characterisation of the traced symmetric monoidal category freely generated over a monoidal signature. In Sect. 4 we study the interpretation of networks in $\mathbf{FinVect}_k$, and, in particular, the interpretations needed for our completeness results. These are presented in the concluding Sect. 5, which also gives a completeness theorem for interpretations with finite fields (Theorem 5), a discussion of some open problems, and a completeness result for compact closed categories (Corollary 5), obtained using the biadjunction of [6] between such categories and traced symmetric monoidal categories.

2 Preliminaries

2.1 Traced Symmetric Monoidal Categories

A *monoidal category* is a category $\mathcal{C}$ equipped with a bifunctor $\otimes : \mathcal{C}^2 \to \mathcal{C}$, an object I and natural isomorphisms $a_{A,B,C} : (A \otimes B) \otimes C \xrightarrow{\sim} A \otimes (B \otimes C)$, $l_A : I \otimes A \xrightarrow{\sim} A$ and $r_A : A \otimes I \xrightarrow{\sim} A$ satisfying standard conditions [7,5]. It is *strict* if these natural isomorphisms are identities. A *symmetric* monoidal category is a monoidal category equipped with a specified natural isomorphism $c_{X,Y} : X \otimes Y \xrightarrow{\sim} Y \otimes X$, again subject to standard axioms. A *trace* on such a symmetric monoidal category is a family of functions:

$$Tr^X_{A,B} : \mathcal{C}(A \otimes X, B \otimes X) \to \mathcal{C}(A, B)$$

subject to the following conditions:

- *tightening* (naturality): $Tr^X_{A',B'}((k \otimes 1_X) \circ f \circ (h \otimes 1_X)) = k \circ Tr^X_{A,B}(f) \circ h$
- *yanking*: $Tr^X_{X,X}(c_{X,X}) = id_X$
- *superposition*: $Tr^X_{C \otimes A, C \otimes B}(id_C \otimes f) = id_C \otimes Tr^X_{A,B}(f)$
- *exchange*:
$$Tr^X_{A,B}(Tr^Y_{A \otimes X, B \otimes X}(f)) = Tr^Y_{A,B}(Tr^X_{A \otimes Y, B \otimes Y}((1_B \otimes c_{X,Y}) \circ f \circ (1_A \otimes c_{Y,X})))$$

where, for ease of presentation, the associativity isomorphisms a have been omitted in the last two conditions. For example, the unabbreviated exchange axiom is:

$$Tr^X_{A,B}(Tr^Y_{A \otimes X, B \otimes X}(f)) =$$
$$Tr^Y_{A,B}(Tr^X_{A \otimes Y, B \otimes Y}($$
$$a^{-1}_{B,Y,X} \circ (1_B \otimes c_{X,Y}) \circ a_{B,X,Y} \circ f \circ a^{-1}_{A,X,Y} \circ (1_A \otimes c_{Y,X}) \circ a_{A,Y,X}))$$

where $f : (A \otimes X) \otimes Y \to (B \otimes X) \otimes Y$. Note that this axiomatisation is not quite the same as the original axiomatisation [6] or another popular formulation (see e.g., [2,3]); however, it is not hard to see that they are all equivalent.[1] A *traced symmetric monoidal category* is a symmetric monoidal category equipped with a (specified) trace.

The following graphical notation for the trace may help the reader. Given $f : A \otimes X \to B \otimes X$, its trace $Tr^X_{A,B}(f) : A \to B$ is shown as a feedback:

---

[1] The *vanishing* condition for the unit $Tr^I(f) = f$ was redundant in the original axiomatisation. The vanishing condition for tensor $Tr^{X \otimes Y}(f) = Tr^X(Tr^Y(f))$ and the *sliding* condition $Tr^X((1 \otimes h) \circ f) = Tr^Y(f \circ (1 \otimes h))$ can all be derived from the axioms presented here.

tightening : $Tr^X_{A',B'}((k \otimes 1_X) \circ f \circ (h \otimes 1_X)) = k \circ Tr^X_{A,B}(f) \circ h$

yanking : $Tr^X_{X,X}(c_{X,X}) = 1_X$

superposition : $Tr^X_{C \otimes A, C \otimes B}(1_C \otimes f) = 1_C \otimes Tr^X_{A,B}(f)$

exchange : $Tr^X_{A,B}(Tr^Y_{A \otimes X, B \otimes X}(f)) = Tr^Y_{A,B}(Tr^X_{A \otimes Y, B \otimes Y}((1_B \otimes c_{X,Y}) \circ f \circ (1_A \otimes c_{Y,X})))$

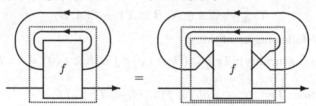

Fig. 1. Axioms for Trace

The above axioms are presented using this notation in Figure 1.

2.2   Finite Dimensional Vector Spaces

Finite dimensional vector spaces over a field k and linear maps form a traced symmetric monoidal category **FinVect**$_k$. The monoidal structure is given by the standard tensor product, and the trace is a natural generalization of the standard 'sum of diagonal elements' trace, sometimes called the 'partial trace'; the trace $Tr^W_{U,V}(f) : U \to V$ of a linear map $f : U \otimes_k W \to V \otimes_k W$ is given by:

$$\left(Tr^W_{U,V}(f)\right)_{i,j} = \Sigma_k f_{i \otimes k, j \otimes k}$$

where i, j run over bases of U and V. If $U = V = k$, we have $Tr^W(f) = \sum_k f_{k,k}$ as expected. If $\{e_1, \ldots, e_n\}$ is a basis of W, this is the same as $\sum_i \langle f(e_i) | e_i \rangle$ where $\langle - | - \rangle$ is the canonical scalar product such that $\langle e_i | e_j \rangle = \delta_{ij}$.

The partial trace is the unique trace for this monoidal structure on $\mathbf{FinVect}_k$. This is because $\mathbf{FinVect}_k$ is compact closed, and every compact closed category has a unique trace with respect to its monoidal structure.

3   Cyclic Networks

We present a theory of cyclic networks similar to the theory of cyclic sharing graphs given in [3].

3.1   Sorts and Signatures

We introduce a notion of multisorted signature suitable for interpretation over monoidal categories. If S is our set of sorts we call elements of S^*, the set of finite sequences of sorts, *arities*. Given such an arity v, we write $|v|$ for its length and v_i for its i-th component (for $1 \leq i \leq |v|$).

Definition 1. *An S-sorted signature is a triple (F, ar_{in}, ar_{out}) where F is a set whose elements are called* function symbols, *and where $ar_{in}, ar_{out} : F \to S^*$ are mappings assigning to each function symbol f two arities: an* input arity $ar_{in}(f)$ *and an* output arity $ar_{out}(f)$.

We may refer to a signature by the set F alone, leaving the arity functions implicit.

Definition 2. *We define $F_\bullet$ to be the extension of F with additional function symbols $\bullet_s$ for each sort $s \in S$, with $ar_{in}(\bullet_s) = ar_{out}(\bullet_s) = \varepsilon$.*

The function symbol $\bullet_s$ will be used to represent the trivial cycle of sort s (the trace of the identity at s).

3.2   Networks

Definition 3. *Let F be an S-sorted signature. A network from v to w in S^* over F is a tuple N of the form (X, φ, π), where:*

- X *is a finite set (of nodes)*
- φ *is a function from X to $F_\bullet$ (the labelling function, assigning a function symbol to each node)*
- π *is a bijection between*

$$O_N = \{\langle x, i \rangle \mid x \in X, 1 \leq i \leq |ar_{out}(\varphi(x))|\} \cup \{j \mid 1 \leq j \leq |v|\}$$

and

$$D_N = \{\langle x, i \rangle \mid x \in X, 1 \leq i \leq |ar_{in}(\varphi(x))|\} \cup \{j \mid 1 \leq j \leq |w|\}$$

such that the following constraints on arities are satisfied:

- $\pi\langle x, i\rangle = \langle y, j\rangle$ *implies* $ar_{out}(\varphi(x))_i = ar_{in}(\varphi(y))_j$
- $\pi\langle x, i\rangle = j$ *implies* $ar_{out}(\varphi(x))_i = w_j$
- $\pi(i) = \langle y, j\rangle$ *implies* $v_i = ar_{in}(\varphi(y))_j$
- $\pi(i) = j$ *implies* $v_i = w_j$

We say that v and w are the input *and* output *arities of the network, and write* $N : v \to w$.

It may help the reader to think of O as the set of ports from which flow originates and D as the set of ports to which flow goes. The function π then shows how the ports are linked.

Example 1. Let $S = \{A, B\}$ be the set of sorts. We consider the following signature (F, ar_{in}, ar_{out}) on S, where $F = \{f, g\}$ and:

$$ar_{in}(f) = AB \quad ar_{out}(f) = AA$$
$$ar_{in}(g) = A \quad ar_{out}(g) = B$$

Then, for instance, $(\{f, g, a\}, \varphi, \pi) : A \to A$ with $\varphi(f) = f, \varphi(g) = g, \varphi(a) = \bullet_A$ and:

$$\pi\langle f, 1\rangle = 1$$
$$\pi\langle f, 2\rangle = \langle g, 1\rangle$$
$$\pi\langle g, 1\rangle = \langle f, 2\rangle$$
$$\pi(1) \quad = \langle f, 1\rangle$$

is a network which may be pictured as follows:

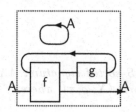

3.3 Homomorphisms

Definition 4. *Let $N = (X, \varphi, \pi) : v \to w$ and $N' = (X', \varphi', \pi') : v \to w$ be networks with the same input and output arities. A homomorphism from N to N' is given by a function $f : X \to X'$ such that:*

- $\varphi'(f(x)) = \varphi(x)$
- $\pi\langle x, i\rangle = \langle y, j\rangle$ *implies* $\pi'\langle f(x), i\rangle = \langle f(y), j\rangle$
- $\pi\langle x, i\rangle = j$ *implies* $\pi'\langle f(x), i\rangle = j$
- $\pi(i) = \langle y, j\rangle$ *implies* $\pi'(i) = \langle f(y), j\rangle$
- $\pi(i) = j$ *implies* $\pi'(i) = j$

The first condition just says that f does not change the function symbol assigned to each node. The other four requirements are equivalent to the commutation of the following diagram:

$$\begin{array}{ccc} O_N & \xrightarrow{\pi} & D_N \\ f^O \downarrow & & \downarrow f^D \\ O_{N'} & \xrightarrow{\pi'} & D_{N'} \end{array}$$

where f^O and f^D send $\langle x, i \rangle$ to $\langle f(x), i \rangle$ and j to j.

We evidently have a category with objects the networks of given input and output arities and morphisms the homomorphisms. Since, as one easily sees, the inverse of a bijective homomorphism is also a homomorphism, the isomorphisms are the bijective homomorphisms. Note that we deal with trivial cycles as nodes and hence homomorphisms must send trivial cycles to trivial cycles.

3.4   Interpretations in Traced Categories

Let us fix a traced symmetric monoidal category $\mathcal{C}$. We are mainly interested in the case of finite dimensional vector spaces and linear maps over a field, but it is natural to state the general case, and necessary if we want to say something about the classifying category built from networks.

Definition 5. *Let F be an S-sorted signature. Then an* interpretation μ *of F in $\mathcal{C}$ consists of the following data:*

- *an object $[\![s]\!]^\mu$ of $\mathcal{C}$ for each sort $s \in S$*
- *an arrow $[\![f]\!]^\mu : [\![ar_{in}(f)]\!]^\mu \to [\![ar_{out}(f)]\!]^\mu$ for each function symbol $f \in F$, while for $\bullet_s$ we put $[\![\bullet_s]\!]^\mu = Tr^{[\![s]\!]^\mu}(id_{[\![s]\!]^\mu})$*

where we define the interpretation of arities by $[\![\varepsilon]\!]^\mu = I$ and $[\![sw]\!]^\mu = [\![s]\!]^\mu \otimes [\![w]\!]^\mu$.

Definition 6. *Let F be an S-sorted signature and let μ be an interpretation of F. Then the value $[\![(X, \varphi, \pi)]\!]^\mu : [\![v]\!]^\mu \to [\![w]\!]^\mu$ of a network $(X, \varphi, \pi) : v \to w$ with respect to μ is defined to be the trace of:*

$$\left(\bigotimes_{x \in X} [\![ar_{out}(\varphi(x))]\!]^\mu \right) \otimes [\![v]\!]^\mu \xrightarrow{\hat{\pi}} \left(\bigotimes_{x \in X} [\![ar_{in}(\varphi(x))]\!]^\mu \right) \otimes [\![w]\!]^\mu$$
$$\xrightarrow{(\otimes [\![\varphi(x)]\!]^\mu) \otimes [\![w]\!]^\mu} \left(\bigotimes_{x \in X} [\![ar_{out}(\varphi(x))]\!]^\mu \right) \otimes [\![w]\!]^\mu$$

where $\hat{\pi}$ is the isomorphism induced by π.

Proposition 1. *If two networks are isomorphic, they have the same value.*

3.5   The Traced Monoidal Category of Networks

Fixing an S-sorted signature F, we now define several constructions on networks over F.

Definition 7. – Identity Networks. *The identity network on arity v is defined to be $(\emptyset, \emptyset, id) : v \to v$, where id is the identity permutation.*

– Sequential Composition of Networks. *For networks $N = (X, \varphi, \pi) : v \to w$ and $N' = (X', \varphi', \pi') : w \to u$, their sequential composition $N' \circ N : v \to u$ is the network $(X \uplus X', \varphi \uplus \varphi', \pi'') : v \to u$, where $(\varphi \uplus \varphi')(x) = \varphi(x)$ for $x \in X$ and $(\varphi \uplus \varphi')(y) = \varphi'(y)$ for $y \in X'$, and π'' sends (i) $p \in O_N$ to $\pi'(\pi(p))$ if $\pi(p) \in \mathbb{N}$, otherwise to $\pi(p)$, and (ii) $\langle y, j \rangle \in O_{N'}$ to $\pi'\langle y, j \rangle$.*

– Parallel Composition of Networks. *For networks $N = (X, \varphi, \pi) : v \to w$ and $N' = (X', \varphi', \pi') : v' \to w'$, their parallel composition $N \otimes N' : vv' \to ww'$ is the network $(X \uplus X', \varphi \uplus \varphi', \pi'') : vv' \to ww'$ where (i) $\pi''(p) = \pi(p)$ for $p \in O_N$, (ii) $\pi''(|v| + i) = |w| + \pi'(i)$ $(1 \le i \le |v'|)$ if $\pi'(i) \in \mathbb{N}$, otherwise $\pi''(|v| + i) = \pi'(i)$, and (iii) $\pi''\langle y, i \rangle = |w| + \pi'\langle y, i \rangle$ if $\pi'\langle y, i \rangle \in \mathbb{N}$, otherwise $\pi''\langle y, i \rangle = \pi'\langle y, i \rangle$.*

– Symmetry Networks. *The symmetry network on arities v and w is defined to be $(\emptyset, \emptyset, c_{|v|,|w|}) : vw \to wv$ where $c_{m,n}(i) = i + n$ for $1 \le i \le m$ and $c_{m,n}(m + i) = i$ for $1 \le i \le n$.*

– Traces of Networks. *The trace $Tr^s_{v,w}(N) : v \to w$ of $N = (X, \varphi, \pi) : vs \to ws$ is the network:*

   • $(X \uplus \{a\}, \varphi', \pi') : v \to w$ *if $\pi(|v| + 1) = |w| + 1$, where $\varphi'(x) = \varphi(x)$ for $x \in X$ and $\varphi'(a) = \bullet_s$, and $\pi' = \pi \setminus \{\langle |v| + 1, |w| + 1 \rangle\}$.*

   • $(X, \varphi, \pi') : v \to w$ *if $\pi(|v| + 1) \ne |w| + 1$, where $\pi'(p) = \pi(|v| + 1)$ if $p = \pi^{-1}(|w| + 1)$ and $\pi'(p) = \pi(p)$, otherwise..*

This definition is extended to non-primitive arities by setting $Tr^\varepsilon_{v,w}(N) = N$ for $N : v \to w$ and $Tr^{su}_{v,w}(N) = Tr^s_{v,w}(Tr^u_{vs,ws}(N))$ for $N : vsu \to wsu$.

Lemma 1. *The constructions above are well-defined on equivalence classes of networks up to network isomorphism.*

We can now introduce the traced symmetric monoidal category $\mathbf{Net}_{(S,F)}$. Its objects are the arities (elements of S^*) and an arrow from v to w is an equivalence class of networks over F with input arity v and output arity w, up to network isomorphism. Composition is given by sequential composition, and the identity arrows by the identity networks. The tensor of two objects is their concatenation and the tensor of two arrows is given by parallel composition; the symmetry maps are given by the symmetry networks. Finally, trace is given by the trace on networks. Using the above lemma it is now straightforward to show:

Proposition 2. $\mathbf{Net}_{(S,F)}$ *forms a traced strict symmetric monoidal category.*

3.6 $\mathbf{Net}_{(S,F)}$ as a Classifying Category

Just as in traditional functorial model theory, it is not hard to see that giving an interpretation of an S-sorted signature F in a traced symmetric monoidal category $\mathcal{C}$ is equivalent to giving a structure-preserving functor (traced functor) from $\mathbf{Net}_{(S,F)}$ to $\mathcal{C}$. This observation can be strengthened to be an equivalence of the category $\mathbf{Mod}((S, F), \mathcal{C})$ of interpretations of F in $\mathcal{C}$ and the category

TrMon(Net$_{(S,F)}$, $\mathcal{C}$) of traced functors from **Net**$_{(S,F)}$ to $\mathcal{C}$ and monoidal natural transformations, where we define a morphism between interpretations μ and μ' to be a family of arrows $h_s : [\![s]\!]^\mu \to [\![s]\!]^{\mu'}$ which commutes with the interpretations of function symbols, that is, for f with $ar_{in}(f) = s_1 \ldots s_m$ and $ar_{out}(f) = t_1 \ldots t_n$, the following diagram commutes:

$$
\begin{array}{ccc}
[\![s_1]\!]^\mu \otimes \left(\cdots \otimes [\![s_m]\!]^\mu \right) & \xrightarrow{\;[\![f]\!]^\mu\;} & [\![t_1]\!]^\mu \otimes \left(\cdots \otimes [\![t_n]\!]^\mu \right) \\
\Big\downarrow {\scriptstyle h_{s_1} \otimes (\cdots \otimes h_{s_m})} & & \Big\downarrow {\scriptstyle h_{t_1} \otimes (\cdots \otimes h_{t_n})} \\
[\![s_1]\!]^{\mu'} \otimes \left(\cdots \otimes [\![s_m]\!]^{\mu'} \right) & \xrightarrow[\;[\![f]\!]^{\mu'}\;]{} & [\![t_1]\!]^{\mu'} \otimes \left(\cdots \otimes [\![t_n]\!]^{\mu'} \right)
\end{array}
$$

Proposition 3. *There is an equivalence of categories:*

$$\mathbf{Mod}((S,F), \mathcal{C}) \simeq \mathbf{TrMon}(\mathbf{Net}_{(S,F)}, \mathcal{C})$$

Proof (Outline). Given an interpretation in a traced (possibly non-strict) symmetric monoidal category $\mathcal{C}$, we can extend it to a strong traced functor from **Net**$_{(S,F)}$ to $\mathcal{C}$. This also sends morphisms between interpretations to monoidal natural transformations, and we obtain a fully faithful functor from **Mod** $((S,F), \mathcal{C})$ to **TrMon(Net**$_{(S,F)}$, $\mathcal{C}$). In addition, given a strong traced functor from **Net**$_{(S,F)}$, we can construct an isomorphic strong traced functor which comes from an interpretation. □

4 Networks, Homomorphisms and Interpretations in Finite Dimensional Vector Spaces

We have seen that to give a strict traced functor from **Net**$_{(S,F)}$ to a traced symmetric monoidal category $\mathcal{C}$ is to give an interpretation of the signature (S,F) in $\mathcal{C}$. We are particularly interested in interpretations in **FinVect**$_k$, for various fields k; we call such interpretations interpretations *over* k. Proposition 1 gives us the soundness of such interpretations:

If two networks are isomorphic, they have the same value for all interpretations over any field k.

Our aim is to establish the converse when k has characteristic 0:

If two networks have the same value under all interpretations over k then they are isomorphic.

To this end a number of simplifying assumptions will prove convenient:

– We consider only the single-sorted case. This will involve no loss of generality, due to the following: any signature F has an associated single-sorted signature F_o obtained by identifying all its sorts; any network N over F then has an associated network N_o over F_o; and for any networks $N, N' : u \to v$ over F, if N_o and N'_o are isomorphic, so are N and N'. In the single-sorted case we identify arities with non-negative integers and write $\bullet$ for the (unique) function symbol for trivial cycles.

- In the single-sorted case, we consider only closed networks, those with no inputs and outputs and so of the form $N : 0 \to 0$. We will later reduce the case of non-closed networks to that of closed ones: introducing extra (dummy) function symbols $f_m : 0 \to m$ and $f^n : n \to 0$ for all $m, n > 0$, one has that two networks $N, N' : m \to n$ are isomorphic if and only if their compositions with (the networks consisting of) f_m and f^n are isomorphic.
- Finally, we consider only non-empty networks without trivial cycles, i.e., those which do not contain any $\bullet$-labelled node. The more general case will not present significant additional difficulties.

So, in the rest of this section, by a network we mean, unless otherwise stated, a non-empty closed network without trivial cycles over a single-sorted signature.

4.1   Basic Facts about Networks and Homomorphisms

We recall the definition of parallel composition (Definition 7) for *closed* networks $N = (X, \varphi, \pi)$ and $N' = (X', \varphi', \pi')$. The network $N \otimes N'$ is $(N \uplus N', \varphi'', \pi'')$ where:

- $\varphi''(x) = \varphi(x)$ for $x \in X$ and $\varphi''(y) = \varphi'(y)$ for $y \in X'$,
- $\pi''\langle x, i \rangle = \pi\langle x, i \rangle$ for $x \in X$ and $\pi''\langle y, i \rangle = \pi'\langle y, i \rangle$ for $y \in X'$.

For closed networks, parallel composition $N \otimes N'$ and sequential composition $N \circ N'$ agree. We also note that $N \otimes N'$ is the coproduct of N and N' in the category of networks and homomorphisms.

Definition 8. *Let x and x' be nodes in a network $N = (X, \varphi, \pi)$. They are directly connected, written $x \sim y$, if either $\pi\langle x, i \rangle = \langle x', j \rangle$ or $\pi\langle x', i \rangle = \langle x, j \rangle$, for some i and j. Connectedness (of nodes) is the equivalence relation generated by $\sim$.*

A non-empty equivalence class of nodes with respect to connectedness is called a connected component. *A network is* connected *if any two of its nodes are connected, i.e., if it is itself a connected component.*

In the following, we may refer to a network just by its set of nodes, leaving φ and π implicit. This convention is helpful as we are interested in decomposing a network into its connected components. We notice that a connected component is itself a (connected) network when equipped with the restrictions of φ and π. Each network X can be decomposed as:

$$X \cong X_1 \otimes \cdots \otimes X_n$$

where the X_i are the connected components of X.

We need some information on homomorphisms and connectedness. First, they clearly preserve connection, and so connectedness. Next:

Lemma 2. *Let $f : X \to Y$ be a homomorphism, and suppose that we have $f(x) = y \sim y'$. Then there is an x' such that $x \sim x'$ and $f(x') = y'$.*

We then have the following proposition:

Proposition 4. *Let $f : X \to Y$ be a homomorphism. For each connected component C of X, the image $f(C) \subseteq Y$ is a connected component of Y.*

Corollary 1. *Let $f : X \to Y$ be a homomorphism. If Y is connected, then f is a surjection.*

The following immediate consequence will be important later.

Corollary 2. *Let $f : X \to Y$ be a homomorphism and suppose that Y is connected and $|X| = |Y|$. Then f is an isomorphism.*

Lemma 3. *Let $f, g : X \to Y$ be homomorphisms. Suppose that $f(x) = g(x)$ and $x \sim x'$. Then $f(x') = g(x')$.*

This yields:

Proposition 5. *Let $f, g : X \to Y$ be homomorphisms. If X is connected and $f(x) = g(x)$ for some $x \in X$, then $f = g$.*

The following upper bound on the number of homomorphisms is a direct consequence of this proposition.

Corollary 3. *Let X and Y be networks, and suppose that X is connected. Then $|\hom(X, Y)| \le |Y|$.*

Proposition 6. *Let $f : X \to Y$ be a homomorphism. Then, for any $y \sim y'$ in Y, $|f^{-1}(y)| = |f^{-1}(y')|$.*

Proof. We may suppose, without loss of generality, that for some i and j, $\pi_Y \langle y, i \rangle = \langle y', j \rangle$. Then we may define a bijection $\theta : f^{-1}(y) \cong f^{-1}(y')$ by $\theta(x) = (\pi_X \langle x, i \rangle)_1$; its inverse is given by $\theta^{-1}(x') = (\pi_X^{-1} \langle x', j \rangle)_1$. $\qquad\square$

The following corollary is then immediate:

Corollary 4. *If $f : X \to Y$ is a homomorphism and Y is connected, then $|X|$ is a multiple of $|Y|$.*

4.2 Interpretations over a Field k

An interpretation μ of a (one-sorted) signature over a field k is specified by a vector space V^μ and a linear map $[\![f]\!]^\mu : [\![ar_{in}(f)]\!]^\mu \to [\![ar_{out}(f)]\!]^\mu$ for each function symbol f, where $[\![m]\!]^\mu = \underbrace{V^\mu \otimes \cdots \otimes V^\mu}_{m}$. Let X be a closed network over this signature, possibly empty or with trivial cycles. Its value with respect to the interpretation μ is then the trace of:

$$\bigotimes_{x \in X} [\![ar_{out}(\varphi(x))]\!]^\mu \xrightarrow{\hat\pi} \bigotimes_{x \in X} [\![ar_{in}(\varphi(x))]\!]^\mu \xrightarrow{\otimes_{x \in X} [\![\varphi(x)]\!]^\mu} \bigotimes_{x \in X} [\![ar_{out}(\varphi(x))]\!]^\mu$$

where $\hat{\pi}$ is the linear map induced by π, and for $\bullet$ we put $[\![\bullet]\!]^\mu = \dim V^\mu$.

Note that for any two closed networks X, Y over this signature we have that $[\![X \otimes Y]\!]^\mu = [\![X]\!]^\mu [\![Y]\!]^\mu$. It follows that the value of a network X with t trivial cycles and non-trivial connected components $X_1, \ldots, X_n$ is given by:

$$[\![X]\!]^\mu = d^t [\![X_1]\!]^\mu \cdots [\![X_n]\!]^\mu$$

where d is the dimension of the interpretation of the sort by μ.

Definition 9. *Let μ_1, μ_2 be two interpretations. The interpretation $\mu_1 + \mu_2$ is defined by:*

- $V^{\mu_1 + \mu_2} = V^{\mu_1} \oplus V^{\mu_2}$,
- $[\![f]\!]^{\mu_1 + \mu_2} \left(\bigotimes_{1 \leq i \leq ar_{in}(f)} v_i + u_i \right) = [\![f]\!]^{\mu_1} \left(\bigotimes_{1 \leq i \leq ar_{in}(f)} v_i \right) + [\![f]\!]^{\mu_2} \left(\bigotimes_{1 \leq i \leq ar_{in}(f)} u_i \right)$

  where the evident inclusions of $[\![m]\!]^{\mu_1}$ and $[\![m]\!]^{\mu_2}$ in $[\![m]\!]^{\mu_1 + \mu_2}$ have been omitted.

Proposition 7. *Let μ_1, μ_2 be two interpretations. If X is a connected network, then $[\![X]\!]^{\mu_1 + \mu_2} = [\![X]\!]^{\mu_1} + [\![X]\!]^{\mu_2}$.*

Proof. Let

$$m : \bigotimes_{x \in X} \bigotimes_{1 \leq j \leq ar_{out}(\varphi(x))} V^{\mu_1} \oplus V^{\mu_2} \longrightarrow \bigotimes_{x \in X} \bigotimes_{1 \leq j \leq ar_{out}(\varphi(x))} V^{\mu_1} \oplus V^{\mu_2}$$

be the linear map whose trace determines the value of X under $\mu_1 + \mu_2$. Also, let m_1, m_2 be the maps whose trace determines the value of X under μ_1 and μ_2 respectively. Suppose that $v = \bigotimes_{x \in X} \bigotimes_{1 \leq j \leq ar_{out}(\varphi(x))} v_{\langle x,j \rangle}$ is a basis vector such that $\langle v | m(v) \rangle \neq 0$. Since v is assumed to be a basis vector, we have that each $v_{\langle x,j \rangle}$ is either in V^{μ_1} or V^{μ_2}, and is a basis vector of the respective space. We claim that all the $v_{\langle x,j \rangle}$ must lie in the same space. First, we notice that for given x all the $v_{\pi_X^{-1}\langle x,i \rangle}$ for $i < ar_{in}x$ must lie in the same space, for otherwise $[\![\varphi(x)]\!]^{\mu_1 + \mu_2} \left(\bigotimes_i v_{\langle x,i \rangle} \right) = 0$ and hence $m(v) = 0$. Thus each x is associated to either V^{μ_1} or V^{μ_2}, and its directly connected nodes are also associated to the same space. Hence either $v \in V^{\mu_1}$ and $m(v) = m_1(v)$ or $v \in V^{\mu_2}$ and $m(v) = m_2(v)$. As the trace of m is obtained by summing up all such $\langle v | m(v) \rangle$, we have the required result. $\square$

4.3   The Counting Interpretation

Let us fix a field k. We now describe the key part of the proof: given a connected network X we define an interpretation $\mu(X, \lambda)$ over k which, in essence, counts the contribution of each function symbol in the network X.

Definition 10. *Let X be a connected network and $\lambda \in k \backslash \{0\}$ be a non-zero scalar. The interpretation $\mu(X, \lambda)$ is defined as follows:*

- *The (unique) sort 1 is interpreted as the vector space $V^{\mu(X,\lambda)}$ with basis the input ports of X, i.e., the set $\{\langle x, i \rangle \mid 1 \leq i \leq ar_{in}(\varphi(x))\}$. (Hence $\dim V^{\mu(X,\lambda)} = \sum_{x \in X} ar_{in}(\varphi(x))$.)*
- $[\![f]\!]^{\mu(X,\lambda)} : [ar_{in}(f)]^{\mu(X,\lambda)} \to [ar_{out}(f)]^{\mu(X,\lambda)}$ *is given by:*

$$[\![f]\!]^{\mu(X,\lambda)}\left(\bigotimes_{1 \leq i \leq ar_{in}(f)} p_i \right) = \lambda \sum_{\substack{x : \varphi(x) = f \\ p_i = \langle x, i \rangle}} \bigotimes_{1 \leq j \leq ar_{out}(f)} \pi \langle x, j \rangle$$

Notice that if $ar_{in}(f) > 0$ then the sum consists of at most one summand. In this case we have:

$$[\![f]\!]^{\mu(X,\lambda)}\left(\bigotimes_i p_i \right) = \begin{cases} \lambda \bigotimes_j \pi \langle x, j \rangle & \text{if } p_i = \langle x, i \rangle \text{ for all } i \\ 0 & \text{otherwise} \end{cases}$$

That is to say, $[\![f]\!]^{\mu(X,\lambda)}$ is non-zero if it is applied to the input of an f-labelled node in X and in this case returns the output of that node. The semantics of an input-less function symbol (a constant) is λ times the sum over all its outputs occurring in X. We also notice that all function symbols that do not actually occur in X receive zero meaning. If F contains a symbol f with $ar_{in}(f) = ar_{out}(f) = 0$ then, since X is connected, either X does not contain f-labelled nodes at all, hence $[\![f]\!]^{\mu(X,\lambda)} = 0$, or X consists of a single f-labelled node, in which case $V^\mu = k$ and $[\![f]\!]^{\mu(X,\lambda)} = \lambda$.

Theorem 1. *Let X and Y be networks, and assume that X is connected. Then, for any $\lambda \in k \backslash \{0\}$, we have:*

$$[\![Y]\!]^{\mu(X,\lambda)} = \lambda^{|Y|} |\hom(Y, X)|$$

Proof. Recall that $V^{\mu(X,\lambda)}$ is the vector space with basis vectors the input ports of X, i.e., the set $\{\langle x, i \rangle \mid 1 \leq i \leq ar_{in}(\varphi(x))\}$. Let

$$m : \bigotimes_{y \in Y} \bigotimes_{1 \leq j \leq ar_{out}(\varphi(y))} V^{\mu(X,\lambda)} \longrightarrow \bigotimes_{y \in Y} \bigotimes_{1 \leq j \leq ar_{out}(\varphi(y))} V^{\mu(X,\lambda)}$$

be the linear map so that $[\![Y]\!]^{\mu(X,\lambda)} = Tr(m)$. Unfolding the definition yields:

$$m\left(\bigotimes_{y \in Y} \bigotimes_{1 \leq j \leq ar_{out}(\varphi(y))} \langle x_{(y,j)}, i_{(y,j)} \rangle \right) = \lambda^{|Y|} \bigotimes_{y \in Y} \sum_x \bigotimes_{1 \leq j \leq ar_{out}(\varphi(y))} \pi_X \langle x, j \rangle$$

where the sum ranges over those $x \in X$ satisfying $\varphi_X(x) = \varphi_Y(y)$ and also $\langle x_{\pi_Y \langle y, i \rangle}, i_{\pi_Y \langle y, i \rangle} \rangle = \langle x, i \rangle$ for all $1 \leq i \leq ar_{in}(\varphi_Y(y))$.

Now the trace of m equals $\lambda^{|Y|}$ times the number of the basis vectors v of the space $\bigotimes_{y \in Y} \bigotimes_{1 \leq j \leq ar_{out}(\varphi(y))} V^{\mu(X,\lambda)}$ which occur in $m(v)$, i.e., for which $\langle v \mid m(v) \rangle = \lambda^{|Y|}$. We show that these basis vectors are in 1-1 correspondence with homomorphisms from Y to X. If $v = \bigotimes_{y \in Y} \bigotimes_{1 \leq j \leq ar_{out}(\varphi(y))} \langle x_{(y,j)}, i_{(y,j)} \rangle$

satisfies $\langle v \mid m(v) \rangle \neq 0$ then for each y the sum in $m(v)$ must contain a summand corresponding to v. More precisely:

$$
\begin{aligned}
&\forall y \in Y \exists x \in X \\
&\quad \varphi_Y(y) = \varphi_X(x) &\quad (a) \\
&\quad \forall i \ \langle x_{\pi_Y^{-1}\langle y,i\rangle}, i_{\pi_Y^{-1}\langle y,i\rangle}\rangle = \langle x,i\rangle &\quad (b) \\
&\quad \forall j \ \langle x_{\langle y,j\rangle}, i_{\langle y,j\rangle}\rangle = \pi\langle x,j\rangle &\quad (c)
\end{aligned}
$$

As explained above, either X is a singleton set or it does not contain function symbols with neither inputs nor outputs. In each case, we have that for each $y \in Y$ there exists a *unique* $x \in X$ satisfying (b) and (c). In the former case, there is only one x anyway; in the latter case either (b) or (c) is a nonempty conjunction and establishes uniqueness.

We have thus determined a function $f : Y \to X$ such that (b) and (c) hold with x replaced with $f(y)$. We claim that f is a homomorphism. Indeed, if $\pi_Y^{-1}\langle y,i\rangle = \langle y',j\rangle$ then by (b) we have $\langle f(y),i\rangle = \langle x_{\langle y',j\rangle}, i_{\langle y',j\rangle}\rangle$. On the other hand, (c) shows $\langle x_{\langle y',j\rangle}, i_{\langle y',j\rangle}\rangle = \pi_X\langle f(y'),j\rangle$, thus $\langle f(y),i\rangle = \pi_X\langle f(y'),j\rangle$ or $\pi_X^{-1}\langle f(y),i\rangle = \langle f(y'),j\rangle$ establishing homomorphism.

Conversely, if $f : Y \to X$ is a homomorphism, we define a basis vector $v = \bigotimes_{y\in Y} \bigotimes_{1\leq j\leq ar_{out}(\varphi(y))} \langle x_{\langle y,j\rangle}, i_{\langle y,j\rangle}\rangle$ by:

$$
\begin{cases}
x_{\langle y,j\rangle} = f(y') \\
i_{\langle y,j\rangle} = i
\end{cases}
\quad \text{when } \pi_Y\langle y,j\rangle = \langle y',i\rangle \tag{1}
$$

Now, towards showing (a), (b), (c) above, given $y \in Y$ we put $x = f(y)$. Condition (a) follows directly from the homomorphism property; condition (b) is direct from the definition of $\langle x_{\langle y,j\rangle}, i_{\langle y,j\rangle}\rangle$; for condition (c), we assume $\pi_Y\langle y,j\rangle = \langle y',i\rangle$ hence $\pi_X\langle f(y),j\rangle = \langle f(y'),i\rangle = \langle x_{\langle y,j\rangle}, i_{\langle y,j\rangle}\rangle$ using the homomorphism property and the definition of $\langle x_{\langle y,j\rangle}, i_{\langle y,j\rangle}\rangle$.

It is obvious that going back and forth starting with a homomorphism f yields that homomorphism back. To show the converse, assume that we are given a basic vector determined by a family $\{\langle \hat{x}_{\langle y,j\rangle}, \hat{i}_{\langle y,j\rangle}\rangle\}_{\langle y,j\rangle}$. We define a homomorphism $f : Y \to X$ by letting $f(y)$ be the unique x satisfying conditions (a), (b), (c) above. We then define another basic vector $\{\langle x_{\langle y,j\rangle}, i_{\langle y,j\rangle}\rangle\}_{\langle y,j\rangle}$ by (1).

Given $y \in Y$ and $1 \leq j \leq ar_{out}(\varphi(y))$ we have:

$$
\langle \hat{x}_{\langle y,j\rangle}, \hat{i}_{\langle y,j\rangle}\rangle = \pi_X\langle f(y),j\rangle
$$

by condition (b) above. On the other hand, if $\pi_Y\langle y,j\rangle = \langle y',i\rangle$ then:

$$
\pi_X\langle f(y),j\rangle = \langle f(y'),i\rangle = \langle x_{\langle y,j\rangle}, i_{\langle y,j\rangle}\rangle
$$

by the homomorphism property and (1), thus:

$$
\langle \hat{x}_{\langle y,j\rangle}, \hat{i}_{\langle y,j\rangle}\rangle = \langle x_{\langle y,j\rangle}, i_{\langle y,j\rangle}\rangle
$$

as required. □

Theorem 1 and Proposition 7 immediately yield:

Theorem 2. *Let X be a network with connected components $X_1, \ldots, X_n$, let λ_1, $\ldots, \lambda_n$ be non-zero scalars, and let Y be a network with connected components $Y_1, \ldots, Y_m$. Then we have:*

$$[\![Y]\!]^{\sum_{i=1}^{n} \mu(X_i, \lambda_i)} = \prod_{j=1}^{m} \sum_{i=1}^{n} \lambda_i^{|Y_j|} |\hom(Y_j, X_i)|$$

5   Completeness Results

We begin by considering closed networks over a one-sorted signature. In the following definition we assume a standard enumeration of (the isomorphism classes) of the connected non-empty and non-trivial such networks.

Definition 11. *Let X be a closed network over a one-sorted signature, and suppose its non-trivial connected components are $X_1, \ldots, X_n$ ($n \geq 0$), and let $\lambda_1, \ldots, \lambda_n$ be distinct variables (taken from some standard enumeration of variables). Then the interpretation μ_X over $\mathbb{Q}(\lambda_1, \ldots, \lambda_n)$ is given by:*

$$\mu_X = \mu(X_1, \lambda_1) + \cdots + \mu(X_n, \lambda_n) + \zeta_2$$

where ζ_2 is the interpretation interpreting 1 by a two-dimensional space and assigning all function symbols the value 0.

Now if Y is any closed network over the same signature as X, with non-trivial connected components $Y_1, \ldots, Y_m$ and with t trivial cycles, we have by the above remarks on the interpretation of such networks, Proposition 7 and Theorem 2 that:

$$[\![Y]\!]^{\mu_X} = d^t \prod_{j=1}^{m} \sum_{i=1}^{n} \lambda_i^{|Y_j|} |\hom(Y_j, X_i)| \tag{2}$$

where $d \geq 2$ is the dimension of the interpretation of 1 by μ_X. Note that this is a polynomial in $\lambda_1, \ldots, \lambda_n$ with positive integer coefficients, and non-zero in case $n > 0$ and X and Y have the same non-trivial connected components up to isomorphism. Writing $\deg(\lambda_i, [\![Y]\!]^{\mu_X})$ for the largest exponent of λ_i in $[\![Y]\!]^{\mu_X}$, we have:

$$\deg(\lambda_i, [\![Y]\!]^{\mu_X}) = \sum_{j : \hom(Y_j, X_i) \neq \emptyset} |Y_j| \tag{3}$$

where $Y_1, \ldots, Y_m$ are the components of Y.

Lemma 4. *Let X and Y be closed networks over a one-sorted signature, at least one of which has a non-trivial connected component. If*

$$[\![X]\!]^{\mu_X} = [\![Y]\!]^{\mu_X}$$

and

$$[X]^{\mu_Y} = [Y]^{\mu_Y}$$

then X and Y are isomorphic.

Proof. Let $X_1, \ldots, X_m$ and $Y_1, \ldots, Y_n$ be the standard enumerations of the non-trivial connected components of X and Y, respectively.

Let U be a connected component in X or Y. The height of U is defined as the length of the longest sequence of homomorphisms

$$U_0 \xrightarrow{f_0} U_1 \xrightarrow{f_1} U_2 \longrightarrow \ldots \xrightarrow{f_{k-1}} U_k = U$$

where U_i are connected components in X or Y, and none of the f_i are isomorphisms. Notice that the height is well-defined by Corollaries 1 and 2.

The multiplicity of component U in X (or Y) is defined as the number of isomorphic copies of U in X (or Y). We show by course-of-values induction on h that each component of X or Y of height h has the same multiplicity in X and in Y.

So assume that U is a connected component of height h and that components of height less than h have equal multiplicities in X and Y. Let us write x and y for the multiplicity of U in X and Y respectively. By the definition of height we have:

$$\sum_{i:\hom(X_i,U)\neq\emptyset} |X_i| = x|U| + \sum_{i:\hom(X_i,U)\neq\emptyset \wedge \mathrm{height}(X_i)<h} |X_i|$$

and:

$$\sum_{j:\hom(Y_j,U)\neq\emptyset} |Y_i| = y|U| + \sum_{i:\hom(Y_i,U)\neq\emptyset \wedge \mathrm{height}(Y_i)<h} |Y_i|$$

Now, supposing without loss of generality that U occurs in X as X_1, we conclude by equation 3 that:

$$\sum_{i:\hom(X_i,U)\neq\emptyset} |X_i| = \deg(\lambda_1, [X]^{\mu_X}) = \deg(\lambda_1, [Y]^{\mu_X}) = \sum_{j:\hom(Y_j,U)\neq\emptyset} |Y_i|$$

Combining this with the induction hypothesis shows $x|U| = y|U|$, hence $x = y$.

So X and Y have the same non-trivial connected components, up to isomorphism. So, as $[X]^{\mu_X} = [Y]^{\mu_X}$, we see by equation 2 above and the following remark that they have the same number of trivial cycles, which concludes the proof.   □

Theorem 3. *If two networks over a given signature have equal value under all interpretations over fields of the form $\mathbb{Q}(\lambda_1, \ldots, \lambda_n)$ then they are isomorphic.*

Proof. We have already described how the general case can be reduced in turn to that of one-sorted signatures and then to that of closed such networks. The previous lemma deals with all such cases except the trivial one where both networks consist only of trivial cycles.   □

In order to strengthen our completeness result to fields of characteristic 0, we encode polynomials with positive integer coefficients as natural numbers:

Proposition 8. *Let d and C be positive integers. There exist natural numbers $k_1, \ldots, k_n$ such that for any polynomials $p, q \in \mathbb{N}[\lambda_1, \ldots, \lambda_n]$ with total degree less or equal to d and coefficients smaller than C we have:*

$$p = q \iff p(k_1, \ldots, k_n) = q(k_1, \ldots, k_n)$$

We then have:

Theorem 4. *Let k be a field of characteristic 0. If two networks over a given signature have equal value under all interpretations over k then they are isomorphic.*

A number of natural questions arise on considering this theorem. As regards generalisations, we do not know if the corresponding result is true for any field of positive characteristic. Nevertheless, a small refinement of our proof yields the following weaker result:

Theorem 5. *If two networks have equal value under all interpretations over all finite fields then they are isomorphic.*

Proof. For this, one makes use of the fact that for positive integers d, C there always exists a finite field k and $l_1, \ldots, l_n \in k$ such that for any two polynomials $p, q \in \mathbb{N}[\lambda_1, \ldots, \lambda_n]$ with total degree less or equal to d and coefficients smaller than C one has $p = q$ iff $p(l_1, \ldots, l_n) = q(l_1, \ldots, l_n)$ in k, and then simply proceeds as in the proof of Theorem 4, taking a finite field with characteristic large enough so that no undesired cancellations occur. □

One may also ask if Theorem 4 can be strengthened. Perhaps there is a uniform bound on the dimensions of the vector spaces needed for completeness. Alternatively there may be a result similar to those of Statman for the simply typed λ-calculus [13]. This might associate to each network N a bound on the dimensions of the vector spaces needed to decide whether any other network is isomorphic to N; there may even a be single interpretation such that another network is isomorphism to N iff it has the same value as N in that interpretation.

5.1   Completeness for Compact Closed Categories

The category **FinVect**$_k$ is not only traced symmetric monoidal but also compact closed. So it is natural to ask if our completeness result also holds for compact closed categories. This is indeed the case: it is a corollary of the result for the traced case and the structure theorem of Joyal, Street and Verity [6].

As noted before, every compact closed category has a unique trace. The structure theorem says that the forgetful 2-functor from the 2-category **CompCat** of compact closed categories to the 2-category **TrMon** of traced symmetric monoidal categories has a left biadjoint whose unit is fully faithful. More concretely, given a traced symmetric monoidal category $\mathcal{C}$, there is a compact closed category **Int** $\mathcal{C}$ whose objects are pairs of objects of $\mathcal{C}$ and whose arrows from (A^+, A^-) to (B^+, B^-) are the arrows from $A^+ \otimes B^-$ ro $B^+ \otimes A^-$ in $\mathcal{C}$. The identity arrow on (A^+, A^-) is $id_{A^+ \otimes A^-}$, and the composition of $f : (A^+, A^-) \to (B^+, B^-)$ with $g : (B^+, B^-) \to (C^+, C^-)$ is given by:

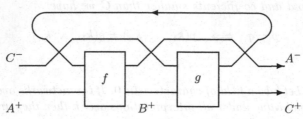

As regards the compact closed structure, the interested reader is referred to [6] (but our symmetric case is much simpler than the original braided case). In the case $\mathcal{C} = \mathbf{Net}_{(S,F)}$, we regard **Int** $\mathbf{Net}_{(S,F)}$ as a category of 'bi-directional networks' modulo isomorphism, where a bi-directional network from (v^+, v^-) to (w^+, w^-) is just a network from $v^+ w^-$ to $w^+ v^-$.

There is a fully faithful strong traced functor $N : \mathcal{C} \to$ **Int** $\mathcal{C}$ sending A to (A, I). Furthermore, for any compact closed category $\mathcal{D}$ and any strong traced functor $\mathcal{F} : \mathcal{C} \to \mathcal{D}$, there is a unique (up to isomorphism) strong monoidal functor $\bar{\mathcal{F}} :$ **Int** $\mathcal{C} \to \mathcal{D}$ such that $\bar{\mathcal{F}} \circ N$ is isomorphic to $\mathcal{F}$; explicitly, $\bar{\mathcal{F}}$ sends (A^+, A^-) to $\mathcal{F}A^+ \otimes (\mathcal{F}A^-)^*$, and $f : (A^+, A^-) \to (B^+, B^-)$ to:

$$
\mathcal{F}A^+ \otimes (\mathcal{F}A^-)^* \xrightarrow{id \otimes \eta \otimes id} \mathcal{F}A^+ \otimes \mathcal{F}B^- \otimes (\mathcal{F}B^-)^* \otimes (\mathcal{F}A^-)^*
$$
$$
\xrightarrow{\simeq} \mathcal{F}(A^+ \otimes B^-) \otimes (\mathcal{F}B^-)^* \otimes (\mathcal{F}A^-)^*
$$
$$
\xrightarrow{(\mathcal{F}f) \otimes id} \mathcal{F}(B^+ \otimes A^-) \otimes (\mathcal{F}B^-)^* \otimes (\mathcal{F}A^-)^*
$$
$$
\xrightarrow{\simeq} \mathcal{F}B^+ \otimes (\mathcal{F}B^-)^* \otimes (\mathcal{F}A^-)^* \otimes \mathcal{F}A^-
$$
$$
\xrightarrow{id \otimes \varepsilon} \mathcal{F}B^+ \otimes (\mathcal{F}B^-)^*
$$

In particular, **TrMon**$(\mathcal{C}, \mathcal{D})$ is equivalent to **CompCat**(**Int** $\mathcal{C}, \mathcal{D}$).

We can routinely define the notion of interpretations of signatures in a compact closed category $\mathcal{D}$ and morphisms between them, but this is the same as giving interpretations of signatures in $\mathcal{D}$ regarded as a traced monoidal category and morphisms between them. From this, we note that **Int** $\mathbf{Net}_{(S,F)}$ is the free compact closed category over the signature (S, F) because, for any compact closed $\mathcal{D}$, we have the following equivalences:

$$
\mathbf{Mod}((S, F), \mathcal{D}) \simeq \mathbf{TrMon}(\mathbf{Net}_{(S,F)}, \mathcal{D}) \simeq \mathbf{CompCat}(\mathbf{Int}\,\mathbf{Net}_{(S,F)}, \mathcal{D})
$$

So we can speak of the value of a bidirectional net given an interpretation of its signature over k, i.e., in $\mathbf{FinVect}_k$: one simply applies the functor obtained from the interpretation by the above chain of equivalences to the isomorphism class of the net.

Corollary 5. *Let k be a field of characteristic 0. If two bidirectional nets have equal value under all interpretations over k then they are isomorphic.*

For the proof, one uses the definition of $\overline{\mathcal{F}}$ to reduce the question to the case of $\mathbf{Net}_{(S,F)}$ and the result is then immediate from Theorem 4.

Acknowledgements

Hasegawa acknowledges the support of the Japanese Ministry of Education, Culture, Sports, Science and Technology, Grant-in-Aid for Young Scientists (B) 17700013. Plotkin acknowledges the support of a Royal Society-Wolfson award.

References

1. Bloom, S., Ésik, Z.: Iteration Theories, EATCS Monographs on Theoretical Computer Science. Springer, Heidelberg (1993)
2. Hasegawa, M.: Recursion from cyclic sharing: traced monoidal categories and models of cyclic lambda calculi. In: de Groote, P., Hindley, J.R. (eds.) TLCA 1997. LNCS, vol. 1210, pp. 196–213. Springer, Heidelberg (1997)
3. Hasegawa, M.: Models of Sharing Graphs: A Categorical Semantics of let and letrec. Distinguished Dissertation Series. Springer, Heidelberg (1999), also available as Ph.D. thesis ECS-LFCS-97-360, University of Edinburgh (1997)
4. Hyland, M., Power, A.J.: Symmetric monoidal sketches and categories of wirings. Electr. Notes Theor. Comput. Sci 100, 31–46 (2004)
5. Joyal, A., Street, R.: Braided tensor categories. Adv. Math. 102, 20–78 (1993)
6. Joyal, A., Street, R., Verity, D.: Traced monoidal categories. Math. Proc. Cambridge Phil. Soc. 119(3), 447–468 (1996)
7. Mac Lane, S.: Categories for the Working Mathematician. Springer, Heidelberg (1971)
8. Pardo, D., Rabinovich, A.M., Trakhtenbrot, B.A.: Synchronous circuits over continuous time: feedback, reliability and completeness. Fundam. Inform. 62(1), 123–137 (2004)
9. Rabinovich, A.M., Trakhtenbrot, B.A.: Nets and data flow interpreters. In: Proc. Fourth Symp. on Logic in Computer Science, pp. 164–174. IEEE Computer Society Press, Washington (1989)
10. Simpson, A.K.: Categorical completeness results for the simply-typed lambda-calculus. In: Dezani-Ciancaglini, M., Plotkin, G. (eds.) TLCA 1995. LNCS, vol. 902, pp. 414–427. Springer, Heidelberg (1995)
11. Simpson, A.K., Plotkin, G.: Complete axioms for categorical fixed-point operators. In: Proc. Fifteenth Symp. on Logic in Computer Science, pp. 30–41. IEEE Computer Society Press, Washington (2000)
12. Soloviev, S.V.: Proof of a conjecture of S. Mac Lane. Ann. Pure Appl. Logic 90, 101–162 (1997)
13. Statman, R.: Completeness, invariance, and definability. J. Symbolic Logic 47, 17–26 (1982)
14. Ştefănescu, G.: Network Algebra. Series in Discrete Mathematics and Theoretical Computer Science. Springer, Heidelberg (2000)
15. Trakhtenbrot, B.A.: On operators, realizable in logical nets. Doklady AN SSSR (Proceedings of the USSR Academy of Sciences) 112(6), 1005–1007 (1957)
16. Trakhtenbrot, B.A.: On the power of compositional proofs for nets: relationships between completeness and modularity. Fundam. Inform. 30(1), 83–95 (1997)
17. Yetter, D.N.: Functorial Knot Theory. World Scientific, Singapore (2001)

Tree Automata over Infinite Alphabets

Michael Kaminski and Tony Tan

Department of Computer Science, Technion – Israel Institute of Technology,
Haifa 32000, Israel
{kaminski,tantony}@cs.technion.ac.il

Dedicated to Boris (Boaz) Trakhtenbrot
on the occasion of his 85th birthday.

Abstract. A number of models of computation on trees labeled with symbols from an *infinite* alphabet is considered. We study closure and decision properties of each of the models and compare their computation power.

1 Introduction

In recent years a new application area for regular word and tree languages has evolved during one of the most important developments in World Wide Web (WWW) – the emergence of the *Extensible Markup Language* (XML). For many purposes, XML documents are modeled as labeled finite trees, where the *finite* set of labels corresponds to the set of element names allowed in the document. Thus, concepts from regular word and tree languages became important in XML research; see [1,8,11,9,12,14,18].

However, this abstraction ignores an important aspect of XML – the presence of *attributes* attached to the leaves of trees. Since attributes may assume values from an *infinite* set, modeling XML documents by trees over a finite alphabet is not adequate in any scenario. Therefore, a more natural way to model XML documents in this setting is to allow, besides the finite set of element labels, an *infinite* set of possible data values. Consequently, there is a need to extend the notion of regular and tree languages in such a way that, on one hand, as many as possible settings involving attributes can be captured and (most of) the desirable properties of the language class are retained on the other.

In this paper we extend *finite-memory automata*; see [2,5,6], to *tree* automata over *infinite alphabets*.

It should be noted, however, that there is a different model of computation over infinite alphabets, called *pebble automata over infinite alphabets*; see [8,10,13]. They are "orthogonal" to finite-memory automata. Also, the *tree walking* automata with pebbles ([8]) naturally extend to infinite alphabets for typechecking purposes. In our opinion, all these automata are inappropriate for modeling XML. This is because the tree languages they accept lack basic decision properties, though, unlike the language considered in this paper, they are closed under *complement*. However, for practical purposes it is very common to

A. Avron et al. (Eds.): Trakhtenbrot/Festschrift, LNCS 4800, pp. 386–423, 2008.

ask whether an XML scheme admits even one document. Since the emptiness problem for pebble automata is undecidable, using these automata for modeling XML documents is, at least, arguable.

In our extension of finite-memory automata to trees *each* head scanning a symbol at the tree node is equipped with a finite number of registers. This number is the same for all heads of the automaton. There are two ways to scan an input tree: *top-down* from the root to the leaves; see [15], and *bottom-up* from the leaves to the root; see [4,17].[1] When moving top-down from a parent node to its children, the automaton *splits* the head and the corresponding set of registers; and when moving bottom-up from the child nodes to their parent node, the automaton replaces the two[2] heads by one and *merges* the corresponding set of registers by forgetting some of their contents. Whereas the definition of a top-down finite-memory tree automata is very natural: the split results in two heads each carrying a set of registers with the same contents, the definition of the bottom-up one is less obvious, because it requires an automaton to merge two sets of registers whose contents may be quite different and even disjoint.

In addition, there are (at least) two possibilities of updating the content of the automaton registers. One possibility is to replace the content of one of the registers with the currently scanned *new* input symbol, as was done in [5,6], and the other is to replace the content of one of the registers with any *new* symbol from the infinite input alphabet, as was done in [2]. That is, in the latter case, the automaton does not necessarily have to arrive to the symbol in order to store it in its registers. Such ability will be referred to as a *nondeterministic reassignment*.

The above two distinctions result in four models of finite-memory tree automata which we consider in our paper. It appears that both top-down and bottom-up models with deterministic reassignment (the first possibility) are not strong enough for modeling XML. The former cannot even accept the tree language consisting of all trees having two different leaves labeled with the same symbol; see Example 2,[3] whereas the latter cannot accept a very simple tree language consisting of all trees whose root label differs from the labels of all other nodes. In fact, unlike in the case of a finite alphabet, the computation powers of these two models are incomparable.

In contrast with the above, the computation models with nondeterministic reassignment, which are stronger than those with the deterministic one, seem to be appropriate for modeling XML. We show that top-down and bottom-up tree automata with nondeterministic reassignment have the same computation power (which indicates that the definition is robust) and are proper extensions

---

[1] Recall that in Computer Science trees grow top-down.

[2] This paper deals with binary trees, only. Since finite branching trees can be encoded by binary trees in a standard manner, our computation models naturally generalize to unranked trees.

[3] For some purposes, an XML document might require certain integrity constraints. For example, it might require that the value at a certain position of a document also occurs (or differs from the value) at some other position.

of the deterministic reassignment models. In particular, the tree languages from the above two examples are accepted by tree automata with nondeterministic reassignment.

Also, we establish some closure and decision properties of tree languages accepted by tree automata with deterministic and nondeterministic reassignment and show how these tree languages are related to *context-free languages over infinite alphabets* introduced in [2]. This relationship is of importance, because the latter are tightly connected to *Document Type Definitions* which define XML documents.

The paper is organized as follows. In the next section we give a rough review of XML concepts. Section 3 contains the notation used throughout this paper. In Sect. 4 and 5 we introduce the tree automata with deterministic and nondeterministic reassignment, respectively. Section 6 deals with decision problems related to our models of computation. In Sect. 7 we establish a relationship between the tree languages defined in this paper and the context-free languages introduced in [2]. Finally, Appendices A and B contain the proof of equivalence of top-down and bottom-up tree automata with nondeterministic reassignment and Appendix C summarizes the closure properties of tree languages accepted by our tree automata.

2   Basic Notions of XML

In this section we briefly sketch the core idea of XML and its connection to tree languages. This is done via an example below. Readers interested in the details are referred to [16] or [19].

Consider the following XML document that displays information about a factory.

```
<factory name = "super">
    <section name = "productions">
        <product id = 011>
            <No.> 1 </No.>
            <label> notebook </label>
        </product>
        <product id = 294>
            <No.> 2 </No.>
            <label> pencil </label>
        </product>
    </section>
    <section name = "advertisement">
        <product id = 011>
            <No.> 1 </No.>
            <label> notebook </label>
        </product>
    </section>
</factory>
```

Like in the case of HTML, the building blocks of XML are elements which are delimited by the start- and end-tags. A start-tag of a `product`-element, for example, is `<product>` and the end-tag is `</product>`. Everything in between `<product>` and `</product>` constitutes a `product`-element.

An element can be nested into another one. For example, the element `<No.>` `1 </No.>` is a subelement of the outer `product`-element. Elements may also have attributes. These are name value pairs separated by the equality sign. For example, `<product id = 011>` indicates that the value of the `id` attribute of that particular `product`-element is 011.

An XML document can be viewed as a tree in a natural way. Fig. 1 below shows the above XML document in form of a tree.

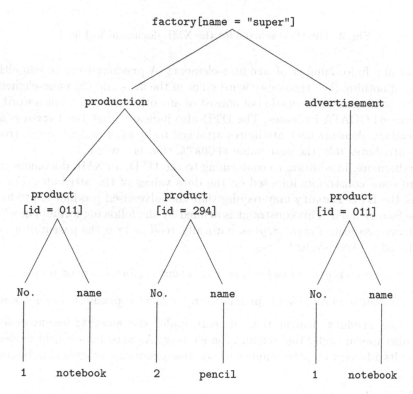

Fig. 1. A tree representation of an XML document

XML documents can be defined by *Document Type Definitions* (DTDs). These are, basically, *extended context-free grammars* (cf. context-free grammars over infinite alphabets in Sect. 7), i.e., context-free grammars with regular expressions at the right-hand sides. For example, the DTD in Fig. 2 defines the scheme of an XML document for a "factory."

This DTD specifies that the element `factory` is the outermost and consists of the `production`- and the `advertisement`-elements. Each of them, in turn,

```
<!DOCTYPE factory [
        <!ELEMENT factory (production, advertisement)>
        <!ELEMENT production (product)*>
        <!ELEMENT advertisement (product)*>
        <!ELEMENT product (No.,name) | ε >
        <!ELEMENT No. (#PCDATA)>
        <!ELEMENT name (#PCDATA)>
        <!ATTLIST factory name #PCDATA>
        <!ATTLIST product id #PCDATA>
]>
```

Fig. 2. The DTD scheme for the XML document in Fig. 1

consists of a finite number of **product**-elements. A **product**-element can either consist of nothing (i.e., the empty word ϵ), or of the **No.**- and the **name**-elements. The **No.**- and **name**-elements do not consist of any other elements, but a word, as the term #PCDATA indicates. The DTD also indicates that the **factory**- and the **product**-elements have attributes attached to it: **name** and **id**, respectively. These attributes take the data value #PCDATA, that is a word.

Furthermore, in addition to conforming to the DTD, an XML document may require some constraints imposed on the data values of the attributes. For example, the DTD **factory** may require that the advertised product is produced by the factory itself. This constraint is defined by the following formula in which the binary predicate $Parent(x, y)$ is, naturally, read as "x is the parent of y" and y is the advertised product.

$$\exists x(Parent(x, y) \wedge x.label = \text{advertisement} \wedge y.label = \text{product}) \rightarrow$$

$$\exists u \exists v(Parent(u, v) \wedge u.label = \text{production} \wedge v.label = \text{product} \wedge v.id = y.id)$$

So, any **product**-element that appears under the **advertisement**-element must also appear under the **production**-element. As **product**-element is identified by its **id**-attribute, the number of possible **product**-elements is unbound.

3   Notation

We fix an infinite alphabet Σ not containing # that is reserved to denote an empty register. For a word $w = w_1 w_2 \cdots w_r$ over $\Sigma \cup \{\#\}$, we define the *content* of w, denoted $[w]$, by $[w] = \{w_j \neq \# : j = 1, 2, \ldots, r\}$. That is, $[w]$ consists of all symbols of Σ which appear in w.

A word $w_1 \cdots w_r$ such that $w_i = w_j$ and $i \neq j$ imply $w_i = \#$ is called an *assignment*. That is, an assignment is a word over $\Sigma \cup \{\#\}$ where each symbol from Σ appears at most one time. Assignments represent the contents of the registers of an automaton: the symbol in the ith register is w_i. If $w_i = \#$, then the ith register is empty. The set of all assignments of length r is denoted by $\Sigma^r_{\neq}$.

A set of words $T \subseteq \{0,1\}^*$ is called a (binary) *tree* if it satisfies the following two conditions.

- For each $n \in T$ and each prefix n' of n, $n' \in T$. That is, T is *prefix closed*.
- If $n \in T$, then either both or none of $\{n0, n1\}$ are in T.

We write $n_1 \preccurlyeq n_2$, if n_1 is a prefix of n_2.

Elements of a tree are called *nodes*. The *root* of the tree is the empty word ϵ. A node n of a tree T is called a *leaf*, if neither of $\{n0, n1\}$ belongs to T. If n is not a leaf, then nodes $n0$ and $n1$ are called the *left* and *right children* of n, respectively, and n is called the *parent* of both $n0$ and $n1$. A node n_1 is an ancestor of n_2, or n_2 is a descendant of n_1, if $n_1 \preccurlyeq n_2$.

The hight of a node n in a tree T is the length of n. The root ϵ is of hight 0. The hight of a tree T is the hight of the longest node of T.

A Σ-*tree* is a map σ from a tree into Σ. The set of all Σ-trees is denoted by $\mathcal{T}(\Sigma)$. We denote by $dom(\sigma) = T$, the domain of $\sigma : T \to \Sigma$.

Finally, let $\sigma : T \to \Sigma$ be a Σ-tree and let $n_1, n_2, \ldots, n_m$ be the list of all leaf nodes of T in the lexicographical order. The Σ-word $\sigma(n_1)\sigma(n_2)\cdots\sigma(n_m)$ is called the *frontier* of σ and is denoted by $\ell(\sigma)$.

4 Tree Automata with Deterministic Reassignment

In this section we define the tree version of finite memory automata (FMA) introduce in [6], both the top-down and bottom-up cases. We start with the top-down case, which is a straightforward extension of the ordinary word FMA.

4.1 Top-Down Finite-Memory Automata with Deterministic Reassignment

Definition 1. (Cf. [6, Definition 1].) *A top-down finite-memory automaton with deterministic reassignment ($\downarrow$-FMA) is a system $A = \langle S, s_0, \boldsymbol{u}, \rho, \mu, F \rangle$, where*

- *S is finite set of states.*
- *$s_0 \in S$ is the initial state.*
- *$\boldsymbol{u} = u_1 u_2 \cdots u_r \in \Sigma^r_{\neq}$ is the initial assignment to the r registers of A.*
- *$\rho : S \to \{1, 2, \ldots, r\}$ is a function from S into $\{1, 2, \ldots, r\}$ called the deterministic reassignment. The intuitive meaning of ρ is as follows. If A is in a state p and the input symbol appears in no register, then the automaton reassigns the $\rho(p)$th register with the input symbol.*
- *$\mu \subseteq S \times \{1, 2, \ldots, r\} \times S^2$ is the transition relation, whose elements are called transitions and are also written in the form*

$$(p, i) \to (p_0, p_1),$$

where $p, p_0, p_1 \in S$ and $i \in \{1, 2, \ldots, r\}$. Intuitively, in a Σ-tree σ the transition $(p, i) \rightarrow (p_0, p_1)$ "applies" at a node n if n is labeled with the state p and the content of the ith register is $\sigma(n)$. Subsequently, the children $n0$ and $n1$ of n are labeled with the states p_0 and p_1, respectively.

- $F \subseteq S \times \{1, 2, \ldots, r\}$ is the set of final relations.

Like in the case of FMA, an actual state of A is a state of S together with the content of all registers. That is, A has infinitely many states which are pairs $(p, \boldsymbol{w})$, where $p \in S$ and $\boldsymbol{w} \in \Sigma^{r\neq}$ is the content of the registers of A. These are called *configurations* of A. The set of all configurations of A is denoted by S^c. The pair $(s_0, \boldsymbol{u})$, denoted s_0^c, is called the *initial* configuration.

The transition relation μ induces the following relation μ^c on $S^c \times \Sigma \times (S^c \times S^c)$, whose elements are written in the form

$$(p, \boldsymbol{w}), \sigma \rightarrow (p_0, \boldsymbol{w}_0), (p_1, \boldsymbol{w}_1),$$

where $p, p_0, p_1 \in S$ and $\boldsymbol{w}, \boldsymbol{w}_0, \boldsymbol{w}_1 \in \Sigma^{r\neq}$.

Let $\boldsymbol{w} = w_1 w_2 \cdots w_r$, $\boldsymbol{w}_0 = w_{0,1} w_{0,2} \cdots w_{0,r}$, and $\boldsymbol{w}_1 = w_{1,1} w_{1,2} \cdots w_{1,r}$. Then $(p, \boldsymbol{w}), \sigma \rightarrow (p_0, \boldsymbol{w}_0), (p_1, \boldsymbol{w}_1)$ belongs to μ^c if and only if the following conditions are satisfied.

- If $\sigma = w_i \in [\boldsymbol{w}]$, then $\boldsymbol{w}_0 = \boldsymbol{w}_1 = \boldsymbol{w}$ and $(p, i, (p_0, p_1)) \in \mu$.
- If $\sigma \notin [\boldsymbol{w}]$, then $w_{0,\rho(p)} = w_{1,\rho(p)} = \sigma$, $w_{0,i} = w_{1,i} = w_i$ for each $i \neq \rho(p)$, and $(p, \rho(p), (p_0, p_1)) \in \mu$.

The set of final relations F defines the set of *final "Σ-relations"* F^c. A pair $((p, \boldsymbol{w}), \sigma) \in F^c$ if the following holds.

- If $\sigma = w_i$, then $(p, i) \in F$; and
- if $\sigma \notin [w_1 w_2 \cdots w_r]$, then $(p, \rho(p)) \in F$.

A *run* of A on a Σ-tree $\sigma : T \rightarrow \Sigma$ is a mapping $R : T \rightarrow S^c$ such that

- $R(\epsilon) = s_0^c$ (recall that $s_0^c = (s_0, \boldsymbol{u})$ is the initial configuration of A) and
- for each non-leaf node $n \in T$, $(R(n), \sigma(n)) \rightarrow (R(n0), R(n1)) \in \mu^c$.

We say that A *accepts* a Σ-tree $\sigma : T \rightarrow \Sigma$, if there exists a run R of A on σ, called an *accepting* run, such that for each leaf n of T, $(R(n), \sigma(n)) \in F^c$. The set of all trees accepted by A is denoted by $L(A)$.

Example 1. Let $A_\epsilon = \langle \{s_0, p\}, s_0, \#\#, \rho, \mu, \{(s_0, 1), (p, 2)\} \rangle$, where

- $\rho(s_0) = 1$, $\rho(p) = 2$, and
- $\mu = \{(s_0, 1, (p, p)), (p, 2, (p, p))\}$.

Then $L(A_\epsilon) = L_\epsilon$, where

$$L_\epsilon = \{\sigma : T \rightarrow \Sigma : \text{ for each } n \in T \setminus \{\epsilon\}, \ \sigma(n) \neq \sigma(\epsilon)\}.$$

For example, an accepting run $R : T \rightarrow \{s_0, p\} \times \Sigma^{2\neq}$ of A_ϵ on a Σ-tree $\sigma : T \rightarrow \Sigma$ such that for each $n \in T \setminus \{\epsilon\}$, $\sigma(n) \neq \sigma(\epsilon)$ is defined by

- $R(\epsilon) = (s_0, \#\#)$,
- $R(0) = (p, \sigma(\epsilon)\#)$,
- $R(1) = (p, \sigma(\epsilon)\#)$, and
- for $n \neq \epsilon$ and $i = 0, 1$, $R(ni) = (p, \sigma(\epsilon)\sigma(n))$.

Example 2. In this example we show that the tree language

$$L_2 = \left\{ \sigma : T \rightarrow \Sigma : \begin{array}{l} \text{there exist two different leaves } n', n'' \in T \\ \text{such that } \sigma(n') = \sigma(n'') \end{array} \right\}$$

is not accepted by a top-down finite-memory automaton.

Indeed, assume to the contrary there exists a top-down tree finite-memory automaton $A = \langle S, s_0, \boldsymbol{u}, \rho, \mu, F \rangle$ that accepts L_2. In particular, A accepts the Σ-tree $\sigma : \{\epsilon, 0, 1\} \rightarrow \Sigma$, where $\sigma(\epsilon) \neq \sigma(0)$ and $\sigma(0) = \sigma(1) \notin [\boldsymbol{u}]$. Let R be an accepting run of A on σ and $R(0) = (q, \boldsymbol{w})$, Then, $\sigma(0) \notin [\boldsymbol{w}]$, implying $(q, \rho(q)) \in F$.

Consider the Σ-tree $\sigma' : \{\epsilon, 0, 1\} \rightarrow \Sigma$, where $\sigma'(\epsilon) = \sigma(\epsilon)$, $\sigma'(0) \neq \sigma(0)$, $\sigma'(0) \notin [\boldsymbol{u}] \cup \{\sigma(\epsilon)\}$, and $\sigma'(1) = \sigma(1)$. Then R is also a run of A on σ'. Since $\sigma'(\epsilon) = \sigma(\epsilon)$, $R(0) = (q, \boldsymbol{w})$. In addition, $\sigma'(0) \notin [\boldsymbol{w}]$ and $(q, \rho(q)) \in F$ imply $\sigma' \in L(A)$. However, this contradicts $L(A) = L_2$.

4.2   Bottom-Up Finite-Memory Automata with Deterministic Reassignment

To define bottom-up automata we need to choose r symbols (to fill r registers at the parent node) out of, possibly, $2r$ symbols stored in the registers at the child nodes. Such a choice is based on the notion of a *type* defined below.

Definition 2. *An r-type is a subset t of $\{1, 2, \ldots, r\} \times \{1, 2, \ldots, r\}$ such that for all $(i_0, i_1), (j_0, j_1) \in t$,*

$$(i_0, i_1) \neq (j_0, j_1) \text{ implies both } i_0 \neq i_1 \text{ and } j_0 \neq j_1.$$

The set of all r-types is denoted by $\mathcal{T}_r$.

For two assignments $\boldsymbol{w}_0 = w_{0,1} w_{0,2} \cdots w_{0,r}$ and $\boldsymbol{w}_1 = w_{1,1} w_{1,2} \cdots w_{1,r}$ we define the type $t(\boldsymbol{w}_0, \boldsymbol{w}_1)$ by

$$t(\boldsymbol{w}_0, \boldsymbol{w}_1) = \{(i', i'') : w_{0,i'} = w_{1,i''}\}.^4$$

A function $f : \{1, \ldots, r\} \rightarrow \{0, 1\} \times \{1, \ldots, r\}$ is said to be a valid selector *for an r-type t if for all $1 \leq i < j \leq r$,*

$$f(i) = (0, i') \text{ and } f(j) = (1, j') \text{ imply } (i', j') \notin t.$$

Intuitively, f is used to select r registers out of $2r$ registers for the assignment at the parent node. If $f(i) = (0, i')$, then the content of the ith register at the parent node comes from the i'th register of the left child. Similarly, if $f(i) = (1, i')$, then

---

[4] That is, the elements of $t(\boldsymbol{w}_0, \boldsymbol{w}_1)$ indicate the registers with the same content. It follows from the defintion of an assignment that the type $t(\boldsymbol{w}_0, \boldsymbol{w}_1)$ is well defined.

the content of the ith register at the parent node comes from the i'th register of the right child. By definition, if f is a valid selector for an r-type t, then the resulting assignment does not have two registers with the same content.

A triple of assignments (u, v, w) is an instance of (t, f), where $t \in \mathcal{T}_r$ and f is a valid selector for t if

- $t = t(u, v)$, and
- $w = w_1 \cdots w_r$ is defined by

$$w_i = \begin{cases} u_j \text{ if } f(i) = (0, j) \\ v_j \text{ if } f(i) = (1, j) \end{cases},$$

where $u = u_1 \cdots u_r$ and $v = v_1 \cdots v_r$.

Intuitively, the assignment w is the result of merging the assignments u and v according to the the the function f. Since f is a valid selector for t, w does not have two registers with the same content.

Definition 3. *A bottom-up finite-memory automaton with deterministic reassignment ($\uparrow$-FMA) is a system $A = \langle S, s_0, u, \rho, \tau, \mu, F \rangle$, where*

- *S is a finite set of* states.
- *$s_0 \in S$ is the* initial *state.*
- *$u = u_1 u_2 \cdots u_r \in \Sigma^{r\neq}$ is the initial* assignment *to the r registers of A.*
- *$\rho : S \to \{1, 2, \ldots, r\}$ is the* deterministic reassignment.
- *$\mu \subseteq (S \times \{1, 2, \ldots, r\})^2 \times S$ is the* transition relation, *whose elements are* transitions *and are also written in the form*

$$(p_0, k_0), (p_1, k_1) \to p,$$

where $p_0, p_1, p \in S$ and $k_0, k_1 \in \{1, 2, \ldots, r\}$. The intuitive meaning of transition $(p_0, k_0), (p_1, k_1) \to p$ is as follows. In a Σ-tree σ it "applies" at nodes $n0$ and $n1$ labeled states p_0 and p_1, respectively, if the content of the k_0th register at node $n0$ is $\sigma(n0)$ and the content of the k_1th register at node $n1$ is $\sigma(n1)$. Then the label of n is p.
- *τ is a* merging relation *whose elements are of the form*

$$((p_0, k_0), (p_1, k_1), t, f),$$

where $p_0, p_1 \in S$ and $k_0, k_1 \in \{1, 2, \ldots, r\}$, $t \in \mathcal{T}_r$, and f is a valid selector for t. The intuitive meaning of an element $((p_0, k_0), (p_1, k_1), t, f)$ of τ is as follows. In a Σ-tree σ it "applies" at nodes $n0$ and $n1$ labeled p_0 and p_1, respectively, if the content of the k_0th register at node $n0$ is $\sigma(n0)$ and the content of the k_1th register at node $n1$ is $\sigma(n1)$ and the type of the assignments at $n0$ and $n1$ is t. Then the assignments are merged according to f.
- *$F \subseteq S \times \{1, 2, \ldots, r\}$ is the set of* final relations.

The transition relation μ induces the following relation μ^c on $(S^c \times \Sigma)^2 \times S^c$ whose elements are also written in the form

$$(p_0, \boldsymbol{w}_0, \sigma_0), (p_1, \boldsymbol{w}_1, \sigma_1) \to (p, \boldsymbol{w}),$$

where $p, p_0, p_1 \in S$ and $\boldsymbol{w}, \boldsymbol{w}_0, \boldsymbol{w}_1 \in \Sigma^{r \neq}$.

Let $\boldsymbol{w} = w_1 w_2 \cdots w_r$, $\boldsymbol{w}_0 = w_{0,1} w_{0,2} \cdots w_{0,r}$, and $\boldsymbol{w}_1 = w_{1,1} w_{1,2} \cdots w_{1,r}$. Then $(p_0, \boldsymbol{w}_0, \sigma_0), (p_1, \boldsymbol{w}_1, \sigma_1) \to (p, \boldsymbol{w})$ belongs to μ^c if and only if there exist $v_0, v_1 \in \Sigma^{r \neq}$, $v_0 = v_{0,1} v_{0,2} \cdots v_{0,r}$ and $v_1 = v_{1,1} v_{1,2} \cdots v_{1,r}$, $(p_0, k_0), (p_1, k_1) \to p \in \mu$, and $((p_0, k_0), (p_1, k_1), t, f) \in \tau$ such that the following conditions are satisfied.

- If $\sigma_0 \in [\boldsymbol{w}_0]$, then $v_0 = \boldsymbol{w}_0$. Otherwise,

$$v_{0,i} = \begin{cases} w_{0,i} & \text{if } i \neq \rho(p_0) \\ \sigma_0 & \text{if } i = \rho(p_0) \end{cases}.$$

- Similarly, if $\sigma_1 \in [\boldsymbol{w}_1]$, then $v_1 = \boldsymbol{w}_1$. Otherwise,

$$v_{1,i} = \begin{cases} w_{1,i} & \text{if } i \neq \rho(p_1) \\ \sigma_1 & \text{if } i = \rho(p_1) \end{cases}.$$

- $v_{0,k_0} = \sigma_0$ and $v_{1,k_1} = \sigma_1$.[5]
- $t = t(\boldsymbol{v}_0, \boldsymbol{v}_1)$.
- $(\boldsymbol{v}_0, \boldsymbol{v}_1, \boldsymbol{w})$ is an instance of (t, f).

A *run* of A on a Σ-tree $\boldsymbol{\sigma} : T \to \Sigma$ is a mapping $R : T \to S^c$ such that

- for each leaf n of T, $R(n) = s_0^c$ (recall that $s_0^c = (s_0, \boldsymbol{u})$ is the initial configuration of A), and
- for each non-leaf node $n \in T$, $(R(n0), \boldsymbol{\sigma}(n0)), (R(n1), \boldsymbol{\sigma}(n1)) \to R(n) \in \mu^c$.

The set of final relations F defines the following set *final "Σ-relations"* F^c. A pair $((p, \boldsymbol{w}), \sigma)$ is in F^c if the following holds.

- If $\sigma = w_i$, then $(p, i) \in F$; and
- if $\sigma \notin [w_1 w_2 \cdots w_r]$, then $(p, \rho(p)) \in F$.

We say that A *accepts* a Σ-tree $\boldsymbol{\sigma} : T \to \Sigma$ if there exists a run R of A on $\boldsymbol{\sigma}$, called an *accepting* run, such that $(R(\epsilon), \boldsymbol{\sigma}(\epsilon)) \in F^c$. The set of all trees accepted by A is denoted by $L(A)$.

Example 3. (Cf. Example 1.) The set of Σ-trees L_ϵ from Example 1 is not accepted by bottom-up finite-memory automata.

Indeed, assume to the contrary L_ϵ is accepted by an r-register bottom-up automaton A with the initial assignment $\boldsymbol{u}$. Consider a tree $\boldsymbol{\sigma} \in L_\epsilon$, where $\boldsymbol{\sigma}(\epsilon) \notin [\boldsymbol{u}]$ and $\boldsymbol{\sigma}(n_1) \neq \boldsymbol{\sigma}(n_2)$, whenever $n_1 \neq n_2$. During the course of computation,

---

[5] Note that if $\sigma_0 \notin [\boldsymbol{w}_0]$ (respectively, $\sigma_1 \notin [\boldsymbol{w}_1]$), then $k_0 = \rho(p_0)$ (respectively, $k_1 = \rho(p_1)$).

till the last step, $\sigma(\epsilon)$ does not appear in the registers of A. When A reaches the root, it reassigns one of the registers with $\sigma(\epsilon)$ and verifies whether one of the final Σ-relations holds.

Assume, in addition, that the range of σ contains more than r different symbols. Therefore, when A reaches the root of the tree, there is a node n such that $\sigma(n)$ is no longer in the registers of A. If we replace $\sigma(\epsilon)$ with $\sigma(n)$, A would still accept the obtained Σ-tree, in contradiction with $L(A) = L_\epsilon$.

Example 4. (Cf. Example 2.) The set of Σ-trees L_2 from Example 2 is accepted by a bottom-up finite-memory automaton that operates as follows. It "guesses" two different nodes, remembers their labels, and carries them up to the node where the paths from the two guessed nodes meet. At this meeting node the automaton checks whether the labels at the two guessed nodes are the same.

5   Tree Automata with Nondeterministic Reassignment

In this section we introduce the notion of tree automata with *nondeterministic reassignment* – both for the top-down and the bottom-up cases. Unlike the deterministic reassignment automata which may only reassign one of their registers with the *current* input symbol, these automata are allowed to reassign a number of registers with arbitrary symbols from Σ. That is, the reassignment function $\rho : S \to 2^{\{1,\dots,r\}}$ mapping the states from S into the power set $2^{\{1,\dots,r\}}$.

5.1   Top-Down Finite-Memory Automata with Nondeterministic Reassignment

Definition 4 below is the top-down tree counterpart of the *infinite-alphabet pushdown automata* introduced in [2] and the *look-ahead finite-memory automata* introduced in [20].

Definition 4. *A top-down finite-memory automaton with nondeterministic reassignment ($\downarrow$-NR-FMA) is a system $A = \langle S, s_0, \boldsymbol{u}, \rho, \mu, F \rangle$, where all components of A, except ρ, are as Definition 1. The nondeterministic reassignment ρ is a function from S into $2^{\{1,2,\dots,r\}}$. The intuitive meaning of ρ is as follows. In state p the automaton may reassign the registers whose indices belong to $\rho(p)$ with any pairwise different symbols of Σ. Of course, these symbols must differ from those in the registers whose indices do not belong to $\rho(p)$.*

The transition relation μ^c on $S^c \times \Sigma \times (S^c \times S^c)$ is defined similarly to that of top-down finite-memory automata. Let $p, p_0, p_1 \in S$ and $\boldsymbol{w}, \boldsymbol{w}_0, \boldsymbol{w}_1 \in \Sigma^{r\neq}$. Then $(p, \boldsymbol{w}), \sigma \to (p_0, \boldsymbol{w}_0), (p_1, \boldsymbol{w}_1)$ belongs to μ^c if and only if the following conditions are satisfied. Let $\boldsymbol{w} = w_1 w_2 \cdots w_r$, $\boldsymbol{w}_0 = w_{0,1} w_{0,2} \cdots w_{0,r}$, and $\boldsymbol{w}_1 = w_{1,1} w_{1,2} \cdots w_{1,r}$. Then

- $\boldsymbol{w}_0 = \boldsymbol{w}_1$,
- for all $i \notin \rho(p)$, $w_{0,i} (= w_{1,i}) = w_i$,

– for some $k = 1, 2, \ldots, r$, $w_{0,k} = \sigma\, (= w_{1,k})$, and
– $(p, k) \to (p_0, p_1) \in \mu$.

To extend F onto $S^c \times \Sigma$ we need one more bit of notation. For two assignments $\boldsymbol{v}, \boldsymbol{w} \in \Sigma^{r \neq}$, $\boldsymbol{v} = v_1 \cdots v_r$ and $\boldsymbol{w} = w_1 \cdots w_r$, and $S \subseteq \{1, \ldots, r\}$, we write $\boldsymbol{v} =_S \boldsymbol{w}$ if for all $i \notin S$, $v_i = w_i$. That is, $\boldsymbol{v}$ and $\boldsymbol{w}$ are equal "modulo" the symbols in the positions in S.

The set of *final* "Σ-relations" F^c is defined as follows. A pair $((p, \boldsymbol{w}), \sigma) \in F^c$ if for some $\boldsymbol{w}' \in \Sigma^{r \neq}$, $\boldsymbol{w}' = w_1' w_2' \cdots w_r'$, such that $\boldsymbol{w} =_{\rho(p)} \boldsymbol{w}'$, $\sigma = w_i'$, and $(p, i) \in F$.

Now a run and acceptance of for $\downarrow$-NR-FMAs are defined exactly as for $\downarrow$-FMAs.

Example 5. (Cf. Example 2.) The set of Σ-trees L_2 from Example 2 is accepted by a $\downarrow$-NR-FMA that operates as follows. In the root of the input it "guesses" the symbol that appears at two different nodes and then (nondeterministically) verifies that the guess is correct.

Proposition 1. *If a set of Σ-trees is accepted by a $\downarrow$-FMA, then it is also accepted by a $\downarrow$-NR-FMA.*

Proof. Given a $\downarrow$-FMA $A = \langle S, s_0, \boldsymbol{u}, \rho, \mu, F \rangle$, consider the following two $\downarrow$-NR-FMAs $A^- = \langle S', s_0^-, \boldsymbol{u}, \rho', \mu', F' \rangle$ and $A^+ = \langle S', s_0^+, \boldsymbol{u}, \rho', \mu', F' \rangle$, where

– $S' = \bigcup\limits_{s \in S} \{s^-, s^+\}$;[6]
– $\rho'(s^-) = \emptyset$ and $\rho'(s^+) = \{\rho(s)\}$, $p \in S$;
– μ' is the union of
  - $\{(s^-, i, s_0^{\mp}, s_1^{\mp}) : (s, i, s_0, s_1) \in \mu\}$ and
  - $\{(s^+, \rho(s), s_0^{\mp}, s_1^{\mp}) : (s, \rho(s), s_0, s_1) \in \mu\}$;
  and
– $F' = \bigcup\limits_{s \in F} \{s^-, s^+\}$.

It can be easily seen that $L(A) = L(A) \cup L(A^+)$. Indeed, both automata are allowed to make a nondeterministic reassignment (according to ρ) only in states of the form s^+, but they have to use it immediately. Consequently, the reassigned symbol must be the current input. Thus, actually, both of them behave like A, except, possibly, the first move at the root of the input Σ-tree.

Since $\downarrow$-NR-FMA languages are closed under union; see Appendix C, the proposition follows.   □

5.2   Bottom-Up Finite-Memory Automata with Nondeterministic Reassignment

Definition 5 below is the "bottom-up" counterpart of Definition 4.

---

[6] That is, S' consists of two copies of S.

Definition 5. *A bottom-up finite-memory automaton with nondeterministic reassignment ($\uparrow$-NR-FMA) is a system $A = \langle S, s_0, \boldsymbol{u}, \rho, \tau, \mu, F \rangle$, where all components of A, but ρ are as Definition 1 and the nondeterministic reassignment ρ is a function from S into $2^{\{1,2,...,r\}}$.*

The relation μ^c on $S^c \times \Sigma \times (S^c \times S^c)$ is defined similarly to that of bottom-up finite-memory automata. The only difference is that each head may reassign nondeterministically a set of their registers before merging. Namely, for $p, p_0, p_1 \in S$ and $\boldsymbol{w}, \boldsymbol{w}_0, \boldsymbol{w}_1 \in \Sigma^{r\neq}$, $(p_0, \boldsymbol{w}_0, \sigma_0), (p_1, \boldsymbol{w}_1, \sigma_1) \to (p, \boldsymbol{w})$ belongs to μ^c if and only if the following holds.

Let $\boldsymbol{w} = w_1 w_2 \cdots w_r$, $\boldsymbol{w}_0 = w_{0,1} w_{0,2} \cdots w_{0,r}$, and $\boldsymbol{w}_1 = w_{1,1} w_{1,2} \cdots w_{1,r}$. Then there exist assignments $\boldsymbol{v}_0 = v_{0,1} \cdots v_{0,r}$ and $\boldsymbol{v}_1 = v_{1,1} \cdots v_{1,r}$, a transition $(p_0, k_0), (p_1, k_1) \to p \in \mu$, and a merging relation $((p_0, k_0), (p_1, k_1), t, f) \in \tau$ such that

- $v_{0,i} = w_{0,i}$ for all $i \notin \rho(p_0)$;
- $v_{1,i} = w_{1,i}$ for all $i \notin \rho(p_1)$;
- $v_{0,k_0} = \sigma_0$ and $v_{1,k_1} = \sigma_1$;
- $t = t(\boldsymbol{v}_0, \boldsymbol{v}_1)$; and
- $(\boldsymbol{v}_0, \boldsymbol{v}_1, \boldsymbol{w})$ is an instance of (t, f).

That is, for the assignment $\boldsymbol{w}$ at the parent node, f selects r out of $2r$ values of the "reassigned" assignments at the child nodes.

The set of final relations F defines the set of *final "Σ-relations"* F^c. A pair $((p, \boldsymbol{w}), \sigma) \in F^c$ if for some $\boldsymbol{w}' \in \Sigma^{r\neq}$, $\boldsymbol{w}' = w_1' w_2' \cdots w_r'$, such that $\boldsymbol{w} =_{\rho(p)} \boldsymbol{w}'$ $\sigma = w_i'$, and $(p, i) \in F$.

Example 6. (Cf. Example 3.) The set of Σ-trees L_ϵ from Example 1 is accepted by a $\uparrow$-NR-FMA that operates as follows. In each leaf of the input the automaton "guesses" the symbol that appears at the root. Then, going down, it verifies that the input symbols are different from those at the leaves and that the "guessed" symbols are the same. Finally, when arriving to the root the automaton verifies that the guess is correct, i.e., the guessed symbol is one that appears at the root.

Proposition 2. *If a set of Σ-trees is accepted by a $\uparrow$-FMA, then it is also accepted by a $\uparrow$-NR-FMA.*

The proof of Proposition 2 is similar to that of Proposition 1 and is omitted.

5.3   The Main Result

We conclude this section with the main result of our paper stating that top-down and bottom-up finite-memory automata with nondeterministic reassignment have the same computation power.

Theorem 1. *A set of Σ-trees is accepted by a $\downarrow$-NR-FMA if and only if it is accepted by a $\uparrow$-NR-FMA. Moreover, the conversions of a $\downarrow$-NR-FMA to its equivalent $\uparrow$-NR-FMA and vice versa are effective.*

The proof of Theorem 1 is long and technical. It is presented in the appendices in the end of this paper.

Corollary 1. *Both $\downarrow$-FMA and $\uparrow$-FMA can be simulated by either of $\downarrow$-NR-FMA or $\uparrow$-NR-FMA.*

Note that by Examples 1, 2, 3, and 4, the inclusions provided by Corollary 1 are proper.

6   Decision Properties

In this section we show that the membership and emptiness problems for $\downarrow$-NR-FMAs are decidable. Thus, by Theorem 1 and Propositions 1 and 2, these problems are decidable for all other models of automata introduced in this paper. We also show that the universality and, consequently, the inclusion problems are undecidable for all models of automata introduced in this paper.

Propositions 3 and 4 below deal with decidability of the membership and emptiness problems. The former asks whether a given $\downarrow$-NR-FMA accepts a given Σ-tree, and the latter asks whether the language of a given $\downarrow$-NR-FMA is empty.

Proposition 3. *The membership problem for $\downarrow$-NR-FMAs is decidable.*

Proof. Let $A = \langle S, s_0, \boldsymbol{u}, \rho, \mu, F \rangle$ be a $\downarrow$-NR-FMA and let $\boldsymbol{\sigma} : T \rightarrow \Sigma$ be a Σ-tree. We contend that $\boldsymbol{\sigma} \in L(A)$ if and only if there is an accepting run of A on $\boldsymbol{\sigma}$ in which the assignment at each node belongs to $\Sigma_0^{r\neq}$, where

$$\Sigma_0 = \boldsymbol{\sigma}(T) \cup [\boldsymbol{u}] \cup \{\#\} \cup \{\theta_1, \theta_2, \ldots, \theta_r\}, \quad \theta_i \notin \boldsymbol{\sigma}(T), \ i = 1, 2, \ldots, r.$$

The "if" direction is immediate, and for the proof of the "only if" direction we just replace the symbols which appear in an accepting run of A on $\boldsymbol{\sigma}$, but do not belong to $\boldsymbol{\sigma}(T) \cup [\boldsymbol{u}] \cup \{\#\}$ with appropriate elements of $\{\theta_1, \theta_2, \ldots, \theta_r\}$.[7]

Therefore, given an input Σ-tree $\boldsymbol{\sigma} : T \rightarrow \Sigma$, we may restrict ourselves to the configurations of A from $S \times \Sigma_0^{r\neq}$, which brings us to an ordinary *finite alphabet* tree automaton. Since the membership problem for the latter is decidable, the proposition follows. $\square$

Proposition 4. *The emptiness problem for $\downarrow$-NR-FMAs is decidable.*

The proof of Proposition 4 is based on Lemma 1 below.

Lemma 1. *Let $A = \langle S, s_0, \boldsymbol{u}, \rho, \mu, F \rangle$ be an r-register $\downarrow$-NR-FMA such that $L(A) \neq \emptyset$ and let $\Sigma_r = \{\theta_1, \theta_2, \ldots, \theta_r\}$ be an r-element subset of Σ that includes $[\boldsymbol{u}]$. Then there is a Σ-tree $\boldsymbol{\sigma} : T \rightarrow \Sigma_r$ in $L(A)$.*

Proof. Let $\boldsymbol{\sigma} : T \rightarrow \Sigma \in L(A)$ and let $R : T \rightarrow S^c$ be an accepting run of A on $\boldsymbol{\sigma}$. To construct a Σ-tree $\boldsymbol{\sigma} : T \rightarrow \Sigma_r$ in $L(A)$ we need the function $I : T \rightarrow \{1, 2, \ldots, r\}$ defined below.

---

[7] Obviously, such symbols can be introduced by reassignment, only.

- For a non-leaf node $n \in T$, if $R(n0) = (p, w_1 w_2 \cdots w_r)$ and $\sigma(n) = w_i$, then $I(n) = i$.
- For a leaf node $n \in T$, if $R(n) = (p, w_1 w_2 \cdots w_r)$ and $w' \in \Sigma^{r \neq}$, $w' = w'_1 w'_2 \cdots w'_r$, is such that $w =_{\rho(p)} w'$, $\sigma = w'_i$, and $(p, i) \in F$, then $I(n) = i$.

Let $u = u_1 \cdots u_r$. We may assume that for each $i = 1, 2, \ldots, r$, $u_i \neq \#$ implies $u_i = \theta_i$. Then a Σ-tree $\sigma_r : T \to \Sigma_r$ satisfying the lemma is defined by $\sigma_r(n) = \theta_{I(n)}$, $n \in T$. This Σ-tree is accepted by the run of A whose state components are the same as of R and that *always* reassigns the i register with Θ_i, $i = 1, 2, \ldots, r$. □

Proof. (of Proposition 4) Let $A = \langle S, s_0, u, \rho, \mu, F \rangle$ be an r-register $\downarrow$-NR-FMA and let $\Sigma_r = \{\theta_1, \theta_2, \ldots, \theta_r\}$ be an r-element subset of Σ that includes $[u]$. It follows from Lemma 1 that $L(A) \neq \emptyset$ if and only if there is a Σ-tree $\sigma : T \to \Sigma_r$ in $L(A)$.

Since on the inputs $\sigma : T \to \Sigma_r$ the configurations of A belong to $S \times \Sigma_r^{r \neq}$, the emptiness of A is reduced to the emptiness of an ordinary *finite alphabet* tree automaton that is is decidable. □

It was shown that in [13] the universality problem of finite-memory automata is undecidable.[8] Consequently, this problem is also undecidable for all above models of tree automata.

Proposition 5. *The universality problem for $\downarrow$-FMAs and $\uparrow$-FMAs is undecidable.*

Corollary 2. *The inclusion problem for $\downarrow$-FMAs and $\uparrow$-FMAs is undecidable.*

7   Context-Free Languages over Infinite Alphabets and Their Relationship with Tree Automata

In this section we recall the definition of *quasi context-free languages* from [2] and show how they are related to the tree languages introduced in this paper.

In short, a *quasi context-free grammar* is a context-free grammar, where each variable carries the same number r of of registers. The terminal alphabet of a grammar G is the set $\{1, 2, \ldots, r\}$. Let V be the set of variables of G. The productions of G are of the form

$$(A, k) \to \alpha_1 \alpha_2 \cdots \alpha_n,$$

where $1 \leq k \leq r$ and $\alpha_i \in V \cup \{1, \ldots, r\}$, $i = 1, 2, \ldots, n$. The above production allows us

- to replace the content of the kth register carried by A with any symbol of Σ that differs from the symbols stored in the other registers,[9] and

---

[8] That is, is undecidable whether a given finite-memory automaton accepts Σ^*.

[9] Actually, automata with nondeterministic reassignment were motivated by [2].

– to replace A with the word $\beta_1 \cdots \beta_n$, where β_i is the content of the jth register, if $\alpha_i = j$, and is α_i, if α_i is a variable, $i = 1, 2, \ldots, n$.

The language generated by G consists of all words in Σ^* obtained by repeatedly applying the productions of G, starting with the *start* variable. It is called a *quasi context-free* language, cf. DTDs in Sect. 2. The precise definition of quasi context-free grammars and languages is as follows.

Definition 6. *([2, Definition 1]) An* infinite-alphabet context-free grammar *is a system $G = \langle V, \boldsymbol{u}, R, S \rangle$, where*

- V *is a finite set of* variables *disjoint with Σ;*
- $\boldsymbol{u} = u_1 u_2 \cdots u_r \in \Sigma^{r \neq}$ *is the* initial *assignment;*
- $R \subseteq (V \times \{1, 2, \ldots, r\}) \times (V \cup \{1, 2, \ldots, r\})^*$ *is a set of* productions, *whose elements are written in the form $(A, i, \boldsymbol{a})$ as $(A, i) \to \boldsymbol{a}$, where $A \in V$, $i = 1, 2, \ldots, r$, and $\boldsymbol{a} \in (V \cup \{1, 2, \ldots, r\})^*$; and*
- $S \in V$ *is the* start *variable.*

For $A \in V$, $\boldsymbol{w} = w_1 w_2 \cdots w_r \in \Sigma^{r \neq}$, and $\boldsymbol{X} = X_1 X_2 \cdots X_n \in (\Sigma \cup (V \times \Sigma^{r \neq}))^*$, we write $(A, \boldsymbol{w}) \Rightarrow \boldsymbol{X}$ if there exist a production $(A, i) \to \boldsymbol{a} \in R$, $\boldsymbol{a} = a_1 a_2 \cdots a_n \in (V \cup \{1, 2, \ldots, r\})^*$, and a symbol $\sigma \notin [\boldsymbol{w}] \setminus \{w_i\}$ such that the condition below is satisfied.

Let $\boldsymbol{w}' \in \Sigma^{r \neq}$ be obtained from $\boldsymbol{w}$ by replacing w_i with σ. Then, for $j = 1, 2, \ldots, n$ the following holds.

- If $a_j = k$ for some $k = 1, 2, \ldots, r$, then $X_j = w'_k$.
- If $a_j = B$ for some $B \in V$, then $X_j = (B, \boldsymbol{w}')$.

For two words $\boldsymbol{X}$ and $\boldsymbol{Y}$ over $\Sigma \cup (V \times \Sigma^{r \neq})$, we write $\boldsymbol{X} \Rightarrow \boldsymbol{Y}$ if there exist words $\boldsymbol{X}_1$, $\boldsymbol{X}_2$, and $\boldsymbol{X}_3$ over $\Sigma \cup (V \times \Sigma^{r \neq})$ and $(A, \boldsymbol{w}) \in V \times \Sigma^{r \neq}$, such that $\boldsymbol{X} = \boldsymbol{X}_1 (A, \boldsymbol{w}) \boldsymbol{X}_2$, $\boldsymbol{Y} = \boldsymbol{X}_1 \boldsymbol{X}_3 \boldsymbol{X}_2$, and $(A, \boldsymbol{w}) \Rightarrow \boldsymbol{X}_3$.

As usual, the reflexive and transitive closure of $\Rightarrow$ is denoted by $\Rightarrow^*$. The language $L(G)$ generated by G is defined by $L(G) = \{\sigma \in \Sigma^* : (S, \boldsymbol{u}) \Rightarrow^* \sigma\}$ and is referred to as a *quasi-context-free* language.

Example 7. Let G be a 1-register grammar with the set of variables $V = \{S\}$, the *initial assignment* #, and the following two production.

$$(S, 1) \to 1S1 \,|\, \epsilon.$$

Then $L(G) = \{\sigma\sigma^R \mid \sigma \in \Sigma^*\}$.[10] For example, the word $\sigma_1 \sigma_2 \sigma_3 \sigma_3 \sigma_2 \sigma_1$ is derived as follows.

$$(S, \#) \Rightarrow \sigma_1(S, \sigma_1)\sigma_1 \Rightarrow \sigma_1\sigma_2(S, \sigma_2)\sigma_2\sigma_1$$
$$\Rightarrow \sigma_1\sigma_2\sigma_3(S, \sigma_3)\sigma_3\sigma_2\sigma_1 \Rightarrow \sigma_1\sigma_2\sigma_3\sigma_3\sigma_2\sigma_1.$$

---

[10] As usual, σ^R is the *reversal* of σ.

We end this paper with the theorem below that relates quasi context-free languages to the tree languages introduced in this paper. Recall that for a Σ-tree $\sigma : T \rightarrow \Sigma$, the *frontier* $\ell(\sigma)$ of σ is the word $\sigma(n_1)\sigma(n_2)\cdots\sigma(n_m)$, where $n_1, n_2, \ldots, n_m$ is the list of all leaf nodes of T in the lexicographical order. Below the set frontiers of all elements of a set of Σ-trees L is denoted by $\ell(L)$:
$\ell(L) = \{\ell(\sigma) : \sigma \in L\}$.

Theorem 2. *Let L be a tree language accepted by a top-down (or bottom-up) finite-memory automaton A with a deterministic (or non-deterministic) reassignment. Then $\ell(L)$ is quasi-context-free language.*

Conversely, for every quasi-context-free language L, there exists a top-down (or bottom-up) finite-memory automata A with deterministic (or non-deterministic) reassignment such that $\ell(L(A)) = L$.

We omit the proof that is quite straightforward. For example, any Σ- tree σ accepted by a $\downarrow$-FMA A, after an appropriate modification, can be thought of as a derivation tree of the word $\ell(\sigma)$. Conversely, given a quasi context-free grammar G, we may assume that all derivation trees of the words in $L(G)$ are binary.[11] Therefore, the set of productions of G can be thought of as the set of transition of a $\downarrow$-FMA A, implying that $\ell(L(A))$ is exactly the language generated by G.

Acknowledgment

This research was supported by the Jewish communities of Germany research fund, by the Technion vice-president fund for the promotion of research at the Technion, and by a grant from the Software Technology Laboratory (STL) in the Department of Computer Science, the Technion. In addition, the work of the second author was supported by the Raphael and Miriam Mishan Fellowship.

References

1. Bex, G.J., Maneth, S., Neven, F.: A formal model for an expressive fragment of XSLT. Information and System 27(1), 21–39 (2002)
2. Cheng, E.Y.C., Kaminski, M.: Context-free languages over infinite alphabets. Acta Informatica 35, 245–267 (1998)
3. Comon, H., et al.: Tree Automata Techniques and Applications (2005), http://www.grappa.univ-lille3.fr/tata/
4. Doner, J.E.: Tree acceptors and some of their applications. Journal of Computer and System Sciences 4, 406–451 (1970)
5. Kaminski, M., Francez, N.: Finite-memory automata. In: Proceedings of the 31th Annual IEEE Symposium on Foundations of Computer Science, pp. 683–688. IEEE Computer Society Press, Los Alamitos (1990)

---

[11] It is well known that any tree can be converted to a binary tree that preserves the order of the leaves.

6. Kaminski, M., Francez, N.: Finite-memory automata. Theoretical Computer Science 138, 329–363 (1994)
7. Kaminski, M., Tan, T.: Regular expressions for languages over infinite alphabets. Fundamenta Informaticae 69, 301–318 (2006)
8. Milo, T., Suciu, D., Vianu, V.: Type checking for XML transformers. Journal of Computer and System Sciences 66, 66–97 (2003)
9. Neven, F., Schwentick, T.: Expressive and efficient pattern languages for tree-structured data. In: Proceedings of the Nineteenth International Symposium on Principles of Database Systems, pp. 145–156. ACM Press, New York (2000)
10. Neven, F., Schwentick, T., Vianu, V.: Towards regular languages over infinite alphabets. In: Sgall, J., Pultr, A., Kolman, P. (eds.) MFCS 2001. LNCS, vol. 2136, pp. 560–572. Springer, Heidelberg (2001)
11. Neven, F.: Automata, logic and XML. In: Bradfield, J.C. (ed.) CSL 2002 and EACSL 2002. LNCS, vol. 2471, pp. 2–26. Springer, Heidelberg (2002)
12. Neven, F., Schwentick, T.: Query automata on finite trees. Theoretical Computer Science 275, 633–674 (2002)
13. Neven, F., Schwentick, T., Vianu, V.: Finite state machines for strings over infinite alphabets. ACM Transactions on Computational Logic 5, 403–435 (2004)
14. Papakonstantinou, Y., Vianu, V.: DTD inference for views of XML data. In: Proceedings of the Twentieth International Symposium on Principles of Database Systems, pp. 35–46. ACM Press, New York (2001)
15. Rabin, M.: Decidability of second order theories and automata on infinite trees. Transactions of the American Mathematical Society 141, 1–35 (1969)
16. Ray, E.: Learning XML. O'Reilly & Associates, Inc, Sebastopol (2001)
17. Thatcher, J., Wright, J.: Generalized finite automata theory. Mathematical System Theory 2, 57–81 (1968)
18. Vianu, V.: A web odyssey: from Codd to XML. In: Proceedings of the 20th International Symposium on Principles of Database Systems, pp. 1–15. ACM Press, New York (2001)
19. XML Core Working Group: Extensible Markup Language (XML). World Wide Web Consortium, http://www.w3.org/XML/
20. Zeitlin, D.: Look-ahead finite-memory automata. Master's thesis, Department of Computer Science, Technion - Israel Institute of Technology (2006)

A   Proof of the "only if" Part of Theorem 1

For an r-register $\downarrow$-NR-FMA $A = \langle S, s_0, \boldsymbol{u}, \rho, \mu, F \rangle$ we construct an r-register $\uparrow$-NR-FMA $\widetilde{A} = \langle \widetilde{S}, \widetilde{s}_0, \widetilde{\boldsymbol{u}}, \widetilde{\rho}, \widetilde{\tau}, \widetilde{\mu}, \widetilde{F} \rangle$ such that $L(A) = L(\widetilde{A})$.

Similarly to the proof of [7, Lemma 5.1] it can be shown that, without loss of generality, the following assumptions hold.

- $\boldsymbol{u} = \#^{r-m}\theta_1 \cdots \theta_m$, where $\theta_1, \ldots, \theta_m \in \Sigma$, and
- only the first $r - m$ registers of A can be reassigned, i.e, the range of ρ is a subset of $\{1, 2, \ldots, r - m\}$.

We precede the formal description of $\widetilde{A}$ with a general intuitive explanation. One would expect the construction to be just the transition reversing, i.e., a transition $(p, k) \to (p_0, p_1)$ of A to become a "transition" $((p_0, p_1), k) \to p$ of $\widetilde{A}$.

This is indeed almost so, but with the following modification. Since transitions of a $\uparrow$-NR-FMA merge two heads and depend on two input symbols from Σ, we combine two transitions of A into one "reversed" transition of $\widetilde{A}$. That is, two transitions $(p_0, k_0) \rightarrow (p_{00}, p_{01})$ and $(p_1, k_1) \rightarrow (p_{10}, p_{11})$ of A are combined into one transition $((p_{00}, p_{01}), k_0), ((p_{10}, p_{11}), k_1) \rightarrow (p_0, p_1)$ of $\widetilde{A}$ and "moved" one level up, as illustrated in Fig. 3 and 4 below. Fig. 3 shows an application of two top-down transitions at two nodes (sharing the same parent node) labeled σ_0 and σ_1. Fig. 4 shows their reversal bottom-up transition applied at the same two nodes, but in the converse direction.

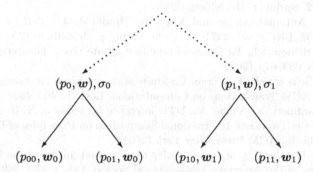

Fig. 3. An application of two transitions of A

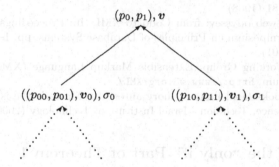

Fig. 4. Reversing and combining two transitions of A

Note that assignments w_0 and w_1 equal to w "modulo" the symbols in the positions belonging to $\rho(p_0)$ and $\rho(p_1)$, respectively; and except for the symbols at the positions in $\rho(p_0) \cap \rho(p_1)$, w can be recovered from w_0 and w_1. The symbols at the positions in $\rho(p_0) \cap \rho(p_1)$ can be "guessed" by a nondeterministic reassignment of the "reversal" automaton $\widetilde{A}$. Thus, for $p_0, p_1 \in S$, we let $\widetilde{\rho}((p_0, p_1))$ be $\rho(p_0) \cap \rho(p_1)$. In fact, the reversed transitions in $\widetilde{A}$ yield the assignments v, v_0, v_1 such that $v =_{\rho(p_0) \cap \rho(p_1)} w$, $v_0 =_{\rho(p_{00}) \cap \rho(p_{01})} w_0$ and $v_1 =_{\rho(p_{10}) \cap \rho(p_{11})} w_1$.

To define the transition relation $\widetilde{\mu}$ and the merging relation $\widetilde{\tau}$ we need the following definition and the corresponding auxiliary result.

Let t be an r-type and let f be a valid selector for t. The pair (t, f) is called an *inverse structure associated with a pair of states* (p_0, p_1) if the following conditions are satisfied.

- $(i, i) \in t$ for all $i \in \{1, \ldots, r\} - (\rho(p_0) \cup \rho(p_1))$.

- $f(i) = \begin{cases} (0, i) \text{ or } (1, i) & \text{for } i \in \{1, \ldots, r\} \setminus (\rho(p_0) \cup \rho(p_1)) \\ (1, i) & \text{for } i \in \rho(p_0) \setminus \rho(p_1) \\ (0, i) & \text{for } i \in \rho(p_1) \setminus \rho(p_0) \end{cases}$

Note that the value of f on the elements of $\rho(p_0) \cap \rho(p_1)$ is arbitrary. The set of all inverse structures associated with a pair of states (p_0, p_1) will be denoted $I(p_0, p_1)$.

Lemma 2. *Let (t, f) be an inverse structure associated with a pair of states (p_0, p_1) and let $\boldsymbol{v}_0, \boldsymbol{v}_1, \boldsymbol{v}, \boldsymbol{w} \in \Sigma^{r\neq}$ be such that $\boldsymbol{v}_0 =_{\rho(p_0)} \boldsymbol{w}$ and $\boldsymbol{v}_1 =_{\rho(p_1)} \boldsymbol{w}$. Then $(\boldsymbol{v}_0, \boldsymbol{v}_1, \boldsymbol{v})$ is an instance of (t, f) if and only if $\boldsymbol{v} =_{\rho(p_0) \cap \rho(p_1)} \boldsymbol{w}$.*

Moreover, if $\boldsymbol{v} =_{\rho(p_0) \cap \rho(p_1)} \boldsymbol{w}$ and $(\boldsymbol{v}_0, \boldsymbol{v}_1, \boldsymbol{v})$ is an instance of (t, f), then $\boldsymbol{v}_0 =_{\rho(p_0)} \boldsymbol{w}$ and $\boldsymbol{v}_1 =_{\rho(p_1)} \boldsymbol{w}$.

Proof. Let $\boldsymbol{v}_0 = v_{0,1} \cdots v_{0,r}$, $\boldsymbol{v}_1 = v_{1,1} \cdots v_{1,r}$, $\boldsymbol{v} = v_1 \cdots v_r$, and $\boldsymbol{w} = w_1 \cdots w_r$. It immediately follows from the definition that $(\boldsymbol{v}_0, \boldsymbol{v}_1, \boldsymbol{v})$ is an instance of (t, f) if and only if

$$v_i = \begin{cases} v_{0,i} = w_i & \text{if } i \notin \rho(p_1) \\ v_{1,i} = w_i & \text{if } i \notin \rho(p_0) \end{cases}, \quad i = 1, \ldots, r,$$

which is equivalent to $\boldsymbol{v} =_{\rho(p_0) \cap \rho(p_1)} \boldsymbol{w}$.

Now we prove the second part. Let $\boldsymbol{v} =_{\rho(p_0) \cap \rho(p_1)} \boldsymbol{w}$. In particular, for all $i \notin \rho(p_0)$, $w_i = v_i$. If $(\boldsymbol{v}_0, \boldsymbol{v}_1, \boldsymbol{v})$ is an instance of (t, f), by the definition of (t, f), $v_i = v_{0,i}$. Therefore, $\boldsymbol{w} =_{\rho(p_0)} \boldsymbol{v}_0$. In a similar way we can show that $\boldsymbol{w} =_{\rho(p_1)} \boldsymbol{v}_1$. $\square$

Now we are ready to define the desired $\uparrow$-NR-FMA $\widetilde{A} = \langle \widetilde{S}, \widetilde{s}_0, \widetilde{\boldsymbol{u}}, \widetilde{\rho}, \widetilde{\tau}, \widetilde{\mu}, \widetilde{F} \rangle$.

- $\widetilde{S} = S \times S \cup \{\widetilde{s}_0\}$, where $\widetilde{s}_0$ is a new state.
- The initial state of $\widetilde{A}$ is $\widetilde{s}_0$.
- $\widetilde{\boldsymbol{u}} = \#^{r-m}\theta_1 \cdots \theta_m$.
- $\widetilde{\rho}(\widetilde{s}_0) = \{1, \ldots, r - m\}$, and
  $\widetilde{\rho}((p_0, p_1)) = \rho(p_0) \cap \rho(p_1)$, for all $(p_0, p_1) \in \widetilde{S}$.
- $\widetilde{\mu} = \widetilde{\mu}_1 \cup \widetilde{\mu}_2 \cup \widetilde{\mu}_3 \cup \widetilde{\mu}_4$, where
  - $\widetilde{\mu}_1 = \{(\widetilde{s}_0, k_0), (\widetilde{s}_0, k_1) \to (p_0, p_1) : (p_0, k_0), (p_1, k_1) \in F\}$;
  - $\widetilde{\mu}_2 = \{(\widetilde{s}_0, k_0), ((p_{10}, p_{11}), k_1) \to (p_0, p_1) :$
    $(p_0, k_0) \in F$ and $(p_1, k_1) \to (p_{10}, p_{11}) \in \mu\}$;
  - $\widetilde{\mu}_3 = \{((p_{00}, p_{01}), k_0), (\widetilde{s}_0, k_1) \to (p_0, p_1) :$
    $(p_0, k_0) \to (p_{00}, p_{01}) \in \mu$ and $(p_1, k_1) \in F\}$;
  - $\widetilde{\mu}_4 = \{((p_{00}, p_{01}), k_0), ((p_{10}, p_{11}), k_1) \to (p_0, p_1) :$
    $(p_0, k_0) \to (p_{00}, p_{01}), (p_1, k_1) \to (p_{10}, p_{11}) \in \mu\}$.
- $\widetilde{\tau} = \widetilde{\tau}_1 \cup \widetilde{\tau}_2 \cup \widetilde{\tau}_3 \cup \widetilde{\tau}_4$, where

- $\tilde{\tau}_1 = \{((\tilde{s}_0, k_0), (\tilde{s}_0, k_1), t, f) :$
  for some $p_0, p_1 \in S$, $(p_0, k_0), (p_1, k_1) \in F$ and $(t, f) \in I(p_0, p_1)\}$;[12]
- $\tilde{\tau}_2 = \{((\tilde{s}_0, k_0), ((p_{1,0}, p_{1,1}), k_1), t, f) :$
  for some $p_0, p_1 \in S$, $(p_0, k_0) \in F$ and $(p_1, k_1) \to (p_{1,0}, p_{1,1}) \in \mu$ and
  $$(t, f) \in I(p_0, p_1)\};$$
- $\tilde{\tau}_3 = \{(((p_{0,0}, p_{0,1}), k_0), (\tilde{s}_0, k_1), t, f) :$
  for some $p_0, p_1 \in S$, $(p_0, k_0) \to (p_{0,0}, p_{0,1}) \in \mu$ and $(p_1, k_1) \in F$ and
  $$(t, f) \in I(p_0, p_1)\};$$
- $\tilde{\tau}_4 = \{(((p_{00}, p_{01}), k_0), ((p_{10}, p_{11}), k_1), t, f) :$
  for some $p_0, p_1 \in S$, $(p_0, k_0) \to (p_{00}, p_{01})$, $(p_1, k_1) \to (p_{10}, p_{11}) \in \mu$ and
  $$(t, f) \in I(p_0, p_1)\}.$$

- $\widetilde{F} = \{((q_0, q_1), k) : (s_0, k) \to (q_0, q_1) \in \mu\}.$

The proof of the equality $L(\widetilde{A}) = L(A)$ is based on Lemma 3 below. Roughly speaking, Lemma 3 is the formal description of the construction in Fig. 3 and 4. It shows how an accepting run of A can be "reversed" into an accepting run of $\widetilde{A}$, or more precisely, it shows how transitions from μ^c are converted into transitions from $\tilde{\mu}^c$, and vice versa.

Lemma 3 consists of four parts corresponding to the type of nodes on which transitions take place. Its part (i) shows how an accepting run of A can be "reversed" into an accepting run of $\widetilde{A}$ at the leaf nodes and vice versa. Part (ii) shows how an accepting run of A can be "reversed" into an accepting run of $\widetilde{A}$ when one of the two sibling nodes is a leaf and the other is an interior node. Part (iii) of the lemma settles the case of the interior nodes. Finally, part (iv) of Lemma 3 deals with the case of the root node ϵ.

Lemma 3

(i) (See Fig. 5 and 6.) If $((p_0, \boldsymbol{w}), \sigma_0), ((p_1, \boldsymbol{w}), \sigma_1) \in F^c$, then there is an assignment $\boldsymbol{v} =_{\tilde{\rho}(p_0, p_1)} \boldsymbol{w}$ such that

$$((\tilde{s}_0, \tilde{\boldsymbol{u}}), \sigma_0), ((\tilde{s}_0, \tilde{\boldsymbol{u}}), \sigma_1) \to ((p_0, p_1), \boldsymbol{v}) \in \tilde{\mu}^c.$$

Conversely, if

$$((\tilde{s}_0, \tilde{\boldsymbol{u}}), \sigma_0), ((\tilde{s}_0, \tilde{\boldsymbol{u}}), \sigma_1) \to ((p_0, p_1), \boldsymbol{v}) \in \tilde{\mu}^c.$$

and $\boldsymbol{w} =_{\tilde{\rho}(p_0, p_1)} \boldsymbol{v}$, *then* $((p_0, \boldsymbol{w}), \sigma_0), ((p_1, \boldsymbol{w}), \sigma_1) \in F^c$.

(ii) (a) (See Fig. 7 and 8.) If $((p_0, \boldsymbol{w}), \sigma_0) \in F^c$,

$$((p_1, \boldsymbol{w}), \sigma_1) \to (p_{10}, \boldsymbol{w}_1), (p_{11}, \boldsymbol{w}_1) \in \mu^c$$

and $\boldsymbol{v}_1 =_{\tilde{\rho}(p_{10}, p_{11})} \boldsymbol{w}_1$, *then there is an assignment* $\boldsymbol{v} =_{\tilde{\rho}(p_0, p_1)} \boldsymbol{w}$ *such that*

$$((\tilde{s}_0, \tilde{\boldsymbol{u}}), \sigma_0), (((p_{10}, p_{11}), \boldsymbol{v}_1), \sigma_1) \to ((p_0, p_1), \boldsymbol{v}) \in \tilde{\mu}^c.$$

---

[12] Recall that $I(p_0, p_1)$ denotes the set of all inverse structures associated with the pair of states (p_0, p_1).

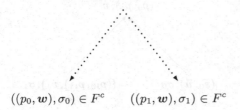

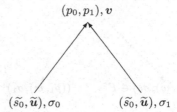

Fig. 5. Application of two final relations of A

Fig. 6. Reversing and combining two final relations of A

Conversely, if

$$((\tilde{s}_0, \tilde{u}), \sigma_0), (((p_{10}, p_{11}), v_1), \sigma_1) \to ((p_0, p_1), v) \in \tilde{\mu}^c$$

and $w =_{\tilde{\rho}(p_0, p_1)} v$, *then* $((p_0, w), \sigma_0) \in F^c$ *and there is an assignment* $w_1 =_{\tilde{\rho}(p_{10}, p_{11})} v_1$ *such that*

$$((p_1, w), \sigma_1) \to (p_{10}, w_1), (p_{11}, w_1) \in \mu^c.$$

(b) *If*

$$((p_0, w), \sigma_0) \to (p_{00}, w_0), (p_{01}, w_0) \in \mu^c,$$

$((p_1, w), \sigma_1) \in F^c$ *and* $v_0 =_{\tilde{\rho}(p_{00}, p_{01})} w_0$, *then there is an assignment* $v =_{\tilde{\rho}(p_0, p_1)} w$ *such that*

$$(((p_{00}, p_{01}), v_0), \sigma_0), ((\tilde{s}_0, \tilde{u}), \sigma_1) \to ((p_0, p_1), v) \in \tilde{\mu}^c.$$

Conversely, if

$$(((p_{00}, p_{01}), v_0), \sigma_0), ((\tilde{s}_0, \tilde{u}), \sigma_1) \to ((p_0, p_1), v) \in \tilde{\mu}^c$$

and $w =_{\tilde{\rho}(p_0, p_1)} v$, *then there is an assignment* $w_0 =_{\tilde{\rho}(p_{00}, p_{01})} v_0$ *such that*

$$((p_0, w), \sigma_0) \to (p_{00}, w_0), (p_{01}, w_0) \in \mu^c$$

and $((p_1, w), \sigma_1) \in F^c$.

(*iii*) *(See Fig. 3 and 4.) If*

$$((p_0, w), \sigma_0) \to (p_{00}, w_0), (p_{01}, w_0) \in \mu^c,$$

$$((p_1, w), \sigma_1) \to (p_{10}, w_1), (p_{11}, w_1) \in \mu^c,$$

$v_0 =_{\tilde{\rho}(p_{00}, p_{01})} w_0$ *and* $v_1 =_{\tilde{\rho}(p_{10}, p_{11})} w_1$, *then there is an assignment* $v =_{\tilde{\rho}(p_0, p_1)} w$ *such that*

$$(((p_{00}, p_{01}), v_0), \sigma_0), (((p_{10}, p_{11}), v_1), \sigma_1) \to ((p_0, p_1), v) \in \tilde{\mu}^c.$$

Conversely, if

$$(((p_{00}, p_{01}), v_0), \sigma_0), (((p_{10}, p_{11}), v_1), \sigma_1) \to ((p_0, p_1), v) \in \tilde{\mu}^c$$

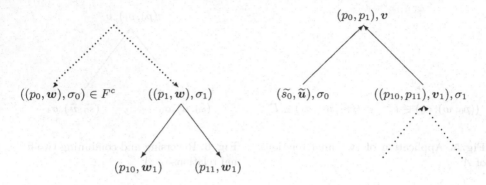

Fig. 7. An application of a final relation and a transitions of A

Fig. 8. Reversing a final relation and a transition of A into a transition of $\widetilde{A}$

and $w =_{\widetilde{\rho}(p_0,p_1)} v$, then there are assignments $w_0 =_{\widetilde{\rho}(p_{00},p_{01})} v_0$ and $w_1 =_{\widetilde{\rho}(p_{10},p_{11})} v_1$ such that

$$((p_0, w), \sigma_0) \to (p_{00}, w_0), (p_{01}, w_0) \in \mu^c$$

and

$$((p_1, w), \sigma_1) \to (p_{10}, w_1), (p_{11}, w_1) \in \mu^c.$$

(iv) (See Fig. 9 and 10.) If

$$((s_0, u), \sigma) \to (p_0, w), (p_1, w) \in \mu^c$$

and $v =_{\widetilde{\rho}(p_0,p_1)} w$, then $(((p_0, p_1), v), \sigma) \in \widetilde{F}^c$.
Conversely, if $(((p_0, p_1), v), \sigma) \in \widetilde{F}^c$, then there is an assignment $w = \widetilde{\rho}(p_0,p_1) v$ such that

$$((s_0, u), \sigma) \to (p_0, w), (p_1, w) \in \mu^c.$$

We postpone the proof of the lemma to the end of this appendix and prove the "only if" part of Theorem 1 first.

Proof. (of the "only if" part of Theorem 1.) We prove that $L(A) = L(\widetilde{A})$ by showing how to convert an accepting run of A on a Σ-tree $\sigma : T \to \Sigma$, into an accepting run of $\widetilde{A}$ on σ, and vice versa.

For an accepting run $R : T \to S^c$, $R(n) = (p_n, w_n)$, $n \in T$, of A on σ we construct an accepting run $\widetilde{R} : T \to S^c$ of $\widetilde{A}$ on σ bottom-up, i.e., from the leaves to the root, by induction, as follows.

By definition, for a leaf node $n \in T$, $\widetilde{R}(n) = \widetilde{s}_0^{\,c} = (\widetilde{s}_0, \widetilde{u})$, and for an interior node $n \in T$, $\widetilde{R}(n) = ((p_{n0}, p_{n1}), v_n)$, where the assignment $v_n =_{\widetilde{\rho}(p_{n0},p_{n1})} w_{n0}(= w_{n1})$ is defined as follows.[13]

---

[13] Recall that by definition of μ^c, $w_{n0} = w_{n1}$.

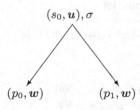

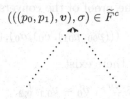

Fig. 9. An application of a transition of A at the root node

Fig. 10. Reversing a transition of A at the root node into a final relation of $\widetilde{A}$

- If both children of n are leaf nodes, then v_n is provided by part (i) of Lemma 3.
- If one child of n is a leaf node and the other is an interior node, then v_n is provided by part (ii) of Lemma 3.
- Finally, if both children of n are interior nodes, then v_n is provided by part (iii) of Lemma 3.

Now, by part (iv) of Lemma 3, $\widetilde{R}$ is an accepting run of $\widetilde{A}$ on σ.

The converse direction can shown in a similar manner. That is, an accepting run R of A on σ is constructed from an accepting run $\widetilde{R}$ of $\widetilde{A}$ by applying the converse direction of Lemma 3.   □

It remains to prove Lemma 3.

Proof. (of Lemma 3) We will prove only part (iii) of the lemma. The proofs of the other parts are very similar.

Let

$$(p_0, \boldsymbol{w}), \sigma_0 \rightarrow (p_{00}, \boldsymbol{w}_0), (p_{01}, \boldsymbol{w}_0) \in \mu^c$$

and

$$(p_1, \boldsymbol{w}), \sigma_1 \rightarrow (p_{10}, \boldsymbol{w}_1), (p_{11}, \boldsymbol{w}_1) \in \mu^c,$$

$\boldsymbol{w} = w_1 \cdots w_r$, $\boldsymbol{w}_0 = w_{0,1} \cdots w_{0,r}$ and $\boldsymbol{w}_1 = w_{1,1} \cdots w_{1,r}$. That is,

- $(p_0, k_0) \rightarrow (p_{00}, p_{01}) \in \mu$ and $w_{0,k_0} = \sigma_0$;
- $(p_1, k_1) \rightarrow (p_{10}, p_{11}) \in \mu$ and $w_{1,k_1} = \sigma_1$.

Then, by the definition of $\widetilde{\mu}_4$,

$$((p_{00}, p_{01}), k_0), ((p_{10}, p_{11}), k_1) \rightarrow (p_0, p_1) \in \widetilde{\mu}_4$$

and

$$(((p_{00}, p_{01}), k_0), ((p_{10}, p_{11}), k_1), t, f) \in \widetilde{\tau}_4,$$

where (t, f) is an inverse structure of (p_0, p_1); see the definition of $\widetilde{\mu}$ and $\widetilde{\tau}$.

Let $\boldsymbol{v}_0 =_{\widetilde{\rho}(p_{00}, p_{01})} \boldsymbol{w}_0$, $\boldsymbol{v}_1 =_{\widetilde{\rho}(p_{10}, p_{11})} \boldsymbol{w}_1$, and let an assignment $\boldsymbol{v}$ be such that $(\boldsymbol{w}_0, \boldsymbol{w}_1, \boldsymbol{v})$ is an instance of (t, f). Then

$$(((p_{00}, p_{01}), \boldsymbol{v}_0), \sigma_0), (((p_{10}, p_{11}), \boldsymbol{v}_1), \sigma_1) \rightarrow ((p_0, p_1), \boldsymbol{v}) \in \widetilde{\mu}^c.$$

Since $\boldsymbol{w}_0 =_{\rho(p_0)} \boldsymbol{w}$ and $\boldsymbol{w}_1 =_{\rho(p_1)} \boldsymbol{w}$, by the first part of Lemma 2, $\boldsymbol{v} =_{\widetilde{\rho}(p_0, p_1)} \boldsymbol{w}$.[14]

---

[14] Recall that $\widetilde{\rho}(p_0, p_1) = \rho(p_0) \cap \rho(p_1)$.

For the proof of the converse part of the lemma, let

$$(((p_{00}, p_{01}), \boldsymbol{v}_0), \sigma_0), (((p_{10}, p_{11}), \boldsymbol{v}_1), \sigma_1) \rightarrow ((p_0, p_1), \boldsymbol{v}) \in \widetilde{\mu}^c.$$

That is, there exist

- $\boldsymbol{v}'_0 \in \Sigma^{r\neq}$, $\boldsymbol{v}'_0 = v'_{0,1} v'_{0,2} \cdots v'_{0,r}$, such that $\boldsymbol{v}'_0 =_{\widetilde{\rho}(p_{00}, p_{01})} \boldsymbol{v}_0$;
- $\boldsymbol{v}'_1 \in \Sigma^{r\neq}$, $\boldsymbol{v}'_1 = v'_{1,1} v'_{1,2} \cdots v'_{1,r}$, such that $\boldsymbol{v}'_1 =_{\widetilde{\rho}(p_{10}, p_{11})} \boldsymbol{v}_1$;
- $((p_{00}, p_{01}), k_0), ((p_{10}, p_{11}), k_1) \rightarrow (p_0, p_1) \in \widetilde{\mu}$, where $v'_{0,k_0} = \sigma_0$ and $v'_{1,k_1} = \sigma_1$; and
- $((p_{00}, p_{01}), k_0, (p_{10}, p_{11}), k_1, t, f) \in \widetilde{\tau}$, where $(\boldsymbol{v}'_0, \boldsymbol{v}'_1, \boldsymbol{v})$ is an instance of (t, f).

By the definition of $\widetilde{\mu}$, both transitions $(p_0, k_0) \rightarrow (p_{00}, p_{01})$ and $(p_1, k_1) \rightarrow (p_{10}, p_{11})$ are in μ; and (t, f) is an inverse structure of (p_0, p_1).

Let $\boldsymbol{w}$ be an assignment such that $\boldsymbol{w} =_{\widetilde{\rho}(p_0, p_1)} \boldsymbol{v}$. Since $\boldsymbol{w} =_{\widetilde{\rho}(p_0, p_1)} \boldsymbol{v}$ and $(\boldsymbol{v}'_0, \boldsymbol{v}'_1, \boldsymbol{v})$ is an instance of (t, f), by the second part of Lemma 2, $\boldsymbol{w} =_{\rho(p_0)} \boldsymbol{v}'_0$ and $\boldsymbol{w} =_{\rho(p_1)} \boldsymbol{v}'_1$. Therefore, we can put $\boldsymbol{w}_0 = \boldsymbol{v}'_0$ and $\boldsymbol{w}_1 = \boldsymbol{v}'_1$, implying

$$(p_0, \boldsymbol{w}), \sigma_0 \rightarrow (p_{00}, \boldsymbol{w}_0), (p_{01}, \boldsymbol{w}_0) \in \mu^c$$

and

$$(p_1, \boldsymbol{w}), \sigma_1 \rightarrow (p_{10}, \boldsymbol{w}_1), (p_{11}, \boldsymbol{w}_1) \in \mu^c.$$

Since $\boldsymbol{v}'_0 =_{\widetilde{\rho}(p_{00}, p_{01})} \boldsymbol{v}_0$ and $\boldsymbol{v}'_1 =_{\widetilde{\rho}(p_{10}, p_{11})} \boldsymbol{v}_1$, the converse part of the lemma follows. $\qquad\square$

B Proof of the "if" Part of Theorem 1

For an r-register $\uparrow$-NR-FMA $A = \langle S, s_0, \boldsymbol{u}, \rho, \tau, \mu, F \rangle$ we construct a $2r$-register $\downarrow$-NR-FMA $\widetilde{A} = \langle \widetilde{S}, \widetilde{s}_0, \widetilde{\boldsymbol{u}}, \widetilde{\rho}, \widetilde{\mu}, \widetilde{F} \rangle$ such that $L(A) = L(\widetilde{A})$.

Like in the previous proof, we assume that

- $\boldsymbol{u} = \#^{r-m} \theta_1 \cdots \theta_m$, where $\theta_1, \ldots, \theta_m \in \Sigma$, and
- only the first $r - m$ registers of A can be reassigned, i.e, the range of ρ is a subset of $\{1, 2, \ldots, r - m\}$.

We precede the formal description of $\widetilde{A}$ with a general intuitive explanation. One would expect the construction to resemble the reversing the classical automata, i.e., a transition $(p_0, k_0), (p_1, k_1) \rightarrow p$ of A to become the "transition" $(p, k_0, k_1) \rightarrow (p_0, p_1)$ of $\widetilde{A}$. This is indeed almost so, but with the following modification. Dually to the construction in Appendix A, reversing of a bottom-up transition $(p_0, k_0), (p_1, k_1) \rightarrow p$ results in two top-down transitions $((p, 0), k_0) \rightarrow (p_0, 0), (p_0, 1)$ and $((p, 1), k_1) \rightarrow (p_1, 0), (p_1, 1)$ at the lower level. The state components 0 and 1 indicate the child nodes of the parent node, where these transitions are applied: 0 indicates the left child and 1 indicates the right one.

One half of the $2r$ registers of $\widetilde{A}$, the *main* registers, is intended to contain the corresponding assignment of A at the parent node, while the other half is intended to "recover" the symbols forgotten in the merging. To identify the (r out of $2r$) main registers, the states of $\widetilde{A}$ are equipped with a pointer function

$$\pi : \{1, 2, \ldots, r\} \to \{1, 2, \ldots, 2r\},$$

where the value $\pi(i)$ is the index of the register of $\widetilde{A}$ containing the symbol stored in ith register of A. These pointers are also used to mimic the merging relation. That is, the pointers at the child nodes and the pointers at the parent node are defined in such a way that mimics the type and the valid selector used to merge the assignments at the child nodes.

More precisely,

$$\widetilde{S} = S \times \{0, 1\} \times \Pi_r^2 \times \mathcal{T}_r \times \mathcal{F}_r \cup \{\widetilde{s}_0\},$$

where $\widetilde{s}_0$ is a *new* state (the initial state of $\widetilde{A}$), Π_r is the set of all injective functions from $\{1, \ldots, r\}$ into $\{1, \ldots, 2r\}$, and $\mathcal{F}_r$ is the set of functions from $\{1, \ldots, r\}$ into $\{0, 1\} \times \{1, \ldots, r\}$. A bottom-up transition $(p_0, k_0), (p_1, k_1) \to p$ and a corresponding "merging attribute" $((p_0, k_0), (p_1, k_1), t, f)$ are simulated by two top-down transitions

$$((p, 0, \pi, \pi_0, t, f), \pi_0(k_0)) \to (p_0, 0, \pi_0, \pi_0', t_0, f_0), (p_0, 1, \pi_0, \pi_0'', t_0, f_0)$$

and

$$((p, 1, \pi, \pi_1, t, f), \pi_1(k_1)) \to (p_1, 0, \pi_1, \pi_1', t_1, f_1), (p_1, 1, \pi_1, \pi_1'', t_1, f_1),$$

where

1. $\pi_0(i) = \pi_1(j)$ implies $(i, j) \in t$, and
2. $\pi(j) = \begin{cases} \pi_0(i) \text{ if } f(j) = (0, i) \\ \pi_1(i) \text{ if } f(j) = (1, i) \end{cases}$

A triple of pointers (π_0, π_1, π) satisfying the above conditions 1 and 2 is said to *comply with* the pair (t, f).

The pointers π_0 and π_1 in the states $(p, 0, \pi, \pi_0, t, f)$ and $(p, 1, \pi, \pi_1, t, f)$, respectively, point at the assignments of A at the corresponding child nodes, and the pointer π points at the assignment A at the parent node.

The registers whose indices lie outside of the ranges of π_0 and π_1 are intended to contain the forgotten symbols which are recovered by a non-deterministic reassignment. That is, we define

$$\widetilde{\rho}(p, 0, \pi, \pi_0, t, f) =$$

$$\{\pi_0(i) : f(j) = (0, i) \text{ and } j \in \rho(s)\} \cup \Big(\{1, \ldots, 2r\} \setminus \mathrm{Range}(\pi_0)\Big)$$

and

$$\widetilde{\rho}(p, 1, \pi, \pi_1, t, f) =$$
$$\{\pi_1(i) : f(j) = (1, i) \text{ and } j \in \rho(s)\} \cup \Big(\{1, \ldots, 2r\} \setminus \text{Range}(\pi_1)\Big).$$

The above description of $\widetilde{A}$ is illustrated in Fig. 11 and 12 below. In Fig. 11 w, w_0, and w_1 are the assignments at the nodes labeled with the states p, p_0, and p_1, respectively,[15] in a run of A. Fig. 12 shows the reversing of the transition in Fig. 11. The intended meaning of the states in Fig. 12 (that depicts the corresponding run of $\widetilde{A}$) is that $\pi_0(v_0) =_{\rho(p_0)} w_0$, $\pi_1(v_1) =_{\rho(p_1)} w_1$, $\pi(v) =_{\rho(p)} w$,[16] and t and f are the type and the valid selector applied in the merging transition of A that results in w.

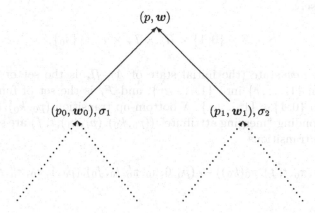

Fig. 11. An application of a bottom-up transition of A

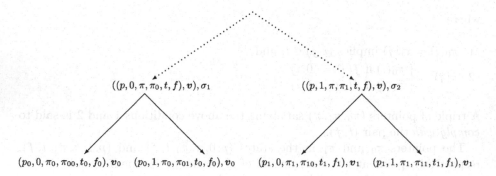

Fig. 12. Reversing and "splitting" the transition of A in Fig. 11

We proceed with the formal description of $\widetilde{A} = \langle \widetilde{S}, \widetilde{s}_0, \widetilde{u}, \widetilde{\rho}, \widetilde{\mu}, \widetilde{F} \rangle$ (that has $2r$ registers). Even though some of the components of $\widetilde{A}$ have been define earlier, we list them one more time fore the sake of continuity.

---

[15] In particular, w results in merging w_0 and w_1 after reassignment.

[16] $\pi(w_1 \cdots w_r)$ denotes $w_{\pi(1)} \cdots w_{\pi(r)}$, etc..

- $\widetilde{S} = S \times \{0,1\} \times \Pi_r^2 \times \mathcal{T}_r \times \mathcal{F}_r \cup \{\widetilde{s}_0\}$, where $\widetilde{s}_0$ is a new state.
- $\widetilde{s}_0$ is the initial state.
- $\widetilde{u} = \#^{r-m}\theta_1 \cdots \theta_m \#^r$.
- The reassignment $\widetilde{\rho}$ is defined as follows.

$$\widetilde{\rho}(\widetilde{s}_0) = \{1, \ldots, r-m\} \cup \{r+1, \ldots, 2r\},$$

and for each $p \in S$,
$\widetilde{\rho}(p, 0, \pi, \pi_0, t, f) =$

$$\{\pi_0(i) : f(j) = (0, i) \text{ and } j \in \rho(p)\} \cup \Big(\{1, \ldots, 2r\} \setminus \text{Range}(\pi_0)\Big)$$

and
$\widetilde{\rho}(p, 1, \pi, \pi_1, t, f) =$

$$\{\pi_1(i) : f(j) = (1, i) \text{ and } j \in \rho(p)\} \cup \Big(\{1, \ldots, 2r\} \setminus \text{Range}(\pi_1)\Big).$$

- The transition relation $\widetilde{\mu}$ consists of the following transitions.
  - For each $(p, k) \in F$ and all π_0, π_1 such that (π_0, π_1, π_{id})[17] complies with (t, f) it contains

$$(\widetilde{s}_0, k) \to (p, 0, \pi_{id}, \pi_0, t, f), (p, 1, \pi_{id}, \pi_1, t, f);$$

  and
  - for each $(p_0, k_0), (p_1, k_1) \to p \in \mu$, each $((p_0, k_0), (p_1, k_1), t, f) \in \tau$, and all $\pi, \pi_0, \pi_1, \pi_{00}, \pi_{01}, \pi_{10}, \pi_{11} \in \Pi_r$, $t_0, t_1 \in \mathcal{T}_r$, and $f_0, f_1 \in \mathcal{F}_r$ such that (π_0, π_1, π) complies with (t, f), $(\pi_{00}, \pi_{01}, \pi_0)$ complies with (t_0, f_0), and $(\pi_{10}, \pi_{11}, \pi_1)$ complies with (t_1, f_1), it contains both

$$(p, 0, \pi, \pi_0, t, f), \pi_0(k_0) \to (p_0, 0, \pi_0, \pi_{00}, t_0, f_0), (p_1, 1, \pi_0, \pi_{01}, t_0, f_0),$$

  and

$$(p, 1, \pi, \pi_1, t, f), \pi_1(k_1) \to (p_1, 0, \pi_1, \pi_{10}, t_1, f_1), (p_1, 1, \pi_1, \pi_{11}, t_1, f_1).$$

- Finally, $\widetilde{F}$ is defined as follows. For each $(s_0, k_0), (s_0, k_1) \to p \in \mu$ (that starts from the initial state s_0) each $((s_0, k_0), (s_0, k_1), t, f) \in \tau$, and all $\pi, \pi_0, \pi_1 \in \Pi_r$ such that (π_0, π_1, π) complies with (t, f), it contains

$$((p, 0, \pi, \pi_0, t, f), \pi_0(k_0))$$

  and

$$((p, 1, \pi, \pi_1, t, f), \pi_1(k_1)).$$

The proof of the equality $L(\widetilde{A}) = L(A)$ is based on Lemma 4 below that is a formalization of Fig. 11 and 12. It shows how an accepting run of A can be "reversed" into an accepting run of $\widetilde{A}$, or more precisely, it shows how transitions from μ^c are converted into transitions from $\widetilde{\mu}^c$, and vice versa.

---

[17] Here π_{id} denotes the identity function on $\{1, 2, \ldots, r\}$. That is, $\pi_{id}(i) = i$, for all $i = 1, \ldots, r$.

Lemma 4 consists of four parts corresponding to the type of nodes on which transitions take place. Its part (i) deals with the case of the root node ϵ. Part (ii) of the lemma settles the case of the interior nodes. Part (iii) shows how an accepting run of A can be "reversed" into an accepting run of $\widetilde{A}$ when one of the two sibling nodes is a leaf and the other is an interior node. Finally, part (iv) of Lemma 3 shows how an accepting run of A can be "reversed" into an accepting run of $\widetilde{A}$ at the leaf nodes and vice versa.

Lemma 4

(i) (See Fig. 13 and 14.) *For all pointers* π_0, π_1, *all types* t, *all valid selectors* f *for* t *such that* (π_0, π_1, π_{id}) *comply with* (t, f), *and all final relations* $((p, \boldsymbol{w}), \sigma) \in F^c$, *there is an assignment* $\boldsymbol{v}$ *such that* $\pi_{id}(\boldsymbol{v}) =_{\rho(p)} \boldsymbol{w}$ *and*

$$((\widetilde{s}_0, \widetilde{\boldsymbol{u}}), \sigma) \to ((p, 0, \pi_{id}, \pi_0, t, f), \boldsymbol{v}), ((p, 1, \pi_{id}, \pi_1, t, f), \boldsymbol{v}) \in \widetilde{\mu}^c.$$

Conversely, for all

$$((\widetilde{s}_0, \widetilde{\boldsymbol{u}}), \sigma) \to ((p, 0, \pi_{id}, \pi_0, t, f), \boldsymbol{v}), ((p, 1, \pi_{id}, \pi_1, t, f), \boldsymbol{v}) \in \widetilde{\mu}^c$$

and all assignments $\boldsymbol{w}$ *such that* $\boldsymbol{w} =_{\rho(p)} \pi_{id}(\boldsymbol{v})$, $((p, \boldsymbol{w}), \sigma) \in F^c$.

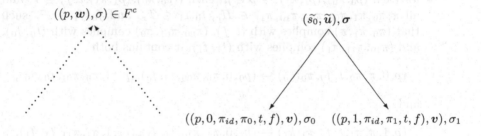

Fig. 13. An application of a final relation of A at the root node

Fig. 14. Reversing the final relation of A at the root node in Fig. 13

(ii) (See Fig. 11 and 12.) *For all*

$$(p_0, \boldsymbol{w}_0), \sigma_0, (p_1, \boldsymbol{w}_1), \sigma_1 \to (p, \boldsymbol{w}) \in \mu^r,$$

all types t *and all valid selectors* f *for* t *which yield* $\boldsymbol{w}$ *in the above transition, all assignments* $\boldsymbol{v}$ *such that* $\pi(\boldsymbol{v}) =_{\rho(p)} \boldsymbol{w}$, *all pointers* π, π_0, π_1, $\pi_{00}, \pi_{01}, \pi_{10}, \pi_{11}$, *and all types* t_0 *and* t_1 *and valid selectors* f_0 *and* f_1 *for* t_0 *and* t_1, *respectively, such that* (π_0, π_1, π) *comply with* (t, f), $(\pi_{00}, \pi_{01}, \pi_0)$ *comply with* (t_0, f_0), *and* $(\pi_{10}, \pi_{11}, \pi_1)$ *comply with* (t_1, f_1), *there are assignments* $\boldsymbol{v}_0$ *and* $\boldsymbol{v}_1$ *such that* $\pi_0(\boldsymbol{v}_0) =_{\rho(p_0)} \boldsymbol{w}_0$, $\pi_1(\boldsymbol{v}_1) =_{\rho(p_1)} \boldsymbol{w}_1$, *and both transitions*

$(((p, 0, \pi, \pi_0, t, f), \boldsymbol{v}), \sigma_0) \rightarrow$

$$((p_0, 0, \pi_0, \pi_{00}, t_0, f_0), \boldsymbol{v}_0, (p_0, 1, \pi_0, \pi_{01}, t_0, f_0), \boldsymbol{v}_0)$$

and

$(((p, 1, \pi, \pi_1, t, f), \boldsymbol{v}), \sigma_1) \rightarrow$

$$((p_1, 0, \pi_1, \pi_{10}, t_1, f_1), \boldsymbol{v}_1, (p_1, 1, \pi_1, \pi_{11}, t_1, f_1), \boldsymbol{v}_1)$$

are in $\widetilde{\mu}^c$.

Conversely, for all transitions
$(((p, 0, \pi, \pi_0, t, f), \boldsymbol{v}), \sigma_0) \rightarrow$

$$((p_0, 0, \pi_0, \pi_{00}, t_0, f_0), \boldsymbol{v}_0, (p_0, 1, \pi_0, \pi_{01}, t_0, f_0), \boldsymbol{v}_0)$$

and

$(((p, 1, \pi, \pi_1, t, f), \boldsymbol{v}), \sigma_1) \rightarrow$

$$((p_1, 0, \pi_1, \pi_{10}, t_1, f_1), \boldsymbol{v}_1, (p_1, 1, \pi_1, \pi_{11}, t_1, f_1), \boldsymbol{v}_1)$$

in $\widetilde{\mu}^c$ and all assignments $\boldsymbol{w}_0$ and $\boldsymbol{w}_1$ such that $\pi_0(\boldsymbol{v}_0) =_{\rho(p_0)} \boldsymbol{w}_0$ and $\pi_1(\boldsymbol{v}_1) =_{\rho(p_1)} \boldsymbol{w}_1$, there is an assignment $\boldsymbol{w}$ such that $\pi(\boldsymbol{v}) =_{\rho(p)} \boldsymbol{w}$ and

$$(p_0, \boldsymbol{w}_0), \sigma_0, (p_1, \boldsymbol{w}_1), \sigma_1 \rightarrow (p, \boldsymbol{w}) \in \mu^c,$$

(*iii*) (*a*) (See Fig. 15 and 16.) *For all transitions*

$$((p_0, \boldsymbol{w}_0), \sigma_0), ((s_0, \boldsymbol{u}), \sigma_1) \rightarrow (p, \boldsymbol{w}) \in \mu^c,$$

all types t and all valid selectors f for t which yield $\boldsymbol{w}$ in the above transition, all assignments $\boldsymbol{v}$ such that $\pi(\boldsymbol{v}) =_{\rho(p)} \boldsymbol{w}$, all pointers π, π_0, π_1, π_{00}, π_{01}, and all types t_0 and valid selectors f_0 for t_0 such that (π_0, π_1, π) comply with (t, f) and $(\pi_{00}, \pi_{01}, \pi_0)$ comply with (t_0, f_0), there is an assignment $\boldsymbol{v}_0$ such that $\pi_0(\boldsymbol{v}_0) =_{\rho(p_0)} \boldsymbol{w}_0$,

$$(((p, 1, \pi, \pi_1, t, f), \boldsymbol{v}), \sigma_1) \in \widetilde{F}^c,$$

and
$(((p, 0, \pi, \pi_0, t, f), \boldsymbol{v}), \sigma_0) \rightarrow$
$$((p_0, 0, \pi_0, \pi_{00}, t_0, f_0), \boldsymbol{v}_0, (p_0, 1, \pi_0, \pi_{01}, t_0, f_0), \boldsymbol{v}_0) \in \widetilde{\mu}^c.$$

Conversely, for all

$$(((p, 1, \pi, \pi_1, t, f), \boldsymbol{v}), \sigma_1) \in \widetilde{F}^c,$$

and

$((p, 0, \pi, \pi_0, t, f), \boldsymbol{v}), \sigma_0 \rightarrow$
$$((p_0, 0, \pi_0, \pi_{00}, t_0, f_0), \boldsymbol{v}_0, (p_0, 1, \pi_0, \pi_{01}, t_0, f_0), \boldsymbol{v}_0 \in \widetilde{\mu}^c,$$

and all assignment w_0 such that $w_0 =_{\rho(p_0)} \pi_0(v_0)$, there is an assignment w such that $w =_{\rho(p)} \pi(v)$ and

$$((p_0, w_0), \sigma_0), ((s_0, u), \sigma_1) \to (p, w) \in \mu^c.$$

(b) For all transitions

$$((s_0, u), \sigma_0), ((p_1, w_1), \sigma_1) \to (p, w) \in \mu^c,$$

all types t and all valid selectors f for t which yield w in the above transition, all assignments v such that $\pi(v) =_{\rho(p)} w$ and for all pointers π, π_0, π_1, π_{10}, π_{11}, ad all types t_1 and valid selectors f_1 for t_1 such that (π_0, π_1, π) comply with (t, f) and $(\pi_{10}, \pi_{11}, \pi_1)$ comply with (t_1, f_1), there is an assignment v_1 such that $\pi_1(v_1) =_{\rho(p_1)} w_1$,

$$(((p, 0, \pi, \pi_0, t, f), v), \sigma_0) \in \widetilde{F}^c$$

and

$$((p, 1, \pi, \pi_1, t, f), v), \sigma_1 \to$$
$$(p_1, 0, \pi_1, \pi_{10}, t_1, f_1), v_1, (p_1, 1, \pi_1, \pi_{11}, t_1, f_1), v_1 \in \widetilde{\mu}^c.$$

Conversely, for all

$$((p, 1, \pi, \pi_1, t, f), v), \sigma_1 \to$$
$$(p_1, 0, \pi_1, \pi_{10}, t_1, f_1), v_1, (p_1, 1, \pi_1, \pi_{11}, t_1, f_1), v_1 \in \widetilde{\mu}^c,$$

and

$$(((p, 0, \pi, \pi_0, t, f), v), \sigma_0) \in \widetilde{F}^c,$$

and all assignments w_1 such that $w_1 =_{\rho(p_1)} \pi_1(v_1)$, there is an assignment w such that $w =_{\rho(p)} \pi(v)$ and

$$((s_0, u), \sigma_0), ((p_1, w_1), \sigma_1) \to (p, w) \in \mu^c.$$

(iv) (See Fig. 17 and 18.) For all

$$((s_0, u), \sigma_0), ((s_0, u), \sigma_1) \to (p, w) \in \mu^c,$$

all types t and all valid selectors f for t which yield w in the above transition, all assignments v such that $\pi(v) =_{\rho(p)} w$, and for all pointers π, π_0, π_1 such that (π_0, π_1, π) comply with (t, f),

$$(((p, 0, \pi, \pi_0, t, f), v), \sigma_0) \in \widetilde{F}^c$$

and

$$(((p, 1, \pi, \pi_1, t, f), v), \sigma_1) \in \widetilde{F}^c.$$

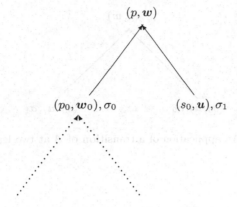

Fig. 15. An application of a transition of A at a leaf node

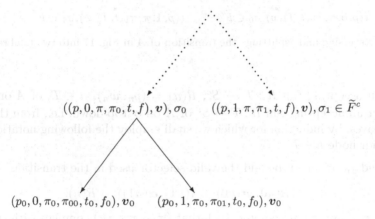

Fig. 16. Reversing and "splitting" the transition of A in Fig. 15 into a transition and a final relation

Conversely, for all pairs of final relations

$$(((p, 0, \pi, \pi_0, t, f), \boldsymbol{v}), \sigma_0), (((p, 1, \pi, \pi_1, t, f), \boldsymbol{v}), \sigma_1) \in \widetilde{F^c},$$

there is an assignment $\boldsymbol{w}$ such that $\boldsymbol{w} =_{\rho(p)} \pi(\boldsymbol{v})$ and

$$((s_0, \boldsymbol{u}), \sigma_0), ((s_0, \boldsymbol{u}), \sigma_1) \to (p, \boldsymbol{w}) \in \mu^c.$$

We postpone the proof of the lemmas to the end of this appendix and prove the "if" part of Theorem 1 first.

Proof. (of the "if" part of Theorem 1.) We prove that $L(A) = L(\widetilde{A})$ by showing how to convert an accepting run of A on a Σ-tree $\boldsymbol{\sigma} : T \to \Sigma$, into an accepting run of $\widetilde{A}$ on $\boldsymbol{\sigma}$, and vice versa.

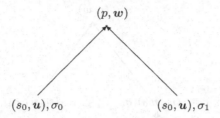

Fig. 17. An application of a transition of A at two leaf nodes

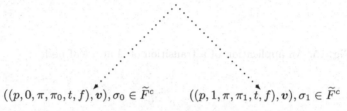

$((p, 0, \pi, \pi_0, t, f), \boldsymbol{v}), \sigma_0 \in \widetilde{F}^c$ $((p, 1, \pi, \pi_1, t, f), \boldsymbol{v}), \sigma_1 \in \widetilde{F}^c$

Fig. 18. Reversing and "splitting" the transition of A in Fig. 17 into two final relations of $\widetilde{A}$

For an accepting run $R : T \to S^c$, $R(n) = (p_n, \boldsymbol{w}_n)$, $n \in T$, of A on $\boldsymbol{\sigma}$ we construct an accepting run $\widetilde{R} : T \to S^c$ of $\widetilde{A}$ on $\boldsymbol{\sigma}$ top-down, i.e., from the root to the leaves, by induction for which we shall employ the following notation. For an interior node $n \in T$,

- t_n and f_n are the type and the valid selector used in the transition

$$(R(n0), \boldsymbol{\sigma}(n0)), (R(n1), \boldsymbol{\sigma}(n1)) \to R(n);$$

- π_{n0}, π_{n1}, and π_n are pointers such that $(\pi_{n0}, \pi_{n1}, \pi_n)$ complies with (t_n, f_n); and
- $\pi_\epsilon = \pi_{id}$.

By definition, for the root node ϵ, $\widetilde{R}(\epsilon) = \widetilde{s_0}^c = (\widetilde{s_0}, \widetilde{\boldsymbol{u}})$, and assume that $\widetilde{R}(n)$ has been constructed for a non-leaf node $n \in T$. Then

- $\widetilde{R}(n0) = ((p_n, 0, \pi_n, \pi_{n0}, t_n, f_n), \boldsymbol{v}_n)$ and
- $\widetilde{R}(n1) = ((p_n, 1, \pi_n, \pi_{n1}, t_n, f_n), \boldsymbol{v}_n)$,

where $\pi_{id}(\boldsymbol{v}_n) =_{\rho(p_n)} \boldsymbol{w}_n$ is defined below..

- The assignment $\boldsymbol{v}_\epsilon$ is provided by part (i) of Lemma 4, and
- for non-root interior node n, the assignment $\boldsymbol{v}_n$ is provided by part (ii) of Lemma 4.

By parts (iii) and (iv) of Lemma 4, $\widetilde{R}$ is an accepting run of $\widetilde{A}$ on $\boldsymbol{\sigma}$.

The converse direction can shown in a similar manner. That is, an accepting run R of A on $\boldsymbol{\sigma}$ is constructed from an accepting run $\widetilde{R}$ of $\widetilde{A}$ by applying the converse direction of Lemma 4.   □

It remains to prove Lemma 4.

Proof. (of Lemma 4.) We will prove only part (ii) of the lemma. The proofs of the other parts are very similar.

Let

$$(p_0, \boldsymbol{w}_0), \sigma_0, (p_1, \boldsymbol{w}_1), \sigma_1 \rightarrow (p, \boldsymbol{w}) \in \mu^c.$$

That is, there exist

- assignments $\boldsymbol{w}'_0 = w'_{0,1} w'_{0,2} \cdots w'_{0,r}$ and $\boldsymbol{w}'_1 = w'_{1,1} w'_{1,2} \cdots w'_{1,r}$ such that $\boldsymbol{w}'_0 =_{\rho(p_0)} \boldsymbol{w}_0$ and $\boldsymbol{w}'_1 =_{\rho(p_1)} \boldsymbol{w}_1$;
- a transition $(p_0, k_0), (p_1, k_1) \rightarrow p \in \mu$ such that $w'_{0,k_0} = \sigma_0$ and $w'_{1,k_1} = \sigma_1$; and
- a merging relation $((p_0, k_0), (p_1, k_1), t, f) \in \tau$ such that $(\boldsymbol{w}'_0, \boldsymbol{w}'_1, \boldsymbol{w})$ is an instance of (t, f).

Let $\pi_0, \pi_1, \pi, \pi_{00}, \pi_{01}, \pi_0, \pi_{10}, \pi_{11}, \pi_1$ be pointers and t_0, t_1 and f_0, f_1 be types and the corresponding valid selectors such that (π_0, π_1, π) complies with (t, f), $(\pi_{00}, \pi_{01}, \pi_0)$ complies with (t_0, f_0), and $(\pi_{10}, \pi_{11}, \pi_1)$ complies with (t_1, f_1).

By definition, $\tilde{\mu}$ contains both

$$(p, 0, \pi, \pi_0, t, f), \pi_0(k_0) \rightarrow (p, 0, \pi_0, \pi_{00}, t_0, f_0), (p, 1, \pi_0, \pi_{01}, t_0, f_0)$$

and

$$(p, 0, \pi, \pi_1, t, f), \pi_1(k_1) \rightarrow (p, 0, \pi_1, \pi_{10}, t_1, f_1), (p, 1, \pi_1, \pi_{11}, t_1, f_1)$$

Let $\boldsymbol{v}$ be an assignment such that $\pi(\boldsymbol{v}) =_{\rho(p)} \boldsymbol{w}$. By the definition of the reassignments $\tilde{\rho}((p, 0, \pi, \pi_0, t, f))$ and $\tilde{\rho}((p, 1, \pi, \pi_0, t, f))$, there are assignments

- $\boldsymbol{v}_0 =_{\tilde{\rho}((p,0,\pi,\pi_0,t,f))} \boldsymbol{v}$ such that $\pi_0(\boldsymbol{v}_0) = \boldsymbol{w}'_0$, and
- $\boldsymbol{v}_1 =_{\tilde{\rho}((p,1,\pi,\pi_0,t,f))} \boldsymbol{v}$ such that $\pi_1(\boldsymbol{v}_1) = \boldsymbol{w}'_1$.

Since (π_0, π_1, π) comply with (t, f), $\pi(\boldsymbol{v}_0) = \boldsymbol{w} = \pi(\boldsymbol{v}_1)$. Thus, both

$$((p, 0, \pi, \pi_0, t, f), \boldsymbol{v}), \sigma_0 \rightarrow (p_0, 0, \pi_0, \pi_{00}, t_0, f_0), \boldsymbol{v}_0, (p_0, 1, \pi_0, \pi_{01}, t_0, f_0), \boldsymbol{v}_0$$

and

$$((p, 1, \pi, \pi_1, t, f), \boldsymbol{v}), \sigma_1 \rightarrow (p_1, 0, \pi_1, \pi_{10}, t_1, f_1), \boldsymbol{v}_1, (p_1, 1, \pi_1, \pi_{11}, t_1, f_1), \boldsymbol{v}_1$$

are in $\tilde{\mu}^c$.

For the proof of converse part of the lemma, let

$$((p, 0, \pi, \pi_0, t, f), \boldsymbol{v}), \sigma_0 \rightarrow (p_0, 0, \pi_0, \pi_{00}, t_0, f_0), \boldsymbol{v}_0, (p_0, 1, \pi_0, \pi_{01}, t_0, f_0), \boldsymbol{v}_0 \in \tilde{\mu}^c$$

and

$$((p, 1, \pi, \pi_1, t, f), \boldsymbol{v}), \sigma_1 \rightarrow (p_1, 0, \pi_1, \pi_{10}, t_1, f_1), \boldsymbol{v}_1, (p_1, 1, \pi_1, \pi_{11}, t_1, f_1), \boldsymbol{v}_1 \in \tilde{\mu}^c$$

where $v_0 =_{\widetilde{\rho}((p,0,\pi,\pi_0,t,f))} v$, $v_1 =_{\widetilde{\rho}((p,1,\pi,\pi_1,t,f))} v$, and (π_0, π_1, π), $(\pi_{00}, \pi_{01}, \pi_0)$, and $(\pi_{10}, \pi_{11}, \pi_1)$ comply with (t, f), (t_0, f_0), and (t_1, f_1), respectively.

Therefore, by definition, there are transitions

$$(p, 0, \pi, \pi_0, t, f), \pi_0(k_0) \to (p_0, 0, \pi_0, \pi_{00}, t_0, f_0), (p_0, 1, \pi_0, \pi_{01}, t_0, f_0) \in \widetilde{\mu}$$

and

$$(p, 1, \pi, \pi_1, t, f), \pi_1(k_1) \to (p_1, 0, \pi_1, \pi_{10}, t_1, f_1), (p_1, 1, \pi_1, \pi_{11}, t_1, f_1) \in \widetilde{\mu}$$

such that $v_{0,\pi_0(k_0)} = \sigma_0$ and $v_{1,\pi_1(k_1)} = \sigma_1$.

By the definition of $\widetilde{\mu}$,

$$(p_0, k_0), (p_1, k_1) \to p \in \mu$$

and

$$((p_0, k_0), (p_1, k_1), t, f) \in \tau.$$

Let w_0, w_1 be assignments such that $w_0 =_{\rho(p_0)} \pi_0(v)$ and $w_1 =_{\rho(p_1)} \pi_1(v)$, and let denote $w_0' = \pi_0(v)$ and $w_1' = \pi_1(v)$.

Since $w_{0,k_0} = v_{\pi_0(k_0)} = \sigma_0$ and $w_{1,k_1} = v_{\pi_1(k_1)} = \sigma_1$,

$$(p_0, w_0), \sigma_0, (p_1, w_1), \sigma_1 \to (p, w) \in \mu^c$$

where w is an assignment such that (w_0', w_1', w) is an instance of (t, f).

Thus, the proof will be complete if we show that $\pi(v) =_{\rho(p)} w$. Let $i \notin \rho(p)$. Since (π_0, π_1, π) complies with (t, f), and (w_0', w_1', w) is an instance of (t, f),

$$v_{\pi(i)} = \begin{cases} v_{\pi_0(j)} = v_{0,\pi_0(j)} = w_{0,j}' = w_i, \text{ where } f(j) = (0, i) \\ v_{\pi_1(j)} = v_{1,\pi_1(j)} = w_{1,j}' = w_i, \text{ where } f(j) = (1, i) \end{cases}$$

and the converse part of the lemma follows.   □

C   Closure Properties

In this section we establish some basic closure properties of the tree languages defined defined by $\downarrow$-NR-FMAs. The proofs are pretty standard and can be easily modified for all other models of tree automata introduced in this paper.

Let $A_i = \langle S_i, s_{0,i}, u_i, \rho_i, \mu_i, F_i \rangle$, $i = 1, 2$, be $\downarrow$-FMAs. Without loss of generality, we assume that A_1 and A_2 possess the following properties.

1. The sets of states S_1 and S_2 are disjoint.
2. Both the initial states $s_{0,1}$ and $s_{0,2}$ are not accessible from the other states of the corresponding automaton. That is, there is no transition of the form $(q, k) \to (s_{0,i}, q')$ or $(q, k) \to (q', s_{0,i})$, $i = 1, 2$.
3. Both automata have the same initial assignment of the form $\theta_1 \cdots \theta_m \#^r$, and registers $1, \ldots, m$ are never reset.[18] This property can be verified similarly to the proof of [7, Lemma 5.1].

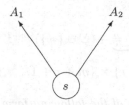

Fig. 19. The diagram of A

Closure under union. We construct an automaton A that accepts $L(A_1) \cup L(A_2)$ as the "union" of A_1 and A_2. The initial state of A is a new state s. ¿From this state A can simulate either of A_1 and A_2, as illustrated in Fig. 19.

The initial assignment of A is $\theta_1 \cdots \theta_m \# ^{2r}$. Except for the input symbols from $\{\theta_1, \theta_2, \ldots, \theta_m\}$, A_1 is simulated by the registers $(m+1)$ to $(m+r)$, while A_2 is simulated by the last r registers. The transitions of A_2 are renamed from $(q, k) \rightarrow (q_0, q_1)$ to $(q, k+r) \rightarrow (q_0, q_1)$ for all $k = m+1, m+2, \cdots, m+r$. For the inputs from $\{\theta_1, \theta_2, \ldots, \theta_m\}$, both A_1 and A_2 use the first m registers. Finally, $\rho(s) = \rho(s_1) \cup \{k+r : k \in \rho(s_2)\}$.

Closure under concatenation.[19] An automaton A accepting $L(A_1)L(A_2)$ results in "extending" A_1 with A_2. That is, A starts by simulating A_1 until it reaches a final relation. Then it continues by simulating A_2. This is achieved by changing all the final relations $(q, k) \in F_1$ to $(q, k) \rightarrow (s_{0,2}, s_{0,2})$. The set of final relations of A is F_2.

Closure under Kleene star.[20] The construction is similar to the concatenation case. To accept $L(A_1)^*$, the automaton A simulate A_1 a number of times, which is achieved by adding the transition $(q, k) \rightarrow (s_{0,1}, s_{0,1})$ to μ_1 for each final relation $(q, k) \in F_1$ and setting $\rho(s_{0,1}) = \{m+1, \ldots, m+r\}$.

Closure under intersection. The construction here is a bit more involved. The basic idea is to use the (equivalent) tree automata model similar to the M-FMA introduced in [6, Sect. 3]. These automata are allowed to "consult" a number of registers at the same time.

Definition 7. (Cf. [20, Definition 5. pp. 20-22].) *A top-down M-finite-memory automaton ($\downarrow$-M-FMA) is a system $A = \langle S, s_0, \boldsymbol{u}, \rho, \mu, F \rangle$, where*

---

[18] Consequently, only registers $m+1, \ldots, m+r$ may be reset.

[19] The concatenation of two tree languages L_1 and L_2 is defined by extending each leaf in every tree from L_1 with two child nodes in each of which a tree from L_2 is rooted. In terms of the corresponding definition in [3, Sect. 2.2, p. 52], we can view this as first extending each leaf in every tree from L_1 with two children labeled with a new symbol $\square$ and then applying $\cdot_\square$.

[20] For a tree language L, L^* is the collection of all iterated concatenations of $L \cup \{\square\}$, where $\square$ is the "empty tree," i.e., a single node labeled with the symbol $\square$. In terms of the corresponding definition in [3, Sect. 2.2, p. 54], $L^* = L^{*,\square} \left(= \bigcup_{n \geq 0} L^{n,\square} \right)$.

- S is the finite set of states.
- s_0 is the initial state.
- $\boldsymbol{u} = \theta_1 \cdots \theta_m \underbrace{\# \cdots \#}_{r} \underbrace{\# \cdots \#}_{r} \in (\Sigma \cup \{\#\})^{(m+2r)}$ is the initial assignment.

- $\rho : S \to \{m+1, \ldots, m+r\} \times \{m+r+1, \ldots, m+2r\}$ is the reassignment function.
- μ is the set of transitions of the following form
  - $(p, k) \to (p_0, p_1) \in S \times \{1, \ldots, m\} \times S \times S$,
  - $(p, (k_0, k_1)) \to (p_0, p_1) \in S \times \{m+1, \ldots, m+r\} \times \{m+r+1, \ldots, m+2r\} \times S \times S$.
- F is the set of final relations of the following form
  - $(p, k) \in S \times \{1, \ldots, m\}$,
  - $(p, (k_0, k_1)) \in S \times \{m+1, \ldots, m+r\} \times \{m+r+1, \ldots, m+2r\}$.

Similarly, the transition relation μ induces the following relation μ^c which is defined as follows. $(p, \boldsymbol{w}), \sigma \to (p_0, \boldsymbol{w}'), (p_1, \boldsymbol{w}')$ belongs to μ^c if and only if the following conditions are satisfied. $\boldsymbol{w}' = w'_1 \cdots w'_{m+2r}$, where $w'_i = w_i$, for all $i \notin \{i, j : (i, j) \in \rho(p)\}$ and

1. If $\sigma \in \{\theta_1, \ldots, \theta_m\}$, then $\boldsymbol{w}'_k = \theta$ and $(p, k, (p_0, p_1)) \in \mu$.
2. If $\sigma \notin \{\theta_1, \ldots, \theta_m\}$, then $\sigma = w'_{k_0} = w'_{k_1}$, for some $k_0 < k_1$ and the triple $(p, (k_0, k_1), (p_0, p_1))$ belongs to μ.

The relation F^c is defined as follows. A pair $((p, \boldsymbol{w}), \sigma) \in F^c$ if the following holds.

- If $\sigma \in \{\theta_1, \ldots, \theta_m\}$, then $\boldsymbol{w}'_k = \theta$ and $(p, k, (p_0, p_1)) \in F$.
- If $\sigma \notin \{\theta_1, \ldots, \theta_m\}$, then $\sigma = w'_{k_0} = w'_{k_1}$, where $m < k_0 < k_1$ and $(p, (k_0, k_1), (p_0, p_1))$ belong to F.

Proposition 6. *For each $\downarrow$-M-NR-FMA A there exists an $\downarrow$-NR-FMA A' such that $L(A) = L(A')$.*

Proof. We convert A into a standard $(m + r^2)$-register $\downarrow$-NR-FMA A' whose initial assignment is $\theta_1 \cdots \theta_m \#^{r^2}$.

Let $\varphi : \{m+1, \ldots, m+r\} \times \{m+r+1, \ldots, m+2r\} \to \{m+1, \ldots, m+r^2\}$ be a one-to-one function. The simulation of A by A' is done by referring each pair of registers (k_0, k_1) to a single register $\varphi(k, k')$ of A'. Formally, $A' = \langle S', s'_0, \boldsymbol{u}', \rho', \mu', F'' \rangle$ is defined as follows.

- $S' = S$ and $s'_0 = s_0$.
- $\boldsymbol{u}' = \theta_1 \cdots \theta_m \#^{r^2}$.
- $\rho'(q_1, q_2) = \varphi(\rho(q_1) \times \{m+r+1, \ldots, m+2r\} \cup \{m+1, \ldots, m+r\} \times \rho(q_2))$.
- μ' consists of the following two types of transitions:
  - for every $((q_1, q_2)(k_1, k_2)) \to ((q'_1, q''_1)(q_2, q''_2)) \in \mu$ it contains

$$((q, q'), \varphi((k, k'))) \to ((q_0, q'_0), (q_1, q'_1)),$$

- for every $((q_1, q_2), k) \rightarrow ((q_1', q_2''), (q_2, q_2'')) \in \mu$, it contains

$$((q_1, q_2), k) \rightarrow ((q_1', q_2''), (q_2, q_2''))$$

  itself.
- F' consists of the following two types of relations:
  - for every $((q_1, q_2), (k_1, k_2)) \in F$ it contains $((q_1, q_2), \varphi(k_1, k_2))$, and
  - for every $((q_1, q_2), k) \in F$ it contains $((q_1, q_2), k)$ itself.   □

Now we construct an ↓-M-FMA A that accepts $L(A_1) \cap L(A_2)$ by simultaneously simulating A_1 and A_2. This is done by defining A as the product of A_1 and A_2.

The precise description of A is as follows. Like in the case of the closure under union, the initial assignment of A is $\theta_1 \cdots \theta_m \#^{2r}$ and the automata A_1 and A_2 are simulated on the registers $\{1, \ldots, m+r\}$ and $\{1, \ldots, m\} \cup \{m+r+1 \ldots, m+2r\}$, respectively. That is, $A = \langle S, s_0, \boldsymbol{u}, \rho, \mu, F \rangle$ is defined as follows.

- $S = S_1 \times S_2$.
- $s_0 = (s_{0,1}, s_{0,2})$.
- $\boldsymbol{u} = \theta_1 \cdots \theta_m \#^{2r}$.
- $\rho(q_1, q_2) = \{\rho_1(q_1), \rho_2(q_2) + r\}$.
- μ consists of the following two types of transitions:
  - for every $(q_1, k_1) \rightarrow (q_1', q_1'') \in \mu_1$ and every $(q_2, k_2) \rightarrow (q_2', q_2'') \in \mu_2$ such that $m \leq k_1, k_2 \leq m + r$, it contains $((q_1, q_2), (k_1, k_2 + r)) \rightarrow (q_1', q_2')(q_1'', q_1'')$, and
  - for every $(q_1, k) \rightarrow (q_1', q_1'') \in \mu_1$ and every $(q_2, k) \rightarrow (q_2', q_2'') \in \mu_2$ such that $1 \leq k \leq m$, it contains $((q_1, q_2), k) \rightarrow (q_1', q_2')(q_1'', q_1'')$.
- F consists of the following two types of relations:
  - for every $(q_1, k_1) \in F_1$ and every $(q_2, k_2) \in F_2$ such that $m \leq k_1, k_2 \leq m + r$, it contains $((q_1, q_2), (k_1, k_2 + r))$, and
  - for every $(q_1, k) \in F_1$ and every $(q_2, k) \in F_2$ such that $1 \leq k \leq m$, it contains $((q_1, q_2), k)$.

Connectives in Cumulative Logics*

Daniel Lehmann

School of Computer Science and Engineering, Hebrew University,
Jerusalem 91904, Israel
lehmann@cs.huji.ac.il

It is a great pleasure for me to present those reflections to the Festschrift of
Boaz (Boris) Trakhtenbrot, who, with great constancy, manifested his interest
in my explorations in quantic and other exotic logics and whose
encouragements have been most appreciated.[1]

Abstract. Cumulative logics are studied in an abstract setting, i.e.,
without connectives, very much in the spirit of Makinson's [11] early
work. A powerful representation theorem characterizes those logics by
choice functions that satisfy a weakening of Sen's property α, in the spirit
of the author's [9]. The representation results obtained are surprisingly
smooth: in the completeness part the choice function may be defined on
any set of worlds, not only definable sets and no definability-preservation
property is required in the soundness part. For abstract cumulative logics, proper conjunction and negation may be defined. Contrary to the
situation studied in [9] no proper disjunction seems to be definable in
general. The cumulative relations of [8] that satisfy some weakening of
the consistency preservation property all define cumulative logics with
a proper negation. Quantum Logics, as defined by [3] are such cumulative logics but the negation defined by orthogonal complement does not
provide a proper negation.

1 Introduction

The study of nonmonotonic logics on a language without connectives was started
in [9], where logics related to the *preferential* system of [8] have been studied.
The present paper pursues this approach by studying less powerful nonmonotonic
logics related to the *cumulative* and *loop-cumulative* systems there. It also asks
the question: which are the natural connectives in those logics?

One reason for being interested in such logics is the remarkable presentation
of Quantum Logic as a cumulative, even loop-cumulative, logic proposed by Engesser and Gabbay in [3]. In [1], Birkhoff and von Neumann suggested that the

---

* This work was partially supported by the Jean and Helene Alfassa fund for research
in Artificial Intelligence.

[1] A preliminary version of the present work was written in May 2002 and then revised in August 2002. It has been presented in an invited lecture to the 2002
ASL European Summer meeting in Muenster, Germany. It appeared as Leib-
niz Center for Research in Computer Science TR-2002-28 and is available at
http://arxiv.org/pdf/cs/0205079.

A. Avron et al. (Eds.): Trakhtenbrot/Festschrift, LNCS 4800, pp. 424–440, 2008.
© Springer-Verlag Berlin Heidelberg 2008

logic of quantum mechanics be isomorphic to the algebra of closed subspaces of Hilbert spaces, under "set product" (i.e., intersection), "closed linear sum", and "orthogonal complement". Many researchers studied the properties of those operations and their results are reviewed in [2]. Recently, Engesser and Gabbay [3] proposed a very different and most intriguing connection between Logics and Quantum mechanics. For them every quantum state defines a consequence relation. They showed that those consequence relations are nonmonotonic. They also showed they satisfy cumulativity, the focus of early studies of nonmonotonicity, in particular [11,8,6,12,4,5]. Whereas Engesser and Gabbay assume a language closed under the propositional connectives (as did Birkhoff and von Neumann), even though those connectives are not at all classical, the purpose of this paper is to study and try to characterize the consequence operations presented by Quantum mechanics before any connectives are defined, in the style of the author's [9] and in the tradition of Makinson's early work. Since Quantum Logics fail, in general, to satisfy two of the properties assumed there, representation results for larger families than those of [9] are needed. Such results will be developed first. For the conservative extension results to be proven below, models closer to the cumulative models of [8] or of [12] could have been used. The models presented here and their tight link with the failure of Coherence have been preferred both for their intrinsic interest and for compatibility with [9].

All in all, the connectives proposed in the present work are probably not the right ones for Quantum Physics. I think that quantum negation must be interpreted as orthogonal complementation and, most importantly, that quantum conjunction should not be a commutative operation interpreted as intersection, as proposed by Birkhoff and Von Neumann but should be some non-commutative operation. Such ideas are developed in [10].

2 C-logics

2.1 Definition

The framework is the one presented in [9]. Let $\mathcal{L}$ be any non-empty set. The elements of $\mathcal{L}$ should be viewed as propositions or formulas and $\mathcal{L}$ is therefore a language. At present no structure is assumed on $\mathcal{L}$ and its elements are therefore to be taken as atomic propositions. Let $\mathcal{C} : 2^{\mathcal{L}} \longrightarrow 2^{\mathcal{L}}$.

Definition 1. *The operation $\mathcal{C}$ is said to be a C-logic iff it satisfies the two following properties.*

$$\text{Inclusion} \quad \forall A \subseteq \mathcal{L}, \ A \subseteq \mathcal{C}(A),$$

$$\text{Cumulativity} \quad \forall A, B \subseteq \mathcal{L}, \ A \subseteq B \subseteq \mathcal{C}(A) \Rightarrow \mathcal{C}(A) = \mathcal{C}(B).$$

Note that neither Monotonicity nor Supraclassicality is assumed.

2.2 Properties

Lemma 1 (Makinson). *An operation C is a C-logic iff it satisfies Inclusion,*

$$\text{\textbf{Idempotence}} \quad \forall A \subseteq \mathcal{L}, \;\; C(C(A)) = C(A)$$

and

$$\text{\textbf{Cautious Monotonicity}} \quad \forall A, B \subseteq \mathcal{L} \; A \subseteq B \subseteq C(A) \Rightarrow C(A) \subseteq C(B)$$

Proof. Let us prove, first, that a C-logic satisfies Idempotence. By Inclusion $A \subseteq C(A) \subseteq C(A)$, therefore, by Cumulativity: $C(A) = C(C(A))$. Assume, now that C satisfies Inclusion, Idempotence and Cautious Monotonicity. Let $A \subseteq B \subseteq C(A)$. By Cautious Monotonicity, we have $C(A) \subseteq C(B)$. Therefore, we have $B \subseteq C(A) \subseteq C(B)$. By Cautious Monotonicity again, we have: $C(B) \subseteq C(C(A))$. By Idempotence, then, we conclude $C(B) \subseteq C(A)$ and therefore $C(B) = C(A)$.   □

Lemma 2 (Makinson). *An operation C is a C-logic iff it satisfies Inclusion and*

$$\text{\textbf{2-Loop}} \quad A \subseteq C(B), B \subseteq C(A) \Rightarrow C(A) = C(B).$$

Proof. Assume C is a C-logic and $A \subseteq C(B)$. By Inclusion, we have $B \subseteq A \cup B \subseteq C(B)$ and by Cumulativity: $C(B) = C(A \cup B)$. Similarly $B \subseteq C(A)$ implies $C(A) = C(A \cup B)$.

Assume now that C satisfies Inclusion and 2-Loop, and that $A \subseteq B \subseteq C(A)$. By Inclusion: $A \subseteq B \subseteq C(B)$. By 2-Loop, then, we have: $C(A) = C(B)$.   □

The finer study of C-logics relies, as for monotonic logics, on the notions of a consistent set and of a theory.

Definition 2. *A set $A \subseteq \mathcal{L}$ is said to be* consistent *iff $C(A) \neq \mathcal{L}$. A set A for which $C(A) = \mathcal{L}$ is said to be* inconsistent.

The following follows from Idempotence.

Lemma 3. *A set A is consistent iff $C(A)$ is.*

Lemma 4. *If $A \subseteq B$ and A is inconsistent, so is B.*

Proof. miuns .1em If $C(A) = \mathcal{L}$, we have $A \subseteq B \subseteq C(A)$ and, by Cumulativity, $C(B) = C(A) = \mathcal{L}$.   □

Definition 3. *A set $A \subseteq \mathcal{L}$ is said to be* maximal consistent *iff it is consistent and any strict superset $B \supset A$ is inconsistent.*

Definition 4. *A set $T \subseteq \mathcal{L}$ is said to be a* theory *iff $C(T) = T$.*

The following is obvious (by Inclusion).

Lemma 5. *There is only one inconsistent theory, namely $\mathcal{L}$.*

Lemma 6. *Any maximal consistent set A is a theory.*

Proof. By Inclusion $A \subseteq \mathcal{C}(A)$. By Lemma 3 the set $\mathcal{C}(A)$ is consistent and by maximality: $A = \mathcal{C}(A)$. ☐

Notation: *Given any C-logic $\mathcal{C}$ let us define $Cn_\mathcal{C} : 2^\mathcal{L} \longrightarrow 2^\mathcal{L}$ by:*

$$Cn_\mathcal{C}(A) = \bigcap_{T \supseteq A, \ T \text{ a theory}} T.$$

When $\mathcal{C}$ is clear from the context, we shall simply write Cn. Note that $Cn(A) = \bigcap_{B \supseteq A} \mathcal{C}(B)$. Note also that Cn does not in this paper denote classical consequence as it often does. The following follows from Lemma 5.

Lemma 7

$$Cn(A) = \bigcap_{T \supseteq A, \ T \text{ a consistent theory}} T.$$

Lemma 8. $A \subseteq Cn(A) \subseteq \mathcal{C}(A)$.

Proof. By the definition of Cn and the fact that $\mathcal{C}(A)$ is a theory that includes A (Idempotence and Inclusion). ☐

Lemma 9. *A set $A \subseteq \mathcal{L}$ is inconsistent iff $Cn(A) = \mathcal{L}$.*

Proof. If $Cn(A) = \mathcal{L}$, then, by Lemma 8, $\mathcal{C}(A) = \mathcal{L}$. If A is inconsistent, then, by Lemma 4, there is no consistent theory that includes A and, by Lemma 7, $Cn(A) = \mathcal{L}$. ☐

Lemma 10. $\mathcal{C}(A) = Cn(\mathcal{C}(A)) = \mathcal{C}(Cn(A))$.

Proof. By Lemma 8, we have $\mathcal{C}(A) \subseteq Cn(\mathcal{C}(A)) \subseteq \mathcal{C}(\mathcal{C}(A))$. By Idempotence, then, the first equality is proved. By Lemma 8 and Cumulativity, we have $\mathcal{C}(A) = \mathcal{C}(Cn(A))$. ☐

Corollary 1. *For any theory T, $Cn(T) = T$.*

Proof. $Cn(T) = Cn(\mathcal{C}(T)) = \mathcal{C}(T) = T$. ☐

Lemma 11. *The operation Cn is monotonic, i.e., if $A \subseteq B$, then $Cn(A) \subseteq Cn(B)$ and also idempotent, i.e., $Cn(Cn(A)) = Cn(A)$.*

Proof. Monotonicity follows immediately from the definition of Cn. For Idempotence, notice that, by Monotonicity and Corollary 1, any theory T that includes A also includes $Cn(A)$: $Cn(A) \subseteq Cn(T) = T$. ☐

2.3   f-Models

Assume $\mathcal{M}$ is a set (of worlds), about which no assumption is made, and $\models \subseteq \mathcal{M} \times \mathcal{L}$ is a (satisfaction) binary relation (nothing assumed either). For any set

$A \subseteq \mathcal{L}$, we shall denote by $\widehat{A}$ or by $\mathrm{Mod}(A)$ the set of all worlds that satisfy all elements of A:

$$\widehat{A} = \mathrm{Mod}(A) = \{x \in \mathcal{M} \mid x \models a, \forall a \in A\}.$$

For typographical reasons we shall use both notations, sometimes even in the same formula. For any set of worlds $X \subseteq \mathcal{M}$, we shall denote by $\overline{X}$ the set of all formulas that are satisfied in all elements of X:

$$\overline{X} = \{a \in \mathcal{L} \mid x \models a, \forall x \in X\}.$$

The following are easily proven, for any $A, B \subseteq \mathcal{L}$, $X, Y \subseteq \mathcal{M}$: they amount to the fact that the operations $X \mapsto \overline{X}$ and $A \mapsto \widehat{A}$ form a Galois connection.

$$A \subseteq \overline{\widehat{A}} \quad , \quad X \subseteq \widehat{\overline{X}}$$

$$\widehat{A \cup B} = \widehat{A} \cap \widehat{B} \quad , \quad \overline{X \cup Y} = \overline{X} \cap \overline{Y}$$

$$A \subseteq B \Rightarrow \widehat{B} \subseteq \widehat{A} \quad , \quad X \subseteq Y \Rightarrow \overline{Y} \subseteq \overline{X}$$

$$A \subseteq B \Rightarrow \overline{A} \subseteq \overline{B} \quad , \quad X \subseteq Y \Rightarrow \widehat{X} \subseteq \widehat{Y}$$

$$\widehat{A} = \widehat{\overline{\widehat{A}}} \quad , \quad \overline{X} = \overline{\widehat{\overline{X}}}$$

The last technical notion that will be needed is that of a definable set of worlds. It will be used in the completeness proof below, but not in Definition 6.

Definition 5. *A set X of worlds is said to be definable iff either one of the two following equivalent conditions holds:*

1. $\exists A \subseteq \mathcal{L}$ such that $X = \widehat{A}$, or
2. $X = \widehat{\overline{X}}$.

The set of all definable subsets of X will be denoted by D_X.

The proof of the equivalence of the two propositions above is obvious.

Lemma 12. *If X and Y are definable sets of worlds, then their intersection $X \cap Y$ is also definable.*

Proof. By the remarks above: if $X = \widehat{A}$ and $Y = \widehat{B}$, $X \cap Y = \widehat{A} \cap \widehat{B} = \widehat{A \cup B}$. □

Definition 6. *A choice function on $\mathcal{M}$ is a function $f : 2^{\mathcal{M}} \to 2^{\mathcal{M}}$.*

Note that f is defined on arbitrary sets of worlds, not only on definable sets as in [9]. In the same vein, we do not require here that the image by f of a definable set be definable as was necessary in the corresponding soundness result of [9].

Definition 7. *A triplet $\langle \mathcal{M}, \models, f \rangle$ is an f-model (for language $\mathcal{L}$) iff $\models$ is a binary relation on $\mathcal{M} \times \mathcal{L}$ and f is a choice function $\mathcal{M}$ that satisfies, for any sets X, Y:*

Contraction $f(X) \subseteq X$

and

f-Cumulativity $f(X) \subseteq Y \subseteq X \Rightarrow f(Y) = f(X).$

Definition 8. *An f-model is said to be a* restricted *f-model iff its choice function f also satisfies, for any set X:*

$$\text{f-Consistency} \quad f(X) = \emptyset \Rightarrow X = \emptyset.$$

2.4   Properties of f-Models

This section makes clear the relation between f-models and the models of [9]. It will not be used in the sequel. There one considered choice functions satisfying Contraction,

$$\textbf{Coherence} \quad X \subseteq Y \Rightarrow X \cap f(Y) \subseteq f(X)$$

and

$$\textbf{f-Monotonicity} \quad f(X) \subseteq Y \subseteq X \Rightarrow f(Y) \subseteq f(X).$$

Lemma 13. *Any function f that satisfies Inclusion, Coherence and f-Monotonicity satisfies f-Cumulativity.*

Proof. Assume $f(X) \subseteq Y \subseteq X$, we must show that $f(X) \subseteq f(Y)$ (the opposite inclusion is guaranteed by f-Monotonicity). By Coherence: $Y \cap f(X) \subseteq f(Y)$. $\square$

2.5   Soundness

Theorem 1. *Let $\langle \mathcal{M}, \models, f \rangle$ be an f-model and the operation $\mathcal{C}$ be such that:*

$$\mathcal{C}(A) = \overline{f(\widehat{A})}. \tag{1}$$

Then $\mathcal{C}$ is a C-logic.

Note that if $\mathcal{C}$ is defined as above, then $\overline{\overline{A}} \subseteq \mathcal{C}(A)$.

Proof. By Contraction $f(\widehat{A}) \subseteq \widehat{A}$ and therefore $\overline{\overline{A}} \subseteq \overline{f(\widehat{A})}$. But $A \subseteq \overline{A}$. We have proved Inclusion.

Assume now $A \subseteq B \subseteq \mathcal{C}(A)$. We have: $\widehat{\mathcal{C}(A)} \subseteq \widehat{B} \subseteq \widehat{A}$ and also $\widehat{\mathcal{C}(A)} = \widehat{\overline{f(\widehat{A})}}$. But $f(\widehat{A}) \subseteq \overline{f(\widehat{A})}$. We have: $f(\widehat{A}) \subseteq \widehat{B} \subseteq \widehat{A}$. By f-Cumulativity, then: $f(\widehat{A}) = f(\widehat{B})$ and $\mathcal{C}(A) = \mathcal{C}(B)$. $\square$

2.6   Representation

Theorem 2. *If $\mathcal{C}$ is a C-logic, then there is a restricted f-model $\langle \mathcal{M}, \models, f \rangle$ such that $\mathcal{C}(A) = \overline{f(\widehat{A})}$.*

Notice that, comparing to Theorem 1 we are getting f-Consistency for free: given any f-model, there is a restricted f-model that defines the same consequence operation. No purely semantic derivation of this result is currently known.

Proof. For $\mathcal{M}$ take all *consistent* theories of $\mathcal{C}$. Set $T \models a$ iff $a \in T$. It follows that, by Lemma 7, for any $A \subseteq \mathcal{L}$,

$$\overline{A} = \mathcal{C}n(A).$$

We must now define a choice function f. Consider some $X \subseteq \mathcal{M}$. We distinguish two cases. If there is some $A \subseteq \mathcal{L}$ such that $\widehat{\mathcal{C}(A)} \subseteq X \subseteq \widehat{A}$, we let $f(X) = \widehat{\mathcal{C}(A)}$, and otherwise we let $f(X) = X$.

The correctness of the definition above relies on the fact that, in the first case, $f(X)$ does not depend on the A chosen. This will be proved now and, then, we shall return to the proof of Theorem 2.

Lemma 14. *If* $\widehat{\mathcal{C}(A)} \subseteq X \subseteq \widehat{A}$ *and* $\widehat{\mathcal{C}(B)} \subseteq X \subseteq \widehat{B}$, *then* $\mathcal{C}(A) = \mathcal{C}(B)$.

Proof. It is enough to prove that, under the assumptions above, $\mathcal{C}(A) = \mathcal{C}(\mathcal{C}n(A,B))$. We have $X \subseteq \widehat{A} \cap \widehat{B} = \widehat{A \cup B}$. Therefore $\mathcal{C}n(A,B) \subseteq \overline{X} \subseteq \mathcal{C}n(\mathcal{C}(A)) = \mathcal{C}(A)$, by Lemma 10. Then $A \subseteq \mathcal{C}n(A,B) \subseteq \mathcal{C}(A)$ and, by Cumulativity, $\mathcal{C}(A) = \mathcal{C}(\mathcal{C}n(A,B))$. □

We return now to the proof of Theorem 2. Notice that, for any $A \subseteq \mathcal{L}$, $f(\widehat{A}) = \widehat{\mathcal{C}(A)}$. Therefore $\overline{f(\widehat{A})} = \mathcal{C}n(\mathcal{C}(A)) = \mathcal{C}(A)$ by Lemma 10. We shall now show that f satisfies Contraction, f-Cumulativity and f-Consistency. If $f(X) = \widehat{\mathcal{C}(A)} \subseteq X$, Contraction is clear. If $f(X) = X$, Contraction is also clear. For f-Cumulativity, assume $f(X) \subseteq Y \subseteq X$. If $f(X) = X$, then $Y = X$ and $f(Y) = f(X)$. Otherwise, $\widehat{\mathcal{C}(A)} \subseteq Y \subseteq X \subseteq \widehat{A}$ and $f(Y) = \widehat{\mathcal{C}(A)} = f(X)$. For f-Consistency, assume that $f(X) = \emptyset$. If $f(X) = X$, then $X = \emptyset$. If $f(X) = \widehat{\mathcal{C}(A)} \subseteq X \subseteq \widehat{A}$, then $\widehat{\mathcal{C}(A)} = \emptyset$ and there is no consistent theory that contains $\mathcal{C}(A)$. We conclude that $\mathcal{C}(A)$ is inconsistent and therefore A is inconsistent by Lemma 3, no consistent theory contains A by Lemma 4 and $\widehat{A} = \emptyset$. We conclude that $X = \emptyset$. □

2.7   Connectives in C-logics

Conjunction and Negation. We shall show that C-logics admit a classical conjunction and a classical negation. Let us assume now, for the remainder of this section, that the language $\mathcal{L}$ is closed under a binary connective written $\wedge$ and a unary connective written $\neg$.

Theorem 3. *If* $\langle \mathcal{M}, \models, f \rangle$ *is a restricted f-model that behaves classically with respect to* $\wedge$ *and* $\neg$, *i.e., for any* $m \in \mathcal{M}$,

 − $m \models a \wedge b$ *iff* $m \models a$ *and* $m \models b$
 − $m \models \neg a$ *iff* $m \not\models a$,

then the inference operation defined by the f-model satisfies:

 − $\wedge$-**R** $\mathcal{C}(A, a \wedge b) = \mathcal{C}(A, a, b)$
 − $\neg$-**R1** $\mathcal{C}(A, a, \neg a) = \mathcal{L}$
 − $\neg$-**R2** *if* $\mathcal{C}(A, \neg a) = \mathcal{L}$, *then* $a \in \mathcal{C}(A)$,

where $C(A, a)$ denotes $C(A \cup \{a\})$.

Proof. The first property follows from the fact that

$$\text{Mod}(A \cup \{a \wedge b\}) = \text{Mod}(A \cup \{a\} \cup \{b\}).$$

For the second property notice that $\text{Mod}(A \cup \{a\} \cup \{\neg a\}) = \emptyset$ implies, by Contraction, that $f(\text{Mod}(A \cup \{a\} \cup \{\neg a\})) = \emptyset$. For the third property, since no element of $\mathcal{M}$ satisfies both a and $\neg a$, if $C(A, \neg a) = \mathcal{L}$, there is no m that satisfies $C(A, \neg a)$ and $f(\text{Mod}(A \cup \neg a)) = \emptyset$. Since the model is a restricted f-model, $\text{Mod}(A \cup \neg a) = \emptyset$. Therefore every m that satisfies A also satisfies a and $a \in C(A)$. □

The reader should notice that it is claimed that, if $C(A, \neg a) = \mathcal{L}$, then any m satisfying A also satisfies a, but it is *not* claimed that, under this hypothesis, $a \in Cn(A)$. Indeed Cn is defined via the theories of C and the relation of those to the elements of $\mathcal{M}$ is not straightforward. The reader should also note that a similar result (Equation 8.5) was obtained in [9] only assuming Coherence. Here Coherence is not required, f-Consistency is required in its place. The reader may notice that, in the presence of all other properties, $\neg$-R2 is equivalent to: if $a \in C(A, \neg a)$ then $a \in C(A)$.

The following theorem shows the converse of Theorem 3. It requires a compactness assumption. We shall, then, assume that C satisfies the following:

Weak Compactness $C(A) = \mathcal{L} \Rightarrow \exists$ a finite $B \subseteq A$ such that $C(B) = \mathcal{L}$.

In a monotonic setting and in the presence of a proper negation Weak Compactness implies Compactness, i.e., if $a \in C(A)$, there is some finite subset B of A such that $a \in C(B)$, but this is not the case in a nonmonotonic setting.

Theorem 4. *If C satisfies Weak Compactness, Inclusion, Cumulativity, $\wedge$-R, $\neg$-R1 and $\neg$-R2, then there is a restricted f-model that behaves classically with respect to $\wedge$ and $\neg$ such that $C(A) = f(\widehat{A})$.*

Before presenting a proof of Theorem 4, three lemmas are needed.

Lemma 15. *Assume C satisfies Cumulativity, Weak Compactness, $\neg$-R1 and $\neg$-R2. If $a \notin C(A)$, there is a maximal consistent set $B \supseteq A$ such that $a \notin B$.*

Proof. By $\neg$-R2, $A \cup \{\neg a\}$ is consistent. By Weak Compactness, Cumulativity and Zorn's lemma, there is a maximal consistent set B that contains it. This B does not contain a by $\neg$-R1. □

Lemma 16. *Assume C satisfies Inclusion, Cumulativity, $\wedge$-R, $\neg$-R1 and $\neg$-R2. If A is a maximal consistent set, then*

- *$a \wedge b \in A$ iff $a \in A$ and $b \in A$,*
- *$\neg a \in A$ iff $a \notin A$.*

Proof. By ∧-R, $a \wedge b \in \mathcal{C}(A)$ iff $a \in \mathcal{C}(A)$ and $b \in \mathcal{C}(A)$, but, by Lemma 6, A is a theory. If $\neg a \in A$, then $a \notin A$ since A is consistent, by $\neg$-R1. If $\neg a \notin A$, then by the maximality of A, $\mathcal{C}(A, \neg a) = \mathcal{L}$ and, by $\neg$-R2, $a \in \mathcal{C}(A)$, but A is a theory.   □

Lemma 17. *Assume $\mathcal{C}$ satisfies Inclusion, Cumulativity, Weak Compactness, $\neg$-R1 and $\neg$-R2. Then*

$$Cn(A) = \bigcap_{B \supseteq A, B \text{ maximal consistent}} B.$$

Proof. The left-hand side is a subset of the right-hand side by Lemmas 7 and 6. But if $a \notin Cn(A)$, then there is a theory T such that $a \notin T = \mathcal{C}(T)$ and, by Lemma 15, there is a maximal consistent B that includes T but does not contain a.   □

Let us now proceed to the proof of Theorem 4.

Proof. We modify the construction of Theorem 2, by considering not all consistent theories but only maximal consistent sets. Those maximal consistent sets are theories and behave classically for ∧ and ¬ by Lemma 16. By Lemma 17, for any $A \subseteq \mathcal{L}$,

$$\widetilde{A} = Cn(A).$$

The remainder of the proof is unchanged.   □

We may now show that, if one considers only tautologies or entailments, propositional nonmonotonic cumulative logic is not weaker than (and therefore exactly the same as) monotonic logic. Section 2.7 will show that this equivalence between cumulative and monotonic logic does not hold if one considers proof-theoretic, inferential properties of the logics. In the following theorem, we consider a propositional language in which negation and conjunction are considered basic and other connectives are defined in the usual classical way.

Theorem 5. *Let $\mathcal{L}$ be a propositional calculus (negation and conjunction basic, other connectives defined classically) and $a, b \in \mathcal{L}$. The following propositions are equivalent.*

1. *a logically implies b, i.e., $a \models b$,*
2. *for every operation $\mathcal{C}$ that satisfies Inclusion, Idempotence, Monotonicity, Weak Compactness and the rules ∧-R, $\neg$-R1 and $\neg$-R2 above: $b \in \mathcal{C}(a)$,*
3. *for every operation $\mathcal{C}$ that satisfies Inclusion, Cumulativity, Weak Compactness and the rules ∧-R, $\neg$-R1 and $\neg$-R2 above: $b \in \mathcal{C}(a)$,*
4. *for every such $\mathcal{C}$ and for any $A \subseteq \mathcal{L}$: $b \in \mathcal{C}(A, a)$,*
5. *for every such $\mathcal{C}$: $\mathcal{C}(a, \neg b) = \mathcal{L}$.*

Proof. Property 5 implies 4, since, by Cumulativity, $\mathcal{C}(a, \neg b) = \mathcal{L}$ implies that we have $\mathcal{C}(A, a, \neg b) = \mathcal{L}$, and, by the rule $\neg$-R2: $b \in \mathcal{C}(A, a)$. Property 4 obviously implies 3, that implies 2 since Monotonicity and Idempotence imply Cumulativity. Property 2 implies 1 since the operation $\mathcal{C}_\models$ defined by $\mathcal{C}_\models(A) = \{b \mid A \models b\}$ satisfies all the conditions of 2.

The only non-trivial part of the proof is that 1 implies 5. Assume $a \models b$ and $\mathcal{C}$ satisfies Inclusion, Cumulativity, Weak Compactness and the rules $\wedge$-R, $\neg$-R1 and $\neg$-R2. By Theorem 4, there is a set $\mathcal{M}$, a satisfaction relation $\models$ that behaves classically with respect to $\wedge$ and $\neg$ and a choice function satisfying Contraction, f-Cumulativity and f-Consistency such that $\mathcal{C}(a, \neg b) = \overline{f(\widehat{\{a\}} \cap \widehat{\{\neg b\}})}$. But, by assumption $\widehat{\{a\}} \cap \widehat{\{\neg b\}} = \emptyset$. By Contraction, then $\mathcal{C}(a, \neg b) = \overline{\emptyset} = \mathcal{L}$. $\qquad\square$

Theorem 5 shows that the proof theory of the semantically-classical conjunction and negation in a nonmonotonic setting is the same as in a monotonic setting. The following shows, that, in yet another sense, C-logics admit a proper conjunction and a proper negation: one may conservatively extend any C-logic on a set of atomic propositions to a language closed under conjunction and negation. It is customary to consider Introduction-Elimination rules, such as $\wedge$-R, $\neg$-R1 and $\neg$-R2 as definitions of the connectives. Hacking [7, Section VII] discusses this idea and proposes that, to be considered as bona fide definitions of the connectives, the rules must be such that they ensure that any legal logic on a small language may be conservatively extended to a legal logic on the language extended by closure under the connective.

Theorem 6. *Let P be an arbitrary set of atomic propositions and $\mathcal{C}$ a C-logic over P. Let $\mathcal{L}$ be the closure of P under $\wedge$ and $\neg$. Then, there exists a C-logic $\mathcal{C}\prime$ on $\mathcal{L}$ that satisfies $\wedge$-R, $\neg$-R1 and $\neg$-R2, such that, for any $A \subseteq P$, $\mathcal{C}(A) = P \cap \mathcal{C}'(A)$.*

Proof. By Theorem 2, there is a restricted f-model on P $\langle \mathcal{M}, \models, f \rangle$ such that $\mathcal{C}(A) = \overline{f(\widehat{A})}$. Let us now extend $\models$ to $\mathcal{L}$ by $m \models a \wedge b$ iff $m \models a$ and $m \models b$ and $m \models \neg a$ iff $m \not\models a$. We claim that $\langle \mathcal{M}, \models, f \rangle$ is now a restricted f-model on $\mathcal{L}$, whose satisfaction relation $\models$ behaves classically for $\neg$ and $\wedge$. Indeed, the properties required from f do not involve the satisfaction relation at all, they deal with subsets of $\mathcal{M}$ exclusively. Let us define, for any $A \subseteq \mathcal{L}$, $\mathcal{C}'(A) = \overline{f(\widehat{A})}$. By Theorem 1, $\mathcal{C}'$ is a C-logic. By Theorem 3 it satisfies $\wedge$-R, $\neg$-R1 and $\neg$-R2. It is left to us to see that $\mathcal{C}(A) = P \cap \mathcal{C}'(A)$, for any $A \subseteq P$. This follows straightforwardly from the fact that both $\mathcal{C}(A)$ and $\mathcal{C}'(A)$ are the sets of formulas (the former of P, the latter of $\mathcal{L}$) satisfied by all members of the set $f(\widehat{A})$. $\qquad\square$

Disjunction. We have seen that any C-logic admits classical negation and conjunction. The reader may think that this implies that it also admits a classical disjunction defined as $a \vee b = \neg(\neg a \wedge \neg b)$. Indeed it is the case that, if we define disjunction in this way one of the basic properties of disjunction is satisfied:

$$\vee\text{-R1} \quad a \in \mathcal{C}(A) \implies a \vee b \in \mathcal{C}(A) \text{ and } b \vee a \in \mathcal{C}(A).$$

But the other fundamental property of disjunction does not hold.

$$\vee\text{-R2} \quad \mathcal{C}(A, a) \cap \mathcal{C}(A, b) \subseteq \mathcal{C}(A, a \vee b).$$

Theorem 7 of [9] shows that if f satisfies Coherence, the logic $\mathcal{C}$ defined by 1 can be extended conservatively to a propositional language such that disjunction behaves as properly, i.e.:

– if $a \in \mathcal{C}(A)$ then $a \vee b \in \mathcal{C}(A)$,
– if $a \in \mathcal{C}(A)$ then $b \vee a \in \mathcal{C}(A)$ and
– if $c \in \mathcal{C}(A, a)$ and $c \in \mathcal{C}(A, b)$ then $c \in \mathcal{C}(A, a \vee b)$.

The author conjectures a proper disjunction cannot always be defined if f satisfies only f-Cumulativity.

2.8 Connection with Previous Work

Theorem 7. *Let $\mathcal{L}$ be a propositional calculus and $\mathcal{C}$ an operation that satisfies Weak-Compactness, Inclusion, Cumulativity, $\wedge$-R, $\neg$-R1 and $\neg$-R2. Define a binary relation among propositions by: $a \hspace{-0.3em}\sim\hspace{-0.3em} b$ iff $b \in \mathcal{C}(a)$. Then, the relation $\hspace{-0.3em}\sim\hspace{-0.3em}$ is a cumulative relation in the sense of [8].*

Proof. We shall show that $\hspace{-0.3em}\sim\hspace{-0.3em}$ satisfies Left Logical Equivalence, Right Weakening, Reflexivity, Cut and Cautious Monotonicity. For Left-Logical-Equivalence, suppose $\models a \leftrightarrow a'$. By Theorem 5, $a' \in \mathcal{C}(a)$ and, by Cumulativity, $\mathcal{C}(a) = \mathcal{C}(a, a')$. But, similarly, exchanging a and a': $\mathcal{C}(a') = \mathcal{C}(a, a')$ and $\mathcal{C}(a) = \mathcal{C}(a')$. For Right Weakening, by Theorem 5 $b \models b'$ implies $b' \in \mathcal{C}(a, b)$. If $a \hspace{-0.3em}\sim\hspace{-0.3em} b$, by Cumulativity $\mathcal{C}(a) = \mathcal{C}(a, b)$ and $a \hspace{-0.3em}\sim\hspace{-0.3em} b'$. Reflexivity follows from Inclusion. Cut and Cautious Monotonicity together are equivalent to: if $a \hspace{-0.3em}\sim\hspace{-0.3em} b$, then $a \wedge b \hspace{-0.3em}\sim\hspace{-0.3em} c$ iff $a \hspace{-0.3em}\sim\hspace{-0.3em} c$. Assume $b \in \mathcal{C}(a)$, then, by Cumulativity, $\mathcal{C}(a) = \mathcal{C}(a, b)$. □

The converse cannot be true since it is easy to see that any cumulative relation defined as above satisfies: if $a \wedge \neg b \hspace{-0.3em}\sim\hspace{-0.3em}$ **false** then $a \hspace{-0.3em}\sim\hspace{-0.3em} b$. But this last property is not satisfied by all cumulative relations. It is satisfied, though, by all cumulative relations that satisfy the *consistency-preservation* property favored by Makinson, i.e., if $a \hspace{-0.3em}\sim\hspace{-0.3em}$ **false** then $a \models$ **false**. The author conjectures the following holds by the results of [5].

Conjecture 1. If $\hspace{-0.3em}\sim\hspace{-0.3em}$ is a cumulative relation, that satisfies

$$a \wedge \neg b \hspace{-0.3em}\sim\hspace{-0.3em} \textbf{false} \Rightarrow a \hspace{-0.3em}\sim\hspace{-0.3em} b,$$

then there is an operation $\mathcal{C}$ that satisfies Weak-Compactness, Inclusion, Cumulativity, $\wedge$-R, $\neg$-R1 and $\neg$-R2 such that $b \in \mathcal{C}(a)$ iff $a \hspace{-0.3em}\sim\hspace{-0.3em} b$.

3 L-logics

A sub-family of C-logics will be defined now. It corresponds to the cumulative with loop (CL) relations of [8].

Definition 9. *The operation $\mathcal{C}$ is said to be an L-logic iff it satisfies the following two properties.*

Inclusion $\forall A \subseteq \mathcal{L}$, $A \subseteq \mathcal{C}(A)$,

Loop $\forall n \forall i = 0, \ldots, n-1$ modulo n $A_i \subseteq \mathcal{C}(A_{i+1}) \Rightarrow \mathcal{C}(A_0) = \mathcal{C}(A_1)$.

The assumption of Loop is: $A_0 \subseteq \mathcal{C}(A_1)$, $A_1 \subseteq \mathcal{C}(A_2)$, ..., $A_{n-1} \subseteq \mathcal{C}(A_0)$. The conclusion could equivalently have been: $\mathcal{C}(A_i) = \mathcal{C}(A_j)$ for any $i, j = 0, \ldots, n-1$. Notice that for $n = 2$, the condition Loop is the condition 2-Loop of Lemma 2. Therefore any L-logic is a C-logic. The characteristic property of L-logics is embedded in the relation to be defined now.

Definition 10. *Let T and S be theories. Let us define $T \leq S$ iff there exists a set $A \subseteq S$ such that $\mathcal{C}(A) = T$.*

The following holds without any assumption on $\mathcal{C}$.

Lemma 18. *The relation $\leq$ is reflexive. If T, S are two theories such that $T \subseteq S$, then $T \leq S$.*

Proof. $S \subseteq S$ and $\mathcal{C}(S) = S$ imply $S \leq S$. $T \subseteq S$ and $\mathcal{C}(T) = T$ imply $T \leq S$. $\square$

The next lemma holds only for L-logics. Notice that, even for L-logics, the relation $\leq$ is not transitive in general.

Lemma 19. *If $\mathcal{C}$ is an L-logic, and $T_0 \leq T_1$, ..., $T_{n-1} \leq T_0$, then $T_0 = T_1 = \ldots = T_{n-1}$.*

Proof. Assume $T_0 \leq T_1$, ..., $T_{n-1} \leq T_0$. There are $A_i \subseteq T_{i+1}$ such that $\mathcal{C}(A_i) = T_i$. Therefore $A_{i-1} \subseteq \mathcal{C}(A_i)$ and by Loop $\mathcal{C}(A_i) = \mathcal{C}(A_j)$. $\square$

In particular, the relation $\leq$ is anti-symmetric for L-logics (in fact for C-logics).

Definition 11. *Let T and S be theories. Let us define $T < S$ iff $T \leq S$ and $S \not\leq T$, or equivalently (for C-logics) $T \leq S$ and $T \neq S$. Let $<^+$ be the transitive closure of $<$.*

Lemma 20. *If $\mathcal{C}$ is an L-logic, then the relation $<^+$ is irreflexive and therefore a strict partial order.*

Proof. By Lemma 19. $\square$

4 The Case of Quantum Logic

4.1 Quantum Consequence Operations

Birkhoff and von Neumann [1] framed Quantum Logics in Hilbert style, i.e., as a set of valid propositions in propositional calculus. Engesser and Gabbay [3] proposed to view Quantum Logics in a different light: as a consequence relation describing what can be deduced from what. They assume a language closed under the propositional connectives, but their definition makes perfect sense and is very rich even on a language that contains only atomic propositions. This is, in this paper's view, a major step taken by Engesser and Gabbay since Birkhoff and von Neumann's framework does not allow any interesting consideration in the absence of connectives. The setting proposed by Engesser and Gabbay allows us to discuss first the nature of Quantum Deduction without any need to posit connectives, and then to consider the proof-theoretic and semantics properties of connectives one at a time.

Assume a Hilbert space $\mathcal{H}$ and an element $h \in \mathcal{H}$ are given. Assume also a non-empty set (language) $\mathcal{L}$ of closed subspaces of $\mathcal{H}$ is given. Thus, the elements of $\mathcal{L}$, the atomic propositions are closed subspaces of $\mathcal{H}$. For every proposition $a \in \mathcal{L}$, we shall denote by a_p the projection on the subspace a: for every $x \in \mathcal{H}$, $a_p(x)$ is the element of a closest to x. For every set of propositions: $A \subseteq \mathcal{L}$, $A^* \stackrel{\text{def}}{=} \bigcap_{a \in A} a$ and A_p^* will denote the projection on A^*, i.e., on the intersection of all the elements of A.

Definition 12 (Engesser-Gabbay). *Let $\mathcal{C}_h : 2^{\mathcal{L}} \longrightarrow 2^{\mathcal{L}}$ be defined by:*

$$b \in \mathcal{C}_h(A) \text{ iff } A_p^*(h) \in b. \tag{2}$$

Theorem 8. *The operation $\mathcal{C}_h$ defined above is an L-logic.*

Engesser and Gabbay essentially noticed already that $\mathcal{C}_h$ is a C-logic. From now on, we shall write $\mathcal{C}$ for $\mathcal{C}_h$ when no confusion can arise. We need a lemma.

Lemma 21. *If $B \subseteq \mathcal{C}(A)$, then $A_p^*(h) = (A^* \cap B^*)_p(h)$ and $d(h, A^*) \geq d(h, B^*)$.*

Proof. For any $b \in B$, $A_p^*(h) \in b$. Therefore $A_p^*(h) \in B^*$.   □

Let us now prove Theorem 8

Proof. Indeed, $A_p^* h \in A^*$ and therefore, for any $a \in A$, $A_p h \in a$, and we have shown Inclusion.

Assume $A \subseteq B \subseteq \mathcal{C}(A)$. By Lemma 21, we have $A_p^*(h) = (A^* \cap B^*)_p(h)$, but $B^* \subseteq A^*$ and $A_p^*(h) = B_p^*(h)$. Therefore $\mathcal{C}(A) = \mathcal{C}(B)$ and we have shown Cumulativity.

$$A_1 \subseteq \mathcal{C}(A_0), A_2 \subseteq \mathcal{C}(A_1), \ldots, A_0 \subseteq \mathcal{C}(A_n) \Rightarrow \mathcal{C}(A_0) = \mathcal{C}(A_1)$$

For Loop, assume $A_i \subseteq \mathcal{C}(A_{i+1})$, for $i = 0, \ldots, n-1 \pmod{n}$. By Lemma 21, $d(h, A_{i+1}) \geq d(h, A_i)$ and therefore all those distances are equal: $d(h, A_0) = d(h, A_1)$, $A_{0p}^*(h) = A_{1p}^*(h)$ and $\mathcal{C}(A_0) = \mathcal{C}(A_1)$.   □

4.2 Open Question

Do the four properties above characterize those consequence operations presentable by Hilbert spaces? Or are there other properties shared by those operations presentable by Hilbert spaces that do not follow from the above? The answer to this question is not known.

Engesser and Gabbay show that any operation presentable by a Hilbert space admits an *internalizing* connective, i.e. a connective $\leadsto$ such that $b \in \mathcal{C}(a)$ iff $a \leadsto b \in \mathcal{C}(\emptyset)$. This could be understood as a sign that Hilbert space logics have some special quality. It should not. Any C-logic can be extended by such an internalizing connective.

Consider any C-logic $\mathcal{C}$. By Theorem 2, there is a restricted f-model that defines $\mathcal{C}$. We shall now extend the language and close it under a new connective $\leadsto$ and build an f-model on the extended language. This is done by defining, by induction on the structure of the propositions, both the satisfaction relation and the extension of $\mathcal{C}$ to the new language as below. For any $m \in \mathcal{M}$, $m \models a \leadsto b$ iff $b \in \mathcal{C}(a)$. Then $\mathcal{C}$ is the C-logic defined by the f-model on the extended language. Notice that either all elements of $\mathcal{M}$ satisfy $a \leadsto b$ or none of them does. This definition by induction is legal since the function f is always the same and we still have an f-model. The $\mathcal{C}$ defined by this model on the extended language is a C-logic and a conservative extension of the original $\mathcal{C}$.

Suppose now that $a \leadsto b \in \mathcal{C}(\emptyset)$. This means that every element of $f(\mathcal{M})$ satisfies $a \leadsto b$. Since the model is a restricted model, $f(\mathcal{M})$ is not empty, there is some element of $\mathcal{M}$ that satisfies $a \leadsto b$ and $b \in \mathcal{C}(a)$. Suppose now that $b \in \mathcal{C}(a)$, then every element of $\mathcal{C}$ satisfies $a \leadsto b$ and $a \leadsto b \in \mathcal{C}(\emptyset)$.

4.3   Connectives

Conjunction. Conjunction is unproblematic. Even infinite conjunctions are easily defined. If A is a set of propositions, and each $a \in A$ is associated with some closed subspace a^*, we may associate the proposition $\bigwedge_{a \in A} a$ with the closed subspace $\bigcap_{a \in A} a^*$, i.e., A^* and the rule $\wedge$-R is validated: $\mathcal{C}(A, B) = \mathcal{C}(\bigwedge a, B)$.

Negation. The situation for negation is most intriguing. By Theorem 8 any operation $\mathcal{C}$ presented as a Quantum Logic is an L-logic, therefore a C-logic. Theorem 6 shows that C-logics admit a negation satisfying ¬-R1 and ¬-R2. We therefore expect Quantum Logics to admit such a negation. But the treatment of negation proposed by Birkhoff and von Neumann and later used by Engesser and Gabbay does not do the job in the following sense. Suppose we define $(\neg a)^* = (a^*)^\perp$ where $\perp$ denotes the orthogonal complement. It is easy to see that ¬-R1 is satisfied since the intersection of a subspace and its orthogonal complement is $\{0\}$, but ¬-R2 is not satisfied. Consider for example three generic (not parallel and not orthogonal) one-dimensional subspaces (lines through the origin) a, b and c in the real plane. Let h be any non-zero vector of c. The intersection of a and $b^\perp$ is $\{0\}$ and therefore $\mathcal{C}(a, \neg b) = \mathcal{L}$. But $b \notin \mathcal{C}(a)$ since the projection of h on a is not in b. This failure of ¬-R2, which is the principle of proof by contradiction, was in fact already noted or guessed by Birkhoff and von Neumann. In section 17, p. 837, they compare Quantum Logics with other non-classical logics introduced on introspective or philosophical grounds, such as intuitionistic logic. They note that even though "logicians have usually assumed that properties of negation were the ones least able to withstand a critical analysis, the study of (quantum) mechanics points to the distributive identities as the weakest link in the algebra of logic." And they conclude: "our conclusion agrees perhaps more with those critiques of logic, which find most objectionable the assumption that to deduce an absurdity from the conjunction of a and not b, justifies one in inferring that a implies b". This paper's conclusions agree only in part, and will be presented below.

If the $\bot$ operator of Birkhoff and von Neumann does not do the job, one may ask whether some other operator would. The following shows that no operator on closed subspaces of a Hilbert space provides a suitable negation. M. Magidor and A. Nissimov helped here.

Theorem 9. *There is no operation n on closed subspaces that guarantees that if one defines*

$$(\neg a)^* = n(a^*)$$

the operation $\mathcal{C}_h$ defined in Definition 12 satisfies the properties $\neg$-R1 and $\neg$-R2 for any h.

Proof. First, if $\mathcal{C}_h$ satisfies $\neg$-R1 for any h, one easily sees that it must be the case that for any $Y = b^*$, $Y \cap n(Y) = \{0\}$. Consider now, some strict subspace $Y = b^*$ of $\mathcal{H}$. Since $\mathcal{H}$ cannot be the union of two disjoint non trivial subspaces, there must be some non-zero element $x \in \mathcal{H}$ that is not in the union $Y \cup n(Y)$. Let us choose $h = x$. Let X be the one-dimensional subspace defined by x and assume that $X = a^*$. Notice that $X \cap Y = X \cap n(Y) = \{0\}$. By the latter, $\mathcal{C}(a, \neg b) = \mathcal{L}$. By $\neg$-R2, we must have $b \in \mathcal{C}(a)$, i.e., $X_p(h) \in Y$. But $X_p(h) = x$ and $x \notin Y$. $\square$

Disjunction. A proper disjunction should satisfy $\vee$-R1 and $\vee$-R2 defined in Section 2.7. We have seen that C-logics do not always support such a disjunction. It is left to be seen whether Quantum Logics support such a disjunction. In any C-logic that satisfies $\wedge$-R, $\vee$-R1 and $\vee$-R2, the distributive equality holds, in the sense that $\mathcal{C}(A, a \wedge (b \vee c)) = \mathcal{C}(A, (a \wedge b) \vee (a \wedge c))$. The only Quantum Logics that admit a proper disjunction are therefore those Quantum Logics that support the distributive law. This is a very limited family.

5 Conclusions and Future Work

Quantum Logics are nonmonotonic logics as noticed by Engesser and Gabbay, they are also very respectable nonmonotonic logics since they are L-logics. It is indeed surprising that Quantum Logics come to satisfy formal properties designed with a completely different intention: to describe properties "introduced on introspective grounds" and intended to describe disciplined "jumping to conclusions".

The study of C-logics is unexpectedly smooth and attractive. The basic intuition behind the cumulative relations of [8] is confirmed: cumulative relations yield classical connectives but the disjunction (that may be defined as usual from negation and conjunction) does not behave proof-theoretically as a proper disjunction should. The section on L-logics is less interesting. I am not sure where it leads. The results are straightforward translations from [8] and L-logics do not seem to behave in any better way with respect to connectives than C-logics. The reason why this may be interesting is that Quantum Logics are not only C-logics but also L-logics. But two main questions about Quantum Logics are left open: can all L-logics be presented as Quantum Logics or do Quantum

Logics satisfy additional properties? What is the meaning for Quantum Logics of the classical negation and conjunction that can be defined for any C-logic? Intersection of closed subspaces provides a perfect semantics for the conjunction natural in C-logic, but it may not be the right conjunction for Quantum Physics. Orthogonal complement does not provide a suitable semantics for negation, but Theorem 6 ensures there is a respectable negation. It seems doubtful that one could find a suitable corresponding operation among closed subspaces of Hilbert spaces that would enable us to associate a closed subspace to the negation of a closed subspace. This probably means that one cannot assume that the negation of an observable is an observable. But must we insist that the negation of an observable be observable? Couldn't negation mean something about what we know and not about the world? Disjunction is probably incompatible with Quantum Logics altogether.

Acknowledgments

I am most grateful to Kurt Engesser for helping me through the geometry of Hilbert spaces. Some of the examples used in the paper are his. David Makinson and Alexander Bochman provided important feedback. An anonymous reader's thoughtful comments helped improve the presentation.

References

1. Birkhoff, G., von Neumann, J.: The logic of quantum mechanics. Annals of Mathematics 37, 823–843 (1936)
2. Dalla Chiara, M.L.: Quantum logic. In: Gabbay, D.M., Guenthner, F. (eds.) Handbook of Philosophical Logic, 2nd edn., vol. 6, pp. 129–228. Kluwer, Dordrecht (2001), http://www.philos.unifi.it/persone/dallachiara.htm
3. Engesser, K., Gabbay, D.M.: Quantum logic, Hilbert space, revision theory. Artificial Intelligence 136(1), 61–100 (2002)
4. Freund, M., Lehmann, D.: Nonmonotonic inference operations. Bulletin of the IGPL 1(1), 23–68 (1993), Max-Planck-Institut für Informatik, Saarbrücken, Germany
5. Freund, M., Lehmann, D.: Nonmonotonic reasoning: from finitary relations to infinitary inference operations. Studia Logica 53(2), 161–201 (1994)
6. Freund, M., Lehmann, D., Makinson, D.: Canonical extensions to the infinite case of finitary nonmonotonic inference operations. In: Brewka, G., Freitag, H. (eds.) Workshop on Nomonotonic Reasoning, Sankt Augustin, FRG, December 1989, vol. (443), pp. 133–138 (1989), Arbeitspapiere der GMD no. 443
7. Hacking, I.: What is logic? The Journal of Philosophy 76(6), 285–319 (1979)
8. Kraus, S., Lehmann, D., Magidor, M.: Nonmonotonic reasoning, preferential models and cumulative logics. Artificial Intelligence 44(1–2), 167–207 (2021)
9. Lehmann, D.: Nonmonotonic logics and semantics. Journal of Logic and Computation 11(2), 229–256 (2001) CoRR: cs.AI/0202018

10. Lehmann, D.: A presentation of quantum logic based on an and then connective. Journal of Logic and Computation (to appear, 2007), doi:10.1093/logcom/exm054
11. Makinson, D.: General theory of cumulative inference. In: Reinfrank, M., et al. (eds.) Non-Monotonic Reasoning 1988. LNCS, vol. 346, pp. 1–18. Springer, Heidelberg (1988)
12. Makinson, D.: General patterns in nonmonotonic reasoning. In: Gabbay, D.M., Hogger, C.J., Robinson, J.A. (eds.) Handbook of Logic in Artificial Intelligence and Logic Programming, Nonmonotonic and Uncertain Reasoning, vol. 3, pp. 35–110. Oxford University Press, Oxford (1994)

Reasoning in Dynamic Logic about Program Termination

Daniel Leivant*

Computer Science Department, Indiana University,
Bloomington, IN 47405, USA
leivant@cs.indiana.edu

For Boris Trakhtenbrot, a grandmaster and a friend.

Abstract. Total correctness assertions (TCAs) have long been considered a natural formalization of successful program termination. However, research dating back to the 1980s suggests that validity of TCAs is a notion of limited interest; we corroborate this by proving compactness and Herbrand properties for the valid TCAs, defining in passing a new sound, complete, and syntax-directed deductive system for TCAs.

It follows that proving TCAs whose truth depends on underlying inductive data-types is impossible in logics of programs that are sound for all structures, such as Dynamic Logic (DL) based on Segerberg-Pratt's PDL, even when augmented with powerful first-order theories like Peano Arithmetic. The Convergence Rule of [6] bypasses this difficulty, but is methodologically and conceptually problematic, in addition to being unsound for general validity. We propose instead to bind variables to inductive data via DL's box operator, leading to an alternative formalization of termination assertions, which we dub *Inductive TCA (ITCA)*. We show that validity of ITCAs is directly reducible to validity of *partial* correctness assertions, confirming the foundational importance of the latter.

1 Dynamic Logic

1.1 Correctness Assertions

The aim of practical verification of imperative programs is to prove that given programs terminate for inputs of interest, with output fulfilling desired properties. First-order Dynamic Logic (DL) provides a convenient framework for specifying such requirements (see e.g. [6,7]).

To focus on the essentials, we refer to the simplest non-trivial imperative programming language, namely regular programs. Given a first-order vocabulary V, we admit two atomic programs: *assignments* $x := \mathbf{t}$ ($\mathbf{t}$ a V-term), and *tests* $?\chi$ (χ a first-order V-formula).[1] Compound programs are generated using

---

* Research partially supported by NSF grant CCR-0105651. Preliminary version appeared as [11].

[1] Implementable programs use only quantifier free tests. Conversely, we could allow tests over DL-formulas, so called "rich tests"; we mention in the sequel particular forms of rich tests.

A. Avron et al. (Eds.): Trakhtenbrot/Festschrift, LNCS 4800, pp. 441–456, 2008.

composition, union, and Kleene's $*$ (nondeterministic iteration). Guarded itera-
tive programs ("while" programs) are then definable: $\textbf{skip} \equiv ?\top$, $\textbf{abort} \equiv ?\bot$,
$(\textbf{if } \chi \textbf{ then } \alpha \textbf{ else } \beta) \quad \equiv \quad (?\chi; \alpha) \cup (?\neg\chi; \beta)$, and $(\textbf{while } \chi \textbf{ do } \alpha) \quad \equiv$
$(?\chi; \alpha)^*; (?\neg\chi)$.

Given a V-structure $\mathcal{S}$, the operational semantics over $\mathcal{S}$ of programs α is
defined by a straightforward recurrence on the complexity of α (see e.g. [7]).

The *formulas of Dynamic Logic* over a vocabulary V are generated from the
atomic V-formulas (including equations) using propositional connectives, quantifiers, and the modal operators: if α is a program (over V) and φ a formula then
$[\alpha]\varphi$ is a formula, intended to express that φ is true at any state reached by a
completed execution of α. The dual operator, $\langle\alpha\rangle\varphi$, can be defined as $\neg[\alpha]\neg\varphi$.

Within DL one can formulate two basic forms of program correctness. A *partial
correctness assertion* (PCA) is a formula of the form $\varphi \to [\alpha]\psi$, abbreviated
as $\varphi[\alpha]\psi$, stating that every terminating execution of program α leads to a
state verifying the first-order post-condition ψ, provided the initial state verifies
the first-order pre-condition φ. Dually, a *total correctness assertion* (TCA) is
a formula $\varphi \to \langle\alpha\rangle\psi$, abbreviated as $\varphi\langle\alpha\rangle\psi$, stating that *some* execution of
program α terminates in a state that verifies the post-condition ψ, provided the
initial state verifies the pre-condition φ.

1.2 Segerberg's Axiomatization

One natural deductive formalism for DL, which we dub *Segerberg's Dynamic
Logic* (**SDL**), is obtained by merging Segerberg's axioms for propositional dynamic logic [16,7] with Floyd-Hoare's Assignment rule and first-order logic. (This
is closely related to the formalism 14.12 of [7], with the Convergence Rule omitted.)

Of course, the presence of programs calls for some caution with inference rules
involving variables. We consider an occurrence of a variable x to be *bound* not
only by quantifiers but also by programs: x is bound in $[\alpha]\varphi$ if it is assigned-to
in α. The operation $\{t/x\}$, of substituting a term t for all free occurrences of x,
is *legal* for a formula φ if x has no free occurrence in φ where a variable in t is
bound. In such cases, the definition of $\{t/x\}\varphi$ proceeds as usual, by recurrence
on α and φ.

A deductive calculus for DL is now obtained by augmenting a deductive calculus for first-order logic (such as Gentzen's natural deduction or sequential
calculus) with the following rules and axioms.

I Axiom-templates and rules for programs in general. These are the rules of the
rudimentary modal logic **K**:

$$\text{GENERALIZATION:}^2 \qquad \frac{\Gamma \vdash \varphi}{\Gamma \vdash [\alpha]\varphi} \qquad \text{(no free variables in } \Gamma)$$

$$\text{BOX:} \qquad [\alpha](\varphi \to \psi) \to ([\alpha]\varphi \to [\alpha]\psi)$$

---

[2] Note that the box operator is not applied to Γ in the consequence, since we insist
that Γ has no free variables.

II Axiom-templates defining the intended meaning of atomic programs in terms of first-order logic:

$$\text{ASSIGNMENT: } [x := t]\varphi(x) \leftrightarrow \varphi(t) \ (x \text{ not assigned-to in } \varphi)$$

$$\text{TEST: } [?\chi]\varphi \leftrightarrow (\chi \rightarrow \varphi)$$

III Syntax directed rules for program constructs.

$$\text{COMPOSITION: } [\alpha;\beta]\varphi \leftrightarrow [\alpha][\beta]\varphi$$

$$\text{BRANCHING: } [\alpha \cup \beta]\varphi \leftrightarrow [\alpha]\varphi \wedge [\beta]\varphi$$

$$\text{ITERATION: } [\alpha^*]\varphi \leftrightarrow \varphi \wedge [\alpha][\alpha^*]\varphi$$

IV Induction for $*$:

$$[\alpha^*](\varphi \rightarrow [\alpha]\varphi) \rightarrow (\varphi \rightarrow [\alpha^*]\varphi)$$

It is easy to formulate a variant of the deductive calculus above for reasoning about guarded iterative programs, rather than regular programs.

Clearly, **SDL** is sound for general validity: for all vocabularies V, if a DL V-formula φ is provable in **SDL**, then it is true in every V-structure $\mathcal{S}$ and environment therein.

2   Validity of TCAs

From the Compactness Theorem for first-order logic it follows that inductive data-types, such as the natural numbers, lists, or strings over an alphabet, cannot be delineated by a first-order theory. In particular, no first-order theory defines, in all structures, the denotations of the numerals $\mathbf{0}, \mathbf{s(0)}, \ldots$. The validity of a TCA in all structures must therefore refer also to "non-standard" data, an ominous requirement that all but trivializes the concept. This is reflected in the existence of sound and complete axiomatizations of the valid TCAs [13,15], in contrast to the set of valid partial-correctness assertions, which is not RE. In fact, validity of TCAs is directly reducible to validity of first-order formulas [9]. Using the latter, we provide here further evidence of the triviality of TCA validity, by proving compactness and Herbrand-style properties for it.

2.1   Explicit Rendition of Dynamic Logic

As observed in [9], the operational semantics of regular programs α can be defined explicitly within an extension of first-order logic with relational variables and quantification over them. To avoid costly pedantry, we posit that all formulas and programs under discussion have free variables among $\bar{x} = x_1 \ldots x_k$. We then write $\varphi[\bar{t}]$, where $\bar{t} = (t_1 \ldots t_k)$ are terms, for the result of simultaneously substituting $t_1 \ldots t_k$ for all free occurrences of $x_1 \ldots x_k$, respectively.

For each program α we define a formula M_α, with free variables $\bar{u} = u_1 \ldots u_k$ and $\bar{v} = v_1 \ldots v_k$ (all variables distinct), with the following property. For every V-structure $\mathcal{S}$ and environment η therein, $\mathcal{S}, \eta \models M_\alpha(\bar{u}, \bar{v})$ iff there is an execution of α starting in environment $\eta[\bar{x} := \eta\bar{u}]$ and terminating in environment $\eta[\bar{x} := \eta\bar{v}]$.

$$M_\alpha(\bar{u}, \bar{v}) \equiv v_i = t(\bar{u}) \wedge (\wedge_{j \neq i} v_j = u_j) \qquad \text{for } \alpha \equiv (x_i := t(\bar{x}))$$

$$M_{?\chi}(\bar{u}, \bar{v}) \equiv \chi(\bar{u}) \wedge \bar{u} = \bar{v}$$
$$\text{where } \alpha \equiv ?\chi(\bar{x})$$

$$M_{\alpha;\beta}(\bar{u}, \bar{v}) \equiv \exists\bar{w}.\, M_\alpha(\bar{u}, \bar{w}) \wedge M_\beta(\bar{w}, \bar{v})$$

$$M_{\alpha \cup \beta}(\bar{u}, \bar{v}) \equiv M_\alpha(\bar{u}, \bar{w}) \vee M_\beta(\bar{w}, \bar{v})$$

$$M_{\alpha^*}(\bar{u}, \bar{v}) \equiv \forall Q.\, Q(\bar{u}) \wedge \mathrm{Cl}_\alpha[Q] \rightarrow Q(\bar{v})$$
$$\text{where}$$
$$\mathrm{Cl}_\alpha[Q] \equiv \forall \bar{z}, \bar{w}.\, Q(\bar{z}) \wedge M_\alpha(\bar{z}, \bar{w}) \rightarrow Q(\bar{w})$$

We use the explicit rendition above of program semantics to render DL formulas φ by second-order formulas $\varphi^\sharp$.

$$\varphi^\sharp \equiv_{\mathrm{df}} \varphi \qquad \text{for } \varphi \text{ atomic}$$

$$(\varphi \wedge \psi)^\sharp \equiv_{\mathrm{df}} \varphi^\sharp \wedge \psi^\sharp \quad \text{and similarly for other connectives}$$

$$(\forall y.\varphi)^\sharp \equiv_{\mathrm{df}} \forall y.(\varphi^\sharp) \quad \text{and similarly for } \exists$$

$$([\alpha]\varphi)^\sharp \equiv_{\mathrm{df}} \forall \bar{u}.\, M_\alpha(\bar{x}, \bar{u}) \rightarrow \{\bar{u}/\bar{x}\}(\varphi^\sharp)$$

$$(\langle\alpha\rangle\varphi)^\sharp \equiv_{\mathrm{df}} \exists \bar{u}.\, M_\alpha(\bar{x}, \bar{u}) \wedge \{\bar{u}/\bar{x}\}(\varphi^\sharp)$$

Proposition 1. *For every DL formula φ, $\models \varphi \leftrightarrow \varphi^\sharp$. That is, for every structure $\mathcal{S}$ and environment η therein, $\mathcal{S}, \eta \models \varphi$ iff $\mathcal{S}, \eta \models \varphi^\sharp$.*

Note that if $*$ does not occur in φ, then $\varphi^\sharp$ is first-order.

2.2   Completeness of First-Order Logic for Convergence Assertions

Call a second-order formula φ *relationally-universal* if every second-order $\forall$ in φ occurs positively, and every second-order $\exists$ occurs negatively.[3] Call a DL formula φ a *convergence assertion* if for every program α with $*$, every modal operator $\langle\alpha\rangle$ occurs positively, and every $[\alpha]$ occurs negatively. For example, the conjunction of any number of TCAs and negations of PCA's is a convergence assertion.

---

[3] Recall that a position in a formula is *positive* if it is in the negative scope of an even number of implications and negations.

Lemma 1. *If a DL formula φ is a convergence assertion, then $\varphi^\sharp$ is a relationally-universal formula semantically equivalent to φ.*

Note: It is easy to see that Lemma 1 remains true even if convergence assertions are allowed as tests in programs (this more general form is used in [15]). The discussion has to be modified simply by (1) Defining by simultaneous recurrence *convergence programs,* that allow such tests, and the more general notion of *convergence assertions,* that refer to such programs; (2) Defining by simultaneous recurrence on programs and formulas the interpretation M_α of convergence programs α, and the interpretation $\varphi^\sharp$ of convergence assertions φ.

Lemma 2. *Every relationally-universal formula φ can be mapped effectively, in logarithmic space, to a first-order formula φ° such that $\models \varphi$ iff $\models \varphi^\circ$.*

Proof. Given a relationally-universal formula φ, convert it first into a prenex formula φ'. This can clearly be done within logarithmic space, and all relational $\forall$ in φ' occur positively, as can be proved by a trivial induction on φ.

Consider the following choice principle $\mathbf{C}_{01}$ (from objects to relations):

$$\forall x \exists R\, \varphi(x, R) \;\rightarrow\; \exists Q \forall x\, \varphi(x, Q_x) \tag{1}$$

Here the arity of Q is $1 +$ the arity of R, and $Q_x(\bar{\mathbf{t}})$ stands for $Q(x, \bar{\mathbf{t}})$. Note that the backward implication for (1) is trivial. Taking the dual of (1), we obtain the schema

$$\exists x \forall R\, \psi(x, R) \;\leftrightarrow\; \forall Q \exists x\, \psi(x, Q_x) \tag{2}$$

Using (2), the relational prenex formula φ' can be converted into a semantically equivalent formula φ'' of the form $\overrightarrow{\mathbf{Q}}\, \varphi^\circ$, where $\overrightarrow{\mathbf{Q}}$ is a block of universal relational quantifiers, and φ° is first-order. Clearly, φ° is valid iff φ'' is valid, i.e. iff φ is valid.

It is easy to see that φ° can be computed from φ', and therfore from φ, in logarithmic space. □

Corollary 1. *Every convergence assertion φ can be mapped effectively, in logarithmic space, to a first-order formula $(\varphi^\sharp)^\circ$, such that $\models \varphi$ iff $\models \varphi^\circ$.*

Proof. By Lemma 1 $\varphi^\sharp$ is semantically equivalent to φ, and by Lemma 2 $\models \varphi^\sharp$ iff $\models (\varphi^\sharp)^\circ$. □

2.3   Positive Results for TCAs

Corollary 1 makes it possible to apply the rich meta-theory of first-order logic to convergence assertions. Since for first-order formulas validity is equivalent to provability, we have

Theorem 1. *The set of valid convergence-assertions is recursively enumerable.*

Theorem 1 was first proved in [13], using a different method. The proof above has the merit of exposing clearly the direct role of first-order logic, and of being applicable to all imperative program constructs whose semantics is definable by universal relational quantification, as is the case for virtually all constructs. The first-order nature of total correctness was rediscovered, in a different guise, in [14].

A related result in [13] establishes a connection between program semantics and TCA's:

Theorem 2. *If programs α and β differ semantically then they have difference total-correctness theories, i.e. there are first-order formulas φ and ψ such that one of the two TCAs $\varphi\langle\alpha\rangle\psi$ and $\varphi\langle\beta\rangle\psi$ is valid, whereas the other is not.*

Proof. Suppose α and β differ semantically: there is a structure $\mathcal{S}$ such that one of the two programs, say α, maps some $\mathcal{S}$-environment η to environment η', and β does not. Let $\eta = \eta_0,\ \eta_1,\ \ldots,\ \eta_n = \eta'$ be the sequence of environments that form the execution-trace of α on $\mathcal{S}$ starting with η. Each η_{i+1} is either identical to η_i (with α performing a successful test χ_i in environment η_i), or else η_{i+1} is obtained from η_i by an assignment. Thus, the values taken by the variables $x_1 \ldots x_k$ present are all expressible in terms of the initial values: for each environment η_i and each variable x_j there is a term $\mathbf{t}_{ij}$ with $\eta_i(x_j) = \eta(\mathbf{t}_{ij}(\bar{x}))$.

For each one of the tests χ_i above, let $\chi_i' \equiv \chi_i(\bar{\mathbf{t}}_i)$, where $\bar{\mathbf{t}}_i \equiv (\mathbf{t}_{i1} \ldots \mathbf{t}_{ik})$. Let φ be the conjunction of the formulas χ_i', and $\bar{v} = v_1 \ldots v_k$ be fresh variables. Then the TCA

$$(\varphi \wedge (\bar{x} = \bar{v}))\ \langle\alpha\rangle\ (\bar{x} = \bar{\mathbf{t}}_n(\bar{v}))$$

is valid, whereas the TCA

$$(\varphi \wedge (\bar{x} = \bar{v}))\ \langle\beta\rangle\ (\bar{x} = \bar{\mathbf{t}}_n(\bar{v}))$$

fails in structure $\mathcal{S}$ and environment η. $\qquad\square$

2.4   Degenerative Properties of TCAs

The expressiveness of first-order logic for TCAs, and the triviality of Theorem 2, are warning signs that TCAs are of limited interest. We show now that the validity of a TCA never reflects the intended semantics of the iteration operator $*$, confirming the limited interest of TCAs.

Given a program α and $n \geqslant 0$, $\alpha{\upharpoonright}n$ will stand for the program obtained from α by interpreting $*$ as iteration up to n times; that is:

- For α atomic $\alpha{\upharpoonright}n$ is α.

- $(\alpha; \beta){\upharpoonright}n$ is $(\alpha{\upharpoonright}n); (\beta{\upharpoonright}n)$.

- $(\alpha \cup \beta){\upharpoonright}n$ is $(\alpha{\upharpoonright}n) \cup (\beta{\upharpoonright}n)$.

- $(\alpha^*){\upharpoonright}n$ is $\cup_{i \leqslant n}(\alpha{\upharpoonright}n)^i$.

For DL formulas φ we define $\varphi{\upharpoonright}n$ by a similar interpretation of $*$, that is: $([\alpha]\varphi){\upharpoonright}n \equiv_{\mathrm{df}} [\alpha{\upharpoonright}n](\varphi{\upharpoonright}n)$, $(\langle\alpha\rangle\varphi){\upharpoonright}n \equiv_{\mathrm{df}} \langle\alpha{\upharpoonright}n\rangle(\varphi{\upharpoonright}n)$, $(\varphi \wedge \varphi'){\upharpoonright}n \equiv_{\mathrm{df}} (\varphi{\upharpoonright}n) \wedge (\varphi'{\upharpoonright}n)$, etc. By a straightforward induction on the complexity of programs α and formulas φ, we obtain

Lemma 3. *1.* $\models \langle\alpha\lceil n\rangle\psi \to \langle\alpha\rangle\psi$, *for all formulas ψ.*

2. Let S be a structure and η an environment therein. If $S, \eta \models \langle\alpha\rangle\varphi$, where φ is first-order, then for some $n \geqslant 0$, $S, \eta \models \langle\alpha\lceil n\rangle\varphi$.
More generally, if φ is a convergence assertion, and $S, \eta \models \varphi$, then $S, \eta \models \varphi\lceil n$ for some $n \geqslant 0$.

3. If φ is a convergence assertion then

$$\models (\varphi\lceil n) \to (\varphi\lceil m) \quad \text{for all } m \geqslant n \qquad \text{and} \qquad \models (\varphi\lceil n) \to \varphi$$

Theorem 3. *Let φ be a convergence assertion. If $\models \varphi$, then $\models (\varphi\lceil n)$ for some $n \geqslant 0$.*

Proof. Towards contradiction, assume that $\models \varphi$, but $\not\models (\varphi\lceil n)$ for all n. Let Γ be the first-order theory consisting of $(\varphi)^\sharp)^\circ$ as well as all formulas $\neg(\varphi\lceil n)^\sharp$, $n \geqslant 0$ (recall the definition of φ° from Corollary 1, and of $\psi^\sharp$ from Proposition 1). Note that $(\varphi\lceil n)^\sharp$ is always first-order, because $\varphi\lceil n$ is $*$-free.

By assumption, for each n there is a model of $\neg(\varphi\lceil n)$, which by Proposition 1 is a model of $\neg(\varphi\lceil n)^\sharp$, and by Corollary 1 must also be a model of $(\varphi^\sharp)^\circ$, since we assume that φ is valid. By Lemma 3(3) it follows that every finite sub-theory of Γ has a model. So, by the Compactness Theorem for first-order logic, Γ is true in some interpretation (S, η).

Since $S, \eta \models (\varphi)^\sharp)^\circ$, we have, by Corollary 1, $S, \eta \models \varphi$, and so, by Lemma 3(2), $S, \eta \models (\varphi\lceil n)$ for some n. But $S, \eta \models \Gamma$, and in particular $S, \eta \models \neg(\varphi\lceil n)^\sharp$, which by Proposition 1 implies that $S, \eta \models \neg(\varphi\lceil n)$, a contradiction.   □

Note. Theorem 3 implies that if $\varphi\langle\alpha^*\rangle\psi$ is valid then so is $\varphi\langle\cup_{i \leqslant n}\alpha^i\rangle\psi$ for some n, not that $\varphi\langle\alpha^n\rangle\psi$ is valid for some n. For example, consider

$$\alpha \quad \equiv \quad (?(x = 0); x := \mathbf{s}(x)) \cup (?(x \neq 0); x := \mathbf{0})$$

Clearly, $\langle\alpha^*\rangle x = \mathbf{s}(\mathbf{0})$ is valid. But in a structure where $\mathbf{0} \neq \mathbf{s}(\mathbf{0})$, there is no n for which $\langle\alpha^n\rangle (x = \mathbf{s}(\mathbf{0}))$ holds true for all values for x. Thus $(\mathbf{0} \neq \mathbf{s}(\mathbf{0})) \langle\alpha^*\rangle (x = \mathbf{s}(\mathbf{0})$ is valid, but for each $n \geqslant 0$ $(\mathbf{0} \neq \mathbf{s}(\mathbf{0}) \langle\alpha^n\rangle (x = \mathbf{s}(\mathbf{0}))$ is not valid.   □

Theorem 3 shows that the validity of TCAs fails to reflect the infinite nature of iteration. Combining this with the proof of Theorem 2, we conclude further:

Theorem 4. *If a TCA $\varphi\langle\alpha\rangle\psi$ is valid, then there is a finite set T of k-ary vectors of terms such that (with fresh variables $\bar{v} = (v_1 \ldots v_k)$)*

$$\models (\varphi \wedge \bar{x} = \bar{v}) \langle\alpha\rangle (\psi \wedge (\bigvee_{\bar{t} \in T} \bar{x} = \bar{t}(\bar{v})))$$

2.5   A Syntax Directed Deductive System for TCAs

The recursive enumerability of the set of valid TCAs begs for a sound and complete deductive calculus. A complete deductive system for TCAs over **while** programs was indeed given by Meyer and Halpern [12]. Schmitt [15] gave a formalism for proving all convergence assertions, but there the proof of a TCA may use convergence assertions more general than TCAs.[4] We give a more syntax-directed deductive system momentarily.

Since all the deductive systems mentioned have axioms and rules that are derived in **SDL**, we trivially obtain

Theorem 5. SDL *is sound and complete for validity of TCAs.*

The emphasis in the Theorem's statement on TCAs is essential, of course: the set of valid PCAs is not RE, and so **SDL** is not complete for validity of PCAs. (See [10] for related results.)

We now proceed to give a deductive calculus **TC**, whose simplicity corresponds to the inherent triviality of TCAs. The calculus **TC** refers only to first-order formulas and to TCAs, and is sound and complete for validity. Our completeness proof uses only basic properties of first-order logic.

As is the case for Hoare's Logic for PCAs, **TC** refers both to modal formulas (here TCAs) and first-order formulas. The inference rules, to be added to those for classical first-order logic, are as follows.

ASSIGNMENT: $\qquad (\{t/x\}\varphi) \langle x := t \rangle \, \varphi$

TEST: $\qquad (\xi \wedge \varphi) \, \langle ?\xi \rangle \, \varphi$

COMPOSITION: $\qquad \dfrac{\varphi \, \langle \alpha \rangle \, \chi \qquad \chi \, \langle \beta \rangle \, \psi}{\varphi \, \langle \alpha; \beta \rangle \, \psi}$

BRANCHING: $\qquad \dfrac{\varphi_0 \, \langle \alpha_0 \rangle \psi \qquad \varphi_1 \, \langle \alpha_1 \rangle \psi}{(\varphi_0 \vee \varphi_1) \, \langle \alpha_0 \cup \alpha_1 \rangle \, \psi}$

ITERATION: $\qquad \dfrac{\{ \, \varphi_i \, \langle \alpha^i \rangle \, \psi \, \}_{i<n}}{(\vee_{i<n}\varphi_i) \, \langle \alpha^* \rangle \, \psi}$

PRE-CONSEQUENCE: $\qquad \dfrac{\varphi \rightarrow \varphi' \qquad \varphi' \, \langle \alpha \rangle \, \psi}{\varphi \, \langle \alpha \rangle \, \psi}$

The Iteration Rule as stated above is somewhat unusual, in that the number of premises (n) is not fixed. The reader who finds this approach objectionable may prefer the following two rules, from which the ITERATION Rule above is easily derivable.

---

[4] The proof in [15] erroneously invokes Keisler's Model Existence Theorem for $\mathbf{L}_{\omega_1\omega}$ for formulas with parameters, but can be rephrased to correctly use Keisler's Theorem for sentences only [Personal communication with Peter Schmitt].

$$\text{EXACT-ITERATION:} \qquad \frac{\varphi \langle \alpha^i \rangle \, \psi}{\varphi \langle \alpha^* \rangle \, \psi}$$

$$\text{DISJUNCTION:} \qquad \frac{\varphi_0 \langle \alpha \rangle \, \psi \qquad \varphi_1 \langle \alpha \rangle \, \psi}{(\varphi_0 \vee \varphi_1) \langle \alpha \rangle \, \psi}$$

The *soundness* of the calculus above is obvious. Its semantic completeness is also straightforward:

Theorem 6. *If a TCA $\varphi\langle\alpha\rangle\psi$ is valid, then it is provable in* **TC**.

Proof. We proceed by induction on α. Assume that $\varphi\langle\alpha\rangle\psi$ is valid.

- If α is an assignment $x := \mathbf{t}$, then $\models \varphi \rightarrow \{\mathbf{t}/x\}\psi$, and by the ASSIGNMENT Rule $\{\mathbf{t}/x\}\psi \, \langle x := \mathbf{t}\rangle\psi$. So $\varphi\langle\alpha\rangle\psi$ follows by PRE-CONSEQUENCE.

- If α is a test $?\chi$, then $\varphi \rightarrow (\chi \wedge \psi)$ is valid, whereas $(\chi \wedge \psi) \langle \alpha \rangle \, \psi$ holds by the TEST Rule. So $\varphi\langle\alpha\rangle\psi$ holds by PRE-CONSEQUENCE.

- Suppose α is $\beta; \gamma$. By Theorem 3 we have, for some n,

$$\models \varphi \rightarrow \langle\beta{\restriction}n\rangle \langle\gamma{\restriction}n\rangle \, \psi$$

whence, by Proposition 1 and Lemma 3(1)

$$\models \varphi \, \langle\beta\rangle \, (\langle\gamma{\restriction}n\rangle \, \psi)^\sharp$$

and so by IH

$$\vdash \varphi \, \langle\beta\rangle \, (\langle\gamma{\restriction}n\rangle \, \psi)^\sharp \qquad\qquad (3)$$

Also,

$$\models (\langle\gamma{\restriction}n\rangle \, \psi)^\sharp \, \langle\gamma{\restriction}n\rangle \, \psi$$

whence, by Lemma 3,

$$\models (\langle\gamma{\restriction}n\rangle \, \psi)^\sharp \, \langle\gamma\rangle \, \psi$$

which by IH implies

$$\vdash (\langle\gamma{\restriction}n\rangle \, \psi)^\sharp \, \langle\gamma\rangle \, \psi$$

Combining the latter with (3) and using the COMPOSITION Rule, we obtain $\vdash \varphi \, \langle\beta; \gamma\rangle \, \psi$,

- If α is $\beta \cup \gamma$, then, again by Theorem 3, there is an n for which

$$\models \varphi \rightarrow (\langle\beta{\restriction}n\rangle \, \psi)^\sharp \vee (\langle\gamma{\restriction}n\rangle \, \psi)^\sharp,$$

$$\models (\langle\beta{\restriction}n\rangle \, \psi)^\sharp \, \langle\beta\rangle \, \psi,$$

$$\text{and} \ \models (\langle\gamma{\restriction}n\rangle \, \psi)^\sharp \, \langle\gamma\rangle \, \psi.$$

By IH the latter two TCAs are provable, and so by BRANCHING and PRE-CONSEQUENCE we obtain $\vdash \varphi \, \langle\beta \cup \gamma\rangle \, \psi$.

• Finally, consider the case where α is β^*. By Theorem 3 we have, for a sufficiently large n,

$$\models \varphi \rightarrow \langle \cup_{i \leqslant n} (\beta \upharpoonright n)^i \rangle \, \psi$$

and so

$$\models \varphi \rightarrow \bigvee_{i \leqslant n} (\langle (\beta \upharpoonright n)^i \rangle \, \psi)^\sharp \tag{4}$$

Also, for each $i \leqslant n$, the TCA

$$(\langle (\beta \upharpoonright n)^i \rangle \, \psi)^\sharp \, \langle \beta^i \rangle \, \psi$$

is valid, and therefore provable, by IH, repeated i times. Thus, by the the ITERATION Rule,

$$\vdash (\langle (\beta \upharpoonright n)^i \rangle \, \psi)^\sharp \, \langle \beta^* \rangle \, \psi$$

Using (4), we conclude by ORE-CONSEQUENCE that $\vdash \varphi \, \langle \beta^* \rangle \, \psi$. $\square$

3 Convergence Assertions for Inductive Data

3.1 Unprovable Convergence Assertions

The first-order nature of TCA validity shows that TCAs, understood as logical statements, do not capture the infinitary nature of inductive data, such as the natural numbers. That is, there is an unavoidable gap between the semantics of program convergence, which is anchored in the standard natural numbers, and the semantics of data in the background theory, which might include non-standard elements. This is illustrated by very simple TCAs, such as

$$P \rightarrow \langle (x := \mathbf{p}(x))^* \rangle \, (x{=}0) \tag{5}$$

where P is some finite axiomatization of arithmetic,[5] that includes the definition of $\mathbf{p}$ as cut-off predecessor: $\mathbf{p}(\mathbf{0}) = \mathbf{0}$, $\mathbf{p}(\mathbf{s}x) = x$. Since there are non-standard models of P, with elements that are not denotations of numerals, the TCA (5) cannot be valid.[6] It follows that (5), although trivially true in the intended structure, cannot be proved in any DL formalism which is sound for all structures, such as **SDL**.

This remains true even if we augment **SDL** with axioms for inductive data, such as Peano Arithmetic, since the semantics of program convergence will remain different from the semantics of counting in the grafted first-order theory.

Of course, (5) can be proved by induction on the formula

$$\varphi(n) \quad \equiv \quad (x = n) \rightarrow \langle \, (x := \mathbf{p}(x))^* \rangle \, (x{=}0)$$

---

[5] Take, for example, Peano Arithmetic with induction up to some fixed level Σ_n in the arithmetical hierarchy.

[6] The same argument applies to the entailment $P \models \langle \, (x := x{-}1)^* \rangle \, (x{=}0)$, where P is an *infinite* theory, say full Peano Arithmetic, or even the set of all sentences true in the standard structure of the natural numbers.

but φ is not a first-order formula, and so this instance of induction is not part of the background theory. As long as the background theory has no direct access to modal formulas, it cannot be used to derive TCAs whose truth depends on the inductive data considered.

To prove convergence assertions whose truth does depend on the inductive nature of underlying data we may: (a) Abandon the separation between logic and inductive data; this route was followed by Harel [6]; or (b) Consider, as we do here, more general formal renditions of convergence, which account directly for inductive data within the framework of DL itself.

3.2   The Convergence Rule

The termination of imperative programs is commonly proved by the Variance Method: one attaches to each instance of a looping construct in the program (such as a **while** loop or a recursive procedure) a parameter ranging over the field A of a well-founded relation $\succ$, and shows that each cycle reduces that parameter under $\succ$:

$$\frac{\vdash \varphi \wedge a = x \rightarrow \langle \alpha \rangle \varphi \wedge a \succ x}{\vdash \varphi \wedge A(x) \rightarrow \langle \alpha^* \rangle \varphi} \qquad (a \text{ not assigned-to in } \alpha)$$

Taking $\succ$ to be the natural order on $\mathbb{N}$, the Variance Rule yields the Convergence Rule of [6]:

$$\frac{\vdash \varphi(\mathbf{s}x) \rightarrow \langle \alpha \rangle \varphi(x)}{\vdash \varphi(x) \wedge N(x) \rightarrow \langle \alpha^* \rangle \varphi(0)} \qquad \begin{array}{l}(x \text{ not assigned-to in } \alpha \\ N \text{ interpreted as } \mathbb{N})\end{array}$$

Note that this rule fuses the interpretation of counting in the background theory and in the program semantics, thus forcing the numeric variables to range precisely over the natural numbers. In particular, the rule is not sound only for structures in which N is interpreted as $\mathbb{N}$, structures dubbed *arithmetical* in [6].

The rationale of [6] for the Convergence Rule was ostensibly to establish a completeness property for DL, analogous to Cook's Relative Completeness Theorem for Hoare-style logics. However, Cook's notion of relative completeness is itself problematic, and the arithmetic completeness of [6] faces additional pitfalls. One is the soundness of the rule only for a special class of structures, itself not first-order axiomatizable.

Also, whereas Hoare's Logic for PCAs is based on a formal separation between rules for programs (Hoare's rules) and rules for data (the background theory), the essential feature of the Convergence Rule is that it fuses the two. When programs and data are fused, and programs and their semantics are codable by data (as is the case in arithmetic structures), the very rationale for factoring out rules for programs from axioms for data is weakened, and one might arguably reason directly about programs in a first-order theory, as done for example in [1].

Interestingly, Hajek [5] was so taken aback by the unsoundness of the Convergence Rule, that he mistakenly took it to be an unintended error, which he proposed to correct by adopting a weaker variant, in which the inductive formula φ in the statement of the rule above has no free variable occurring in the program α, whether assigned to or not. However, since **SDL** is complete for sound TCAs (Theorem 5), and Hajek's variant is sound for general validity, adding it to **SDL** will yield no new TCAs, regardless of the background theory.

Needless to say, proving program termination by the Variance Method is of immense practical importance. Our contention is that the method is a mathematical tool (referring to particular structures, i.e. well-orderings) rather than a logical principle.

3.3   Inductive TCAs

Interestingly, the machinery of DL itself provides the means to refer to inductive data, such as the natural numbers. Thus program termination over $\mathbb{N}$ can be rendered directly by a suitable generalization of TCAs.

DL is particularly suitable for enforcing variables to assume values in a given inductively generated algebra, such as $\mathbb{N}$. Let $N(x)$ be the program $x := 0;\ (x := \mathbf{s}(x))^*$ (with $\mathbf{s}$ denoting the successor function); then the modal operator $[N(x)]$ is semantically equivalent to $\forall x \in \mathbb{N}$. More precisely, a formula $[N(x)]\varphi$ is true in a structure $\mathcal{S}$ and an environment η therein iff the formula $\forall x \in N.\ \varphi$ is true in that environment, where N is the set of denotations in $\mathcal{S}$ of the numerals $\mathbf{0}, \mathbf{s}(\mathbf{0}), \mathbf{s}(\mathbf{s}(\mathbf{0})), \ldots, \mathbf{s}^{[n]}(\mathbf{0}), \ldots$.

We can therefore state that a TCA $\varphi\langle\alpha\rangle\psi$ is true whenever a variable x assumes initially the denotation of a numeral, by the DL formula

$$[N(x)](\varphi \rightarrow \langle\alpha\rangle\psi).$$

The program $N(x)$ is nondeterministic, but by conceding some brevity and elegance we can use instead a deterministic **while** program (using a fresh variable y):

$$y := 0;\ \textbf{while } y \neq x \textbf{ do } y := \mathbf{s}(y)$$

If the axioms defining the predecessor function are added to the background theory, yet another alternative is

$$y := x;\ \textbf{while } y \neq 0 \textbf{ do } y := \mathbf{p}(y)$$

Similar programs can be given for any inductively generated algebra. For example, the set Σ^* of words over a finite alphabet Σ can be identified with the free algebra generated from the constant ε, denoting the empty word, and, for each $a \in \Sigma$, a unary function identifiers a. For example, $\{0, 1\}^*$ is generated from ε and unary $\mathbf{0}$ and $\mathbf{1}$. (Confusion with the constants 0 and 1 can be avoided by using instead $\mathbf{s}_0$ and $\mathbf{s}_1$ as function identifiers.) A word such as 011 is represented in the algebra as $\mathbf{0}(\mathbf{1}(\mathbf{1}(\varepsilon)))$.

Let

$$W_{\{0,1\}}(x) \quad \equiv \quad x := \varepsilon;\ ((x := \mathbf{0}(x)) \cup (x := \mathbf{1}(x)))^*$$

Then $[W_{\{0,1\}}(x)]\ \varphi$ is true in a structure $\mathcal{S}$ and environment η therein exactly when φ is true for all denotations of terms representing $\{0, 1\}^*$. A deterministic program defining $\{0,1\}^*$ must rely here on a destructor function **p**. Adding to the background theory its defining equations,

$$pfp(\epsilon) = \epsilon \quad pfp(\mathbf{0}(x)) = x \quad pfp(\mathbf{1}(x)) = x$$

the defining program is similar to the last one above for $\mathbb{N}$.

Analogous programs can be defined for any free algebra, even if they are multi-sorted. For example, to have x range over the algebra of lists over $\mathbb{N}$, with Λ denoting NIL and **c** denoting **cons**, we use the program

$$L_N(x) \equiv$$
$$x := \Lambda; \ (y := \mathbf{0}; \ (y := \mathbf{s}(y))^*; \ x := \mathbf{c}(y,x))^*$$

We define *inductive total correctness assertions (ITCA)* to be the DL formulas of the form $[\alpha](\varphi \rightarrow \langle \beta \rangle \psi)$, where φ and ψ are first-order. Such formulas can express a variety of program properties. For example, $\forall n \in \mathbb{N}. \langle \alpha^n \rangle \varphi$ is expressed, using a fresh variable x, by $[N(n)] \langle x := \mathbf{0}; \ \text{while} \ (x \neq n) \ \text{do} \ (\alpha; \ x := \mathbf{s}(x)) \rangle \varphi$ (compare [6, §6.4].)

3.4 Reduction of ITCAs to Partial Correctness Assertions

Surprisingly, the validity of each ITCA is equivalent to the validity of a certain *partial* correctness assertion. This observation attests to the foundational importance of PCAs.

Theorem 7. *Every ITCA φ can be effectively converted, in logarithmic space, to a PCA whose validity is equivalent to the validity of φ.*

Proof. Given an ITCA $[\alpha](\varphi \rightarrow \langle \beta \rangle \psi)$, its $\sharp$-translation is (modulo trivial variations) of the form

$$\forall \bar{v}.M_\alpha(\bar{x}, \bar{v}) \wedge \varphi(\bar{v}) \rightarrow (\exists \bar{w}. \ M_\beta(\bar{v}, \bar{w}) \wedge \psi(\bar{w})) \tag{6}$$

We use the choice principle $\mathbf{C}_{01}$ defined above to rewrite $M_\beta(\bar{v}, \bar{w})$ as a relationally universal formula $\forall \bar{Q}.M_\beta^0(\bar{Q}, \bar{v}, \bar{w})$. So (6) is equivalent to

$$\forall \bar{v}.M_\alpha(\bar{x}, \bar{v}) \wedge \varphi(\bar{v}) \rightarrow \exists \bar{w} \ \forall \bar{Q}. \ M_\beta^0(\bar{Q}, \bar{v}, \bar{w}) \wedge \psi(\bar{w})$$

which by $\mathbf{C}_{01}$ is equivalent to

$$\forall \bar{R} \ \forall \bar{v} \ M_\alpha(\bar{x}, \bar{v}) \wedge \varphi(\bar{v}) \rightarrow \exists \bar{w} \ M_\beta^0(\bar{R}_{\bar{w}}, \bar{v}, \bar{w}) \wedge \psi(\bar{w})$$

The validity of the latter formula is equivalent to the validity of

$$\forall \bar{v} \ M_\alpha(\bar{x}, \bar{v}) \wedge \varphi(\bar{v}) \rightarrow \exists \bar{w} \ M_\beta^0(\bar{R}_{\bar{w}}, \bar{v}, \bar{w}) \wedge \psi(\bar{w})$$

i.e. of the PCA

$$[\alpha] \ (\varphi \rightarrow \exists \bar{w} \ M_\beta^0(\bar{R}_{\bar{w}}, \bar{x}, \bar{w}) \wedge \psi(\bar{w})) \qquad \square$$

3.5   ITCAs and the Variance Method

Hoare's logic is conveniently conveyed as a system of program annotation, which Variance Method supplements [3], as e.g. in the annotation of a **while** loop:

$$\{x \in \mathbb{N}\}$$
$$\{\varphi\}$$
$$\textbf{while } \chi \textbf{ do}$$
$$\{x = \mathbf{s}(n)\}$$
$$\alpha$$
$$\{x = n\}$$
$$\textbf{end}$$
$$\{\psi\}$$

Note that this format can be viewed as the display of a DL proof of the ITCA $[N(x)](\varphi \rightarrow \langle \alpha \rangle \psi)$.

The use of the Variance Method for an arbitrary well-ordering $\succ$, over a field A, is captured in DL analogously, as an admission of the $\succ$-induction schema

$$(\forall y \prec x. \varphi(y) \rightarrow \varphi(x)) \rightarrow (\forall x \in A)\varphi(x)$$

where φ is itself a TCA. However, here the well-foundedness of $\succ$ is declared generically, without reference to an inductive process, and so ITCAs are of no direct relevance.

4   Summary and Conclusions

Logics for imperative programs have been studied decades ago, and yet some fundamental issues have not been completely addressed to date. In this paper we considered the issue of expressing and proving the correct termination of simple imperative programs. Using the second-order definition of program semantics as a core unifying method, we reproved the recursive enumerability of valid TCAs and their separation ability, and proved additional degenerative properties, namely compactness and a Herbrand-like theorem. Indeed, using this method, TCA validity is seen clearly to be a first-order concept. We exhibited in passing the first sound, complete, and syntax-directed deductive system for TCAs.

The degenerative properties of TCA validity can be traced to its first-order nature, and the major limitative results about classical first-order logic. Among these limitations is the failure to axiomatize inductive data-types, such as the natural numbers. Since TCA of interest are those whose truth depends on an underlying inductive data-type, it follows that they can not be proved in any semantically-sound logic of programs, such as Segerberg's Dynamic Logic. We defined here an extension of TCA's, the *inductive* TCAs (ITCA), which express directly, within the framework of DL itself, termination properties for inductive data.

ITCAs have additional applications, to be discussed elsewhere. Notably, they suggest a general notion of program provability in DL. A unary function f over $\mathbb{N}$ is *provable* in a given deductive formalism $\mathbf{D}$ for DL, if there is a program α that computes f, for which $\mathbf{D} \vdash [N(x)]\langle\alpha\rangle \top$. (It is straightforward to generalize this notion to functions of arbitrary arity, and over arbitrary inductive data types.) This notion of provable termination is independent of any coding and auxiliary concepts, of the kind invoked by definitions of "provable recursive functions" for formalization of arithmetic (see e.g. [8]). These functions are precisely the provably recursive functions of Peano Arithmetic. Allowing "rich tests" in programs (i.e. tests that are themselves DL formulas, rather than first-order formulas) does not make a difference. This result is not surprising, in view of known relations between DL and Peano Arithmetic [2,4], but neither is it trivial, because those results focus on the interpretation of DL in Peano Arithmetic, whereas we need to focus on the dual interpretation. The characterization is of interest also because it establishes an inherently computational bridge between logics of programs and traditional Proof Theory. It also suggests a panoply of interesting questions concerning the proof theoretic calibration of various deductive calculi for Dynamic Logic. For example, we believe that the provable functions of DL with Induction restricted to TCAs are precisely the primitive recursive functions, and that further restricting Induction to first-order (i.e. program-free) formulas characterizes precisely the Kalmar-elementary functions.

References

1. Andreka, H., Nemeti, I., Sain, I.: A complete logic for reasoning about programs via nonstandard model theory, Parts I and II. Theoretical Computer Science 17, 193–212, 259–278 (1982)
2. Bergstra, J.A., Tucker, J.V.: Hoare's Logic and Peano's Arithmetic. Theoretical Computer Science 22, 265–284 (1983)
3. Gries, D. (ed.): The science of programming. Springer, Berlin (1981)
4. Hajek, P.: Arithmetical interpretations of Dynamic Logic. Journal of Symbolic Logic 48, 704–713 (1983)
5. Hajek, P.: A simple dynamic logic. Theoretical Computer Science 46, 239–259 (1986)
6. Harel, D.: First-Order Dynamic Logic. LNCS, vol. 68. Springer, Heidelberg (1979)
7. Harel, D., Kozen, D., Tiuryn, J.: Dynamic Logic. MIT Press, Cambridge (2000)
8. Kreisel, G.: Survey of proof theory. Journal of symbolic Logic 33, 321–388 (1968)
9. Leivant, D.: Logical and mathematical reasoning about imperative programs. In: Conference Record of the Twelfth Annual Symposium on Principles of Programming Languages, pp. 132–140. ACM, New York (1985)
10. Leivant, D.: Partial corretness assertions provable in dynamic logics. In: Walukiewicz, I. (ed.) FOSSACS 2004. LNCS, vol. 2987, Springer, Heidelberg (2004)
11. Leivant, D.: Proving termination assertions in dynamic logics. In: Nineteenth Symposium on Logic in Computer Science, pp. 89–99. IEEE Computer Society Press, Washington (2004)
12. Meyer, A., Halpern, J.: Axiomatic definition of programming languages: a theoretical assessment. Journal of the ACM 29, 555–576 (1982)

13. Meyer, A., Mitchell, J.: Termination assertions for recursive programs: completeness and axiomatic definability. Information and Control 56, 112–138 (1983)
14. Sain, I.: Total correctness in nonstandard logics of programs. Theoretical Computer Science 50, 285–321 (1987)
15. Schmitt, P.H.: Diamond formulas: A fragment of Dynamic Logic with recursive enumerable validity problem. Information and Computation 61, 147–158 (1984)
16. Segerberg, K.: A completeness theorem in the modal logic of programs (preliminary report). Notics of the American Mathematical Society 24(6), A–552 (1977)

The Grace of Quadratic Norms: Some Examples

Leonid A. Levin*

Computer Science Department, Boston University,
111 Cummington St., Boston, MA 02215, USA

> My token tribute to the anniversary of
> Boaz Trakhtenbrot and to his major role
> in developing computer theory in Russia.

Abstract. Here I share a few notes I used in various course lectures, talks, etc. Some may be just calculations that in the textbooks are more complicated, scattered, or less specific; others may be simple observations I found useful or curious.

1 Nemirovski Estimate of Mean of Arbitrary Distributions with Bounded Variance

The popular Chernoff bounds assume severe restrictions on distribution: it must be cut-off, or vanish exponentially, etc. In [3], an equally simple bound uses no conditions at all beyond independence and known bound on variance. It is not widely used because it is not explained anywhere with very explicit computation. I offer this summary:

Assume independent variables x_i with the same unknown mean m and known lower bounds b_i on inverses $1/v_i$ of variance. We estimate m with $< 2^{-k}$ chance of error exceeding ε. This requires $\sum b_i$ of about $12k/\varepsilon^2$.

First, we normalize x_i to set $\varepsilon = 1$, spread them into $2k - 1$ groups, and in each group j take an average X_j, weighted in proportion to b_i. The inverse variance bounds B_j for X_j are additive and we assure $B_j \geq (\sqrt{2} + 1)^2 = b$.

By Chebyshev's inequality, X_j deviate from m to each side by ≥ 1 with probability $\leq 1/(b+1)$. (We assume equality: the general case follows by modifying the distribution.) Their median then deviates from m by ≥ 1 with probability

$$
P \leq 2 \sum_{i=0}^{k-1} \binom{2k-1}{i} \frac{b^i}{(b+1)^{2k-1}} < \frac{2(b+1)}{(b+1)^{2k}} \binom{2k-1}{k} \sum_{i=0}^{k-1} b^i = Q \ .
$$

Now, $n! = (n/e)^n \sqrt{2\pi n + \theta_n}$, $\pi/3 < \theta_n \leq e^2 - 2\pi$, and $\binom{2k}{k} < 4^k/\sqrt{\pi k}$. So,

$$
P < Q = \frac{b+1}{(b+1)^{2k}} \binom{2k}{k} \frac{b^k - 1}{b - 1} < \left(\frac{4b}{(b+1)^2} \right)^k \frac{b+1}{(b-1)\sqrt{\pi k}} = 2^{-k} \sqrt{\frac{2}{\pi k}} \ .
$$

* Supported by NSF grant 0311411.

A. Avron et al. (Eds.): Trakhtenbrot/Festschrift, LNCS 4800, pp. 457–459, 2008.
© Springer-Verlag Berlin Heidelberg 2008

2 Leftover Hash Lemma

The following lemma is often useful to convert a stream of symbols with absolutely unknown (except for a lower bound on its entropy) distribution into a source of perfectly uniform random bits $b \in Z_2 = \{0, 1\}$.

The version I give is close to that in [2], though some aspects are closer to that from [1]. Unlike [1], I do not restrict hash functions to be linear and do not guarantee polynomial reductions, i.e. I forfeit the case when the unpredictability of the source has computational, rather than truly random, nature.

However, like [1], I restrict hash functions only in probability of collisions, not requiring pairwise uniform distribution.

Let G be a probability distribution on Z_2^n with Renyi entropy $-\log \sum_x G^2(x) \geq m$. Let $f_h(x) \in Z_2^k$, $h \in Z_2^t$, $x \in Z_2^n$ be a hash function family in the sense that for each $x, y \neq x$ the fraction of h with $f_h(x) = f_h(y)$ is $\leq 2^{-k} + 2^{-m}$.

Let U^t be the uniform probability distribution on Z_2^t and $s = m - k - 1$. Consider a distribution $P(h, a) = 2^{-t} G(f_h^{-1}(a))$ generated by identity and f from $U^t \otimes G$. Let $\mathbf{L}_1(P, Q) = \sum_z |P(z) - Q(z)|$ be the $\mathbf{L}_1$ distance between distributions P and $Q = U^i$, $i = t + k$. It never exceeds their $\mathbf{L}_2$ distance

$$\mathbf{L}_2(P, Q) = \sqrt{2^i \sum_z (P(z) - Q(z))^2} .$$

Lemma 1 (Leftover Hash Lemma)

$$\mathbf{L}_1(P, U^i) \leq \mathbf{L}_2(P, U^i) < 2^{-s/2} .$$

Note that h must be uniformly distributed but can be reused for many different x. These x need to be independent only of h, not of each other as long as they have $\geq m$ entropy in the distribution conditional on all their predecessors.

Proof

$$(\mathbf{L}_2(P, U))^2 = 2^i \sum_{h,a} P(h, a)^2 + 2^i \sum_z (2^{-2i} - 2P(z)2^{-i}) = 2^i \sum_{h,a} P(h, a)^2 - 1$$

$$= -1 + 2^i \sum_{x,y} G(x)G(y)2^{-2t} \sum_a \|\{h : f_h(x) = f_h(y) = a\}\|$$

$$= -1 + 2^{k-t} \sum_{x,y} G(x)G(y)\|\{h : f_h(x) = f_h(y)\}\|$$

$$= -1 + 2^{k-t} \left(\sum_x G(x)^2 2^t + \sum_{x,y \neq x} G(x)G(y)\|\{h : f_h(x) = f_h(y)\}\| \right)$$

$$\leq -1 + 2^k 2^{-m} + 2^{k-t}(1 - 2^{-m})2^t(2^{-k} + 2^{-m}) < 2^{-s} . \qquad \square$$

3   Disputed Ballots and Poll Instabilities

Here is another curious example of advantages of quadratic norms.

The ever-vigilant struggle of major parties for the heart of the median voter makes many elections quite tight. Add the Electoral College system of the US Presidential elections and the history may hang on a small number of ballots in one state. The problem is not in the randomness of the outcome. In fact, chance brings a sort of fair power sharing unplagued with indecision: either party wins sometimes, but the country always has one leader. If a close race must be settled by dice, so be it. But the dice must be trusty and immune to manipulation!

Alas, this is not what our systems assure. Of course, old democratic traditions help avoiding outrages endangering younger democracies, such as Ukraine. Yet, we do not want parties to compete on tricks that may decide the elections: appointing partisan election officials or judges, easing voter access in sympathetic districts, etc. Better to make the randomness of the outcome explicit, giving each candidate a chance depending on his/her share of the vote. It is easy to implement the lottery in an infallible way, the issue is how its chance should depend on the share of votes.

In contrast to the present one, the system should avoid any big jump from a small change in the number of votes. Yet, chance should not be proportional to the share of votes. Otherwise each voter may vote for himself, rendering election of a random person. The present system encourages voters to consolidate around candidates acceptable to many others. The 'jumpless' system should preserve this feature. This can be done by using a non-linear function: say the chance in the post-poll lottery be proportional to the *squared* number of votes. In other words, a voter has one vote per each person he agrees with.[1] Consider for instance an 8-way race where the percents of votes are 60, 25, 10, 1, 1, 1, 1, 1. The leader's chance will be 5/6, his main rival's 1/7, the third party candidate's 1/43 and the combined chance of the five 'protest' runners 1/866.

This system would force major parties to determine the most popular candidate via some sort of primaries, and will almost exclude marginal runners. However it would have no discontinuity rendering any small change in the vote distribution irrelevant. The system would preserve an element of chance, but would be resistant to manipulation.

References

1. Goldreich, O., Levin, L.A.: A Hard-Core Predicate for any One-way Function. Section 5. In: STOC (1989)
2. Hastad, J., Impagliazzo, R., Levin, L.A., Luby, M.: A Pseudorandom Generator from any One-way Function. Section 4.5. SICOMP 28(4) (1999)
3. Nemirovsky, A.S., Yudin, D.B.: Problem Complexity and Method Efficiency in Optimization. Wiley, New York (1983)

---

[1] The dependence of lottery odds on the share of votes may be sharper. Yet, it must be smooth to minimize the effects of manipulation. Even (trusty) noise alone, e.g., discarding a randomly chosen half of the votes, can "smooth" the system a little.

Nested Petri Nets
for Adaptive Process Modeling*

Irina A. Lomazova

Program Systems Institute of the Russian Academy of Science,
Pereslavl-Zalessky, 152020, Russia
irina@lomazova.pereslavl.ru

To my teacher B.A. Trakhtenbrot with sincere gratitude.

Abstract. We consider Nested Petri nets (NP-nets), i.e. Petri nets in
which tokens can be Petri nets themselves. To increase flexibility and
give tools for modeling adaptive processes we extend this formalism by
allowing operations on net tokens. We prove decidability of some crucial
for verification problems and thus show that, in spite of very flexible
structure, NP-nets maintain "good" properties of ordinary Petri nets.

1 Introduction

Petri nets are a well-known formalism for modeling concurrent and distributed
systems of less than Turing power. Limited expressiveness of Petri nets is more a
merit, than a demerit, for many constitutive behavioral properties are decidable
for Petri nets.

In this paper we consider Nested Petri nets (NP-nets) [12,13,14,15,16,17] as
a tool for adaptive modeling of concurrent and distributed processes. Informally
speaking Nested Petri nets are Petri nets with net tokens, which may be generated, transferred, copied and removed as usual Petri net tokens. Net tokens may
autonomously fire its inner transitions, thus changing their own marking. Net
tokens may synchronize with one another (horizontal synchronization). There is
also a mechanism of synchronizing transition firings in two adjacent levels (vertical synchronization). Vertical synchronization means simultaneous firing of a
system net together with all net tokens "involved" in this firing.

The idea of net tokens being Petri nets goes back to R. Valk [20], and "nets in
nets" approach is extensively studied in the Petri net literature [1,10,11,18,21].
In all these works, the goal was to extend the expressive power or the expressive
comfort of Petri nets, and, as a rule, this leads to Turing expressibility and
undecidability of almost all interesting properties.

For NP-nets it was proved, that though they are strictly more expressive
than usual Petri nets (reachability is undecidable for NP-nets), they are still less

---

* This research was partly supported by the Russian Foundation for Basic Research
(grants 06-01-00106 and 07-01-00702).

A. Avron et al. (Eds.): Trakhtenbrot/Festschrift, LNCS 4800, pp. 460–474, 2008.

expressive than Turing machines and such problems as Termination, Coverability and some others are decidable for them.

Here we extend nested Petri nets with operations on net tokens. Now in NP-nets transition firing may transform involved net tokens and built new net tokens from the former ones by applying special net operations. This approach was already used in [4,5] for adaptive modeling of workflow processes. Nested workflow nets were defined there and decidability of soundness (proper termination – a crucial property for workflow modeling) was established.

Now we consider in a certain sense more general situation – adaptive modeling for arbitrary processes, when net tokens represent some subprocesses and a system net controls their execution. Extending NP-nets with net operations gives rich facilities and flexibility in adaptive modeling. With net operations it is possible to change subprocess structure in runtime. Choosing a set of net operations may depend on application field. Here, having in mind that net tokens are subprocesses, we restrict ourselves to net tokens being nets with two distinguished places: initial (source) place with no input arcs and final (sink) place with no output arcs. For such nets operations similar to Process algebra operations can be naturally defined.

The idea of controlled modification of token nets is also considered for high-level net and rule (HLNR) systems in [8,2]. Unlike our approach that easily supports arbitrary (but limited) nesting level and synchronization between different levels of a nested net as well as between different net tokens, this work considers nesting of depth one only. Moreover, [8] carries structural modification of P/T token nets by means of rule tokens, whereas our approach uses predefined and well-known operations, such as sequential and parallel composition.

Our goal was to extend NP-nets by self-modifying facilities, retaining decidability results. In this paper we prove that the Coverability and the Termination problems are decidable for NP-nets with net operations.

2   Petri Nets

By $\mathbb{N}$ we denote the set of natural numbers.

A Petri net is a bipartite graph $N = (P, T, F)$ with a finite set of nodes $P \cup T$, where $P \cap T = \emptyset$, and an incidence function (flow relation) $F : (P \times T) \cup (T \times P) \rightarrow \mathbb{N}$, describing arc multiplicity. Nodes from P are called *places*, they correspond to local states of the system. Nodes from T are called *transitions* and correspond to actions or events. A *marking* in a Petri net is a function $M : P \rightarrow \mathbb{N}$, mapping each place to some natural number (possibly zero). Thus a marking may be considered as a multiset over the set of places. Pictorially, P-elements are represented by circles, T-elements by boxes, and the flow relation F by directed arcs. Places may carry tokens represented by filled circles. A current marking M is designated by putting $M(p)$ tokens into each place $p \in P$.

A toy example of a Petri net, modeling some assumed research laboratory, is shown in Fig. 1. Here black tokens are researchers, who can be either at home (place p_1) or at work (place p_2). Being at work two researchers can produce a project (a token in the place p_3). So, transitions t_1 and t_2 correspond to going home and to the laboratory correspondingly. The transition t_3 models discussion; t_3 "consumes" two researchers at work and produces a project, returning tokens from p_2 back.

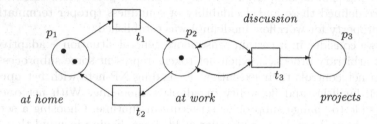

Fig. 1. Ordinary Petri net

For a transition $t \in T$ an arc (x, t) is called an *input arc*, and an arc (t, x) – an *output arc*; the *preset* $^{\bullet}t$ and the *postset* $t^{\bullet}$ are defined as the multisets over P such that $^{\bullet}t(p) = F(p, t)$ and $t^{\bullet}(p) = F(t, p)$ for each $p \in P$. A transition $t \in T$ is *enabled* in a marking M iff $\forall p \in P \; M(p) \geq F(p, t)$. An enabled transition t may *fire* yielding a new marking $M' =_{\text{def}} M - {}^{\bullet}t + t^{\bullet}$, i.e. $M'(p) = M(p) - F(p, t) + F(t, p)$ for each $p \in P$ (denoted $M \xrightarrow{t} M'$).

Being a helpful and efficient formalism for modeling and analysis of concurrent and distributed system ordinary Petri nets have serious limitations connected with the size of modeled systems. Large unstructured Petri nets are difficult to understand and to deal with. To diminish a Petri net size and to make a model more readable high-level Petri nets were introduced in the early eighties [19].

High-level Petri nets have the following characteristic properties:

- tokens in a high-level Petri net marking are not indistinguishable black dots, but individual objects;
- transitions may fire in different modes depending on consumed tokens.

Thus, tokens in high-level Petri nets may be of different types and a marking maps a place not just to a number of tokens, but to a multiset of tokens. To retain good Petri net features (e.g. decidability of some important semantic properties) it is usually supposed that the number of token types (colors) is finite. Arc labels are also changed. In high-level Petri nets arcs are labeled not with natural numbers indicating arc multiplicity, but with expressions depending on variables. Different variables binding define different modes of transition firings. Thus a variable is binded to a token of some type, and an expression get a multiset of tokens as its value.

Figure 2 shows an example of high-level Petri net *CPN*, extending the example in Fig. 1. Now researchers are represented by individual tokens A, B and C. Each of them, as in the previous example, can be either at home or at work. A variable x in a cycle of going home from work and back may be bound with any

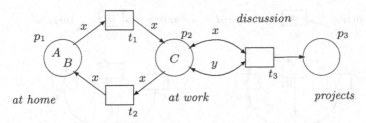

Fig. 2. High-level Petri net *CPN*

of these three values. Each binding generates some mode of firing for transitions t_1 and t_2. Thus now t_1 represents actually three transitions- one for each of the participants A, B and C. In the initial marking workers A and B are at home, and C is at work. After firing of t_1 with the binding $x := A$ workers A and C will be at work and could produce a project via discussion.

A popular class of high-level Petri nets are Colored Petri nets (CPNs) of Jensen [9]. In CPN types are called colors, so in a marking places contain colored tokens. It is well known that high-level Petri nets can be modeled by ordinary Petri nets provided the number of colors or types of individual tokens is finite. That means that for each high-level Petri net an ordinary Petri net with the equivalent behavior can be constructed. Such an "unfolding" may lead to a high growth of the net size.

3   Nested Petri Nets

In *nested Petri nets (NP-nets)* [12,13,16] tokens may be Petri nets themselves. A NP-net consists of a *system net* and *element nets*. Marked element nets are *net tokens*.

We illustrate NP-nets by further extending our example with a research laboratory. Now a research worker will be modeled by a Petri net (an element net) with three local states: passive (tied), active (rested) and inspired (ready to produce a project). Fig. 3 shows two almost similar element nets EN_1 and EN_2 for two types of researchers. Both in EN_1 and EN_2 transitions u_1, u_2, u_3, and u_4 correspondingly represent changing states from passive to active, from active to inspired and so on. The only difference in these two nets – labels for the transition u_2: α in EN_1 and $\overline{\alpha}$ in EN_2. Labels α and $\overline{\alpha}$ are for horizontal synchronization, i.e. synchronization between two net tokens residing in one system place. Label $\overline{\alpha}$ is complementary to α and vice versa. A transition with a label for horizontal synchronization may fire only simultaneously with some transition marked by the complementary label in some other net token in the same system place.

A system net *SN* in our example almost coincides with the high-level Petri net *CPN* in Fig. 2. The only syntactical difference – transition t_3 is labeled by $\overline{\lambda}$ – complementary to the label for u_3 in element nets. These labels are for vertical synchronization between system and element nets.

The main difference between the high-level Petri net *CPN* in Fig. 2 and the NP-net *NPN* shown in Fig. 4 concerns their markings. Tokens in places p_1 and

464   I.A. Lomazova

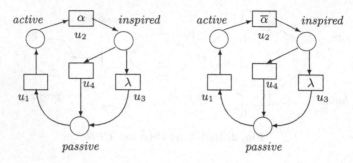

Fig. 3. Element nets EN_1 and EN_2

p_2 in the system net SN are copies of element nets EN_1 or EN_2 with their own markings. Tokens in p_3 (projects) are usual black dots, i.e. atom (not net) tokens. Thus element nets serve as token description for a system net. In our example we have one net token with the label α (the element net EN_1) and two net tokens with the label $\overline{\alpha}$ (the element net EN_2). So, here new ideas can't be generated without the worker EN_1 (a "creative" researcher).

A possible initial marking for NP-net NPN_1 is schematically shown in Fig. 4. It corresponds to the situation, when one tired creative worker is at home, and two other participants are at work. One of them is active, and another is ready to write a project.

Firing rules for transitions are defined as follows. An unlabeled transition both in element and system nets may fire *autonomously*. Autonomous transition firing in a system net is executed according to the usual rules for high-level Petri nets, where net tokens are considered to be common atomic tokens (autonomous system transition firing does not change net token markings). We say that a net token is *involved* in some transition firing, if it is consumed or produced by this firing. Net tokens in our example are ordinary Petri nets. Unlabeled transitions in net tokens may fire autonomously according to the usual rules for ordinary Petri nets, after that the net token (with a new marking) remains in the same position in its system net as before.

So, in our example an autonomous transition u_1 may fire in the net token residing in p_1, changing the state of the creative researcher from passive to active. After that this researcher may move to work (autonomous system transition firing). Note, that these two transitions may fire in any order. Some other transition firings are also possible in the initial state, e.g. active researcher may go home.

In nested Petri nets net tokens may also synchronize with one another (horizontal synchronization) and with the system net (vertical synchronization). Thus in our example after becoming active and coming to work the creative worker may discuss synchronously with another active worker (both being in the same place p_2), after that both net tokens involved in it come to the inspired state, i.e. two transitions with complementary labels for horizontal synchronization in two net tokens residing in the same system place may fire simultaneously.

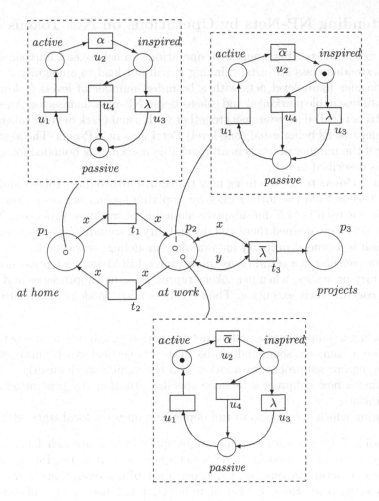

Fig. 4. NP-net NPN_1

In our example labels $\overline{\lambda}$ in the system net and λ in net tokens are for vertical synchronization. It means that transition t_3 in the system net may fire only simultaneously with transitions marked by complementary labels in all involved in its firing net tokens. So, in our example t_3 may fire only when there are two inspired net tokens in the place p_3, and after this firing two net tokens involved in it change to the passive state.

It was proved in [13,17] that, in contrast to high-level Petri nets, NP-nets are strictly more expressive than ordinary Petri nets, but they are still less than Turing power. The Reachability problem (to decide whether a given target marking is reachable from an initial one) being decidable for ordinary Petri nets is undecidable for NP-nets. However, the Coverability problem (to decide whether a marking, containing a given target marking, is reachable from an initial one) is decidable for NP-nets. The same is true about the Termination.

4   Extending NP-Nets by Operations on Net Tokens

Here we extend nested Petri nets with operations on net tokens. Further we also call the extended class NP-nets, thinking it will not lead to ambiguity.

We consider many-level nets with a bounded number of levels (taking into account all reachable markings) and tokens being NP-nets themselves. An upper-layer Petri net (called a system net) together with usual black or color tokens may contain net tokens being usual (one-level) Petri nets or NP-nets. The algorithm checking if the number of levels in all reachable markings is bounded for a given NP-net is described in [14].

Now in NP-nets transition firing may transform involved net tokens and build new net tokens from the former ones by applying special net operations. This approach was used in [4,5] for adaptive modeling of workflow processes. Nested workflow nets were defined there and decidability of soundness (proper termination, which is a crucial property for workflow modeling) was established.

Here we consider in a certain sense more general situation – adaptive modeling for arbitrary processes, when net tokens represent some subprocesses and a system net controls their execution. Then NP-nets allow modeling such situations as:

- canceling a subprocess (a system net transition simply removes a net token in case it comes to some undesirable state via vertical synchronization);
- changing one subprocess by another (via the similar mechanism);
- starting a new subprocess in some specific situation (by generating a new net token);
- choosing which subprocess to run depending on some local state, etc.

Extending NP-nets with net operations gives even more rich facilities and flexibility in adaptive modeling. With net operations it is possible to change subprocess structure in runtime. Choosing a set of net operations may depend on application field. Here, having in mind that net tokens are subprocesses, we restrict ourselves to net tokens being nets with two distinguished places: initial(source) place with no input arcs and final (sink) place with no output arcs. For such nets we can define operations similar to operations in Process algebra.

Let α, β be two marked nets (net tokens) with distinguished initial and final places. We define the following net operations (cf. Fig. 5):

1. $(\alpha.\beta)$ – sequential composition of nets (merging of the final place of α with the initial place of β).
2. $(\alpha|\beta)$ – OR-composition (choice composition) of nets (merging two initial places into one initial place and two final places into one final place).
3. $(\alpha\|\beta)$ – AND-composition (parallel composition) of nets (combining α and β by adding two new transitions – "initial" and "final" – as it is shown in Fig. 5).
4. $refine_t(\alpha, \beta)$ – in a net α refining transition t by net β with the same input and output sets of places.

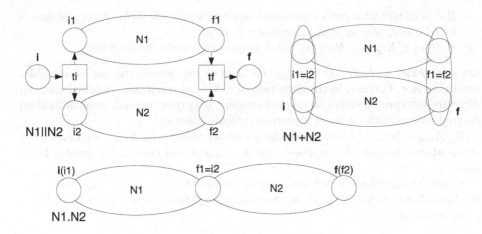

Fig. 5. Parallel composition $N1\|N_2$, Choice $N_1 + N_2$, Sequential composition $N1 \cdot N_2$

These operations allow for example to add a new subprocess as an alternative to some other subprocess, or to increase reliability by running two similar subprocesses in parallel, or to insert some additional check and so on.

5   Definitions of Extended NP-Nets

Here we give a definition of NP-nets extended with operations on net tokens. A NP-net is defined as a tuple of several net components with one designated component called a system net. Other net components are called element nets. First of all, from now on we suppose all element nets to have one initial and one final places.

A net component is a colored Petri net. Together with usual types (colors) of high-level Petri nets in NP-nets we use net types. That is, we consider each element net EN_i as a type τ_i and a associate the set of elements (values) with it. Elements of type τ_i are marked nets of the form (EN_i, M). As usual, variables and constants in arc expressions are supposed to have some fixed types. Binding functions map variables to values of corresponding types.

In arc expressions we use *variables* from $Var = \{v, \ldots\}$ and *constants* from $Con = \{c, \ldots\}$. We define $Expr_{in}$ to be a language of input expressions (over $Var \cup Con$) and $Expr_{out}$ to be a language of output expressions as follows. Note, that $Expr_{in} \subset Expr_{out}$.

Definition 1. *Let* $Atom = Var \cup Con$.

1. *An atom* $a \in Atom$ *is an expression in* $Expr_{in}$.
2. *If* $e_1, e_2 \in Expr_{in}$ *are expressions, then* $(e_1 + e_2)$ *is also an expression in* $Expr_{in}$.

3. *If e is constructed from atoms with net types with the help of net operations listed above, then e is an expression in $Expr_{out}$.*
4. *If $e_1, e_2 \in Expr_{out}$, then $(e_1 + e_2)$ is also an expression in $Expr_{out}$.*

Let $e \in Expr = Expr_{in} \cup Expr_{out}$. By $Var(e)$ we denote the set of variables occurring in e. Further, in the definition of a NP-net constants, expressions from *Expr* are interpreted either as marked element nets (net tokens), or as individual colored tokens without inner structure (atomic tokens).

By $A_{net} = \{\alpha, \ldots,\}$ we denote the set of net tokens, by A_{atom} we denote the set of atomic tokens. We suppose, that $A_{atom} \neq \emptyset$ and contains at least a black dot.

Some of transitions in NP-nets may be marked by synchronization labels: we distinguish labels for vertical and horizontal synchronization, each label has a complementary one.

Definition 2. *A nested Petri net (NP-net) is a tuple $NPN = (Expr, Lab, SN, EN_1, \ldots, EN_k, \mathcal{U})$ with*

1. *Expr is a language of expressions defined above.*
2. *Lab is a set of labels defined above.*
3. *$SN, EN_1, \ldots, EN_k$ $(k \geq 1)$ is a finite number of net components, where SN is called a system net, and $EN_i (1 \geq i \geq k)$ are element nets of NPN.*
4. *$\mathcal{U} = (A, \mathcal{I})$ is a model for Expr consisting of a universe A and interpretation function $\mathcal{I}$, where $A = A_{net} \cup A_{atom}$, and A_{net} is the set of marked element nets (net tokens), A_{atom} is a set of colored tokens.*
   Interpretation function $\mathcal{I} : Con \rightarrow A$ interprets constants as net or atomic tokens. Net operations are interpreted in the natural way (as described above) and "+" designates the multiset addition.

Here a net component is defined as a tuple (N, W, Λ), where:

- *$N = (P, T, F)$ is a net with a set of places P, a set of transitions T, and a flow relation F.*
- *W is a function mapping an arc $(x, y) \in F$ to an expression from Expr, so that an input to a transition arc gets an expression from $Expr_{in}$, and an output arc – from $Expr_{out}$. Moreover, W must satisfy the following restrictions (called arc expression restrictions):*
  - *there are no net constants (with values in A_{net}) in input arc expressions;*
  - *every net variable has not more than one occurrence in each input arc expression;*
  - *for every two expressions $W(p_1, t)$ and $W(p_2, t)$ ascribed to two input arcs of the same transition t the set $Var(W(p_1, t)) \cap Var(W(p_2, t))$ does not contain net variables;*
  - *for each variable v in an output arc expression $W(t, q)$ there should be at least one input arc expression $W(p, t)$ containing this variable.*
- *Λ is a partial function of transition labeling assigning labels from Lab to transitions in a net component. Note that a system net SN can't have labels for upper synchronization.*

So, an NP-net consists of a system net, and a set of element nets, which define structure of net tokens. A system net is a high-level net with a special language of arc expressions. A language model $\mathcal{U}$ is specified by element nets. A model universe contains marked element nets along with individual tokens. Note, that tokens in a net token can themselves be marked nets. Thus an NP-net can have several layers. In addition some transitions in system and element nets are marked by special labels for synchronization of transition firings.

We now come to defining NP-net behavior.

Let $NPN = (Expr, Lab, SN, EN_1, \ldots, EN_k, \mathcal{U})$ be a NP-net. A marking M in NPN is a function mapping each place $p \in P$ in SN to some (possibly empty) multiset $M(p)$ of tokens from A. Thus a marking in a NP-net is defined as a marking of its system net.

Let t be a transition in SN, ${}^\bullet t = \{p_1, \ldots, p_i\}$, $t^\bullet = \{q_1, \ldots, q_j\}$ be sets of its pre- and post-elements. Then $W(t) = \{W(p_1, t), \ldots, W(p_i, t), W(t, q_1), \ldots, W(t, q_j)\}$ will denote a set of all arc expressions adjacent to t. A *binding* of t is a function b assigning a value $b(v)$ (of the corresponding type) from A to each variable v occurring in some expression in $W(t)$. Obviously, given a binding of t, a value $\theta(b) = b(\theta)$ of any expression $\theta \in W(t)$ can be computed.

A transition t in SN is *enabled* in a marking M w.r.t. a binding b iff $\forall p \in {}^\bullet t : W(p, t)(b) \subseteq M(p)$, i. e. each input place p adjacent to t contains a multiset value of input arc label $W(p, t)$.

The enabled transition fires yielding a new marking M', write $M \to M'$, such that for all places p, $M'(p) = (M(p) \setminus W(p, t)(b)) \cup W(t, p)(b)$.

For net tokens from A_{net}, which serve as variable values in input arc expressions from $W(t)$, we say, that they are *involved* in the firing of t. (They are removed from input places and may be brought to output places of t or used as building material for new net tokens).

There are three kinds of steps in a NP-net NPN.

An autonomous step. Let t be a transition without synchronization labels in a system net or in a net token. Then an autonomous step is a firing of t according to the usual rules, when all tokens are considered as if they were atomic ones. An autonomous step does not change inner markings of net tokens, but it can transfer or remove some of them. Also new net tokens may evolve as a result of such a step.

A horizontal synchronization step. Let t_1, t_2 be two transitions labeled with complementary labels for horizontal synchronization in two net tokens residing in one place in some marking. If both t_1 and t_2 are enabled in this marking then they may fire simultaneously changing their markings according to the usual rules and remaining in the same outward place.

A vertical synchronization step. Let t be a transition labeled λ for lower synchronization in system net SN or in some net token, let t be enabled in a marking M w.r.t. a binding b and let $\alpha_1, \ldots, \alpha_k \in A_{net}$ be net tokens involved in this firing of t. Then t can fire provided in each α_i $(1 \leq i \leq k)$ a transition labeled by the upper synchronization label λ is also enabled. A vertical synchronization step includes two stages: first, firing of transitions in all net tokens involved in the firing of t and then, firing of t in the system net w.r.t. binding b.

We say marking M' is directly reachable from marking M and write $M \to M'$ if there is a step in NPN leading from M to M'. A run of NP-net NPN is a sequence of markings $M_0 \to M_1 \to \ldots$ successively reachable from the initial marking M_0.

6   NP-Nets as Well-Structured Transition Systems

Our former decidability results [13,14] for NP-nets without operations on net tokens were mostly based on the theory of Well-Structured Transition Systems (WSTS) [3]. Now we consider nested nets extended with operations on net tokens. To show that such extended nets are still WSTS is a challenge we deal with here.

Recall that a transition system is a pair $\mathcal{S} = \langle S, \to \rangle$, where S is an abstract set of states (or markings), and $\to \subseteq S \times S$ is any transition relation. For a transition system $\mathcal{S} = \langle S, \to \rangle$ we write $Succ(s)$ for the set $\{s' \in S \mid s \to s'\}$ of immediate successors of s. $\mathcal{S}$ is finitely branching if all $Succ(s)$ are finite.

A *quasi-ordering* is any reflexive and transitive relation $\leq$ (over some set X). A *well-quasi-ordering* (a wqo) is any quasi-ordering $\leq$ such that, for any infinite sequence $x_0, x_1, x_2, \ldots$, in X, there exist indices i, j with $i < j$ and $x_i \leq x_j$. Note, that if $\leq$ is a wqo then any infinite sequence contains an infinite increasing subsequence: $x_{i_0} \leq x_{i_1} \leq x_{i_2} \ldots$.

Definition 3. *A transition system $\Sigma = \langle S, \to \rangle$ is well-structured (a WSTS) iff there is an ordering $\leq \subseteq S \times S$ on states, such that $\leq$ is a wqo, and $\leq$ is "compatible" with $\to$, where "compatible" means that whenever $s_1 \leq t_1$, and a transition $s_1 \to s_2$ exists, there also has to exist a transition $t_1 \to t_2$, such that $s_2 \leq t_2$.*

We define a wqo on markings for NP-nets and show that together with the step relation $\to$ on states it forms WSTS.

Let $EN^1, \ldots, EN^k$ be element nets, and let $Nets = Nets(EN^1, \ldots, EN^k)$ be the set of all nets obtained from $EN^1, \ldots, EN^k$ by net operations from $\{., +, ||, refine_t\}$. Let also $A = A_{atom} \cup A_{net}$ be set of tokens as defined before. Then by $MNets(A)$ we denote the set of nets from $Nets$, marked with tokens from A. We suppose that $MNets(A)$ contains also an empty net ϵ, s.t. $\alpha.\epsilon = \epsilon.\alpha = \alpha + \epsilon = \epsilon + \alpha = \alpha||\epsilon = \epsilon||\alpha = \alpha$, $refine_t(\epsilon, \alpha) = \epsilon$, and $refine_t(\alpha, \epsilon)$ is not defined. We now define an embedding relation $\preceq_{emb}$ on nets from $MNets$.

Definition 4. *The embedding relation $\preceq_{emb} \subseteq MNets \times MNets$ is defined as a transitive closure of the following relation: for each $(N, M), (N, M'), (N_1, M_1), (N_2, M_2) \in MNets$ we have*

1. $\epsilon \preceq_{emb} (N, M)$.
2. $(N, M) \preceq_{emb} (N, M')$ *iff for every place* $p \in N$ *there exists a multiset injection* $j_p : M(p) \to M'(p)$, *such that for each token* $\alpha \in M(p)$ *we have* $\alpha = j_p(\alpha)$ *for an atomic token and* $\alpha \preceq j_p(\alpha)$ *for a net token. In words, it means that for each place* p *in* N *a multiset* $M'(p)$ *of tokens can be reduced to* $M(p)$ *by removing some tokens and/or by replacing some of net tokens by smaller (w.r.t.* $\preceq_{emb}$) *ones.*
3. *For any operation* $* \in \{., +, ||, refine_t\}$ *if* $(N_1, M_1) \preceq_{emb} (N_2, M_2)$, *then* $(N_1, M_1) * (N, M) \preceq_{emb} (N_2, M_2) * (N, M)$, *and* $(N, M) * (N_1, M_1) \preceq_{emb} (N, M) * (N_2, M_2)$.

Lemma 1. *The relation* $\preceq_{emb}$ *on MNets(A) is an ordering relation, and* $\preceq_{emb}$ *is a wqo provided the set of element nets and the set* A_{atom} *of atomic tokens are finite.*

The proof of this lemma is based on inductive application of Higman lemma [6] and is omitted here.

Lemma 2. *Let* (N, M) *be a NP-net,* $\Re(N, M)$ *– the set of all reachable states for* (N, M). *Then* $\langle \Re(N, M), \to, \preceq_{emb} \rangle$ *is a WSTS.*

Proof (Sketch). We are to show, that (1) the relation $\preceq_{emb}$ on the set $\Re(N, M)$ is a wqo, and (2) the compatibility property, which can be represented by the following diagram, where M_1 and M_2 are two system net markings, t, t' – some transition firings:

$$M_1 \quad \preceq_{emb} \quad M_2$$
$$t \downarrow \qquad\qquad \downarrow t'$$
$$M_1' \quad \preceq_{emb} \quad M_2'$$

(1) Let R be a binary relation on X and define a relation R^* on vectors in X^n by $(x_1, \ldots, x_n) R^* (y_1, \ldots, y_n)$ iff $\forall 1 \leq i \leq n : x_i R y_i$. It's easy to check that if a binary relation R on X is a wqo, then the relation R^* on X^n induced by R is also a wqo. Let us fix some linear ordering $p_1, p_2, \ldots, p_n$ on places in a system net. Then markings in the system net may be represented as vectors $M(p_1), M(p_2), \ldots, M(p_n)$. So, a proof of the theorem can be reduced to the case, when a system net has only one place.

Then, since a set of atom token colors for a concrete net is fixed and finite, by the same property of wqo a proof can be reduced to the case when there are atom tokens of only one type, i.e. black dot tokens (or tokens of some other certain color). By Lemma 1 the embedding on net tokens is a wqo. Finally, the statement follows from Higman lemma [6].

(2) Checking compatibility is a technical work, which we omit here.

As is well known ([3]), each WSTS has a finite coverability tree. So the next theorem immediately follows from the last lemma.

Theorem 1. *For each NP-net a finite coverability tree can be effectively constructed.*

Indeed, since for NP-nets the ordering relation is decidable, reachability tree is finitely branching, and the finite set of all states, directly reachable from a given state, can be effectively computed, a finite coverability tree can be effectivly constructed for any NP-net.

7   Decidability Results for NP-Nets

Now we come to decidability of semantic properties for NP-nets.

A net terminates if there exists no infinite execution (*Termination Problem*). The *Control-State Maintainability Problem* is to decide, given an initial marking M and a finite set $\{M_1, M_2, \ldots, M_n\}$ of markings, whether there exists a computation starting from M with all its inner markings covering (not less than w.r.t. some ordering) one of the M_i's. The dual problem, called the *Inevitability Problem*, is to decide whether all computations starting from M eventually visit a state not covering one of the M_i's, e.g. for Petri nets we can ask whether a given place will eventually be emptied.

It was proved in [3] that the Termination problem, the Control-state maintainability problem, and the Inevitability problem are decidable for WSTSs with (1) decidable $\leq$, and (2) effective $Succ(s)$, where $Succ(s)$ is the set of all states, directly reachable from s. This allows us to obtain the following decidability results:

Theorem 2. *Termination, the control-state maintainability problem and the inevitability problem (w.r.t. $\preceq_{emb}$) are decidable for NP-nets.*

Remark 1. Solving termination, the control-state maintainability and the inevitability problems for a NP-net can be done by building a coverability tree, which turns to be finite for NP-nets.

The *Coverability problem* is to decide, given an initial marking M and a target marking M', whether a marking covering M' is reachable from M. To prove decidability of the Coverability problem for NP-nets we use the so called set-saturation method for WSTSs [3].

Let $LTS = \langle S, \rightarrow, \leq \rangle$ be a WSTS. For $X \subseteq S$ the set $\uparrow X =_{\mathrm{def}} \{y \in S \mid (\exists x \in X) : y \geq x\}$ is upword-closed. A basis of an upword-closed I is a set I^b such that $I = \cup_{x \in I^b} \uparrow x$. It is known, that for a wqo any upword-closed set has a finite basis. By $Pred(s)$ we denote the set $\{s' \in S \mid s \rightarrow s'\}$ of immediate predecessors of s. For $X \subseteq S$ respectively we have $Pred(X) =_{\mathrm{def}} \cup_{s \in X} Pred(s)$.

Let C be an upword-closed set. Consider the sequence $C_0 \subseteq C_1 \subseteq \ldots$, where $C_0 =_{\mathrm{def}} C$, $C_{i+1} =_{\mathrm{def}} C_i \cup Pred(C_i)$, and denote $Pred^*(C) =_{\mathrm{def}} \cup_{i \geq 0} C_i$. Then to solve the Coverability problem is to check that $M \in Pred^*(\uparrow M_0)$ for a given initial state M_0 and a target set M. A WSTS has effective pred-basis if there exists an algorithm accepting any state $s \in S$ and returning $pb(s$ – a finite basis of $\uparrow Pred(\uparrow s)$.

It was also proved in [3], that The Coverability problem is decidable for WSTS's with (1) decidable $\leq$, and (2) effective pred-basis.

Lemma 3. *Any NP-net has effective pred-basis.*

The proof of this lemma is rather technical and extends the proof for NP-nets without net operations in [16]. As a consequence of this lemma and the state-saturation method we get the following

Theorem 3. *The Coverability problem is decidable for NP-nets.*

8 Conclusion

In this paper we extended NP-nets by net operations, providing new facilities for adaptive modeling. Now NP-nets allow to model not only canceling or recovering a subprocess, but also to change its structure in a run time.

We have proved, that the Termination problem, as well as the Coverability problem (which can be considered as a weak analogue of Reachability) are decidable for extended NP-nets. Thus being a rather expressive and flexible modeling formalism, NP-nets with net operations retain "good" properties of usual Petri nets.

References

1. Biberstein, O., Buchs, D., Guelfi, N.: Object-Oriented Nets with Algebraic Specifications: The CO-OPN/2 Formalism. In: Agha, G.A., De Cindio, F., Rozenberg, G. (eds.) APN 2001. LNCS, vol. 2001, pp. 73–130. Springer, Heidelberg (2001)
2. Ehrig, H., Padberg, J.: Graph grammars and Petri net transformations. In: Desel, J., Reisig, W., Rozenberg, G. (eds.) ACPN 2003. LNCS, vol. 3098, pp. 496–536. Springer, Heidelberg (2004)
3. Finkel, A., Schnoebelen, Ph.: Well-structured transition systems everywhere! Theoretical Computer Science 256(1-2), 63–92 (2001)
4. van Hee, K., et al.: Nested Nets for Adaptive Systems. In: Donatelli, S., Thiagarajan, P.S. (eds.) ICATPN 2006. LNCS, vol. 4024, pp. 241–260. Springer, Heidelberg (2006)
5. van Hee, K., et al.: Checking Properties of Adaptive Workflow Nets. In: Concurrency, Specification and Programming, Informatik-Bericht 206, vol. 1, pp. 92–103. Humboldt-Universitat zu Berlin, Berlin (2006)
6. Higman, G.: Ordering by divisibility in Abstract Algebra. Proc. London Math. Soc. 3(2), 326–336 (1952)
7. Hoffman, K.: Run time modification of algebraic high level nets and algebraic higher order nets using folding and unfolding construction. In: Hommel, G. (ed.) Proceedings of the 3rd Internation Workshop Communication Based Systems, pp. 55–72. Kluwer Academic Publishers, Dordrecht (2000)
8. Hoffmann, K., Ehrig, H., Mossakowski, T.: High-level nets with nets and rules as tokens. In: Ciardo, G., Darondeau, P. (eds.) ICATPN 2005. LNCS, vol. 3536, pp. 268–288. Springer, Heidelberg (2005)
9. Jensen, K.: Coloured Petri Nets - Basic Concepts, Analysis Methods and Practical. Springer, Heidelberg (1992)

10. Köhler, M., Rölke, H.: Reference and value semantics are equivalent for ordinary object petri nets. In: Ciardo, G., Darondeau, P. (eds.) ICATPN 2005. LNCS, vol. 3536, pp. 309–328. Springer, Heidelberg (2005)
11. Lakos, C.: From coloured Petri nets to object Petri nets. In: DeMichelis, G., Díaz, M. (eds.) ICATPN 1995. LNCS, vol. 935, pp. 278–297. Springer, Heidelberg (1995)
12. Lomazova, I.A.: Modeling Multi-Agent Dynamic Systems with Nested Petri Nets. In: Program Systems: Theoretical Foundations and Applications, pp. 143–156. Fizmatlit, Moscow, Nauka (1999) (in Russian)
13. Lomazova, I.A.: Nested Petri nets — a Formalism for Specification and Verification of Multi-Agent Distributed Systems. Fundam. Inform. 43(1-4), 195–214 (2000)
14. Lomazova, I.A.: Nested Petri nets: Multi-level and recursive systems. Fundam. Inform. 47(3-4), 283–293 (2001)
15. Lomazova, I.A.: Modeling dynamic objects in distributed systems with nested Petri nets. Fundam. Inform. 51(1-2), 121–133 (2002)
16. Lomazova, I.A. (ed.): Nested Petri nets: modeling and analysis of distributed systems with object structure. Nauchny Mir, Moscow (2004) (in Russian)
17. Lomazova, I.A., Schnoebelen, P.: Some decidability results for nested Petri nets. In: Bjorner, D., Broy, M., Zamulin, A.V. (eds.) PSI 1999. LNCS, vol. 1755, pp. 208–220. Springer, Heidelberg (2000)
18. Moldt, D., Wienberg, F.: Multi-Agent-Systems Based on Coloured Petri Nets. In: Azéma, P., Balbo, G. (eds.) ICATPN 1997. LNCS, vol. 1248, pp. 82–101. Springer, Heidelberg (1997)
19. Smith, E.: Principles of high-level net theory. In: Reisig, W., Rozenberg, G. (eds.) APN 1998. LNCS, vol. 1491, pp. 174–210. Springer, Heidelberg (1998)
20. Valk, R.: Nets in computer organization. In: Brauer, W., Reisig, W., Rozenberg, G. (eds.) APN 1986. LNCS, vol. 255, pp. 218–233. Springer, Heidelberg (1987)
21. Valk, R.: Object Petri nets: Using the nets-within-nets paradigm. In: Desel, J., Reisig, W., Rozenberg, G. (eds.) ACPN 2003. LNCS, vol. 3098, pp. 819–848. Springer, Heidelberg (2004)

Checking Temporal Properties
of Discrete, Timed and Continuous Behaviors

Oded Maler[1], Dejan Nickovic[1], and Amir Pnueli[2,3]

[1] Verimag, 2 Av. de Vignate, 38610 Gières, France
Dejan.Nickovic@imag.fr, Oded.Maler@imag.fr
[2] Weizmann Institute of Science, Rehovot 76100, Israel
[3] New York University, 251 Mercer St. New York, NY 10012, USA
Amir.Pnueli@cs.nyu.edu

Words, even infinite words, have their limits.
Dedicated to B.A. Trakhtenbrot on his 85th Birthday.

Abstract. We survey some of the problems associated with checking
whether a *given* behavior (a sequence, a Boolean signal or a continuous
signal) satisfies a property specified in an appropriate temporal logic and
describe two such monitoring algorithms for the real-time logic MITL.

1   Introduction

This paper is concerned with the following problem:

> Given a temporal property φ how does one check that a given behavior ξ
> satisfies it.

Within this paper we assume that the behavior to be checked is produced by a
model of a dynamical system S, although some of the techniques are applicable
to behaviors generated by real physical systems. Unlike formal verification which
aims at showing that *all* behaviors generated by S satisfy φ, here S is used to
generate *one behavior at a time* and can thus be viewed as a *black box*. This setting has been studied extensively in recent years both in the context of digital
hardware, under the names of "dynamic" verification, or assertion checking as
well as for software, where it is referred to as *runtime verification* [15,39]. We
will use the term *monitoring*. In this framework the question of *coverage*, that is,
finding a finite number of test cases whose behavior will guarantee overall correctness, is delegated outside the scope of the property monitor. This approach
can be used when the system model is too large to be verified formally. It is also
applicable when the "model" in question is nothing but a hardly-formalizable
simulation program, as is often the case in electrical simulation of circuits. On the
other hand, the explicit presentation of ξ itself, rather than using the generating
model S, raises new problems.

Most of the work described in this paper has been performed within the
European project PROSYD[1] with the purpose of extending some ingredients

---

[1] IST-2003-507219 PROSYD (Property-Based System Design).

A. Avron et al. (Eds.): Trakhtenbrot/Festschrift, LNCS 4800, pp. 475–505, 2008.
© Springer-Verlag Berlin Heidelberg 2008

of verification methodology from digital (discrete) to analog (continuous and hybrid) systems. Consequently, we treat systems and behaviors described at three different levels of abstraction (discrete, timed and continuous). Hence we find it useful to start with a generic model of a dynamical system defined over an abstract state space which evolves in an abstract time domain, see also [28,29]. The three models used in the paper are obtained as special instances of this model.

States and Behaviors. A model S of a system is defined over a set $V = \{x_1, \ldots x_n\}$ of *state variables*, each ranging over a domain X_i. The *state space* of the system is thus $X = X_1 \times \cdots \times X_n$. The system evolves over a time domain T which is a linearly-ordered set. A *behavior* of the system is a function from the time domain to the state space, $\xi : T \to X$. We consider *complete* behaviors, where ξ is defined all over T, as well as *partial* behaviors where ξ is defined only on a downward-closed subset of T, that is, some interval of the form $[0, r)$. We use the notation $\xi[t_1, t_2]$ for the restriction of ξ to the interval $[t_1, t_2]$ and let $\xi[t] = \bot$ when $t \geq r$. We denote the set of all possible (complete and partial) behaviors over a set X by X^*.[2]

Systems. The dynamics of a system S is defined via a rule of the form $x' = f(x, u)$ which determines the future state x' as a function of the current state x and current input $u \in U$. As mentioned earlier, we do not have access to f and our interaction with the model is restricted to stimulating it with an input $\nu \in U^*$ and then observing and checking the generated behavior $\xi \in X^*$.

Properties. Regardless of the formalism used to express it, a property φ defines a subset L_φ of X^*. A property monitor is a device or algorithm for deciding whether a given behavior ξ satisfies φ (denoted by $\xi \models \varphi$) or, equivalently, whether $\xi \in L_\varphi$.

The paper starts with properties of discrete (digital) systems, a well-studied and mature domain, where some of the problems associated with monitoring (non-causality of the specification formalism, satisfiability by finite traces, online vs. offline) are already manifested. We then move to *timed* discrete systems, whose behaviors can be viewed as *continuous-time Boolean signals*, which raise a lot of new issues such as sampling, event detection, variability bounds, etc. Most of the paper will investigate monitoring at this level of abstraction where we made some original contributions. Finally we move to continuous (analog) signals which, in addition to dense time, admit also *numerical real values*. Although for many types of properties (and in particular those expressible in our *signal temporal logic* [37,31]) checking continuous properties can be reduced to checking timed properties, there are further issues, such as approximation errors, raised by the continuous domain and by the manner in which signals are generated by numerical simulators.

---

[2] For discrete time behaviors, it is common to use X^* for finite behaviors and X^ω for infinite ones, but these distinctions are less meaningful when we come to continuous behaviors.

2   Discrete (Digital) Systems: Properties

Discrete models are used for modeling digital hardware (at gate level and above) as well as software. At this level of abstraction the set $\mathbb{N}$ of natural numbers is taken as the underlying time domain. In this case the difference between $\xi[t]$ and $\xi[t+1]$ reflects the changes in state variables that took place in the system within one clock cycle (hardware) or one program step (software).[3] The state space of digital systems is often viewed as the set $\mathbb{B}^n$ of Boolean n-bit vectors.[4] Behaviors are, hence, n-dimensional Boolean *sequences* generated by system models which are essentially finite automata (transition systems) which can be encoded in a variety of formalisms such as systems of Boolean equations with primed variables or unit delays, hardware description languages at various levels of abstraction, programming languages, etc.

Semantically speaking, a property is a subset of the set of all sequences (also known in computer science as a *formal language*) indicating the behaviors that we allow the system to have. Such subsets can be defined syntactically using a variety of formalisms such as logical formulae, regular expressions or automata that accept them. In this paper we focus on *temporal logic* [35,36] which can be viewed as a useful syntactic sugar for the first-order fragment of the monadic logic of order [40]. This section does not present new results but is rather a synthetic survey of the state-of-the-art which can serve as an entry point to the vast literature and which, we feel, is a pre-requisite for understanding the timed and continuous extensions.

2.1   Temporal Logic (Future)

The temporal logic of linear time (LTL) is perhaps the most popular property specification formalism. In a nutshell it is a language for specifying certain relationships between values of the state variables at *different time instants*, that is, at different positions in the sequence. For example, we may require that whenever $x_1 = 1$ at position t then $x_2 = 0$ at position $t + 3$. A property monitor is thus a device that observes sequences and checks whether they satisfy all such relationships. We repeat briefly some standard definitions concerning the syntax and semantics of LTL. By *semantics* we mean the rules according to which a sequence is declared as satisfying or violating a formula φ.

The syntax of LTL is given by the following grammar:

$$\varphi := p \mid \neg\varphi \mid \varphi_1 \vee \varphi_2 \mid \bigcirc \varphi \mid \varphi_1 \mathcal{U} \varphi_2,$$

where p belongs to a set $P = \{p_1, \ldots, p_n\}$ of propositions indicating values of the corresponding state variable. The basic temporal operators are *next* ($\bigcirc$),

---

[3] We mention here the existence and usefulness of *asynchronous* (*event triggered* rather than *time triggered*) systems and models, where the interpretation of a step is different.

[4] In software, as well as in high-level models of hardware, systems may include state variables ranging over larger domains such as bounded and unbounded numerical variables or dynamically-varying data structures such as queues and trees, but, at least in the hardware context, those can be encoded by bit vectors.

which specifies what should hold in the next step and *until* ($\mathcal{U}$), which requires φ_1 to hold until φ_2 becomes true, without bounding the temporal distance to this becoming. From these basic LTL operators one can derive other standard Boolean operators as well as temporal operators such as *eventually* ($\Diamond$) and *always* ($\square$):

$$\Diamond\varphi = \mathrm{T}\,\mathcal{U}\varphi \quad \text{and} \quad \square\varphi = \neg\Diamond\neg\varphi.$$

Models of LTL are *Boolean sequences* of the form $\xi : \mathbb{N} \to \mathbb{B}^n$. We also use p to denote the sequence obtained by projecting a sequence ξ on the dimension corresponding to p. The satisfaction relation $(\xi, t) \models \varphi$, indicating that sequence ξ satisfies φ starting from position t, is defined inductively as follows:

$$
\begin{aligned}
(\xi,t) &\models p && \leftrightarrow p[t] = 1 \\
(\xi,t) &\models \neg\varphi && \leftrightarrow (\xi,t) \not\models \varphi \\
(\xi,t) &\models \varphi_1 \vee \varphi_2 && \leftrightarrow (\xi,t) \models \varphi_1 \text{ or } (\xi,t) \models \varphi_2 \\
(\xi,t) &\models \bigcirc\varphi && \leftrightarrow (\xi,t+1) \models \varphi \\
(\xi,t) &\models \varphi_1\mathcal{U}\varphi_2 && \leftrightarrow \exists t' \geq t\ (\xi,t') \models \varphi_2 \text{ and } \forall t'' \in [t,t'), (\xi,t'') \models \varphi_1 \\[1ex]
(\xi,t) &\models \Diamond\varphi && \leftrightarrow \exists t' \geq t\ (\xi,t') \models \varphi \\
(\xi,t) &\models \square\varphi && \leftrightarrow \forall t' \geq t\ (\xi,t') \models \varphi
\end{aligned}
$$

A sequence ξ satisfies φ, denoted by $\xi \models \varphi$, iff $(\xi,0) \models \varphi$.

2.2   Temporal Logic (Past)

The past fragment of LTL is defined by a syntax similar to the future fragment where the *next* and *until* operators are replaced by *previously* ($\ominus$) and *since* ($\mathcal{S}$). As with future LTL, useful derived operators are *sometime in the past* $\diamondsuit$ and *always in the past* $\boxdot$ defined as

$$\diamondsuit\varphi = \mathrm{T}\,\mathcal{S}\varphi \quad \text{and} \quad \boxdot\varphi = \neg\diamondsuit\neg\varphi$$

Their semantics is given by

$$
\begin{aligned}
(\xi,t) &\models \ominus\varphi && \leftrightarrow t = 0 \text{ or } (\xi,t-1) \models \varphi \\
(\xi,t) &\models \varphi_1\mathcal{S}\varphi_2 && \leftrightarrow \exists t' \in [0,t]\ (\xi,t') \models \varphi_2 \text{ and } \forall t'' \in (t',t], (\xi,t'') \models \varphi_1 \\[1ex]
(\xi,t) &\models \diamondsuit\varphi && \leftrightarrow \exists t' \in [0,t](\xi,t') \models \varphi \\
(\xi,t) &\models \boxdot\varphi && \leftrightarrow \forall t' \in [0,t]\ (\xi,t') \models \varphi
\end{aligned}
$$

A *finite* sequence satisfies a past property φ if it satisfies it from the last position "backwards", that is, $\xi \models \varphi$ if $(\xi, |\xi|) \models \varphi$.

3   Discrete Systems: Checking Temporal Properties

We describe here the fundamental problems associated with checking temporal properties as well as the common approaches for tackling them. These are problems that exist already in the simplest model of Boolean sequences and are propagated, with additional complications to the timed and continuous domains.

3.1   Causality and Non-determinism

A major difficulty in checking properties expressed in future LTL is due to the *non-causal* definition of the satisfaction relation. To see what this means it might be helpful to look at the definition of LTL semantics as a procedure which is *recursive* on both the structure of φ and on the sequential structure of ξ. This procedure is called initially with φ and with $\xi[0]$ as arguments because we want to determine the satisfiability of φ from position zero. Then the semantic rules "call" the procedure recursively with sub formulae of φ and with further positions of ξ. In other words, the satisfiability of φ at time t may depend on the value of ξ at some *future* time instant $t' \geq t$. Even worse, some temporal operators refer to future time instants in a *quantified* manner, for example, requiring some p to hold in *all future time instants*. The satisfiability of such a property may sometime be determined only at infinity, that is, "after" we can be sure that no instance of $\neg p$ is observed.

Note that for past LTL, the recursion goes *backward* in time and the satisfaction of a past formula φ by a sequence ξ at position t is determined according to the values of ξ at the interval $[0,t]$ and in this sense, past LTL is causal. However it has been argued that the futuristic specification style is more natural for humans. The past fragment of LTL admits an immediate translation to deterministic automata and a simple monitoring procedure [16] based on this observation.

The "classical" theoretical scheme for using LTL in formal verification is based on translating a formula φ into a non-deterministic automaton over infinite sequences (an ω-automaton) $\mathcal{A}_\varphi$ that accepts exactly the sequences that satisfy it. The non determinism is needed to compensate for the non causality: the automaton has to "guess" at time t whether future observations at some $t' > t$ will render φ satisfied at t, and split the computation into two paths according to these predictions. A path that made a wrong prediction will be aborted later, either within a finite number of steps (if the guess is falsified by some observation) or via the ω-acceptance condition (if the falsification is due to non-occurrence of an event at infinity). Satisfiability of the formula can thus be determined by checking whether the ω-language accepted by $\mathcal{A}_\varphi$ is not empty. This reduces to checking the existence of an accepting cycle in $\mathcal{A}_\varphi$ which is reachable from an initial state. Verification is achieved by checking whether S may generate an infinite behavior rejected by $\mathcal{A}_\varphi$ (or accepted by $\mathcal{A}_{\neg\varphi}$). It should be noted that simplified procedures have been developed and implemented when the property in question belongs to a subclass of LTL, such as safety.

3.2   Evaluating Incomplete Behaviors

In monitoring we do not exploit the model S that generates the sequences, but rather observe sequences as they come. The major problem here, with respect to the standard semantics of LTL which is defined over *complete infinite sequences*, is the impossibility to observe infinite sequences in finite time.[5] Hence, the extension of LTL semantics to *incomplete behaviors* is a major issue in monitoring.

After having observed a finite sequence ξ we can be in one of the following three basic situations with respect to a property φ:

1. All possible infinite completions of ξ satisfy φ. Such a situation may happen, for example, when φ is $\Diamond p$ and p occurs in ξ. In this case we say that ξ *positively determines* φ.
2. All possible infinite completions of ξ violate φ. For example when φ is $\Box \neg p$ and p occurs in ξ. In this case we say that ξ *negatively determines* φ.
3. Some possible completions of ξ do satisfy φ and some others violate it. For example, any sequence where p has not occurred has extensions that satisfy, as well as extensions that violate, formulae such as $\Diamond p$ or $\Box \neg p$. In this case we say that ξ is *undecided*.

It should be noted that the "undecided" category can be refined according to both methodological, quantitative, and logical considerations. One might want to distinguish, for example, between "not yet violated" (in the case of $\Box \neg p$) and "not yet satisfied" (in the case of $\Diamond p$). The quantitative aspects enter the picture as well because the longer we observe a sequence ξ free of p, the more we tend to believe in the satisfaction of $\Box \neg p$, although the doubt will always remain. On the other hand, the satisfaction of a formula like $\bigcirc^k p$, a shorthand for $\bigcirc(\bigcirc(\ldots \bigcirc p)\ldots))$, although undecided for sequences shorter than k, will be revealed in finite time. The most general type of answer concerning the satisfiability of φ by a finite sequence ξ would be to give exactly the set of completions of ξ that will make it satisfy φ, defined as

$$\xi \backslash \varphi = \{\xi' : \xi \cdot \xi' \models \varphi\}.$$

Positive and negative determination correspond, respectively, to the special cases where $\xi \backslash \varphi = X^*$ and $\xi \backslash \varphi = \emptyset$. This "residual" language can be computed syntactically as the left quotient ("derivative") of φ by ξ.

In certain situations we would like to give a decisive answer at the end of the sequence. In the case of positive and negative determination we can reply with a yes/no answer. More general rules for assigning semantics to every finite sequence have been proposed [27,11]. Let us consider some sub-classes of LTL formulae for which such a finitary semantics clearly makes sense. The simplest among those is bounded-LTL, where the only temporal operator is *next* and where satisfiability of a formula φ at time 0 is always determined by the values of the sequence up to some $t \leq k$, with k being a constant depending on φ. Note that this class is

---

[5] To be more precise, there are some classes of infinite sequences such as the *ultimately-periodic* ones, that admit a finite representation and an easily-checkable satisfiability, however we work under the assumption that we do not have much control over the type of sequences provided by the simulator and hence we have to treat arbitrary finite sequences. It is worth noting that if S is input-deterministic then an ultimately-periodic input induces an ultimately-periodic behavior.

not as useless as it might seem: one can use "syntactic sugar" operators such as $\Box_{[0,r]}\varphi$ as shorthand for $\bigwedge_{i=0}^{r-1}(\bigcirc^i\varphi)$. The implication for monitoring is that every *sufficiently-long* sequence is determined with respect to such formulae (see also [25]).

The next class is the class of *safety* properties[6] where the only quantification of the time variable is *universal* as in $\Box\varphi$. It is not hard to see that ω-languages corresponding to such formulae consist of infinite words that *do not have a prefix* in some finitary language. While monitoring a finite sequence ξ relative to such a formula, we can be in either of the following two situations. Either such a prefix has been observed and hence any continuation of ξ will be rejected and ξ can be declared as violating, or no such prefix has been observed but nothing prevents its occurrence in the future and ξ is undecided. A similar and dual situation holds for eventually property such as $\Diamond\varphi$ that quantify existentially over time, and where an occurrence of a finitary prefix satisfying φ renders the sequence accepted.

With respect to these sub-classes one can adopt the following policy: interpret any quantification $Qt, Q \in \{\forall, \exists\}$ as $Qt \leq |\xi|$ and hence a safety that has not been violated during the lifetime of ξ is considered as satisfied, and an eventuality not fulfilled by that time is interpreted as violated. This principle may be extended to more complex formulae that involve nesting of temporal operators but in this case the interpretation seems less intuitive.

Let us remark that although models of *past* LTL are finite sequences, the problem of undecided sequences still exists. Consider for example the property $\boxminus p$. As soon as $\neg p$ is observed, we can say the the formula is negatively determined and need not wait for the rest of the sequence. On the other hand, as long as $\neg p$ has not been observed, although the prefix satisfies the property we cannot give conclusive results until the "official" end of the sequence, because $\neg p$ may always be observed in the next instant. Hence the treatment of past properties is not much different from future ones, except for the simpler construction of the corresponding automaton

Naturally many solutions have been proposed to this problem in the context of monitoring and runtime verification and we mention few. The work of [1] concerning the FoCs property checker of IBM, as well as those of [22] are restricted to safety (prefix-closed) or eventuality properties and report violation when it occurs. On the other hand, the approach of giving the residual language is proposed in [23] and [41] in the context of timed properties. A systematic study of the possible adaptation of LTL semantics to finite sequences ("truncated paths") is presented in [11]. This semantics has been adopted by the semiconductor industry standard *property specification language* PSL [10].

Our approach to monitoring is invariant under all these semantical choices. As a minimal requirement for being used, the chosen semantics should associate with every formula φ a function $\Omega_\varphi : X^* \to D$ which maps all finite sequences

---

[6] To be more precise safety properties can be written as positive Boolean combinations of formulae of the form $\Box\varphi$ where φ is a past property, and eventuality properties are negations of safety properties.

into a domain D that contains $\mathbb{B}$ (satisfied/violated) and is augmented with some additional values for undecided formulae.

3.3   Offline and Online Monitoring

In this section we discuss different forms of interaction between the mechanism that generates behaviors and the mechanism that checks whether they satisfy a given property. The behaviors are generated by some kind of a *simulator* that computes states sequentially. Without loss of generality we may assume that the systems we are interested in are *not reverse-deterministic* and, hence, the natural way to generate behaviors is from the past to the future. One may think of three basic modes of interaction (see Fig. 1):

1. *Offline*: The behaviors are completely generated by the simulator before the checking procedure starts. The behaviors are kept in a file which can be read by the monitor in either direction.
2. *Passive Online*: The simulator and the checker run in parallel, with the latter observing the behaviors progressively.
3. *Active Online*: There is a feed-back loop between the generator and the monitor so that the latter may influence the choice of inputs and, hence, the subsequent values of ξ. Such "adaptive" test generation may steer the system toward early detection of satisfaction or violation, and is outside the scope of this paper.

Each behavior is a finite sequence ξ, whose satisfiability value with respect to φ is defined via $\Omega_\varphi(\xi)$ regardless of the checking method. However there are some practical reasons to prefer one method over the other. First, to save time, we would like the checking procedure to reach the most refined conclusions as soon as possible. In the offline setting this will only reduce checking time, while in the online setting the effects of early detection of satisfaction/violation can be much more significant. This is because in certain systems (analog circuits is a notorious example) simulation time is *very long* and if the monitor can abort a simulation once its satisfiability is decided, one can save a lot of time.

The difference between online and offline is, of course, much more significant in situations where monitoring is done with respect to a *physical device*, not its simulated model. We discuss briefly several instances of this situation. The first is when chips are tested after fabrication by injecting real signals to their ports and observing the outcome. Here, the response time of the tester is very important and early (online) detection of violation can have economic importance. In other circumstances we may be monitoring a system which is already up and running. One may think of the supervision of a complex safety-critical plant where the monitoring software should alert the operator about dangerous developments that manifest themselves by property violation or by progress toward such violations. Such a situation calls for online monitoring, although offline monitoring can be used for "post mortem" analysis, for example, analyzing the "black box" after an airplane crash. Monitoring can be used for diagnosis and improvement of non-critical systems as well. For example analyzing whether the behavior of

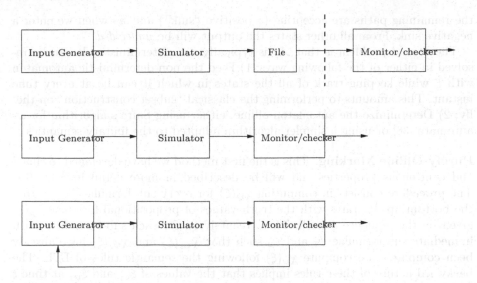

Fig. 1. Offline, passive online and active online modes of interaction between a test generator and a checker

an organization satisfies some specifications concerning the business rules of the enterprise, e.g. "every request if treated within a week". Such an application of monitoring can be done offline by inspecting transaction logs in the enterprise data base.

In the sequel we describe three basic methods for checking satisfaction of LTL formulae by sequences.

The Automaton-Based Method. This is an online-oriented approach that follows the principles used in formal verification. To monitor a property φ we first construct the automaton $\mathcal{A}_\varphi$ that accepts exactly the sequences satisfying φ and then let it read every sequence ξ as it is generated. There is a vast literature concerning the construction of automata from LTL formulae [43] and monitoring does not depend too much on the choice of the translation algorithm. We have, however a preference for the compositional construction, presented in [19] and extended for timed systems in [33]. For each sub-formula ψ of φ, this procedure constructs a sequence $\chi_\psi(\xi)$ indicating the satisfaction of ψ over time, that is $\chi_\psi(\xi)$ has value 1 at t iff $(\xi, t) \models \psi$.

There are two major problems that need to be tackled while employing this method. The first problem is that the natural automaton for φ will be an automaton over *infinite sequences*. This automaton needs to be transformed, via a suitable definition of acceptance conditions, into an automaton over finite sequences that realizes the chosen finitary semantics, as discussed in the previous section. For example, if our satisfiability domain consists of *yes*, *no* and *undecided*, we will output *yes* as soon as the automaton enters a state from which all the remaining paths are accepting (a positive "sink") and *no* when we enter a negative sink. From all other states the output will be *undecided*.

The second problem is that $\mathcal{A}_\varphi$ is typically non-deterministic. It can be resolved in either of the following ways: 1) Feed the non-deterministic automaton with ξ while keeping track of all the states in which it can be at every time instant. This amounts to performing the classical "subset construction" on-the-fly; 2) Determinize the automaton offline, either using Safra's algorithm for ω-automata [38] or using a simpler algorithm adapted to the finitary semantics.

Purely-Offline Marking. This is the first method we have developed to timed and continuous properties and will be described in more detail in Sect. 6.1. The procedure consists in computing $\chi_\psi(\xi)$ for every sub-formula ψ of φ from the bottom up. It starts with the truth values of propositional formulae $\chi_p(\xi)$ given by the sequence ξ itself. Then, recursively, for each sub-formula ψ with immediate sub-formulae ψ_1 and ψ_2 such that $\chi_{\psi_1}(\xi)$ and $\chi_{\psi_2}(\xi)$ have already been computed, we compute $\chi_\psi(\xi)$ following the semantic rules of LTL. The backward nature of these rules implies that the values of ξ_{ψ_1} and ξ_{ψ_2} at time t will "propagate" to values of ξ_ψ at some $t' \leq t$. The satisfaction function χ_φ for the main formula is computed at the end.

Incremental Marking. This approach combines the simplicity of the offline procedure with the advantages of online monitoring in terms of early detection of violation or satisfaction. After observing a prefix of the sequence $\xi[0, t_1]$ we apply the offline procedure. If, as a result, $\chi_\varphi(\xi)$ is determined at time zero we are done. Otherwise we observe a new segment $\xi[t_1, t_2]$ and then apply the same procedure based on $\xi[0, t_2]$.

A more efficient implementation of this procedure need not start the computation from scratch each time a new segment is observed. It will be often the case that $\chi_\psi(\xi)$ for some sub-formulae ψ is already determined for some subset of $[0, t_1]$ based on $\xi[0, t_1]$. In this case we only need to propagate upwards the new information obtained from $\xi[t_1, t_2]$, combined, possibly, with some additional residual information from the previous segment that was not sufficient for determination in the previous iterations. This procedure will be described in more algorithmic detail in Sect. 6.2.

The choice of the granularity (length of segments) in which this procedure is invoked depends on trade-offs between the computational cost and the importance of early detection.

4   The Timed Level of Abstraction

Coming to export the specification, testing and verification framework from the digital to the analog world, one faces two major conceptual and technical problems [30].

1. The state variables range over subsets of the set of *real numbers* that represent physical magnitudes such as voltage or current;

2. The systems evolve over a *physical* time scale modeled by the real numbers and not over a *logical* time scale defined by a central clock or by events.

Mathematically speaking, the behaviors that should be specified and checked are *signals*, function from $\mathbb{R}_{\geq 0}$ to $\mathbb{R}^n$ rather than *sequences* from $\mathbb{N}$ to $\mathbb{B}^n$ or to some other finite domain. The first problem for monitoring is the problem of how to represent a signal defined over the real time axis inside the computer, given that it is a function defined over an infinite (and non-countable) domain. The very same problem is encountered, of course, by numerical simulators that produce such signals.

Based on our conviction that the dense time problem is more profound than the infinite-state problem we use the following approach. Using a finite number of predicates over the continuous state space, analog signals are transformed into Boolean ones and are checked against properties expressed in a real-time temporal logic whose atomic propositions correspond to those predicates. This allows us to tackle the problem of dense time in isolation. Aspects specific to the continuous state space are discussed in Sect. 7. Note that one can naturally combine these predicates with genuine Boolean propositions to specify properties of hybrid systems (mixed-signal systems in the circuit jargon).

Handling an infinite state space, such as the continuum, using finite formulae is a fundamental mathematical problem. In finite domains one can characterize every individual state by a distinct formula. For example, there is a bijection between $\mathbb{B}^n$ and the set of Boolean terms over $\{p_1, \ldots, p_n\}$ which has one literal for each p_i. The common way to speak of subsets of infinite sets such as $\mathbb{R}^n$ is via *predicates*, functions from $\mathbb{R}^n$ to $\mathbb{B}$, for example inequalities of the form $x_i < d$.

We thus adopt the following approach. Let $\mu_1, \ldots, \mu_m$ be m predicates of the form $\mu : \mathbb{R}^n \to \mathbb{B}$. These predicates define a mapping $M : \mathbb{R}^n \to \mathbb{B}^m$ assigning to every real point a Boolean vector indicating the predicates it satisfies. Applying this mapping in a pointwise fashion to an analog signal $\xi : \mathbb{R}_{\geq 0} \to \mathbb{R}^n$ we obtain a Boolean signal $M(\xi) = \xi' : \mathbb{R}_{\geq 0} \to \mathbb{B}^m$ describing the evolution over time of the truth values of these predicates with respect to ξ (see Fig. 2). Events such as *rising* and *falling* in the Boolean signal correspond to some *qualitative* changes in the analog signal, for example threshold crossing of some continuous variable. This is an intermediate level of abstraction where we can observe the *temporal distance* between such events and need to confront the problems introduced by the dense time domain. Timed formalisms such as real-time temporal logics or timed automata are tailored for modeling, specification, verification and monitoring at this level of abstraction, which in addition to its applicability to analog circuits, is also very useful to model phenomena such as delays in digital circuits and execution times of software and, in fact, anything in life that can be modeled as a process where some time has to elapse between its initiation and termination.[7]

---

[7] It is a pity that the study and utilization of timed models outside academic "formal methods" circles is so negligible compared to their vast, almost universal, domain of application.

$s = x_1 \| x_2$

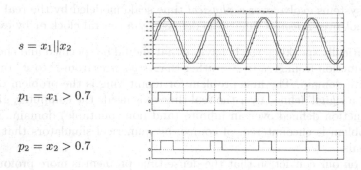

$p_1 = x_1 > 0.7$

$p_2 = x_2 > 0.7$

Fig. 2. A 2-dimensional continuous signal and the 2-dimensional Boolean signal obtained from it via the predicates $x_1 > 0.7$ and $x_2 > 0.7$

4.1 Dense-Time Signals: Representation

The major problems in handling Boolean signals by computerized tools are due to the properties of the time domain. In digital systems we have the *discrete order* $(\mathbb{N}, <)$, which means that there is a relation (successor) that generates the whole order relation. In other words, for every t and t' such that $t < t'$, there is a finite positive k such that $t' = Suc^k(t)$. This also implies that whenever we put a bound r on the range of the time variable, the set $\{t : 0 \le t \le r\}$ is *finite* and every behavior defined on the interval $[0, r]$ can be represented by a finite set $\{\xi[0], \xi[1], \ldots, \xi[r]\}$.

The dense order $(\mathbb{R}, <)$ does not admit such a property, and for every $t < t'$ one can find t'' such that $t < t'' < t'$. This implies that in order to specify a dense-time signal, even if restricted to a bounded time interval $[0, r]$, one might need to specify an *infinite* set of values. For arbitrary analog signals the only way to provide these values throughout the entire interval is via analytic expressions such as $\xi[t] = \sin(t)$. Otherwise an analog signal can only be partially represented by its values at a finite subset of the time domain consisting of *sampling points* (more on that in Sect. 7). As for Boolean signals, let us note that functions from $\mathbb{R}_+$ to $\mathbb{B}$ can be rather weird objects, potentially switching between 0 and 1 infinitely many times in a bounded interval of time (the so-called Zeno phenomenon).[8] From now on we restrict our attention to non-Zeno Boolean signals.

A non-Zeno Boolean signal ξ defined over an interval $[0, r)$ decomposes naturally into a finite sequence of intervals $I_0, I_1, \ldots, I_k$ such that $I_0 = [0, t_1)$, $I_i = [t_i, t_{i+1})$, $I_k = [t_{k-1}, r)$, the value of ξ is constant in every interval, and $\xi(I_{i+1}) = \neg\xi(I_i)$. The set of intervals, together with the value at $t = 0$ determine the value of ξ at *any* point and can serve as a basis for checking properties relative to ξ.

---

[8] Such Zeno signals can be obtained from analog signals via Booleanization: just consider a signal representing a damped oscillation around zero and its Boolean image via the predicate $x < 0$.

4.2   Dense-Time Signals: Properties

The temporal operators of LTL are of two types. The *next* operator is bounded and quantitative. It specifies something that should happen within the very next step or, if used iteratively, within a bounded number of steps. The *until* operator and its derivatives are unbounded and qualitative, requiring that something should or should not hold at some unspecified future instant. The latter properties are not affected seriously from the passage to dense time, while quantitative operators need to be redefined. To start with, the *next* operator which specifies at t what should hold at the *smallest* t' such the $t < t'$ becomes meaningless. Instead one has to use operators that specify at t what should hold at time $t + d$ or during the interval $t \oplus [a, b] = [t + a, t + b]$. Many temporal logics over such metric time have been proposed and studied [21,4,17,18] and we will focus on the logic MITL, which is a natural adaptation of LTL to dense time [3].

Dense time also has an influence on the different monitoring procedures. As we shall see, the offline procedure based on marking the truth values of sub-formulae over time, can be rather easily adapted to signals. However the online approaches are more problematic. Consider the approach based on translating a formula into an automaton that accepts its models. The appropriate automaton will be a timed automaton, which reads signals continuously and uses auxiliary clock variables to measure times since the occurrence of certain events. Automata corresponding to MITL formulae are, more often than not, non-deterministic, a feature that, in a discrete-time framework, can be resolved using subset construction, either offline or on the fly. Dense non-determinism is another story as the automaton may stay during an interval in a state q while at any moment during the interval it may take a transition to q', thus spawning uncountably-many runs of the automaton. The impossibility of an offline determinization of timed automata is a well-known fact in the domain, but in Sect. 6.3 we will mention some remedies to this problem.

We can now move to more detailed definitions of signals and their corresponding temporal logics, followed by the description of their monitoring algorithms.

5   Boolean Signals and Their Temporal Logics

5.1   Signals

Two basic semantic domains can be used to describe timed behaviors. *Time-event sequences* consist of instantaneous events separated by time durations while discrete-valued *signals* are functions from time to some discrete domain. The reader may consult the introduction to [6] for more details on the algebraic characterization of these domains. In this work we use Boolean signals as the semantic domain, which is the natural choice, both for the logic MITL and the circuit application domain.

Let the time domain T be the set $\mathbb{R}_{\geq 0}$ of non-negative real numbers. A Boolean signal is a function $\xi : T \to \mathbb{B}^n$. We use $\xi[t]$ for the value of the signal at time t and the notation $\sigma_1^{t_1} \cdot \sigma_2^{t_2} \cdots$ for a signal whose value is σ_1 at the interval $[0, t_1)$,

σ_2 in the interval $[t_1, t_1 + t_2)$, etc. A signal whose value is defined only on an interval $[0, r)$ is called finite and of *metric*[9] length r (denoted by $|\xi| = r$). The restriction of a signal to length d is defined as

$$\xi' = \langle \xi \rangle_d \text{ iff } \xi'[t] = \begin{cases} \xi[t] & \text{if } t < d \\ \bot & \text{otherwise} \end{cases}$$

For the sake of simplicity we restrict ourselves to *left-closed right-open* signal segments and to timed modalities that use only closed intervals. As a consequence we exclude signals with *punctual* "intervals" which are meaningless in the algebraic definition of signals [6,5]. The more general case was treated in [3].

Different Boolean signals can be combined and separated using the standard operations of *pairing* and *projection* defined as

$$\xi_1 \parallel \xi_2 = \xi_{12} \text{ if } \forall t \, \xi_{12}[t] = (\xi_1[t], \xi_2[t])$$
$$\xi_1 = \pi_1(\xi_{12}) \quad \xi_2 = \pi_2(\xi_{12})$$

In particular, $\pi_p(\xi)$ will denote the projection of ξ on the dimension that corresponds to proposition p.

Any Boolean operation OP can be "lifted" to an operation on signals as

$$\xi = \text{OP}(\xi_1, \xi_2) \text{ iff } \forall t \, \xi[t] = \text{OP}(\xi_1[t], \xi_2[t])$$

When we apply operations on signals of different lengths we use the convention

$$\text{OP}(v, \bot) = \text{OP}(\bot, v) = \bot$$

which guarantees that if $\xi = \text{OP}(\xi_1, \xi_2)$ then $|\xi| = \min(|\xi_1|, |\xi_2|)$.

Any reasonable Boolean signal can be represented using a countable number of intervals. An *interval covering* of a given interval $I = [0, r)$ is a sequence $\mathcal{I} = I_1, I_2 \dots$ of left-closed right-open intervals such that $\bigcup I_i = I$ and $I_i \cap I_j = \emptyset$ for every $i \neq j$. An interval covering $\mathcal{I}'$ is said to *refine* $\mathcal{I}$, denoted by $\mathcal{I}' \prec \mathcal{I}$ if $\forall I' \in \mathcal{I}' \, \exists I \in \mathcal{I}$ such that $I' \subseteq I$.

An interval covering $\mathcal{I}$ is said to be *consistent* with a signal ξ if $\xi[t] = \xi[t']$ for every t, t' belonging to the same interval I_i. In that case we can use the notation $\xi(I_i)$. Clearly, if $\mathcal{I}$ is consistent with ξ, so is any $\mathcal{I}' \prec \mathcal{I}$. We restrict ourselves to signals of *finite variability*, that is, signals admitting a finite consistent interval covering. We denote by $\mathcal{I}_\xi$ the *minimal* interval covering consistent with a finite variability signal ξ. The set of positive intervals of ξ is $\mathcal{I}_\xi^+ = \{I \in \mathcal{I}_\xi : \xi(I) = 1\}$ and the set of negative intervals is $\mathcal{I}_\xi^- = \mathcal{I}_\xi - \mathcal{I}_\xi^+$.

A signal ξ is said to be *unitary* if $\mathcal{I}_\xi^+$ is a singleton. Any finite-variability signal ξ over a bounded interval can be decomposed into a union of k unitary signals such that $\xi = \xi^1 \vee \dots \vee \xi^k$, see Fig. 3.

The *concatenation* $\xi = \xi_1 \cdot \xi_2$ of two signals ξ_1 and ξ_2 defined over the intervals $[0, r_1)$ and $[0, r_2)$ respectively is a signal over $[0, r_1 + r_2)$ defined as:

$$\xi[t] = \begin{cases} \xi_1[t] & \text{if } t < r_1 \\ \xi_2[t - r_1] & \text{otherwise} \end{cases}$$

---

[9] To distinguish it from the *logical* length which corresponds to the number of state changes.

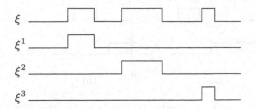

Fig. 3. A signal ξ and its unitary decomposition (ξ^1, ξ^2, ξ^3)

The *d-suffix* of a signal ξ is the signal $\xi' = d\backslash\xi$ obtained from ξ by removing the prefix $\langle\xi\rangle_d$ from ξ, that is,

$$\xi'[t] = \xi[t + d] \text{ for every } t \in [0, |\xi| - d).$$

The *Minkowski sum* and *difference* of two sets P_1 and P_2 are defined as

$$P_1 \oplus P_2 = \{x_1 + x_2 : x_1 \in P_1, x_2 \in P_2\}$$
$$P_1 \ominus P_2 = \{x_1 - x_2 : x_1 \in P_1, x_2 \in P_2\}.$$

Of particular interest are the applications of these operations to one-dimensional sets consisting of elements of the time domain T:

$$\{t\} \oplus [a, b] = [t + a, t + b], \qquad [m, n) \oplus [a, b] = [m + a, n + b)$$

$$\{t\} \ominus [a, b] = [t - b, t - a], \qquad [m, n) \ominus [a, b] - [m - b, n - a)$$

The operation that will be used for computing the satisfiability of a formula whose major operator is a bounded temporal operator is the operation of *back shifting*.

Definition 1 (Back Shifting). *The $[a, b]$-back-shifting of a Boolean signal ξ', denoted by $\xi = \text{SHIFT}_{[a,b]}(\xi')$, is a signal ξ such thet for every t, $\xi[t] = 1$ iff there exists $t' \in t \oplus [a, b]$ such that $\xi'[t'] = 1$.*

The resemblance of this definition to the semantics of the $\Diamond_{[a,b]}$ operator (to be defined in Sect. 5.2) is not a coincidence. If $\varphi = \Diamond_{[a,b]}\varphi'$ then the respective satisfiability signals of φ and φ' satisfy $\chi_\varphi = \text{SHIFT}_{[a,b]}(\chi_{\varphi'})$. This operation is easy to compute on a representation based on an interval covering of the signals. When ξ' is a unitary signal with $\mathcal{I}_{\xi'}^+ = \{I'\}$, the result of back shifting is the unitary signal ξ with $\mathcal{I}_\xi^+ = \{I\}$ where $I = I' \ominus [a, b] \cap T$ (the intersection with T is needed to remove negative values, see Fig. 4).

5.2   Real-Time Temporal Logic

The syntax of MITL is defined by the grammar

$$\varphi := p \mid \neg\varphi \mid \varphi_1 \vee \varphi_2 \mid \varphi_1 \mathcal{U}_{[a,b]}\varphi_2 \mid \varphi_1 \mathcal{U}\varphi_2$$

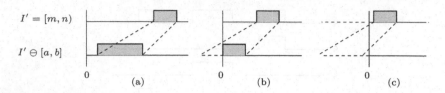

Fig. 4. Three instances of back shifting $I = [m, n) \ominus [a, b]$: (a) $I = [m - b, n - a)$; (b) $I = [0, n - a]$ because $m - b < 0$; (c) $I = \emptyset$ because $n - a < 0$

where p belongs to a set $P = \{p_1, \ldots, p_n\}$ of propositions and $b > a \geq 0$ are rational numbers.[10] From basic MITL operators one can derive other standard Boolean and temporal operators, in particular the time-constrained *eventually* and *always* operators:

$$\Diamond_{[a,b]}\varphi = \mathrm{T} \, \mathcal{U}_{[a,b]}\varphi \quad \text{and} \quad \Box_{[a,b]}\varphi = \neg\Diamond_{[a,b]}\neg\varphi$$

We interpret MITL over n-dimensional Boolean signals and define the satisfiability relation similarly to LTL.

$$
\begin{aligned}
(\xi, t) &\models p &&\leftrightarrow p[t] = \mathrm{T} \\
(\xi, t) &\models \neg\varphi &&\leftrightarrow (\xi, t) \not\models \varphi \\
(\xi, t) &\models \varphi_1 \vee \varphi_2 &&\leftrightarrow (\xi, t) \models \varphi_1 \text{ or } (\xi, t) \models \varphi_2 \\
(\xi, t) &\models \varphi_1 \mathcal{U} \varphi_2 &&\leftrightarrow \exists t' \geq t \; (\xi, t') \models \varphi_2 \text{ and} \\
& &&\quad \forall t'' \in [t, t'], (\xi, t'') \models \varphi_1 \\
(\xi, t) &\models \varphi_1 \mathcal{U}_{[a,b]} \varphi_2 &&\leftrightarrow \exists t' \in t \oplus [a, b] \; (\xi, t') \models \varphi_2 \text{ and} \\
& &&\quad \forall t'' \in [t, t'], (\xi, t'') \models \varphi_1 \\
\\
(\xi, t) &\models \Diamond_{[a,b]}\varphi &&\leftrightarrow \exists t' \in t \oplus [a, b] \; (\xi, t') \models \varphi \\
(\xi, t) &\models \Box_{[a,b]}\varphi &&\leftrightarrow \forall t' \in t \oplus [a, b] \; (\xi, t') \models \varphi
\end{aligned}
$$

The past version of MITL is obtained by replacing the $\mathcal{U}_{[a,b]}$ operator by the *since* operator $\mathcal{S}_{[a,b]}$, from which one can derive the time-constrained *sometime in the past* ($\Diamondblack$) and *always in the past* ($\boxminus$), operators. The semantics of the past operators is defined as

$$
\begin{aligned}
(\xi, t) &\models \varphi_1 \mathcal{S}_{[a,b]} \varphi_2 &&\leftrightarrow \exists t' \in t \ominus [a, b] \; (\xi, t') \models \varphi_2 \text{ and} \\
& &&\quad \forall t'' \in [t', t], (\xi, t'') \models \varphi_1
\end{aligned}
$$

$$
\begin{aligned}
(\xi, t) &\models \Diamondblack_{[a,b]}\varphi &&\leftrightarrow \exists t' \in t \ominus [a, b] \; (\xi, t') \models \varphi \\
(\xi, t) &\models \boxminus_{[a,b]}\varphi &&\leftrightarrow \forall t' \in t \ominus [a, b] \; (\xi, t') \models \varphi
\end{aligned}
$$

In this paper we focus on the more difficult future fragment of MITL.

6   Checking Timed Properties

In this section we describe two procedures for checking MITL properties:

---

[10] In fact, it is sufficient to consider integer constants.

1. An offline marking procedure that propagates truth values upwards from propositions via super-formulae up to the main formula. This procedure has been first presented in [31].
2. An incremental marking procedure that updates the marking each time a new segment of the signal is observed. This procedure is described in [37].

A central notion in all these algorithms is that of the *satisfaction signal* $\xi' = \chi_\varphi(\xi)$ associated with a formula φ and a signal ξ. In this signal $\xi'[t] = 1$ whenever $(\xi, t) \models \varphi$. We remind the reader that due to non-causality the value of $\xi'[t]$ is not necessarily known at time t, that is, after observing $\xi[t]$, and may depend on future values of ξ. Whenever the identity of ξ is clear from the context, we will use the shorthand notation χ_φ.

6.1 Offline Marking

This algorithm [31] works as follows. It has as input a formula φ and an n-dimensional Boolean signal of length r. For every sub-formula ψ of φ it computes its satisfiability signal $\chi_\psi(\xi)$. To simplify the discussion we restrict the presentation to a bounded version of MITL where the unbounded *until* is not used. Hence we have properties that are fully determined if the signal is sufficiently long. In the case where the signal is too short the output is *undecided*, denoted by $\perp$. The procedure is recursive on the structure (parse tree) of the formula. It goes down until the propositional variables whose values are determined directly by ξ, and then propagates values as it comes up from the recursion. We will use OP$_1$ and OP$_2$ for arbitrary unary and binary logical or temporal operators. As a preparation for the incremental version, we do not pass ξ and χ_φ as input or output parameters but rather store them in global data structures.

Algorithm 1. OFFLINEMITL

input : an MITL Formula φ

switch φ **do**
 case p
  | $\chi_\varphi := \pi_p(\xi)$;
 end
 case OP$_1(\varphi_1)$
  OFFLINEMITL (φ_1);
  $\chi_\varphi := $ COMBINE(OP$_1, \varphi_1$);
 end
 case OP$_2(\varphi_1, \varphi_2)$
  OFFLINEMITL (φ_1);
  OFFLINEMITL (φ_2);
  $\chi_\varphi := $ COMBINE $($OP$_2, \chi_{\varphi_1}, \chi_{\varphi_2})$;
 end
end

Most of the work in this algorithm is done by the COMBINE function which for $\varphi = $ OP$_2(\varphi_1, \varphi_2)$ computes χ_φ from the signals χ_{φ_1} and χ_{φ_2}, which may differ

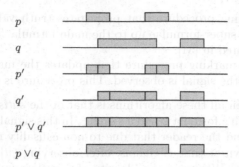

Fig. 5. To compute $p \vee q$ we first refine the interval covering to obtain the semantically-equivalent representations p' and q'. We then perform interval-wise operations to obtain $p' \vee q'$ and then merge adjacent positive intervals

in length. We describe briefly how this function works for each of the operators, with a sufficient detail to understand how it operates on the representation of the input and output signals by their sets of positive intervals. For the sake of readability we omit the description of various mundane optimizations.

$\chi_\varphi := \textbf{Combine}(\neg, \chi_{\varphi_1})$ The negation is computed by simply changing the Boolean value of each minimal interval in the representation of χ_{φ_1}.

$\chi_\varphi := \textbf{Combine}(\vee, \chi_{\varphi_1}, \chi_{\varphi_2})$ For the disjunction we first construct a refined interval covering $\mathcal{I} = \{I_1, \ldots, I_k\}$ for $\chi_{\varphi_1} \| \chi_{\varphi_2}$ so that the mutual values of both signals become uniform in every interval. Then we compute the disjunction interval-wise, that is, $\varphi(I_i) = \varphi_1(I_i) \vee \varphi_2(I_i)$. Finally we merge adjacent intervals having the same Boolean value to obtain the minimal interval covering $\mathcal{I}_{\chi_\varphi}$. This procedure is illustrated in Fig. 5.

$\chi_\varphi := \textbf{Combine}(\Diamond_{[a,b]}, \chi_{\varphi_1})$ This is the most important part of our procedure which computes $\chi_\varphi := \text{SHIFT}_{[a,b]}(\xi_{\varphi_1})$. For every positive interval $I \in \mathcal{I}^+{}_{\varphi_1}$ we compute its back shifting $I \ominus [a,b] \cap T$ and insert it to $\mathcal{I}^+{}_\varphi$. Overlapping positive intervals in $\mathcal{I}^+{}_\varphi$ are merged to obtain a minimal consistent interval covering. In the process, all the negative intervals shorter than $b - a$ disappear.[11]

$\chi_\varphi := \textbf{Combine}(\mathcal{U}_{[a,b]}, \chi_{\varphi_1}, \chi_{\varphi_2})$ The implementation of the timed *until* operator is based on the equivalence $\varphi_1 \mathcal{U}_{[a,b]} \varphi_2 \leftrightarrow (\Diamond_{[a,b]}(\varphi_1 \wedge \varphi_2)) \wedge \varphi_1$ when χ_{φ_1} is a unitary signal. This is because for a unitary signal, if φ_1 holds at t_1 and at t_2 it must hold during the whole interval. This does not hold for arbitrary signals, see Fig. 6. In order to treat the general case where χ_{φ_1} is a non-unitary signal we first need to decompose it into the unitary signals $\chi_{\varphi_1}^1, \ldots, \chi_{\varphi_1}^k$ and then compute

$$\chi_\varphi^i = (\text{SHIFT}_{[a,b]}(\chi_{\varphi_1}^i \wedge \chi_{\varphi_2})) \wedge \chi_{\varphi_1}^i$$

---

[11] This procedure can be viewed alternatively as shifting the *negative* intervals by $[b, a]$.

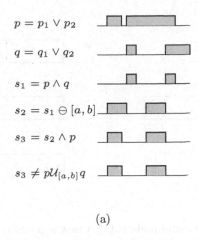

$$p = p_1 \lor p_2$$

$$q = q_1 \lor q_2$$

$$s_1 = p \land q$$

$$s_2 = s_1 \ominus [a, b]$$

$$s_3 = s_2 \land p$$

$$s_3 \neq p\,\mathcal{U}_{[a,b]}\, q$$

(a)

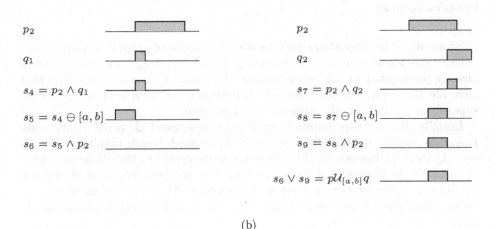

$$p_2$$

$$q_1$$

$$s_4 = p_2 \land q_1$$

$$s_5 = s_4 \ominus [a, b]$$

$$s_6 = s_5 \land p_2$$

$$p_2$$

$$q_2$$

$$s_7 = p_2 \land q_2$$

$$s_8 = s_7 \ominus [a, b]$$

$$s_9 = s_8 \land p_2$$

$$s_6 \lor s_9 = p\,\mathcal{U}_{[a,b]}\, q$$

(b)

Fig. 6. Computing satisfiability of $p\,\mathcal{U}_{[a,b]}\,q$ via the satisfiability of $\Diamond_{[a,b]}(q \land p) \land p$. (a) wrong results obtained with non-unitary signals; (b) correct results obtained with a unitary decomposition $p = p_1 \lor p_2$ and $q = q_1 \lor q_2$. The computation with p_1 is omitted as it has an empty intersection with q

for each $i \in [1, k]$. Finally we recompose the resulting signals as

$$\chi_\varphi = \bigvee_{i=1}^{k} \chi_\varphi^i.$$

6.2 Incremental Marking

Incremental marking is performed using a kind of piecewise-online procedure invoked each time a new segment of ξ, denoted by Δ_ξ, is observed. For each

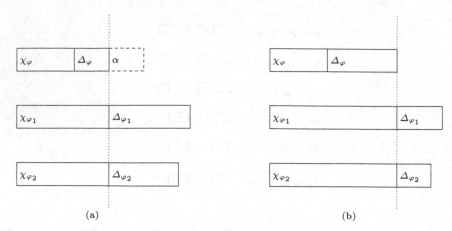

Fig. 7. A step in an incremental update: (a) A new segment α for φ is computed from Δ_{φ_1} and Δ_{φ_2}; (b) α is appended to Δ_φ and the endpoints of χ_{φ_1} and χ_{φ_1} are shifted forward accordingly

sub-formula ψ the algorithm stores its already-computed satisfaction signal partitioned into a concatenation of two signals $\chi_\psi \cdot \Delta_\psi$ with χ_ψ consisting of values already propagated to the super-formula of ψ, and Δ_ψ consists of values that have already been computed but which have not yet been propagated to the super-formula and can still influence its satisfaction.

Initially all signals are empty. Each time a new segment Δ_ξ is read, a recursive procedure similar to the offline procedure is invoked, which updates every χ_ψ and Δ_ψ from the bottom up. The difference with respect to the offline algorithm is that only the segments of the signal that have not been propagated upwards participate in the update of their super-formulae. This may result in a lot of saving when the signal is very long, as has been demonstrated empirically in [37].

As an illustration consider $\varphi = \text{OP}(\varphi_1, \varphi_2)$ and the corresponding truth signals of Fig. 7-(a). Before the update we always have $|\chi_\varphi \cdot \Delta_\varphi| = |\chi_{\varphi_1}| = |\chi_{\varphi_2}|$: the parts Δ_{φ_1} and Δ_{φ_2} that may still affect φ are those that start at the point from which the satisfaction of φ is still unknown. We apply the COMBINE procedure on Δ_{φ_1} and Δ_{φ_2} to obtain a new (possibly empty) segment α of Δ_φ. This segment is appended to Δ_φ in order to be propagated upwards, but before that we need to shift the borderline between χ_{φ_1} and Δ_{φ_1} (as well as between χ_{φ_2} and Δ_{φ_2}) in order to reflect the update of Δ_φ. The procedure is described in Algorithm 2.

6.3   Monitoring Using Timed Automata

Our contribution to the automaton-based approach for checking timed properties will be described elsewhere and we mention the relevant results briefly. In [32] we have shown how to build deterministic timed automata from past MITL properties and gave an alternative proof of the impossibility to do so for future MITL. The difference in dererminizability between the past and future

---

Algorithm 2. INC-OFFLINE-MITL

---

input : an MITL Formula φ and an increment Δ_ξ of a signal

switch φ **do**

    case p

    | $\Delta_\varphi := \Delta_\varphi \cdot \pi_p(\Delta_\xi)$;

    end

    case OP$_1(\varphi_1)$

    | INC-OFFLINE-MITL (φ_1, Δ_ξ);

    | $\alpha :=$ COMBINE(OP$_1, \Delta_{\varphi_1}$);

    | $d := |\alpha|$;

    | $\Delta_\varphi := \Delta_\varphi \cdot \alpha$;

    | $\chi_{\varphi_1} := \chi_{\varphi_1} \cdot \langle \Delta_{\varphi_1} \rangle_d$;

    | $\Delta_{\varphi_1} := d \backslash \Delta_{\varphi_1}$

    end

    case OP$_2(\varphi_1, \varphi_2)$

    | INC-OFFLINE-MITL (φ_1, Δ_ξ);

    | INC-OFFLINE-MITL (φ_2, Δ_ξ);

    | $\alpha :=$ COMBINE(OP$_2, \Delta_{\varphi_1}, \Delta_{\varphi_2}$);

    | $d := |\alpha|$;

    | $\Delta_\varphi := \Delta_\varphi \cdot \alpha$;

    | $\chi_{\varphi_1} := \chi_{\varphi_1} \cdot \langle \Delta_{\varphi_1} \rangle_d$;

    | $\Delta_{\varphi_1} := d \backslash \Delta_{\varphi_1}$;

    | $\chi_{\varphi_2} := \chi_{\varphi_2} \cdot \langle \Delta_{\varphi_2} \rangle_d$;

    | $\Delta_{\varphi_2} := d \backslash \Delta_{\varphi_2}$

    end

end

---

fragments turned out to be a syntactical accident *not related* to the difference in the causality between past and future (note that the logic MTL, admitting *punctual* modalities such as $\Diamond_d$ is non-deterministic in both directions). The reason is that interval-based modalities, when they point backwards, erase the effect of small fluctuations in Boolean signals, see [32]. In [33] we adapted the compositional construction of non-deterministic automata from LTL [19] to MITL. Finally we have shown in [34] how to construct deterministic timed automata for the *bounded* fragment of the more general logic MTL under *bounded variability* assumptions. More technical details concerning the techniques used can be found in those papers. We mention the works of [42,24] and [12,13,14] which inspired part of our work.

7 Continuous Signals

The algorithms developed for dense-time Boolean signals, provide a solid basis for monitoring continuous signals when the properties belong to the *signal temporal logic* (STL) [31,37] which is nothing but MITL, parameterized by a set of numerical predicates playing the role of atomic propositions. For such properties, each continuous signal is transformed, via the numerical predicates appearing in

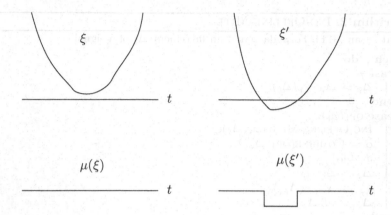

Fig. 8. Two signals which are close from a continuous point of view, one satisfying the property $\Box(x > 0)$ and one violating it

the property, into a Boolean signal which is checked against the MITL "skeleton" of the formula. In the rest of this section we discuss technical problems related to the applicability of the "Booleanization" procedure.

As we have seen, non-Zeno Boolean signals, albeit the fact that they are defined over dense time domain, admit an *exact finite representation* via the switching points that define their *true* and *false* intervals. This is no longer the case for continuous signals where we have a contrast between the *ideal mathematical object*, consisting of an uncountable number of pairs $(t, \xi[t])$ with t ranging over some interval $[0, r) \subseteq \mathbb{R}_{\geq 0}$, and any *finite* representation which consists of a collection of such pairs, with t restricted to range over a finite set of *sampling points*. The values of ξ at sampling points t_1 and t_2 may, at most, impose some constraints on the values of ξ inside the interval (t_1, t_2). Such constraints can be based on the dynamics of the generating system and the manner in which the numerical simulator produces the signal values at the sampling points. Numerical analysis is a very mature domain with a lot of accumulated experience concerning tradeoffs between accuracy and computation time. Its major premise is that given a model of the system as a continuous dynamical system defined by a differential equation[12], one can improve the quality of a discrete-time approximation of its behavior by employing denser sets of sampling points and more sophisticated numerical integration procedures.[13]

In order to speak quantitatively about the approximation of a signal by another we need the concept of a *distance/metric* imposed on the space of continuous signals. A metric is a function that assigns to two signals ξ_1 and ξ_2 a non-negative value $\rho(\xi_1, \xi_2)$ which indicates how they resemble each other. Using metrics one can express the "convergence" of a numerical integration scheme as

---

[12] It is worth noting that some models used for rapid simulation of transistor networks cannot always be viewed as continuous dynamical systems in the classical mathematical sense.

[13] For systems which are stable the quality can be improved indefinitely.

the condition that $\lim_{d \to 0} \rho(\xi, \xi_d) = 0$ where ξ is the ideal mathematical signal and ξ_d is its numerical approximation using an integration step d.

Metrics and norms for continuous signals are used extensively in circuit design, control and signal processing. There are, however, major problems concerned with their application to property monitoring due to the incompatibility between the continuous nature of the signals and the discrete nature of $\{0, 1\}$-properties, a phenomenon which is best illustrated using the following simple example. Consider the property $\Box(x > 0)$ and an ideal mathematical signal ξ that satisfies the property but which passes very close to zero at some points. We can easily transform ξ into a signal ξ' which is *very close* to ξ under any reasonable continuous metric, but according to the metric induced by the property, these signals are as distant as can be: one of them satisfies the property and the other violates it (see Fig. 8).

Moreover, if the sojourn time of a signal below zero is short, an arbitrary shift in the sampling can make the monitor miss the zero-crossing event and declare the signal as satisfying (see Fig. 9). In this sense properties are not *robust* as small variations in the signal may lead to large variations in its property satisfaction. Let us mention some interesting ideas due to P. Caspi [20] concerning new metrics for bridging the gap between the continuous and the discrete points of view. Such metrics are expressible, by the way, in STL [37].

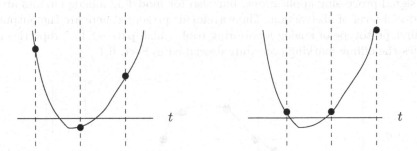

Fig. 9. Shifting the sampling points, zero crossing can be missed

The abovementioned issues can be handled pragmatically in our context, without waiting for a completely-satisfactory theoretical solution to this fundamental problem. The following assumptions facilitate the monitoring of sampled continuous signals against STL properties, passing through the timed abstraction:

1. *Sufficiently-dense sampling*: the simulator detects every change in the truth value of any of the predicates appearing in the formula at a sufficient accuracy. This way the positive intervals of all the Boolean signals that correspond to these predicates are determined. This requirement imposes some level of sophistication on the simulator that has to perform several back-and-forth iterations to locate the time instances where a threshold crossing occurs. Many simulation tools used in industry have already such event-detection features. A survey of the treatment of discontinuous phenomena by numerical simulators can be found in [26].

2. *Bounded variability*: some restrictive assumptions can be made about the values of the signal between two sampling points t_1 and t_2. For example one may assume that ξ is monotone so that if $\xi[t_1] \leq \xi[t_2]$ then $\xi[t_1'] \leq \xi[t_2']$ for every t_1' and t_2' such that $t_1 < t_1' < t_2' < t_2$. An alternative condition could be a condition a-la Lipschitz: $|\xi[t_2] - \xi[t_1]| \leq K|t_2 - t_1|$. Such conditions guarantee that the signal does not get wild between the sampling points, otherwise property checking based on these values is useless.

Under such assumptions every continuous signal which is given by a discrete-time representation, based on sufficiently-dense sampling, induces a well-defined Boolean signal ready for MITL monitoring. Let us add at this point a general remark that the standards of exactness and exhaustiveness as maintained in discrete verification cannot and should not be exported to the continuous domain, and even if we are not guaranteed that all events are detected, we can compensate for that by using safety margins in the predicates and properties.

8 Monitoring STL Properties

In this section we illustrate the monitoring of STL properties against signals produced by the numerical simulator Matlab/Simulink, used mainly for control and signal-processing applications, but also for modeling analog circuits at the functional level of abstraction. The waveforms presented here are the output of our first prototype of analog monitoring tool, which parses STL properties and applies the offline marking procedure described in Sect. 6.1.

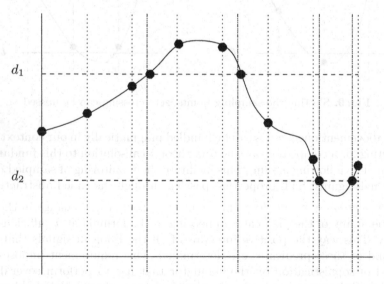

Fig. 10. Sufficiently-dense sampling with respect to the two thresholds d_1 and d_2. The set of sampling points consists of a uniform grid augmented with the threshold-crossing points

$s = x_1 \| x_2$

$p_1 = x_1 > 0.7$

$p_2 = x_2 > 0.7$

$\Diamond_{[3,5]} p_2$

$p_1 \rightarrow \Diamond_{[3,5]} p_2$

$\Box_{[0,300]} (p_1 \rightarrow \Diamond_{[3,5]} p_2)$

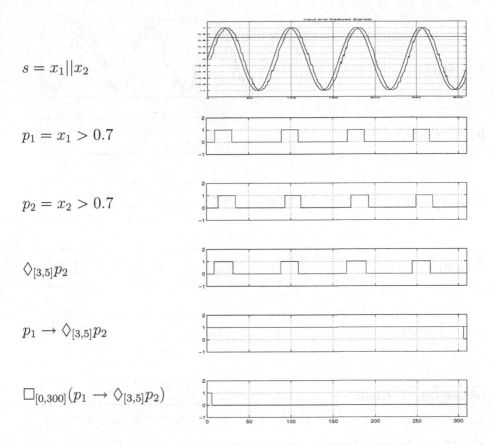

Fig. 11. A 2-dimensional signal satisfying the property $\Box_{[0,300]} ((x_1 > 0.7) \Rightarrow \Diamond_{[3,5]} (x_2 > 0.7))$. Boolean signals correspond to the evolution of the truth values of sub-formulae over time

8.1 Following a Reference Signal

As a first example consider the property

$$\varphi_1 : \quad \Box_{[0,300]} ((x_1 > 0.7) \Rightarrow \Diamond_{[3,5]} (x_2 > 0.7))$$

which requires that whenever x_1 crosses the threshold 0.7, so does x_2 within $t \in [3,5]$ time units. We fix x_1 to be the sinusoid

$$x_1[t] = \sin(\omega t),$$

and let x_2 be a signal generated by

$$x_2[t] = \sin(\omega(t + d)) + \theta$$

where d is a random delay ranging in $[3,5]$ degrees and θ is an additive random noise. The marking procedure is illustrated in Fig. 11. The Boolean signals corresponding to the atomic propositions p_1 and p_2 are derived from the sampled

$s = x_1 \| x_2$

$p_1 = x_1 > 0.7$

$p_2 = x_2 > 0.7$

$\Diamond_{[3,5]} p_2$

$p_1 \to \Diamond_{[3,5]} p_2$

$\Box_{[0,300]}\left(p_1 \to \Diamond_{[3,5]} p_2\right)$

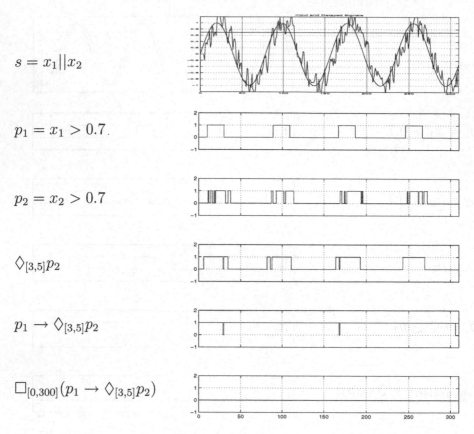

Fig. 12. A 2-dimensional signal violating the property $\Box_{[0,300]}((x_1 > 0.7) \Rightarrow \Diamond_{[3,5]}(x_2 > 0.7))$

analog signal. From there the truth values of the sub-formulae $\Diamond_{[3,5]}(x_2 > 0.7)$, $(x_1 > 0.7) \Rightarrow \Diamond_{[3,5]}(x_2 > 0.7)$ are marked as intermediate steps toward the marking of φ_1 which is satisfied in this example. In Fig. 12 we apply the same procedure to check φ_1 against a signal in which x_2 was generated with a much larger additive noise $\theta \in [-0.5, 0.5]$. The fluctuations in the value of x_2 are reflected in the Boolean abstraction p_2 and lead to a violation of the property at some points where $x_1 > 0.7$ is not followed by $x_2 > 0.7$ within the pre-specified delay.

8.2 Stabilizability

The second example is a very typical stabilizability property used extensively in control and signal processing. The system in question is supposed to maintain a controlled variable y around a fixed level despite disturbances x coming from the outside world. The actual system used to generate this example is a water-level controller for a nuclear plant. The disturbances come from changes in the

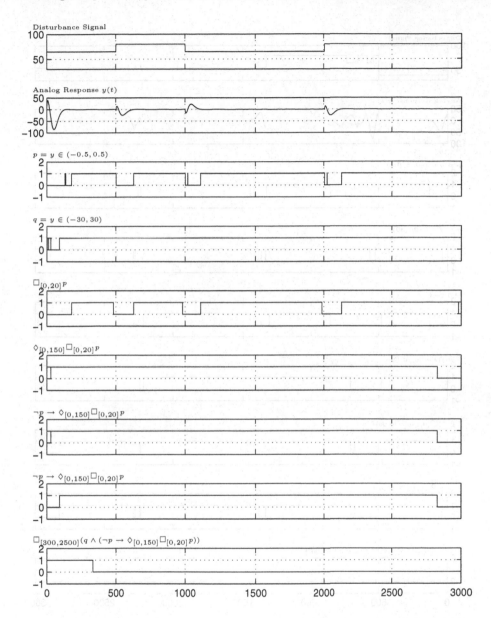

Fig. 13. A disturbance signal and an analog response y satisfying the stabilizability property $\Box_{[300,2500]}((|y| \leq 30) \wedge ((|y| > 0.5) \Rightarrow \Diamond_{[0,150]}\Box_{[0,20]}(|y| \leq 0.5)))$

system load that trigger changes in the operations of the reactor which, in turn, influences the water level, see [9]. Other instances of the same type of problem may occur when the voltage of a circuit has to be kept constant despite variations in the current due to changes in the circuit workload.

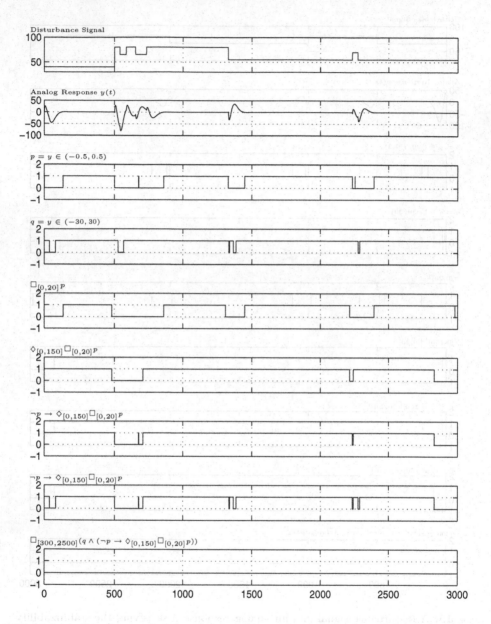

Fig. 14. A disturbance signal and an analog response y violating the stabilizability property $\square_{[300,2500]}((|y| \leq 30) \wedge ((|y| > 0.5) \Rightarrow \Diamond_{[0,150]}\square_{[0,20]}(|y| \leq 0.5)))$

We want y to stay always in the interval $[-30, 30]$ (except, possibly, for an initialization period of duration 300) and if, due to a disturbance, it goes outside the interval $[-0.5, 0.5]$, it should return to it within 150 time units and stay there for at least 20 time units. The whole property is

$$\varphi_2: \quad \square_{[300,2500]}((|y| \leq 30) \wedge ((|y| > 0.5) \Rightarrow \Diamond_{[0,150]}\square_{[0,20]}(|y| \leq 0.5))).$$

The results of applying our offline monitoring procedure to this formula appear in Figures 13 and 14. When the disturbance is well-behaving, the property is verified, while when the disturbance changes too fast, the property is violated both by over-shooting below -30 and by taking more than 150 time units to return to $[-0.5, 0.5]$.

9 Conclusions

Motivated by the exportation of some ingredients of formal verification technology toward analog circuits and continuous systems in general, we embarked on the development of a monitoring procedure for temporal properties of continuous signals. During the process we have gained better understanding of temporal satisfiability in general as well as of the relation between real-time temporal logics and timed automata. The ideas presented in this paper have been implemented into an *analog monitoring tool* AMT [37] that has been applied to real-life case studies.

References

1. Abarbanel, Y., et al.: FoCs: Automatic Generation of Simulation Checkers from Formal Specifications. In: Emerson, E.A., Sistla, A.P. (eds.) CAV 2000. LNCS, vol. 1855, pp. 538–542. Springer, Heidelberg (2000)
2. Alur, R., Dill, D.L.: A Theory of Timed Automata. Theoretical Computer Science 126, 183–235 (1994)
3. Alur, R., Feder, T., Henzinger, T.A.: The Benefits of Relaxing Punctuality. Journal of the ACM 43, 116–146 (1996)
4. Alur, R., Henzinger, T.A.: Logics and Models of Real-Time: A Survey. In: Huizing, C., et al. (eds.) REX 1991. LNCS, vol. 600, pp. 74–106. Springer, Heidelberg (1992)
5. Asarin, E.: Challenges in Timed Languages. Bulletin of EATCS 83 (2004)
6. Asarin, E., Caspi, P., Maler, O.: Timed Regular Expressions. The Journal of the ACM 49, 172–206 (2002)
7. Beer, I., et al.: The Temporal Logic Sugar. In: Berry, G., Comon, H., Finkel, A. (eds.) CAV 2001. LNCS, vol. 2102, Springer, Heidelberg (2001)
8. Bensalem, S., et al.: Testing Conformance of Real-time Applications with Automatic Generation of Observers. In: RV 2004 (2004)
9. Donzé, A.: Etude d'un Modèle de Contrôleur Hybride. Master's thesis, INPG (2003)
10. Eisner, C., Fisman, D.: A Practical Introduction to PSL. Springer, Heidelberg (2006)
11. Eisner, C., et al.: Reasoning with Temporal Logic on Truncated Paths. In: Hunt Jr., W.A., Somenzi, F. (eds.) CAV 2003. LNCS, vol. 2725, pp. 27–39. Springer, Heidelberg (2003)

12. Geilen, M.C.W., Dams, D.R.: An On-the-fly Tableau Construction for a Real-time Temporal Logic. In: Joseph, M. (ed.) FTRTFT 2000. LNCS, vol. 1926, pp. 276–290. Springer, Heidelberg (2000)
13. Geilen, M.C.W.: Formal Techniques for Verification of Complex Real-time Systems, PhD thesis, Eindhoven University of Technology (2002)
14. Geilen, M.C.W.: An Improved On-the-fly Tableau Construction for a Real-time Temporal Logic. In: Hunt Jr., W.A., Somenzi, F. (eds.) CAV 2003. LNCS, vol. 2725, pp. 394–406. Springer, Heidelberg (2003)
15. Havelund, K., Rosu, G. (eds.): Runtime Verification RV 2002. ENTCS 70(4) (2002)
16. Havelund, K., Rosu, G.: Synthesizing Monitors for Safety Properties. In: Katoen, J.-P., Stevens, P. (eds.) ETAPS 2002 and TACAS 2002. LNCS, vol. 2280, pp. 342–356. Springer, Heidelberg (2002)
17. Henzinger, T.A.: It's about Time: Real-time Logics Reviewed. In: Sangiorgi, D., de Simone, R. (eds.) CONCUR 1998. LNCS, vol. 1466, pp. 439–454. Springer, Heidelberg (1998)
18. Hirshfeld, Y., Rabinovich, A.: Logics for Real Time: Decidability and Complexity. Fundamenta Informaticae 62, 1–28 (2004)
19. Kesten, Y., Pnueli, A.: A Compositional Approach to CTL* Verification. Theoretical Computer Science 331, 397–428 (2005)
20. Kossentini C., Caspi, P.: Approximation, Sampling and Voting in Hybrid Computing Systems, HSCC (to appear, 2006)
21. Koymans, R.: Specifying Real-time Properties with with Metric Temporal Logic. Real-time Systems, 255–299 (1990)
22. Kim, M., et al.: Monitoring, Checking, and Steering of Real-time Systems. RV 2002, ENTCS 70(4) (2002)
23. Kristoffersen, K.J., Pedersen, C., Andersen, H.R.: Runtime Verification of Timed LTL using Disjunctive Normalized Equation Systems. RV 2003 ENTCS 89(2) (2003)
24. Krichen, M., Tripakis, S.: Black-box Conformance Testing for Real-time Systems. In: Graf, S., Mounier, L. (eds.) SPIN 2004. LNCS, vol. 2989, pp. 109–126. Springer, Heidelberg (2004)
25. Kupferman, O., Vardi, M.Y.: On Bounded Specifications. In: Nieuwenhuis, R., Voronkov, A. (eds.) LPAR 2001. LNCS (LNAI), vol. 2250, pp. 24–38. Springer, Heidelberg (2001)
26. Mosterman, P.J.: An Overview of Hybrid Simulation Phenomena and their Support by Simulation Packages. In: Vaandrager, F.W., van Schuppen, J.H. (eds.) HSCC 1999. LNCS, vol. 1569, pp. 165–177. Springer, Heidelberg (1999)
27. Lichtenstein, O., Pnueli, A., Zuck, L.D.: The Glory of the Past. In: LCTES 2000. LNCS, pp. 196–218 (1985)
28. Maler, O.: A Unified Approach for Studying Discrete and Continuous Dynamical Systems. In: CDC, pp. 2083–2088. IEEE, Los Alamitos (1998)
29. Maler, O.: Control from Computer Science. Annual Reviews in Control 26, 175–187 (2002)
30. Maler, O.: Analog Circuit Verification: a State of an Art. ENTCS 153, 3–7 (2006)
31. Maler, O., Nickovic, D.: Monitoring Temporal Properties of Continuous Signals. In: Lakhnech, Y., Yovine, S. (eds.) FORMATS 2004 and FTRTFT 2004. LNCS, vol. 3253, pp. 152–166. Springer, Heidelberg (2004)
32. Maler, O., Nickovic, D., Pnueli, A.: Real Time Temporal Logic: Past, Present, Future. In: Pettersson, P., Yi, W. (eds.) FORMATS 2005. LNCS, vol. 3829, pp. 2–16. Springer, Heidelberg (2005)

33. Maler, O., Nickovic, D., Pnueli, A.: From MITL to Timed Automata. In: Asarin, E., Bouyer, P. (eds.) FORMATS 2006. LNCS, vol. 4202, pp. 274–289. Springer, Heidelberg (2006)
34. Maler, O., Nickovic, D., Pnueli, A.: On Synthesizing Controllers from Bounded-Response Properties. In: Damm, W., Hermanns, H. (eds.) CAV 2007. LNCS, vol. 4590, pp. 95–107. Springer, Heidelberg (2007)
35. Manna, Z., Pnueli, A.: The Temporal Logic of Reactive and Concurrent Systems - Specification. Springer, Heidelberg (1992)
36. Manna, Z., Pnueli, A.: Temporal Verification of Reactive Systems: Safety. Springer, Heidelberg (1995)
37. Nickovic, D., Maler, O.: AMT: A Property-Based Monitoring Tool for Analog Systems. In: Raskin, J.-F., Thiagarajan, P.S. (eds.) FORMATS 2007. LNCS, vol. 4763, pp. 304–319. Springer, Heidelberg (2007)
38. Safra, S.: On the Complexity of ω-Automata. In: FOCS 1988, pp. 319–327 (1988)
39. Sokolsky, O., Viswanathan, M.(eds.): Runtime Verification RV 2003. ENTCS 89(2) (2003)
40. Trakhtenbrot, B.A.: Finite Automata and the Logic of One-place Predicates. DAN SSSR 140 (1961)
41. Thati, P., Rosu, G.: Monitoring Algorithms for Metric Temporal Logic Specifications. RV (2004)
42. Tripakis, S.: Fault Diagnosis for Timed Automata. In: Damm, W., Olderog, E.-R. (eds.) FTRTFT 2002. LNCS, vol. 2469, pp. 205–224. Springer, Heidelberg (2002)
43. Vardi, M.Y., Wolper, P.: An Automata-theoretic Approach to Automatic Program Verification. In: LICS 1986, pp. 322–331. IEEE, Los Alamitos (1986)

Token-Free Petri Nets

Antoni Mazurkiewicz

Institute of Computer Science of PAS, Warsaw

This paper is dedicated to Boaz Trakhtenbrot
on the occasion of the 85th anniversary of his birthday,
with a deep gratitude for many inspiring
discussions. Thank you, Boaz, for all of them!

Abstract. In the paper a modification of classical Petri nets is defined.
The fundamental notions of nets, as places, transitions, and firing rules
are retained. In this way the composition properties of classical nets as
well as their full expressive power is preserved. However, while preserving basic notions of Petri nets theory, the presented modification offers
much more freedom in defining data types and the way of place contents
transformations. It turns out that such a modification leads to a simplification of the net description (neither tokens nor arrows are necessary
anymore) and yet strengthening its expressive power.

1 Introduction

For more than 40 years Petri nets [3] serve as an efficient formal model of concurrent behavior of complex discrete systems. There exists a rich bibliography of
books and works devoted to this model and many applications of nets have been
already created. The model is extremely simple: it uses three basic concepts, of
places, of transitions, and of a flow relation. The behavior of a net is represented
by changing distribution of tokens situated in the net places, according to some
simple rules (so-called firing rules). Non-negative integers play essential role in
the description of tokens distribution, indicating the number of tokens assigned
to nets places. Transitions determine the way of changing the distribution, taking off a number of tokens from entry places and putting a number of tokens in
exit places of an active transition. Formally, to any transition some operations
on numbers stored in places are assigned and therefore the behavior of nets is described by means of a simple arithmetic with adding and subtracting operations
on non-negative integers.

Nets became attractive for several reasons, namely because:

- simplicity of description,
- demonstrativeness, mainly due to their graphic representation,
- a deep insight into concurrency phenomena,
- facility of applications to basic concurrent systems.

A. Avron et al. (Eds.): Trakhtenbrot/Festschrift, LNCS 4800, pp. 506–520, 2008.

Among various versions of Petri nets three of them are worth of a special attention, namely condition/event nets, elementary net systems, and place/transition nets. In each of them tokens play slightly different parts. Condition/event nets explain the background of concurrency; tokens mark conditions holding and in this way describe the whole state of the system described. Elementary net systems are tools for modeling concurrent computations and tokens mark points of distributed control, indicating instantaneous stage of computation. Place/transition nets are tools for complex systems description; tokens describe overall configurations of the system. The last type of nets will constitute the basis for designing token-free nets discussed in the paper.

The paper is organized as follows: First, some facts from the formal language theory will be recalled or defined; they will serve for defining the behavior of discussed nets. Next, token-free nets will be defined and some properties of their behavior will be given. Having defined token-free nets, their composition properties are considered; in particular, it is shown that token-free nets can be viewed as the composition of one-place nets (automata). To have a handy tool for proving general properties of token-free nets, their canonical representation is introduced. The last part of the paper concerns adjusting net specifications to restrict their behavior (e.g. to avoid possible deadlocks).

2   Preliminaries

The reader is assumed to be acquainted with basic notion from automata theory, language theory and Petri nets theory. To fix the notation, let us recall here some of them. Let Σ be an alphabet, i.e. a finite set of symbols. Any finite sequence of symbols from Σ is called a string over Σ. The empty string is denoted by ϵ. The set of all strings over Σ is denoted by Σ^*. If u, v are strings, uv denotes their concatenation (composition). Any set $L \subseteq \Sigma^*$ is a language over Σ. Let L', L'' be languages; then $L'L'' = \{w'w'' \mid w' \in L', w'' \in L''\}$. If it causes no ambiguity, we shall not distinguish between a singleton set and its element, one-symbol string and its single symbol, and between a one-string language and its single element. In particular, aL denotes the language $\{aw \mid w \in L\}$, for any symbol a and language L. For any strings u, w we say that u is a prefix of w and write $u \leq w$, if there is v with $w = uv$ (if $v \neq \epsilon$, u is said to be a *proper* prefix of w). Language L is prefix-closed, if any prefix of any string in L is in L as well. Let $A \subseteq \Sigma$, $w \in \Sigma^*$; projection $\pi_A(w)$ of w onto A is defined recursively; for any symbol a $\pi_A(a) = a$, if $a \in A$, or $\pi_A(a) = \epsilon$, otherwise; then $\pi_A(\epsilon) = \epsilon$ and $\pi_A(uv) = \pi_A(u)\pi_A(v)$, for all $u, v \in \Sigma^*$. In other words, $\pi_A(w)$ is the effect of erasing in w all symbols not belonging to A. If L is a language over Σ and $A \subseteq \Sigma$, then $\pi_A(L) = \{\pi_A(w) \mid w \in L\}$ is the projection of L onto A. Let L be a language, w be a string; the language $\{u \mid wu \in L\}$ is called the *left quotient* of L by w (or the *continuation* of w in L) and is denoted by L/w.

3   Token-Free Nets

Let us discuss advantages offered by nets. Without no doubt, they are intuition appealing and enforcing intended specification. By their very nature, they enforce natural numbers as a tool for system state description. Natural numbers form a basic formal concept, easy to understand and to manipulate: operations on numbers are simple and easy. The cardinality of their set is countable, by definition. Operations in nets are limited to (multiple) successor/predecessor operations; they have an obvious meaning and need no additional explanations.

However, some of positive features of nets create also their weakness. Namely, the simplicity of the model makes difficult some complex situations description; the formalism being well suited to describe basic phenomena may turn out to be difficult for some real applications, e.g. relations $\{(n, m)\}$ (replace n with m), $\{(0, 0)\}$ (test for "0"), $\{(n, n + 1) \mid n > m\}$ (increase n by 1 provided $n > m$) are not admissible in classical PT nets formalism. Integers may turn out to be too primitive structures for dealing with more subtle objects; enlarging the net structure to the real word situations could make the description difficult to grasp. Finally, the intuition appealing graphical representation may be dangerous for rigorous treatment and formal reasoning. This is the reason why a constantly growing number of different extensions of original nets to the so-called higher level ones has been invented, as nets with multiple arcs, different capacity of places, self loops, inhibitor arcs, colored nets, etc. All of them are aiming to overcome difficulties mentioned above, but they did not improve the situation radically. In this paper a very modest modification of the original net concept is introduced. Namely, it is proposed to:

- retain notions of places and transitions;
- replace tokens with arbitrary objects (consequently, accepting various types of places);
- retain notions of enabling and firing (suitably modified);
- extend flow relation to an arbitrary relation (consequently, replace in nets arrows by edges).

These modifications, though simple and straightforward, yet cover many features introduced by higher level nets and lead to the token-free nets concept defined below.

Definition 1. Any token-free net $(P, T, G; S, F, \sigma^0)$ is defined by its *structure* (P, T, G) and its *specification* (S, F, σ^0), where

$$
\begin{array}{ll}
P & \text{is a finite non-empty set, (of } \textit{places}) , \\
T & \text{is a finite set(of } \textit{transitions}), P \cap T = \emptyset, \\
G \subseteq P \times T, & (\textit{influence} \text{ relation}).
\end{array}
$$

If $(p, t) \in G$, place p is said to be a *neighbor* of transition t, and transition t to be a *neighbor* of place p. Pair (p, t) is an *edge* of the net. The specification (S, F, σ^0) of a net is the triple:

S is a set (*of (individual) values*),
$F : G \longrightarrow 2^{S \times S}$,   (*transition assignment*),
$\sigma^0 : P \longrightarrow S$,   (*initial valuation*).

Any mapping $\sigma : P \longrightarrow S$ assigning to place $p \in P$ its value $\sigma(p) \in S$ is called a *valuation* of net N. F is a mapping which to each pair $(p, t) \in G$ assigns a binary relation in S, denoted by $F_t(p)$ or $F(p, t)$, and called the *transition relation* for edge (p, t). Relation $F_t(p)$ is intended to define the way of transforming the value assigned to p by execution of t.

Token-free nets can be represented by diagrams similar to those used in the classical net theory, with circles representing places, boxes representing transitions, and edges joining places with transitions that are in the influence relation. An example of a graphical representation of net structure (P, T, G) with $P = \{p, q, r, t\}, T = \{a, b, c, d\}, G = \{(p, a), (q, a), (q, c), (q, d), (r, b), (r, c), (t, c),$ $(t, d)\}$, is given in Fig. 1.

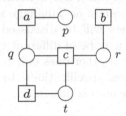

Fig. 1. Token-free net structure

From now on, whenever the context indicates the type of a net, the adjective "token-free" will be omitted; if necessary, original nets will be referred to as "classical nets".

Definition 2. Let $N = (P, T, G; S, F, \sigma^0)$ be a net, $p \in P, t \in T$. Set

$$P_t = \{p \mid (p, t) \in G\}, \qquad T_p = \{t \mid (p, t) \in G\}.$$

Say that p *enables* t (or that t is enabled by p) at valuation σ, if $\sigma(p)$ is in the domain of $F(p, t)$. Let σ', σ'' be two valuations of N, $t \in T$ be a transition of N; say that t transforms in N valuation σ' into σ'' (or that t can *fire* at valuation σ') and write $\sigma' \xrightarrow{t}_N \sigma''$, if the following equivalence holds:

$$\sigma' \xrightarrow{t}_N \sigma'' \iff \begin{vmatrix} (\sigma'(p), \sigma''(p)) \in F_t(p), & \text{if } p \in P_t, \\ \sigma''(p) = \sigma'(p), & \text{if } p \notin P_t. \end{vmatrix}$$

From the above equivalence it follows that $\sigma' \xrightarrow{t}_N \sigma''$ implies that each $p \in P_t$ enables t at σ'. In such a case say that t is *enabled* at σ'. Extend relation $\xrightarrow{t}_N$ to $\xrightarrow{w}_N$ for $w \in T^*$ in the standard way:

$$\sigma' \xrightarrow{w}_N \sigma'' \iff \begin{vmatrix} \sigma' = \sigma'', & \text{if } w = \epsilon, \\ \exists \sigma : \sigma' \xrightarrow{u}_N \sigma \xrightarrow{v}_N \sigma'', & \text{if } w = uv. \end{vmatrix}$$

Then

$$T(N) = \{w \mid \exists \sigma : \sigma^0 \xrightarrow{w}_N \sigma\}, \quad V(N) = \{\sigma \mid \exists w : \sigma^0 \xrightarrow{w}_N \sigma\}$$

are called respectively the (sequential) *behavior* and *reachability set* of net N. Elements of $T(N)$ are called, traditionally, *firing sequences* of N.

Thus, execution of transition t at valuation σ results in replacing value $\sigma'(p)$ of place p with value $\sigma''(p)$ provided $(\sigma'(p), \sigma''(p)) \in F(p, t)$. From the behavior definition it follows that a transition can be executed only if values assigned to all its neighbor places enable it; a place valuation can be changed only if at least one of neighbor transitions is enabled and then precisely one of them is executed. Therefore, transition execution depends solely on values assigned to its neighbor places; the way the value assigned to a place is transformed depends exclusively on the transition being executed and does not depend on values of other neighbors of the transition. It results in a strong locality of net actions.

A non-classical example of token-free nets are 'string nets', with strings over alphabet Σ are individual values, or 'language nets', with languages over Σ as the values. This type of nets will be discussed later on. Observe that values assigned to different places may be of different types, e.g. values of some places can be strings, while those of other ones can be integers. It makes possible to deal with places with a mixed specifications. In Table 1 a comparison between classical nets and token-free ones is presented.

Table 1. Classical and token-free nets comparison

	CLASSICAL NETS	TOKEN-FREE NETS
Places	P	P
Transitions	T	T
Structure relation	$P \times T \cup T \times P$	$P \times T$
Type	Directed	Undirected
Values	Natural numbers	Any domain element
Represented by	Tokens	—
Configurations	Markings	Valuations
Basic operations	Successor/predecessor	Any
Assigned to	Arrows	Edges
Preconditions	Included in the structure	Domains of relations

Classical place/transition nets are particular cases of token-free nets, in which values of all places are non-negative integers, and relations are defined either by $F_t(p) = \{n, n+k) \mid n+k \le c\}$ (then pair (p, t) is represented by an arc leading from t to p, or by $F_t(p) = \{n, n - k) \mid k \leq n\}$ (then the arc leads from p to t). Arc of both kinds are labeled by k, the *multiplicity* of arcs, and with places labeled with c, the capacity of p). As it follows from the classical net theory [4], the behavior of such nets agrees with the definition given above. Behavior of token-free nets will be of the primary interest through the whole paper.

Below are some examples of what we can gain by introducing token-free nets.

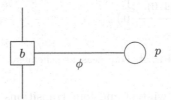

Fig. 2. Waiting for a condition

Example 1. Let $\phi = \{(x, x) \mid \rho(x)\}$ be the transition relation assigned to the edge (p, t) (Fig. 2). Then transition t can be executed only if value of p meets condition ρ.

Example 2. Let numbers 1, 2, 3 ,4 represent traffic lights: 1 – green, 2 – green &yellow, 3 – red, 4 – red; let $\oplus = \{(1, 2), (2, 3), (3, 4), (4, 1)\}$ be the transition relation assigned to edges (p, c), (q, c) and $f = \{(1, 1)\}$ be the transition relation assigned to edges (p, a) and (q, b) (Fig. 3). Places p, q represent traffic lights for directions SN and EW, respectively. Transition c corresponds to the device controlling lights. Transitions a and b indicates changing traffic lights. Then the scheme in Fig. 3 models a traffic light system on crossroad with two directions: NS and EW. The behavior of this scheme is the prefix closure of language $a^*(ccb^*cca^*)^*$; a^* represent stream of vehicles in NS direction, b^* in EW direction. For traffic direction p the sequence of light colors is 1, 2, 3, 4) (i.e. green,green&yellow, red,red), for q it is 3, 4, 1, 2 i.e (red, red, green, green&yellow).

Example 3. In this example two different token-free nets representing well-known "four seasons scheme" are presented. The left-hand scheme is a faithful translation of original classical scheme, with four places corresponding to four seasons

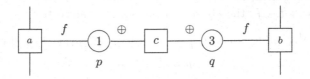

Fig. 3. Traffic lights token-free net

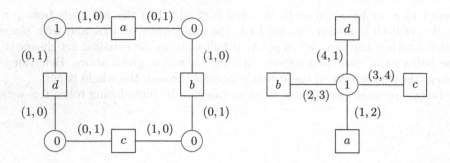

Fig. 4. Four seasons schemes

(spring, summer, fall, and winter) and four transitions terminating one season
and starting the other. In the token-free version edges are labeled with singleton relations (containing one pair of values each; braces around singletons are
omitted to simplify the notation). The right-hand scheme is a token-free net
with only one place which can be valuated with four numbers 1, 2, 3, and 4,
corresponding to four seasons mentioned above. Both schemes have the same
behavior, namely prefix closure of language $(abcd)^*$.

4   Behavior Properties

Let $N = (P, T, G, S, F, \sigma^0)$ be an net fixed for the rest of this section. Recall
that $T_p = \{t \mid (p, t) \in G\}$ and F is a mapping assigning to each pair $(p, t) \in G$ a
transition relation in $S \times S$. Extend F to $F^* : P \times T^* \longrightarrow 2^{S \times S}$ defining F^* for
each $p \in P$, $w \in T^*$ as follows:

$$(s', s'') \in F^*(p, w) \Leftrightarrow \begin{vmatrix} s' = s'', & \text{if } w = \epsilon, \\ (s', s'') \in F(p, t), & \text{if } (p, t) \in G, \\ \exists s : (s', s) \in F^*(p, u') \wedge \\ (s, s'') \in F^*(p, u''), & \text{if } w = u'u''. \end{vmatrix}$$

Lemma 1. *Let* $\pi_p : T^* \longrightarrow T_p^*$ *be the projection,* σ', σ'' *be valuations. Then*

$$\sigma' \xrightarrow{w}_N \sigma'' \Leftrightarrow \forall p \in P : (\sigma'(p), \sigma''(p)) \in F^*(p, \pi_p(w)).$$

Proof. If $w = \epsilon$, equivalence $\sigma' \xrightarrow{\epsilon}_N \sigma'' \Leftrightarrow \forall p \in P : (\sigma'(p), \sigma''(p)) \in F^*(p, \pi_p(\epsilon))$
is obvious. If $w = t \in T$, the above equivalence is reduced to

$$\sigma' \xrightarrow{t}_N \sigma'' \Leftrightarrow \forall p \in P : p \in P_t \wedge (\sigma'(p), \sigma''(p)) \in F(p, t) \vee$$
$$p \notin P_t \wedge \sigma'(p) = \sigma''(p),$$

which follows directly from the definition of transition. If $w = u'u''$, the equiv-
alence $\sigma' \xrightarrow{u'u''}_N \sigma'' \Leftrightarrow \forall p \in P : (\sigma'(p), \sigma''(p)) \in F^*(p, \pi_p(u'u''))$ follows by
induction from the extension definition and from properties of projections. $\square$

Corollary 1. *For any net N with set of places P:*

$$w \in T(N) \iff \forall p \in P : \pi_p(w) \in \pi_p(T(N)).$$

Call set $L_p(N) = \{w \mid (\sigma^0(p), \sigma(p)) \in F^*(p, \pi_p(w))\}$ the *local behavior* of N. From Lemma 1 it follows that $L_p(N) = \pi_p(T(N))$ for all $p \in P$.

Let $N = (P, T, G, S, F, \sigma^0)$ be a net, let $L_p \subseteq T_p^*$ for each $p \in P$. By the *composition* of languages L_p we understand the language

$$\underset{p \in P}{\&} L_p = \{w \in T^* \mid \bigwedge_{p \in P} (\pi_p(w) \in L_p)\}.$$

Proposition 1 follows directly from the above definition.

Proposition 1. *The behavior of any net is the composition of all its local behaviors.*

Definition 3. Let $N = (P, T, G; S, F, \sigma^0)$ be a token-free net, $T_p = \{t \mid (p, t) \in G\}$, $\pi_p : T^* \longrightarrow T_p^*$ be projection. Strings $w', w'' \in T^*$ are said to be *Shields' equivalent*, if $\forall p : \pi_p(w') = \pi_p(w'')$. Shields' equivalence, as depending on net N, is denoted by $\equiv_N$; however, if the net is known, the subscript N will be omitted. Two languages L', L'' are Shields' equivalent, if $\forall \in P : \pi_p(L') = \pi_p(L'')$. The Shields' closure $\langle L \rangle$ of language L is the prefix closure of the set $\{w \mid \exists u \in L : w \equiv u\}$. A language $L \subseteq T^*$ is Shields' closed, if $L = \langle L \rangle$.

Intuitively speaking, Shields' equivalence of strings (or languages) means that they look like identical from each local point of view of the net.

Lemma 2. *Let $(P, T, G; S, F, \sigma^0)$ be a token-free net, $L \subseteq T^*, w \in T^*$. Then $w \in \langle L \rangle \iff \forall p \in P : \pi_p(w) \in \pi_p(L)$.*

Proof. $w \in \langle L \rangle$ implies $\exists u : w \equiv u \in L$ which in turn implies $\pi_p(w) = \pi_p(u) \in \pi_p(L)$ for all $p \in P$. Conversely, $\forall p \in P : \pi_p(w) \in \pi_p(L)$ implies $\exists u \in L : \forall p \in P : \pi_p(w) \in \pi_p(u)$ which in turn implies $w \in \langle L \rangle$. $\square$

Theorem 1. *The behavior of any token-free net is Shields' closed.*

Proof. Let M be the behavior of net N. Prefix closedness of M follows directly from definition. To prove that M is Shields' closed, let $w \in M$ and $u \equiv w$. Since $w \in M$, there is σ such that $\sigma^0 \xrightarrow{w}_N \sigma$. Then, by Lemma 1, $\forall p \in P : (\sigma^0(p), \sigma(p)) \in F^*(p, \pi_p(w))$. Since $w \equiv u$, $\pi_p(w) \iff \pi_p(u)$, we have also $\forall p \in P : (\sigma^0(p), \sigma(p)) \in F^*(p, \pi_p(u))$. It means $\sigma^0 \xrightarrow{u}_N \sigma$, i.e. $u \in M$. $\square$

Language $L \subseteq T^*$ is *linear*, if $\forall u, v \in L \implies v \leq u \lor u \leq v$ (recall that $u \leq v$ means u is a prefix of v). Shields' closure of any linear subset of $T(N)$ is called a (single) *run* of net N.

Proposition 2. *Let $N = (P, T, G; S, F, \sigma^0)$ be a token-free net, $T_p = \{t \mid (p, t) \in G\}$, $\pi_p : T^* \longrightarrow T_p^*$ be projection, R be a run of a net. Then $\pi_p(R)$ is linear for any $p \in P$.*

Proof. It follows directly from the monotonicity of projections: $u \le v \Rightarrow \pi_p(u) \le \pi_p(v)$ for all $u, v \in T^*, p \in P$. □

Proposition 3. *With denotations of the previous proposition, let $R = \boldsymbol{T}(N)$ and let $\pi_p(R)$ be linear for all $p \in P$. Then R is a (single) run of N.*

Proof. To prove the proposition, let L be maximal linear subset of R; then $\forall p \in P : \pi_p(L) = \pi_p(R)$, since otherwise either L would be not maximal, or $\pi_p(R)$ would be not linear for some $p \in P$; hence $R = \langle L \rangle$. □

5 Composition Properties

Token-free nets enjoy a composition property similar to that which holds for classical Petri nets [2].

Definition 4. Let $N_i = (P_i, T_i, G_i; S_i, F_i)$ be token-free nets, for each $i = 1, 2, \ldots, n$ such that $P_i \cap P_j = \emptyset = P_i \cap T_j$ for all $i \ne j$. Then the system

$$(\bigcup_{i=1}^{n} P_i, \bigcup_{i=1}^{n} T_i, \bigcup_{i=1}^{n} G_i; \bigcup_{i=1}^{n} S_i, F, \sigma^0)$$

such that $\sigma^0(p) = \sigma_i^0(p), F(p, t) = F_i(p, t)$ for each $p \in P_i, i = 1, 2, \ldots, n$ and $t \in \bigcup_{i=1}^{n} T_i$, is a token-free net, called the *composition* of nets N_i and denoted by $\bigcup_{i=1}^{n} N_i$.

Observe that due to assumed disjointness of P_i the above definition is correct. Sets T_i need not be disjoint; actually, transitions common to a number of composition components establish the only link binding them. The following theorem expresses the main composition property of nets, similar to that valid for classical ones (see e.g. [1,2]).

Theorem 2. *Let $N_i = (P_i, T_i, G_i; S_i, F_i)$ be token-free nets, $(i = 1, 2, \ldots, n)$, such that $P_i \cap P_j = \emptyset = P_i \cap T_j$ for all $i \ne j$. Then*

$$\boldsymbol{T}(\bigcup_{i=1}^{n} N_i) = \underset{i=1}{\overset{n}{\&}} \boldsymbol{T}(N_i).$$

Proof. Denote $\bigcup_{i=1}^{n} N_i$ by N and similarly $\bigcup_{i=1}^{n} P_i, \bigcup_{i=1}^{n} T_i, \bigcup_{i=1}^{n} G_i$ by P, T, G, respectively. Let $\pi_i : T^* \longrightarrow T_i^*$ be the projection mapping. We have to prove that

$$w \in \boldsymbol{T}(N) \Leftrightarrow \forall i : \pi_i(w) \in \boldsymbol{T}(N_i).$$

By Lemma 1 $\forall i : \pi_i(w) \in \boldsymbol{T}(N_i) \Leftrightarrow \forall i : \forall p \in P_i : \pi_p(\pi_i(w)) \in \pi_p(\boldsymbol{T}(N_i))$. Since P_i are pairwise disjoint and $P = P_1 \cup \cdots \cup P_n$, by definition of N we have

$$\forall i : \forall p \in P_i : \pi_p(\pi_i(w)) \in \pi_p(\boldsymbol{T}(N_i)) \Leftrightarrow \forall p \in P : \pi_p(w) \in \pi_p(\boldsymbol{T}(N))$$

and again by Lemma 1 $\forall p \in P : \pi_p(w) \in \pi_p(\boldsymbol{T}(N)) \Leftrightarrow w \in \boldsymbol{T}(N)$. That is, $\forall i : \pi_i(w) \in \boldsymbol{T}(N_i) \Leftrightarrow w \in \boldsymbol{T}(N)$. It ends the proof. □

It is worthwhile to note that the composition property holds in effect of a strong locality of transitions executions that make possible changing a place value independently of values assigned to other places. Extreme form of nets composition (or decomposition) is the following.

Definition 5. By an *automaton* we understand here any ordered quadruple $A = (S, \Sigma, \delta, s^0)$ such that S is a set (of *states* of A), Σ is a finite set (of *symbols* of A), $\delta : \Sigma \longrightarrow 2^{S \times S}$ is the *transition relation* of A, and $s^0 \in S$ is the *initial state* of A. Automaton A is *finite*, if set S is finite. Let $\delta^* : \Sigma^* \longrightarrow 2^{S \times S}$ be an extension of δ defined recursively for all strings in Σ^*:

$$(s', s'') \in \delta^*(\epsilon) \Leftrightarrow s' = s'',$$
$$(s', s'') \in \delta^*(wt) \Leftrightarrow \exists s : (s', s) \in \delta^*(w) \wedge (s, s'') \in \delta(t).$$

The set $L(A) = \{w \mid \exists s : (s^0, s) \in \delta^*\}$ is the language *accepted* by automaton A and string $w \in \Sigma^*$ is said to be *accepted* by automaton A, if $w \in L(A)$.

Let $N = (P, T, G; S, F, \sigma^0)$ be a token-free net. To each place $p \in P$ assign finite automaton A_p defined by equality $A_p = (S_p, \Sigma_p, \delta_p, s_p^0)$, where $S_p = S, \Sigma_p = T_p, \delta_p(t) = F_p(t)$, and $s_p^0 = \sigma^0(p)$, called the *local automaton* of N. Conversely, given a finite family of local automata $A_i = (S_i, \Sigma_i, F_i, s_i^0)$, $i \in I$, with $\delta_i(t) \subseteq S_i \times S_i$ for each $t \in T_i$ and $s_i^0 \in S_i$, their composition $\&_{i \in I} A_i$ can be defined as net $(I, T, G; S, F, \sigma^0)$, such that $T = \bigcup_{i \in I} \Sigma_i, G = \{(i, t) \mid i \in I, t \in \Sigma_i\}$, $S = \bigcup_{i \in I} S_i$, and $\sigma^0(i) = s_i^0$ for all $i \in I$.

Corollary 2. *Any token-free net is the composition of its local automata.*

Proof. It is a simple consequence of the Theorem 2.   □

The following theorem says that any net can be reduced to a single automaton.

Theorem 3. *Any token-free net is equivalent to a single automaton.*

Proof. Let $N = (P_N, T_N, G_N; S_N, F_N, \sigma_N^0)$ be a token-free net, $A_N = (S_A, \Sigma_A, \delta_A, s_A^0)$ be an automaton defined by the following equalities

$$S_A = \{\sigma \mid \sigma : P_N \longrightarrow S_N\};$$
$$\Sigma_A = T_N;$$
$$\delta_A(t) = \{(\sigma', \sigma'') \mid \sigma' \xrightarrow{t}_N \sigma''\};$$
$$s_A^0 = \sigma_N^0.$$

Then $T(N) = L(A_N)$. Indeed $(\sigma', \sigma'') \in \delta_A(t) \Leftrightarrow \sigma' \xrightarrow{t}_N \sigma_1''$; from here by an easy induction, we get $(\sigma', \sigma'') \in \delta_A(w) \Leftrightarrow \sigma' \xrightarrow{w}_N \sigma_1''$ for all $w \in T_N^*$, and, in consequence, $T(N) = L(A_N)$.   □

6   Canonical Representation

Token-free nets offer much of freedom in choosing valuations of places suitable for a variety of intended models. However, if behavior of nets defined by transition sequences is of primary interest, one can chose a specific valuation making possible to represent behavior of an arbitrary net. In other words, if such a specification exists, it is sufficient to express all nets possibilities. In this section we shall give such a specification. Recall that $L/u = \{w \mid uw \in L\}$ for any string u and language L (the left quotient of L by u).

Definition 6. Token-free net $N = (P, T, G; S, F, \sigma^0)$ is *canonical*, if there exists non-empty language $M^0 \subseteq T^*$ such that for all $p \in P, t \in T$

$$S = 2^{T^*},$$
$$F(p, t) = \{(X, Y) \mid X, Y \subseteq T^*, \emptyset \neq Y = X/t\},$$
$$\sigma^0(p) = \pi_p(M^0).$$

Canonical net $(P, T, G; S, F, \sigma^0)$ is then fully determined by its structure (P, T, F) and language M^0; such a net will be denoted by $(P, T, G; M^0)$ and said to be *initialized* with M^0.

From this definition it follows that in canonical net transition t is enabled at valuation σ if and only if languages assigned (by σ) to all places influenced by t contain some strings beginning with t.

Lemma 3. *For any prefix closed language L and string w, $L/w \neq \emptyset \Leftrightarrow w \in L$.*

Proof. Let L be prefix closed. If $L/w \neq \emptyset$, then there is u s.t. $wu \in L$; since L is prefix closed, $w \in L$. Conversely, if $w \in L$, then $\{\epsilon\} \subseteq L/w =\neq \emptyset$.   □

Theorem 4. *For any canonical net C initialized with M^0: $\boldsymbol{T}(C) = \langle M^0 \rangle$.*

Proof. By the definition of the behavior it follows $w \in \boldsymbol{T}(C) \Leftrightarrow \exists \sigma : \sigma^0 \overset{w}{\rightarrow} \sigma$. In case of canonical net it means $w \in \boldsymbol{T}(C) \Leftrightarrow \exists M \neq \emptyset : M^0 \overset{w}{\rightarrow} M$. By the definition of canonical nets and Lemma 1 $M_0 \overset{w}{\rightarrow} M \Leftrightarrow \forall p \in P : (\pi_p(M^0), \pi_p(M)) \in F^*(p, \pi_p(w))$. From definition of canonical nets the right-hand side of this equivalence means $\pi_p(M) = \pi_p(M^0)/\pi_p(w)$. Since M is non-empty, $\forall p \in P : \pi_p(M^0)/\pi_p(w) \neq \emptyset$. by Lemma 3 the last condition holds if and only if $\forall p \in P : \pi_p(w) \in \pi_p(M^0)$; by Lemma 2 it holds if and only if $w \in \langle M^0 \rangle$. It ends the proof.   □

Example 4. Consider nets represented graphically in Fig. 5, a classical net (a) with the prefixes of string *adcbacd* in its behavior and its canonical version (b) with initial valuation σ^0 shown in the first row of Table 2 such that $\sigma^0(p) = w_p = \pi_p(adcbacd)$) for $p = 1, 2, 3, 4, 5$ (the braces around singletons are omitted).

The effect of execution of sequence *adcbacd* is shown in Table 2 below; columns correspond to places of the net, each row displays valuation of the net after

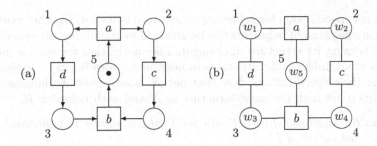

Fig. 5. Classical net (a) and token-free canonical net (b)

Table 2. Behavior of a canonical net

i	1	2	3	4	5
w_i	adad	acac	dbd	cbc	aba
a	dad	cac	dbd	cbc	ba
d	ad	cac	bd	cbc	ba
c	ad	ac	bd	bc	ba
b	ad	ac	d	c	a
a	d	c	d	c	
c	d		d		
d					

execution of transition given in the leftmost column. Observe that the order of transition execution is defined by their enabling conditions that replace classical arrows.

Let $N_i = (P_i, T_i, G_i; S_i, F_i, \sigma_i^0), (i = 1, 2)$ be token-free nets. Say that N_1, N_2 are *equivalent*, if $T(N_1) = T(N_2)$, and *strongly equivalent*, if they are equivalent and, moreover, their structures are identical: $(P_1, T_1, G_1) = (P_2, T_2, G_2)$. The following theorem gives a possibility of proving behavior properties of nets by proving them for canonical nets exclusively.

Theorem 5. *Any token-free net is strongly equivalent to a canonical net.*

Proof. Let $N = (P, T, G; S, F, \sigma^0)$ be a token-free net. Then, by Theorem 4, the behavior $T(C)$ of canonical net $C = (P, T, G; T(N))$ is equal to $\langle T(N) \rangle$. Since by Theorem 1 the behavior $T(N)$ is Shields' closed, $T(C) = T(N)$. Nets N and C have the same structure, hence they are strongly equivalent. □

7   Specification Adjusting

In Sect. 5 operations on token-free nets that preserve their behaviors were considered; in the present section the inverse question is addressed, namely how to modify a net specification for achieving an assumed behavior. In other words, the question is what kind of behaviors can be attained by modifying the specification of a net, keeping its structure unchanged. The motivation for such a question is how to avoid unfavorable situations in nets, as e.g. deadlocks or livelocks, by modifying their specifications. Say that net N is *adjustable* to language R, if there exists a net with the same structure as N and with behavior R.

Theorem 6. *Any token-free net with set T of transitions is adjustable to any Shields' closed subset of T^*.*

Proof. It is an immediate consequence of Theorem 4, initializing with R a canonical net with the same structure as N; since R is Shields' closed, $\langle R \rangle = R$.   □

Corollary 3. *Any token-free net with T as the set of transitions is adjustable to T^*; any net is adjustable to $\{\epsilon\}$.*

In between of these two extremities there is a broad variety of nets with all possible behaviors, up to Shields' equivalence. The question is, and in fact it was a motivation of this work, does there exist a net specification eliminating from its behavior all terminated strings and preserving all extendable ones (a string in a language is extendable, if it is a proper prefix of another string in this language, and is terminated, if otherwise). In general, such a specification does not exist, as it is shown in Example 5.

Example 5. Consider net N given in Fig. 6 specified with classical symbols. Its behavior is the language $\langle(a(bfek \cup dhcg))^*a(bfcg \cup ekdh)\rangle$. Projection of this language on T_p and T_q are respectively $(a(b \cup d))^*$ and $(a(c \cup e))^*$. Thus, Shields' closure of the behavior would contain strings from $\langle a(b \cup d)(c \cup e)\rangle$; if we want to have $\langle a(be \cup dc)\rangle$, we must accept in the behavior string $\langle a(bc \cup de)\rangle$ as well, which leads to a deadlock. Informally, free decisions made in separate places cannot be coordinated. In this case the set of all extendable strings of $\boldsymbol{T}(N)$ is not Shields' closed.

However, one can easily adjust this net to perform alternate executions: $\langle(abefkadhcg)^*\rangle$, enabling execution of each transition. This can be made by

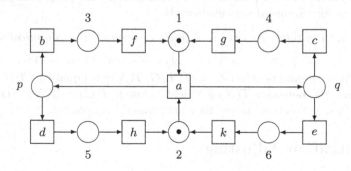

Fig. 6. Net with deadlock

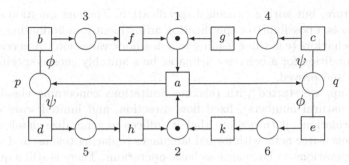

Fig. 7. Adjusted net

replacing classical arrows leading from places p and q with "token-free edges" and assigning to them singleton relations $\phi = (1,2)$ and $\psi = (3,0)$, as shown in Fig. 7.

8   Conclusions

In the paper a generalization of Petri Nets, a broadly known formal tool for concurrent systems description, is proposed. The great impact of the Petri Nets theory on the whole concurrency theory is due to a proper choice of description primitives and their mutual interplay. These primitives are places, transitions, and the way transitions executions transform contents of their neighbor places. Using these notions one can describe basic phenomena of concurrency, as independency, conflict, process, cooperation, and many others. However, in classical net theory there are notions that seem to be of secondary importance, or even debatable, hence stimulating different proposals of net modifications. These are notions of flow direction, tokens, marking, capacities, and others. In Sect. 3, to separate relevant notions from irrelevant ones, and at the same time to simplify net description, the token-free nets have been introduced, with simplified structure and arbitrary, hence flexible, specification. First, it has been shown that this modification does not cause substantial changes in the behavior description (Sect. 4). It turns out that due to a strong locality of transformations (inherited from classical nets), the proposed modification does not destroy the composition property. Moreover, it is shown in Sect. 5 that token-free nets can be viewed as a composition of classical automata. On the other hand, it is also shown that any token-free net is equivalent (from the point of view of behavior) to a single place net, i.e. to an automaton, at the cost of possible lost of independency of transitions. While in the first part of the paper the behavior of different nets with various specifications were discussed, in the last part of it the nets with a fixed structure, but various specifications are considered. First, a canonical specification of token-free nets has been defined. It has been proved that canonical specification enjoys a universal property: any token-free net with an arbitrary specification can be simulated by a canonical net with the same behavior and the same structure, but with a canonical specification. The last question discussed in the paper is a possibility of specification adjustment for eliminating some unfavorable behaviors (e.g. those leading to a deadlock valuation). A necessary and sufficient condition for a behavior definable be a suitably chosen specification is formulated and proved.

To sum up, we started with relaxing limitations concerning classical nets, as tokens (natural numbers), fixed flow direction, and limited ways of marking transformations; it turned out that it suffices to generalize classical nets to canonical token-free nets, with formal languages replacing tokens, and with continuation operations on languages as basic operations. There is still a quest for a satisfactory concurrency notion; the sufficient condition for concurrency of some actions is independency of their resources (as e.g. formulated in [2,5]), but it is not a necessary condition: any net can be replaced by one-place net with the same behavior, hence a net without separate resources.

References

1. Diekert, V., Rozenberg, G.: Book of Traces. World Scientific Pub. Co, Singapore (1995)
2. Mazurkiewicz, A.: Semantics of Concurrent Systems: A Modular Fixed Point Trace Approach, Instituut voor Toegepaste Wiskunde en Informatica, Rijksuniversiteit Leiden, TR-84-19 (1984)
3. Petri, C.A.: Concepts of Net Theory. In: Proc. of MFCS 1973, High Tatras, Math. Institute of Slovak Academy of Sciences, pp. 137–146 (1973)
4. Reisig, W.: Petri Nets. In: EATCS Monographs on Theoretical Computer Science, Springer, Heidelberg (1985)
5. Shields, M.W.: Non-sequential behaviour, part I. Int. Report CSR-120-82, Dept. of Computer Science, University of Edinburgh (1979)

Proof Search Tree and Cut Elimination

Grigori Mints

Stanford University, Stanford CA 94305, USA
mints@csli.stanford.edu

This paper is dedicated to B. Trakhtenbrot whose influence and support I felt
for many years.

Abstract. A new cut elimination method is obtained here by "proof
mining" (unwinding) from the following non-effective proof that begins
with extracting an infinite branch $\mathcal{B}$ when the canonical search tree $\mathcal{T}$
for a given formula E of first order logic is not finite. The branch $\mathcal{B}$
determines a semivaluation so that $\mathcal{B} \models \bar{E}$ and (*) every semivaluation
can be extended to a total valuation. Since for every derivation d of E
and every model $\mathcal{M}$, $\mathcal{M} \models E$, this provides a contradiction showing
that $\mathcal{T}$ is finite, $\exists l(\mathcal{T} < l)$. A primitive recursive function $L(d)$ such that
$\mathcal{T} < L(d)$ is obtained using instead of (*) the statement: For every r, if
the canonical search tree $\mathcal{T}^{r+1}$ with cuts of complexity $r + 1$ is finite,
then $\mathcal{T}^r$ is finite.

In our proof the reduction of $(r+1)$-cuts does not introduce new r-cuts
but preserves only one of the branches.

1 Introduction

Normalization of derivations by eliminating the cut rule

$$\frac{\Gamma \to C \quad C, \Gamma \Rightarrow A}{\Gamma \Rightarrow A} \; cut$$

is one of the main tools of proof theory. Continuing work done in [4] we obtain
a new cut elimination method by "proof mining" (unwinding) from a familiar
non-effective proof consisting of four parts.

1. If the canonical search tree without cut $\mathcal{T}$ for a given formula E of first order
   logic is not finite, then there exists an infinite branch $\mathcal{B}$ of $\mathcal{T}$.
2. The branch $\mathcal{B}$ determines a semivaluation (partial model for subformulas of
   E) so that $\mathcal{B} \models \bar{E}$.
3. Every semivaluation can be extended to a total valuation (model).
4. For every derivation d of E and every model $\mathcal{M}$, $\mathcal{M} \models E$.

This is a contradiction showing that $\mathcal{T}$ is finite, $\exists l(\mathcal{T} < l)$. A primitive recursive
function $L(d)$ such that $\mathcal{T} < L(d)$ is obtained in the present paper after replacing
the statement 3 by

A. Avron et al. (Eds.): Trakhtenbrot/Festschrift, LNCS 4800, pp. 521–536, 2008.

(3') For every r, if the canonical search tree T^{r+1} with cuts of rank $r+1$ is finite, then T^r is finite.

By rank we mean the maximal depth of logic connectives in a formula:

$$\mathrm{rk}(A\&B) = \mathrm{rk}(A \vee B) = \max\{\mathrm{rk}(A), \mathrm{rk}(B)\} + 1; \ \ rk(QxA) = rk(A) + 1$$

$\mathrm{rk}(A) = 0$ for atomic A.

We consider formulas in the negative normal form, that is with negations pushed down to atomic formulas. Our proof of (3') provides a reduction of $(r+1)$-cuts that (unlike familiar Gentzen's reduction) does not introduce new r-cuts but preserves only one of the branches, depending however of the truth-values of subformulas in the "current" node of the tree T^r. This node is changed in the course of reduction.

To save notation we assume that the endformula of an original derivation d is a Σ_1^0 sentence of first order logic,

$$E = \exists \mathbf{x} M(\mathbf{x}), \qquad \mathbf{x} := x_1, \ldots x_p,$$

with a quantifier free M, but given derivation d as well as proof search trees may contain formulas of an arbitrary complexity.

T^r is a canonical proof search tree for E with cuts over formulas of rank $\leq r$ beginning with a quantifier. In particular

$T := T^0$ is a proof search tree with $\exists$ rules instantiated by terms in H (the Herbrand universe for E) and **cuts over all atomic formulas with terms in** H. We wish to prove:

If d is a derivation of E, then $T < l$ for some l.

It is assumed that the eigenvariables for the $\forall$ inference (that can occur for $r > 0$) in a proof search tree is uniquely determined by the formula introduced by the rule. In this way the eigenvariables in T^r for different r but the same principal formula are the same.

The step 3 in the non-effective proof of cut-elimination consists in an *extension of a semivaluation* to all formulas of complexity r and hence to a total valuation. This is proved by induction on the rank r of the semivaluation with a trivial base $r = 0$ and the induction step:

$$\exists f Sem(f, r) \rightarrow \exists f Sem(f, r + 1) \tag{1.1}$$

where $Sem(f, r)$ means that f is a semivaluation defined for all formulas of complexity r.

As noticed in [2], given derivation $d : E$ with cuts of rank R provides a bound l such that $T^R < l$: this l is a level in T^R where all rules present in d had been already applied. In particular for $R = 0$ any derivation with only atomic cuts provides a bound for T.

As G.Kreisel pointed out [1], semivaluations are closely related to infinite branches of the proof search tree. By König's Lemma,

$$\exists f Sem(f, r) \iff T^r \text{ is infinite,}$$

therefore (1.1) can be converted into the implication

$$\exists l(T^{r+1} < l) \to \exists l(T^r < l) \qquad (1.2)$$

of Σ_1^0-formulas. A familiar proof of this implication consists of converting it back to (1.1) and using arithmetical comprehension.

There are (at least) two ways of replacing this with a primitive recursive proof of (1.2). One way is to note that a standard Gentzen-style cut reduction applied to all cuts of complexity $r + 1$ in T^{r+1} leads to a derivation d of complexity r that provides a simple bound for T^r.

This argument led to the algorithm for bounding T discussed in [4] : eliminate all non-atomic cuts, then restrict T as above, completely bypassing (1.1).

Our present method relies on a non-effective proof of (1.2) which is closer to extension of a semivaluation f of complexity r to a semivaluation, say g of complexity $r+1$. If T^{r+1}, or any other derivation $d : E$ of complexity $r+1$ is finite, and g is given, assumption $g(E) = \bot$ leads to a contradiction by computing the truth values of all formulas and sequent S in T^{r+1} and proving $g(S) = \top$. The values of g for the formulas of complexity $\leq r$ with parameters from T^r are the same as the given values of f.

Instead of using comprehension to determine the values under g of remaining formulas (of complexity $r + 1$), consider which of these values are needed at the beginning under the "depth first" strategy: in the uppermost leftmost $r + 1$-cut try to compute the left hand side premise $\exists x A(x)$ first:

$$
\begin{array}{c}
\dfrac{\Delta, A(t_i)}{\Delta, \exists x A(x)} \vdots b \\[2ex]
\vdots \qquad \dfrac{\Gamma, \bar{A}(b)}{\Gamma, \forall x \bar{A}(x)} \forall \\[2ex]
\dfrac{\Gamma, \exists x A(x) \qquad \Gamma, \forall x \bar{A}(x)}{\Gamma, E}
\end{array}
\qquad (1.3)
$$

The values $f(A(t_i))$ are all defined, and it is possible to act as if

$$g(\exists x A(x)) = \max_i f(A(t_i))$$

for all $A(t_i)$ *in a given derivation*. This allows to prune the cut (1.3) retaining only one of the premises depending on this value of $g(\exists x A(x))$. At the end, all $r + 1$-cuts are removed and (1.2) is established.

We present a non-effective proof of (1.2) modified in this way for the case $r = 0$ in Section 3. The same proof works for arbitrary r and by conservativeness of Konig's Lemma over primitive recursive arithmetic PRA its unwinding provides a primitive recursive cut elimination algorithm (as pointed out by U. Kohlenbach). Section 4 presents our unwinding leading to a new cut elimination procedure for first order logic described in Section 5. Exposition in Section 5 is independent from the rest of the paper.

Discussions with G. Kreisel, S. Feferman and especially U. Kohlenbach helped to clarify the goal of this work and the statements of results and proofs.

2   Preliminaries

2.1   Tree Notation

Let's recall some notation concerning finite sequences of natural numbers. We use a, b, c as variables for *binary finite sequences*

a $=< a_0, \ldots, a_n >$ where $a_i \in \{0, 1\}$, $lth(a) := n + 1$, $(a)_i := a_i$.

Concatenation $*$:

$< a_0, \ldots, a_n > * < b_0, \ldots, b_m > := < a_0, \ldots, a_n, b_0, \ldots, b_m >$.

$<>$ is the empty sequence with $lth(<>) = 0$. a $\subseteq$ b : $\Longleftrightarrow$ $\exists cb = a * c$;

a $<$ b iff a *lexicographically strictly precedes* b, that is situated strictly to the left in the tree of all finite sequences:

a $\subset$ b or for some $j < lth(a)$, $(a)_i = (b)_i$ for all $i < j$, and $(a)_j < (b)_j$.

Consider a primitive recursive tree $\mathcal{T}$ of *binary sequences* with the root $<>$:

$$b \in \mathcal{T} \& a \subseteq b \to a \in \mathcal{T}; \qquad a \in \mathcal{T} \to (\forall i < lth(a))(a)_i \leq 1$$

$\mathcal{T}_a$ is the subtree of $\mathcal{T}$ with the root a: $\{b \in \mathcal{T} : a \subseteq b\}$.

In fact we use labeled trees. $\mathcal{T}(a) = 0$ means a $\notin \mathcal{T}$, while $\mathcal{T}(a) \neq 0$ means that a $\in \mathcal{T}$ and contains some additional information. A node a $\in \mathcal{T}$ is a *leaf* if b $\supset$ a implies b $\notin \mathcal{T}$. In this case all branches of $\mathcal{T}$ through a are *closed*.

$$\mathcal{T} < l := (\forall a : lth(a) = l)(a \notin \mathcal{T}); \qquad \mathcal{T} > l := (\exists a : lth(a) = l + 1)(a \in \mathcal{T})$$

and similar bounded formulas with replacement of $<, >$ by $\leq, \geq$.

2.2   Tait Calculus; Canonical Proof Trees

We consider first order formulas in positive normal form (negations only at atomic formulas). Negation $\bar{A}$ of a formula A is defined in a standard way by de-Morgan rules. Derivable objects are *sequents*, that is multisets of formulas.

Axioms: $A, \bar{A}, \Gamma$

Inference Rules:

$$\frac{A, \Gamma \quad B, \Gamma}{A \& B, \Gamma} \ \& \qquad \frac{A, B, \Gamma}{A \vee B, \Gamma} \ \vee \qquad \frac{M(t), \exists x B(x), \Gamma}{\exists x B(x), \Gamma} \ \exists$$

$$\frac{B(a), \Gamma}{\forall x B(x), \Gamma} \ \forall \qquad \frac{C, \Gamma \quad \bar{C}, \Gamma}{\Gamma} \ cut$$

The *eigenvariable* a in $\forall$ inference should be fresh. The term t in the rule $\exists$ is called *the term* of that rule.

Definition 1. *The Herbrand Universe H of a Σ_1^0-formula E consists of all terms generated from constants and free variables occurring in M by function symbols occurring in M. If the initial supply is empty, add a new constant.*

For a given formula $E = \exists x_1 \ldots x_p M$ list all p-tuples of terms of the Herbrand universe H in a sequence

$$\mathbf{t}_1, \ldots, \mathbf{t}_i, \ldots \tag{2.1}$$

We assume also some Godel enumeration $Gn(A)$ of all terms and formulas A. The *canonical proof search tree* $\mathcal{T}$ for a sentence $E = \exists \mathbf{x} M(\mathbf{x})$ is constructed by bottom-up application of the rules &, $\vee$ (first) and $\exists$ and atomic cut when &, $\vee$ are not applicable.

$\mathcal{T}$ assigns sequents to nodes a of the tree of finite binary sequences. To express a $\notin \mathcal{T}$ (when a is situated over an axiom or the second premise of a one-premise inference rule) we write $\mathcal{T}(a) = 0$.

$\mathcal{T}(<>)$ contains sequent E. If $\mathcal{T}(a)$ is already constructed and is not an axiom (:=*closed* ' node or branch), then it is extended preserving all existing formulas. Principal formulas of the propositional rules are preserved (for bookkeeping). If all branches of $\mathcal{T}$ are closed, then the whole *tree is closed*.

The following *fairness conditions* are assumed. There exists a primitive recursive function L_0 such that for each a $\in \mathcal{T}$ and every non-closed b $\supseteq$ a, b $\in \mathcal{T}$ with $lth(b) \geq lth(a) + L_0(a)$

1. If $C\&D \in \mathcal{T}(a)$ then $C \in \mathcal{T}(b)$ or $D \in \mathcal{T}(b)$,
2. If $C \vee D \in \mathcal{T}(a)$ then $C \in \mathcal{T}(b)$ and $D \in \mathcal{T}(b)$,
3. $B(\mathbf{t}_i) \in \mathcal{T}(b)$ for every $i \leq lth(a)$,
4. For every atomic formula A over H with $Gn(A) < lth(a)$ either $A \in \mathcal{T}(b)$ or $\bar{A} \in \mathcal{T}(b)$.

For $r > 0$ let H^r be the Herbrand universe for $\mathcal{T}^r$ generated by the functions (including constants) in E from the eigenvariables (chosen in a standard way as above) for all $\forall$-formulas of rank $\leq r$.

We say that a formula A *agrees* with a node a $\in \mathcal{T}^r$ if A contains free only eigenvariables of the $\forall$ rules situated under a.

The canonical proof search tree $\mathcal{T}^r$ of complexity $r > 0$ is defined similarly to $\mathcal{T}$, but now cuts are applied to atomic formulas and formulas of complexity $\leq r$ beginning with quantifier that agree with a given node.

The following *fairness conditions* are assumed. There exists a primitive recursive function L_r such that for each a $\in \mathcal{T}^r$ and every non-closed b $\supseteq$ a, b $\in \mathcal{T}^r$ with $lth(b) \geq lth(a) + L_r(a)$

1. If $C\&D \in \mathcal{T}(a)$ then $C \in \mathcal{T}(b)$ or $D \in \mathcal{T}(b)$,
2. If $C \vee D \in \mathcal{T}(a)$ then $C \in \mathcal{T}(b)$ and $D \in \mathcal{T}(b)$,
3. If $\exists y C(y) \in \mathcal{T}(a)$, then $C(t) \in \mathcal{T}(b)$ for every term t with $Gn(t) \leq lth(a)$,
4. If $\forall y C(y) \in \mathcal{T}(a)$, then $C(t) \in \mathcal{T}(b)$ for some term t,
5. For every every formula A over H^r with $Gn(A) < lth(a)$ that agrees with a and is atomic or begins with a quantifier, either $A \in \mathcal{T}(b)$ or $\bar{A} \in \mathcal{T}(b)$.

3 Cuts of Rank 1: Modified Non-effective Proof

Let us modify the proof of (1.2) for $r = 0$ restated as follows:

Lemma 1. $\mathcal{T}$ *is infinite* $\Rightarrow \mathcal{T}^1$ *is infinite.*

Proof. Suppose $\mathcal{T}$ is infinite. Then every non-closed branch $\mathcal{B}$ of $\mathcal{T}$ (existing by Koenig's Lemma) is a *countermodel* for E: $\mathcal{B} \models \bar{E}$.

Write $\mathcal{B} \models^+ A$ (A is decided by $\mathcal{B}$) to express that A is propositionally implied by the values of formulas present in the branch $\mathcal{B}$: there are $A_1, \ldots A_k \in \mathcal{T}(\mathrm{b})$ for some $\mathrm{b} \in \mathcal{B}$ such that $\bar{A}_1 \& \ldots \& \bar{A}_k \to A$ is a tautology. Note that for every quantifier free formula A with parameters in H (the Herbrand universe of E).

$$\mathcal{B} \models^+ A \text{ or } \qquad \mathcal{B} \models^+ \bar{A},$$

since all atomic formulas in A are decided by cuts in every branch of $\mathcal{T}$.

Consider a new inference rule:

$$\frac{\mathcal{B} \models^+ A \quad \bar{A}, \Gamma}{\Gamma} \; \mathcal{B}\text{-cut}$$

for a quantifier free A with all terms in the Herbrand universe H.

Define *extended derivations* as ones using $\mathcal{B}$-cut and cuts of rank 1 in addition to ordinary cut free rules.

Assume there exists an extended derivation d of E and prove $\mathcal{B} \models E$ using induction on the number of cuts in d. This implies $\mathcal{T}$ is finite by contradiction with $\mathcal{B} \not\models E$.

Induction base. No cuts. Induction on d using the fact that $\mathcal{B}$ is defined for all needed subformulas.

Induction step.

1. For some $\exists$ inference in (1.3) one has $\mathcal{B} \models^+ A(t_i)$. Replace (1.3) by the rule

$$\frac{\mathcal{B} \models^+ A(t_i) \quad \overset{\vdots t_i}{\bar{A}(t_i), \Delta}}{\Delta} \; \mathcal{B}\text{-cut}$$

   where the right hand side premise is obtained by substitution of t_i for b. Now apply IH.

2. For some particular cut in (1.3) and for all $\exists$ inferences as above $\mathcal{B} \models \bar{A}(t_i)$. Replace the cut (1.3) by its left branch, erasing all formulas $\exists x A$ traceable to this cut and replacing corresponding $\exists$ inference by a $\mathcal{B}$-cut with the premise $\mathcal{B} \models \bar{A}(t_i)$. Now IH is applicable.

Let us assume the only free variables of the terms t_i are the eigenvariables of the $\forall$ inferences situated below. Then if the case 1 does not obtain, the situation in the case 2 always occurs for one of the cuts (1.3), namely for the uppermost cut in the leftmost (with respect to $r + 1$-cuts) branch. Indeed, $r + 1$-eigenvariables do not occur in the conclusions of $\exists$ rules in that branch. This concludes the proof. □

4   Reduction of Rank

We present a combinatorial proof of (1.2) obtained by "unwinding" the proof for complexity 1 in the previous subsection.

We use finite objects similar to $\mathcal{B}$-derivations. For a node $a \in \mathcal{T}^r$ consider a rule

$$\frac{A \in \mathcal{T}^r(a) \quad A, \Gamma}{\Gamma} \text{ a-cut}$$

for a formula A agreeing with a and use a-*derivations* using this rule. For comparison with $\mathcal{B}$-cut recall that $\mathcal{B} \models \bar{A}$ for $A \in \mathcal{T}(a), a \in \mathcal{B}$.

Assume also that every free variable of the term t in $\exists$ rule either occurs free in the conclusion or is an eigenvariable of a rule occurring below. This can be achieved by replacing redundant free variables by a constant 0. (Recall however remarks by G. Kreisel on essential use of such "redundant" variables in unwinding of mathematical proofs).

4.1   A Bound for $\mathcal{T}^r$

Eliminating $r + 1$-Cuts Till the end of the subsection 4.1

$$d : \Gamma, E$$

denotes an a-derivation of complexity $r + 1$, with Γ agreeing with a, complexity of $\Gamma \leq r$, $a \in \mathcal{T}^r$ with eigenvariables for $\forall$-formulas of the complexity $\leq r$ chosen in a standard way.

At the beginning $d = \mathcal{T}^r$, at the end all $r + 1$-cuts are replaced by a-rules for a suitable a that changes in the process.

Definition 2. *A formula A is decided by* a *if $A \in \mathcal{T}(a)$ (i.e. A is explicitly false in* a*) or $\bar{A} \in \mathcal{T}(a)$ (i.e. A is explicitly true in* a*).*

Consider an $r + 1$-cut in d:

$$\frac{\Delta, \exists x A(x) \quad \Delta, \forall x \bar{A}(x)}{\Delta, E} \text{ cut} \qquad (4.1)$$

Such a cut is *leftmost*, if there are no $\forall$-premises of $r + 1$-cuts below it, that is it is in the left branch of every $r + 1$-cut situated below. Choose a leftmost cut (4.1) such that there are no $r + 1$-cuts above it. List all side formulas of $\exists$ rules traceable to the formula $\exists x A(x)$.

$$A(t_1), \ldots, A(t_m) \qquad (4.2)$$

Note that the terms t_i do not contain eigenvariables of the $\forall$ rules of complexity $r + 1$, so $t_i \in H^r$.

Lemma 2. *Let $a \in \mathcal{T}^r$, $d : \Gamma, E$ be a-derivation, and all formulas $A(t_i)$ in (4.2) be decided by* a*.*

Then the cut (4.1) can be replaced by a-cuts, leading to a new a-derivation of Γ, E.

Proof. Consider possible cases.

1. At least one of the side formulas $A(t_i)$ is "true" in a: $\bar{A}(t_i) \in T(a)$. Then we can replace (4.1) by a-cuts. Delete the left branch, replace all formulas traceable to $\forall x \bar{A}(x)$ by $\bar{A}(t_i)$ and substitute the eigenvariable of rules introducing $\forall x \bar{A}(x)$ by t_i. Both (4.1) and all such $\forall$ rules become a-cuts:

$$\frac{\bar{A}(t_i) \in T^r(a) \quad \bar{A}(t_i), \Sigma}{\Sigma} \tag{4.3}$$

2. All side formulas $A(t_i)$ are "false" in a: $A(t_i) \in T^r(a)$. Then the formula $\exists x A$ is redundant. Delete the right branch of the cut (4.1). From the left branch delete all formulas traceable to $\exists x A(x)$. Then $\exists$ rules introducing this formula become a-cuts:

$$\frac{A(t_i) \in T^r(a) \quad A(t_i), \Sigma}{\Sigma} \tag{4.4}$$

$\square$

We say that the transformation in Lemma 2 *reduces d* to a new derivation. We show that d can be (primitive recursively in all parameters) reduced to derivations without $(r+1)$-cuts by climbing up T^r.

Lemma 3. *Let* $a \in T^r$, $d : \Gamma, E$ *be* a-*derivation. Then there exists a level* $l \geq lth(a)$ *such that every node* $b \supset a$, $b \in T^r$ *decides all formulas in (4.2).*

Proof. Formulas $A(t_i)$ contain only eigenvariables of the rules of complexity $\leq r$, hence they contain only terms in H^r. Using the fairness function L_r find a level $l \geq lth(a)$ such that all these formulas appear (possibly negated) by the level l.

$\square$

For $a \in T^r$ let $L^{prop}(a)$ be the first level $\geq lth(a)$ of T^r saturated with respect to propositional rules appled to all formulas in $T^r(a)$.

Lemma 4. *Let* $a \in T^r$, $d : \Gamma$ *be an* a-*derivation,* $\Gamma \supseteq T^r(a)$ *and let d consist of* a-*cuts and propositional rules. Then*

$$T_a^r < L^{prop}(a)$$

that is the restriction of T_a^r *to this level is a derivation.*

Proof. Let $A_1, \ldots, A_n$ be all a-cut formulas in d. Then

$$T^r(a) \supseteq \Gamma, A_1, \ldots A_n \tag{4.5}$$

by the proviso in the a-cut. Now use induction on d. Induction base is obvious, the case of a-cut in the induction step follows from (4.5). Consider a propositional rule, say

$$\frac{A, \Gamma \quad B, \Gamma}{A \& B, \Gamma}$$

There is a level $l \leq L^{prop}(a)$ such that every non-closed node $b \supseteq a$, $lth(b) = l$ contains A or B and $L^{prop}(b) \leq L^{prop}(a)$. Now apply IH.

$\square$

For an arbitrary finite sequence $A_1, \ldots, A_m$ of formulas of complexity $\leq r$ with parameters in H^r that agree with a let $L^E(\mathrm{a}, A_1, \ldots, A_m, d)$ be the first level $l \geq lth(\mathrm{a})$ containing $A_1, \ldots, A_m$ up to negation and saturated with respect to all rules appled to all formulas in $\mathcal{T}^r(\mathrm{a})$, $A_1, \ldots A_m$ in d. The latter condition means in particular that if $A_i = \exists x B(x)$ is instantiated by a term t in d and $A_i \in \mathcal{T}^r(\mathrm{b})$ with $\mathrm{b} \supseteq \mathrm{a}$, $lth(\mathrm{b}) \geq l$, then $B(t) \in \mathcal{T}^r(\mathrm{b})$.

Lemma 5. *Let* $\mathrm{a} \in \mathcal{T}^r$, $d : \Gamma$ *be an* a-*derivation of complexity* r, $\Gamma \supseteq \mathcal{T}^r(\mathrm{a})$. *Then*

$$\mathcal{T}_\mathrm{a}^r < L^E(\mathrm{a}, A_1, \ldots, A_m, d)$$

where $A_1, \ldots, A_m$ *is the complete list of the principal and side formulas of quantifier rules in* d.

Proof. Like in the previous Lemma, with all quantifier rules treated using the new bound. □

Lemma 6. *There is a primitive recursive function* L_1 *such that* $\mathcal{T}^r < L_1(d)$ *for every* $d : E$ *of complexity* $r + 1$ *with the standard choice of eigenvariables.*

Proof. Combine the previous Lemmata. □

The original derivation $d : E$ may fail to satisfy the standardness condition for eigenvariables. This condition can be enforced by renaming eigenvariables and deleting redundant formulas and branches. We use fairness properties of $\mathcal{T}^r$ instead.

Theorem 1. *There is a primitive recursive function* L_* *such that* $\mathcal{T}^r < L_*(d)$ *for every* $d : E$ *of complexity* r.

Proof. Rename eigenvariables in d into standard eigenvariables. This invalidates some of the $\forall$ rules in d by violating the proviso for eigenvariables. However Lemma 5 is still applicable to the resulting figure. □

Theorem 2. *There is a primitive recursive function* L *such that* $\mathcal{T} < L(d)$ *for every* $d : E$.

Proof. Iterate the previous theorem. □

Let's see what happens if $d : E$ is not a derivation but an infinite figure constructed by inference rules of predicate logic. In this case the proofs we gave in Lemma 3 and Lemma 2 do not go through for several reasons. First, it is possible that the given branch $\mathcal{B}$ or the leftmost branch (with respect to $r + 1$-cuts) contains infinitely many $r + 1$-cuts so that every $\exists$-formula is instantiated by a term containing $r + 1$-eigenvariable. Second, even if there is only finite number of $r + 1$-cuts, the search through all instance $A(t_i)$ of $\exists x A(x)$ can be infinite. This agrees with the fact that finding the branch in $\mathcal{T}^{r+1}$ is not in general recursive (in a given branch of $\mathcal{T}^r$).

4.2 Operation $\Gamma \cdot d$

Assume that weakening is among our rules:

$$\frac{\Gamma}{\Gamma, A} \; W$$

Definition 3. *The operation $\Gamma \cdot d$ adds the finite set Γ of formulas to all sequents in a derivation $d : \Delta$ and prunes all rules that become redundant.*

We give a detailed definition by induction on d. Unless stated otherwise, the rules are simply augmented by Γ:

$$\frac{d' \quad d''}{\frac{\Delta' \quad \Delta''}{d : \; \Delta}} \quad \text{is defined to be} \quad \frac{\Gamma \cdot d' \quad \Gamma \cdot d''}{\frac{\Gamma \cdot \Delta' \quad \Gamma \cdot \Delta''}{d : \; \Gamma \cdot \Delta}}$$

and similarly for the one-premise rules. Let's list all exceptional cases.

1. *If Γ, Δ is an axiom (that is contains $A, \bar{A}$ for some A) then $\Gamma \cdot d$ is an axiom.*
2. *If $A \in \Gamma$ that is $\Gamma = \Gamma, A$ (since sequents are finite sets), and d ends in a cut*

$$\frac{d' \qquad d''}{\Delta, A \quad \Delta, \bar{A}}{d : \; \Delta}$$

   then

$$\Gamma \cdot d := \frac{\Gamma \cdot d'}{\Gamma, \Delta, A}$$

   We abbreviate this definition to "d is replaced by $\Gamma \cdot d'$". If $\bar{A} \in \Gamma$ the cut is replaced by $\Gamma \cdot d''$.
3. *Let $A \in \Gamma$. If d ends in the &-rule introducing $A \& B$*

$$\frac{d'}{\Delta, A \quad \Delta, B}{d : \; \Delta, A \& B}$$

   then d is replaced by $\Gamma \cdot d'$ and a weakening:

$$\frac{\Gamma \cdot d' : \; \Gamma, \Delta, A}{\Gamma \cdot d : \; \Gamma, A, \Delta, A \& B} \; W$$

   A cut over $A \& B$ is relaced by $(\Gamma \cdot)$ the left premise. Similarly when $B \in \Gamma$.
4. *If $A, B \in \Gamma$, then $\vee$-rules introducing $A \vee B$ and $A \vee B$-cuts are similarly replaced.*
5. *If $A[t] \in \Gamma$ then an $\exists$-rule with the side formula $A[t]$ is deleted:*

$$\frac{d' : \; \Delta, A[t], \exists x A[x]}{d : \; \Delta, \exists x A[x]} \qquad \Gamma \cdot d' : \Gamma, \Delta, A[t], \exists x A[x]$$

6. If $A[t] \in \Gamma$ for some t, then $\forall x A$-cuts are replaced by the left premise and $\forall$-rule

$$\frac{d'[a]}{\frac{\Delta, A[a]}{\Delta, \forall x A[x]}} \quad \text{is replaced by} \quad \frac{\Gamma \cdot d'[t]}{\frac{\Gamma, \Delta, A[t]}{\Gamma, A[t], \Delta, \forall x A[x]}} W$$

7. If $\bar{A}, \bar{B} \in \Gamma$, then $A \& B$-cuts are replaced by axiom-like derivations:

$$\frac{axiom \quad axiom}{\frac{\Gamma, \bar{A}, \bar{B}, A \quad \Gamma, \bar{A}, \bar{B}, B}{\Gamma, \bar{A}, \bar{B}, A \& B}}$$

8. If $\bar{A}, \in \Gamma$ or $\bar{B}, \in \Gamma$, then $A \vee B$-cuts are similarly replaced by axiom-like derivations.

9. For a cut

$$d'$$
$$\vdots$$
$$\frac{\Sigma_n, \exists x A[x], A[t_n]}{\Sigma_n, \exists x A[x]}$$
$$\vdots$$
$$\frac{\Sigma_1, \exists x A[x], A[t_1]}{\Sigma_1, \exists x A[x]}$$
$$\vdots$$
$$\frac{d_0 : \Delta, \exists x A[x] \quad d_1 : \Delta, \forall x \bar{A}[x]}{\Delta} \tag{4.6}$$

the reduction is similar. If $\bar{A}[t] \in \Gamma$ for some t, then the cut is replaced (as above in $\forall x A$-case) by a weakening and $\Gamma \cdot d_1$.

If $A[t_1], \ldots, A[t_n] \in \Gamma$ for all side formulas of $\exists$-rules traceable to the cut formula $\exists x A[x]$ then the right-hand side of the cut is pruned, that is d is replaced by a weakening and $\Gamma \cdot d_0$.

So in the latter case the result is

$$\Gamma d'$$
$$\vdots$$
$$\Gamma, \Sigma_n, \exists x A[x]$$
$$\vdots$$
$$\Gamma, \Sigma_1, \exists x A[x]$$
$$\vdots$$
$$\Gamma, \Delta$$

Definition 4. *A finite set of formulas Γ agrees with a derivation d if for every $\forall$-inference in d*

$$\frac{\Delta, A[b]}{\Delta, \forall x A[x],}$$

if the eigenvariable b occurs in Γ, then $A[b]$ or $\exists x \bar{A}[x]$ is a member of Γ.

Lemma 7. *If Γ agrees with d then $\Gamma \cdot d$ is a derivation.*

Proof. Check all rules. Two things can go wrong: the proviso for eigenvariables in $\forall$ rules and pruning of cuts when suitable subformulas of the cut formula are present in Γ (up to negation). Since Γ agrees with d, each $\forall$ rule stays correct or is pruned.

A cut over a formula $C = A \odot B$, $\odot \in \{\&, \vee\}$ is pruned if Γ contains $\Gamma' \subseteq \{\pm A, \pm B\}$ such that the $\neg\Gamma'$ decides C, that is tautologically implies $\bar{C}$ or C. In the first case all rules traceable to C in the derivation of the l.h.s. premise of the cut are pruned after adding Γ, hence C becomes redundant in this premise and can be introduced by a weakening. If $\neg\Gamma' \models C$, then Γ', C has an axiom-like derivation. $\qquad\square$

Lemma 8. *Let $d : E$ and let $a \in T^r$ be a non-closed node such that*

$$T^r(a) \cdot d \text{ contains cuts of rank } r + 1 \ .$$

Then there is an $L > lev(a)$ such that for every non-closed $b \in T^r$, $lev(b) = L$

$$T^r(b) \cdot (T^r(a) \cdot d) \text{ contains fewer cuts of rank } r + 1 \text{ than } T^r(a) \cdot d$$

The function $L = L(a, d)$ is primitive recursive.

Proof. Take the leftmost uppermost cut of rank $r + 1$ in $T^r(a) \cdot d$:

$$\frac{\Sigma, C \quad \Sigma, \bar{C}}{\Sigma}$$

List all $\forall$-inferences of rank $\leq r$ in $T^r(a) \cdot d$. Using fairness function for T^r find level $L_0 \geq lev(a)$ such that all main formulas $\forall z F[z]$ of such inferences are decided by every non-closed node $b \supseteq a$ of level $\geq L_0$, and in the case $\forall z F[z] \in T^r(b)$ also $F[a] \in T^r(b)$ for the eigenvariable a. This guarantees that $T^r(b)$ agrees with $T^r(a) \cdot d$.

Now increase L to guarantee all immediate subformulas of C are decided in b. This is problematic in the case $C = \exists x A[x]$. In that case we require all side formulas $A[t_1], \ldots, A[t_n]$ of $\exists$-rules traceable to C are decided. If at least one of $\neg A[t_i]$ is present, the C-cut is replaced by the right premise, otherwise by the left premise of the cut. $\qquad\square$

A cut reduction step is applied to an assignment of derivations of rank $\leq r + 1$ to non-overlapping non-empty nodes $a_i \in T^r$:

$$(a_1, d_1; \ldots; a_p, d_p); \qquad lev(a_i) = l; \ rk(d_i) \leq r + 1$$

Define

$$R(0, d) := (\emptyset, d)$$

Assume $R(n, d) = (a_1, d_1; \ldots; a_p, d_p)$.

Assume that for $i < s$ we have $rk(T^r(a_i) \cdot d_i) \leq r$, while for $i \geq s$, $rk(T^r(a_i) \cdot d_i) = r + 1$ (up to the order of a_i). For every $i \geq s$ write down all nodes

$$b_{i,1}, \ldots, b_{i,q_i}$$

of the level $L(a_i, d_i)$ with $rk(T^r(b_{i,j}) \cdot d_i) = r + 1$, denote $d_{i,j} := T^r(b_{i,j}) \cdot d_i$ and define

$$R(n+1, d) := (a_1, d_1; \ldots; a_{s-1}, d_{s-1}; b_{s,1}, d_{s,1}; \ldots; b_{s,q_s}, d_{s,q_s}; \ldots; b_{p,q_p}, d_{p,q_p})$$

Lemma 9. $R(n, d)$ *is a derivation.*

Theorem 3. *Let* $rk(d) \leq r + 1$ *and* d *contains* n *cuts of rank* $r + 1$. *Then* $rk(R(n, d)) \leq r$.

Proof. Induction on n. If $n = 0$, then $rk(d) = 0$. If $n > 0$, $\qquad\qquad$ □

4.3   Elimination of Atomic Cuts

On one view (for example for natural deduction) atomic cuts are just an illusion. More seriously, standard Gentzen cut elimination is simply substitution.

Let's give an argument closer to proof-mining.

Since the endformula is assumed to be existential $E = \exists x M(\mathbf{x})$, the derivation $d : E$ with atomic cuts contains only instances from the Herbrand universe of E

$$M(\mathbf{t}_1), \ldots, M(\mathbf{t}_n)$$

Wait till they appear in the search tree without any cuts and are analyzed. Assume that by this moment some node b of the proof search tree is not closed. Then it determines a countermodel for $M(\mathbf{t}_1), \ldots, M(\mathbf{t}_n)$ a contradiction.

More detailed proof: for $\Gamma = T(b)$, the derivation $\Gamma \cdot d$ is empty: all propositional rules are pruned. Hence it is an axiom, that is Γ is an axiom.

5   A Variant of Ordinary Cut Elimination

The complete search in the proof search tree T^r, especially the presence of all cuts of the rank $\leq r$ seemed to be essential for the possibility to prune cuts of rank $r + 1$ from the given derivation d. However in fact we need only rank-r cuts that decide instances $A[t]$ of rank-$r + 1$ cut formulas $\exists x A[x]$. $r + 1$ denotes here the maximal complexity of cuts in given derivation.

In this section a new cut-elimination algorithm is presented. It is obtained from the algorithm described in previous sections by cleaning out and changing some details and is independent of the previous sections.

It turns out that the subformulas needed for cut elimination are almost present in d itself, so that cut elimination algorithm to be described in the present section can work with d only, without need for T^r.

5.1 Propositional Cut

As a warm-up note that propositional cuts are easily eliminated (i. e., replaced by cuts of smaller complexity) by a familiar trick exactly corresponding to computation of the truth-value of $A\&B$ from the given values of A, B. A cut

$$\frac{\dot{\Sigma}, A \quad \dot{\Sigma}, B}{\Sigma, A\&B} \quad \frac{\Pi, \dot{\bar{A}}, \bar{B}}{\Pi, \bar{A} \vee \bar{B}}$$

$$\frac{\overset{\vdots}{\Delta, \dot{A}\&B} \quad \overset{\vdots}{\Delta, \dot{\bar{A}} \vee \bar{B}}}{d : \Delta} \; Cut_{A\&B}$$

is replaced by cuts over A and B. Adding formulas $\pm A, \pm B$ to the derivation d allows to choose one of the branches in the cut $C_{A\&B}$. If both $\bar{A}$ and $\bar{B}$ are added, then the r.h.s. of the derivation d is retained after replacing $\bar{A} \vee \bar{B}$ by $\bar{A}, \bar{B}$.

In all remaining cases one of A, B (say A) without negation is retained. Then the l.h.s. in $Cut_{A\&B}$ is transformed by replacing $A\&B$ by A and choosing the l.h.s. in the corresponding &-rules. The result of eliminating $C_{A\&B}$ looks as follows:

$$\begin{array}{cc} & \overset{\vdots}{\Sigma, \dot{\bar{A}}, B} \quad \overset{\vdots}{\Pi, \dot{\bar{A}}, \bar{B}} \\ \overset{\vdots}{\Sigma, A} & \\ \overset{\vdots}{} & \frac{\overset{\vdots}{\Delta, \bar{A}, B} \quad \overset{\vdots}{\Delta, \bar{A}, \bar{B}}}{\Delta, \bar{A}} \; Cut_B \\ \frac{\overset{\vdots}{\Delta, A}}{\Delta} & Cut_A \end{array}$$

5.2 Quantifier Cut

Consider now the remaining case of the $r + 1$-cut with an existential cut formula. Choose an uppermost leftmost $r + 1$-cut C, that is assume that there are neither $r + 1$-cuts over C nor $r + 1$-cuts having C over their r.h.s. premise. List all $\exists$-inferences traceable to the cut formula $\exists x A(x)$ of the cut C:

$$d'$$
$$\vdots$$
$$\frac{\Sigma_n, \exists x A[x], A[t_n]}{\Sigma_n, \exists x A[x]}$$

$$\vdots$$
$$\frac{\Sigma_1, \exists x A[x], A[t_1]}{\Sigma_1, \exists x A[x]} \qquad d''[x]$$
$$\vdots \qquad\qquad \frac{\Delta, \bar{A}[x]}{}$$
$$\frac{d_0 : \ \Delta, \exists x A[x] \quad \Delta, \forall x \bar{A}[x]}{d : \Delta} \; C \qquad\qquad (5.1)$$

We try to replace this with the figure:

$$\frac{\tilde{d}_0}{\Delta, A[t_1], \ldots, A[t_n] \quad \Delta, \bar{A}[t_1] \quad \ldots \Delta, \bar{A}[t_n]}{\Delta} \tag{5.2}$$

ending in n cuts of complexity r, where $\tilde{d}_0 = (A[t_1], \ldots, A[t_n]) \cdot d_0$ is the result of adding formulas $A[t_1], \ldots, A[t_n]$ to d_0 (Section 4.2). Unfortunately $\tilde{d}$ is not in general a derivation. If some of $A[t_i]$ contains an eigenvariable of a $\forall$-rule in d_0, that rule is invalidated by adding $A[t_i]$. In the cut reduction from the previous section such a rule is pruned from d_0 by formulas added to Δ, since their main and side formulas are decided by cuts in T^r. Now we simulate the same effect by pushing such $\forall$-inferences down from d_0 instead.

List all principal formulas of $\forall$ rules in d_0 such that their eigenvariable occurs in one of the $t_1, \ldots, t_n$, and inductively eliminate them beginning from the lowermost ones by adding r-cuts from below to the derivation d. The inductive parameter (of the cut C) is the number of different $\forall$-formulas to be eliminated from the derivation ending in C. Note that by the choice of C and the assumption that the endformula of the whole derivation is purely existential (Σ_1^0) all $\forall$-formulas have complexity $\leq r$.

Suppose some of these $\forall$-formulas has been eliminated resulting in the derivation e. Take in e one of the remaining formulas with one of the lowermost $\forall$ rules to be eliminated:

$$
\begin{array}{c}
\vdots \\
\Theta, B[b] \\
\hline
\Theta, \forall y B[y] \\
\vdots \\
\Omega \\
\vdots \\
e : \Sigma
\end{array}
$$

Then e is replaced by

$$
\begin{array}{cc}
\begin{array}{c}
\vdots \\
\Theta, B[b] \\
\vdots \\
\Omega, B[b] \\
\vdots \\
\Sigma, B[b] \\
\hline
\Sigma, \forall y B[y]
\end{array}
&
\begin{array}{c}
\text{axiom} \\
\overline{\Theta, \forall y A[y], \exists y \bar{B}[y]} \\
\vdots \\
\Omega, \exists y \bar{B}[y] \\
\vdots \\
\Sigma, \exists y \bar{B}[y]
\end{array}
\end{array}
$$
$$\frac{}{\Sigma} \; Cut_{\forall y B(y)}$$

Here the left hand side is obtained by adding $B[b]$ to all sequents in e between Σ and the premise of the $\forall$-rule introducing $\forall x A[x]$. This rule is deleted.

The right hand side is obtained by adding $\exists x \bar{B}[x]$ to all sequents in e between Σ and the conclusion of the $\forall$-rule introducing $\forall x A[x]$. This rule is again deleted. In this way the new figure is a derivation with fewer main $\forall$-formulas.

The new figure contains an $r + 1$-cut at each side. Since $\forall$-rules introducing $\forall y B(y)$ are eliminated from the derivation d, each of these two $r + 1$-cuts has smaller induction parameter.

After elimination of all such $\forall$-rules the cut reduction in (5.2) can be applied.

References

1. Kreisel, G., Mints, G., Simpson, S.: The Use of Abstract Language in Elementary Metamathematics. Lecture Notes in Mathematics 253, 38–131 (1975)
2. Mints, G.: The Universality of the Canonical Tree. Soviet Math. Dokl. 14, 527–532 (1976)
3. Mints, G.: E-theorems. J. Soviet Math. 8, 323–329 (1977)
4. Mints, G.: Unwinding a Non-effective Cut Elimination Proof. In: Grigoriev, D., Harrison, J., Hirsch, E.A. (eds.) CSR 2006. LNCS, vol. 3967, pp. 259–269. Springer, Heidelberg (2006)

Symbolic Verification Method for Definite Iterations over Tuples of Altered Data Structures and Its Application to Pointer Programs[*]

Valery Nepomniaschy

A.P. Ershov Institute of Informatics Systems, Russian Academy of Sciences,
Siberian Division, 6, Lavrentiev Ave., Novosibirsk 630090, Russia
vnep@iis.nsk.su

This paper is dedicated to Boris Trakhtenbrot,
my teacher, on the occasion of his 85th anniversary.
His lectures on axiomatic semantics of complex program
constructs have introduced me into the program verification field.

Abstract. The symbolic method for verifying definite iterations over hierarchical data structures without loop invariants is extended to allow tuples of altered data structures and the termination statement which contains a condition depending on variables modified by the iteration body. Transformations of these generalized iterations to the standard ones are proposed and justified. A technique for generating verification conditions is described. The generalization of the symbolic verification method allows us to apply it to pointer programs. As a case study, programs over doubly-linked lists are considered. A program that merges in-place ordered doubly-linked lists is verified by the symbolic method without loop invariants.

1 Introduction

The axiomatic approach to program verification is based on the Hoare method and consists of the following stages: annotating a program by pre-, post-conditions and loop invariants; generating verification conditions with the help of proof rules and proving the verification conditions [6]. The loop invariant synthesis is an important problem [8] which is far from being solved [27]. In the functional approach to program verification, loops are annotated by functions expressing the loop effect [10]. However, the synthesis of such functions remains a difficult problem in most cases [14]. Attempts to extract loop invariants from programs by using special tools [3] are found to be successful only for quite simple kinds of invariants.

---

[*] This work was partly supported by Russian Foundation for Basic Research under grant 04-01-00114.

A. Avron et al. (Eds.): Trakhtenbrot/Festschrift, LNCS 4800, pp. 537–554, 2008.

Two ways for overcoming the difficulties of program verification are considered. In papers on code certification [16,29], verification of simple but important program properties is considered in the framework of the axiomatic approach. However, the invariant synthesis problem imposes an essential limitation on the practical use of this method [29]. Another way for overcoming the difficulty is to use loops of a special form which allows us to simplify the verification process. The special form of loops called simple loops is proposed in [1]. The simple loops are similar to for-loops in that they contain the only control variable called a loop parameter and, also, its alteration in a given finite domain does not depend on other variables modified by the loop body. Therefore, the simple loops are definite iterations because of their termination for all values of input variables. Although the reduction of for-loops to $while$-loops is often used for verification, attempts to use the specific character of for-loops in the framework of the axiomatic approach should be noted [4,5,7]. In the framework of the functional approach, a general form of the definite iteration as an iteration over all elements of an arbitrary structure has been proposed in [28], where spreading of such iterations in practical programming has been justified.

We developed a symbolic method integrating axiomatic and functional approaches in [17,18] for verifying for-loops with the statement of assignment to array elements as the loop body . We extended the method in [21,22] to the definite iterations over data structures without restrictions on the iteration bodies. This method is based on using a replacement operation that represents the loop effect in a symbolic functional form and allows us to express the loop invariant. The symbolic method uses a special technique for proving verification conditions containing the replacement operation. Along with data structures defined in [28], the symbolic method allows us to use hierarchical data structures that are constructed from given structures [22]. The symbolic method is more appropriate for so-called flat loops [1] which have no embedded loops. Moreover, the symbolic verification method is applied to a new kind of definite iterations called iterations over tuples of data structures [23]. These iterations allow loops with several input data structures to be represented compactly and naturally, and to be used as flat loops. This method has been successfully applied to verification of programs over arrays and files [22,23]. However, the attempt to apply the method to pointer programs has revealed the following problems. First, the data structures can be modified by the iteration body, and, therefore, the loops are not simple. Second, the iteration body can include the termination statement containing a condition that depends on variables modified by the iteration body. In [24] the symbolic verification method has been extended to definite iterations over altered data structures and applied to verification of pointer programs with one input data structure.

The purpose of this paper is to extend the symbolic method to definite iterations with several input altered data structures and to apply it to pointer program verification. An outline of standard definite iterations over hierarchical data structures is given in Sect. 2. Definite iterations over tuples of unaltered data structures that extend the iterations from [23] are described in Sect. 3, where their reduction to the standard iterations is justified. Similar results for the iterations extended by the termination statement are represented in Sect. 4. In Sect. 5, definite iterations over tuples of altered data structures are introduced and their reduction to the iterations considered in Sects. 3 and 4 is described. A technique of generating verification conditions for standard definite iterations, as well as for pointer programs, is given in Sect. 6. Application of the symbolic method to verification of programs over doubly-linked lists is presented in Sect. 7, where we consider a program that merges two lists. Advantages and prospects of the symbolic verification method are discussed in Sect. 8.

2 Definite Iteration over Hierarchical Data Structures

We introduce the following notation. Let $\{s_1, \ldots, s_n\}$ be a multiset consisting of elements $s_1, \ldots, s_n$, $U_1 - U_2$ be the difference of multisets U_1 and U_2, $U_1 \bigcup U_2$ be the union of these multisets, and $|U|$ be the power of a finite multiset U. Let $[v_1, \ldots, v_m]$ denote a vector consisting of elements v_i $(1 \leq i \leq m)$ and $\emptyset$ denote the empty vector. A concatenation operation $con(V_1, V_2)$ is defined in the usual fashion for vectors V_1 and V_2. For a function $f(x)$ we assume that

$$f^0(x) = x, f^i(x) = f(f^{i-1}(x)) \, (i = 1, 2, \ldots).$$

Let us remind the notion of a data structure [28]. Let $memb(S)$ be a finite multiset of elements of a structure S, $empty(S)$ be a predicate "$memb(S)$ is empty", $choo(S)$ be a function which returns an element of $memb(S)$, $rest(S)$ be a function which returns a structure S' such that $memb(S') = memb(S) - \{choo(S)\}$. The functions $choo(S)$ and $rest(S)$ will be undefined if and only if $empty(S)$. This definition, which abstracts from the way of determination of the functions $choo(S)$ and $rest(S)$, is quite flexible. For example, if a tree is defined as a data structure, a tree traversal method is fixed. So, such different traversal methods result in different data structures.

Let us remind a definition of useful functions related to the structure S [21]. Let $vec(S)$ denote a vector $[s_1, \ldots, s_n]$ such that $s_i = choo(rest^{i-1}(S))$ $(i = 1, \ldots, n)$ in the case of $\neg empty(S)$ and $memb(S) = \{s_1, \ldots, s_n\}$. The vector $vec(S)$ is empty if $empty(S)$. The function $vec(S)$ defines such an unfolding of the structure S that gives uniquely its use. Structures S_1 and S_2 are called equivalent $(S_1 = S_2)$, when $vec(S_1) = vec(S_2)$. The functions $head(S)$ and $last(S)$ will be undefined in the case of $empty(S)$. A function $head(S)$ returns a structure such that $vec(head(S)) = [s_1, \ldots, s_{n-1}]$ if $vec(S) = [s_1, \ldots, s_n]$ and $n \geq 2$. If $n = 1$, then $empty(head(S))$. Let $last(S)$ be a partial function such that $last(S) = s_n$ if $vec(S) = [s_1, \ldots, s_n]$. Let $str(s)$ denote a structure S which contains the only element s. A concatenation operation $con(S_1, S_2)$ is defined in [21] so that $con(vec(S_1), vec(S_2)) = vec(con(S_1, S_2))$. Let $con(s, S) = con(str(s), S)$ and $con(S, s) = con(S, str(s))$. In the case of $\neg empty(S)$, $con(choo(S), rest(S)) = con(head(S), last(S)) = S$. Moreover, the property $head(rest(S)) = rest(head(S))$ provided $\neg empty(rest(S))$ results from [21].

Along with the data structures, we use hierarchical data structures. Let us determine the rules for construction of a hierarchical structure S from given structures $S_1, \ldots, S_m$ [22]. We will use $T(S_1, \ldots, S_m)$ to denote a term constructed from data structures S_i $(i = 1, \ldots, m)$ with the help of the functions $choo, last, rest, head, str, con$. For a term T which represents a data structure, we denote the function $|memb(T)|$ by $lng(T)$. The function can be calculated by the following rules: $lng(S_i) = |memb(S_i)|$, $lng(con(T_1, T_2)) = lng(T_1) + lng(T_2)$, $lng(rest(T)) = lng(head(T)) = lng(T) - 1$, $lng(str(s)) = 1$. Let a hierarchical data structure $S = STR(S_1, \ldots, S_m)$ be defined by the functions $choo(S)$ and $rest(S)$ constructed with the help of conditional $if - then - else$, superposition and Boolean operations from the following components:

— terms not containing $S_1, \ldots, S_m$;
— the predicate $empty(S_i)$ and the functions $choo(S_i), rest(S_i), last(S_i)$, $head(S_i)$ $(i = 1, \ldots, m)$;
— terms of the form $STR(T_1, \ldots, T_m)$ such that $\sum_{i=1}^{m} lng(T_i) < \sum_{i=1}^{m} lng(S_i)$;
— an undefined element ω.

Note that the undefined value ω of the functions $choo(S)$ and $rest(S)$ means $empty(S)$.

Let us suppose that the iteration body consists of a sequence of assignment and conditional statements. The iteration body is represented as the vector assignment statement $v := body(v, x)$, where x is the iteration parameter, v is a vector of other variables, $body(v, x)$ is a vector of conditional expressions constructed with the help of the operation $if - then - else$. Such a representation is formed by a sequence of suitable substitutions which replace both conditional statements by conditional expressions and a sequence of assignment statements by one vector assignment statement.

Let us consider a definite iteration of the form

$$\textbf{for } x \textbf{ in } S \textbf{ do } v := body\,(v, x) \textbf{ end} \tag{1}$$

where S is a data structure that may be hierarchical and the iteration body $v := body(v, x)$ does not change the structure S. The result of this iteration is an initial value v_0 of the vector v if $empty(S)$. Assume that $\neg empty(S)$ and $vec(S) = [s_1, \ldots, s_n]$. Then the iteration body iterates sequentially for x defined as $s_1, \ldots, s_n$.

3   Iterations over Tuples of Unaltered Data Structures

To define a definite iteration over a tuple of data structures $S_1, \ldots, S_m$ (possibly hierarchical), we will use a function $sel(x_1, \ldots, x_m)$ for selection of one structure from the structures $S_1, \ldots, S_m$, where $x_i \in memb(S_i) \bigcup \{\omega\}$ $(i = 1, \ldots, m)$. The function $sel(x_1, \ldots, x_m)$ returns an integer j $(1 \leq j \leq m)$ such that $x_j \neq \omega$, and also $sel(\omega, \ldots, \omega)$ is undefined. In the case of $x_j \neq \omega$ and $x_i = \omega$ for all $i \neq j$ $(i = 1, \ldots, m)$, $sel(x_1, \ldots, x_m) = j$ and the definition of the function sel can be omitted.

Let us consider the iteration over the tuple of structures $S_1, \ldots, S_m$ of the form

$$\textbf{for } x_1 \textbf{ in } S_1, \ldots, x_m \textbf{ in } S_m \textbf{ do } t := sel(x_1, \ldots, x_m); v := body\,(v, x_t, t) \textbf{ end} \tag{2}$$

where v is a data vector ($x_i \notin v$ for all $i = 1, \ldots, m$) and the iteration body does not change the structures $S_1, \ldots, S_m$. If $empty(S_i)$ for each $i = (1, \ldots, m)$, then the iteration result is an initial value v_0 of the vector v. Otherwise, we assume $x_i = choo(S_i)$ for each $i = 1, \ldots, m$ and $t = sel(x_1, \ldots, x_m)$, where $choo(S_i) = \omega$ provided $empty(S_i)$. A new value v_1 of the vector v is defined so that $v_1 = body(v_0, x_t, t)$. The structure S_t is replaced by the structure $rest(S_t)$, and the other structures S_i ($i \neq t$) are not changed. The process is applied to v_1 and the resulted structures until all structures become empty. The resulted value v_d ($d = \sum_{i=1}^{m} |\, memb(S_i)\, |$) of the vector v is assumed to be the result of iteration (2).

It should be noted that the function $body(v, x_t, t)$ does not depend on variables x_i for $i \neq t$ resulting in simplification of the reduction of iteration (2) to iteration (1). Such variables x_i can be doubled by variables from the vector v.

Here the purpose is to reduce iteration (2) to iteration (1) with the help of hierarchical structures. We introduce the following notation in order to define a hierarchical structure $S = STR(S_1, \ldots, S_m)$ from the structures $S_1, \ldots, S_m$ and the function $sel(x_1, \ldots, x_m)$. Let $EMPTY = (empty(S_1) \wedge \ldots \wedge empty(S_m))$, $t_1 = sel(choo(S_1), \ldots, choo(S_m))$, $REST = STR(r(S_1), \ldots, r(S_m))$ provided $\neg EMPTY$, $r(S_{t_1}) = rest(S_{t_1})$ and $r(S_i) = S_i$ for $i \neq t_1$. Then $(choo(S), rest(S)) = \textbf{if } EMPTY \textbf{ then } (\omega, \omega)$ $\textbf{else } ((choo(S_{t_1}), t_1), REST)$. Notice that this definition is consistent with the definition of hierarchical structures from Sect. 2, since the quantifiers bounded by the set $\{1, \ldots, m\}$ can be expressed by applying conjunction or disjunction m times, and $empty(S) \equiv (choo(S) = \omega)$.

Theorem 1. *Iteration (2) is equivalent to the iteration*

$$\textbf{for } (x, \tau) \textbf{ in } S \textbf{ do } v := body\,(v, x, \tau) \textbf{ end} \tag{3}$$

Proof. We will use induction on $d = 0, 1, 2, \ldots$. If $d = 0$, then $empty(S_i)$ for each $i = 1, \ldots, m$, and, therefore, Theorem 1 holds. Let us suppose $d > 0$ and iterations (2) and (3) are equivalent if $\sum_{i=1}^{m} |\, memb(S_i)\, | = d - 1$. It is evident that iteration (2) is equivalent to the program

$$v := body(v, choo(S_{t_1}), t_1);$$
$$\textbf{for } x_1 \textbf{ in } r(S_1), \ldots, x_m \textbf{ in } r(S_m) \tag{4}$$
$$\textbf{do } t := sel(x_1, \ldots, x_m); v := body\,(v, x_t, t) \textbf{ end}.$$

By the inductive hypothesis, program (4) is equivalent to the program

$$v := body(v, choo(S_{t_1}), t_1); \textbf{ for } (x, \tau) \textbf{ in } REST \textbf{ do } v := body\,(v, x, \tau) \textbf{ end} \tag{5}$$

It remains to notice that program (5) is equivalent to iteration (3) according to $rest(S) = REST, choo(S) = (choo(S_{t_1}), t_1)$ and $S = con(choo(S), rest(S))$ provided $d > 0$. □

We introduce the following notation in order to formulate useful properties of the hierarchical structure $S = STR(S_1, \ldots, S_m)$. We will use $memb_1(S)$ to denote the multiset of the first components of elements of $memb(S)$. Let $vec_i(S)$ be the subsequence of $vec(S)$ which consists of all elements (r, i) for a suitable r, and $vec_{i1}(S)$ be the sequence of the first components of elements of $vec_i(S)(i = 1, \ldots, m)$.

Theorem 2

2.1. $memb_1(S) = \bigcup_{i=1}^m memb(S_i)$.

2.2. $vec_{i1}(S) = vec(S_i)$ for each $i = 1, \ldots, m$.

Proof. We will use induction on $d = 0, 1, 2, \ldots$. If $d = 0$, then $empty(S)$, $empty(S_i)$ for each $i = 1, \ldots, m$ and Theorem 2 holds. Let us suppose $d > 0$ and Theorem 2 holds for $\sum_{i=1}^m \mid memb(S_i) \mid = d - 1$.

2.1. By the induction hypothesis, $memb_1(S) = memb_1(con(choo(S), rest(S))) = \{choo(S_{t_1})\} \bigcup memb(rest(S_{t_1})) \bigcup \bigcup_{i \neq t_1} memb(S_i) = \bigcup_{i=1}^m memb(S_i)$ according to $memb(S_{t_1}) = \{choo(S_{t_1})\} \bigcup memb(rest(S_{t_1}))$.

2.2. By the induction hypothesis, $vec_{i1}(S) = vec_{i1}(con(choo(S), rest(S))) = $ **if** $i = t_1$ **then** $con(choo(S_i), vec(rest(S_i)))$ **else** $vec(S_i) = vec(S_i)$ according to $vec_{i1}(str(choo(S)))$ is the empty sequence and $vec_{i1}(rest(S)) = vec(S_i)$, when $i \neq t_1$. $\qquad\qquad\square$

In the case of $S = STR(S_1, S_2)$, $head(S)$ and $last(S)$ can be expressed in terms of $head(S_i)$ and $last(S_i)$ $(i = 1, 2)$ as follows.

Theorem 3. *If* $S = STR(S_1, S_2)$ *and* $\neg empty(S)$, *then* $\neg empty(S_1)$, $head(S) = STR(head(S_1), S_2)$ *and* $last(S) = (last(S_1), 1)$ *or* $\neg empty(S_2)$, $head(S) = STR(S_1, head(S_2))$ *and* $last(S) = (last(S_2), 2)$.

Proof. For definiteness, we suppose $t_1 = 1$. We will use induction on $d = 1, 2, \ldots$. If $d = 1$, then $\neg empty(S_1)$, $empty(head(S_1))$, $empty(S_2)$, $empty(head(S))$, $last(S) = (last(S_1), 1)$ and therefore Theorem 3 holds. In the case of $d > 1$, $head(S) = head(con(choo(S), rest(S))) = con(choo(S), head(rest(S))) = con(choo(S), head(STR(rest(S_1), S_2)))$ and $last(S) = last(rest(S))$. By the induction hypothesis for the structure $rest(S) = STR(rest(S_1), S_2)$, two cases are possible.

1. $\neg empty(rest(S_1))$ and $head(STR(rest(S_1), S_2)) = STR(head(rest(S_1)), S_2)$. Let us denote the structure $STR(head(S_1), S_2)$ by HS. Then, $head(S) = con(choo(S), STR(head(rest(S_1)), S_2))$, $HS = con(choo(HS), rest(HS)) = con(choo(HS), STR(rest(head(S_1)), S_2))$. From $\neg empty(head(S))$, $choo(HS) = choo(S)$ and $rest(head(S_1)) = head(rest(S_1))$, it follows that $head(S) = HS$. Moreover, $last(rest(S)) = (last(rest(S_1)), 1) = (last(S_1), 1)$.

2. $\neg empty(S_2)$ and $head(STR(rest(S_1), S_2)) = STR(rest(S_1), head(S_2))$. Let us denote the structure $STR(S_1, head(S_2))$ by SH. Then, $choo(S_2) = choo(head(S_2))$ provided $\neg empty(head(S_2))$, $head(S) = con(choo(S), STR(rest(S_1), head(S_2)))$, $SH = con(choo(SH), rest(SH))$, $rest(SH) = $

$STR(rest(S_1), head(S_2))$. From $choo(S) = choo(SH) = (choo(S_1), 1)$ it follows that $head(S) = SH$. Moreover, $last(rest(S)) = last(S_2)$.   □

Claim 1. *There exists a structure $S = STR(S_1, S_2, S_3)$ for which Theorem 3 does not hold.*

Proof. Let $memb(S_i) = \{a_i\}$ $(i = 1, 2, 3)$, where $a_1 < a_2 < a_3$. We define the function $sel(x_1, x_2, x_3)$ as **if** $x_1 \neq \omega \wedge x_2 \neq \omega \wedge x_3 \neq \omega$ **then** 1 **else if** $x_i = \omega \wedge x_j \neq \omega \wedge x_l \neq \omega$ **then** $max(j, l)$, where i, j, l are different. Then $vec(S) = [(a_1, 1)(a_3, 3)(a_2, 2)]$, $vec(head(S)) = [(a_1, 1)(a_3, 3)]$, $last(S) = (a_2, 2) = (last(S_2), 2)$. Therefore, $vec(STR(S_1, head(S_2), S_3)) = [(a_3, 3)(a_1, 1)] \neq vec(head(S))$.   □

4   Iterations with Termination Statement

Let us consider the following definite iteration over a tuple of unaltered data structures with a body including the termination statement EXIT:

> **for** x_1 **in** $S_1, \ldots, x_m$ **in** S_m **do** $t = sel(x_1, \ldots, x_m), v := body(v, x_t, t)$;
>
> **if** $cond(v, x_1, \ldots, x_m)$ **then** $EXIT$ **end**   (6)

where $x_i \notin v$ $(i = 1, \ldots, m)$, and the iteration body does not change the structures S_j $(j = 1, \ldots, m)$.

Let us define operational semantics of iteration (6) for an initial value v_0 of the vector v. Let $t_1 = sel(choo(S_1), \ldots, choo(S_m))$, $v_1 = body(v_0, choo(S_{t_1}), t_1)$, $b_1 = cond(v_1, choo(S_1), \ldots, choo(S_m))$. If $empty(S_i)$ for all $i = 1, \ldots, m$, then v_0 is the result of iteration (6). Otherwise, v_1 is the result of iteration (6) in the case of $b_1 = true$. If $b_1 = false$, then this process is continued with $v = v_1$ and the structure $rest(S_{t_1})$ instead of S_{t_1}, when the other structures S_i $(i \neq t_1)$ are not changed. A value v_l of the vector v that results from this process is the result of iteration (6).

Here the purpose is to reduce iteration (6) to iteration (1) with the help of hierarchical structures. We impose such restrictions on the body of iteration (6) that allows us to replace the termination condition $cond(v, x_1, \ldots, x_m)$ by $cond(v_0, x_1, \ldots, x_m)$ for an initial value v_0 of the vector v.

We introduce the following notation:
$\overline{S} = [S_1, \ldots, S_m], f(\overline{S}) = [f(S_1), \ldots, f(S_m)]$ for an unary function f,
$vec_0(S_i) = $ **if** $\neg empty(S_i)$ **then** $vec(S_i)$ **else** $\emptyset, t_j = sel(choo(vec_{j-1}(\overline{S})))$,
$vec_j(S_i) = $ **if** $i \neq t_j$ **then** $vec_{j-1}(S_i)$ **else** $rest(vec_{j-1}(S_i))$,
where $i = 1, \ldots, m, j = 1, \ldots, d, d = \sum_{i=1}^{m} | memb(S_i) |$.

The function $body(v, x_t, t)$ preserves the condition $cond(v, x_1, \ldots x_m)$ with respect to the structures $S_1, \ldots, S_m$ and the function $sel(x_1, \ldots, x_m)$ if $cond(body(v, x_t', t), x_1, \ldots, x_m) = cond(v, x_1, \ldots, x_m)$ for all $v, x_1, \ldots, x_m, t, x_t'$ for which there exist numbers l, l_0 such that $l_0 \leq l, x_i = choo(vec_l(S_i))$, $x_i' = choo(vec_{l_0}(S_i)), t = sel(x_1', \ldots, x_m')$ $(i = 1, \ldots, m)$.

Lemma 1. *If the function $body(v, x_t, t)$ preserves the condition $cond(v, x_1, \ldots, x_m)$ with respect to the structures $S_1, \ldots, S_m$ and the function $sel(x_1, \ldots, x_m)$, then iteration (6) with an initial value v_0 of the vector v is equivalent to the iteration*

$$\textbf{for } x_1 \textbf{ in } S_1, \ldots, x_m \textbf{ in } S_m \textbf{ do } t := sel(x_1, \ldots, x_m); v := body(v, x_t, t); \quad (7)$$
$$\textbf{if } cond(v_0, x_1, \ldots, x_m) \textbf{ then } EXIT \textbf{ end}$$

Proof. Lemma 1 is evident in the case of $d = 1$, when iteration (6) is terminated after the first step. Let $d > 1$, $v_i = body(v_{i-1}, choo(vec_{i-1}(S_{t_i})), t_i)$, $b_i = cond(v_i, choo(vec_{i-1}(\overline{S})))$ $(i = 1, \ldots, d-1)$. Let us define a number k $(1 \leq k \leq d-1)$ as follows. If $b_i = false$ for all $i = 1, \ldots, d-1$, then we suppose $k = d - 1$. Otherwise, there exists the only number k such that $b_k = true$ and in the case of $k > 1$ $b_i = false$ for all $i = 1, \ldots, k-1$. Lemma 1 follows immediately from the condition

$$\forall j(1 \leq j \leq k \rightarrow b_j = cond(v_0, choo(vec_{j-1}(\overline{S})))). \quad (8)$$

This condition results from the following more general condition for $i = j$:

$$\forall j(1 \leq j \leq k \rightarrow \forall i(1 \leq i \leq j \rightarrow cond(v_i, choo(vec_{j-1}(\overline{S}))) = \quad (9)$$
$$cond(v_0, choo(vec_{j-1}(\overline{S}))))).$$

To prove condition (9), we use induction on $i = 1, \ldots, j$. The function $body$ (v, x_t, t) preserves the condition $cond(v, x_1, \ldots, x_m)$ for $v = v_0, l = j - 1, l_0 = 0$ and therefore condition (9) holds for $i = 1$. Let us suppose $1 < i < j$. The function $body(v, x_t, t)$ preserves the condition $cond(v, x_1, \ldots, x_m)$ for $v = v_i, l = j - 1, l_0 = i$ and therefore $cond(v_{i+1}, choo(vec_{j-1}(\overline{S}))) = cond(v_i, choo(vec_{j-1}(\overline{S})))$. It remains to apply the inductive hypothesis. □

Let us define a hierarchical structure $T_0 = STR(S_1, \ldots, S_m)$ with the help of the function $sel(x_1, \ldots x_m)$ and the condition $cond(v_0, x_1, \ldots, x_m)$ as follows. $(choo(T_0), rest(T_0)) = \textbf{if } EMPTY \textbf{ then } (\omega, \omega) \textbf{ else } ((choo(S_{t_1}), t_1),$ $\textbf{if } b_0 \textbf{ then } \omega \textbf{ else } REST)$, where $b_0 = cond(v_0, choo(\overline{S}))$, $EMPTY = (empty(S_1) \wedge \ldots \wedge empty(S_m))$, $REST = STR(r(S_1), \ldots, r(S_m))$, $r(S_{t_1}) = rest(S_{t_1})$ and $r(S_i) = S_i$ for $i \neq t_1$.

Theorem 4. *If the function $body(v, x_t, t)$ preserves the condition $cond(v, x_1, \ldots, x_m)$ with respect to the structures $S_1 \ldots, S_m$ and the function $sel(x_1, \ldots, x_m)$, then iteration (6) with an initial value v_0 of the vector v is equivalent to the iteration*

$$\textbf{for } (x, \tau) \textbf{ in } T_0 \textbf{ do } v := body(v, x, \tau) \textbf{ end} \quad (10)$$

Proof. . It follows from Lemma 1 that it is sufficient to prove the equivalence of iterations (7) and (10) for $v = v_0$. For this we will use induction on $d = 0, 1, \ldots$. If $d = 0$, then $EMPTY$ and the equivalence is evident. Let us suppose that

$d > 0$ and the equivalence holds for $d - 1$. Two cases are possible.
1. $b_0 = true$. Then iteration (7) is equivalent to the statement
$v := body(v, choo(S_{t_1}), t_1)$ and the structure T_0 consists of one element
$choo(T_0) = (choo(S_{t_1}), t_1)$. Therefore, iterations (7) and (10) are equivalent.
2. $b_0 = false$. Then iteration (7) is equivalent to the program

$$v := body(v, choo(S_{t_1}), t_1); \text{ for } x_1 \text{ in } r(S_1), \ldots, x_m \text{ in } r(S_m) \text{ do}$$
$$t := sel(x_1, \ldots, x_m); v := body(v, x_t, t);$$
$$\text{if } cond(v_0, x_1, \ldots, x_m) \text{ then } EXIT \text{ end} \tag{11}$$

By the inductive hypothesis, program (11) is equivalent to the program

$$v := body(v, choo(S_{t_1}), t_1); \text{ for } (x, \tau) \text{ in } T_1 \text{ do } v := body(v, x, \tau) \text{ end} \tag{12}$$

where $T_1 = STR(r(S_1), \ldots, r(S_m))$. It follows from $choo(T_0) = (choo(S_{t_1}), t_1)$
and $REST = T_1$ that program (12) is equivalent to iteration (10).   □

Thus, introduction of the vector v_0 which represents initial values of variables
of the vector v allows us to reduce iteration (6) to more simple iteration (10)
for the mentioned restrictions on the termination condition $cond(v, x_1, \ldots, x_m)$.
If this condition does not depend on variables from v, then Theorem 4 can be
essentially simplified as follows.

Corollary 1. *The iteration*
for x_1 in $S_1, \ldots, x_m$ in S_m do $t := sel(x_1, \ldots, x_m); v := body(v, x_t, t);$
if $cond(x_1, \ldots, x_m)$ then $EXIT$ end
is equivalent to the iteration
for (x, τ) in T do $v := body(v, x, \tau)$ end
where $(choo(T), rest(T)) = $ if $EMPTY$ then (ω, ω) else $((choo(S_{t_1}), t_1),$
if $cond(choo(\overline{S}))$ then ω else $REST)$.

5   Iterations over Tuples of Altered Data Structures

The definite iteration over tuples of altered data structures has the form (2),
where the structure S_i can depend on variables from the vector v
($i = 1, \ldots, m$). Let w_i denote a vector consisting of all variables on which the
structure $S_i = S_i(w_i)$ depends ($i = 1, \ldots, m$). If S_i does not depend on variables
from v, then w_i is the empty vector and it can be omitted. It should be noted
that the vector w_i can consist of one variable S_i when, for example, S_i is a linear
list which can be changed by an iteration body. Let $Init$ denote a set consisting
of admissible initializations of variables from v. The set $Init$ can depend on a
program containing iteration (2).

Let v_j (w_{ij}, respectively) denote a vector consisting of the values of variables
from v (w_i, respectively). Let us say that v_j extends w_{ij} ($v_j \supset w_{ij}$) if, for each
variable y from the vector w_i and its value $y_j \in w_{ij}$, the property $y_j \in v_j$ holds.

Let us define operational semantics of iteration (2) for a vector v_0 consisting of the initial values of variables from v such that $v_0 \in Init$. Let $S_{i0} = S_i(w_{i0})$ provided $w_{i0} \subset v_0$ $(i = 1, \ldots, m)$, $d = \sum_{i=1}^{m} | memb(S_{i0}) |$.

We introduce the following notation :

$vec_0(S_{i0}) = $ **if** $\neg empty(S_{i0})$ **then** $vec(S_{i0})$ **else** $\emptyset$,

$t_j = sel(choo(vec_{j-1}(S_{10})), \ldots, choo(vec_{j-1}(S_{m0})))$,

$vec_j(S_{i0}) = $ **if** $i \neq t_j$ **then** $vec_{j-1}(S_{i0})$ **else** $rest(vec_{j-1}(S_{i0}))$,

$VEC_{j0} = [vec_j(S_{10}), \ldots, vec_j(S_{m0})]$,

$v_j = body(v_{j-1}, choo(vec_{j-1}(S_{t_j0})), t_j)(j = 1, \ldots, d)$.

An unrestricted variation of the structures S_i by the body of iteration (2) can result in an infinite iterative process. To provide finiteness of the iteration, we impose a restriction RTR1 on iteration (2) such that, at the j-th step of the iterative process, the iteration body does not change VEC_{j0} for $j = 1, \ldots, d-1$. Therefore, after the j-th step of the iterative process, the vector of undelivered elements of the structure S_i coincides with $vec_j(S_{i0})$, when $v = v_{j-1}$ and $x_i = choo(vec_{j-1}(S_{i0}))$ $(i = 1, \ldots, m)$.

The result of the iterative process is defined to be v_d. The following claim follows immediately from operational semantics of iteration (2).

Claim 2. *Iteration (2) with an initial value $v_0 \in Init$ of the vector v, provided $S_i = S_i(w_i)$ and $w_{i0} \subset v_0$ $(i = 1, \ldots, m)$, is equivalent to the program*

$$w_1 := w_{10}; \ldots w_m := w_{m0}; \textbf{ for } x_1 \textbf{ in } S_1(w_{10}), \ldots, x_m \textbf{ in } S_m(w_{m0})$$
$$\textbf{do } t := sel(x_1, \ldots, x_m); v := body(v, x_t, t) \textbf{ end} \tag{13}$$

where, in the case of $w_k = \emptyset$, the statement $w_k := w_{k0}$ is omitted in (13) and $S_k(w_{k0})$ is replaced by S_k $(k = 1, \ldots, m)$.

Let us define operational semantics of iteration (6) for a vector $v = v_0 \in Init$, when the structure S_i depends on variables from the vector w_i $(i = 1, \ldots, m)$. Let $b_j = cond(v_j, choo(vec_{j-1}(S_{10})), \ldots, choo(vec_{j-1}(S_{m0})))$ $(j = 1, \ldots, d-1)$. We impose a restriction RTR2 on iteration (6) such that on at the j-th step of the iterative process the iteration body does not change VEC_{j0}, when $\neg b_1 \wedge \ldots \wedge \neg b_j$ $(j = 1, \ldots, d-1)$. v_d is defined as the result of the iterative process in the case of $d \leq 1$ or $\neg b_1 \wedge \ldots \wedge \neg b_{d-1}$ for $d > 1$. If $b_1 = true$, then v_1 is defined as the result of the iterative process. Otherwise, there exists j such that $2 \leq j < d \wedge \neg b_1 \wedge \ldots \wedge \neg b_{j-1} \wedge b_j$. In this case v_j is defined to be the result of the iterative process.

The following claim follows immediately from operational semantics of iteration (6).

Claim 3. *Iteration (6) with an initial value $v_0 \in Init$ of the vector v, provided $S_i = S_i(w_i)$ and $w_{i0} \subset v_0$ $(i = 1, \ldots, m)$, is equivalent to the program*

$$w_1 := w_{10}; \ldots w_m := w_{m0}; \textbf{ for } x_1 \textbf{ in } S_1(w_{10}), \ldots, x_m \textbf{ in } S_m(w_{m0}) \textbf{ do}$$
$$t := sel(x_1, \ldots, x_m); v := body(v, x_t, t);$$
$$\textbf{if } cond(v, x_1, \ldots, x_m) \textbf{ then } EXIT \textbf{ end} \tag{14}$$

where, in the case of $w_k = \emptyset$, the statement $w_k := w_{k0}$ is omitted in (14) and S_k (w_{k0}) is replaced by S_k ($k = 1, \ldots, m$).

6   Generation of Verification Conditions

Let us remind the definition of the replacement operation $rep(v, S, body)$ which presents the effect of iteration (1) [21]. Let its result for $v = v_0$ be a vector v_n such that $n = 0$ provided $empty(S)$, and $v_i = body(v_{i-1}, s_i)$ for all $i = 1, \ldots, n$ provided $\neg empty(S)$ and $vec(S) = [s_1, \ldots, s_n]$. The following theorem presents useful properties of the replacement operation [23].

Theorem 5
5.1. Iteration (1) is equivalent to the multiple assignment statement

$$v := rep(v, S, body).$$

5.2. $rep(v, con(S_1, S_2), body) = rep(rep(v, S_1, body), S_2, body)$.
5.3. $rep(v, str(s), body) = body(v, s)$.

Corollary 2
2.1. $\neg empty(S) \rightarrow rep(v, S, body) = body(rep(v, head(S), body), last(S))$.
2.2. $\neg empty(S) \rightarrow rep(v, S, body) = rep(body(v, choo(S)), rest(S), body)$.

The replacement operation allows us to formulate the following proof rule without invariants for iteration (1). Let $R(y \leftarrow exp)$ be the result of substitution of an expression exp for all occurrences of a variable y into a formula R. Let $R(vec \leftarrow vexp)$ denote the result of a synchronous substitution of the components of an expression vector $vexp$ for all occurrences of the corresponding components of a vector vec into a formula R.

rl1. $\{P\}prog\{Q(v \leftarrow rep(v, S, body))\} \vdash$
        $\{P\}prog;$ **for** x **in** S **do** $v := body(v, x)$ **end** $\{Q\}$

where P is a pre-condition, Q is a post-condition which does not depend on the iteration parameter x, $prog$ is a program fragment, and $\{P\}$ $prog$ $\{Q\}$ denotes partial correctness of a program $prog$ with respect to P and Q.

The following corollary is evident from Theorem 5.1.

Corollary 3. *The proof rule rl1 is derived in the standard system of proof rules for usual statements including the multiple assignment statement.*

Projections of vectors $body(v, x)$ and $rep(v, S, body)$ on a variable y from the vector v are denoted by $body_y(v, x)$ and $rep_y(v, S, body)$, respectively.

Let us consider Pascal pointer programs. We will use the method from [11] to describe axiomatic semantics of these programs. With a pointer type we associate a section of a heap which presents a computer memory. Let L be a set of heap elements to which pointers can refer. An element to which a pointer p

refers is denoted by $p\uparrow$ in programs or by $\subset p\supset$ in specifications, or by $L\subset p\supset$ in specifications when it belongs to the set L. We will denote the predicate $\subset p\supset\in L$ by $pnto(L,p)$. Let $upd(L,\subset p\supset,e)$ be a set resulted from the set L by replacing the element $L\subset p\supset$ with the value of the expression e. In the case when the set L consists of records with the field k, we use $upd(L,\subset p\supset,k,e)$ to denote a set resulted from the set L by replacing the field $L\subset p\supset .k$ of the record $L\subset p\supset$ with the value of the expression e. We consider $upd(L,\subset q\supset,(k_1,\ldots,k_n),(e_1,\ldots,e_n))$ to be a shorthand for $upd(\ldots(upd(L,\subset q\supset,k_1,e_1),\ldots),\subset q\supset,k_n,e_n)$ in the case of different fields $k_1,\ldots,k_n$, and when the expressions $e_1,\ldots,e_n$ do not depend on $\subset q\supset$.

To generate verification conditions for programs which contain statements $q\uparrow:=e$, $q\uparrow.k:=e$, $new(p)$, $dispose(r)$, we use their equivalent forms:
$L:=upd(L,\subset q\supset,e)$ when $pnto(L,q)$, $L:=upd(L,\subset q\supset,k,e)$ when $pnto(L,q)$,
$L:=L\bigcup\{\subset p\supset\}$ when $\neg pnto(L,p)$, $L:=L-\{\subset r\supset\}$ when $pnto(L,r)$,
respectively.

The following proof rules, where P is a pre-condition, Q is a post-condition, *prog* is a program fragment, result from these equivalent representations.

rl2. $\{P\}\,prog\,\{Q(L\leftarrow upd(L,\subset q\supset,e))\}\vdash\{P\}\,prog;q\uparrow:=e\,\{Q\}$
  when $pnto(L,q)$.

rl3. $\{P\}\,prog\,\{Q(L\leftarrow upd(L,\subset q\supset,k,e))\}\vdash\{P\}\,prog;q\uparrow.k:=e\,\{Q\}$
  when $pnto(L,q)$.

rl4. $\{P\}\,prog\,\{Q(L\leftarrow L\cup\{\subset p\supset\})\}\vdash\{P\}\,prog;new(p)\,\{Q\}$
  when $\neg pnto(L,p)$.

rl5. $\{P\}\,prog\,\{Q(L\leftarrow L-\{\subset r\supset\})\}\vdash\{P\}\,prog;dispose(r)\,\{Q\}$
  when $pnto(L,r)$.

7   Case Study: Iterations over Doubly-Linked Lists

7.1   Specification Means

In this section we assume that a set L that forms a doubly-linked list, consists of records with the fields *key*, *next* and *prev*. The *key* field contains an integer (possible, zero) that serves as an identification name for an element. The *next* and *prev* fields contain a pointer or nil.

The predicate $reach_n(L,r,q)$ ($reach_p(L,r,q)$, respectively) means that the element $\subset q\supset$ is reached from the element $\subset r\supset$ in the set L via pointers from the *next* (*prev*, respectively) field. Let $root_n(L)$ ($root_p(L)$, respectively) be a pointer to the head element of the set L with respect to n-reachability (p-reachability, respectively), i.e. such an element from which all other elements of the set L can be reached via pointers from the *next* (*prev*, respectively) field. Let $l=last_n(L)$ ($l=last_p(L)$, respectively) be such an element of the set L

that the field $l.next$ ($l.prev$, respectively) contains nil or a pointer to an element which does not belong to the set L.

The predicate $dset(L)$ means that the set L is doubly-linked, i.e. there exist pointers $root_n(L)$ and $root_p(L)$, as well as elements $last_n(L)$ and $last_p(L)$, such that $last_p(L) = \subset root_n(L) \supset$ and $last_n(L) = \subset root_p(L) \supset$. Notice that there exist the only pointers $root_n(L)$ and $root_p(L)$, as well as the only elements $last_n(L)$ and $last_p(L)$ for the doubly-linked set L. A doubly-linked set L can be considered as a structure L such that $choo(L) = \subset root_n(L) \supset$ and $rest(L)$ results from the set L by removing the element $choo(L)$.

The predicate $dlist(L)$ means that the set L is a doubly-linked list, i.e. $dset(L)$ and $last_n(L).next = last_p(L).prev = nil$. The other useful kind of doubly-linked sets, so-called semilists, is defined by the predicate $dpset(L)$ which means $dset(L)$ and $last_n(L).next = nil$.

Let us define several useful operations over doubly-linked sets. A doubly-linked set which contains the only element l is denoted by $dset(l)$. Let us consider disjoint doubly-linked sets L_1 and L_2 such that $\neg pnto(L_1, last_n(L_2).next)$ and $\neg pnto(L_2, last_p(L_1).prev)$. We define their concatenation as a doubly-linked set $L = con(L_1, L_2)$ such that $L = L_1' \cup L_2'$, where L_1' (L_2', respectively) results from L_1 (L_2, respectively) by placing the pointer $root_n(L_2)$ ($root_p(L_1)$, respectively) into the field $last_n(L_1).next$ ($last_p(L_2).prev$, respectively). Let us extend the definition of $con(L_1, L_2)$ such that $con(L_1, L_2) = L_i$, where i=1 if L_2 is the empty set $\emptyset$, and i=2 if $L_1 = \emptyset$. We consider $con(L, l)$ and $con(l, L)$ to be a short form for $con(L, dset(l))$ and $con(dset(l), L)$, respectively. A set $con(con(L_1, L_2), L_3)$ is denoted by $con(L_1, L_2, L_3)$. It should be noted that for doubly-linked sets L_1 and L_2, the set $L = con(L_1, L_2)$ is doubly-linked, and, moreover, $dpset(L)$ in the case of $dpset(L_2)$.

A sequence which is the projection of the doubly-linked set L on the key field in the direction given by pointers in the $next$ field is denoted by $L.key$. In the case of the empty set L, let $L.key$ be the empty sequence.

For a sequence seq of different integers, we denote by $sord(seq)$ a predicate whose value is true, if the sequence seq has been sorted in the order $<$, and false otherwise. Let $set(seq)$ be the set of all elements of the sequence seq.

7.2 Merging Ordered Doubly-Linked Lists

The following annotated program prog1 merges in-place ordered doubly-linked lists L_1 and L_2 into an ordered list L, where the sets of keys of elements of L_1 and L_2 are disjoint.

$\{P\}z := nil; y_1 := root_n(L_1); y_2 := root_n(L_2);$
for x_1 **in** L_1, x_2 **in** L_2 **do** $t := sel(x_1, x_2);$
if $z \neq nil$ **then begin** $z\uparrow.next := y_t; y_t\uparrow.prev := z$ **end**;
$z := y_t; y_t := x_t.next;$ **if** $x_1 = \omega \vee x_2 = \omega$ **then** $EXIT$ **end** $\{Q\}$,
where $sel(x_1, x_2) = $ **if** $x_1.key < x_2.key$ **then** 1 **else** 2,
$y_t = $ **if** $t = 1$ **then** y_1 **else** y_2,
$P : L_1 = L_{10} \wedge L_2 = L_{20} \wedge dlist(L_{10}) \wedge dlist(L_{20}) \wedge L = L_{10} \cup L_{20} \wedge sord(L_{10}.key) \wedge$
$sord(L_{20}.key) \wedge set(L_{10}.key) \cap set(L_{20}.key) = \emptyset,$

$Q : dlist(L) \wedge sord(L.key) \wedge set(L.key) = set(L_{10}.key) \cup set(L_{20}.key)$.
It should be noted that the program prog1 has variables $L_1, L_2, L, y_1, y_2, z, x_1, x_2$, t, and the elements $z{\uparrow}$ and $y_t{\uparrow}$ can be written in the form $L \subset z \supset$ and $L \subset y_t \supset$, respectively. Moreover, y_i is a pointer to a scanned element of the list L_i ($i = 1, 2$), and z is a pointer to an element of the set L which has been selected by means of the function sel at the previous step of the iterative process.

The iteration that is contained in prog1 is considered for the initialization $Init$:
$L_1 = L_{10}, L_2 = L_{20}, L = L_{10} \cup L_{20}, z = nil, y_1 = root_n(L_{10}), y_2 = root_n(L_{20})$,
where $dlist(L_{10})$ and $dlist(L_{20})$ hold because of true precondition P.

Claim 4. *The restriction RTR2 holds for the iteration from prog1 under the condition of the initialization $Init$.*

Proof. Let $v = v_0 \in Init$. We will use the induction on $j = 1, \ldots, d - 1$. If $j = 1$, then $z = nil$ and the set $L = L_{10} \cup L_{20}$ does not change after the first step of the iterative process. Therefore, the restriction RTR2 holds for $j = 1$. Let us suppose that $j > 1$ and $\neg b_1 \wedge \ldots \wedge \neg b_j$. Two elements of the set L $L \subset z \supset$ and $L \subset y_t \supset$ can be changed at the j-th step of the iterative process. Detecting that the element $L \subset z \supset$ has been selected at the (j-1)-th step and the element $L \subset y_t \supset$ has been selected at the j-th step, we see that these elements do not belong to VEC_{j0}. Claim 4 follows from this and the inductive hypothesis. □

By Claim 3, the program prog1 is equivalent to the following program prog2 for the initialization $Init$:

$\{P\}z := nil; y_1 := root_n(L_1); y_2 := root_n(L_2); L_1 := L_{10}, L_2 := L_{20}$;
for x_1 **in** L_{10}, x_2 **in** L_{20} **do** $t := sel(x_1, x_2)$;
if $z \neq nil$ **then begin** $z{\uparrow}.next := y_t; y_t{\uparrow}.prev := z$ **end**;
$z := y_t; y_t := x_t.next;$ **if** $x_1 = \omega \vee x_2 = \omega$ **then** $EXIT$ **end** $\{Q\}$.

By Corollary 1, the program prog2 is equivalent to the program prog3:
$\{P\}z := nil; y_1 := root_n(L_1); y_2 := root_n(L_2); L_1 := L_{10}, L_2 := L_{20}$;
for (x, τ) **in** S **do if** $z \neq nil$ **then begin** $z{\uparrow}.next := y_\tau; y_\tau{\uparrow}.prev := z$ **end**;
$z := y_\tau; y_\tau := x.next$ **end** $\{Q\}$,
where the hierarchical structure $S = STR(L_{10}, L_{20})$ is defined as follows.
$(choo(S), rest(S)) =$ **if** $EMPTY$ **then** (ω, ω) **else** $((choo(L_{t_10}), t_1)$,
if b_1 **then** ω **else** $REST)$,
where $t_1 = sel(choo(L_{10}), choo(L_{20})), b_1 = (choo(L_{10}) = \omega \vee choo(L_{20}) = \omega)$,
$REST =$ **if** $t_1 = 1$ **then** $STR(rest(L_{10}), L_{20})$ **else** $STR(L_{10}, rest(L_{20}))$.

The body of the iteration that is contained in prog3 can be written in the form $(L, y_1, y_2, z) := body(L, y_1, y_2, z, x, \tau)$, where $body(L, y_1, y_2, z, x, \tau) =$
(**if** $z \neq nil$ **then** $upd(upd(L, \subset z \supset, next, y_\tau), \subset y_\tau \supset, prev, z)$ **else** L,
if $\tau = 1$ **then** $x.next$ **else** y_1, **if** $\tau = 1$ **then** y_2 **else** $x.next, y_\tau)$.

The following verification condition VC is generated from prog3 with the help of the proof rule rl1.
$VC : P(L, L_1, L_2, L_{10}, L_{20}) \rightarrow Q(L', L_{10}, L_{20})$,
where $L' = rep_L((L, root_n(L_1), root_n(L_2), nil), S, body)$.

In order to prove the condition VC by induction, we will replace doubly-linked lists by semilists. The verification condition VC immediately follows from the property

$prop(L, L_1, L_2, L_{10}, L_{20}) = (P'(L, L_1, L_2, L_{10}, L_{20}) \rightarrow Q'(L', L_{10}, L_{20}))$,

where $P'(L, L_1, L_2, L_{10}, L_{20}) : L_1 = L_{10} \wedge L_2 = L_{20} \wedge dpset(L_{10}) \wedge dpset(L_{20}) \wedge L = L_{10} \cup L_{20} \wedge \neg pnto(L_{20}, choo(L_{10}).prev) \wedge \neg pnto(L_{10}, choo(L_{20}).prev) \wedge sord(L_{10}.key) \wedge sord(L_{20}.key) \wedge set(L_{10}.key) \cap set(L_{20}.key) = \emptyset$,

$Q'(L, L_{10}, L_{20}) : dpset(L) \wedge sord(L.key) \wedge choo(L).key = choo(L_{t_10}).key \wedge choo(L).prev = choo(L_{t_10}).prev \wedge set(L.key) = set(L_{10}.key) \cup set(L_{20}.key)$.

Claim 5. *The property $prop(L, L_1, L_2, L_{10}, L_{20})$ holds.*

Proof. Let us suppose $t_1 = 1$. We will use induction on $k = |memb(S)| \ (k \geq 2)$. If $k = 2$, then $|memb(L_{10})| = 1$ and $L' = con(L_{10}, L_{20})$. Indeed, the structure $STR(rest(L_{10}), L_{20})$ consists of the only element $(choo(L_{20}), 2)$ and $L' = rep_L((L_{10} \cup L_{20}, root_n(L_{10}), root_n(L_{20}), nil), S, body) = upd(upd(L_{10} \cup L_{20}, \subset root_n(L_{10}) \supset, next, root_n(L_{20})), \subset root_n(L_{20}) \supset, prev, root_n(L_{10}))$ by Theorem 5.3 and Corollary 2.2. Therefore, from $P'(L, L_1, L_2)$ it follows that $dpset(con(L_{10}, L_{20}))$ and $sord(con(L_{10}, L_{20}).key)$. $Q'(L', L_{10}, L_{20})$ follows immediately from this.

Let us suppose that $k > 2$ and $prop(L, L_1, L_2, L_{10}, L_{20})$ holds for $|memb(S)| < k$. Two cases are possible.

1. $t_2 = 1$. Then $L' = \{choo(L_{10})\} \cup L''$, where $L'' = rep_L((rest(L_{10}) \cup L_{20}, root_n(rest(L_{10})), root_n(L_{20}), nil), STR(rest(L_{10}), L_{20}), body) = rep_L((rest(L_{10}) \cup L_{20}, choo(rest(L_{10})).next, root_n(L_{20}), root_n(rest(L_{10}))), STR(rest^2(L_{10}), L_{20}), body)$. By the inductive hypothesis for L'', it follows from $P'(L'', L_1, L_2, rest(L_{10}), L_{20})$ that $Q'(L'', rest(L_{10}), L_{20})$. Therefore, $L' = con(choo(L_{10}), L'')$. $Q'(L', L_{10}, L_{20})$ follows from this, $sord(L''.key)$ and $choo(L_{10}).key < choo(rest(L_{10})).key = \subset root_n(L'') \supset.key$.

2. $t_2 = 2$. Then $L' = \{upd(choo(L_{10}), next, root_n(L_{20}))\} \cup upd(L'', \subset root_n(L'') \supset, prev, root_n(L_{10}))$, where $L'' = rep_L((rest(L_{10}) \cup L_{20}, root_n(rest(L_{10})), root_n(L_{20}), nil), STR(rest(L_{10}), L_{20}), body) = rep_L((rest(L_{10}) \cup L_{20}, choo(L_{10}).next, choo(L_{20}).next, root(L_{20})), STR(rest(L_{10}), rest(L_{20})), body)$. By the inductive hypothesis for L'', it follows from $P'(L'', L_1, L_2, rest(L_{10}), L_{20})$ that $Q'(L'', rest(L_{10}), L_{20})$. Therefore, $L' = con(choo(L_{10}), L'')$. $Q'(L', L_{10}, L_{20})$ follows from this, $sord(L''.key)$, $choo(L_{10}).key < choo(L_{20}).key$ and $choo(L'') = choo(L_{20})$. $\square$

8   Conclusion

A generalization of the symbolic verification method for definite iterations over tuples of data structures to allow modification of data structures by the iteration body and termination of the iteration under a condition is described in this paper. The generalization extends application domains of the symbolic method since generalized iterations allow us to represent a new important case of while-loops and to apply the method to verification of pointer programs with several input data structures.

Restrictions RTR1 and RTR2 imposed on iterations over tuples of altered data structures allow us to change only the current and previously processed values of iteration parameters and to retain the important property of termination of the iterations. The idea of reduction of the iterations to the iterations over tuples of unaltered data structures by means of introducing special variables that store initial values of variables from the iteration body was found to be fruitful as demonstrated by Claims 2,3 and by the example in Sect. 7.2.

Instead of loop invariants, the symbolic method uses properties of the replacement operation which, as a rule, are simpler than the invariants. To represent the invariants, new notions related to a specific character of verified programs are often required. The proof of verification conditions including the replacement operation does not require introducing such notions. Instead, the symbolic method uses properties of both hierarchical data structures and the replacement operation that are expressed by Theorems 2, 3, 5 and Corollary 2.

Verification of pointer programs has been considered in [2,9,12,24,25,26] in the framework of axiomatic approach. It should be noted that the symbolic method has been applied to verification of two programs over linear singly-linked lists for elimination of elements with zero keys and for a search of an element with reordering [24]. Interesting examples of pointer program verification, which include a program for in-place merging ordered singly-linked lists, have been given in [2], where a verification method based on the method proposed in [15] has been developed. To verify this program, a special list representation and a complicated loop invariant are used in [2]. An application of the tool Isabelle/HOL to pointer program verification using the method [2] has been described in [12]. Verification of a program over doubly-linked lists for elimination of elements with zero keys has been presented in [26], where a Hoare-like logic oriented to pointer program verification has been proposed. This logic has been formalized in [25] as the separation logic. The symbolic verification method has two advantages as compared with [2] and [26], since it does not use both loop invariants and special list representations. For Hoare-style verification of pointer programs, decidable logics and simulation of data structures including doubly-linked lists have been adapted in [9]. Such a new verification method uses loop and simulation invariants. Correctness proofs of some routines over singly-linked lists have been considered in [13] as a case study of a reliable library of object-oriented components.

The symbolic verification method is promising for applications. In [19,20], we described a tool SPECTRUM which uses the symbolic method for verification of linear algebra programs. It is suggested to develop a tool for program verification by the symbolic method and apply it to pointer program verification.

References

1. Abd-El-Hafiz, S.K., Basili, V.R.: A knowledge - based approach to the analysis of loops. IEEE Trans. of Software Eng. 22(5), 339–360 (1996)
2. Bornat, R.: Proving pointer programs in Hoare logic. In: Backhouse, R., Oliveira, J.N. (eds.) MPC 2000. LNCS, vol. 1837, pp. 102–126. Springer, Heidelberg (2000)
3. Ernst, M.D., et al.: Dynamically discovering likely program invariants to support program evolution. IEEE Trans. of Software Eng. 27(2), 99–123 (2001)
4. Gries, D., Gehani, N.: Some ideas on data types in high-level languages. Comm. ACM 20(6), 414–420 (1977)
5. Hehner, E.C.R., Gravell, A.M.: Refinement semantics and loop rules. In: Woodcock, J.C.P., Davies, J., Wing, J.M. (eds.) FM 1999. LNCS, vol. 1709, pp. 1497–1510. Springer, Heidelberg (1999)
6. Hoare, C.A.R.: An axiomatic basis of computer programming. Comm. ACM 12(10), 576–580 (1969)
7. Hoare, C.A.R.: A note on the for statement. BIT 12(3), 334–341 (1972)
8. Hoare, C.A.R.: The verifying compiler: a grand challenge for computing research. In: Broy, M., Zamulin, A.V. (eds.) PSI 2003. LNCS, vol. 2890, pp. 1–12. Springer, Heidelberg (2004)
9. Immerman, N., et al.: Verification via structure si+mulation. In: Alur, R., Peled, D.A. (eds.) CAV 2004. LNCS, vol. 3114, pp. 281–294. Springer, Heidelberg (2004)
10. Linger, R.C., Mills, H.D., Witt, B.I.: Structured programming: theory and practice. Addison Wesley, Reading (1979)
11. Luckham, D.C., Suzuki, N.: Verification of array, record and pointer operations in Pascal. ACM Trans. on Programming Languages and Systems 1(2), 226–244 (1979)
12. Mehta, F., Nipkow, T.: Proving pointer programs in higher-order logic. In: Baader, F. (ed.) CADE 2003. LNCS (LNAI), vol. 2741, pp. 121–135. Springer, Heidelberg (2003)
13. Meyer, B.: Towards practical proofs of class correctness. In: Bert, D., P. Bowen, J., King, S. (eds.) ZB 2003. LNCS, vol. 2651, pp. 359–387. Springer, Heidelberg (2003)
14. Mills, H.D.: Structured programming: retrospect and prospect. IEEE Software 3(6), 58–67 (1986)
15. Morris, J.M.: A general axiom of assignment, Assignment and linked data structures. Lecture Notes of Intern. Summer School on Theoretical foundations of programming methodology, D. Reidel, pp. 25–41 (1982)
16. Necula, G.C.: Proof-carrying code. In: Proc. 24th Annual ACM Symposium on Principles of Programming Languages, pp. 106–119. ACM Press, New York (1997)
17. Nepomniaschy, V.A.: Loop invariant elimination in program verification. Programming and Computer Software 3, 129–137 (1985) (English translation of Russian Journal "Programmirovanie")
18. Nepomniaschy, V.A.: On problem–oriented program verification. Programming and Computer Software 1, 1–9 (1986)
19. Nepomniaschy, V.A., Sulimov, A.A.: Problem-oriented means of program specification and verification in project SPECTRUM. In: Miola, A. (ed.) DISCO 1993. LNCS, vol. 722, pp. 374–378. Springer, Heidelberg (1993)
20. Nepomniaschy, V.A., Sulimov, A.A.: Problem-oriented verification system and its application to linear algebra programs. Theoretical Computer Science 119, 173–185 (1993)

21. Nepomniaschy, V.A.: Symbolic verification method for definite iteration over data structures. Information Processing Letters 69, 207–213 (1999)
22. Nepomniaschy, V.A.: Verification of definite iteration over hierarchical data structures. In: Finance, J.-P. (ed.) ETAPS 1999 and FASE 1999. LNCS, vol. 1577, pp. 176–187. Springer, Heidelberg (1999)
23. Nepomniaschy, V.A.: Verification of definite iteration over tuples of data structures. Programming and Computer Software 1, 1–10 (2002)
24. Nepomniaschy, V.A.: Symbolic verification method for definite iteration over altered data structures. Programming and Computer Software 1, 1–12 (2005)
25. O'Hearn, P., Reynolds, J., Yang, H.: Local reasoning about programs that alter data structures. In: Fribourg, L. (ed.) CSL 2001 and EACSL 2001. LNCS, vol. 2142, pp. 1–19. Springer, Heidelberg (2001)
26. Reynolds, J.C.: Reasoning about shared mutable data structure. In: Proc. Symp. in celebration of the work of C.A.R. Hoare, Oxford, pp. 1–22 (1999)
27. Stark, J., Ireland, A.: Invariant discovery via failed proof attempts. In: Flener, P. (ed.) LOPSTR 1998. LNCS, vol. 1559, pp. 271–288. Springer, Heidelberg (1999)
28. Stavely, A.M.: Verifying definite iteration over data structures. IEEE Trans. of Software Eng. 21(6), 506–514 (1995)
29. Whalen, M., Schumann, J., Fischer, B.: Synthesizing certified code. In: Eriksson, L.-H., Lindsay, P.A. (eds.) FME 2002. LNCS, vol. 2391, pp. 431–450. Springer, Heidelberg (2002)

Categories of Elementary Sets over Algebras and Categories of Elementary Algebraic Knowledge

Boris Plotkin[1] and Tatjana Plotkin[2]

[1] Department of Mathematics, Hebrew University, Jerusalem, Israel
borisov©math.huji.ac.il
[2] Department of Computer Science, Bar-Ilan University, Ramat Gan, Israel
plotkin©macs.biu.ac.il

To our dear friend B.A.Trahtenbrot on the occasion of his birthday.

Abstract. For every variety of algebras Θ and every algebra H in Θ we consider the category of algebraic sets $K_\Theta(H)$ in Θ over H. We consider also the category of elementary sets $LK_\Theta(H)$. The latter category is associated with a geometrical approach to the First Order Logic over algebras. It is also related to the category of elementary knowledge about algebra H. Grounding on these categories we formally introduce and study the intuitive notions of coincidence of algebraic geometries over algebras H_1 and H_2 from Θ, of coincidence of logics over H_1 and H_2, and of coincidence of the corresponding knowledge. This paper is a survey of ideas stimulated by this approach.

1 Preliminaries

1.1 Varieties of Algebras, Free Algebras

A *variety of algebras* is a class of algebras defined by a signature (system of symbols of operations Ω) and a set of identities for these operations. For example, the variety of semigroups is given by the operation of multiplication and the associativity identity. There are also various varieties of groups, of associative algebras, of Lie algebras, and many others. In every variety Θ there exists the free algebra $W = W(X)$ over a set X. This means that X is a subset in W and for every algebra $H \in \Theta$ and every mapping $\nu : X \to H$ this ν is uniquely extended up to a homomorphism $\mu : W \to H$. Thus, we have a commutative diagram

$$X \xrightarrow{\;id\;} W$$
$$\nu \searrow \quad \downarrow \mu$$
$$H$$

For example, if Θ is the variety of commutative associative algebras over a field P, $X = \{x_1, \ldots, x_n\}$, then the algebra of polynomials $P[x_1, \ldots, x_n]$ is the free in Θ algebra over X. Free semigroups and free monoids are well known and have lots of applications.

A. Avron et al. (Eds.): Trakhtenbrot/Festschrift, LNCS 4800, pp. 555–570, 2008.

The notion of category is quite popular. For every variety of algebras Θ define an important category Θ^0. Its objects are free in Θ algebras $W = W(X)$, where X runs all finite subsets in some infinite universum X^0. Morphisms in Θ^0 are arbitrary homomorphisms $s : W(X) \to W(Y)$.

1.2 Affine Spaces in Θ over H

For every algebra H denote by $H^{(n)}$ the n-th cartesian degree of the set H. Its elements are the points $a = (a_1, \ldots, a_n)$, $a_i \in H$. Let, further, $X = \{x_1, \ldots, x_n\}$ and take the set H^X of mappings $\nu : X \to H$. We have the canonical bijections $H^X \to H^{(n)}$ and $Hom(W, H) \to H^{(n)}$. A point $(\mu(x_1), \ldots, \mu(x_n))$, $\mu(x_i) = a_i = \nu(x_i)$ corresponds to an element $\mu : W \to H$. Each of the sets $H^{(n)}$, H^X, and $Hom(W, H)$ we treat as the affine space for the given H and X. In particular, a homomorphism $\mu : W \to H$ is a point of the affine space $Hom(W, H)$.

Let us define the category of affine spaces $K_\Theta^0(H)$. Its objects are the sets of homomorphisms $Hom(W, H)$ and morphisms are of the form

$$\tilde{s} : Hom(W_1, H) \to Hom(W_2, H),$$

where $s : W_2 \to W_1$ is a morphism in Θ^0 and the mapping $\tilde{s}$ is given by the rule $\tilde{s}(\nu) = \nu s : W_2 \to H$ for $\nu : W_1 \to H$. We have a contravariant functor $\Theta^0 \to K_\Theta^0(H)$ which implies a duality of the categories if and only if the identities of the algebra H determine the whole variety Θ, i.e., $Var(H) = \Theta$.

1.3 Algebraic Sets and Elementary Sets

Let us take $W = W(X)$ and consider equations of the form $w = w'$, where $w, w' \in W$. These w and w' are the terms in W, $w = w(x_1, \ldots, x_n)$, $w' = w'(x_1, \ldots, x_n)$, $X = \{x_1, \ldots, x_n\}$. A point $a = (a_1, \ldots, a_n) \in H^{(n)}$ is a solution of $w = w'$ in the algebra H if $w(a_1, \ldots, a_n) = w'(a_1, \ldots, a_n)$. Similarly, a point $\mu \in Hom(W, H)$ is a solution of $w = w'$ if $w(\mu(x_1), \ldots, \mu(x_n)) = w^\mu = w'(\mu(x_1), \ldots, \mu(x_n)) = w'^\mu$. The equality $w^\mu = w'^\mu$ means that the pair (w, w') belongs to the kernel of the homomorphism μ, denoted by $Ker(\mu)$. The kernel $Ker(\mu)$ is a congruence of the algebra W, and the quotient algebra $W/Ker(\mu)$ can also be considered.

Let now T be a system of equations in W. Set

$$T_H' = A = \{\mu : W \to H \mid T \subset Ker(\mu)\}$$

for every $H \in \Theta$. Here A is the set of points, satisfying every equation in T. Such sets A are called algebraic sets.

We consider also elementary sets as subsets in the corresponding $Hom(W, H)$. In order to do that let us treat every equation $w = w'$ as a formula which is an equality in First Order Logic (FOL), written as $w \equiv w'$. The equality $w \equiv w'$ is an element of a special algebra of formulas $\Phi = \Phi(X)$ (see Subsect. 2.6 for the definition). We also view an arbitrary formula $u \in \Phi$ as an equation and look for its solutions in the affine space $Hom(W(X), H)$. The value of a formula

$Val_H(u) = Val_H^X(u)$ is defined to be a subset in $Hom(W(X), H)$. In particular, $Val_H(w \equiv w')$ is the set of all points $\mu : W \to H$ with $(w, w') \in Ker(\mu)$. Later we will see how to determine $Val_H(u)$ for an arbitrary $u \in \Phi$. A point μ is the solution of an "equation" u if $\mu \in Val_H(u)$.

Let us define the logical kernel $LKer(\mu)$ of a point $\mu : W \to H$. A formula $u \in \Phi(X)$ belongs to $LKer(\mu)$ if and only if $\mu \in Val_H(u)$. The usual kernel $Ker(\mu)$ is the set of all (w, w') with $w \equiv w' \in LKer(\mu)$.

Note that the algebra Φ is a Boolean algebra. Recall the definition of a filter of Boolean algebra. Proceed from Φ and a set $T \in \Phi$. T is a filter if

1. $0 \notin T$,
2. $u_1, u_2 \in T$ implies $u_1 \wedge u_2 \in T$,
3. $u \in T$ and $u < v$ imply $v \in T$.

It is obvious that the kernel $LKer(\mu)$ is a filter. This fact is very important. Define an elementary set in $Hom(W, H)$ as follows. Let T be an arbitrary subset in Φ. We set:

$$T_H^L = A = \{\mu : W \to H | T \subset LKer(\mu)\}.$$

If μ is a solution of every "equation" $u \in T$, then $\mu \in A$. The following is also true:

$$T_H^L = \bigcap_{u \in T} Val_H(u) = A.$$

Each A of such a form is called an elementary set. It is clear, that every algebraic set is an elementary set as well.

1.4   Categories of Algebraic and Elementary Sets

Define first the category $Set_\Theta(H)$ of affine sets over an algebra H. Its objects are of the form (X, A), where A is an arbitrary subset in the affine space $Hom(W(X), H)$. The morphisms are

$$[s] : (X, A) \to (Y, B).$$

Here $s : W(Y) \to W(X)$ is a morphism in Θ^0. The corresponding morphism $\tilde{s} : Hom(W(X), H) \to Hom(W(Y), H)$ should be coordinated with A and B by the condition: if $\nu \in A$, then $\tilde{s}(\nu) \in B$. Then the induced mapping $[s] : A \to B$ we consider as a morphism $(X, A) \to (Y, B)$.

In every category a part of its objects determines a full subcategory. Thus, the category $K_\Theta(H)$ of algebraic sets over H is a full subcategory in $Set_\Theta(H)$. We take here objects (X, A), where A is an algebraic set in $Hom(W(X), H)$. If we take for A the elementary sets, then we are coming up with the category of elementary sets $LK_\Theta(H)$. The category $K_\Theta(H)$ is a full subcategory in $LK_\Theta(H)$. We consider $K_\Theta(H)$ as a main algebra-geometrical (AG)-invariant of the algebra H, while $LK_\Theta(H)$ is a main logic-geometrical (LG)-invariant of the same H. We will work with these invariants.

1.5   Homomorphisms and Filters of Boolean Algebras

We recall here the known facts about homomorphisms and filters of Boolean algebras. Let $\alpha : B_1 \to B_2$ be a homomorphism of Boolean algebras. Its kernel $Ker(\alpha)$ is a congruence in B_1, defined by the condition $bKer(\mu)b' \Leftrightarrow \alpha(b) = \alpha(b')$. Cosets in B_1 correspond to the congruence $Ker(\mu)$. The coset $[1] = T$ is a filter, determining the whole congruence $Ker(\alpha)$. The general rule is as follows. If T is a filter in B, then the corresponding congruence ρ in B is given by: $b\rho b' \Leftrightarrow (b \to b') \wedge (b' \to b) \in T$. We consider the quotient algebra B/ρ also denoted by B/T. Here the filter T serves as the kernel of the natural homomorphism. We will use these facts in the sequel.

Let us pass to some structures of algebraic logic, in particular, to the definition of the algebra of formulas $\Phi(X)$.

2   Halmos Categories and Halmos Algebras

Polyadic Halmos algebras and cylindric Tarski algebras are the main structures of algebraic logic [3], [2]. They are used to be defined for infinite set of variables X^0. For our purposes we need to explore another situation when we take the set of all finite subsets X of X^0 instead of one infinite X^0. This, in particular, leads to Halmos categories and special multisorted Halmos algebras. From now on X denotes a finite set.

2.1   Extended Boolean Algebras

Note first that in Algebraic Logic (AL) quantifiers are treated as operations on Boolean algebras. Let B be a Boolean algebra. Its existential quantifier is a mapping $\exists : B \to B$ with the conditions:

1. $\exists 0 = 0$,
2. $\exists a > a$,
3. $\exists(a \wedge \exists b) = \exists a \wedge \exists b$.

The universal quantifier $\forall : B \to B$ is defined dually:

1. $\forall 1 = 1$,
2. $\forall a < a$,
3. $\forall(a \vee \forall b) = \forall a \vee \forall b$.

Here 0 and 1 are zero and unit of the algebra B and a, b are arbitrary elements of B. The quantifiers $\exists$ and $\forall$ are coordinated in the usual way: $\overline{\exists a} = \forall \overline{a}, \overline{\forall a} = \exists \overline{a}$.

Let Θ and $W = W(X) \in \Theta$ be fixed and B be a Boolean algebra. We call B an extended Boolean algebra in Θ over $W(X)$, if

1. There are defined quantifiers $\exists x$ for all $x \in X$ in B with $\exists x \exists y = \exists y \exists x$ for all $x, y \in X$.

2. To every equality $w \equiv w'$, $w, w' \in W$ it corresponds a constant in B, denoted by $w \equiv w'$. Here,
   (a) $w \equiv w$ is the unit of the algebra B.
   (b) For every n-ary operation $\omega \in \Omega$ we have

$$w_1 \equiv w'_1 \wedge \ldots \wedge w_n \equiv w'_n < w_1 \ldots w_n \omega \equiv w'_1 \ldots w'_n \omega.$$

We can consider the variety of such algebras for the given Θ and $W = W(X)$.

2.2 Example

Take an affine space $Hom(W(X), H)$ and let the algebra $Bool(W(X), H) = Sub(Hom(W(X), H))$ be the Boolean algebra of subsets A in $Hom(W(X), H)$. Let us define quantifiers $\exists x$, $x \in X$ on $Bool(W(X), H)$. We set $\mu \in \exists x A$ if and only if there exists $\nu \in A$ such that $\mu(y) = \nu(y)$ for every $y \in X$, $y \neq x$.

Every equality $w \equiv w'$, $w, w' \in W$ is implemented on this algebra as

$$Val_H^X(w \equiv w') = \{\mu : W \to H | (w, w') \in Ker(\mu)\}.$$

As a result we have an extended algebra $Bool(W(X), H)$ in Θ over $W(X)$.

Consider, further, the category $Hal_\Theta(H)$ of extended Boolean algebras for the given $H \in \Theta$. Its morphisms are of the form

$$s_* : Bool(W(X), H) \to Bool(W(Y), H),$$

where $s : W(X) \to W(Y)$ is a morphism in Θ^0. Let us define the transition from s to s_*. We have

$$\tilde{s} : Hom(W(Y), H) \to Hom(W(X), H).$$

Let A be a subset in $Hom(W(X), H)$. We set $s_* A = \tilde{s}^{-1} A$. The map s_* is a homomorphism of Boolean algebras, but, in general, not a homomorphism of extended Boolean algebras.

We have a covariant functor $\Theta^0 \to Hal_\Theta(H)$.

2.3 Halmos Categories

A category Υ is a Halmos category if:

1. Every its object has the form $\Upsilon(X)$, and this object is an extended Boolean algebra in Θ over $W(X)$.
2. Morphisms are of the form $s_* : \Upsilon(X) \to \Upsilon(Y)$, where $s : W(X) \to W(Y)$ are morphisms in Θ^0, s_* are homomorphisms of Boolean algebras and the transition $s \to s_*$ is given by a covariant functor $\Theta^0 \to \Upsilon$.
3. There are identities controlling the interaction of morphisms with quantifiers and equalities. See [8] for the explicit list of identities or [7] for more details.

The category $Hal_\Theta(H)$ is an example of the Halmos category. Another important example is the category of formulas Hal_Θ^0 of the algebras of formulas $Hal_\Theta^0(X) = \Phi(X)$. This category plays in logical geometry the same role as the category Θ^0 does in AG.

2.4   Multisorted Algebras

We will use multisorted algebras in order to define the notion of Halmos algebras. One-sorted algebras are algebras with one domain. In multisorted algebras there are many domains. They are written as $G = (G_i, i \in \Gamma)$, where Γ is a set of sorts, which can be infinite. Categories are often related to multisorted algebras [4].

Every operation ω in G has a specific type $\tau = \tau(\omega)$. In the one-sorted case it is the arity of an operation. In the multisorted case we have $\tau = (i_1, \ldots, i_n; j)$ and a mapping $\omega : G_{i_1} \times \ldots \times G_{i_n} \to G_j$. Morphisms of multisorted algebras are of the form $\mu = (\mu_i, i \in \Gamma) : G \to G'$, where $\mu_i : G_i \to G'_i$ are the mappings and μ is naturally correlated with the operations ω.

Subalgebras, quotient algebras, and cartesian products of multisorted algebras are defined in the usual way. Hence, one can define varieties of multisorted algebras with the given domain Γ and signature Ω. In every such a variety there exist free algebras, determined by multisorted sets.

2.5   Halmos Algebras

We deal with multisorted Halmos algebras, associated with Halmos categories. We first describe the signature. Take $L_X = \{\vee, \wedge, ^-, \exists x, x \in X, M_X\}$ for every X. Here M_X is a set of all equalities over the algebra $W = W(X)$. We add all possible $s : W(X) \to W(Y)$ to L_X, treating them as symbols of unary operations. Denote the new signature by L_Θ. Denote by Γ the set of all finite subsets of the infinite set X^0.

Consider further algebras $\Upsilon = (\Upsilon_X, X \in \Gamma)$. Every Υ_X is an algebra in the signature L_X and an unary operation (mapping) $s_* : \Upsilon_X \to \Upsilon_Y$ corresponds to every $s : W(X) \to W(Y)$. We call an algebra Υ in the signature L_Θ a Halmos algebra, if

1. Every Υ_X is an extended Boolean algebra in the signature L_X.
2. Every mapping $s_* : \Upsilon_X \to \Upsilon_Y$ is coordinated with the Boolean operations and is a homomorphism of Boolean algebras.
3. The identities, controlling interaction of operations s_* with quantifiers and equalities are the same as in the definition of Halmos categories.

It is clear now that each Halmos category Υ can be viewed as a Halmos algebra and vice versa. In particular, this relates to $Hal_\Theta(H)$.

2.6   Categories and Algebras of Formulas

Denote by $M = (M_X, X \in \Gamma)$ a multisorted set with the components M_X.

Take the absolutely free algebra $\Upsilon^0 = (\Upsilon^0_X, X \in \Gamma)$ over M in the signature L_Θ. Elements of each Υ^0_X are FOL formulas which are inductively constructed from the equalities using the signature L_Θ. So, Υ^0 is a multisorted algebra of pure FOL formulas.

Denote by Hal_Θ the variety of Γ-sorted Halmos algebras in the signature L_Θ. Denote by Hal^0_Θ the free algebra of this variety over the multisorted set of equalities M. The same M determines the homomorphism $\pi = (\pi_X, X \in \Gamma)$: $\Upsilon^0 \to Hal^0_\Theta$. If $u \in \Upsilon^0_X$, then the image $u^\pi x = \bar{u}$ in $Hal^0_\Theta(X)$ is viewed as a compressed formula.

Setting $Hal^0_\Theta(X) = \Phi(X)$ we get the wanted algebra of compressed formulas. This is an extended Boolean algebra.

Recall that the Halmos algebra of formulas Hal^0_Θ is also a Halmos category. As it was mentioned above, the category Hal^0_Θ plays in logical geometry the same role as the category Θ^0 does in universal algebraic geometry. We have a covariant functor $\Theta^0 \to Hal^0_\Theta$.

2.7 Value of a Formula

The value $Val^X_H(w \equiv w')$ corresponds to each equality $w \equiv w'$, $w, w' \in W(X)$. This determines a mapping $Val_H : M \to Hal_\Theta(H)$ which is uniquely extended up to homomorphisms $Val^0_H : \Upsilon^0 \to Hal_\Theta(H)$ and $Val_H : Hal^0_\Theta \to Hal_\Theta(H)$. For every $X \in \Gamma$ we have a commutative diagram

$$
\begin{array}{ccc}
\Upsilon^0 & \xrightarrow{\ Val^{0X}_H\ } & Bool(W(X), H) \\
& \searrow{\scriptstyle \pi_X} \quad \nearrow{\scriptstyle Val^X_H} & \\
& \Phi(X) &
\end{array}
$$

Thus, for every $u \in \Upsilon^0_X$ and the corresponding $\bar{u} \in \Phi(X)$ we have the values $Val^{0X}_H(u) = Val^X_H(\bar{u})$.

We call formulas u and v in Υ^0_X semantically equivalent if $Val^0_\Theta(u) = Val^0_\Theta(v)$ for every algebra $H \in \Theta$. It is proved in [7] that

1. Formulas u and v are semantically equivalent if and only if $u^{\pi x} = \bar{u} = \bar{v} = v^{\pi x}$.

2. The variety Hal_Θ is generated by all algebras $Hal_\Theta(H)$, where $H \in \Theta$.

The second proposition motivates the definition of the variety Hal_Θ as a variety, determined by common identities of all $H \in \Theta$.

Let us make a remark on the kernel of the homomorphism Val_H. We have $Ker(Val_H) = Th(H) = (Th^X(H), X \in \Gamma)$. Here $Th(H) = (Th^X(H), X \in \Gamma)$ is the elementary theory of the algebra H, i.e., the set of formulas $u \in Th^X(H)$ such that $Val^X_H(u) = Hom(W(X), H)$. This is the necessary information from algebraic logic.

3   Categories and Bases of Elementary Algebraic Knowledge

This section complements the paper [8]. We will repeat some material from [8] with some changes in approach and notation. The main distinction is that the subject of knowledge in [8] was a model, consisting of an algebra $H \in \Theta$ and some set of relations implemented in H, while here we deal only with the algebra H without additional relations. Besides, in [8] the emphasis was made on finite H while here H is an arbitrary algebra.

3.1 The Category $Knl_\Theta(H)$

Denote the category of knowledge about H by $Knl_\Theta(H)$. The form of its objects is (X, T, A). Here $X \in \Gamma$, $T \subset \Phi(X)$ and $A = T_H^L$ is an elementary set in $Hom(W(X), H)$. We consider T as a description of knowledge and A as its content.

Let us define morphisms $(X, T_1, A) \to (Y, T_2, B)$ in $Knl_\Theta(H)$. Proceeding from some $s : W(X) \to W(Y)$, we have $s_* : \Phi(X) \to \Phi(Y)$. Restrict ourselves with s satisfying $s_* u \in T_2$ if $u \in T_1$. Show that a morphism $[s] : (Y, B) \to (X, A)$ corresponds to every s of such kind. For the given s we have $\tilde{s} : Hom(W(Y), H) \to Hom(W(X), H)$. We need to check that if $\nu \in B$, then $\tilde{s}(\nu) \in A$.

Let $\nu \in B = T_{2H}^L$ be given. For every $v \in T_2$ we have $\nu \in Val_H^X(v)$. Condition $\tilde{s}(\nu) \in A$ means that $\tilde{s}(\nu) \in Val_H^X(u)$ for every $u \in T_1$. Using the fact that Val_H is a homomorphism and, hence, it commutes with s_*, we observe that the latter condition is equal to

$$\nu \in s_* Val_H^X(u) = Val_H^X(s_* u).$$

Since $s_* u = v \in T_2$, the inclusion above holds.

We have proved that if $\nu \in B$, then $\tilde{s}(\nu) \in A$. This gives us the required $[s] : (Y, B) \to (X, A)$. Denote the corresponding morphism in the category of knowledge by

$$(s, [s]) : (X, T_1, A) \to (Y, T_2, B).$$

The product of morphisms is defined componentwise, i.e., $[s_1 s_2] = [s_2][s_1]$.

Now we want to relate the category of knowledge about the algebra H to two other categories. The second one is the category of elementary sets $LK_\Theta(H)$. Having in mind that we consider every set of formulas T as a logical description of knowledge, let us define the first category called the category LD_Θ of sets of formulas. The objects of the category LD_Θ are of the form (X, T), where $X \in \Gamma$, and T is a set of formulas in $\Phi(X)$. Morphisms $s_* : (X, T_1) \to (Y, T_2)$ are defined like in the category $Knl_\Theta(H)$. These categories are connected by the functor

$$Ct_H : LD_\Theta \to LK_\Theta(H).$$

This functor transforms a description of knowledge into its content by the rule: $Ct_H(T) = T_H^L = A$. Correspondingly, $Ct_H(s_*) = [s] : (Y, B) \to (X, A)$ is assigned to a morphism $s_* : (X, T_1) \to (Y, T_2)$. The functor Ct_H is contravariant. The category of knowledge $Knl_\Theta(H)$ and the pair of categories LD_Θ and $LK_\Theta(H)$ equipped with the functor Ct_H are uniquely related. Both can be treated as a knowledge base about the algebra H.

3.2 Isomorphism of Categories of Knowledge

Let H_1, H_2 be two algebras in Θ. Isomorphism of the categories $Knl_\Theta(H_1)$ and $Knl_\Theta(H_2)$ can be represented by a commutative diagram

$$LD_\Theta \xrightarrow{\ \tau\ } LD_\Theta$$

$$Ct_{H_1} \Big\downarrow \qquad\qquad \Big\downarrow Ct_{H_2}$$

$$LK_\Theta(H_1) \xrightarrow{\ F\ } LK_\Theta(H_2)$$

Here τ is an automorphism of the category LD_Θ while F is the isomorphism of the categories of elementary sets, induced by τ. It is an equivalence of the corresponding databases as well. The explanations will follow in the next subsection.

3.3   Inner Automorphism of the Category LD_Θ

We give here some background and explanations from category theory [5]. Let $\varphi_1, \varphi_2 : C_1 \to C_2$ be two functors of categories. Consider an isomorphism $s :$ $\varphi_1 \to \varphi_2$ of functors. It means that for every object A of the category C_1 there is an isomorphism of objects $s_A : \varphi_1(A) \to \varphi_2(A)$ in the category C_2. Besides, for every morphism $\nu : A \to B$ in C_1 there is a commutative diagram

$$\varphi_1(A) \xrightarrow{\ s_A\ } \varphi_2(A)$$

$$\varphi_1(\nu) \Big\downarrow \qquad\qquad \Big\downarrow \varphi_2(\nu)$$

$$\varphi_1(B) \xrightarrow{\ s_B\ } \varphi_2(B)$$

If φ_1 and φ_2 are isomorphic, then we write $\varphi_1 \approx \varphi_2$.

Recall that categories C_1 and C_2 are called equivalent, if there exists a pair of functors $\varphi : C_1 \to C_2$ and $\psi : C_2 \to C_1$ such that $\psi\varphi \approx 1_{C_1}$ and $\varphi\psi \approx 1_{C_2}$. Here 1_C is the identity functor of the category C. So, any equivalence of categories is determined by a pair of functors (φ, ψ). If $\psi\varphi = 1_{C_1}$ and $\varphi\psi = 1_{C_2}$, then $\psi = \varphi^{-1}$ and the equivalence of categories is converted into their isomorphism, given by the invertible functor φ.

If, further, $C_1 = C_2 = C$, then φ is an automorphism of the category C. Suppose that φ satisfies $\varphi \approx 1_C$. Let an isomorphism $s : 1_C \to \varphi$ be given. Then for the object A we have an isomorphism $s_A : A \to \varphi(A)$, and for the morphism ν we have

$$A \xrightarrow{\ s_A\ } \varphi(A)$$

$$\nu \Big\downarrow \qquad\qquad \Big\downarrow \varphi(\nu)$$

$$B \xrightarrow{\ s_B\ } \varphi(B)$$

This diagram gives $\varphi(\nu) = s_B \nu s_A^{-1} : \varphi(A) \to \varphi(B)$. Such an automorphism φ is called inner. See details in [6].

Let us return to the mentioned diagram

$$LD_\Theta \xrightarrow{\ \tau\ } LD_\Theta$$

$$Ct_{H_1} \Big\downarrow \qquad\qquad \Big\downarrow Ct_{H_2}$$

$$LK_\Theta(H_1) \xrightarrow{\ F\ } LK_\Theta(H_2)$$

There are two main problems related to this diagram. The first one is to figure out the structure of automorphisms τ of the category LD_Θ subject to the variety Θ. The second one is to determine relations between the algebras H_1 and H_2 which yield the existence of an isomorphism F, induced by some τ.

Consider the situation when τ is an inner automorphism. Note first of all that every automorphism φ of the category Θ^0 induces an automorphism φ^* of the category Hal_Θ^0.

Indeed, let $W = W(X)$ be an object in Θ^0 and $\varphi(W) = W(Y)$. Then we set $\varphi^*(\Phi(X)) = \Phi(Y)$. Every morphism of the category Hal_Θ^0 has the form $s_* : \Phi(X) \to \Phi(Y)$, where $s : W(X) \to W(Y)$ is a morphism in Θ^0. Here we set $\varphi^*(s_*) = \varphi(s)_* : \varphi^*(\Phi(X)) \to \varphi^*(\Phi(Y))$. Now φ^* is an automorphism of the category Hal_Θ^0.

Let now τ be an inner automorphism of the category Hal_Θ^0 and $\sigma : 1_{Hal_\Theta^0} \to \tau$ be the corresponding isomorphism of functors. For every $\Phi = \Phi(X)$ we have an isomorphism $\sigma_\Phi : \Phi \to \tau(\Phi)$. Here, if $\tau(\Phi) = \Phi(Y)$, then $\sigma_\Phi = s_*$, where $s : W(X) \to W(Y)$ is an isomorphism in Θ^0.

Let us make the next step. Proceed from $\sigma_\Phi : \Phi(X) \to \tau(\Phi(Y))$. It is an isomorphism in the category Hal_Θ^0 and, consequently, an isomorphism of Boolean algebras, in particular, it is a bijection. Let, further, $T \subset \Phi(X)$. Take $T^* = \{\sigma_\Phi(u) | u \in T\} \subset \Phi(Y)$. So, in the category LD_Θ we have the objects (X, T) and (Y, T^*).

Let us construct an automorphism $\bar{\tau}$ of the category LD_Θ by the automorphism τ of the category Hal_Θ^0. For every object (X, T) of the category LD_Θ take $\bar{\tau}(X, T) = (Y, T^*)$. This determines the bijection on the objects of the category LD_Θ. Further we need to define $\bar{\sigma} : 1_{LD_\Theta} \to \bar{\tau}$ by $\sigma : 1_{Hal_\Theta^0} \to \tau$. For every (X, T) we should define an isomorphism $\bar{\sigma}_{(X,T)} : (X, T) \to \bar{\tau}(X, T) = (Y, T^*)$ in LD_Θ. Let us return to $\sigma_\Phi : \Phi(X) \to \Phi(Y)$. This σ_Φ induces a bijection $T \to T^*$. This means that σ_Φ is an isomorphism $(X, T) \to (Y, T^*)$ as well. Denote it by $\bar{\sigma}_{(X,T)}$. Let now $\eta : (X_1, T_1) \to (X_2, T_2)$ be a morphism in LD_Θ. We set $\bar{\tau}(\eta) = \bar{\sigma}_{(X_2,T_2)} \eta \bar{\sigma}_{(X_1,T_1)}^{-1} : \bar{\tau}(X_1, T_1) \to \bar{\tau}(X_2, T_2)$. This determines the automorphism $\bar{\tau}$ with the isomorphism of functors $\bar{\sigma} : 1_{LD_\Theta} \to \bar{\tau}$. $\bar{\tau}$ is also inner. We say that an automorphism τ of the category LD_Θ is the restriction of the automorphism τ of the category Hal_Θ^0. It is easy to show that if $\varphi : \Theta^0 \to \Theta^0$ is an inner automorphism of the category Θ^0, then the corresponding $\varphi^* : Hal_\Theta^0 \to Hal_\Theta^0$ is also an inner automorphism. Hence, we could proceed from $\tau = \varphi^*$. Then the inner automorphism $\bar{\tau}$ of the category LD_Θ is induced by an inner $\varphi : \Theta^0 \to \Theta^0$. It remains the general problem of investigation of automorphisms of the categories Hal_Θ^0 and LD_Θ. The other general problem was already mentioned: for which H_1 and H_2 we have an isomorphism $F : LK_\Theta(H_1) \to LK_\Theta(H_2)$, induced by some τ. For τ we can take inner automorphisms and even the identical τ. Some related remarks will be given in the next section, where we return to logical geometry.

4 Logically-Geometrical Equivalence of Algebras

4.1 The Main Galois Correspondence in Algebraic Geometry and Logical Geometry

Let us start with algebraic geometry and suppose that Θ, $W = W(X)$ in Θ^0, $H \in \Theta$, and $Hom(W, H)$ are given. The Galois correspondence between sets A in $Hom(W, H)$ and binary relations, i.e., a system of equations T in W, is defined by:

$$A = T'_H = \{\mu : W \to H | T \subset Ker(\mu)\} = \bigcap_{(w,w') \in T} Val^X_H(w \equiv w').$$

$$T = A'_H = \bigcap_{\mu \in A} Ker(\mu) = \{(w, w') | A \subset Val^X_H(w \equiv w')\}.$$

In logical geometry we replace $Ker(\mu)$ by the logical kernel $LKer(\mu)$ and instead of systems of equations we consider arbitrary subsets in the algebra of formulas $\Phi = \Phi(X)$. Earlier we had:

$$A = T^L_H = \{\mu : W \to H | T \subset LKer(\mu)\} = \bigcap_{u \in T} Val^X_H(u).$$

In the opposite direction:

$$T = A^L_H = \bigcap_{\mu \in A} LKer(\mu) = \{u | A \subset Val^X_H(u), u \in \Phi(X)\}.$$

Here T is a filter in Φ, called an H-closed filter. We can speak of closures A''_H, $A^{LL}_H \subset A''_H$, T''_H, and T^{LL}_H.

Note that in the case of algebraic geometry we used representation of a system of equations as a system of equalities in $\Phi(X)$.

The formulas above actually determine the Galois correspondence. For example, for the transitions $A \to A^L_H$ and $T \to T^L_H$ it means that $A_1 \subset A_2$ implies $A^L_{2H} \subset A^L_{1H}$ and $T_1 \subset T_2$ implies $T^L_{2H} \subset T^L_{1H}$. Besides, $A \subset A^{LL}_H$, $T \subset T^{LL}_H$.

4.2 Infinitary Logic

For the fixed $\Phi = \Phi(X)$ and for $T \subset \Phi$ consider formulas of the form $(\bigwedge_{u \in T} u) \to v$, written also as $T \to v$, $v \in \Phi$. If the set T is infinite, then it is a formula of infinitary logic. It follows from the definitions that the inclusion $v \in T^{LL}_H$ takes place if and only if the formula $T \to v$ holds in H. In the case of algebraic geometry the latter formula can be rewritten as a quasi-identity (possibly infinitary):

$$\bigwedge_{(w,w') \in T} w \equiv w' \to w_0 \equiv w'_0.$$

If this quasi-identity holds in H, then $w_0 \equiv w'_0 \in T''_H$.

4.3   Coordinate Algebras

Let $A \subset Hom(W, H)$ be an algebraic set, and $T = A'_H$ be an H-closed congruence in W. Then W/T is called the coordinate algebra for H. It is an algebra in the variety Θ. Denote by $C_\Theta(H)$ the category of such coordinate algebras. We have a duality $K_\Theta(H) \to C_\Theta(H)$.

Let, further, $A \subset Hom(W, H)$ be an elementary set and $T = A_H^L$. We have a Boolean algebra $\Phi(X)/T$, which we call the coordinate algebra for A. Denote the category of such coordinate Boolean algebras by $LC_\Theta(H)$. The transition $(X, A) \to \Phi(X)/A_H^L$ determines a contravariant functor $LK_\Theta(H) \to LC_\Theta(H)$. It seems that in the general case this functor is not a duality. Duality takes place if H is an algebra of constants of the corresponding logic. See details in [7].

Consider a particular case $A = Hom(W(X), H)$. Then $A_H^L = Th^X(H)$, and the coordinate algebra for A is the Boolean algebra $\Phi(X)/Th^X(H)$.

4.4   Lattices of Elementary Sets

It can be proved that all elementary sets in $Hom(W, H)$ constitute a lattice which is a sublattice in the lattice of all subsets in the given affine space. This lattice is antiisomorphic to the lattice of all H-closed filters of the corresponding algebra $\Phi = \Phi(X)$. The same is not true for algebraic sets, i.e., they do not constitute a sublattice in the lattice of all subsets of the space $Hom(W, H)$. Indeed, if A and B are algebraic sets, their union $A \cup B$ is not necessarily an algebraic set, but it is an elementary set.

We will be interested in isomorphisms $F : LK_\Theta(H_1) \to LK_\Theta(H_2)$. We call such an isomorphism correct if it is naturally correlated with the lattice operations (see [6]). We regard algebras H_1 and H_2 to have the same logic, if the categories $LK_\Theta(H_1)$ and $LK_\Theta(H_2)$ are correctly isomorphic. We can also proceed from correct equivalence of these categories.

The following subsection is the main one in this section.

4.5   *LG*-Equivalence of Algebras

Definition 1. *Algebras H_1 and H_2 in Θ are called LG-equivalent, if for every $X \in \Gamma$ and every T in $\Phi(X)$ we have $T_{H_1}^{LL} = T_{H_2}^{LL}$.*

Recall (see Subsect. 4.1) that H_1 and H_2 are AG-equivalent, if for every X and every T in $W = W(X)$ we have $T_{H_1}'' = T_{H_2}''$.

It follows from the definitions, that

1. If H_1 and H_2 are LG-equivalent, then they are AG-equivalent.
2. If H_1 and H_2 are LG-equivalent, then they are elementary equivalent in the sense of Tarski, i.e., the elementary theories $Th(H_1)$ and $Th(H_2)$ coincide.

The opposite statement is not true. Therefore LG-equivalence is more strict than elementary equivalence.

Now let us prove that

Proposition 1. *If H_1 and H_2 are LG-equivalent, then the categories $LK_\Theta(H_1)$ and $LK_\Theta(H_2)$ are correctly isomorphic.*

Remark 1. In the case of algebraic geometry (AG case) such a proposition trivially follows from the duality of $LK_\Theta(H)$ and $LC_\Theta(H)$. The LG situation is not so trivial.

Proof. Let (X, A) be an object in $LK_\Theta(H_1)$ and H_1 and H_2 be LG-equivalent. We set: $F(X, A) = (X, B)$, where $B = (A_{H_1}^L)_{H_2}^L$. Here F determines a bijection on the objects of the category.

Take a morphism $[s] = [s]_{H_1} : (X, A_1) \to (X, A_2)$ in $LK_\Theta(H_1)$. We have $s : W(Y) \to W(X)$ and $\tilde{s} : Hom(W(X), H_1) \to Hom(W(Y), H_1)$. If $\nu \in A_1$, then $\tilde{s}(\nu) \in A_2$. We say that s is admissible for A_1 and A_2. The straightforward check shows that the same s is admissible for B_1 and B_2. The same is true in the opposite direction. This, in particular, gives a morphism $[s]_{H_2} : (X, B_1) \to (Y, B_2)$ in $LK_\Theta(H_2)$. We need to prove that there is a one-to-one correspondence between $[s]_{H_1}$ and $[s]_{H_2}$. The last means, that if $[s]_{H_1} = [s_1]_{H_1} = [s_2]_{H_1}$, then $[s]_{H_2} = [s_1]_{H_2} = [s_2]_{H_2}$, and vice versa.

Let $[s_1]_{H_1} = [s_2]_{H_1}$. This equality means that for every $\nu \in A_1 = A$ we have $\tilde{s}_1(\nu) = \tilde{s}_2(\nu)$; $\nu s_1 = \nu s_2$. Then for every $w \in W(Y)$ we have $\nu s_1 w = \nu s_2 w$. This gives $\nu \in Val_{H_1}^X(s_1 w \equiv s_2 w)$, and $A \in Val_{H_1}^X(s_1 w \equiv s_2 w)$. Therefore, $(s_1 w \equiv s_2 w) \in A_{H_1}^L = T = B_{H_2}^L$, where $B = B_1$. Now the inclusion $(s_1 w \equiv s_2 w) \in B_{H_2}^L$ gives $B \subset Val_{H_2}^X(s_1 w \equiv s_2 w)$. For every $\mu \in B$ we have $\mu s_1 w \equiv \mu s_2 w$. It is true for every $w \in W(Y)$, and $\mu s_1 = \mu s_2$, $\tilde{s}_1(\mu) = \tilde{s}_2(\mu)$ for every $\mu \in B$. Thus, $[s_1]_{H_2} = [s_2]_{H_2}$. The opposite direction is checked similarly.

We set $F([s_1]_{H_1} = [s]_{H_2}$, which gives an isomorphism $F : LK_\Theta(H_1) \to LK_\Theta(H_2)$. This isomorphism is correct.

Let us make some general remarks on relations between morphisms and Galois transitions used in the first part of the proof. Take $s : W(X) \to W(Y)$ and, correspondingly, $s_* : \Phi(X) \to \Phi(Y)$. For $T \subset \Phi(Y)$ we set $u \in s_* T$ for $s_* u \in T$. For $T \subset \Phi(X)$ we have $s^* T = \{s_* u | u \in T\}$. Further, $\tilde{s} : Hom(W(Y), H) \to Hom(W(X), H)$. Take $B = s_* A = \tilde{s}^{-1} A$ for $A \subset Hom(W(X), H)$. For $B \subset Hom(W(Y), H)$ we have $s^* B = \{\tilde{s}(\mu) | \mu \in B\}$.

We have the properties:

1. If $T \subset \Phi(X)$, then $(s^* T)_H^L = s_* T_H^L$.
2. If $B \subset Hom(W(Y), H)$, then $(s^* B)_H^L = s_* B_H^L$.
3. If $A \subset Hom(W(X), H)$, then $s^* A_H^L \subset (s_* A)_H^L$.

We view these properties as rules of behavior of elementary sets under the moves of affine spaces. The first rule implies that if A is an elementary set, then so is $s_* A$. Note also that if $u \in \Phi(X)$ and $u \in Th^X(H)$, then $s_* u \in (Th^Y)(H)$. Indeed, $Val_H^Y(s_* u) = s_* Val_H^X(u) = s_* 1 = 1 = Hom(W(Y), H)$ follows from $Val_H^X(u) = 1 = Hom(W(X), H)$.

Let us return to the diagram

$$LD_\Theta \xrightarrow{\ \tau\ } LD_\Theta$$
$$\scriptstyle Ct_{H_1}\Big\downarrow \qquad\qquad \Big\downarrow\scriptstyle Ct_{H_2}$$
$$LK_\Theta(H_1) \xrightarrow{\ F\ } LK_\Theta(H_2)$$

It is important to notice that the algebra H does not participate in the category LD_Θ and the upper arrow is not concerned with H. Automorphisms τ are also free from H. On the other hand, it is natural to ask for which H_1 and H_2 with the isomorphism $F : LK_\Theta(H_1) \to LK_\Theta(H_2)$ there exists a suitable τ.

Consider separately the case of $\tau = 1$ with the diagram

$$LK_\Theta(H_1) \xrightarrow{\ F\ } LK_\Theta(H_2)$$
$$\scriptstyle Ct_{H_1}\searrow \qquad \swarrow\scriptstyle Ct_{H_2}$$
$$LD_\Theta$$

Take an object (X, u), i.e., T consists of one element u. Then $Val^X_{H_2}(u) = F(Val^X_{H_1}(u))$. Here, if $u \in Th^X(H_1)$, but does not belong to $Th^X(H_2)$, then $F(Hom(W(X), H))$ is not in $Hom(W(X), H_2)$. This means that the corresponding F is not correct. It is quite possible that some other τ provides correctness of F. Thus, in order to get a correct F, we should assume that the algebras H_1 and H_2 are elementary equivalent. Unlike LG-equivalence that implies the needed isomorphism of categories, the elementary equivalence does not provide this. We mention here the following

Proposition 2. *Algebras H_1 and H_2 are elementary equivalent if and only if for every $X \in \Gamma$ and finite $T \subset \Phi(X)$ we have $T^{LL}_{H_1} = T^{LL}_{H_2}$. This imparts some "geometrical" meaning to the idea of elementary equivalence.*

4.6   Noetherianity

Here we consider noetherian conditions providing coincidence of LG-equivalence and elementary equivalence. Let us give some definitions.

1. An algebra $H \in \Theta$ is LG-noetherian, if every elementary set over H can be determined by one formula.
2. An algebra H is strongly LG-noetherian, if in every T there exists a finite part T_0 such that $T^L_H = T^L_{0H}$.
3. An algebra H is weakly LG-noetherian, if for every formula $T \to v$, holding in H, there exists a finite part T_0 in T such that $T_0 \to v$ holds in H.

The following proposition takes place:

Proposition 3. *If H_1 and H_2 satisfy one of the noetherianity conditions, then they are LG-equivalent if and only if they are elementary equivalent.*

Concluding this section on elementary sets let us make one more remark. For the given algebra H take its group of automorphisms $Aut(H)$. This group acts naturally in each $Hom(W, H)$.

Theorem 1. *Every elementary set over H is invariant under the action of the group Aut(H).*

5 Problems

In this section we formulate several natural problems joined by the idea of *LG*-equivalence.

Problem 1. Whether there exists an infinite *LG*-noetherian algebra H?

Problem 2. Whether there exists an infinite weakly *LG*-noetherian algebra H?

Problem 3. Whether there exists an infinite weakly *LG*-noetherian Abelian group H?

Problem 4. Construct examples of non-isomorphic *LG*-equivalent algebras.

Remark 2. Constructions of model theory imply that examples of such kind exist. This problem mainly asks what is the situation for the specific varieties Θ and $H \in \Theta$.

Problem 5. Let L_1 and L_2 be two extensions of a field P. Is it true that *LG*-equivalence implies isomorphism?

Remark 3. In the general case this problem has a negative solution (see [1] and references therein). The question makes sense if we impose additional conditions on the extensions L_1 and L_2.

Problem 6. Are there elementary equivalent but not *LG*-equivalent extensions L_1 and L_2?

Problem 7. Consider the *LG*-theory of Abelian groups.

6 Concluding Remarks

In conclusion, we remind the reader that in AG and LG along with the categories $K_\Theta(H)$ and $LK_\Theta(H)$ we consider the categories of coordinate algebras $C_\Theta(H)$ and $LC_\Theta(H)$. Also, along with the category $LC_\Theta(H)$ we consider a close (similar) category LD_Θ which is not related to the algebra H but induced by the transition to the category of knowledge $Knl_\Theta(H)$. This gives us a new insight on the general problem of similarity of logics for different H_1 and H_2. We also speak of the sameness of knowledge about H_1 and H_2.

Note that *LG*-equivalence often implies isomorphism. For example, it is true if H_1 and H_2 are free groups F_n and F_m (Zlil Sela). But we do not know what is the situation if H_1 is a free group $F(X)$ and H_2 is an arbitrary group, *LG*-equivalent to H_1. Whether it is true that H_1 and H_2 are isomorphic? If yes, this will mean that the free group $F(X)$ is *LG*-separated from all other groups.

Acknowledgements

We are grateful to Professor B. Zilber for the very useful discussions and for letting us know the paper [1] which is strongly related to the problems formulated above.

References

1. Grossberg, R.: Classification theory for abstract elementary classes. Logic and Algebra. In: Zhang, Y. (ed.) Contemporary Mathematics vol. 302, pp. 165–204 AMS (2002)
2. Halmos, P.R.: Algebraic logic. New York (1969)
3. Henkin, L., Monk, J.D., Tarski, A.: Cylindric Algebras. North-Holland Publ. Co., Amsterdam (1985)
4. Higgins, P.J.: Algebras with a scheme of operators. Math. Nachr. 27, 115–132 (1963)
5. Mac Lane, S.: Categories for the Working Mathematician. Springer, Heidelberg (1971)
6. Plotkin, B.: Varieties of algebras and algebraic varieties. Israel Math. Journal 96(2), 511–522 (1996)
7. Plotkin, B.: Algebraic geometry in First Order Logic. Sovremennaja Matematika and Applications 22, 16–62 (2004) Journal of Math. Sciences 137(5), 5049–5097 (2006), http://arxiv.org/abs/math.GM/0312485
8. Plotkin, B., Plotkin, T.: An Algebraic Approach to Knowledge Base Models Informational Equivalence. Acta Applicandae Mathematicae 89(1–3), 109–134 (2005), http://arxiv.org/abs/math.GM/0312428

Selection and Uniformization Problems in the Monadic Theory of Ordinals: A Survey

Alexander Rabinovich and Amit Shomrat

Sackler Faculty of Exact Sciences, Tel Aviv University, Israel 69978
{rabinoa,shomrata}@post.tau.ac.il

Dedicated with deepest appreciation and respect to Boris Abramovich Trakhtenbrot whose inspiration as a teacher, a researcher and a role model has been guiding us and many others for many years.

Abstract. A formula $\psi(Y)$ is a *selector* for a formula $\varphi(Y)$ in a structure $\mathcal{M}$ if there exists a unique Y that satisfies ψ in $\mathcal{M}$ and this Y also satisfies φ. A formula $\psi(X, Y)$ *uniformizes* a formula $\varphi(X, Y)$ in a structure $\mathcal{M}$ if for every X there exists a unique Y such that $\psi(X, Y)$ holds in $\mathcal{M}$ and for this Y, $\varphi(X, Y)$ also holds in $\mathcal{M}$. In this paper we survey some fundamental algorithmic questions and recent results regarding selection and uniformization, when the formulas ψ and φ are formulas of the monadic logic of order and the structure $\mathcal{M} = (\alpha, <)$ is an ordinal α equipped with its natural order. A natural generalization of the Church problem to ordinals is obtained when some additional requirements are imposed on the uniformizing formula $\psi(X, Y)$. We present what is known regarding this generalization of Church's problem.

1 Introduction

The aim of this paper is to survey recent results on the selection and uniformization problems for monadic (second-order) logic of order. These results are well-known for the standard discrete time model of natural numbers. When selection and uniformization problems are considered over countable ordinals new and interesting phenomena appear. Our exposition focusses on methodological issues rather than providing the technical details. No proofs are offered, though we sometimes indicate the main ideas of the proofs.

1.1 Selection

Definition 1 (Selection). *Let $\varphi(Y)$, $\psi(Y)$ be formulas and $\mathcal{M}$ a structure. We say that ψ selects (or, is a selector for) φ in $\mathcal{M}$ iff:*

1. *either both formulas are not satisfied in $\mathcal{M}$, or*
2. *ψ defines in $\mathcal{M}$ a unique P and this P satisfies φ in $\mathcal{M}$.*

We say that ψ selects φ over a class $\mathcal{C}$ of structures iff ψ selects φ in every $\mathcal{M} \in \mathcal{C}$.

A. Avron et al. (Eds.): Trakhtenbrot/Festschrift, LNCS 4800, pp. 571–588, 2008.

Generally, once some logic $\mathcal{L}$ and a class $\mathcal{C}$ of structures for $\mathcal{L}$ have been fixed, three basic questions may be raised concerning selection for $\mathcal{L}$-formulas over $\mathcal{C}$. First,

(1) Selection property: Does every $\mathcal{L}$-formula have a selector over $\mathcal{C}$?

When this is the case we shall say that $\mathcal{C}$ has the *selection property* (with respect to $\mathcal{L}$-formulas). When $\mathcal{C}$ lacks the selection property, the following algorithmic question naturally arises:

(2) Deciding selectability: Can we decide, given an $\mathcal{L}$-formula φ, whether it has a selector over $\mathcal{C}$?

Finally, whether or not $\mathcal{C}$ has the selection property, it seems interesting to ask:

(3) Synthesis of a selector: If φ has selectors over $\mathcal{C}$, can one be computed for it?

When both Questions (2) and (3) are answered affirmatively, we say that the *selection problem* over $\mathcal{C}$ is solvable.

We consider the above questions when the logic $\mathcal{L}$ is either the *second-order monadic logic of order* (MLO) or its first-order fragment and $\mathcal{C} = \{(\alpha, <)\}$ for some ordinal α or $\mathcal{C}$ is a class of countable ordinals.

MLO extends first-order logic by allowing quantification over *subsets* of the domain. The binary relation symbol '$<$' is its only non-logical constant. Since our structures are ordinals, we shall assume that '$<$' is interpreted as a well-order of the domain. In short, MLO uses first order variables s, t, ... interpreted as elements and monadic second-order variables X, Y, ... interpreted as subsets of domain. The atomic formulas are $s < t$ and $t \in X$; all other formulas are built from the atomic ones by applying Boolean connectives and quantifiers $\forall, \exists$ for both kinds of variables. An MLO formula is *first-order* if it does not use quantification over set variables.

An MLO formula $\varphi(Y)$ defines in an ordinal α the family of sets which satisfy $\varphi(Y)$ in α. If this family is non-empty, then a selector $\psi(Y)$ for φ defines one set from this family.

MLO plays a very important role in mathematical logic and computer science. The fundamental connection between this logic and automata was discovered independently by Büchi, Elgot and Trakhtenbrot [1, 6, 20–22] and the logic was proved to be decidable over the class of finite chains. Büchi [2] proved the decidability of MLO in $(\omega, <)$ and later [4] that the monadic theory of every ordinal $\leq \omega_1$ is decidable. Shelah [18] showed that the MLO-theory of any ordinal $\alpha < \omega_2$ is decidable. Rabin proved that the MLO theory of the full binary tree $T_2 := (D, <, \text{Left}, \text{Right})$ is decidable [13, 14]. Here D is the set all finite strings over $\{0, 1\}$; the relation symbol '$<$' is interpreted as the prefix relation and the unary predicate 'Left' (respectively, 'Right') is interpreted as the set of strings whose last symbol is '0' (respectively, '1').

The Rabin basis theorem states that if $T_2 \models \exists Z \varphi(Z)$ then there is a regular subset $S \subseteq D$ such that $T_2 \models \varphi(S)$. Since a subset of T_2 is regular iff it is definable, the Rabin basis theorem can be restated as following: the full binary tree has the selection property.

1.2   Uniformization

A *uniformizer* for a binary relation R is a function $f \subseteq R$ such that $\mathrm{dom}(f) = \mathrm{dom}(R)$. That every binary relation has a uniformizer is a statement equivalent to the Axiom of Choice. Existence of a uniformizer becomes mathematically interesting when we place certain restrictions on the uniformizing function f. A *uniformization context* is a pair $\langle \mathcal{R}, \mathcal{F} \rangle$, where $\mathcal{R}$ is a class of binary relations and $\mathcal{F}$ a class of functions. We call $\mathcal{R}$ the *challenge* class and $\mathcal{F}$ the *response* class of the context. Given such a pair, one may ask whether a particular (resp. every) $R \in \mathcal{R}$ has a uniformizer $f \in \mathcal{F}$.[1]

This paper focuses on two uniformization contexts – *definable* and *causal* uniformization – in which the challenge class $\mathcal{R}$ is taken to be the class of relations definable in some ordinal α, when α is viewed as a structure for MLO. Our aim is to survey the current state of research into these two contexts, to report recent developments, and to indicate what seem to us the most interesting questions still left open.

1.3   Definable Uniformization

The first uniformization context we explore we call *definable uniformization* (in the literature, this is referred to simply as "uniformization"; see, for instance, [8] and [11]). Here, the response class $\mathcal{F}$ is taken to be the class of functions $\mathcal{P}(\alpha) \to \mathcal{P}(\alpha)$ definable in the ordinal α (or, strictly speaking, in the structure $(\alpha, <)$ where α is equipped with its natural order):

Definition 2 (Definable uniformization). *Let $\varphi(X,Y)$, $\psi(X,Y)$ be formulas and $\mathcal{M}$ a structure. Say that ψ uniformizes (or, is a uniformizer for) φ in $\mathcal{M}$ iff:*

1. $\mathcal{M} \models \forall X \exists^{\leq 1} Y \, \psi(X,Y)$,
2. $\mathcal{M} \models \forall X \forall Y (\psi(X,Y) \to \varphi(X,Y))$, *and*
3. $\mathcal{M} \models \forall X (\exists Y \varphi(X,Y) \to \exists Y \psi(X,Y))$.

Here "$\exists^{\leq 1} Y \ldots$" stands for "there exists at most one…".
We say that ψ uniformizes φ over a class $\mathcal{C}$ of structures iff ψ uniformizes φ in every $\mathcal{M} \in \mathcal{C}$.

Note that $\varphi(Y)$ has a selector if and only if $X = X \wedge \varphi(Y)$ has a definable uniformizer. Thus, selection is a special case of uniformization. Accordingly, Questions (1)–(3) above can be generalized to the latter case.

(1′) Uniformization property: Does every formula have a uniformizer over $\mathcal{C}$?
(2′) Decidability of uniformization: Can we decide, given a formula φ, whether it has a uniformizer over $\mathcal{C}$?
(3′) Synthesis of a uniformizer: If φ has uniformizers over $\mathcal{C}$, can one be computed for it?

---

[1] Some famous examples are found in descriptive set theory, where one proves, for instance, that for every Π^1_1 relation there is a Π^1_1 uniformizer.

Again, when both Questions $(2')$ and $(3')$ are answered affirmatively, we say that the *uniformization problem* over $\mathcal{C}$ is solvable.

Gurevich and Shelah [8] proved that the full binary tree does not have the uniformization property. In [11], Lifsches and Shelah characterize the trees which have the uniformization property.[2] For ordinals they show:

Theorem 3. *An ordinal α has the uniformization property iff $\alpha < \omega^\omega$.*

This answers Question $(1')$ for ordinals. Question $(3')$ was answered in the affirmative when $\mathcal{C} = \{(\alpha, <)\}$ for $\alpha < \omega^\omega$. However, for an ordinal $\alpha \geq \omega^\omega$, Questions $(2')$ and $(3')$ remain open.

In Section 10 we consider a restricted version of uniformization problem which we call bounded uniformization, and show that the bounded uniformization problem is solvable in every ordinal $\leq \omega_1$.

1.4   Church Uniformization

The second uniformization context we look at is that of *causal uniformization*, better known as the *Church uniformization*. While definable uniformization makes sense in any structure, causal uniformization is only relevant in a linear order.

Definition 4 (Causal operator). *Let $(A, <)$ be a linear order and $f : \mathcal{P}(A) \to \mathcal{P}(A)$. We call f causal iff for all $P, P' \subseteq A$ and $\alpha \in A$,*

$$\text{if } P \cap [0, \alpha] = P' \cap (-\infty, \alpha], \text{ then } f(P) \cap (-\infty, \alpha] = f(P') \cap (-\infty, \alpha].$$

That is, if P and P' agree up to and including α, then so do $f(P)$ and $f(P')$.

When discussing causal uniformization, we fix some ordinal α. Again, we take as challenge class $\mathcal{R}$ the class of relations definable in $(\alpha, <)$. Note that in MLO variables range over subsets of the domain. Thus, relations definable in $(\alpha, <)$ are relations on $\mathcal{P}(\alpha)$. It therefore makes sense to ask, whether a definable relation can b e uniformized by a causal function. Accordingly, in causal uniformization the response class $\mathcal{F}$ consists of all causal functions $f : \mathcal{P}(\alpha) \to \mathcal{P}(\alpha)$ (whether definable or not).

We speak of a causal function f uniformizing a formula φ in $(\alpha, <)$, meaning that f uniformizes the *relation* defined by φ in $(\alpha, <)$. In this context, Question $(1')$ above becomes the question whether any formula φ has a causal uniformizer in $(\alpha, <)$. For any $\alpha \geq 2$, the answer is easily seen to be negative. For example, the formula saying "if $X = \varnothing$, then $Y = $ All; otherwise, $Y = \varnothing$" has no causal uniformizer in $(\alpha, <)$ for any $\alpha \geq 2$. Question $(2')$ is already more interesting.

Definition 5 (Church uniformization problem). *Let α be an ordinal. Given a formula $\varphi(X, Y)$, decide whether there is a causal uniformizer for φ in $(\alpha, <)$.*

---

[2] A *tree* for them is a poset $(T, <)$ such that for every $a \in T$, $\{b \in T \mid b \leq a\}$ is a linear order.

Church [5] was the first to formulate this problem for the case $\alpha = \omega$. Some restricted versions of this problem were solved by Church and Trakhtenbrot. Church's Problem for ω was solved by Büchi and Landweber [3] building on McNaughton's game-theoretical interpretation of this problem [12]. Under this game-theoretical interpretation the causal operators correspond to the strategies of the players.

It would seem that Question $(3')$ is irrelevant to causal uniformization. In what sense can we speak of *computing* a general causal uniformizer? Already in ω, there are $2^{\aleph_0}$ such operators. It becomes relevant once more, when we examine the uniformization context where the response class consists of all operators $f : \mathcal{P}(\alpha) \to \mathcal{P}(\alpha)$ which are *both definable and causal*.

(1″) Does every formula which has a causal uniformizer in $(\alpha, <)$ also has a *definable* causal uniformizer?

When this fails, we have an analogue of Question $(2')$.

(2″) Can we decide, given a formula φ, whether it has a definable causal uniformizer in $(\alpha, <)$?

And, of course,

(3″) If φ has definable causal uniformizers in $(\alpha, <)$, can we compute a formula defining one?

Question $(1'')$ will be answered here for all ordinals. Questions $(2'')$ and $(3'')$ are yet unsolved for $\alpha \geq \omega^\omega$.

Note that $\varphi(Y)$ has a selector if and only if $X = X \wedge \varphi(Y)$ has a definable causal uniformizer. Thus, selection is also a special case of definable causal uniformization. Indeed, looking at selection would turn out to be the key for answering Question $(1'')$.

1.5   The Structure of the Paper

In Sect. 2, we fix our notations and terminology. We also recall some fundamental theorems about the monadic theories of countable ordinals. Section 3 surveys the selection property in an ordinals. In Sect. 4 the selection problem in an ordinal are considered. Section 5 investigates the selection property and the selection problem over classes of countable ordinals. The logic considered in Sect. 3–5 is MLO, while in Sect. 6 we consider the first-order fragment of MLO and other logics with expressive power between first-order MLO and MLO. In Sect. 7 we assign to each formula φ a *selection degree* which measures "how difficult it is to select φ". We show that in a countable ordinal all non-selectable formulas share the same degree.

If a structure $\mathcal{M}$ lacks the selection property, it is natural to ask whether there is a finite expansion of $\mathcal{M}$ which has the selection property. This question is investigated for a countable ordinal in Sect. 8. In Sect. 9 the Church uniformization problem for countable ordinals is considered. In Sect. 10 we treat a restricted version of the definable uniformization problem. Finally, Sect. 11 contains some open problems.

As mentioned above, this paper offers no proofs and only occasionally indicates their main ingredients. For results having to do with selection in a particular ordinal (Sects. 3, 4, 6 and 7) proofs can be found in [16]. Sections 5, 8 and 10 are covered in [17]. All results having to do with Church (= causal) uniformization (Sect. 9) are in [15]. It is perhaps worth mentioning that almost all proofs relay on what is known as the "composition method" (originating in [7] and adapted and ingeniously applied to the case of MLO in [18]). In [16] use is also made of Büchi's translation of MLO-formulas into automata over ordinal words (see, for instance, [4]).

From now on, "uniformization" *simpliciter* would mean "definable uniformization". When we intend to refer to causal uniformization, this would be stated explicitly.

Finally, for the sake of notational simplicity, we state our results for formulas $\varphi(X, Y)$ with free variables X and Y. All results generalize in a straightforward manner to formulas $\varphi(\bar{X}, \bar{Y})$ where $\bar{X}$ and $\bar{Y}$ are finite tuples of (distinct) variables.[3]

2 Preliminaries

2.1 Notations

We use n, k, l, m, p for natural numbers, $\alpha, \beta, \gamma, \delta, \zeta$ for ordinals. The set of natural numbers is $\omega := \{0, 1, 2, \ldots\}$. ω_1 is the first uncountable ordinal. We write $\alpha + \beta$, $\alpha\beta$, α^β for the sum, multiplication and exponentiation, respectively, of ordinals α and β.

We use standard notation for sub-intervals of a chain: if $(A, <)$ is a chain and $b < a$ are in A, we write $(b, a) := \{c \in A \mid b < c < a\}$, $[b, a) := (b, a) \cup \{b\}$, etc.

2.2 MLO

The vocabulary of MLO consists of first-order variables $t_0, t_1, t_2, \ldots$ interpreted as elements of the domain and monadic second-order variable $X_0, X_1, \ldots$ interpreted as subsets of the domain. The atomic formulas are $t_i < t_j$ and $t_i \in X_j$; the MLO formulas are built from atomic ones by applying Boolean connectives and quantifiers $\forall, \exists$ for both kinds of variables. An MLO formula is first-order if it does not use quantification over set variables; note however, that such formula may contain free set variables.

The *quantifier depth* of a formula φ is denoted by $\mathrm{qd}(\varphi)$.

We use lower case letters $s, t, \ldots$ to denote the first-order variables and upper case letters $X, Y, \ldots$ to denote second-order set variables.

---

[3] To fit the general notion of uniformization, the relation defined by $\varphi(\bar{X}, \bar{Y})$ must be thought of as consisting of pairs $(\bar{P}, \bar{Q})$ of tuples of subsets of the domain, where $\mathrm{lg}(\bar{P}) = \mathrm{lg}(\bar{X})$ and $\mathrm{lg}(\bar{Q}) = \mathrm{lg}(\bar{Y})$.

A *structure* is a tuple $\mathcal{M} := (A, <, \bar{a}, \bar{P})$ where: A is a non-empty set, $<$ is a binary relation on A, and $\bar{a} := \langle a_0, \ldots, a_{m-1} \rangle$ (respectively, $\bar{P} := \langle P_0, \ldots, P_{l-1} \rangle$) is a *finite* tuple of elements (respectively, sub*sets*) of A.

Suppose φ is a formula with free-variables among $t_0, \ldots, t_{m-1}, X_0, \ldots, X_{l-1}$. We define the relation $\mathcal{M} \models \varphi$ (read: $\mathcal{M}$ *satisfies* φ) as usual.

The *monadic theory* of $\mathcal{M}$, $\mathrm{MTh}(\mathcal{M})$, is the set of all formulas satisfied by $\mathcal{M}$. When $\bar{a}$ and $\bar{P}$ are the empty tuple (as is most often the case for us), $\mathrm{MTh}(\mathcal{M})$ is a set of *sentences*.

2.3   The Monadic Theory of Countable Ordinals

Büchi (for instance [4]) has shown that there is a *finite* amount of data concerning any ordinal $\leq \omega_1$ which determines its monadic theory:

Theorem 6. *Let* $\alpha \in [1, \omega_1]$. *Write* $\alpha = \omega^\omega \beta + \zeta$ *where* $\zeta < \omega^\omega$ *(this can be done in a unique way). Then the monadic theory of* $(\alpha, <)$ *is determined by:*

1. *whether* α *is countable or* $\alpha = \omega_1$,
2. *whether* $\alpha < \omega^\omega$, *and*
3. ζ.

We can associate with every $\alpha \leq \omega_1$ a finite *code* which holds the data required in the previous theorem. This is clear with respect to (1) and (2). As for (3), if $\zeta \neq 0$, write

$$\zeta = \sum_{i \leq n} \omega^{n-i} \cdot a_{n-i}, \text{ where } n, a_i \in \omega \text{ for } i \leq n \text{ and } a_n \neq 0$$

(this, too, can be done in a unique way), and let the sequence $\langle a_n, \ldots, a_0 \rangle$ encode ζ. The following is then implicit in [4]:

Theorem 7. *There is an algorithm that, given a sentence* φ *and the* code *of an* $\alpha \in [1, \omega_1]$, *determines whether* $(\alpha, <) \models \varphi$.

Agreement In this paper, whenever we say that an algorithm is "given an ordinal..." or "returns an ordinal...", we mean the *code* of the ordinal.

3   The Selection Property in an Ordinal

In [11], Lifsches and Shelah characterize the trees which have the uniformization property.[4] For ordinals they show that an ordinal α has the uniformization property iff $\alpha < \omega^\omega$. It follows, in particular, that $\alpha < \omega^\omega$ has the selection property. On the other hand, it does not immediately follow that all ordinals above ω^ω lack the selection property. Indeed, the selection property is known *not* to imply the uniformization property. As mentioned in Sect. 1, Rabin proved that the full binary tree has the selection property[14], while Gurevich and Shelah proved that the full binary tree lacks the uniformization property [8]. But, in fact, for selection, too, we have:

---

[4] A *tree* for them is a poset $(T, <)$ such that for every $a \in T$, $\{b \in T \mid b \leq a\}$ is a linear order.

Proposition 8 (Selection property). *An ordinal α has the selection property iff $\alpha < \omega^\omega$.*

The proof that in any ordinal $\alpha \geq \omega^\omega$ there are non-selectable formulas, reduces to the cases $\alpha = \omega^\omega$ and $\alpha = \omega_1$. The key to handling these, in turn, is the notion of a *periodic* subset.

If $(A, <, P)$ is a structure and $D \subseteq A$, we write $(A, <, P)_{\restriction D}$ for the *restriction* of $(A, <)$ to D, that is, $(A, <, P)_{\restriction D} := (D, <, P \cap D)$.

Definition 9. *Let $\alpha \in \{\omega^\omega, \omega_1\}$ and $P \subseteq \alpha$. We say that P is* periodic *iff there are $\alpha_0, \alpha_1 < \alpha$ and $P_1 \subseteq \alpha_1$ such that $(\alpha, <, P)_{\restriction [\alpha_0, \alpha)}$ is the "concatenation" of α copies of $(\alpha_1, <, P_1)$, i.e., for every $\beta < \alpha$,*

$$(\alpha, <, P)_{\restriction [\alpha_0 + \alpha_1 \beta, \alpha_0 + \alpha_1(\beta+1))} \text{ is isomorphic to } (\alpha_1, <, P_1).$$

The notion of a periodic subset enters our discussion through the following lemma.

Lemma 10. *Let $\alpha \in \{\omega^\omega, \omega_1\}$. Any definable subset of α is periodic.*

Now, no unbounded ω-sequence in ω^ω is periodic (in fact, an unbounded periodic subset of ω^ω has order-type ω^ω). Note that there is a formula $\theta_{\omega\,\mathrm{ub}}(Y)$ that in every countable limit ordinal α defines the set of all unbounded ω-sequences in α. This formula $\theta_{\omega\,\mathrm{ub}}(Y)$ is the conjunction of the following two formulas:

$$\text{"Y is unbounded"}: \forall t_1 \exists t_2 (t_2 > t_1 \wedge t_2 \in Y), \text{ and}$$
$$\text{"no point is a limit point of Y"}:$$
$$\forall t_1 \exists t_2 \Big(t_1 > 0 \rightarrow \big(t_2 < t_1 \wedge \forall t_3 (t_2 < t_3 < t_1 \rightarrow t_3 \notin Y) \big) \Big).$$

Therefore,

Corollary 11. *The formula $\theta_{\omega\,\mathrm{ub}}(Y)$ saying "Y is an unbounded ω-sequence" has no selector in $(\omega^\omega, <)$.*

To handle ω_1 recall the following definitions:

Definition 12 (Clubs and stationary sets)

1. *Let $C \subseteq \omega_1$. C is called:*
   closed iff for every limit $\beta < \omega_1$, if $\sup(C \cap \beta) = \beta$, then $\beta \in C$.[5]
   a club iff C is closed and unbounded in ω_1.
2. *$S \subseteq \omega_1$ is called* stationary *iff for every club $C \subseteq \omega_1$, $S \cap C \neq \emptyset$.*

Note that being a club and being stationary are definable properties of a subset of ω_1 and that ω_1 itself is definable. It is also easy to show that any unbounded periodic subset of ω_1 contains a club. From this and Lemma 10, one derives:

Corollary 13. *Let $\theta_{stat}(Y)$ say: "Both $Y \cap \omega_1$ and $\omega_1 \setminus Y$ are stationary in ω_1". Then θ_{stat} has no selector in $(\alpha, <)$ for every $\alpha \geq \omega_1$.*

---

[5] That is, C is closed under taking sup.

Note that if two ordinal have the same monadic theory, then ψ selects φ in the first ordinal iff ψ selects φ in the second. By Theorem 6, ω^ω and $\omega^\omega\beta$ have the same monadic theory for every countable ordinal $\beta > 0$. Therefore, $\theta_{\omega\,\mathrm{ub}}(Y)$ is not selectable in $\omega^\omega\beta$ for every countable $\beta > 0$.

For a countable ordinal $\alpha > \omega^\omega$, a formula $\psi_\alpha(Y)$ which is unselectable in α can be constructed as follows. Write $\alpha = \omega^\omega\beta + \zeta$ where $\zeta < \omega^\omega$. If $\zeta = 0$, then $\theta_{\omega\,\mathrm{ub}}(Y)$ is not selectable in α. Otherwise note that since $0 < \zeta < \omega^\omega$ there is a formula $\Psi(t)$ such that $\alpha \models \Psi(\mu)$ iff $\mu = \omega^\omega\beta$. Hence, $\omega^\omega\beta$ is definable in α. The formula $\psi_\alpha(Y)$ saying "Y is unbounded ω-sequence in the interval $[0, \omega^\omega\beta)$" is unselectable in α.

Note that in Corollary 13 a formula not *selectable* in every $\alpha \geq \omega_1$ was presented. We sketched a construction of a formula ψ_α not selectable in a countable ordinal $\alpha \geq \omega^\omega$. However, ψ_α depends on (the code of) α. Is there a single formula not selectable in *every* $\alpha \in [\omega^\omega, \omega_1)$? The answer turns out to be negative:

Proposition 14. *For every $n \in \omega$, we can compute $\xi(n) < \omega^\omega$ such that for every formula $\varphi(Y)$ with $\mathrm{qd}(\varphi) \leq n$ and $\zeta \in [\xi(n), \omega^\omega)$, φ is selectable in $\omega^\omega + \zeta$.*

4   The Selection Problem in an Ordinal $\alpha \leq \omega_1$

In [11], issues of decidability and computability are not discussed. However, from the proof of Proposition 6.1 there, one can extract an algorithm as follows (a detailed proof is given in [17]):

Proposition 15 (Uniformization below ω^ω). *There is an algorithm that, given (the code of) an ordinal α and $\varphi(X, Y)$, computes a $\psi(X, Y)$ that uniformizes φ in α.*

In the case of *selection*, we are able to go beyond ω^ω.

Proposition 16 (Solvability of the selection problem). *There exists an algorithm that, given $\alpha \in [\omega^\omega, \omega_1]$ and a formula $\varphi(Y)$, decides whether φ has a selector in $(\alpha, <)$, and if so, constructs one for it.*

Roughly speaking, the proof breaks into three steps. One shows that:

1. If $\alpha \in \{\omega^\omega, \omega_1\}$, then any formula $\varphi(Y)$ satisfied by a periodic predicate in $(\alpha, <)$ is selectable in $(\alpha, <)$.

By Lemma 10, this means that being satisfied by a periodic predicate (or not being satisfied at all) is a necessary and sufficient condition for selectability in these ordinals.

2. It is decidable whether φ is satisfied by a periodic predicate.
3. Selection in any countable ordinal is reducible to the case of ω^ω.

The full uniformization problem turns out to be trickier. There is currently no proof of the solvability (or insolvability) of the uniformization problem in $(\omega^\omega, <)$. A restricted case of this problem is treated in Sect. 10.

5 Selection over Classes of Countable Ordinals

Here we discuss the selection property and problem over classes of countable ordinals.

5.1 The Selection Property for Subclasses of ω^ω

By Proposition 8, any class of ordinals which has an $\alpha \geq \omega^\omega$ as a member does not have the selection property. What can we say about subclasses of ω^ω? It turns out that there is a simple combinatorial criterion for a class $C \subseteq \omega^\omega$ to have the selection property.

Notation (Trace). Let $0 \neq \alpha < \omega^\omega$. Write $\alpha = \omega^{n_r} a_r + \omega^{n_{r-1}} a_{r-1} + \cdots + \omega^{n_0} a_0$ where $r \in \omega$ and $n_r > n_{r-1} > \ldots > n_0$ and a_i (for $i \leq r$) are positive integers (this presentation is unique). Let $\mathrm{trace}(\alpha) := \{n_r, \ldots, n_0\}$.

Proposition 17. A class $C \subseteq \omega^\omega$ has the selection property iff

$$\forall p \in \omega \exists N(p) \in \omega \forall \alpha \in C \left(\alpha > \omega^{p+N(p)} \rightarrow [p, p+N(p)] \cap \mathrm{trace}(\alpha) \neq \varnothing \right).$$

If in addition, $N(p)$ is computable form p, then selectors are computable over C.

Therefore, $\{\omega^k \mid k \in \omega\}$ does not have the selection property. On the other hand, both of the following classes have the selection property and selectors are computable over them (for both, let $N(p) := 0$ for all $p \in \omega$):

1. $\{\omega, \omega^2 + \omega, \omega^3 + \omega^2 + \omega, \ldots\}$.
2. The class of $\alpha < \omega^\omega$ whose trace is a prefix of ω, that is, such that $\mathrm{trace}(\alpha) = \{0, 1, \ldots, n-1\}$ for some $n \in \omega$.

Note that the first of these classes has order-type ω while the second has order-type ω^ω.

5.2 The Selection Problem over Definable Classes of Countable Ordinals

In [17] we proved that the selection problem is solvable over every MLO definable class of countable ordinals. Thus, given a formula $\varphi(Y)$, we may decide, for instance, whether φ has a selector over the class of all countable ordinals, of countable limit ordinals, etc. In fact, something slightly more general holds.

Proposition 18. There is an algorithm that, given formulas $\pi(t)$ and $\varphi(Y)$ and an ordinal $\delta \leq \omega_1$:

1. decides whether φ has a selector over the class definable by π in $(\delta, <)$, namely over $\{(\alpha, <) \mid \alpha \in \delta \setminus 1 \wedge (\delta, <) \models \pi(\alpha)\}$, and
2. if a selector exists constructs it.

This is indeed more general. For example, ω^ω is not a definable ordinal, but $\{\omega^\omega\}$ *is* definable in $(\omega^\omega + \zeta, <)$ for any $\zeta < \omega^\omega$. The proof of the last proposition is based on a reduction of this problem to the bounded uniformization problem discussed in Sect. 10.

Note that Proposition 17 provides sufficient and necessary conditions for the selection property. However, the definability conditions of Proposition 18 are sufficient but are not necessary conditions for solvability of the selection problem over a class of countable ordinals. The class $\{\omega^k \mid k \in \omega\}$ is not definable (in any ordinal), however, the selection problem over this class is solvable.

6 Selection between First-Order and Second-Order Logics

Let φ be an MLO-formula. Recall that φ is a *first-order formula* iff all quantifiers appearing in φ are first-order quantifiers. Note, however, that φ can contain second-order free variables which range over subsets of the domain.[6] Let us call the set of first-order MLO formulas the *first-order fragment* of MLO. Then all results concerning the selection property and selection problem in *countable* ordinals carry through from MLO to its first-order fragment. This follows from:

Proposition 19. *If α is a countable ordinal and $\varphi(Y)$ is an MLO formula selectable in $(\alpha, <)$, then there is a first-order $\chi(Y)$ that selects φ in $(\alpha, <)$. Furthermore, χ is computable from α and φ.*

From the last proposition, we can infer a little more.

Let $\mathcal{L}_2$ and $\mathcal{L}_1$ be logics. We say that a structure $\mathcal{M}$ has the $\mathcal{L}_2 - \mathcal{L}_1$ *selection property* iff for every $\mathcal{L}_2$-formula φ there is an $\mathcal{L}_1$-formula such that ψ selects φ in $\mathcal{M}$. We say that the $\mathcal{L}_2 - \mathcal{L}_1$ *selection problem* for a structure $\mathcal{M}$ is *solvable* iff there is an algorithm which for every $\varphi \in \mathcal{L}_2$ decides whether there is $\psi \in \mathcal{L}_1$ which selects φ in $\mathcal{M}$, and if so, constructs such a ψ. Using this terminology Proposition 19 can be rephrased as "The MLO $-$ FOMLO selection problem is solvable for ordinal $\alpha \leq \omega_1$". More generally,

Corollary 20. *Let $\mathcal{L}_1$ and $\mathcal{L}_2$ be logics such that:*

1. *For every first-order ϕ, there is an $\mathcal{L}_1$-formula Λ equivalent to it.*
2. *For every $\mathcal{L}_2$-formula Λ, there is an MLO formula φ equivalent to it.*

Then a countable ordinal α has the $\mathcal{L}_2 - \mathcal{L}_1$ selection property iff $\alpha < \omega^\omega$.

If furthermore, in (1) Λ is computable from ϕ and in (2) φ is computable from Λ, then the selection problem in $(\alpha, <)$ is solvable for all $\alpha \leq \omega_1$.

---

[6] This is significant. For instance, there is a first-order $\phi(Y)$ such that the only subset of ω satisfying ϕ in $(\omega, <)$ is the set of even numbers. On the other hand, there is no first-order formula $\phi(y)$, with y an *individual* variable, such that for any $n \in \omega$, $(\omega, <) \models \phi(n)$ iff n is even.

A famous example of a logic $\mathcal{L}$ as in the corollary is *weak* MLO (WMLO), where the second-order quantifiers range over *finite* subsets of the domain. Therefore, MLO − WMLO, WMLO − WMLO and WMLO − FOMLO selection problems are solvable for every $\alpha \leq \omega_1$.

When we turn to ω_1, the first-order fragment of MLO no longer behaves like full MLO.

Proposition 21. $(\omega_1, <)$ *has the* FO *order selection property, but not the* MLO *selection property.*

In fact, an interesting dichotomy holds. Let $\phi(Y)$ be first-order.

1. If ϕ is selectable in $(\omega^\omega, <)$, then ϕ is also selectable in $(\omega_1, <)$, and we can compute for it a (first-order) selector that works in both;
2. If ϕ is not selectable in $(\omega^\omega, <)$, then it is not even *satisfied* in $(\omega_1, <)$ (hence, is trivially selectable).

7   Selection Degrees

We know that the formula $\theta_{\omega\,\mathrm{ub}}(Y)$ saying "Y is an unbounded ω-sequence" has no selector in $(\omega^\omega, <)$. Now, let us look at the formula $\theta_{\omega^2\,\mathrm{ub}}$ saying "Y is unbounded and of order type ω^2". It is immediate from Lemma 10, that $\theta_{\omega^2\,\mathrm{ub}}$, too, has no selector in $(\omega^\omega, <)$. But are there any other interesting relations between these two formulas? Can we say, for instance, that $\theta_{\omega^2\,\mathrm{ub}}$ is even "harder" to select than $\theta_{\omega\,\mathrm{ub}}$ (whatever that might mean)? Or, perhaps the other way round?

To turn this admittedly vague question into a mathematical one, we require a notion of comparing formulas and perhaps an equivalence relation on them. But, as our example shows, semantical equivalence seems not to be the right notion. Note, however, the following. For any unbounded ω^2-sequence $S_2 \subseteq \omega^\omega$, the set of limit points of S_2 (i.e., those $\alpha < \omega^\omega$ such that $\sup(S_2 \cap \alpha) = \alpha$) is an unbounded ω-sequence. Also, this set is *definable* from S_2. On the other hand, given an unbounded ω-sequence $S_1 \subseteq \omega^\omega$, the set $\{\alpha + n \mid \alpha \in S_1, n \in \omega\}$ is an unbounded ω^2-sequence. And, again the latter set is definable from S_1. The example suggests the following definition:

Definition 22 (Reduction). *Let* $\varphi_0(Y)$, $\varphi_1(X)$ *be formulas and* $\mathcal{M}$ *a structure. We say that* φ_0 *is easier than* φ_1 *to select in* $\mathcal{M}$ *(in symbols:* $\varphi_0 \preceq_\mathcal{M} \varphi_1$*) iff there exists a formula* $\psi(X, Y)$ *such that:*

1. *if* φ_1 *is not satisfied in* $\mathcal{M}$*, neither is* φ_0*, and*
2. *if* P *satisfies* φ_1 *in* $\mathcal{M}$*, then* $\psi(P, Y)$ *selects* φ_0 *in* $\mathcal{M}$*, i.e., there is a unique* Q *which satisfies* $\psi(P, Y)$ *in* $\mathcal{M}$ *and this* Q *satisfies* φ_0 *in* $\mathcal{M}$*.*

We call ψ *a reduction of* φ_0 *to* φ_1 *over* $\mathcal{C}$*.*

It is clear that $\preceq_\mathcal{M}$ is a partial preorder on the formulas. The corresponding equivalence classes of $\preceq_\mathcal{M}$ are called *selection degrees* in $\mathcal{M}$.

A formula which has a selector in $\mathcal{M}$ is easier to select than any other formula. A non-selectable formula is never easier than a selectable one. Thus, *the* minimal selection degree is the set of selectable formulas. It turns out that in a countable ordinal, all non-selectable formulas also form a single degree:

Proposition 23. *Every $\alpha \in [\omega^\omega, \omega_1)$ has two selection degrees:*

1. *the class all formulas selectable in $(\alpha, <)$, and*
2. *the class of all non-selectable formulas.*

Furthermore, given α and two non-selectable (in α) formulas φ_0 and φ_1, we can compute a reduction ψ of φ_0 to φ_1.

Phrased somewhat differently, this becomes:

Corollary 24. *Let $\alpha \in [\omega^\omega, \omega_1)$ and $P \subseteq \alpha$. Suppose P satisfies some non-selectable formula in $(\alpha, <)$. Then for every formula $\varphi(Y)$, there is a formula $\psi(X, Y)$ such that $\psi(P, Y)$ selects φ in $(\alpha, <)$.*

For the case $\alpha = \omega^\omega$, the proof proceeds by showing that any formula is easier than the formula $\theta_{\omega \,\mathrm{ub}}(Y)$ which says "Y is an unbounded ω-sequence." Then one shows that, conversely, $\theta_{\omega \,\mathrm{ub}}$ is easier than any non-selectable formula in $(\omega^\omega, <)$. Finally, one reduces every other countable $\alpha \geq \omega^\omega$ to the case $\alpha = \omega^\omega$.

Note that if $\varphi_0(Y)$ and $\varphi_1(X)$ are both satisfiable in $\mathcal{M}$, then φ_0 is easier than φ_1 to select in $\mathcal{M}$ if and only if $\varphi_0(Y) \wedge \varphi_1(X)$ has a uniformizer in $\mathcal{M}$ (indeed, a uniformizer for the latter formula and a reduction of φ_0 to φ_1 are one and the same thing). Thus, Proposition 23 actually solves a special case of the uniformization problem in a countable ordinal, namely, where $\varphi(X, Y)$ has the form $\varphi_0(Y) \wedge \varphi_1(X)$.

8 Labeled Ordinals

Corollary 24 leaves open an interesting question. It tells us that if P satisfies some non-selectable formula in $(\omega^\omega, <)$, then with P as a parameter, we can select all formulas in $(\omega^\omega, <)$. But, the formulas we select using P do not themselves "mention" P. In other words, the proposition does not tell us that $(\omega^\omega, <, P)$ has the selection property.

Let P_ω be an unbounded ω-sequence $\{\omega^k \mid k \in \omega\}$. Then for every $\varphi(Y)$ there is a formula $\psi(X, Y)$ such that $\psi(P_\omega, Y)$ selects $\varphi(Y)$ in ω^ω. However, let $\varphi(X, Y)$ says: "If $x < x'$ are successive elements of X, then $Y \cap [x, x')$ is an ω-sequence unbounded in $[x, x')$". Then it is easy to show $\varphi(P_\omega, Y)$ has no selector in $(\omega^\omega, <, P_\omega)$. But, is this fact an artifact of the specific choice of P? That is, could $(\omega^\omega, <)$ be expanded by finitely many subsets of ω^ω to have the selection property?

Proposition 25. *Let $P := \{\omega, \omega^2 + \omega, \omega^3 + \omega^2 + \omega, \ldots\}$. Then:*

(a) $(\omega^\omega, <, P)$ has the selection property,

(b) *for any formula* $\varphi(X, Y)$, *a selector for* $\varphi(P, Y)$ *in* $(\omega^\omega, <, P)$ *is computable,* and
(c) *the monadic theory of* $(\omega^\omega, <, P)$ *is decidable.*

This proposition can be extended to all $\alpha < \omega^{\omega^2}$.

9   The Church Uniformization Problem

McNaughton [12] observed that the Church uniformization problem can be equivalently phrased in game-theoretic language. This phrasing is easily generalizable to all ordinals.

Definition 26. *For an ordinal* α *and a formula* $\varphi(X, Y)$, *the* McNaughton game $\mathcal{G}^\alpha_\varphi$ *is a game of perfect information of length* α *between two players,* X *and* Y.
  At stage $\beta < \alpha$, X *either accepts or rejects* β; *then,* Y *decides whether to accept or to reject* β.
  For a play π, *we denote by* X_π *(resp.* Y_π*) the set of ordinals* $< \alpha$ *accepted by* X *(resp.* Y*) during the play. Then,*

$$Y \text{ wins } \pi \text{ iff } (\alpha, <) \models \varphi(X_\pi, Y_\pi).$$

What we want to know is: Does either one of X and Y have a *winning strategy* in $\mathcal{G}^\alpha_\varphi$? If so, which of them? That is, can X choose his moves so that, whatever way Y responds we have $\neg\varphi(X_\pi, Y_\pi)$? Or can Y respond to X's moves in a way that ensures the opposite?
  Since at stage $\beta < \alpha$, Y has access only to $X_\pi \cap [0, \beta]$, a winning strategy for Y is one and the same thing as a causal uniformizer for φ. Thus, we may rephrase Definition 5 as follows.

Definition 27 (Game version of the Church uniformization problem).
Let α *be an ordinal. Given a formula* $\varphi(X, Y)$, *decide whether* Y *has a winning strategy in* $\mathcal{G}^\alpha_\varphi$.

In their seminal [3], Büchi and Landweber prove the decidability of the Church uniformization problem in $(\omega, <)$. While in defining the problem, we did not require that the winning strategy (= causal uniformizer) be definable, Büchi and Landweber have shown that in the case of $(\omega, <)$ we can indeed restrict ourselves to definable winning strategies (compare Question $(1'')$ in Sect. 1).

Theorem 28 (Büchi and Landweber [3]). *Let* $\varphi(X, Y)$ *be a formula. Then:*

- Determinacy: *One of the players has a winning strategy in the game* $\mathcal{G}^\omega_\varphi$.
- Decidability: *It is decidable which of the players has a winning strategy.*
- Definable strategy: *The player who has a winning strategy, also has a definable winning strategy.*
- Synthesis algorithm: *We can compute a formula* $\psi(X, Y)$ *that defines (in* $(\omega, <)$*) a winning strategy for the winning player in* $\mathcal{G}^\omega_\varphi$.

It seems that Büchi and Landweber believed their theorem would generalize to all countable ordinals. Indeed, after stating the theorem just quoted they write:

> "We hope to present elsewhere an extension of [the theorem] from ω to any countable ordinal."

But, from Proposition 8 it follows that for every $\alpha \geq \omega^\omega$ there are formulas φ such that Y wins $\mathcal{G}_\varphi^\alpha$, but has no *definable* winning strategy. Indeed, fix any $\alpha \geq \omega^\omega$. Pick a formula $\varphi'(Y)$ not selectable in $(\alpha, <)$ and let $\varphi(X, Y)$ denote $X = X \wedge \varphi'(Y)$. If $\psi(X, Y)$ defined a winning strategy for Y in $\mathcal{G}_\varphi^\alpha$, then (say) $\exists X(X = \varnothing \wedge \psi(X, Y))$ would select φ in $(\alpha, <)$, which is impossible. On the other hand, Y does win this game: she simply plays some fixed $P \subseteq \alpha$ which satisfies φ in $(\alpha, <)$ (ignoring X's moves).

The Büchi-Landweber Theorem in its entirety generalizes to ordinals smaller than ω^ω. Its determinacy and decidability clauses generalize to all countable ordinals. Thus,

Theorem 29. *Let α be a countable ordinal, $\varphi(X, Y)$ a formula.*

- Determinacy: *One of the players has a winning strategy in the game $\mathcal{G}_\varphi^\alpha$.*
- Decidability: *It is decidable which of the players has a winning strategy.*
- Definable strategy: *If $\alpha < \omega^\omega$, then the player who has a winning strategy, also has a definable (in $(\alpha, <)$) winning strategy. For every $\alpha \geq \omega^\omega$, there is a formula for which this fails.*
- Synthesis algorithm: *If $\alpha < \omega^\omega$, we can compute a formula $\psi(X, Y)$ that defines a winning strategy for the winning player in $\mathcal{G}_\varphi^\alpha$.*

A proof of this theorem can be found in [15]. It uses the composition method to reduce games of every countable length to games of length ω.

Finally, for uncountable ordinals the situation changes radically. Let $\varphi_{spl}(X, Y)$ say: "X is stationary, $Y \subseteq X$ and both Y and $X \setminus Y$ are stationary" (recall Definition 12). Then it follows immediately from [10] that each of the following statements is consistent with ZFC:

1. None of the players has a winning strategy in $\mathcal{G}_{\varphi_{spl}}^{\omega_1}$.
2. Y has a winning strategy in $\mathcal{G}_{\varphi_{spl}}^{\omega_1}$.
3. X has a winning strategy in $\mathcal{G}_{\varphi_{spl}}^{\omega_1}$.

In other words, ZFC can hardly tell us anything concerning this game. On the other hand, S. Shelah (private communication) tells us he believes it should be possible to prove:

Conjecture 30. It is consistent with ZFC that $\mathcal{G}_\varphi^{\omega_1}$ is determined for *every* formula φ.

10   The Bounded Uniformization Problem

As mentioned above, the uniformization problem in $(\omega^\omega, <)$ has not so far been solved (or shown to be undecidable). The task of constructing a uniformizer is intuitively harder than that of constructing a selector in that a uniformizer must *respond* to a given subset substituted for the domain variable X with an appropriate subset to be substituted for the image variable Y; it must (uniformly) answer a variety of challenges. In selection X simply does not appear in the formula. Put more abstractly, its *variability* has been reduced to zero. A natural move therefore, when X does appear in the formula, is to place various restrictions on the subsets of the domain substituted for it. One restriction which comes to mind is to consider formulas $\varphi(t, Y)$ where the t is an *first-order* variable, i.e. ranges over elements of the domain. Once we show the solvability of the uniformization problem for such formulas, our next step may be to allow X to range only over *finite* subsets of the domain, or perhaps over sets of order-type ω, etc. These examples are generalized by the following proposition.

Proposition 31 (Solvability of δ-bounded uniformization). *There is an algorithm that, given ordinals $\alpha \in [\omega^\omega, \omega_1]$, $\delta < \omega^\omega$ and a formula $\varphi(X, Y)$, decides whether there is a ψ which uniformizes φ in $(\alpha, <)$, when X is restricted to range over subsets of order-type $< \delta$. If such a ψ exists, the algorithm constructs it.*

Roughly speaking, the proof proceeds by a (non-trivial) reduction of this problem to uniformization over the class of ordinals smaller than δ and to selection in $(\omega^\omega, <)$ (or in $(\omega_1, <)$ when $\alpha = \omega_1$). Proposition 15 tells us the former is solvable, while Proposition 16 handles the latter.

11 Open Problems

We end by presenting several questions and conjectures, whose investigation, we believe, represents the next natural step in the exploration of definable and Church uniformization in the monadic theory of ordinals. First, a question already mentioned.

Question 32. Is the uniformization problem in $(\omega^\omega, <)$ solvable?

Next, we saw that for every countable $\alpha \geq \omega^\omega$ there are McNaughton games, where the winner does not have a definable winning strategy. This leads to (compare Questions $(2'')$ and $(3'')$ of Sect. 1):

Conjecture 33. There is an algorithm that, given $\alpha \in [\omega^\omega, \omega_1)$ and a formula $\varphi(X, Y)$, decides whether there is a *definable* winning strategy in $\mathcal{G}_\varphi^\alpha$, and if so, returns a ψ defining one.

Rabinovich ([15]) shows that if the conjecture holds for $\alpha = \omega^\omega$, then it is true.

We have seen that the only stumbling block for selection in $(\omega^\omega, <)$ was selecting an unbounded ω-sequence (recall Corollary 24). We believe an analogous statement may be true concerning definability of a winning strategy for games of length ω^ω:

Conjecture 34. For every formula $\varphi(X, Y)$, there is a formula $\psi(W, X, Y)$ such that for every unbounded ω-sequence $S \subseteq \omega^\omega$, $\psi(S, X, Y)$ defines in $(\omega^\omega, <)$ a winning strategy for the winner of $\mathcal{G}_\varphi^{\omega^\omega}$.

It is possible to extend the definition of selection degrees to the case of uniformization. First, extend MLO by allowing also atomic formulas of the form $F(X,Y)$ where F is a new relation symbol. Call the resulting language MLO^F. Let $\mathcal{M}$ be a structure with domain A. For every $f : \mathcal{P}(A) \to \mathcal{P}(A)$, denote by $\mathcal{M}^f$ the expansion of $\mathcal{M}$ which interprets F as (the graph of) f.

Let $\varphi_0(X,Y)$, $\varphi_1(X,Y)$ be MLO-formulas (that is, where F does not appear). Say that φ_0 is *easier than* φ_1 *to uniformize in* $\mathcal{M}$ if and only if there exists an MLO^F-formula $\psi(X,Y)$ such that for every f which uniformizes (the relation defined by) φ_1 in $\mathcal{M}$, ψ uniformizes φ_0 in $\mathcal{M}^f$. Now continue as in the case of selection to define *uniformization degree*. It is easy to see that this definition generalizes the one given for selection. A natural question is then:

Question 35. What are the uniformization degrees in $(\omega^\omega, <)$?

Of course, there is no reason to limit ourselves to *countable* ordinals.

Question 36. What are the selection/uniformization degrees in $(\omega_1, <)$?

Recall that it was only for notational convenience that we stated our results for formulas $\varphi(Y)$ having only a single free-variable Y. Our discussion carries through to formulas $\varphi(\bar{Y})$ with finitely many free-variables. In particular, so does the definition of selection degrees. Thus, $\varphi_0(Y_0, \ldots, Y_{l-1})$ is easier than $\varphi_1(X_0, \ldots, X_{m-1})$ to select in $\mathcal{M}$ iff there is $\psi(X_0, \ldots, X_{m-1}, Y_0, \ldots, Y_{l-1})$ such that for every m-tuple $\bar{P}$ satisfying φ_1 in $\mathcal{M}$, $\psi(\bar{P}, \bar{Y})$ selects φ_0 in $\mathcal{M}$. This is important to remember when discussing selection degrees in $(\omega_1, <)$. Indeed, for each $n \in \omega \setminus 1$, let $\varphi_n(X_0, \ldots, X_{n-1})$ say "for all $i < j < n$, X_i is a stationary subset of ω_1 and $X_i \cap X_j = \varnothing$." Then it can be shown that every formula $\varphi(\bar{Y})$ is easier than φ_n for some $n \in \omega \setminus 1$. We suspect also (but this is yet to be proven) that the φ_n represent distinct selection degrees in $(\omega_1, <)$ and that, more generally, φ_{n+1} never shares a degree with a formula having only n free-variables. If this is indeed so, then unlike what held true for countable ordinals, not all interesting phenomena having to do with selection in $(\omega_1, <)$ are exhibited by formulas having a single free-variable.

Further open questions in the context of uniformization and selection are suggested in [19].

References

1. Büchi, J.R.: Weak second-order arithmetic and finite automata. Zeit. Math. Logik und Grundl. Math. 6, 66–92 (1960)
2. Büchi, J.R.: On a decision method in restricted second order arithmetic. In: Proc. Int. Congress on Logic, Method, and Philosophy of Science. 1960, pp. 1–12. Stanford University Press (1962)
3. Büchi, J.R., Landweber, L.H.: Solving sequential conditions by finite-state strategies. Trans. Amer. Math. Soc. 138, 295–311 (1969)
4. Büchi, J.R., Siefkes, D.: The Monadic Second-Order Theory of all Countable Ordinals. In: Bloomfield, R.E., Jones, R.B., Marshall, L.S. (eds.) VDM 1988. LNCS, vol. 328, pp. 1–126. Springer, Heidelberg (1988)

5. Church, A.: Logic, arithmetic and automata. In: Proc. Inter. Cong. Math. 1963, Almquist and Wilksells, Uppsala (1963)
6. Elgot, C.: Decision problems of finite-automata design and related arithmetics. Trans. Amer. Math. Soc. 98, 21–51 (1961)
7. Feferman, S., Vaught, R.L.: The first-order properties of products of algebraic systems. Fundamenta Mathematicae 47, 57–103 (1959)
8. Gurevich, Y., Shelah, S.: Rabin's uniformization problem. Jou. of Symbolic Logic 48, 1105–1119 (1983)
9. Gurevich, Y.: Monadic second-order theories. In: Barwise, J., Feferman, S. (eds.) Model-Theoretic Logics, pp. 479–506. Springer, Heidelberg (1985)
10. Larson, P.B., Shelah, S.: The stationary set splitting game. Mathematical Logic Quarterly (to appear)
11. Lifsches, S., Shelah, S.: Uniformization and Skolem functions in the class of trees. Jou. of Symbolic Logic 63(1), 103–127 (1998)
12. McNaughton, R.: Testing and generating infinite sequences by a finite automaton. Information and Control 9, 521–530 (1966)
13. Rabin, M.O.: Decidability of second-order theories and automata on infinite trees. Trans. Amer. Math. Soc. 141, 1–35 (1969)
14. Rabin, M.O.: Automata on infinite objects and Church's problem. Amer. Math. Soc., Providence, RI (1972)
15. Rabinovich, A.: The Church synthesis problem over countable ordinals (submitted)
16. Rabinovich, A., Shomrat, A.: Selection in the monadic theory of a countable ordinal (submitted)
17. Rabinovich, A., Shomrat, A.: Selection over classes of ordinals expanded by monadic predicates (submitted)
18. Shelah, S.: The monadic theory of order. Annals of Math., Ser. 2 102, 379–419 (1975)
19. Shomrat, A.: Uniformization Problems in the Monadic Theory of Countable Ordinals. M.Sc. Thesis,Tel Aviv University (2007)
20. Trakhtenbrot, B.A.: The synthesis of logical nets whose operators are described in terms of one-place predicate calculus. Doklady Akad. Nauk SSSR 118(4), 646–649 (1958)
21. Trakhtenbrot, B.A.: Certain constructions in the logic of one-place predicates. Doklady Akad. Nauk SSSR 138, 320–321 (1961)
22. Trakhtenbrot, B.A.: Finite automata and monadic second order logic. Siberian Math. J 3, 101–131 (1962) Russian; English translation in: AMS Transl. 59, 23–55 (1966)
23. Trakhtenbrot, B.A., Barzdin, Y.M.: Finite Automata. North Holland, Amsterdam (1973)

The Scholten/Dijkstra Pebble Game Played Straightly, Distributedly, Online and Reversed

Wolfgang Reisig

Department of Computer Science, Humboldt-Universität zu Berlin

With pleasure I remember the visit of Boaz at GMD in Bonn in the early 1980s and at TU Munich in late 1980s as well as long discussions with him in Dagstuhl, and several meetings in Tel Aviv. What a rich source of inspiration!

Abstract. The Scholten/Dijkstra "Pebble Game" is re-examined. We show that the algorithm lends itself to a *distributed* as well as an *online* version, and even to a *reversed* variant.

Technically this is achieved by exploiting the local and the reversible nature of Petri Net transitions. Furthermore, these properties allow to retain the verification arguments of the algorithm.

1 Introduction

University Video Communications [1] distributes a "Distinguished Lecture Series" of "Leaders in Computer Science and Electrical Engineering" where in an "Academic Honour Presentation" Edsger W. Dijkstra talked about *Reasoning about programs*. As an example, Dijkstra presents a "Pebble Game" as an example of a nondeterministic algorithm. Gries in [2] refers the problem to Carl Scholten, due to a letter from Dijkstra in fall 1979. Scholten plays the game with black and white beans in a coffee can. Dijkstra models this algorithm as a guarded command program and proves its decisive properties.

Section 2.1 of this paper recalls Dijkstra's oral presentation of the algorithm as well his program. A Petri Net model of the algorithm is given in Sect. 2.2, and verified in Sect. 2.3.

Part 3 of this paper presents three variants that provide more insight into the nature of the algorithm. In particular, a *distributed* and an *online* version exploit the *local* nature of the algorithm's steps. Furthermore, it is shown that the algorithm can be played "backwards", exploiting the *reversible* nature of the algorithm's steps. Interesting enough, the decisive verification arguments remain valid in all three variants.

2 The Algorithm's Basic Version

2.1 Dijkstra's Algorithm and Model

We quote Dijkstra's oral presentation of [1]:

A. Avron et al. (Eds.): Trakhtenbrot/Festschrift, LNCS 4800, pp. 589–595, 2008.

"... a one person game is played with a big urn full of pebbles where each pebble is white or black. We don't start with an empty urn. ...

The rule is that one goes on playing as long as moves are possible. For a move there have to be at least two pebbles in the urn because a move is the following: What one does is: One shakes the urn, and then looks in the opposite direction, puts one hand in the urn, picks up two pebbles, looks at their color and depending on the color of the two pebbles taken out one puts a pebble in the urn again ...

The idea is that if we take out two pebbles of a different color, we put back the white one. However, if we take out two pebbles of equal color, we put a black one into the urn. (If we take out two white ones, we have to have a sufficient supply of black pebbles) ...

Given the initial content of the urn, what can be said about final pebble?"

Figure 1 represents the algorithm as a nondeterministic guarded command program. B and W are the number of white and black pebbles in the initial state.

$$b := B;\ w := W;$$

$$\underline{\mathrm{do}}\ w \geq 1 \wedge b \geq 1 \rightarrow b := b - 1$$
$$\square\quad b \geq 2 \rightarrow$$
$$b := b - 1$$
$$\square\quad w \geq 2 \rightarrow$$
$$w := w - 2;\ b := b + 1$$
$$\underline{\mathrm{od}}$$

Fig. 1. Dijkstra's solution to the pebble game

Dijkstra suggests to annotate this program by assertions, in particular by a loop invariant, thus showing that the final pebble is white if and only if W is odd.

2.2   A Petri Net Model of the Algorithm

To model the algorithm, we start with the pebbles: Initially as well as at any reachable state, the urn contains finitely many white and black pebbles. As a mathematical structure, they form a *finite multiset* (also called a *bag*). Let PEBBLES denote the bag of pebbles initially in the urn.

Next we turn to the urn: In the context of the algorithm, the urn is an item with two properties:

- the urn can contain any bag of white and black pebbles;
- actions can affect the urn, where an action may remove some of the pebbles available in the urn, and may add pebbles to the urn.

A *place* of a (high-level) Petri Net has exactly these properties. So, we model the urn as a Petri Net place.

Finally, we turn to the three actions: Each action is to remove two pebbles with a specific choice of colors from the urn. Occurrence of an action then returns a pebble to the urn. The returned pebbles's color is specified by the rules of the game. This is exactly what a Petri Net transition with corresponding arc inscriptions describes. So, we model each action as a Petri Net transition.

Figure 2 shows the corresponding Petri net. Its steps are the steps of the pebble game and its sequences of steps are the runs of the pebble game. All together, the Petri net of Fig. 2 models the pebble game.

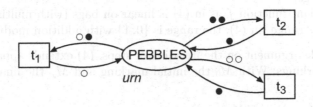

Fig. 2. The basic version of the algorithm

Petri Nets would also allow to model a 2-elementary bag of pebbles be taken out of the urn, instead of two pebbles.

2.3 Verification of the Algorithm

As described by Dijkstra, the algorithm has two decisive properties. Firstly,

$$\text{the algorithm terminates with one pebble, } p, \text{ remaining in the urn.} \qquad (1)$$

Furthermore, the color of p depends only on the number W of white pebbles in PEBBLES:

$$p \text{ is white if and only if } W \text{ is odd} \qquad (2)$$

Property (1) follows immediately from the observation that each occurrence of each transition of the Petri net in Fig. 2 reduces the magnitude of the bag in the urn by one, and that at least one of the transitions is enabled as long as there are at least two pebbles in the urn. As each transition returns a pebble, the process will eventually terminate with one pebble in the urn. Any further formalization of this obvious argument would be a formal overkill.

Proof of property (2) is far less trivial. It is based on a function f that assigns each bag of pebbles one of the numbers 0 or 1. More precisely, if BAG is a bag of pebbles with W white pebbles, let

$$f(BAG) = \begin{cases} 0 & \text{if } W \text{ is even} \\ 1 & \text{if } W \text{ is odd} \end{cases} \qquad (3)$$

In particular, for the one-elementary bags $[\bullet]$ and $[\circ]$, $f([\bullet]) = 0$ and $f([\circ]) = 1$.

The decisive argument is now that f is *stable*: For each step $M \xrightarrow{t} M'$ of the algorithm holds:

$$f(M(urn)) = f(M'(urn)) \tag{4}$$

where $M(urn)$ and $M'(urn)$ represent the token load of the place urn at the markings M and M', respectively. This rises the question of how to prove (4). Petri Net Theory provides the standard technique of *place invariants* to prove this kind of properties.

The place invariant technique exploits two observations:

- the effect of a transition occurrence boils down to the addition and subtraction of bags;
- the invariant function f as in (4) is linear on bags (with multiset addition) and on its range (in (4), the range is $\{0, 1\}$ with addition modulo 2).

By a simple argument on the iteration of steps, (4) extends apparently to all reachable markings: With M_0 the initial marking and M_ω the final marking of urn, we get

$$f(M_\omega(urn)) = f(M_0(urn)). \tag{5}$$

Now it is easy to prove (2): The remaining pebble is white

    iff $M_\omega(urn) = [\circ]$

    iff $f(M_\omega(urn)) = f([\circ]) = 1$

    iff $f(M_0(urn)) = 1$

    iff K is odd.

The intuition behind this proof in fact resembles Dijkstra's loop invariant.

Dijkstra's annotations for the program in Fig. 1 may be mirrored in Petri Nets by help of additional places, where an invariant property corresponds to a constant token load. Verification then reduces to problems of transition enabling, and can be attacked by place invariants just as program verification is based on loop invariants. This is merely a matter of convention and syntactic sugar.

3   Variants of the Algorithm

3.4   A Distributed Version of the Algorithm

Dijkstra assumes *one* player to execute the steps of the algorithms in a sequential order. There is no reason to do so. Many players may concurrently remove pairs of pebbles from the urn. To illustrate the most general case, assume a group of children with each child representing a white or a black pebble. The children are collected in a large circle (representing the urn) on the floor of the kindergarden. Any two children may decide to leave the circle together, with one of them returning, colored according to the rule's game.

How model this behavior? The initial count of n children yields $n(n-1)$ enabled actions, of which no more than $\lfloor n/2 \rfloor$ occur concurrently. Each action occurrence reduces n by one. A sequence of sets of actions would misrepresent this kind of behavior, because nothing in the algorithm enforces lockstep behavior. Sequential subprocesses likewise don't shine up.

From the very beginning, Petri Nets came with the notion of *distributed runs*: A distributed run is a partially ordered set of *transition occurrences*. An occurrence of a transition t is technically represented as an instance of t, with ingoing and outgoing arcs from and to the tokens that are consumed and produced by an occurrence of t. For example, an occurrence of transition t_1 in Fig. 2 reads

$$\begin{array}{c} \circ \\ \bullet \end{array} \searrow \boxed{t_1} \longrightarrow \circ \tag{6}$$

As an example, assume the initial bag PEBBLES of the net in Fig. 2 to contain two white and three black pebbles. Fig. 3 shows three distributed runs with this initial state.

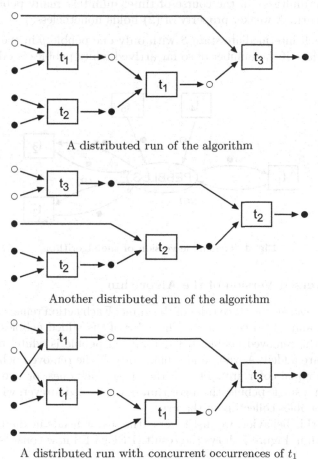

A distributed run of the algorithm

Another distributed run of the algorithm

A distributed run with concurrent occurrences of t_1

Fig. 3. Three distributed runs of the algorithm in Fig. 2

The decisive properties (1) and (2) remain valid in the distributed version. Furthermore, proof of (1) and (2) as given in Sect. 2.3 also apply to the distributed case.

Summing up, the distributed version of the algorithm comes without extra cost. It just employs the notion of distributed runs. This notion is anyway the adequate notion of runs for distributed systems.

3.5   An Online Version of the Algorithm

Both the sequential and the distributed versions of the algorithm assume the initial bag PEBBLE be available before computation starts. There is no reason to do so. The pebbles may be produced elsewhere and be added to the urn *while* the algorithm processes previous pebbles. The urn then serves as an (unbounded) *buffer*. Figure 4 shows this online version of the algorithm.

Property (1) is no longer valid, as we now do not necessarily assume PEBBLES to be a *finite* multiset. In the course of time, infinitely many pebbles may be added to the urn. A weaker property of (2) holds nonetheless:

$$\text{For each intermediate state } S \text{ with only one pebble } p \text{ in the urn,} \atop p \text{ is white if the number of so far arrived white pebbles is odd.} \qquad (7)$$

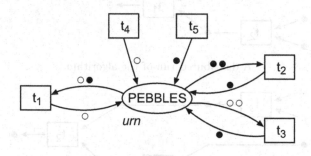

Fig. 4. The online version of the algorithm

3.6   A Reversed Version of the Algorithm

This version reverses the three rules of the game: Each action removes one pebble from the urn and adds two pebbles. The color of the added pebbles depends on the color of the removed pebble: If the removed pebble is white, pebbles with mixed color are added: A white and a black one. If the removed pebble is black, pebbles with equal color are added: Either two black ones, or two white ones. Starting with a single pebble, the algorithm may proceed forever, with unlimited numbers of pebbles collecting in the urn.

To model this behavior, we just reverse the arrow heads in the basic version of the algorithm. Figure 5 shows the result. PEBBLES now consists of only one pebble.

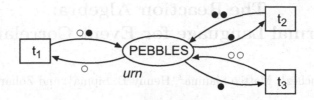

Fig. 5. The reversed version of the algorithm

As an interesting property of this version, in any reachable state S, the number of white pebbles is odd if and only if the color of the initial pebble is white. Proof of this property follows again from the place invariant (4).

One may construct a distributed as well as an online version of the reversed algorithm, in an obvious way.

4 Conclusion

We agree that an adequate invariant decisively supports both intuition and verification of the pebble algorithm. We challenge however that guarded commands was the most adequate modeling technique for this algorithm.

Removing or adding pebbles from or to the urn is *locally confined* to the pebbles affected. Different pairs of pebbles can therefore independently be processed. Both the distributed and the online version of the algorithm exploit locality of the actions.

Removing or adding pebbles from or to the urn are *reversible* actions: Knowing the resulting state and the action that caused the state allows re-tracing the start state. Assignment statements are in general not reversible. The reversed version of the algorithm exploits the reversibility of the actions.

Acknowledgements

H. Völzer brought the pebble game to my attention. G. Goos pointed me at David Gries' book. D. Bjørner gave me some additional hints.

References

1. Dijkstra, E.W.: Reasoning about programs, University Video Communications, Stanford. The Distinguished Lecture Series, Academic Leaders in Computer Science and Electrical Engineering, vol. III (1990)
2. Gries, D.: The Science of Programming, pp. 165–301. Springer, Heidelberg (1981)

The Reaction Algebra:
A Formal Language for Event Correlation*

César Sánchez[1], Matteo Slanina[2], Henny B. Sipma[1], and Zohar Manna[1]

[1] Computer Science Department, Stanford University, Stanford, CA 94305-9025
{cesar,sipma,zm}@CS.Stanford.EDU
[2] Google Inc., 1600 Amphitheatre Pkwy, Mountain View, CA 94043
mslanina@google.com**

To Boaz, pioneer and visionary – in honor of your 85th birthday.

Abstract. Event-pattern reactive programs are small programs that process an input stream of events to detect and act upon given temporal patterns. These programs are used in distributed systems to notify components when they must react.

We present the *reaction algebra*, a declarative language to define finite-state reactions. We prove that the reaction algebra is complete in the following sense: every event-pattern reactive system that can be described and implemented – in any formalism – using finite memory, can also be described in the reaction algebra.

1 Introduction

Interactive computation [6] studies the interaction of computational devices, including reactive and embedded systems, with their (not necessarily computational) environment. The most common approach to study interactive computation is based on machine models such as automata and Turing machines, enriched with output. In this paper we offer a complementary perspective: the *reaction algebra*, a declarative language to describe finite-state reactions. Its relationship to the machine models is similar to the relationship of regular expressions to language acceptors.

The practical motivation for a formalization of event-pattern reactive programs is to offer developers of distributed reactive systems a declarative way to describe temporal reaction patterns that is both formal and practical. The advantage of this design approach is that the interaction between components is made explicit and separate from the application code and can hence be analyzed independently. In addition, the code for pattern detection and reaction can be generated automatically from the event-pattern expressions and can be optimized for different objectives, including minimum processing time per event or smallest footprint.

* This research was supported in part by NSF grants CCR-02-20134, CCR-02-09237, CNS-0411363, CCF-0430102, and CSR- 0615449, and by NAVY/ONR contract N00014-03-1-0939.
** Current affiliation. This work was done while at Stanford University.

A. Avron et al. (Eds.): Trakhtenbrot/Festschrift, LNCS 4800, pp. 596–619, 2008.
© Springer-Verlag Berlin Heidelberg 2008

Event-Patattern Reactive Programming. In recent years the publish/subscribe architecture has become popular in the design of distributed reactive systems. In this architecture, components communicate with each other by events via an event channel. Components publish events to the event channel that may be of interest to other components. Components can also subscribe to the event channel to express interest in receiving certain events. The objectives of the publish/subscribe architecture are flexibility and scalability. Components are loosely coupled and may be added and removed on the fly and activated only when relevant events happen.

Most modern distributed systems are built on a middleware platform, a software layer that hides the heterogeneity of the underlying hardware, offers a uniform interface to the application, and usually provides services that implement common needs. Many middleware platforms provide an event channel that supports the publish/subscribe architecture. There are differences, however, in what kind of subscriptions are supported. Most platforms, including GRYPHON [1], ACE-TAO [24], SIENA [4], and ELVIN [25], support simple "event filtering": components can subscribe with a list of event types and the event channel notifies the component each time an event of one of those types is published. A slightly more expressive mechanism is "event content filtering", in which components in their subscriptions can specify predicates over the data included in the event. Notification, however, is still based on the properties of single events.

A more sophisticated subscription mechanism is "event correlation", which allows subscriptions in the form of temporal patterns. A component is notified only when a sequence of events that satisfies one of the patterns has been published. An implementation of this mechanism must maintain state: it may have to remember events it observed and may even have to store events that may have to be delivered to a component at a later stage. Event correlation is attractive because it separates the interaction logic from the application code and reduces the number of unnecessary notifications. Separation of the interaction logic increases analyzability. It also allows reuse of pattern detection code, thereby simplifying the development of applications. However, providing event correlation as a service requires that it have an intuitive, easy to use description language with a well-defined semantics. The reaction algebra, presented in this paper, aims to provide such a language.

Example 1. Fig. 1 shows an example of a small avionics system. It consists of six components that all communicate with the event channel. The purpose of the system is to control the cockpit's display such that it shows relevant information according to the current mode of operation, in this case *tactical mode* and *navigational mode*. In tactical mode, the Tactical Steering (TS) component collects data from the sensors and publishes events with tactical information to be displayed; in navigational mode the Navigational Steering (NS) component collects the data and performs the calculations. The mode of operation is set by the pilot via the Pilot Control component, which publishes an event to the event channel each time the mode is switched.

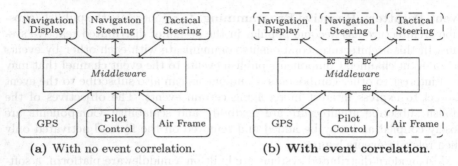

(a) With no event correlation.   (b) **With event correlation**.

Fig. 1. A simple avionics scenario

Without event correlation (Fig. 1(a)) all components receive all events that are published, that is all components are activated by the event channel when an event is published and their application code has to decide whether to react to the event or discard it. This strategy is clearly inefficient. For example, both the TS and NS component need to remember what is the current mode, or alternatively perform superfluous calculations and publish events that will not be used.

With event correlation (Fig.ure 1(b)) the TS component can subscribe with the temporal pattern that specifies that it only wants to be activated when "*an event from* GPS *is received, after an event* MODE=NAVIGATION *is received with no event* MODE=TACTICAL *in between*", and similar for the NS component. In this way neither the TS nor the NS component is activated unnecessarily and no useless events are published. In addition the application code for the NS and TS components is simpler because it does not have to decide whether to react or not.   □

Our first language for event correlation was ECL [21]. It was developed as part of the DARPA PCES project, with implementations integrated in ACE-TAO [24] and FACET [10], the underlying middleware platforms of the Boeing Open Experimental Platform. Next, we proposed PAR [22], a simplified but equally expressive version of ECL. With a formal semantics defined in the style of Plotkin's Structural Operational Semantics [18] in a coalgebraic framework [19], PAR was more suitable for formal analysis. In this paper we further streamline and simplify the presentation resulting in a new language called "the reaction algebra".

We prove that the reaction algebra is at least as expressive as finite-memory machines, that is, that every event-pattern reactive mechanism that can be implemented in finite memory, including Moore and Mealy machines [17,15], can be described by a reaction algebra program (a preliminary short version of the proof appeared in [23]). This result parallels, in the domain of reactive behaviors, the well-known equivalence between regular expressions and finite automata in the field of formal languages [11,14,9] and has equally important implications. Our result is technically more challenging, due to the more complex semantic domain and the determinism of the language. The proof proceeds by constructing a set of formulas, one for each state of the event-pattern machine, and then showing that each formula and its corresponding state are bisimilar. Hence, by coinduction, we can conclude that the observable behaviors are indistinguishable.

Related Work. The main difference between our reaction algebra and other algebraic languages from concurrency theory like CCS [16], CSP [8] and process algebras [2] is that our reaction algebra is a programming language, and therefore it is *deterministic*, while every reasonable concurrency theory models nondeterminism. The reaction algebra resembles synchronous reactive languages such as ESTEREL [3] sharing common features such as immediate reactivity and determinism. There are also some significant differences, however. For instance, every reaction algebra expression has a unique well-defined semantics, while this may not be the case for some syntactically correct ESTEREL programs [26]. Moreover, some correct ESTEREL programs can become incorrect when put in an enclosing context, even if this context corresponds to correct programs on other instantiations. In contrast, every reaction algebra context generates a uniquely defined behavior when instantiated.

Paper Organization. The paper is organized as follows. Section 2 reviews the coalgebraic framework that serves as the semantic domain. Section 3 introduces the reaction algebra, its semantics and some examples of extensions of the basic language. Section 4 shows that reaction algebra expressions can only define regular behaviors while Section 5 shows that they can define all regular behaviors. Finally, Section 6 presents the conclusions.

2 Semantic Domain

Event-pattern reactive programs recognize temporal patterns in an input stream of events and respond by generating output notifications. The reaction algebra enables a declarative specification of these patterns.

2.1 Reactive Machines

We use reactive machines as our model of computation to define the semantics of reaction algebra expressions. Reactive machines resemble finite-state automata: they are state machines over a set of input events. Reactive machines describe behaviors in terms of the output generated after each input event. In addition, to enable compositional definition of languages, reactive machines are equipped with a completion status function that affects reactions to future inputs.

Reactive machines satisfy the following conditions:

- *Determinism and non-blocking*: for every input prefix there is exactly one instantaneous reaction;
- *Causality*: the current output can depend only on past inputs;
- *Immediate reaction*: outputs are generated synchronously with inputs;

Despite these restrictions reactive machines are sufficiently general to model a wide range of reactive formalisms, including message passing systems and I/O automata [13].

Inputs. We assume a set Σ of input events and a finite set *Prop* of predicates over Σ, corresponding to elementary properties of individual events; that is, for all $p \in Prop$, $p \subseteq \Sigma$. An element of $\mathbb{B}(Prop)$, the boolean algebra over *Prop*, is called an *observation*. A *valuation* is an assignment of truth values to all propositions in *Prop*, which is lifted to $\mathbb{B}(Prop)$ in the usual way.

An input event a satisfies an observation $p \in \mathbb{B}(Prop)$, written $a \vDash p$, whenever p is true for all valuations that assign true to the elementary propositions that contain a (i.e., valuations in which for all $q \in Prop$, if $a \in q$ then q is assigned true.) To simplify the presentation in this paper, we assume that Σ is finite and that for every input event a there exists an observation p_a in *Prop*.

Outputs. The output domain of a reactive machine, denoted by $\mathcal{O}$, consists of sets of symbols taken from a finite set Γ. The reason for having sets of symbols rather than single symbols is that reaction algebra expressions can describe multiple patterns to be detected in parallel, each with its own outputs. Outputs of an expression, in that case, are the union of the outputs of the subexpressions. The simplest output, or notification, is a singleton element from Γ. Absence of output is represented by the empty set.

Completion Status. We define a *completion domain* $\mathcal{C} = \{\top, \iota, \bot\}$ containing three completion statuses that intuitively indicate

- $\top$: **success**. The pattern has *just* been observed.
- $\bot$: **failure**. The pattern cannot be observed in any stream that extends the current prefix.
- ι: **incomplete**. More input is needed or the input event processed is not relevant.

All event-pattern behaviors have the property that, once success or failure is declared, any subsequent output will be empty and any completion status will be incomplete.

We now define reactive machines formally:

Definition 1 (Reactive Machine). *A reactive machine over input Σ and output domain $\mathcal{O}$ is a tuple $\mathcal{M} = \langle \Sigma, M, o, \alpha, \partial \rangle$ consisting of a set M of states and three functions defined on an input event and a state:*

- *$o : \Sigma \times M \to \mathcal{O}$, an output function that returns an output notification,*
- *$\alpha : \Sigma \times M \to \mathcal{C}$, a completion function that returns a completion status, and*
- *$\partial : \Sigma \times M \to M$, a derivative function that returns a next state.*

*A machine must satisfy the **silent property**: for every state $m \in M$ and input $a \in \Sigma$, if $\alpha(a, m) \neq \iota$ then $\partial(a, m)$ is silent. A set of states S is silent if, for every state $s \in S$ and input a, $\alpha(a, s) = \iota$, $o(a, s) = \varnothing$ and $\partial(a, s) \in S$. A state is silent if it belongs to some silent set.*

The silent property establishes that a terminated program (or pattern observed) must not exhibit any subsequent behavior, that is, it must not contribute any future outputs.

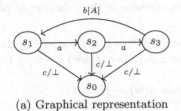

(a) Graphical representation

Σ	a	a	b	a	b	b	a	b	a	c	a	a	b	...
O	$\varnothing$	$\varnothing$	A	$\varnothing$	$\varnothing$	$\varnothing$	$\varnothing$	A	$\varnothing$	$\varnothing$	$\varnothing$	$\varnothing$	$\varnothing$	...
C	ι	ι	ι	ι	ι	ι	ι	ι	ι	$\perp$	ι	ι	ι	...
M	s_2	s_3	s_1	s_2	s_2	s_2	s_3	s_1	s_2	s_0	s_0	s_0	s_0	...

(b) Sample run from initial state s_1

Fig. 2. Example machine $\mathcal{M}$ with a sample evaluation for input "$aababbabacaab\ldots$"

Notation. We will write $o_a m$, $\alpha_a m$, and $\partial_a m$ to stand for $o(a, m)$, $\alpha(a, m)$, and $\partial(a, m)$, respectively. Also, we extend the definitions of α, o, and ∂ to strings of input symbols in the standard way, as $\alpha_{wa}v = \alpha_a \partial_w v$, $o_{wa}v = o_a \partial_w v$, and $\partial_{wa}v = \partial_a \partial_w v$. It is sometimes convenient to use a graphical representation of machines. Nodes are labeled by states. Two nodes, labeled by states $n, m \in M$, are connected by an edge labeled by input event a whenever $\partial_a n = m$. Completion status and outputs are also depicted on the edges, but only if $\alpha_a n \neq \iota$ and $o_a n \neq \varnothing$, respectively. Self-loops with labels ι and $\varnothing$ are not shown.

Example 2. Fig. 2(a) depicts a machine $\mathcal{M}$. Node s_0 is silent since all outgoing edges are self-loops labeled ι and $\varnothing$. The only edge associated with nonempty output connects s_3 to s_1, for which $o_b s_3 = A$. Fig. 2(b) shows the run of $\mathcal{M}$ for input $aababbabacaabb\ldots$, starting from state s_0; below each input symbol appear the output, the completion status, and the next state. □

We use the notions of homomorphism and bisimulation to extract a unique semantics for each state of every machine. Homomorphisms are functions that preserve observable behavior and bisimulations capture whether two behaviors are indistinguishable.

Definition 2 (Homomorphism). *A machine homomorphism from $\mathcal{M}$ to $\mathcal{M}'$ is a function $f : M \to M'$ such that, for all $m \in M$ and $a \in \Sigma$:*

$$o_a m = o'_a f(m),$$
$$\alpha_a m = \alpha'_a f(m) \quad and$$
$$f(\partial_a m) = \partial'_a f(m).$$

Definition 3 (Bisimulation). *A bisimulation between machines $\mathcal{M}$ and $\mathcal{M}'$ is a binary relation $\#$ such that for all $m \in M$, $m' \in M'$ and input symbol a:*

$$if \ m \# m' \ then \ \begin{cases} o_a m = o'_a m', \\ \alpha_a m = \alpha'_a m' \quad and \\ \partial_a m \ \# \ \partial'_a m'. \end{cases}$$

We say that two states m and m' are bisimilar (and we write $m \approx m'$) if there is a bisimulation that relates them.

Example 3. One important instance of a reactive machine is the *machine of all behaviors*, defined as $\mathcal{B}: \langle B, \partial^{\mathcal{B}}, \alpha^{\mathcal{B}}, o^{\mathcal{B}} \rangle$, where

- B is the set of all functions f from input prefixes Σ^+ to $\mathcal{O} \times \mathcal{C}$ satisfying the following silent condition. If $f(w) = \langle o, c \rangle$ for $c \neq \iota$ then $f(wv) = \langle \varnothing, \iota \rangle$ for all input extensions $v \in \Sigma^+$,
- $\partial_a^{\mathcal{B}} f$ of f on input a is the function g such that $g(w) = f(aw)$,
- $o_a^{\mathcal{B}} f$ is the first component of $f(a)$, and
- $\alpha_a^{\mathcal{B}} f$ is the second component of $f(a)$.

It is a routine exercise to check that $\mathcal{B}$ is well defined since the silent condition for machines is implied by the silent condition imposed on the functions in the set B. The elements of B are called "behaviors" or "reactions".   □

In [22] we showed that the definition of a reactive machine (Def. 1) captures a category of coalgebras with a final object. Once Σ and $\mathcal{O}$ are fixed, the machine of all behaviors is final among all machines, i.e., there is exactly one homomorphism (usually denoted $[\![\cdot]\!]_{\mathcal{M}}$ or simply $[\![\cdot]\!]$) from any machine $\mathcal{M}$ into $\mathcal{B}$.

The finality of $\mathcal{B}$ serves two purposes. First, the formal semantics of a language intended to describe event-pattern reactions can be defined by equipping the set of all language expressions with appropriate functions α, o and ∂ (providing that they satisfying the silent condition). By defining these functions, the set of all language expressions becomes a machine. Then, the semantics of an expression φ is obtained by finality as its (unique) homomorphical image $[\![\varphi]\!]$ in $\mathcal{B}$. We call this the principle of definition by corecursion. Second, the finality of $\mathcal{B}$ gives the following principle of proof by coinduction:

Theorem 1 (Coinduction). *If two states m and s from arbitrary machines are bisimilar ($m \approx s$) then they define the same behavior (i.e., $[\![m]\!] = [\![s]\!]$).*

In other words, bisimilarity captures whether two states react in the same way when given the same stream of input symbols.

In Section 5 we use Theorem 1 to show that the behavior of every state of a finite event-pattern machine can be described with a reaction algebra expression.

3   The Reaction Algebra

This section describes the language and semantics of the reaction algebra. We first present in sections 3.1 and 3.2 the syntax and semantics of the basic constructs. These constructs are sufficient to express any behavior that can be represented by a finite reactive machine. In section 3.3 we extend the language with additional constructs that do not increase the expressiveness of the language, but are convenient to describe common patterns that occur in practice.

3.1   Syntax and Informal Semantics

Reaction algebra (RA) expressions are defined inductively according to the following syntax:

$$\alpha ::= p \mid \mathbf{S} \mid \alpha \mid \alpha \mid \alpha \, ; \alpha \mid \mathbf{R} \, \alpha \mid \alpha \triangleright \alpha \mid \overline{\alpha} \mid \alpha \lceil A \rceil \; .$$

The base case is the *simple* expression p that tests whether an input symbol satisfies an observation p from $\mathbb{B}(Prop)$. It ranges over all observations. Compound expressions are constructed with the operators *selection* ($|$), *sequential* composition (;), *repetition* (R), *priority* or *otherwise* operator ($\triangleright$), *complementation* ($\bar{\cdot}$), and *output* operator ($\cdot\lceil\cdot\rceil$). The output A ranges over all output notifications.

Informal Semantics. A RA expression defines a reaction. The execution of a RA expression consists of the processing of input events, one at a time, producing a (possibly empty) output after each event is processed. Informally, the operators behave as follows.

Simple Expression: The expression p declares success when an event is received that matches the observation p; all other events are ignored. No output is generated.

Silent: The expression S does not generate any output and always declares incomplete.

Selection: The expression $x \mid y$ evaluates x and y in parallel, offering each the same events, and generating as output the combination of the subexpressions' outputs. Selection succeeds as soon as one of the branches succeeds and only fails when both branches have failed.

Sequential: Sequential composition, $x \, ; y$, evaluates the first subexpression, and upon successful completion starts the evaluation of the second. If one of them fails, sequential immediately fails. The output generated is that of the currently active subexpression.

Repetition: The expression R x starts by evaluating x, called the *body*. If the evaluation of the body completes with success, it evaluates R x (called the *continuation*) again. If the body fails, repetition declares failure. The output generated is that of the body.

Otherwise: The expression $x \triangleright y$ evaluates x and y in parallel. If x completes first (or at the same time as y), the completion status of $x \triangleright y$ is that of x. Otherwise the completion status is that of y. The output generated is the combination of the subexpressions' outputs.

Negation: The expression $\bar{x}$ behaves as x except that it reverses success with failure and vice-versa. The output generated is the output of the enclosing subexpression.

Output: The expression $x\lceil A\rceil$ evaluates x. Upon successful completion, the output A is generated and combined with any output simultaneously generated by x. The completion status of $x\lceil A\rceil$ is the same as that of x.

3.2   Formal Semantics

The formal semantics of RA expressions is defined by defining the functions α_a, o_a and ∂_a and applying the principle of corecursion, using the finality of the reactive machine of all behaviors $\mathcal{B}$.

The functions are defined inductively, by giving, for each of the operators, the values of α, o, and ∂ on every input symbol, possibly based on the values of the subexpressions. The definitions are presented as rules using the following notation: $x \overset{a}{\leadsto} c$ stands for $\alpha_a x = c$; $x \overset{a}{\to} y$ stands for $\partial_a x = y$ (with $x \overset{a}{\to}_\iota y$ as an abbreviation for both $x \overset{a}{\leadsto} \iota$ and $x \overset{a}{\to} y$); and $x \overset{a}{\Rightarrow} o$ stands for $o_a x = u$.

Simple Expression: The rule (α**Ev$_1$**) captures that a simple expression p succeeds upon receiving an event that satisfies p; (α**Ev$_2$**) and (**Ev**) state that it waits otherwise:

$$(\alpha\mathbf{Ev_1})\ \ p \overset{a}{\leadsto} \top\ \ (\text{if } a \vDash p)$$

$$(\alpha\mathbf{Ev_2})\ \ p \overset{a}{\leadsto} \iota\ \ (\text{if } a \nvDash p) \qquad (\mathbf{Ev})\ \ p \overset{a}{\to} p\ \ (\text{if } a \nvDash p)$$

(**oEv**) states that a simple expression does not generate any output:

$$(\mathbf{oEv})\ \ p \overset{a}{\Rightarrow} \varnothing$$

Silent: The rules for silent: (α**Sil**), (**oSil**) and (**Sil**) establish that the expression S does not generates any observable behavior:

$$(\alpha\mathbf{Sil}):\ \mathsf{S} \overset{a}{\leadsto} \iota \qquad (\mathbf{oSil}):\ \mathsf{S} \overset{a}{\Rightarrow} \varnothing \qquad (\mathbf{Sil}):\ \mathsf{S} \overset{a}{\to} \mathsf{S}$$

We introduce an extra rule that simplifies the definition of many others; it constrains the derivative of an expression that completes to be silent:

$$(\mathbf{GlobalSil})\ \ \frac{x \overset{a}{\not\to} \iota}{x \overset{a}{\to} \mathsf{S}}$$

The rule (**GlobalSil**) guarantees that the derivative of an expression that declares a non silent completion status is the silent expression S. This encompasses the non-silent completion cases for the rest of the operators, and guarantees the silent condition necessary to define a reactive machine.

Selection: The rules for the completion status of selection establish that $x \mid y$ succeeds if either x or y does, and fails only when both x and y fail.

$$(\alpha\mathbf{Sel_1})\ \frac{x \overset{a}{\leadsto} \top}{x \mid y \overset{a}{\leadsto} \top} \qquad \frac{y \overset{a}{\leadsto} \top}{x \mid y \overset{a}{\leadsto} \top} \qquad (\alpha\mathbf{Sel_2})\ \frac{x \overset{a}{\leadsto} \bot \quad y \overset{a}{\leadsto} \bot}{x \mid y \overset{a}{\leadsto} \bot}$$

In every other case, the completion status is incomplete:

$$(\alpha\mathbf{Sel_3})\ \frac{x \overset{a}{\leadsto} \iota \quad y \overset{a}{\not\leadsto} \top}{x \mid y \overset{a}{\leadsto} \iota} \qquad (\alpha\mathbf{Sel_4})\ \frac{x \overset{a}{\not\leadsto} \top \quad y \overset{a}{\leadsto} \iota}{x \mid y \overset{a}{\leadsto} \iota}$$

The output is the combination of the outputs of x and y,

$$(\textbf{oSel}) \ \frac{x \xRightarrow{a} u_1 \qquad y \xRightarrow{a} u_2}{x \mid y \xRightarrow{a} u_1 \cup u_2}$$

and the derivative of a selection is the selection of the derivatives,

$$(\textbf{Sel}_1) \ \frac{x \xrightarrow{a}_\iota x' \qquad y \xrightarrow{a}_\iota y'}{x \mid y \xrightarrow{a} x' \mid y'}$$

unless one of them (not both) fail, in which case the derivative is the derivative of the non-failing subexpression,

$$(\textbf{Sel}_2) \ \frac{x \xrightsquigarrow{a} \bot \qquad y \xrightarrow{a}_\iota y'}{x \mid y \xrightarrow{a} y'} \qquad\qquad (\textbf{Sel}_3) \ \frac{x \xrightarrow{a}_\iota x' \qquad y \xrightsquigarrow{a} \bot}{x \mid y \xrightarrow{a} x'}$$

Sequential: Completion and output of a sequential composition are determined by the first subexpression:

$$(\alpha\textbf{Seq}_1) \ \frac{x \xnrightarrow{a} \bot}{x \,;\, y \xrightsquigarrow{a} \iota} \qquad (\alpha\textbf{Seq}_2) \ \frac{x \xrightsquigarrow{a} \bot}{x \,;\, y \xrightsquigarrow{a} \bot} \qquad (\textbf{oSeq}) \ \frac{x \xRightarrow{a} u}{x \,;\, y \xRightarrow{a} u}$$

The derivative of the sequential composition is given by the two rules:

$$(\textbf{Seq}_1) \ \frac{x \xrightarrow{a}_\iota x'}{x \,;\, y \xrightarrow{a} x' \,;\, y} \qquad\qquad (\textbf{Seq}_2) \ \frac{x \xrightsquigarrow{a} \top}{x \,;\, y \xrightarrow{a} y}$$

Repeat: The rules for completion and output for repeat are:

$$(\alpha\textbf{Rep}_1) \ \frac{x \xnrightarrow{a} \bot}{\text{R}\, x \xrightsquigarrow{a} \iota} \qquad (\alpha\textbf{Rep}_2) \ \frac{x \xrightsquigarrow{a} \bot}{\text{R}\, x \xrightsquigarrow{a} \bot} \qquad (\textbf{oRep}) \ \frac{x \xRightarrow{a} u}{\text{R}\, x \xRightarrow{a} u}$$

The derivative rules state that either the repetition begins if the body succeeds (**Rep₂**), or that the body must be completed first (**Rep₁**):

$$(\textbf{Rep}_1) \ \frac{x \xrightarrow{a}_\iota x'}{\text{R}\, x \xrightarrow{a} x' \,;\, \text{R}\, x} \qquad\qquad (\textbf{Rep}_2) \ \frac{x \xrightsquigarrow{a} \top}{\text{R}\, x \xrightarrow{a} \text{R}\, x}$$

Otherwise: The completion rules for *otherwise* state that $x \triangleright y$ succeeds or fails whenever x does ($\alpha\textbf{Ow}_1$) and in all other cases has the same completion status as y ($\alpha\textbf{Ow}_2$),

$$(\alpha\textbf{Ow}_1) \ \frac{x \xrightsquigarrow{a} c}{x \triangleright y \xrightsquigarrow{a} c} \ c \neq \iota \qquad\qquad (\alpha\textbf{Ow}_2) \ \frac{x \xrightsquigarrow{a} \iota \qquad y \xrightsquigarrow{a} d}{x \triangleright y \xrightsquigarrow{a} d}$$

The outputs of x and y are combined,

$$(\textbf{oOw}) \ \frac{x \xRightarrow{a} u_1 \qquad y \xRightarrow{a} u_2}{x \triangleright y \xRightarrow{a} u_1 \cup u_2}$$

and the derivative of $x \triangleright y$ is the derivative of the subexpressions.

$$(\mathbf{Ow}) \ \frac{x \xrightarrow{a}_\iota x' \quad y \xrightarrow{a}_\iota y'}{x \triangleright y \xrightarrow{a} x' \triangleright y'}$$

Complementation: The completion rules state that success and failure are reversed:

$$(\alpha\mathbf{Neg_1}) \ \frac{x \overset{a}{\rightsquigarrow} \top}{\overline{x} \overset{a}{\rightsquigarrow} \bot} \qquad (\alpha\mathbf{Neg_2}) \ \frac{x \overset{a}{\rightsquigarrow} \iota}{\overline{x} \overset{a}{\rightsquigarrow} \iota} \qquad (\alpha\mathbf{Neg_3}) \ \frac{x \overset{a}{\rightsquigarrow} \bot}{\overline{x} \overset{a}{\rightsquigarrow} \top}$$

and the output and derivative rules reduce output and derivative to those of the subexpression,

$$(\mathbf{oNeg}) \ \frac{x \overset{a}{\Rightarrow} u}{\overline{x} \overset{a}{\Rightarrow} u} \qquad (\mathbf{Neg}) \ \frac{x \xrightarrow{a}_\iota x'}{\overline{x} \xrightarrow{a} \overline{x'}}$$

Output: The completion and derivative rules state that $x\lceil A \rceil$ behaves as x

$$(\alpha\mathbf{Out}) \ \frac{x \overset{a}{\rightsquigarrow} c}{x\lceil A \rceil \overset{a}{\rightsquigarrow} c} \qquad (\mathbf{Out}) \ \frac{x \xrightarrow{a}_\iota x'}{x\lceil A \rceil \xrightarrow{a} x'\lceil A \rceil}$$

and the output rules state that $x\lceil A \rceil$ adds output A to the output of x if x succeeds, and otherwise just produces the output of x,

$$(\mathbf{oOut_1}) \ \frac{x \overset{a}{\Rightarrow} u \quad x \overset{a}{\not\Rightarrow} \top}{x\lceil A \rceil \overset{a}{\Rightarrow} u} \qquad (\mathbf{oOut_2}) \ \frac{x \overset{a}{\Rightarrow} u \quad x \overset{a}{\rightsquigarrow} \top}{x\lceil A \rceil \overset{a}{\Rightarrow} u \cup A}$$

Example 4. The behavior of state s_1 of machine $\mathcal{M}$ in Fig. 2 is described by the expression $\mathbf{R}\left((a\,;a\,;b\lceil A \rceil) \triangleright \overline{c}\right)$. Alternatively, the same behavior is also described by $\left(\mathbf{R}(a\,;a\,;b\lceil A \rceil)\right) \triangleright \overline{c}$. These two expressions can be easily proven equivalent by giving a bisimulation that relates them. □

The following theorem justifies the study of expressiveness up to bisimulation in the reaction algebra:

Theorem 2 ([22]). *Bisimilarity is a reaction algebra congruence. Bisimilarity is the largest reaction algebra congruence that refines output equivalence.*

3.3 Language Extensions

The operators given above are sufficient to describe any behavior that can be represented by a finite reactive machine. For practical applications, however, it is often convenient to have available additional operators that describe common event-pattern behaviors. In this section we introduce some of these additional operators. Several of these operators were specifically requested by Boeing system developers to support the functionality of their Avionics platform. Some of these operators were also included in ECL [21].

The operators presented below do not increase the expressiveness of the language, that is, all of them can be defined in terms of the basic operators defined before. For some of them we will still also give the rules for the ∂, α and o functions, as these functions more directly describe behavior.

Immediate: The *immediate occurrence* of an observation p, written $p!$ can be defined in terms of basic operators as follows:

$$p! \overset{\text{def}}{=} p \triangleright (\neg p)$$

Upon the reception of an input event, it immediately terminates, either succeeding if the event satisfies p or failing otherwise. The immediate reaction is useful to represent transitions in machines. It is easy to see that the elementary observation (the primitive operator in the reaction algebra) can also be defined in terms of immediate reaction, since:

$$p \approx \overline{\text{R} \ (\neg p)!}$$

Two important particular observations are the false observation (satisfied by no event) and the true observation (satisfied by every event). We use **false** to represent the former, and **true** to represent the latter:

$$\textbf{false} \overset{\text{def}}{=} false! \qquad \textbf{true} \overset{\text{def}}{=} true!$$

Note that **false** fails immediately, while **true** succeeds immediately.

Positive and Negative: We define the *positive* and *negative* versions of an expression x as:

$$x^+ \overset{\text{def}}{=} x \mid \overline{x} \qquad x^- \overset{\text{def}}{=} \overline{x^+}$$

An expression differs from its positive and negative versions only in the completion status (x^+ cannot fail, x^- cannot succeed), but not in the instant this termination is produced or in the output generated. The positive and negative operator are both idempotent, they cancel each other, and complementation turns one into the other, as expressed by the following equivalences:

$$(x^+)^+ \approx x^+ \qquad (x^+)^- \approx x^- \qquad \overline{x^+} \approx x^-$$
$$(x^-)^- \approx x^- \qquad (x^-)^+ \approx x^+ \qquad \overline{x^-} \approx x^+$$

More Loops: The repetition construct Rx terminates when x fails. An infinite loop can thus be defined by applying R to the positive version of x:

$$\text{L} \ x \overset{\text{def}}{=} \text{R} \ x^+$$

Note that $\text{S} \approx \text{L} \ \textbf{true}$ and hence can be defined in terms of the other basic constructs. S is the only basic operator that is redundant. We decided to keep S in the set of basic operators for simplicity of the definitions.

Another repetition operator, called *persist*, is useful to represent repeated attempts until success. It first evaluates the body: if the body finishes with success, then *persist* also finishes with success; if the body fails then *persist* restarts the evaluation. Where R x repeats the body while it succeeds, P x persists while it fails.

The defining rules for P are:

$$(\alpha\mathbf{Per_1}) \ \frac{x \overset{a}{\not\rightarrow} \top}{P\,x \overset{a}{\rightsquigarrow} \iota} \qquad (\alpha\mathbf{Per_2}) \ \frac{x \overset{a}{\rightsquigarrow} \top}{P\,x \overset{a}{\rightsquigarrow} \top} \qquad (\mathbf{oPer}) \ \frac{x \overset{a}{\Rightarrow} u}{P\,x \overset{a}{\Rightarrow} u}$$

The derivative rules determine that either the repetition begins if the body succeeds ($\mathbf{Per_1}$), or that the body must be completed first ($\mathbf{Per_2}$):

$$(\mathbf{Per_1}) \ \frac{x \overset{a}{\rightarrow}_\iota x'}{P\,x \overset{a}{\rightarrow} \overline{x'}\,; P\,x} \qquad\qquad (\mathbf{Per_2}) \ \frac{x \overset{a}{\rightsquigarrow} \bot}{P\,x \overset{a}{\rightarrow} P\,x}$$

The following equivalences show that *persist* is the dual of *repetition*:

$$\overline{P\,x} \approx R\,\overline{x} \qquad \overline{R\,x} \approx P\,\overline{x}$$

These duality laws could have been used as an alternative definition of P using only repetition and negation. They also show that Theorem 2 still holds when the basic algebra is enriched with persist.

Persist also provides a more intuitive definition of lazy observation in terms of the immediate observation:

$$p \approx P\,p!$$

Delays: Sometimes it is useful to delay the failing of one expression until some other expression terminates. This can be accomplished with the *waiting for* construct:

$$y\,\mathcal{W}\,x \overset{\text{def}}{=} y \mid x^-$$

If expression y terminates with success, then $y\,\mathcal{W}\,x$ immediately succeeds. If, on the other hand, y fails, then $y\,\mathcal{W}\,x$ waits for x to terminate and then fails.

Accumulation: A pattern commonly occurring in practice is a task that consists of several subtasks executed in parallel that all must succeed before the main task can proceed. This pattern can be described by the *accumulation* operator +: it evaluates its subexpressions in parallel and succeeds when all subexpressions have succeeded and fails as soon as one of them fails, as reflected by the following rules for completion:

$$(\alpha\mathbf{Acc_1}) \ \frac{x \overset{a}{\rightsquigarrow} \bot}{x+y \overset{a}{\rightsquigarrow} \bot} \qquad \frac{y \overset{a}{\rightsquigarrow} \bot}{x+y \overset{a}{\rightsquigarrow} \bot} \qquad (\alpha\mathbf{Acc_2}) \ \frac{x \overset{a}{\rightsquigarrow} \top \qquad y \overset{a}{\rightsquigarrow} \top}{x+y \overset{a}{\rightsquigarrow} \top}$$

In every other case, the completion status is incomplete:

$$(\alpha\mathbf{Acc_3}) \ \frac{x \overset{a}{\rightsquigarrow} \iota \qquad y \overset{a}{\not\rightarrow} \bot}{x+y \overset{a}{\rightsquigarrow} \iota} \qquad (\alpha\mathbf{Acc_4}) \ \frac{x \overset{a}{\not\rightarrow} \bot \qquad y \overset{a}{\rightsquigarrow} \iota}{x+y \overset{a}{\rightsquigarrow} \iota}$$

The output of an accumulation expression is the combination of outputs of its subexpressions:

$$(\mathbf{oAcc}) \ \frac{x \overset{a}{\Rightarrow} u_1 \qquad y \overset{a}{\Rightarrow} u_2}{x+y \overset{a}{\Rightarrow} u_1 \cup u_2}$$

The derivative of an accumulation expression is the accumulation of the derivatives,

$$(\textbf{Acc}_1) \quad \frac{x \xrightarrow{a}_\iota x' \qquad y \xrightarrow{a}_\iota y'}{x + y \xrightarrow{a} x' + y'}$$

unless one of the subexpressions (but not both) succeeds, which case is captured by rules ($\textbf{Acc}_2$) and ($\textbf{Acc}_3$):

$$(\textbf{Acc}_2) \quad \frac{x \overset{a}{\rightsquigarrow} \top \qquad y \xrightarrow{a}_\iota y'}{x + y \xrightarrow{a} y'} \qquad\qquad (\textbf{Acc}_3) \quad \frac{x \xrightarrow{a}_\iota x' \qquad y \overset{a}{\rightsquigarrow} \top}{x + y \xrightarrow{a} x'}$$

Accumulation is the dual of selection, as shown by the following congruences:

$$\overline{x \mid y} \approx \overline{x} + \overline{y}, \qquad \overline{x + y} \approx \overline{x} \mid \overline{y}.$$

which could also have been used as an alternative definition of accumulation from selection and negation.

Parallel: The *parallel* construct is the nonterminating version of accumulation. It executes its subexpressions in parallel without ever terminating, even if all subexpressions terminate

$$x \parallel y \overset{\text{def}}{=} x^+ + y^+ + \textsf{S}$$

The accumulation and parallel operator were two of the operators included in the language ECL [21], but as we show here, they are not necessary, as they can be defined in terms of the basic operators.

Preemption: The construct $x \;\textsf{U}\; y$ (read "try x unless y") allows the occurrence of one pattern (described by y) to preempt further execution of another expression (x). Both expressions are evaluated in parallel. If y completes with success before x then the whole expression fails. We say that y preempts x.

$$(\alpha\textbf{Try}_2) \quad \frac{x \overset{a}{\rightsquigarrow} \iota \qquad y \overset{a}{\rightsquigarrow} \top}{x \;\textsf{U}\; y \overset{a}{\rightsquigarrow} \bot}$$

If x completes no later than y, the completion status is that of x, reflected in the following rules:

$$(\alpha\textbf{Try}_1) \quad \frac{x \overset{a}{\rightsquigarrow} c}{x \;\textsf{U}\; y \overset{a}{\rightsquigarrow} c} \; c \neq \iota \qquad (\alpha\textbf{Try}_3) \quad \frac{x \overset{a}{\rightsquigarrow} \iota \qquad y \overset{a}{\not\rightsquigarrow} \top}{x \;\textsf{U}\; y \overset{a}{\rightsquigarrow} \iota}$$

The output of the try-unless construct is the combination of outputs of the subexpressions

$$(\textbf{oTry}) \quad \frac{x \overset{a}{\Rightarrow} u_1 \qquad y \overset{a}{\Rightarrow} u_2}{x \;\textsf{U}\; y \overset{a}{\Rightarrow} u_1 \cup u_2}$$

The rules for the derivative are:

$$(\textbf{Try}_1) \; \frac{x \xrightarrow{a}_\iota x' \quad y \xrightarrow{a}_\iota y'}{x \; \text{U} \; y \xrightarrow{a} x' \text{U} \, y'} \qquad\qquad (\textbf{Try}_2) \; \frac{x \xrightarrow{a}_\iota x' \quad y \overset{a}{\not\leadsto} \bot}{x \; \text{U} \; y \xrightarrow{a} x'}$$

The try-unless can be defined in terms of previously defined operators, as shown by the following congruence

$$x \; \text{U} \; y \approx x \triangleright (\overline{y} + \text{S})$$

and hence its addition to the language does not increase the expressiveness of the language.

Dual Output: A dual version of the output operator, that generates a notification whenever an expression fails, can be defined by dualizing the rules $(\textbf{oOut}_1)$ and $(\textbf{oOut}_2)$ above:

$$(\textbf{oOutF}_1) \; \frac{x \xrightarrow{a} u \quad x \overset{a}{\not\leadsto} \bot}{x \lfloor A \rfloor \xrightarrow{a} u} \qquad (\textbf{oOutF}_2) \; \frac{x \xrightarrow{a} u \quad x \overset{a}{\leadsto} \bot}{x \lfloor A \rfloor \xrightarrow{a} u \cup A}$$

The rules for completion status and derivative remain the same as for output:

$$(\alpha\textbf{OutF}) \; \frac{x \overset{a}{\leadsto} c}{x \lfloor A \rfloor \overset{a}{\leadsto} c} \qquad (\textbf{Out}) \; \frac{x \xrightarrow{a}_\iota x'}{x \lfloor A \rfloor \xrightarrow{a} x' \lfloor A \rfloor}$$

Duality laws: In the basic reaction algebra, as defined in Section 3.2, enriched with accumulation, persist, and dual output, every expression is equivalent to an expression in negation normal form, that is, an expression in which complementation is applied only to observations. The following congruences, if applied as rewriting rules from left to right, provide a method to calculate the negation normal form of a given reaction algebra expression:

$$\overline{\overline{x}} \approx x$$
$$\overline{x \mid y} \approx \overline{x} + \overline{y} \qquad \overline{x + y} \approx \overline{x} \mid \overline{y}$$
$$\overline{x \triangleright y} \approx \overline{x} \triangleright \overline{y}$$
$$\overline{x \lceil A \rceil} \approx \overline{x} \lfloor A \rfloor \qquad \overline{x \lfloor A \rfloor} \approx \overline{x} \lceil A \rceil$$
$$\overline{\text{R} \, x} \approx \text{P} \, \overline{x} \qquad \overline{\text{P} \, x} \approx \text{R} \, \overline{x}$$

If also the strict operator is included in the language, then complementation can be removed completely, as shown by the following congruence:

$$\overline{p!} \approx (\neg p)! \qquad (\neg\neg p) \approx p.$$

4   Regularity of the Reaction Algebra

A behavior is called *regular* if it can be described by a reactive machine with a finite number of states. We show in this section that the reaction algebra can only express regular behaviors.

Every expression can be decomposed according to its behavior in response to individual observations. Given an expression x, the set of input symbols can be partitioned according to their direct effect on the completion status of x:

$$S(x) = \{a \in \Sigma \mid \alpha_a(x) = \top\}$$
$$F(x) = \{a \in \Sigma \mid \alpha_a(x) = \bot\}$$
$$I(x) = \{a \in \Sigma \mid \alpha_a(x) = \iota\}$$

Also, the one-step reaction of expression x on input a can be defined as:

$$Step_a(x) \overset{\text{def}}{=} p_a! \lceil o_a x \rceil$$

Lemma 1 (Expansion). *Every reaction algebra expression x is equivalent to its expansion with respect to input symbols:*

$$x \approx \Big(\mathop{|}_{a \in S(x)} Step_a x \Big) \rhd \Big(\mathop{|}_{a \in I(x)} Step_a x \, ; \, \partial_a x \Big) \rhd \Big(\overline{ \mathop{|}_{a \in F(x)} Step_a x } \Big).$$

Proof. Let $\varphi(x)$ denote the expansion of expression x. The proof proceeds by showing that the following relation is a bisimulation:

$$R = \{\langle x, \varphi(x)\rangle \mid x \in RA\} \cup \{\langle x, x\rangle \mid x \in RA\} \cup \{\langle x, \mathsf{S} \rhd x \rhd \mathsf{S}\rangle \mid x \in RA\}$$

For an arbitrary input symbol a and pair $\langle x, \varphi(x)\rangle$, both sides produce the same output and completion status in all three possibilities: $a \in S(x)$, $a \in F(x)$ and $a \in I(x)$. In the case that $a \notin I(x)$, then $\partial_a x = \mathsf{S} = \partial_a \varphi(x)$. In the other case, $a \in I(x)$, we have that

$$\partial_a \varphi(x) = \mathsf{S} \rhd \partial_a x \rhd \overline{\mathsf{S}}$$

so $\langle \partial_a \varphi(x), \partial_a x\rangle$ is in R. Hence, R is a bisimulation. $\square$

The fact that RA expressions can be described with finite memory is an easy consequence of the Expansion Lemma and the following proposition.

Proposition 1. *For every reaction algebra expression x, the set of derivatives $\Delta x \overset{\text{def}}{=} \{\partial_w x \mid \text{for some input prefix } w\}$ is finite.*

Proof. The proof proceeds by structural induction on expressions. Clearly, the result holds for observations and the silent expression S since $\Delta p = \{p, \mathsf{S}\}$ and $\Delta \mathsf{S} = \{\mathsf{S}\}$. For selection the derivative is either silent, one of the subterms or a selection of derivatives of the subterms, so the inductive hypothesis can be applied. The same reasoning holds for $\rhd$, complementation, repeat and output. $\square$

Theorem 3. *Every reaction algebra expression is equivalent to a finite machine.*

Proof. Given x we build a machine $\mathcal{M}_x : \langle \Delta x, \alpha, o, \partial \rangle$ by taking Δx as the states. For every state m_y corresponding to expression $y \in \Delta_x$, the functions are defined as:

- $\alpha(m_y)(a) = \alpha(y)(a)$,
- $o(m_y)(a) = o(y)(a)$, and
- $\partial(m_y)(a) = \partial(y)(a)$.

The binary relation $\{\langle y, m_y \rangle \mid$ for $y \in \Delta x\}$ is a bisimulation which, for the particular case of the original expression x, shows that x is equivalent to the corresponding state m_x in $\mathcal{M}_x$. $\square$

5   Expressive Completeness

The converse of Theorem 3 also holds: every state of a finite reactive machine can be described by a reaction algebra expression.

First, we observe that all silent states of a given machine are bisimilar. Therefore, without loss of generality, we assume that the given finite machine has at most one silent state.

We construct a set of reaction algebra expressions, each one capturing the behavior of a state in the machine. The construction proceeds as follows. First, the non-silent states are arbitrarily numbered from 1 to n. We will use v_i to refer to the state indexed i. The silent state, if it exists, receives index $n + 1$ and is denoted by v_{shh}. Then, we incrementally build a set of intermediate formulas whose behavior simulates more and more accurately that of its corresponding state for certain input strings. Finally, using the intermediate formulas we define a set of expressions Φ_i, each one bisimilar to a state v_i.

5.1   Intermediate Formulas

This stage of the construction runs for n rounds. At round k, we build a set of formulas φ_{ij}^k, one for each pair of non-silent states v_i and v_j. The formula φ_{ij}^k approximates the behavior on input prefixes that take from v_i to v_j of the following form:

Definition 4 (Direct Path). *A non-empty input string w is a direct path from state v_1 to state v_2 if $\partial_w v_1 = v_2$ and, for all proper prefixes u of w, $\partial_u v_1 \neq v_2$.*

Direct paths correspond to paths in the graph of the machine that visit the destination node exactly once, at the end of the traverse. The expression φ_{ij}^k captures the behavior of state v_i for direct paths that lead to v_j visiting only states labeled k or less along the way. Upon reaching v_j, φ_{ij}^k completes with success, it fails if a state of index larger than k is reached, and it declares incomplete otherwise. Formally, we classify the set of symbols according to formula φ_{ij}^k as follows:

Definition 5. *Given an index k and nodes v_i and v_j, we partition Σ into:*

- Successful symbols (S_{ij}^k): *symbols a for which $\partial_a v_i = v_j$.*
- Incomplete symbols (I_{ij}^k): *symbols a for which $\partial_a v_i = v_l$, for $l \neq j$ and $l \leq k$.*
- Failing symbols (F_{ij}^k): *symbols a for which $\partial_a v_i = v_l$, for $l \neq j$ and $l > k$.*

Incomplete symbols could, in principle, be extended to direct paths from v_i to v_j (at least no violation of the restriction to visit states labeled k or less has occurred so far). Failing symbols can never be extended to such a path, since a state labeled greater than k (and different from j) is visited.

The correctness of the construction relies on all formulas φ_{ij}^k satisfying the following property, as we will prove at every stage:

Property 1. Let a be an input symbol, and $\partial_a v_i = v_m$ the corresponding derivative (successor state of v_i in the machine):

1.1 if a is an incomplete symbol: $\quad \alpha_a \varphi_{ij}^k = \iota \qquad o_a \varphi_{ij}^k = o_a v_i \qquad \partial_a \varphi_{ij}^k \approx \varphi_{mj}^k,$

1.2 if a is a successful symbol: $\quad \alpha_a \varphi_{ij}^k = \top \qquad o_a \varphi_{ij}^k = o_a v_i \qquad \partial_a \varphi_{ij}^k = \mathsf{S},$

1.3 if a is a failing symbol: $\quad \alpha_a \varphi_{ij}^k = \bot \qquad o_a \varphi_{ij}^k = \varnothing \qquad \partial_a \varphi_{ij}^k = \mathsf{S}.$

Properties 1.1 and 1.2 guarantee that φ_{ij}^k generates the same output as the state v_i for all words in any direct path to v_j that only visit states labeled k or less. Notice that φ_{ij}^k can disagree with state v_i for failing symbols since, in this case, the output of the formula is empty and the output of the state need not be. These properties also establish that the completion status of the formula φ_{ij}^k is success for successful symbols, fail for failing symbols and incomplete for all others. Again, in the case of successful and failing symbols the completion behavior can differ from v_i. Consider, for example, a successful symbol, for which the completion of φ_{ij}^k is $\top$. The corresponding derivative in the machine directly connects v_i to v_j and, since v_j is not the silent state, the completion status is ι. These discrepancies are reduced during the construction as k grows. Eventually, when $k = n$, we have $F_{ij}^n = \varnothing$ and the only discrepancies left are in the completion status.

We now define the formulas φ_{ij}^k inductively:

Base case $(k = 0)$: Let v_i and v_j be two states:

$$\varphi_{ij}^0 \overset{\text{def}}{=} \underset{v_i \xrightarrow{a/\iota \lceil A \rceil} v_j}{\Big|} p_a! \lceil A \rceil.$$

Given an input symbol a, φ_{ij}^0 either immediately succeeds or immediately fails; it succeeds if $\partial_a v_i = v_j$ and fails otherwise. In particular, if there is no input symbol connecting v_i to v_j, then φ_{ij}^0 is equivalent to **false**.

Example 5. For machine $\mathcal{M}$ in Fig. 2(a), where we number states s_1 as 1, s_2 as 2 and s_3 as 3, we obtain:

$$\varphi_{12}^0 = p_a!, \quad \varphi_{31}^0 = p_b! \lceil A \rceil, \quad \varphi_{13}^0 = \textbf{false} \quad \text{and} \quad \varphi_{22}^0 = p_b!$$

Lemma 2. *All formulas φ_{ij}^0 satisfy Property 1.*

Proof. First, Property 1.1 holds vacuously since there are no incomplete symbols in the base case: every given symbol is either successful of failing. If a is a successful symbol, by definition of $p_a!$, φ_{ij}^0 succeeds, and its output coincides with that of state v_i. If, on the other hand, a is a failing symbol, then every branch of the selection fails. Consequently, the completion status of φ_{ij}^0 is $\bot$ and the output is empty.   □

Inductive step $(k > 0)$: We assume that we have defined all the formulas φ_{ij}^{k-1} satisfying Property 1, and proceed to define φ_{ij}^k. First, the particular case where indices j and k are equal is easy: $\varphi_{ik}^k \stackrel{\text{def}}{=} \varphi_{ik}^{k-1}$.

For the following we assume $k \neq j$. There are two kinds of direct paths from v_i to v_j: those that visit v_k and those that do not. We first consider paths that visit state v_k. These paths may loop around v_k (zero, one, or more times), and either keep looping forever or eventually enter a path that visits v_j.

To define a formula that captures this case we make use of φ_{ik}^{k-1}, φ_{kk}^{k-1} and φ_{kj}^{k-1}, previously defined. Note that the formula φ_{kj}^{k-1} must be restarted precisely after φ_{kk}^{k-1} succeeds. This can be achieved with $(\varphi_{kk}^{k-1} * \varphi_{kj}^{k-1})$ using the new binary operator $*$ defined as follows:

$$x * y \stackrel{\text{def}}{=} \big(\mathtt{P}\,(y\,\mathcal{W}\,x)\big) \vartriangleright \mathtt{R}\,x.$$

The $*$ operator is designed to work for sub-formulas such that, for every input, y completes no later than x. This is actually our case: if φ_{kk}^{k-1} completes, then the reached state is indexed k or greater. Consequently, if φ_{kj}^{k-1} has not completed yet, it has to do so at exactly that instant.

Informally, $*$ works as follows. For every input, the output is the combination of that of the subexpressions. For completion, consider all possible cases:

1. y succeeds: regardless of what x does, $y\,\mathcal{W}\,x$ immediately succeeds, and consequently so does the *persist* term $\big(\mathtt{P}\,(y\,\mathcal{W}\,x)\big)$. Therefore, $x * y$ also succeeds.
2. y fails: then, $y\,\mathcal{W}\,x$ waits for x to complete (which can happen at the same time or later). At the point of completion of x, independently of the completion status of x, $y\,\mathcal{W}\,x$ fails, and then the *persist* subexpression restarts. To see what happens with the right branch of $\vartriangleright$, we consider the possible values of x upon completion:
   - x succeeds: $\mathtt{R}\,x$ is restarted, at the same time as the *persist* branch. In other words, the whole formula is restarted at this point. This behavior is used to model a loop around state v_k.
   - x fails: then $\mathtt{R}\,x$ fails, which makes the whole expression fail.

Now, using $*$, we are ready to define the formula that captures the behavior of node v_i for direct paths to v_j that visit v_k:

$$Kleene_{ij}^k \stackrel{\text{def}}{=} \begin{cases} \varphi_{ik}^{k-1}\,;\,(\varphi_{kk}^{k-1} * \varphi_{kj}^{k-1}) & \text{if } i \neq k \\[2mm] \varphi_{kk}^{k-1} * \varphi_{kj}^{k-1} & \text{otherwise} \end{cases}$$

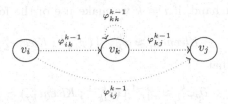

Fig. 3. Direct paths from v_i to v_j, using only nodes indexed k or less classified according to whether v_k is visited. Dotted arrows distinguish paths from edges.

Finally, to complete the definition of φ_{ij}^k we also have to consider the paths that do not visit v_k, captured directly by φ_{ij}^{k-1}, and compose these two cases:

$$\varphi_{ij}^k \overset{def}{=} \varphi_{ij}^{k-1} \mid Kleene_{ij}^k.$$

Lemma 3. *For all nodes v_i, v_j and index k, φ_{ij}^k satisfies Property 1.*

Proof. We proceed by induction on k, with the base case already proved in Lemma 2. For the inductive step we considered the cases for an input symbol a separately:

1. Let a be a successful symbol ($a \in S_{ij}^k$). Then, $\partial_a v_i = v_j$, so a is also a successful symbol for φ_{ij}^{k-1}. Hence, $\alpha_a \varphi_{ij}^{k-1} = \top$ and therefore $\alpha_a \varphi_{ij}^k = \top$ and $\partial_a \varphi_{ij}^k = \mathbf{S}$. Moreover, by inductive hypothesis $o_a \varphi_{ij}^{k-1} = o_a v_i$ so $o_a \varphi_{ij}^k = o_a v_i$. Hence, Property 1.2 holds.

2. Let a be a failing symbol ($a \in F_{ij}^k$). Similar.

3. Let a be an incomplete symbol ($a \in I_{ij}^k$). We consider two cases:

   (a) $v_i \overset{a}{\rightarrow} v_k$. In this case a is in S_{ik}^{k-1} and also in F_{ij}^{k-1}. Consequently,

   $$\alpha_a \varphi_{ij}^k = \alpha_a(Kleene_{ij}^k) = \iota, \quad \text{and} \quad o_a \varphi_{ij}^k = o_a(Kleene_{ij}^k) = o_a v_i,$$

   by inductive hypothesis. Finally, $\partial_a \varphi_{ij}^k = \partial_a Kleene_{ij}^k$. Now, it follows from properties of $*$:

   $$\partial_a Kleene_{ij}^k = (\varphi_{kk}^{k-1} * \varphi_{kj}^{k-1}) = Kleene_{kj}^k = \varphi_{kj}^k.$$

   Then Property 1.1 holds.

   (b) $v_i \overset{a}{\rightarrow} v_l$ with $l < k$. Then, a is also an incomplete symbol for φ_{ij}^{k-1}. Consequently, by inductive hypothesis $\alpha_a \varphi_{ij}^{k-1} = \iota$ and $\alpha_a(Kleene_{ij}^k) = \iota$, and we can conclude that $\alpha_a \varphi_{ij}^k = \iota$. Second, $o_a \varphi_{ij}^k = o_a \varphi_{ij}^{k-1} = o_a v_i$. Finally, if $i \neq k$, then

   $$\begin{aligned} \partial_a \varphi_{ij}^k &= \partial_a \varphi_{ij}^{k-1} \mid \partial_a Kleene_{ij}^k \\ &= \varphi_{lj}^{k-1} \mid (\varphi_{lk}^{k-1} \,;\, Kleene_{kj}^k) \approx \varphi_{lj}^k. \end{aligned}$$

On the other hand, if $i = k$ we make use of the following property of *Kleene*:

$$Kleene_{kj}^k \approx \varphi_{kj}^{k-1} \mid (\varphi_{kk}^{k-1} ; Kleene_{kj}^k),$$

to conclude that

$$\partial_a \varphi_{ij}^k \approx \varphi_{lj}^{k-1} \mid (\varphi_{lk}^{k-1} ; Kleene_{kj}^k) \approx \varphi_{lj}^k.$$

Then, Property 1.1 also holds. □

5.2   Final Formulas

Using the formulas φ_{ij}^n obtained in the last step of the previous stage, we now define formulas Φ_i, one for each non-silent state v_i. The behavior of the silent state v_{shh}, if present, is modeled by the formula S.

First, we need to define variations of the *Kleene* formula to cover the cases of succeeding and failing transitions in the machine. For each state v_i:

$$Kleene_i^\top \stackrel{\text{def}}{=} \varphi_{ii}^n * (\underset{v_i \xrightarrow{a/\top\lceil A\rceil} v_{shh}}{\mid} p_a! \lceil A \rceil) \qquad Kleene_i^\perp \stackrel{\text{def}}{=} \varphi_{ii}^n * (\underset{v_i \xrightarrow{a/\perp\lceil A\rceil} v_{shh}}{\mid} p_a! \lceil A \rceil)$$

The formula $Kleene_i^\top$ captures the behaviors of state v_i for input strings that either loop forever around v_i, or eventually succeed directly from v_i. The formula $Kleene_i^\perp$ works similarly except that it captures behaviors that fail directly from v_i. Note that $Kleene_i^\perp$ succeeds (instead of failing).

Finally, the behavior of v_i is defined by composing all possible paths:

$$\Phi_i \stackrel{\text{def}}{=} \left(Kleene_i^\top \mid \underset{j}{\mid} (\varphi_{ij}^n ; Kleene_j^\top) \right) \triangleright \overline{Kleene_i^\perp \mid \underset{j}{\mid} (\varphi_{ij}^n ; Kleene_j^\perp)}$$

5.3   Proof of Correctness

The correctness of the construction relies on the following lemma:

Lemma 4. *For all states v_i and input symbols a,*
(1) $\alpha_a \Phi_i = \alpha_a v_i$ and $o_a \Phi_i = o_a v_i$.
(2) If $\alpha_a v_i$ is incomplete and $\partial_a v_i = v_l$ then $\partial_a \Phi_i \approx \Phi_l$.

Proof. (1) We proceed by cases:

1. If $v_i \xrightarrow{a/\iota\lceil A\rceil} v_l$, then all the direct branches in $Kleene_i^\top$ and $Kleene_i^\perp$ are not satisfied. Therefore $o_a \Phi_i = \cup_j o_a \varphi_{ij}^n = \cup o_a v_i = o_a v_i$. Moreover, all select branches of both sides of $\triangleright$ are incomplete, so $\alpha_a \Phi_i = \iota = \alpha_a v_i$.
2. If $v_i \xrightarrow{a/\top\lceil A\rceil} v_{shh}$, then $o_a \varphi_{ij}^n = \varnothing$, and $o_a Kleene_i^\top = o_a v_i$ so $o_a \Phi_i = o_a v_i$. Also, $\alpha_a Kleene_i^\top = \alpha_a \Phi_i = \perp = \alpha_a v_i$.
3. The case $v_i \xrightarrow{a/\perp\lceil A\rceil} v_{shh}$ is handled similarly, except that in this case the $Kleene_i^\perp$ succeeds, so $\alpha_a \Phi_i = \perp = \alpha_a v_i$.

(2) For all branches with $j \neq l$, $\partial_a \varphi_{ij}^n \approx \varphi_{lj}^n$, and then $\partial_a (\varphi_{ij}^n ; Kleene_j^\top) \approx (\varphi_{lj}^n ; Kleene_j^\top)$. On the other hand, for $j = l$, since $\alpha_a \varphi_{il}^n = \top$, we have $\partial_a (\varphi_{il}^n ; Kleene_j^\top) = Kleene_l^\top$. Finally, $\partial_a Kleene_i^\top \approx (\varphi_{li}^n ; Kleene_i^\top)$. This holds since all branches inside $Kleene_i^\top$ fail. Hence,

$$\partial_a \Phi_i \approx \left(\frac{\varphi_{li}^n ; Kleene_i^\top \mid Kleene_i^\top \mid \mid_{j \neq l} \varphi_{lj}^n ; Kleene_j^\top}{\triangleright}{\varphi_{li}^n ; Kleene_i^\perp \mid Kleene_i^\perp \mid \mid_{j \neq l} \varphi_{lj}^n ; Kleene_j^\perp} \right)$$

$$\approx \left(\frac{Kleene_l^\top \mid \mid_j \varphi_{lj}^n ; Kleene_j^\top}{\triangleright}{Kleene_l^\perp \mid \mid_j \varphi_{lj}^n ; Kleene_j^\perp} \right) = \Phi_l$$

The reordering of terms in the last step was possible by the commutativity and associativity of the $\mid$ operator.   □

Theorem 4. *Every final formula Φ_i is bisimilar to its corresponding state v_i.*

This is a direct consequence of Lemma 4 and implies that the behavior of state v_i is captured precisely by formula Φ_i. Therefore, every finite graph can be expressed by a reaction algebra expression.

6   Conclusions

We have introduced the reaction algebra as a formal language for interactive computation. While most models of interactive computation start from machine-based formalisms that are "interactive Turing-complete" the reaction algebra is a simple and tractable language that can be enriched to describe more complex behaviors. Our approach can also be interpreted as complementary to most formalisms for the design of reactive systems, like Statecharts [7], which are usually based on machine models. Even though we use machines for the description of the semantics, the main emphasis of our work relies on the study of simple languages to express reactions, and their properties.

The purpose of the reaction algebra is analogous to the role of regular expressions in language acceptors. Where regular expressions aim at easily defining regular sets, the reaction algebra can easily define reactions that can be *efficiently implemented*. Even though for some expressions the smallest finite machine has exponential size, every reaction algebra expression can be evaluated using storage space $O(n)$, performing at most n number of elementary operations per input event. Reaction algebras have been used in practice as an event-pattern reactive programming language; we show in this paper how to extend the basic reaction algebra with new operators.

We have shown that every reactive behavior that can be described and implemented with finite memory can be expressed in RA with a basic set of operators. In addition to its theoretical value, this result has also has practical applications, for example, in the development of compilers and analysis tools. Compilers only need to support the minimal set of constructs, while additional constructs can be reduced to this set by a preprocessor. Similarly, analysis methods need to cover only the basic constructs.

Future work includes: (1) Study whether, unlike regular-expressions (see [5,20,12]), there are equational axiomatizations of the reaction algebra. (2) Construct decision procedures for the problem of equational reasoning of parameterized RA expressions, and for the full first-order case. Efficient solutions will allow the synthesis of reaction algebra expressions and the implementation of behavior-preserving optimizations. (3) Go beyond the finite state case by equipping the reaction algebra with capabilities to store and manipulate data, and study to what extent the expressive power is still complete in some suitable sense, and to what extent the analysis problems are still tractable.

References

1. Aguilera, M.K., et al.: Matching events in a content-based subscription system. In: Symposium on Principles of Distributed Computing, pp. 53–61 (1999)
2. Baeten, J.C.M., Weijland, W.P.: Process Algebra. Cambridge University Press, Cambridge (1990)
3. Berry, G.: Proof, language, and interaction: essays in honour of Robin Milner. In: The foundations of Esterel, pp. 425–454. MIT Press, Cambridge (2000)
4. Carzaniga, A., Rosenblum, D.S., Wolf, A.L.: Design and evaluation of a wide-area event notification service. ACM Transactions on Computer Systems 19(3), 332–383 (2001)
5. Conway, J.H.: Regular algebra and finite machines. Chapman and Hall, Boca Raton (1971)
6. Goldin, D., Smolka, S.A., Wegner, P. (eds.): Interactive Computation: the New Paradigm. Springer, Heidelberg (2006)
7. Harel, D.: Statecharts: A visual formalism for complex systems. Science of Computer Programming 8(3), 231–274 (1987)
8. Hoare, C.A.R.: Communicating Sequential Processes. Prentice-Hall, Englewood Cliffs (1985)
9. Hopcroft, J.E., Ullman, J.D.: Introduction to automata theory, languages and computation. Addison-Wesley, Reading (1979)
10. Hunleth, F., Cytron, R., Gill, C.D.: Building customizable middleware using aspect oriented programming. In: Works. on Advanced Separation of Concerns (OOPSLA 2001) (2001)
11. Kleene, S.C.: Representation of events in nerve nets and finite automata. In: Shannon, C.E., McCarthy, J. (eds.) Automata Studies, vol. 34, pp. 3–41. Princeton University Press, Princeton (1956)
12. Kozen, D.: A completeness theorem for kleene algebras and the algebra of regular events. Information and Computation 110(2), 366–390 (1994)
13. Lynch, N., Tuttle, M.: An introduction to Input/Output automata. CWI-Quarterly 2(3) (1989)

14. McNaughton, R.F., Yamada, H.: Regular expressions and state graphs for automata. IEEE Transactions on Electronic Computers 9, 39–47 (1960)
15. Mealy, G.H.: A method for synthesizing sequential circuits. Bell Systems Technical Journal 34(5), 1045–1079 (1955)
16. Milner, R.: Communication and Concurrency. Prentice-Hall, Englewood Cliffs (1989)
17. Moore, E.F.: Gedanken-Experiments on sequential machines. In: Automata Studies, pp. 129–153 (1956)
18. Plotkin, G.D.: A Structural Approach to Operational Semantics. Technical Report DAIMI FN-19, University of Aarhus (1981)
19. Rutten, J.J.: Automata and coinduction (an exercise in coalgebra). In: CONCUR (1998)
20. Salomaa, A.: Two complete axiom systems for the algebra of regular events. Journal of the ACM 13(1), 158–169 (1966)
21. Sánchez, C., et al.: Event correlation: Language and semantics. In: Alur, R., Lee, I. (eds.) EMSOFT 2003. LNCS, vol. 2855, pp. 323–339. Springer, Heidelberg (2003)
22. Sánchez, C., et al.: Final semantics for Event-Pattern Reactive Programs. In: Fiadeiro, J.L., et al. (eds.) CALCO 2005. LNCS, vol. 3629, pp. 364–378. Springer, Heidelberg (2005)
23. Sánchez, C., et al.: Expressive completeness of an event-pattern reactive programming language. In: Wang, F. (ed.) FORTE 2005. LNCS, vol. 3731, pp. 529–532. Springer, Heidelberg (2005)
24. Schmidt, D., Levine, D., Harrison, T.: The design and performance of a real-time CORBA object event service. In: Proc. of OOPSLA (1997)
25. Segall, B., Arnold, D.: Elvin has left the building: A publish/subscribe notification service with quenching. In: Queensland AUUG Summer Technical Conference, Brisbane, Australia (1997)
26. Tardieu, O.: A deterministic logical semantics for Esterel. In: Workshop on Structural Operational Semantics, SOS (2004)

On Natural Non-dcpo Domains

Vladimir Sazonov

Department of Computer Science, the University of Liverpool,
Liverpool L69 3BX, U.K.
Vladimir.Sazonov@liverpool.ac.uk

Dedicated to my teacher, *Boris Abramovich Trakhtenbrot*,
in his 87th year whose influence on me and help
cannot be overstated.

Abstract. As Dag Normann has recently shown, the fully abstract model for **PCF** of hereditarily sequential functionals is not ω-complete and therefore not continuous in the traditional terminology (in contrast to the old fully abstract continuous dcpo model of Milner). This is also applicable to a wider class of models such as the recently constructed by the author fully abstract (universal) model for **PCF**$^+$ = **PCF** + parallel **if**. Here we will present an outline of a general approach to this kind of "natural" domains which, although being non-dcpos, allow considering "naturally" continuous functions (with respect to existing directed "pointwise", or "natural" least upper bounds) and also have appropriate version of "naturally" algebraic and "naturally" bounded complete "natural" domains. This is the non-dcpo analogue of the well-known concept of Scott domains, or equivalently, the complete f-spaces of Ershov. In fact, the latter version of natural domains, if considered under "natural" Scott topology, exactly corresponds to the class of f-spaces, not necessarily complete.

1   Introduction

The goal of this paper is to present a first brief outline of the so-called "natural" version of domain theory in the general setting, where domains are not necessary directed complete partial orders (dcpos). As Dag Normann has recently shown [6], the fully abstract model of hereditarily-sequential finite type functionals for **PCF** [1,3,5,10][1] is not ω-complete (hence non-dcpo) and therefore not continuous in the traditional terminology. This is also applicable to a potentially wider class of models such as the fully abstract model of (hereditarily) wittingly consistent functionals for **PCF**$^+$ (i.e. **PCF** + parallel **if**) [10]. Note that until the above mentioned negative result in [6] and further positive results in [10] the domain theoretical structure of such models was essentially

---

[1] As to the language **PCF** for sequential finite type functionals see [4,7,9,11]. Note also that the technical part of [10] — the source of considerations of the present paper — is heavily based on [8,9].

A. Avron et al. (Eds.): Trakhtenbrot/Festschrift, LNCS 4800, pp. 620–634, 2008.

unknown. The point of using the term "natural" for these kinds of domains is that in the case of non-dcpos, the ordinary definitions of continuity and finite (algebraic) elements via arbitrary directed least upper bounds (lubs) prove to be inappropriate. A new, restricted concept of "natural" lub is necessary, and it leads to a generalized theory applicable also to non-dcpos. More informally, if some directed least upper bounds do not exist in a partial ordered set D then this can serve as an indication that even some existing least upper bounds can be considered as "unnatural" in a sense. Although "natural" lubs for functional domains can also be characterised technically as "pointwise" (in the well-known sense), using the latter term for the concepts of continuous functions or finite elements as defined in terms of pointwise lubs is, in fact, somewhat misleading. The term "pointwise continuous" is in this sense awkward and of course not intended to be considered as "continuous for each argument value", but rather as "continuous with respect to the pointwise lubs" which is lengthy. Thus, the more neutral and not so technical term "natural" is used instead of "pointwise". Moreover, for general non-functional non-dcpo domains the term "pointwise" does not seem to have the straightforward sense. However we should also note the terminological peculiarity of the term "natural". For example, the existence of "naturally finite but not finite" elements in such "natural" domains is quite possible (see Hypotheses 2.8 in [10] concerning sequential functionals). Although the main idea of the current approach has already appeared in [10], it was applied there only in a special situation of typed non-dcpo models with "natural" understood as (hereditarily) "pointwise". Here our goal is to make the first steps towards a general non-dcpo domain theory of this kind.

2   Natural Domains

A non-empty partially ordered set (poset) $\langle I, \leq \rangle$ is called *directed* if for all $i, j \in I$ there is a $k \in I$ such that $i, j \leq k$. By saying that a (non-empty) family of elements x_i in a poset $\langle D, \sqsubseteq \rangle$ is *directed*, we mean that I, the range of i, is a directed poset, and, moreover, the map $\lambda i.x_i : I \to D$ is *monotonic* in i, that is, $i \leq j \Rightarrow x_i \sqsubseteq x_j$. However in general, if it is not said explicitly or does not follow from the context, x_i may denote a not necessarily directed family. Moreover, we will usually omit mentioning the range I of i, relying on the context. Different subscript parameters i and j may range, in general, over different index sets I and J. As usual $\bigsqcup X$ denotes the ordinary least upper bound (lub) of a subset $X \subseteq D$ in a poset D which may exist or not. That is, this is a partial map $\bigsqcup : 2^D \rightharpoonup D$ with 2^D denoting the powerset of D. If D has a least element, it is denoted as $\bot_D$ or $\bot$ and called *undefined*.

Definition 1

(a) Any poset $\langle D, \sqsubseteq^D \rangle$ (not necessarily a dcpo) is also called a *domain*.

(b) Recall that a *directly complete partial order* (or dcpo domain) is required to be closed under taking directed least upper bounds $\bigsqcup x_i$.[2] (We omit the usual requirement that a dcpo should contain a least element $\bot$.)

---

[2] In general, by $\bigsqcup_i z_i$ we mean $\bigsqcup \{z_i \mid i \in I\}$, and analogously for $\biguplus$ below.

(c) A *natural pre-domain* is a domain D (in general non-dcpo) with a partially defined operator of *natural lub* $\uplus : 2^D \rightharpoonup D$ satisfying the first of the following *four* conditions. It is called a *natural domain* if all these conditions hold:

($\uplus 1$) $\uplus \subseteq \bigsqcup$. That is, for all sets $X \subseteq D$, if $\uplus X$ exists (i.e. X is in the domain of $\uplus$) then $\bigsqcup X$ exists too and $\uplus X = \bigsqcup X$.

($\uplus 2$) If $X \subseteq Y \subseteq D$, $\uplus X$ exists, and Y is upper bounded by $\uplus X$ then $\uplus Y$ exists too (and is equal to $\uplus X$).

($\uplus 3$) $\uplus \{x\}$ exists (and is equal to x).

($\uplus 4$) Let $\{y_{ij}\}_{i \in I, j \in J}$ be an arbitrary non-empty family of elements in D indexed by I and J. Then

$$\uplus_i \uplus_j y_{ij} = \left(\uplus_j \uplus_i y_{ij} = \right) \uplus_{ij} y_{ij} = \uplus_i y_{ii}$$

provided that:

1. Assuming all the required internal natural lubs $\uplus_j y_{ij}$ in $\uplus_i \uplus_j y_{ij}$ and one of the external natural lubs $\uplus_i \uplus_j y_{ij}$ or $\uplus_{ij} y_{ij}$ exist, then both exist and the corresponding equality above holds. (The case of $\uplus_j \uplus_i y_{ij}$ is symmetrical.[3])

2. For the last equality to hold, the family y_{ij} is additionally required to be directed (and monotonic) in each parameter i and j ranging over the same I, and the existence of any natural lub in this equality implies the existence of the other.

The second part of ($\uplus 4$) (directed case) evidently follows also from ($\uplus 1$), ($\uplus 2$), and the following *optional clause* which might be postulated as well.

($\uplus 5$) If $X \subseteq Y \subseteq D$, $\uplus Y$ exists, and X is cofinal with Y (i.e. $\forall y \in Y \exists x \in X$. $y \sqsubseteq x$) then $\uplus X$ exists too (and $= \uplus Y$).

But we will really use only ($\uplus 1$)–($\uplus 4$). Evidently, any pre-domain with unrestricted $\uplus \rightleftharpoons \bigsqcup$ is a natural domain. As an extreme case any *discrete* D with $\sqsubseteq$ coinciding with $=$ and $\uplus \rightleftharpoons \bigsqcup$ is a natural domain. But, as in the case of [10], it may happen that only under a restricted $\uplus \sqsubseteq \bigsqcup$ a natural domain has some additional nice properties such as "natural" algebraicity properties discussed below in Sect. 3. Note that a natural domain is actually a second-order structure $\langle D, \sqsubseteq^D, \uplus^D \rangle$ in contrast to the ordinary dcpo domains represented as a first-order poset $\langle D, \sqsubseteq^D \rangle$ structure.

Definition 2. *Direct product* of natural (pre-) domains $D \times E$ (or more generally, $\prod_{k \in K} D_k$) is defined by letting $\langle x, y \rangle \sqsubseteq^{D \times E} \langle x', y' \rangle$ iff $x \sqsubseteq^D x'$ & $y \sqsubseteq^E y'$, and additionally $\uplus_i \langle x_i, y_i \rangle \rightleftharpoons \langle \uplus_i x_i, \uplus_i y_i \rangle$ for any family $\langle x_i, y_i \rangle$ of elements in $D \times E$ whenever each natural lub $\uplus_i x_i$ and $\uplus_i y_i$ exists.

---

[3] It follows that for the equality $\uplus_i \uplus_j y_{ij} = \uplus_j \uplus_i y_{ij}$ to hold it suffices to require that all the internal and either one of the external natural lubs or the mixed lub $\uplus_{ij} y_{ij}$ exist.

Proposition 1. *The direct product of natural (pre-) domains is a natural (pre-) domain as well.* □

The poset of all monotonic maps $D \to E$ between any domains ordered pointwise ($f \sqsubseteq^{(D \to E)} f' \rightleftharpoons fx \sqsubseteq^E f'x$ for all $x \in D$) is denoted as $(D \to E)$. We will usually omit the superscripts to $\sqsubseteq$.

Definition 3

(a) A monotonic map $f : D \to E$ between natural pre-domains is called *naturally continuous*[4] if $f(\biguplus_i x_i) = \biguplus_i f(x_i)$ for any directed natural lub $\biguplus_i x_i$, assuming it exists (that is, if $\biguplus_i x_i$ exists then $\biguplus_i f(x_i)$ is required to exist and satisfy this equality). The set of all (monotonic and) naturally continuous maps $D \to E$ is denoted as $[D \to E]$.

(b) Given an arbitrary family $f_i : D \to E$ of monotonic maps between natural pre-domains, define a natural lub $f = \biguplus_i f_i : D \to E$ pointwise, as

$$fx \rightleftharpoons \biguplus_i (f_i x),$$

assuming the latter natural lub exists for all x; otherwise $\biguplus_i f_i$ is undefined.

Proposition 2. *For the case of naturally continuous f_i the resulting f in (b) above is a naturally continuous map as well, assuming E is a natural domain.*

Proof. Use the first part of ($\biguplus$4): $f \biguplus_j x_j \rightleftharpoons \biguplus_i (f_i \biguplus_j x_j) = \biguplus_i \biguplus_j (f_i x_j) = \biguplus_j \biguplus_i (f_i x_j) \rightleftharpoons \biguplus_j f x_j$, for x_j directed and having a natural lub (with all other natural lubs evidently existing). □

Moreover, for any non-empty set F of monotonic functions $D \to E$ and a family $f_i \in F$, if the natural lub $\biguplus_i f_i$ exists and is also an element of F then it is denoted as $\biguplus_i^F f_i$; otherwise, $\biguplus_i^F f_i$, is considered as undefined. When defined, $\biguplus_i^F f_i = \bigsqcup_i^F f_i = \bigsqcup_i^{(D \to E)} f_i$. Here $\bigsqcup^F$ denotes the lub relativized to the poset F with the pointwise partial order $\sqsubseteq^{(D \to E)}$ restricted to F. Evidently, $F \subseteq F' \implies \bigsqcup_i^{F'} f_i \sqsubseteq \bigsqcup_i^F f_i$ when both lubs exist. In contrast with $\bigsqcup^F$, the natural lub $\biguplus_i^F f_i = \biguplus_i f_i$ is essentially independent on F, except it is required to be in F. We will omit the superscript F when it is evident from the context. Further, it is easy to show (by pointwise considerations) that

Proposition 3. *For D and E natural pre-domains, any $F \subseteq (D \to E)$ is (trivially) a natural pre-domain under $\biguplus^F$ defined above. It is also a natural domain*

---

[4] Using the adjective 'natural' here and in other definitions below is, in fact, rather annoying. We would be happy to avoid it at all, but we need to distinguish all these 'natural' non-dcpo versions of the ordinary definitions for dcpos relativized to the natural lub $\biguplus$ from similar definitions relativized to the ordinary lub $\bigsqcup$. In principle, if the context is clear, we could omit 'natural', and use this term as well as 'non-natural' only when necessary. Another way is to write '$\biguplus$-continuous' vs. '$\bigsqcup$-continuous', etc. to make the necessary distinctions.

if E is, and, in particular, $(D \to E)$ and $[D \to E]$ are natural domains in this case with $[D \to E]$ closed under (existing, not necessarily directed) natural lubs in $(D \to E)$.

Proof

($\uplus$1) is trivial.

($\uplus$2) For a family of monotonic functions $\{f_j \in F\}_{j \in J}$ and $I \subseteq J$, assume that $\uplus_{i \in I} f_i \in F$ and $f_j \sqsubseteq \uplus_{i \in I} f_i$ for all $j \in J$. It follows that for all $j \in J$ and $x \in D$, $f_j x \sqsubseteq \uplus_{i \in I}(f_i x)$. Therefore, by using ($\uplus$2) for E, $\uplus_{j \in J}(f_j x)$ exists for all x in the natural domain E, and hence $\uplus_{j \in J} f_j$ does exist too in $(D \to E)$ and therefore coincides with $\uplus_{i \in I} f_i \in F$, as required.

($\uplus$3) For any f, $(\uplus\{f\})x = \uplus\{fx\} = fx$. Thus, $\uplus\{f\} = f$, as required.

($\uplus$4) For arbitrary family of functions $f_{ij} \in F$ ($\uplus$4) reduces to the same in E for $y_{ij} = f_{ij}x$ with arbitrary $x \in D$.

1. Indeed, assume all the required internal natural lubs $\uplus_j f_{ij}$ and one of the external natural lubs $\uplus_i \uplus_j f_{ij}$ or $\uplus_{ij} f_{ij}$ exist and belong to F. Then for all $x \in D$ the corresponding assertion holds for $\uplus_j f_{ij}x$ and $\uplus_i \uplus_j f_{ij}x$ or $\uplus_{ij} f_{ij}x$, and therefore $\uplus_i \uplus_j f_{ij}x = \uplus_{ij} f_{ij}x$ in E. This pointwise identity implies both existence of the required natural lubs in F and equality between them $\uplus_i \uplus_j f_{ij} = \uplus_{ij} f_{ij}$.

2. For directed f_{ij}, $i, j \in I$, and one of the natural lubs $\uplus_j f_{ij}$ or $\uplus_j f_{ii}$ existing, we evidently have for all $x \in D$ that $f_{ij}x$ is directed in each parameter i and j, and $\uplus_j f_{ij}x = \uplus_j f_{ii}x$ holds in E, and therefore both the required lubs exist in F and the equality $\uplus_j f_{ij} = \uplus_j f_{ii}$ holds. $\square$

If natural domains D and E are dcpos with $\uplus = \bigsqcup$ then the same holds both for $(D \to E)$ and $[D \to E]$, and the latter domain coincides with that of all (usual) continuous functions with respect to arbitrary directed lubs. This way natural domain theory generalizes that of dcpo domains, and we will see that other important concepts of domain theory over dcpos have natural counterparts in natural domains with all the ordinary considerations extending quite smoothly to the 'natural' non-dcpo case.

These considerations allow us to construct inductively some natural domains of finite type functionals by taking, for each type $\sigma = \alpha \to \beta$, an arbitrary subset $F_{\alpha \to \beta}$ of monotonic (or only naturally continuous) mappings $F_\alpha \to F_\beta$. Of course, we can additionally require that these F_σ are sufficiently closed (say, under λ-definability or sequential computability). This way, for example, the λ-model of hereditarily-sequential finite type functionals can be obtained. E.g. in [10] this was done inductively over level of types with an appropriate definition of sequentially computable functionals in $\mathbb{Q}_{\alpha_1,\dots,\alpha_n \to \text{Basic-Type}} \subseteq (\mathbb{Q}_{\alpha_1}, \dots, \mathbb{Q}_{\alpha_n} \to \mathbb{Q}_{\text{Basic-Type}})$ (over the basic 'flat' domain $\mathbb{Q}_{\text{Basic-Type}} = \mathbf{N}_\perp$). It was proved only a posteriori and quite non-trivially that all sequential functionals are naturally continuous by embeddings: $\mathbb{Q}_{\alpha_1,\dots,\alpha_n \to \text{Basic-Type}} \subseteq [\mathbb{Q}_{\alpha_1}, \dots, \mathbb{Q}_{\alpha_n} \to \mathbb{Q}_{\text{Basic-Type}}]$ and $\mathbb{Q}_{\alpha \to \beta} \hookrightarrow [\mathbb{Q}_\alpha \to \mathbb{Q}_\beta]$, and satisfy further "natural" algebraicity properties discussed in Sect. 3. It was while determining the domain theoretical nature of $\mathbb{Q}_\alpha$ that the idea of natural domains emerged;

and, although it proved to be quite simple, it was unclear at that moment whether anything reasonable could be obtained. What is new here is a general, abstract presentation of natural domains that does not rely, as in [10], on a type structure like that of $\{\mathbb{Q}_\alpha\}$. Unfortunately, it would take too much space to consider here the construction of the λ-model $\{\mathbb{Q}_\alpha\}$ — the source of general considerations of this paper. (See also [1,3,5] where the same model was defined in a different way and where its domain theoretical structure was not described; it was even unknown whether it is different from the older dcpo model of Milner [4] which was shown later by Normann [6].)

Proposition 4. *Let D, E be natural pre-domains and F a natural domain. A two place monotonic function $f : D \times E \to F$ is naturally continuous iff it is so in each argument.*

Proof. "Only if" is trivial (and uses ($\uplus$3) for F). Conversely, for arbitrary directed families x_i and y_i having natural lubs we have

$$f(\underset{i}{\uplus}\langle x_i, y_i\rangle) \rightleftharpoons f((\underset{i}{\uplus} x_i, \underset{i}{\uplus} y_i)) = \underset{i}{\uplus}\underset{j}{\uplus} f(\langle x_i, y_j\rangle) = \underset{ij}{\uplus} f(\langle x_i, y_j\rangle)$$

$$= \underset{i}{\uplus} f(\langle x_i, y_i\rangle) \ ,$$

as required, by applying the natural continuity of f in each argument and using ($\uplus$4) for F. $\qquad\square$

Proposition 5. *There are the natural (in the sense of category theory) order isomorphisms over natural domains preserving additionally in both directions all the existing natural lubs, not necessarily directed[5],*

$$(D \times E \to F) \cong (D \to (E \to F)) \ , \tag{1}$$
$$[D \times E \to F] \cong [D \to [E \to F]]. \tag{2}$$

This makes the class of natural domains with monotonic, resp., naturally continuous morphisms a Cartesian closed category (ccc) in two ways. Moreover, each side of the second isomorphism is a subset of the corresponding side of the first, with embedding making the square diagram commutative.

Proof. Indeed, the isomorphism (1) and its inverse are defined for any $f \in (D \times E \to F)$ and $g \in (D \to (E \to F))$, as usual, by

$$f^* \rightleftharpoons \lambda x.\lambda y.f(x,y) \in (D \to (E \to F)) \ ,$$
$$\hat{g} \rightleftharpoons \lambda(x,y).gxy \in (D \times E \to F) \ .$$

---

[5] And, of course, preserving the ordinary lubs.

Then $\lambda f.f^*$ preserves (in both directions) all the existing natural lubs $(\biguplus_i f_i)^* = \biguplus_i f_i^*$:

$$(\biguplus_i f_i)^* xy \rightleftharpoons (\biguplus_i f_i)(x,y) \rightleftharpoons \biguplus_i f_i(x,y) \rightleftharpoons \biguplus_i ((f_i^*x)y) \rightleftharpoons (\biguplus_i (f_i^*x))y$$

$$\rightleftharpoons ((\biguplus_i f_i^*)x)y \rightleftharpoons (\biguplus_i f_i^*)xy$$

holds for all $x \in D$ and $y \in E$ where if the first natural lub exists then all the others exist too, and conversely. Here we used only the definitions of * and $\biguplus$ for functions. The second isomorphism (2) is just the restriction of the first. For its correctness we should check that f^* (resp. $\hat{g}$) is naturally continuous if f (resp. g) is:

$$f^* \biguplus_i x_i \rightleftharpoons \lambda y.f(\biguplus_i x_i, y) = \lambda y.\biguplus_i f(x_i,y) \rightleftharpoons \biguplus_i \lambda y.f(x_i,y) \rightleftharpoons \biguplus_i f^* x_i$$

by using additionally Proposition 4 in the second equality. Similarly,

$$\hat{g}(\biguplus_i x_i, \biguplus_i y_i) \rightleftharpoons g(\biguplus_i x_i)(\biguplus_i y_i) = \biguplus_i g x_i (\biguplus_i y_i) = \biguplus_i \biguplus_j g x_i y_j$$

$$= \biguplus_i g x_i y_i \rightleftharpoons \biguplus_i \hat{g}(x_i, y_i)$$

by using ($\biguplus 4$) for F. □

Definition 4. An upward closed set U in a natural pre-domain D is called *naturally Scott open* if for all directed families x_i having the natural lub

$$\biguplus_i x_i \in U \Longrightarrow x_i \in U \text{ for some } i.$$

Such subsets constitute the *natural Scott topology* on D.

This is a straightforward generalization of the ordinary *Scott topology* on any poset defined in terms of the usual lub $\bigsqcup$ of directed families. Evidently, each Scott open set (in the standard sense) is naturally Scott open, and therefore the latter sets constitute a T_0-topology.

Proposition 6

(a) *Any natural pre-domain $\langle D, \sqsubseteq^D, \biguplus^D \rangle$ is a T_0-space under its natural Scott topology whose standardly generated partial ordering coincides with the original ordering $\sqsubseteq^D$ on D.*

(b) *Continuous functions between pre-domains defined as preserving the existing natural lubs are also continuous relative to the natural Scott topologies in the domain and co-domain.*

(c) But the converse holds only in the weakened form: continuity of a map f in the sense of natural Scott topologies implies $f(\biguplus_i x_i) = \bigsqcup_i f(x_i)$ for any directed family x_i with existing $\biguplus_i x_i$.[6]

Proof

(a) If $x \sqsubseteq y$ and $x \in U$ for any naturally Scott open $U \subseteq D$ then $y \in U$ because U is upward closed. Conversely, assume $x \not\sqsubseteq y$, and define $U_y \rightleftharpoons \{z \in D \mid z \not\sqsubseteq y\}$. This set is evidently upward closed. Let $\biguplus_i x_i \in U_y$ for a directed family. Then it is impossible that all $x_i \notin U_y$, i.e. $x_i \sqsubseteq y$, because then we should have $\biguplus_i x_i \sqsubseteq y$ — a contradiction. Therefore U_y is a naturally Scott open set (in fact, even Scott open in the standard sense) such that $x \not\sqsubseteq y$, $x \in U_y$ but $y \notin U_y$, as required.

(b) Assume monotonic $f : D \to E$ preserves natural directed lubs and $U \subseteq E$ is naturally Scott open in E. Then $f^{-1}(U)$ is evidently upward closed in D as U is such in E. Further, let $\biguplus_i x_i \in f^{-1}(U)$, i.e. $f(\biguplus_i x_i) = \biguplus_i f(x_i) \in U$ and hence $f(x_i) \in U$ and $x_i \in f^{-1}(U)$ for some i. Therefore $f^{-1}(U)$ is naturally Scott open. That is, f is continuous in the sense of natural Scott topologies in D and E.

(c) Conversely, assume $f : D \to E$ is continuous in the sense of natural Scott topologies in D and E, and $\biguplus_i x_i$ exists in D for a directed family. Let us show that $f(\biguplus_i x_i) = \bigsqcup_i f(x_i)$. The inequality $f(\biguplus_i x_i) \sqsupseteq f(x_i)$ follows by monotonicity of f. Assume y is an upper bound of all $f(x_i)$ in E but $f(\biguplus_i x_i) \not\sqsubseteq y$. Define like above the Scott open set $V_y \rightleftharpoons \{z \in E \mid z \not\sqsubseteq y\}$. Then $f^{-1}(V_y)$ is naturally Scott open containing $\biguplus_i x_i$ and therefore some x_i, implying $f(x_i) \in V_y$, i.e. $f(x_i) \not\sqsubseteq y$ — a contradiction. This means that $f(\biguplus_i x_i) = \bigsqcup_i f(x_i)$. □

3   Naturally Finite Elements

Definition 5. A *naturally finite* element d in a natural pre-domain D is such that for any directed natural lub (assuming it exists) if $d \sqsubseteq \biguplus X$ then $d \sqsubseteq x$ for some x in X. If arbitrary directed lubs $\bigsqcup X$ are considered then d is called just *finite*.

The last part of the definition is most reasonable in the case of dcpos. Otherwise (assuming $\biguplus \neq \bigsqcup$), 'finite' means rather 'non-natural finite'.

Definition 6. A natural pre-domain D is called *naturally (ω-) algebraic* if (it has only countably many naturally finite elements and) each element in D is a natural lub of a (non-empty) directed set of naturally finite elements.

---

[6] In the special case of $\biguplus \rightleftharpoons \bigsqcup$ and standard Scott topologies we have, as usual, the full equivalence of the two notions of continuity of maps with $f(\biguplus_i x_i) = \biguplus_i f(x_i)$. We will see below that the full equivalence of these two notion of continuity holds also for naturally algebraic and naturally bounded complete natural pre-domains.

If D is dcpo with $\biguplus = \bigsqcup$ then the above reduces to the traditional concept of $(\omega\text{-})$ *algebraic dcpo*. It follows, assuming additionally $(\biguplus 2)$, that

$$x = \biguplus \hat{x} \tag{3}$$

where $\hat{x} \rightleftharpoons \{d \sqsubseteq x \mid d \text{ is naturally finite}\}$ for any $x \in D$.

Definition 7. If any two upper bounded elements c, d have least upper bound $c \sqcup d$ in D then D is called *bounded complete*, and it is called *finitely bounded complete* if, in the above, only finite c, d (and therefore $c \sqcup d$) are considered.

This is the traditional definition adapted to the case of an arbitrary poset D. If D is an algebraic dcpo then it is bounded complete iff it is finitely bounded complete. In fact, for dcpos bounded completeness means existence of a lub for any bounded set, not necessarily finite. Algebraic and bounded complete dcpos with least element $\bot$ are also known as *Scott domains* or as the *complete f_0-spaces* of Ershov [2] (or just *Scott-Ershov domains*). For the 'natural', non-dcpo version of these domains we need

Definition 8. A natural pre-domain D is called *naturally bounded complete* if any two naturally finite elements upper bounded in D have a lub (not necessarily natural lub, but evidently naturally finite element).

In such domains any set of the form $\hat{x}$ is evidently directed, if non-empty. (It is indeed non-empty in naturally algebraic pre-domains.)

Proposition 7. *For a naturally algebraic natural domain D the natural lub of an arbitrary family x_i can be represented as*

$$\biguplus_i x_i = \biguplus \bigcup_i \hat{x}_i \tag{4}$$

where both natural lubs either exist or not simultaneously.

Proof. Indeed, let $x_i^0 \sqsubseteq x_i$ denote an arbitrarily chosen naturally finite approximation of x_i, and let j range over naturally finite elements of D. Define $x_{ij} \rightleftharpoons j$ if $j \sqsubseteq x_i$, and $\rightleftharpoons x_i^0$ otherwise. Then $\biguplus_i x_i = \biguplus_i \biguplus \hat{x}_i = \biguplus_i \biguplus_j x_{ij} = \biguplus_{ij} x_{ij} = \biguplus \bigcup_i \hat{x}_i$ by (3) and the first part of $(\biguplus 4)$. $\qquad \square$

Therefore, any naturally algebraic natural domain D is, in fact, defined by the quadruple $\langle D, D^{[\omega]}, \sqsubseteq^D, \mathcal{L} \rangle$ where $D^{[\omega]} \subseteq D$ consists of naturally finite elements in D and $\mathcal{L}$ is the set of all sets of naturally finite elements having a natural lub. Indeed, we can recover $\biguplus_i x_i \rightleftharpoons \bigsqcup \bigcup_i \hat{x}_i$ by (4) whenever $\bigcup_i \hat{x}_i \in \mathcal{L}$. Moreover, in the case of naturally algebraic and naturally bounded complete natural domains D their elements x can be identified, up to the evident order isomorphism, with the *ideals* $\hat{x} \in \mathcal{L}$ (non-empty directed downward closed sets of naturally finite elements ordered under set inclusion and having a natural lub). In particular,

$$x \sqsubseteq y \iff \hat{x} \subseteq \hat{y} . \tag{5}$$

Note 1. The above definition via naturally finite elements and ideals does not always work in practice. Thus, in the real application of this theory to the λ-model of hereditarily sequential finite type functionals $\{\mathbb{Q}_\alpha\}$ [10] we do not have naturally finite elements $\mathbb{Q}_\alpha^{[\omega]} \subseteq \mathbb{Q}_\alpha$ as given. We only have a priori that $\mathbb{Q}_\alpha$ are partial ordered sets with $\perp_\alpha$ and with monotonic application operators $\mathrm{App}_{\alpha,\beta} : \mathbb{Q}_{\alpha\to\beta} \times \mathbb{Q}_\alpha \to \mathbb{Q}_\beta$. That they are, in fact, naturally ω-algebraic, naturally bounded complete natural domains with $\mathrm{App}_{\alpha,\beta}$ naturally continuous requires quite complicated considerations (using appropriate theory of sequential computational strategies) for its proof. Even the fact that the natural (in fact, quite simply defined as pointwise) lub $\biguplus^\alpha$ on $\mathbb{Q}_\alpha$ is fruitful notion to use here was not self-evident at all.

Generalizing the case of dcpos we can improve an appropriate part in Proposition 6 (see also footnote 6):

Proposition 8

(a) For D and E naturally algebraic and naturally bounded complete natural pre-domains, a monotonic map $f : D \to E$ is naturally continuous (in the sense of preserving directed natural lubs) iff for all $x \in D$ and naturally finite $b \sqsubseteq fx$ there exists naturally finite $a \sqsubseteq x$ such that $b \sqsubseteq fa$. This means that natural continuity of functions between such domains is equivalent to topological continuity with respect the natural Scott topology because

(b) Naturally Scott open sets in such domains are exactly arbitrary unions of the upper cones $\breve{a} \rightleftharpoons \{x \mid a \sqsubseteq x\}$ for a naturally finite.

Proof (a) Indeed, for f naturally continuous, $fx = \biguplus f(\hat{x})$, so $b \sqsubseteq fx$ implies $b \sqsubseteq fa$ for some $a \sqsubseteq x$ for naturally finite a, b.

Conversely, assume f satisfies the above b-a-continuity property and $x = \biguplus_i x_i$ be a natural directed lub in D. Let us show that $fx = \biguplus_i fx_i$. The inclusions $fx_i \sqsubseteq fx$ hold by monotonicity of f and imply $\bigcup_i \widehat{fx_i} \subseteq \widehat{fx}$. Now, it suffices to show, by (4) and (5) applied to E, the inverse inclusion $\widehat{fx} \subseteq \bigcup_i \widehat{fx_i}$. Thus, assume $b \sqsubseteq fx$ for a naturally finite b and hence $b \sqsubseteq fa$ for some naturally finite $a \sqsubseteq x = \biguplus_i x_i$ and, therefore, $a \sqsubseteq x_i$ for some i. Then $b \sqsubseteq fa \sqsubseteq fx_i$, as required.

(b) This follows straightforwardly from the definitions of naturally finite elements, naturally Scott open sets, and from the identity $x = \biguplus \hat{x}$ (with $\hat{x}$ directed). $\qquad\square$

Note 2. In fact, it can be shown that naturally algebraic and naturally bounded complete natural (pre-) domains, if considered as topological spaces under the natural Scott topology, are exactly f-spaces of Ershov [2] (i.e. all, not necessary complete f-spaces). But here again we could apply the comments of Note 1. Indeed, $\mathbb{Q}_\alpha$ do not originally appear as f-spaces (represented as in [2] either topologically or order theoretically with finite (or f-) elements as given). This becomes clear only a posteriori, after complicated considerations based, in particular, on the general concept of natural domains (and on a lot of other things). That is why this concept is important in itself.

Further generalizing the traditional dcpo case and working in line with the theory of f-spaces [2], we can show

Proposition 9. *If natural domains D and E are naturally (ω-)algebraic and naturally bounded complete then so are $D \times E$ and $[D \to E]$, assuming additionally in the case of $[D \to E]$ that E contains the least element $\perp_E$. Then such a restricted class of domains with $\perp$ and with naturally continuous morphisms constitute a ccc.*

Proof. For $D \times E$ this is evident. Let us show this for $[D \to E]$. Indeed, let $a_0, \ldots, a_{n-1} \in D$ and $b_0, \ldots, b_{n-1} \in E$ be two arbitrary lists of naturally finite elements satisfying the

Consistency condition: for any $x \in D$ the set $\{b_i \mid a_i \sqsubseteq x, i < n\}$ is upper bounded in E, and hence its lub exists and is naturally finite.

(In general, assume that $a, b, c, d, \ldots$, possibly with subscripts, range over naturally finite elements.) Then define a *tabular* function $\begin{bmatrix} b_0, \ldots, b_{n-1} \\ a_0, \ldots, a_{n-1} \end{bmatrix} \in [D \to E]$ by taking for any $x \in D$

$$\begin{bmatrix} b_0, \ldots, b_{n-1} \\ a_0, \ldots, a_{n-1} \end{bmatrix} x \rightleftharpoons \bigsqcup \{b_i \mid a_i \sqsubseteq x, i < n\} \qquad (6)$$

because this lub does always exist. (Here we use the fact that E contains the least element $\perp_E$ needed to get the lub defined if the set on the right is empty.) In particular, $\begin{bmatrix} b_0, \ldots, b_{n-1} \\ a_0, \ldots, a_{n-1} \end{bmatrix}$ is the least monotonic function $f : D \to E$ for which $b_i \sqsubseteq f a_i$ for all $i < n$, that is,

$$\begin{bmatrix} b_0, \ldots, b_{n-1} \\ a_0, \ldots, a_{n-1} \end{bmatrix} \sqsubseteq f \iff b_i \sqsubseteq f a_i \text{ for all } i < n. \qquad (7)$$

Moreover, this is also a naturally continuous function. Indeed, for any directed family $\{x_k\}_{k \in K}$ in D with the natural lub existing

$$\begin{bmatrix} b_0, \ldots, b_{n-1} \\ a_0, \ldots, a_{n-1} \end{bmatrix} \biguplus_k x_k = \bigsqcup \{b_i \mid a_i \sqsubseteq \biguplus_k x_k\} = \begin{bmatrix} b_0, \ldots, b_{n-1} \\ a_0, \ldots, a_{n-1} \end{bmatrix} x_{k_0}$$

for some $k_0 \in K$ (due directedness of $\{x_k\}_{k \in K}$) so that, in fact, $\begin{bmatrix} b_0, \ldots, b_{n-1} \\ a_0, \ldots, a_{n-1} \end{bmatrix} x_k \sqsubseteq$ $\begin{bmatrix} b_0, \ldots, b_{n-1} \\ a_0, \ldots, a_{n-1} \end{bmatrix} x_{k_0}$ for all $k \in K$ and hence, by ($\biguplus 2$) and ($\biguplus 3$) for E,

$$\begin{bmatrix} b_0, \ldots, b_{n-1} \\ a_0, \ldots, a_{n-1} \end{bmatrix} \biguplus_k x_k = \biguplus_k \begin{bmatrix} b_0, \ldots, b_{n-1} \\ a_0, \ldots, a_{n-1} \end{bmatrix} x_k .$$

It is also follows from (7) that $\begin{bmatrix} b_0, \ldots, b_{n-1} \\ a_0, \ldots, a_{n-1} \end{bmatrix} = \bigsqcup_{i < n} \begin{bmatrix} b_i \\ a_i \end{bmatrix}$. Moreover, this is a naturally finite element in $[D \to E]$. Thus, in the simplest case of $\begin{bmatrix} b \\ a \end{bmatrix}$

$$\begin{bmatrix} b \\ a \end{bmatrix} \sqsubseteq \biguplus_j f_j \overset{(7)}{\iff} b \sqsubseteq \biguplus_j f_j a \iff \exists j. b \sqsubseteq f_j a \iff \exists j. \begin{bmatrix} b \\ a \end{bmatrix} \sqsubseteq f_j$$

for any directed family of naturally continuous functions f_j with $\biguplus_j f_j$ and therefore $\biguplus_j f_j a$ existing. If $\begin{bmatrix} b_0,...,b_{n-1} \\ a_0,...,a_{n-1} \end{bmatrix} \sqsubseteq f$ and $\begin{bmatrix} d_0,...,d_{m-1} \\ c_0,...,c_{m-1} \end{bmatrix} \sqsubseteq f$ then evidently $\begin{bmatrix} b_0,...,b_{n-1} \\ a_0,...,a_{n-1} \end{bmatrix} \sqcup \begin{bmatrix} d_0,...,d_{m-1} \\ c_0,...,c_{m-1} \end{bmatrix} = \begin{bmatrix} b_0,...,b_{n-1},d_0,...,d_{m-1} \\ a_0,...,a_{n-1},c_0,...,c_{m-1} \end{bmatrix} \sqsubseteq f$. Thus, the set $\hat{f}$ of tabular approximations to any monotonic function f is directed. Moreover, any naturally continuous f is, in fact, the natural lub of this set:

$$f = \biguplus \hat{f} = \biguplus \{\varphi \mid \varphi \sqsubseteq f \ \& \ \varphi \text{ tabular}\} \tag{8}$$

because

$$fx = \biguplus \widehat{fx} = \biguplus \{b \mid b \sqsubseteq fx\} = \biguplus \{b \mid \exists \text{ naturally finite } a \sqsubseteq x \ (b \sqsubseteq fa)\}$$
$$= \biguplus \{ \begin{bmatrix} b \\ a \end{bmatrix} x \mid \begin{bmatrix} b \\ a \end{bmatrix} \sqsubseteq f\} = \biguplus \{\varphi x \mid \varphi \sqsubseteq f \ \& \ \varphi \text{ tabular}\}.$$

The last equality holds because, for tabular functions, $\varphi x = \begin{bmatrix} b \\ a \end{bmatrix} x$ for some $\begin{bmatrix} b \\ a \end{bmatrix} \sqsubseteq \varphi$ (where, in accordance with (6), $\begin{bmatrix} b \\ a \end{bmatrix}$ does not necessary is one of the columns of the tabular representation of φ). It also follows from (8) that tabular elements of $[D \to E]$ are exactly the naturally finite ones. Moreover, this domain is naturally $(\omega\text{-})$algebraic and naturally bounded complete. □

Note 3. For any finite list of tabular elements $\varphi_1, \ldots, \varphi_k$ in $[D \to E]$, they are upper bounded in $[D \to E]$ iff the union of tables representing φ_i is consistent in the above sense. This reduces, essentially algorithmically, the problem of upperboundedness for naturally finite elements in $[D \to E]$ to those in D and E. But if we would consider a subset of $F \subseteq [D \to E]$ (say, of sequential or other kind of restricted function(al)s as in [10]) then no such algorithmic reduction for F is possible a priori, even if it is naturally algebraic and naturally bounded complete and its naturally finite elements are represented in the tabular way as above.

4 Semi-formal Considerations on the More General Case of $F \subseteq [D \to E]$ Induced by [4][7]

Here most our of assertions will have a conditional character with intuitively appealing assumptions. Let $F \subseteq (D \to E)$ be an arbitrary natural domain of monotonic functions (for appropriate natural domains D and E). (See Proposition 3. For example, F could consist of naturally continuous sequential function(al)s only.) While postulating the additional requirement of natural continuity and ω-algebraicity property of a function domain F looks quite reasonable from the computational perspective, *the requirement of (natural) bounded completeness might seem questionable in general.* Why should the lub of two (naturally finite)

---

[7] Note that [4] was devoted only to the case of dcpos.

sequential functionals exist at all and be sequentially computable, even if they have a joint upper bound? However the following intuitive, semi-formal and sufficiently general argumentation in favour of natural bounded completeness can be given (and easily formalised for the case of finite type functionals).

The simplest, 'basic' domains D like flat ones may be reasonably postulated to be naturally bounded complete. Also, the greatest lower bound (glb) $x \sqcap y$ of any two elements can be considered computable/natural continuous. (Say, for flat domains we need only conditional **if** and equality $=$ to define $\sqcap$.) Then, assuming that F has the most basic computational closure properties, we can conclude that F is also closed under the naturally continuous operation glb $f \sqcap g = (\lambda x \in D.fx \sqcap gx) \in F$.

Moreover, it seems quite reasonable to assume that the set of naturally finite elements in any 'basic' D is a directed union, $D^{[\omega]} = \bigcup_k D^{[k]}$, of some finite sets $D^{[k]}$ of naturally finite objects which are suitably *finitely restricted* for each k where k (say, $0, 1, 2, \ldots$) may serve as a measure of restriction. For each $D^{[k]} \subseteq D$ we could expect that each $x \in D$ has a best naturally finite lower approximation $x^{[k]} = \Psi^{[k]} x \sqsubseteq x$ from $D^{[k]}$, assuming also $\Psi^{[k]}(\Psi^{[k]} x) = \Psi^{[k]} x$. Thus, $\Psi_D^{[k]} : D \to D$ is just a monotonic projection onto its finite range $D^{[k]}$. It easily follows that the family $\{x^{[k]}\}_k$ is directed for any $x \in D$. Also it is a reasonable assumption that such $\Psi_D^{[k]}$, for the basic domains, are computable and therefore naturally continuous.

Then the fact that each finitely restricted element $x^{[k]}$ is naturally finite can even be deduced as follows: $x^{[k]} \sqsubseteq \biguplus Z$ for a directed set Z implies $x^{[k]} \sqsubseteq \biguplus \{z^{[k]} \mid z \in Z\} = z^{[k]} \sqsubseteq z$ for some z by natural continuity of $\Psi^{[k]}$ and because $D^{[k]}$ is finite.

Further, we could additionally assume that $x = \biguplus_k x^{[k]}$ holds for all x. This implies formally (from our assumptions) that naturally finite and finitely restricted (i.e., of the form $x^{[k]}$) elements in D are the same.

It follows that any two upper bounded finitely restricted elements $d, e \in D^{[k]}$ must have a (not necessarily natural) lub $d \sqcup e$ in D which is also finitely restricted. Indeed, it can be obtained as the greatest lower bound in D of a finite nonempty set:

$$d \sqcup e = \sqcap\{x^{[k]} \mid x \sqsupseteq d, e\}. \tag{9}$$

By induction, given any (not necessary 'basic') naturally ω-algebraic and naturally bounded complete domains D and E with such projections, we should conclude that the composition $\Psi_E^{[k]} \circ f \circ \Psi_D^{[k]}$, denoted as $\Psi_F^{[k]} f$ or $f^{[k]}$ ($f^{[k]} x \rightleftharpoons (fx^{[k]})^{[k]}$), is computable/naturally continuous, assuming $f \in F \subseteq [D \to E]$ is such. Assuming that F has minimal reasonable closure properties, we can conclude that this composition should belong to F as well. But, once all $D^{[k]}$ and $E^{[k]}$ are finite sets consisting only of naturally finite elements, $\Psi_F^{[k]} f$ is just a naturally finite tabular function, which can be reasonably postulated as k-restricted in F, and $\Psi_F^{[k]} : F \to F$ is the corresponding directed family of projections having finite ranges $F^{[k]}$ consisting of some tabular k-restricted functions.

These projections are naturally continuous and, moreover, preserve all existing natural lubs (not necessarily directed) assuming $\Psi_D^{[k]}$ and $\Psi_E^{[k]}$ do:

$$(\Psi_E^{[k]} \circ (\biguplus_i f_i) \circ \Psi_D^{[k]})x = \Psi_E^{[k]}((\biguplus_i f_i)(\Psi_D^{[k]}x)) = \Psi_E^{[k]}(\biguplus_i (f_i(\Psi_D^{[k]}x)))$$

$$= \biguplus_i \Psi_E^{[k]}(f_i(\Psi_D^{[k]}x)) = \biguplus_i ((\Psi_E^{[k]} \circ f_i \circ \Psi_D^{[k]})x) = (\biguplus_i (\Psi_E^{[k]} \circ f_i \circ \Psi_D^{[k]}))x \ .$$

Moreover, having that F consists of only naturally continuous functions, $f = \biguplus_k f^{[k]}$ should hold for all f. Indeed, this follows from the same property in D and E: $fx = f(\biguplus_k x^{[k]}) = \biguplus_k (fx^{[k]}) = \biguplus_k \biguplus_m (fx^{[k]})^{[m]} = \biguplus_k (fx^{[k]})^{[k]} = \biguplus_k (f^{[k]}x)$. Then we can conclude that the tabular functions (of the form $f^{[k]}$ for any $f \in F$) are exactly the naturally finite elements of the natural domain F, and F is naturally ω-algebraic. Finally, having projections $\Psi_F^{[k]}$ and naturally continuous finite glb $\sqcap$ in F (definable by induction like above and therefore existing in F by the natural closure properties), natural bounded completeness of F follows exactly as above in (9) for the case of 'basic' domains.

To define a naturally ω-algebraic and naturally bounded complete natural domain $F \subseteq [D \to E]$, we can fix any (simply) bounded complete set $F^{[\omega]}$ of tabular elements in $[D \to E]$ containing $\perp_{[D \to E]}$, and take F to be any extension of $F^{[\omega]}$ by some (if exists in $[D \to E]$) directed natural lubs of these tabular elements. Then $F^{[\omega]}$ is exactly the set of all naturally finite elements in F. Two extreme versions of F are $F^{[\omega]}$, and the set of all existing directed natural lubs from $F^{[\omega]}$. Besides the fact that this construction looks quite natural in itself, it follows from the above considerations that naturally finite elements in F cannot be anything other than tabular elements, provided there are, as above, directed families of naturally continuous projections $\Psi_D^{[k]}$ and $\Psi_E^{[k]}$ to finite elements such that $x = \biguplus_k x^{[k]}$ and $y = \biguplus_k y^{[k]}$ hold for any $x \in D$ and $y \in E$, and that F is closed under projections $\Psi_F^{[k]}$ defined from $\Psi_D^{[k]}$ and $\Psi_E^{[k]}$.

5 Conclusion

Our presentation is that of the current state of affairs and has the peculiarity that really interesting concrete examples of non-dcpo domains (such as those of hereditarily sequential and wittingly consistent higher type functionals [10]) from which this theory has in fact arisen require too much space to be presented here. The theory is general, but the non-artificial and instructive non-dcpo examples on which it is actually based are rather complicated and in a sense exceptional (dcpo case being more typical and habitual). However we can hope that there will be many more examples where this theory can be used, similarly to the case of dcpos.

One important topic particularly important for applications which was not considered here in depth and which requires further special attention is the possibility of the effective version of naturally algebraic, naturally bounded complete natural domains. Unlike the ordinary dcpo version (Ershov-Scott domains), not everything goes so smoothly here as is noted in connection with the model of hereditarily sequential functionals in Sect. 2.4 of [10]; see also Note 3 above.

Acknowledgments. The author is grateful to Yuri Ershov for a related discussions on f-spaces, to Achim Jung for his comments on the earlier version of presented here non-dcpo domain theory, and to Grant Malcolm for his kind help in polishing the English.

References

1. Abramsky, S., Jagadeesan, R., Malacaria, P.: Full Abstraction for PCF. Information and Computation 163(2), 409–470 (2000)
2. Ershov, Y.L.: Computable functionals of finite types. Algebra and Logic 11(4), 367–437 (1972), http://www.springerlink.com doi: 10.1007/BF02219096 (English Version)
3. Hyland, J.M.E., Ong, C.-H.L.: On Full Abstraction for PCF: I, II, and III. Information and Computation 163, 285–408 (2000)
4. Milner, R.: Fully abstract models of typed λ-calculi. Theoretical Computer Science 4, 1–22 (1977)
5. Nickau, H.: Hereditarily-Sequential Functionals: A Game-Theoretic Approach to Sequentiality. Siegen, PhD Thesis (1996)
6. Normann, D.: On sequential functionals of type 3. Mathematical Structures in Computer Science 16(2), 279–289 (2006)
7. Plotkin, G.: LCF considered as a programming language. Theoretical Computer Science 5, 223–256 (1977)
8. Sazonov, V.Y.: Functionals computable in series and in parallel. Sibirskii Matematicheskii Zhurnal 17(3), 648–672 (1976), http://www.springerlink.com, doi: 10.1007/BF00967869 (English Version)
9. Sazonov, V.Y.: Expressibility of functionals in D.Scott's LCF language. Algebra and Logic 15(3), 308–330 (1976), http://www.springerlink.com, doi: 10.1007/BF01876321 (English Version)
10. Sazonov, V.Y.: Inductive Definition and Domain Theoretic Properties of Fully Abstract Models for PCF and PCF. Logical Methods in Computer Science 3(3:7), 1–50 (2007), http://www.lmcs-online.org
11. Scott, D.S.: A type-theoretical alternative to ISWIM, CUCH, OWHY, Theoretical Computer Science, 121 (1&2), 411–440 Böhm Festschrift (1993) Article has been widely circulated as an unpublished manuscript since (1969)

Church's Problem and a Tour through Automata Theory

Wolfgang Thomas

RWTH Aachen, Lehrstuhl Informatik 7, 52056 Aachen, Germany
thomas@informatik.rwth-aachen.de

Dedicated to Boris A. Trakhtenbrot, pioneer and teacher of automata theory
for generations of researchers, on the occasion of his 85th birthday.

Abstract. Church's Problem, stated fifty years ago, asks for a finite-state machine that realizes the transformation of an infinite sequence α into an infinite sequence β such that a requirement on (α, β), expressed in monadic second-order logic, is satisfied. We explain how three fundamental techniques of automata theory play together in a solution of Church's Problem: Determinization (starting from the subset construction), appearance records (for stratifying acceptance conditions), and reachability analysis (for the solution of games).

1 Introduction

Around 1960, a core of automata theory had been established which led to the first comprehensive expositions, such as the volume *Sequential Machines – Selected Papers* edited by Moore [15] and the monograph [11] of Hopcroft and Ullman. In these early books three essential aspects of automata theory are either underrepresented or missing: the view of automata as transducers (computing functions rather than accepting languages), the use of automata in the study of infinite computations, and the close connection between automata and logic. These directions of study are a focus in the work of B.A. Trakhtenbrot. In the development of automata theory the three aspects often appeared in combination, offered most beautiful results and – as we know today – are highly significant and even indispensable for many applications in the design and analysis of computer systems.

The breakthrough on the relation between automata and logic was the proof of the expressive equivalence between finite automata and weak monadic second-order arithmetic over the natural number ordering, established by Büchi and Elgot (see the joint announcement [3] of 1958 and the two papers [1,8]) and independently by Trakhtenbrot [27] (submitted in July 1957). The Büchi-Elgot-Trakhtenbrot Theorem was extended soon after by Büchi [2] to the full monadic second-order theory of the natural number ordering, together with an expressive model of finite automaton over infinite sequences ("Büchi automaton"). Later Rabin showed how to generalize this theory to cover also infinite trees [20].

A. Avron et al. (Eds.): Trakhtenbrot/Festschrift, LNCS 4800, pp. 635–655, 2008.

The equivalence between formulas of monadic second-order logic and finite automata opened a way to establish algorithms that can test sentences for truth in the standard model of arithmetic. After decades of work on variants of the original question and on improving the efficiency of decision procedures, this approach became the origin of "model-checking", today a vast field which offers techniques for verifying highly nontrivial software and hardware systems.

Regarding the infinite behavior of automata and the use of automata as transducers, a master problem was raised by Church in 1957 [5] (see also [6]). He asked for the synthesis of automata that realize functions over infinite words rather than languages. Church posed the problem whether certain transformations of infinite words that are specified in a system of arithmetic are computable by finite automata (in his words: by circuits):

> Given a requirement which a circuit is to satisfy, we may suppose the requirement expressed in some suitable logistic system which is an extension of restricted recursive arithmetic. The *synthesis problem* is then to find recursion equivalences representing a circuit that satisfies the given requirement (or alternatively, to determine that there is no such circuit). ([5, p.8-9])

Church's Problem was solved by Büchi and Landweber [4] for specifications in monadic second-order logic over $(\mathbb{N}, <)$, building on a fundamental result by McNaughton [13] on the determinization of Büchi automata. These two results, the Büchi-Landweber Theorem and the McNaughton Theorem, are the origin of a field which might be called "synthesis of reactive systems" (rather than "verification"), with the algorithmic theory of infinite games as a core discipline. Today the area attracts much attention – it is concerned with refined studies on Church's synthesis problem and the extension to more general questions (e.g., on infinite stochastic games or multiplayer games).

At a very early stage, it was again Trakhtenbrot who merged the fundamental constructions of the subject in his pioneering monograph with Barzdin [28]. The part due to Trakhtenbrot (namely, Chapters I to III) covers both key results indicated above; it is based on lecture notes of his of 1966, with more material (on the Büchi-Landweber Theorem) added with the translation. It is remarkable to see that the authors call (in the preface to [28]) the Chapters I-III the "old" parts of the theory, while just the Chapters IV and V, which focus on statistical aspects, are mentioned as the "first encouraging steps of a new trend". This judgment was prophetic in the sense that today it is true as it was more than 30 years ago; probably the use of statistical methods will be a key in developing efficient approaches to the present demanding challenges in verification and synthesis. However, for the remainder of this paper, our objective is rather to reflect on the "old" theory. We single out basic ingredients that are relevant to Church's Problem, taking a view as it developed over the past twenty years. We call these methods "determinization", "appearance records", and "reachability analysis". The first two deal with approaches to set up memory structures in finite automata, and the last one is concerned with techniques for exploring transition graphs. The purpose of this paper is to present the integration of these ideas in a solution of Church's problem. Since the details of these constructions are well-known, our exposition focusses on methodological issues rather than offering a full technical treatment[1].

In Sect. 2, we begin with a presentation of Church's Problem. Section 3 briefly discusses the issue of determinization and subset constructions. Determinization is the key construction for transforming Church's Problem into a problem of state-based infinite games, namely into the question of solving so-called "Muller games".

Sections 3 and 4 are the main part of the paper; they are devoted to the solution of Muller games in two stages, following an idea proposed in [24] (as an alternative to the original proof in [4]). The first stage is a stratification of the Muller winning condition; it leads to the so-called "parity condition". We explain how this stratification is obtained by a simple memory structure that we call "appearance records". We present it in two versions (for weak and strong Muller games).

The last step is the solution of parity games (again in their weak and strong version), showing memoryless determinacy of these games. This completes the solution of Church's Problem. We explain (in Sect. 4) that the core of the proof of memoryless determinacy of parity games is provided by (a subtle iteration of) simple reachability tests. In the present game theoretical framework, we deal with alternating reachability.

Of course, this emphasis on three essential constructions just points to some selected central ideas. Automata theory is much too rich to be reducible to these simple principles. For example, we do not touch the large area of automaton minimization. Our exposition is also more motivated by didactic aspects than by claims on practical applicability. For applications in program verification or program synthesis, one often has to find refinements or even alternatives for the basic constructions in order to ensure algorithmically satisfactory solutions.

2   Church's Problem

Let us start with an example. Our objective is to construct a finite automaton that transforms an input stream α of bits into an output stream β of bits such that the following three conditions are satisfied. (We write, e.g., $\alpha(t)$ for the t-th bit of α ($t = 0, 1, \ldots$), and $\exists^{\omega}$ for the quantifier "there exist infinitely many".)

1. $\forall t(\alpha(t) = 1 \quad \rightarrow \quad \beta(t) = 1)$
2. $\neg \exists t \; \beta(t) = \beta(t+1) = 0$
3. $\exists^{\omega} t \; \alpha(t) = 0 \quad \rightarrow \quad \exists^{\omega} t \; \beta(t) = 0$

The desired automaton has to produce the output bit $\beta(t)$ without delay upon receipt of $\alpha(t)$. More specifically, we work with transducers in the format of deterministic Mealy automata. A Mealy automaton has the format $\mathcal{M} = (S, \Sigma, \Gamma, s_0, \delta, \tau)$ where S is the finite set of states, Σ and Γ are the input

---

[1] Readers who want to see a self-contained exposition are referred to the tutorial [25].

alphabet and output alphabet, respectively, s_0 the initial state, $\delta : S \times \Sigma \to S$ the transition function and $\tau : S \times \Sigma \to \Gamma$ the output function. In a graphical presentation we label a transition from p to $\delta(p, a)$ by $a/\tau(p, a)$. The definition of the function $f_{\mathcal{M}} : \Sigma^\omega \to \Gamma^\omega$ computed by $\mathcal{M}$ is then obvious.

For our example, the first two conditions are satisfied easily by producing output 1 at each moment t. But the last condition, which has the form of a fairness constraint, excludes this simple solution; we cannot ignore the zero bits in α. A natural idea is to alternate between outputs 0 and 1 if the inputs are only 0. We arrive at the following procedure:

- for input 1 produce output 1
- for input 0 produce
  - output 1 if last output was 0
  - output 0 if last output was 1

This procedure is executable by the following Mealy automaton. (As initial state we take, for example, the left-hand state.)

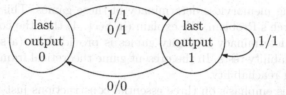

Let us present the specification language and the task of synthesis in more detail. For the formulation of "requirements" we consider the system of monadic second-order logic (MSO) over the successor structure $(\mathbb{N}, +1, <)$, also called S1S (for "second-order theory of one successor") or "sequential calculus". This case was emphasized by Church as an open problem in [6], and today it is understood that "Church's Problem" refers to S1S. In short words, this language uses variables $s, t, \ldots$ for time instances (natural numbers) and variables $X, Y, \ldots$ for sequences. Sequences are identified here with unary predicates over the natural numbers: The bit sequence α is identified with the predicate that holds for t iff $\alpha(t) = 1$; so in S1S one writes $X(t)$ rather than $\alpha(t) = 1$. The atomic formulas are equalities and inequalities between number terms (e.g. $s + 1 + 1 = t$, $s < t$) and formulas $X(\tau)$ with number term τ; the S1S-formulas are built from atomic ones by applying Boolean connectives and the quantifiers $\forall, \exists$ to both kinds of variables. In general, we have S1S-specifications $\varphi(\overline{X}, \overline{Y})$ that speak about *tuples* of predicates (sequences). For an m_1-tuple $\overline{X}$ and an m_2-tuple $\overline{Y}$ this means that the input alphabet under consideration is $\{0, 1\}^{m_1}$ and the output alphabet $\{0, 1\}^{m_2}$. In our explanations and examples we only refer to the case $m_1 = m_2 = 1$.

Church's Problem can now be stated as follows:

Given an S1S-specification $\varphi(\overline{X}, \overline{Y})$, decide whether a Mealy automaton exists that transforms each input sequence $\alpha \in (\{0, 1\}^{m_1})^\omega$ into an output sequence $\beta \in (\{0, 1\}^{m_2})^\omega$ such that $(\mathbb{N}, +1, <) \models \varphi[\alpha, \beta]$ – and if this is the case, construct such an automaton.

Among the many concepts of transformations of sequences, only a very special form is admitted for Church's Problem. Two aspects are relevant, as was at an early stage clarified by Trakhtenbrot [26]: First, the transformation should be "causal" (or: "nonanticipatory"), which means that the output $\beta(t)$ only depends on the prefix $\alpha(0)\ldots\alpha(t)$ of α. (Thus we have a much sharper requirement than continuity in the Cantor space, where $\beta(t)$ is determined by some finite prefix of α, possibly longer than $\alpha(0)\ldots\alpha(t)$.) The second aspect is the computability of the transformation by a finite-state machine (and here we take the above-mentioned format of Mealy automata, to be specific).

As an illustration consider the two transformations T^- and T^+ which "divide by 2", respectively "double" a given sequence α. The transformation T^- maps α to the sequence β that contains every second letter of α (so $\beta(t) = \alpha(2t)$). Clearly T^- is not causal. $T^+(\alpha)$ is defined to be the sequence β which repeats each α-letter once; so we have $\alpha(t) = \beta(2t) = \beta(2t+1)$ for all t. This transformation is causal but not computable by a Mealy automaton; an unbounded memory is needed to store for outputs from time $2t$ onwards the relevant α-segment $\alpha(t)\ldots\alpha(2t)$, for increasing t. Note that we exclude the possibility to produce outputs of length greater than 1 in one step.

Before we enter the solution of Church's Problem in the framework of automata over infinite sequences, it should be mentioned that an alternative approach has been developed by Rabin [21] via automata on infinite trees. Tree automata allow to deal directly with the space of all sequence pairs (α, β) of input- and output-sequences. In the present paper we pursue the "linear" approach as in [4].

3 From Logic to Games

It is useful to study Church's Problem in the framework of infinite games, following an idea that was proposed by McNaughton [12]. A specification φ defines an infinite two-person game between players A and B who contribute the input-, respectively the output-bits in turn. A play of this game is the sequence of pairs $(\alpha(t), \beta(t))$ of bits supplied for $t = 0, 1, \ldots$ by A and B in alternation, and the play $(\alpha(0), \beta(0))\ (\alpha(1), \beta(1))\ (\alpha(2), \beta(2))\ \ldots$ is won by player B iff φ is satisfied by the pair (α, β). So the formula φ serves as a *winning condition* (for player B). A Mealy automaton as presented above defines a *winning strategy* for player B in this game; so we speak of a *finite-state winning strategy*.

In this section the game theoretic form of Church's Problem is developed, in two steps: First, the S1S-specifications are transformed into deterministic automata over infinite words ("ω-automata"), and secondly these automata are converted into arenas of infinite games.

3.1 Determinization and Muller Automata

The first step for solving Church's Problem consists of a transformation of a specification $\varphi(\overline{X}, \overline{Y})$ into a semantically equivalent but "operational" form. The idea is to introduce a finite number of "states" that are visited while a play evolves and at the same time to radically simplify the logical condition to be satisfied. As it turns out, this condition only takes into account which states are visited infinitely often during an infinite play that is built up by players A and B.

This transformation puts Church's Problem into the framework of automata theory. It is remarkable that we do not have any solution of Church's Problem that avoids this transformation at the start – e.g., by an inductive approach of synthesis that is guided by the structure of the original formula φ.

The appropriate model of automaton into which S1S-formulas are to be transformed was introduced by Muller [17] and is called *Muller automaton*. In the present context, a Muller automaton scans deterministically a play $(\alpha(0), \beta(0))$ $(\alpha(1), \beta(1)) \ldots$ as a sequence from $\Sigma = (\{0,1\}^2)^\omega$; the automaton is called equivalent to φ if precisely the plays are accepted that satisfy φ. Its unique run ϱ on a given play between A and B can be viewed as the working of a referee watching the play. The acceptance condition for the run ϱ refers to the "infinity set of ϱ", which is defined as follows (denoting the set of states by Q):

$$\text{Inf}(\varrho) := \{q \in Q \mid \exists^\omega i \; \varrho(i) = q\}$$

The acceptance component of the Muller automaton is a collection $\mathcal{F}$ of state sets, and a run ϱ is declared to be accepting if $\text{Inf}(\varrho)$ belongs to $\mathcal{F}$. Since a set $\text{Inf}(\varrho)$ clearly constitutes a strongly connected subset of the transition graph of the Muller automaton, it suffices to include only strongly connected subsets in $\mathcal{F}$, which we call "accepting loops".

The transformation from S1S to Muller automata can be established by an induction on the construction of S1S-formulas. While the cases of atomic formulas and Boolean operations are straightforward, the quantifier step (without loss of generality regarding the existential second-order quantifier) is of intriguing difficulty. The projection operation involved in an application of the existential quantifier leads immediately to nondeterministic Muller automata, which then have to be determinized.

The determinization problem requires to condense the different runs of a given nondeterministic automaton $\mathcal{A}$ into a single run of a new (deterministic) automaton such that this run allows to decide the existence of a successful run of $\mathcal{A}$. Over a finite word w, acceptance is decided by inspecting the last states of the different $\mathcal{A}$-runs on w. To compute the states reachable by $\mathcal{A}$ via w, it suffices to record the reachable states for each of the prefixes of w. Since the update of this set from one prefix to the next is possible without a reference to previously visited states, the "subset construction" (introduced by Myhill [19] and by Rabin and Scott [22]) suffices, in which a deterministic automaton is built with states that are sets of states of $\mathcal{A}$.

For the determinization of ω-automata one needs additional memory, since repeated visits to certain states on individual runs of the given nondeterministic automaton $\mathcal{A}$ have to be recorded. We consider here the case of nondeterministic Büchi automata (to which the case of nondeterministic Muller automata is easily reduced); a run ϱ of a Büchi automaton is called successful if for infinitely many t the state $\varrho(t)$ belongs to a designated set F of accepting states. To check the existence of a successful run of a Büchi automaton deterministically, there are (at least) three types of appropriate memory structures. Such a structure S should be finite, and it should be usable for the test whether an $\mathcal{A}$-run exists on a given ω-word in which a state from F occurs infinitely often; moreover, the test should involve just the information which memory states of S are visited infinitely often and which only finitely often when processing the ω-word under consideration.

The first idea, pursued by McNaughton in [13] and also in the book [28], is to start new computation threads whenever the nondeterminism of $\mathcal{A}$ requires this, to record whether visits of states in F occur, and to devise a policy of merging runs when they reach the same $\mathcal{A}$-state. A different approach is the construction of Muller and Schupp in [18]. Here a version of the run tree of a nondeterministic Büchi automaton is built up while an input word is scanned. The finite prefixes of the run tree (corresponding to the prefixes of the input word) are compressed so that only finitely many different compressed trees can arise. The subset construction is applied in each step, however dividing the reached states into two parts (consisting of the F-states and non-F-states, respectively), which leads to two son nodes of a leaf of the previous tree. A bound on the height and the width of the trees is realized by a compression of paths without branching and by the deletion of double occurrences of a state. A vertex coloring with three colors serves to keep track of repeated visits to states in F. Finally, the celebrated construction of Safra [23] involves a somewhat sparser use of the subset construction; here a new branch of the run tree is opened only when F-states are encountered (so a new son vertex is only created with F-states). Again, a compression policy ensures that the size of the "Safra trees" stays bounded, and a subtle mechanism serves to record repeated visits to F.

There is no space here to discuss these intriguing constructions in further detail. Even today the subject is not closed; and a major open problem is to devise procedures that substantially reduce (or even minimize) the size of deterministic ω-automata.

3.2 Muller Games

For an analysis of Church's Problem in a game theoretic setting it is useful to distinguish the contribution of bits (in the general case: bit vectors) by the two players A and B. Rather than processing a bit pair $(\alpha(t), \beta(t))$ in one step of the Muller automaton, we introduce two steps, each processing a single bit, and using an intermediate state. Then we have two kinds of states, called A- and B-states. In an A-state, the next bit is to be picked by player A, in a B-state by player B. In a graph theoretical presentation we indicate A-states by boxes and B-states by circles. Thus the transitions of a Muller automaton from a given state are dissolved as follows:

The result is a "game graph". For our example specification above, we obtain the following game graph from a corresponding Muller automaton (the reader should ignore for the moment the boldface notation of some arrows).

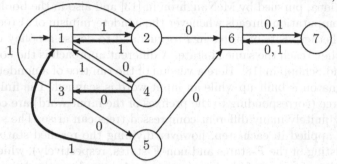

The three conditions of our example formula (Sect. 2) can indeed be captured by this graph, by providing an appropriate list of accepting loops. The first condition requires that a bit 1 chosen by A has to be answered by the bit 1 chosen by B. If this is violated (starting from the initial state 1), state 6 (and hence the loop consisting of states 6 and 7) is entered. The second condition says that player B should not pick two zeroes in succession. If this is violated, we would reach 6 and 7 again. We thus exclude states 6 and 7 from the accepting loops. The third condition on fairness means that if A chooses 0 infinitely often (which happens by going to 4 or 5), then B has to choose 0 infinitely often (which is only possible by going from 4 to 3). Altogether we declare a loop F as accepting if it does not contain 6 or 7 and satisfies $(4 \in F \vee 5 \in F \rightarrow 3 \in F)$.

How should player B pick his bits to ensure that the play visits precisely the states of one of these loops F infinitely often? We have to fix how to move from states 2, 4, 5, 7. From 7 player B has to move to 6 since there is no other choice. The other choices can be fixed as follows: From 2 to 1, from 4 to 3, and from 5 to 1 (see boldface arrows). Then, depending on what Player A does, a play starting in 1 will visit infinitely often the states 1 and 2, or the states 1 to 4, or the states 1, 3, 4, 5, or the states 1 to 5. Each of these loops is accepting.

We see that player B has a winning strategy by fixing his moves as stated above. This winning strategy can be converted into a Mealy automaton when we combine again each pair of two successive moves (by player A and then B) into a single transition. We get an automaton with the states 1 and 3 and the following transitions: From 1 via $\binom{1}{1}$ back to 1, from 1 via $\binom{0}{0}$ to 3, and from 3 via $\binom{0}{1}$ and via $\binom{1}{1}$ back to 1. Up to names of states (and the irrelevant initial state) this is precisely the Mealy automaton mentioned in Sect. 2.

In the remainder of the paper, we shall outline a "solution" of Muller games (and some weaker variants like "'weak Muller games"). By a solution we mean two algorithms: The first decides for each state q (A-state or B-state) whether for plays starting in q player B has a winning strategy, and – in this case – the second algorithm allows to construct a Mealy automaton that executes such a winning strategy. In this analysis we may cancel the labels on the transitions. This is motivated by the fact that the winning condition is formulated in terms of visits of states only, regardless of the labels that are seen while traversing edges. When a winning strategy over the unlabelled game graph is to be constructed, it will be easy to re-introduce the labels and use them for a Mealy automaton as required in the original formulation of Church's Problem.

As a preparation, we now summarize the relevant definitions in some more detail.

3.3   Finite-State Games: The Framework

A *game graph* has the form $G = (Q, Q_A, E)$ where $Q_A \subseteq Q$ and $E \subseteq Q \times Q$ is the transition relation. We assume that $\forall q \in Q : qE \neq \emptyset$ (i.e. $\forall q \exists q' : (q, q') \in E$); so plays cannot end in a deadlock (and hence a subset Q_0 of Q induces again a game graph if from each $q \in Q_0$ there is an edge back to Q_0). We set $Q_B := Q \setminus Q_A$. In this paper edges will always lead from Q_A-states to Q_B-states or conversely. A *play* over G from q is an infinite sequence $\varrho = q_0 q_1 q_2 \ldots$ with $q_0 = q$ and $(q_i, q_{i+1}) \in E$ for $i \geq 0$. We assume that player A chooses the next state from a state in Q_A, and player B from a state in Q_B. The set Q will always be finite in the sequel; so we speak of *finite-state games*.

For the formulation of winning conditions, we add a further item to the game graph, depending on the format of the condition. We use either a collection $\mathcal{F} \subseteq 2^Q$ of sets $R \subseteq Q$, or a coloring $c : Q \to \{0, \ldots, k\}$ for some natural number k. In the special case $c : Q \to \{0, 1\}$ we also consider the subset $F = \{q \in Q \mid c(q) = 1\}$ instead. For a collection $\mathcal{F} \subseteq 2^Q$ we introduce two winning conditions. The first is the Muller winning condition mentioned above; it refers to the set of states visited infinitely often in a play ϱ:

$$\mathrm{Inf}(\varrho) := \{q \in Q \mid \exists^\omega i \; \varrho(i) = q\}$$

Player B wins the play ϱ if $\mathrm{Inf}(\varrho) \in \mathcal{F}$. With these conventions we speak of a *Muller game* $(G, \mathcal{F})$. Another use of a system $\mathcal{F}$ leads to the *weak Muller condition* (also called *Staiger-Wagner condition*). Here we refer to the visited states in a play ("occurrence set"):

$$\mathrm{Occ}(\varrho) := \{q \in Q \mid \exists i \; \varrho(i) = q\}$$

Player B wins a play ϱ according to the weak Muller condition if $\mathrm{Occ}(\varrho) \in \mathcal{F}$. We speak of the *weak Muller game* $(G, \mathcal{F})$. From the example in Sect. 2 we obtain a weak Muller game if we delete the third requirement. The items 1 and 2 are captured over the presented game graph (with the states $1, \ldots, 7$) by the condition that none of the states 6 or 7 is ever visited; this is expressed by the weak Muller condition with the system $\mathcal{F}$ that contains all subsets of $\{1, \ldots, 5\}$.

An important special case of weak Muller games is the *reachability game*, given a set $F \subseteq Q$ of states of the game graph (Q, Q_A, E). The reachability condition for player B is satisfied for a play ϱ if some state of ϱ belongs to F. We speak of the *reachability game* (G, F). One obtains an equivalent weak Muller condition by setting $\mathcal{F} = \{R \subseteq Q \mid R \cap F \neq \emptyset\}$. The reachability game (for player B) yields a game with complemented winning condition for player A, namely to stay in the set $Q \setminus F$ throughout. Such a condition is called a *safety condition*. Taking up our example again, we see that the weak Muller condition mentioned above (covering items 1 and 2 of the requirement) amounts to the safety condition, now for player B, to stay in the set $\{1, \dots, 5\}$ during the whole play.

We now turn to the solution of games, starting with the central concept of strategy. A *strategy for player B from q* is a function $f : Q^+ \to Q$, specifying for any play prefix $q_0 \dots q_k$ with $q_0 = q$ and $q_k \in Q_B$ some vertex $r \in Q$ with $(q_k, r) \in E$ (otherwise the value of f is chosen arbitrarily). A play $\varrho = q_0 q_1 \dots$ from $q_0 = q$ is *played according to strategy f* if for each $q_i \in Q_B$ we have $q_{i+1} = f(q_0 \dots q_i)$. A strategy f for player B from q is called *winning strategy for player B from q* if any play from q which is played according to f is won by player B. In the analogous way, one introduces strategies and winning strategies for player A. We say that A (resp. B) *wins from q* if there is a winning strategy for A (resp. B) from q.

For a game over the graph $G = (Q, Q_A, E)$, the *winning regions of players A and B* are the sets $W_A := \{q \in Q \mid \text{A wins from } q\}$ and $W_B := \{q \in Q \mid \text{B wins from } q\}$. It is obvious that a state cannot belong to both W_A and W_B; so the winning regions W_A, W_B are disjoint. But whether these sets exhaust the whole game graph is a more delicate question. One calls a game *determined* if $W_A \cup W_B = Q$, i.e. from each vertex one of the two players has a winning strategy. Determinacy of infinite games is a central topic in descriptive set theory; with the axiom of choice one can construct games that are not determined. For the games considered in this paper (i.e. games defined in terms of the operators Occ and Inf), determinacy is well-known. Nevertheless we state this claim in the results below, since determinacy is the natural way to show that envisaged winning strategies are complete: In order to show that the domain D of a strategy covers the entire winning region of one player, one verifies that from each state outside D the other player has a winning strategy.

To "solve" a game over the graph $G = (Q, Q_A, E)$ involves two tasks:

1. to decide for each $q \in Q$ whether $q \in W_B$ or $q \in W_A$,
2. and depending on q to construct a suitable winning strategy from q (for player B, respectively A).

For item 2 two kinds of strategies will be employed, the memoryless and the finite-state strategies. A strategy $f : Q^+ \to Q$ is *memoryless* if the value of $f(q_1 \dots q_k)$ only depends on the "current state" q_k. For the definition of finite-state strategies, we first observe that over a finite set Q, a strategy $f : Q^+ \to Q$ can be considered as a word function. We say that f is a *finite-state strategy* if it is computed by a Mealy automaton. In the present context we use the format $\mathcal{S} = (S, Q, Q, s_0, \delta, \tau)$ with state set S, input alphabet Q, output alphabet Q, initial state s_0, transition function $\delta : S \times Q \to S$, and output function $\tau : S \times Q_A \to Q$ for player A (respectively $\tau : S \times Q_B \to Q$ for player B). The *strategy f_S computed by S* is now defined by $f_S(q_0 \cdots q_k) = \tau(\delta^*(s_0, q_0 \cdots q_{k-1}), q_k)$ (where $\delta^*(q, w)$ is the state reached from q after processing the input word w and τ is chosen for the player under consideration).

Now we state the main theorem on weak Muller games and Muller games.

Theorem 1. *Weak Muller games and Muller games are determined, and for a weak Muller game, respectively Muller game $(G, \mathcal{F})$ one can effectively compute the winning regions of the two players, and one can construct, for each state q of G, a finite-state winning strategy from q for the respective winning player.*

The part concerning Muller games is the Büchi-Landweber Theorem and gives the desired solution of Church's Problem. For this, one proceeds as in the previous section, i.e. one transforms a given S1S-formula φ into a Muller automaton $\mathcal{M}$ which is then converted to a game graph G with Muller winning condition. Note that the game graph G inherits an initial state from $\mathcal{M}$. Using the Büchi-Landweber Theorem, one checks whether this initial state belongs to the winning region of player B, and in this case one obtains a Mealy automaton S that realizes a winning strategy from the initial state. The desired finite-state strategy for the original formula φ is then easily constructed as a product automaton from $\mathcal{M}$ and S. Its memory thus combines the state space of the Muller automaton $\mathcal{M}$ with that of the strategy automaton S. It is not yet well understood how these two aspects play together in general. Our example in Sect. 2 illustrates the case that in addition to the states of $\mathcal{M}$ no additional memory is necessary.

3.4 Reachability Games

As a preparatory step for Theorem 1 we solve reachability games. Recall that a reachability game (G, F) involves the winning condition (for player B) that the play should reach somewhere a state of the set F. The solution relies on a simple backward search of the game graph, starting with the set F.

Theorem 2. *A reachability game (G, F) with $G = (Q, Q_A, E)$ and $F \subseteq Q$ is determined, and the winning regions W_A, W_B of players A and B, respectively, are computable, as well as corresponding memoryless winning strategies.*

Proof. We compute, for $i = 0, 1, \ldots$, the vertices from which player B can force a visit in F within i moves. We call this set the i-th "attractor" (for B):

$$\mathrm{Attr}_B^i(F) := \{q \in Q \mid \text{from } q \text{ player B can force a visit of } F \text{ in } \leq i \text{ moves}\}$$

Its computation for increasing i is known from the theory of finite games (and corresponds to the well-known analysis of AND-OR-trees):

$$\mathrm{Attr}_B^0(F) = F,$$
$$\mathrm{Attr}_B^{i+1}(F) = \mathrm{Attr}_B^i(F)$$
$$\cup \{q \in Q_B \mid \exists (q, r) \in E : r \in \mathrm{Attr}_B^i(F)\}$$
$$\cup \{q \in Q_A \mid \forall (q, r) \in E : r \in \mathrm{Attr}_B^i(F)\}$$

So for step $i + 1$ we include a state of Q_B if from it some edge can be chosen into $\text{Attr}_B^i(F)$. We can fix such a choice for each Q_B-state in $\text{Attr}_B^{i+1}(F)$ ($i = 0, 1, \ldots$) in order to build up a memoryless strategy. We include a state in Q_A in $\text{Attr}_B^{i+1}(F)$ if all edges from it lead to $\text{Attr}_B^i(F)$. The sequence $\text{Attr}_B^0(F) \subseteq \text{Attr}_B^1(F) \subseteq \text{Attr}_B^2(F) \subseteq \ldots$ becomes stationary for some index k since Q is finite. Since $k \leq |Q|$ we can define $\text{Attr}_B(F) := \bigcup_{i=0}^{|Q|} \text{Attr}_B^i(F)$.

Later we shall also use the set $\text{Attr}_A(F)$, defined in the analogous way for player A.

With the inductive construction it was explained that $\text{Attr}_B(F) \subseteq W_B$; furthermore we have defined a uniform memoryless winning strategy which can be applied to any state in W_B regardless of the start of the play. (For states in $Q_B \cap F$ the choice of the next state is arbitrary.)

For the converse inclusion $W_B \subseteq \text{Attr}_B(F)$ we have to show that $\text{Attr}_B(F)$ exhausts the winning region W_B. For this, we show that from each state in the complement of $\text{Attr}_B(F)$, player A has a winning strategy (which is again memoryless). It suffices to verify that from any state q in $Q \setminus \text{Attr}_B(F)$ player A can force to stay outside $\text{Attr}_B(F)$ also in the next step. This is checked by a case distinction: If $q \in Q_A$, there must be an edge back into $Q \setminus \text{Attr}_B(F)$, otherwise all edges from q would go to $\text{Attr}_B(F)$ whence q would belong to $\text{Attr}_B(F)$. If $q \in Q_B$, all edges from q must lead to $Q \setminus \text{Attr}_B(F)$, because otherwise there would be an edge to $\text{Attr}_B(F)$ and q would again belong to $\text{Attr}_B(F)$.

4   Appearance Records and Game Simulations

4.1   Appearance Records

For Muller games, both in the weak and the unrestricted form, memoryless strategies are not enough. A simple example illustrates this. Consider the following game graph G and the set $\mathcal{F} = \{\{1, 2, 3\}\}$.

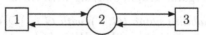

The weak Muller game $(G, \mathcal{F})$ requires for player B to visit all states in order to win. From vertex 2 there is no memoryless winning strategy: Neither the choice to move to 1 nor the choice to move to 3 will ensure to reach each vertex. On the other hand, a one-bit memory will do: When coming back to 2 we should know whether 1 or 3 was visited before, and then we should move to 3, respectively 1 (and maybe do this perpetually from that moment onwards). A general principle derivable from this solution is to "remember where we have been already". This principle corresponds to a simple experience of every-day life: When there is a task ahead consisting of several items, keep a list of what was done already (and thus of what still has to be done).

For the strong Muller game $(G, \mathcal{F})$, both vertices 1 and 3 have to be visited again and again. Clearly it does not suffice just to remember where we have been already: After the visits of 1 and 3 it is necessary to switch from 1 to 3 and back again and again. The natural solution is to "remember where we went last time" – and then to do the choice accordingly, going to the respective "other" vertex.

In the first case (of weak Muller games), we are led to a memory structure that allows to store in an accumulative way the vertices that were already visited in a play. Given a weak Muller game $(G, \mathcal{F})$ with $G = (Q, Q_A, E)$ and $\mathcal{F} \subseteq 2^Q$, we define the transition structure of an automaton $\mathcal{S}$ with the power set 2^Q of Q as its set of states and Q as its input alphabet. Having read the input word $q_1 \cdots q_k$, its state will be $\{q_1, \ldots, q_k\}$. So the initial state is $\emptyset$ and the transition function $\delta : 2^Q \times Q \to 2^Q$ is defined by $\delta(R, p) = R \cup \{p\}$. This memory of subsets of Q with the mentioned update rule is called *appearance record*. We shall show that this memory structure suffices for winning strategies in arbitrary weak Muller games over G. What remains is to fix the output function for $\mathcal{S}$.

Let us now treat the case of (strong) Muller games. Our example above motivates to keep a refined record of the states visited in a play, taking into account the order in which states were "visited last time". A naive way to realize this kind of memory is to arrange the set of visited states in a list where the first entry is the currently visited state q, the second one the state $q' \neq q$ visited last before q, the third one the state q'' different from q, q' visited last before q, q', and so on until the set of previously visited states is exhausted. It will be useful to work with a slight (but essential) refinement of this list structure, which goes back to McNaughton [12] and is today known as *latest appearance record*, short "LAR".

Consider a Muller game $(G, \mathcal{F})$ with $G = (Q, Q_A, E)$ and $Q = \{1, \ldots, n\}$. A LAR is a pair $((i_1, \ldots, i_r), h)$ where the i_j are distinct states from Q and $0 \leq h \leq r$. Following Büchi, we call the index h the *hit* of the LAR. Again we define the transition structure of an automaton $\mathcal{S}$, now with the set of LAR's over Q as its set of states. The initial state is $((\ ), 0)$ (empty list and hit 0). The transition function changes a given LAR upon input $q \in Q$ by listing the new state q at the front: If it was not present in the previous LAR, then it is added (and h is set to be 0); if it occurs in the previous LAR, then it is shifted to the front and the position where it was taken from is the value of h.

We give an example for a set Q of four states, which we name A, B, C, D to avoid confusion with the hit values $1, \ldots, 4$. Also we indicate the hit value h by underlining the h-th position of the corresponding list (if $h > 0$). Suppose a play ϱ starts with the states $A, C, C, D, B, D, C, D, D, \ldots$. Then we obtain the following sequence of LAR's (where we skip the initial LAR $((\ ), 0)$):

Suppose that the play goes on only by states C and D (and both are chosen again and again). Then the states A, B will not be touched anymore, and the hit will assume 2 as maximal value thereafter again and again: It will no more be 3 or 4 (since A, B stay where they are), and it cannot finally stay with value 1 (since then only a single state, namely the leading one of the LAR, would be visited from some point onwards). The maximal hit visited infinitely often thus indicates the cardinality of the set $\mathrm{Inf}(\varrho)$.

Visited state	Reached LAR
A	(A)
C	(CA)
C	$(\underline{C}A)$
D	(DCA)
B	$(BDCA)$
D	$(D\underline{B}CA)$
C	$(CD\underline{B}A)$
D	$(D\underline{C}BA)$
D	$(\underline{D}CBA)$

Let us summarize the definition of the automaton $\mathcal{S} = (S, Q, s_0, \delta)$ which realizes, given a play prefix $i_1 \ldots i_k \in Q^*$, the computation of the resulting LAR. The state set S is the set of LAR's over Q, we have $s_0 = ((\ ), 0)$, and the transition function $\delta : S \times Q \to S$ realizes the update of the LAR as follows: We have $\delta(((i_1 \ldots i_r), h), i) = ((ii_1 \ldots i_r), 0)$ if i does not occur in $(i_1 \ldots i_r)$; otherwise, if $i = i_k$ cancel i from $(i_1 \ldots i_r)$ to obtain $(j_1 \ldots j_{r-1})$ and set $\delta(((i_1 \ldots i_r), h), i) = ((ij_1 \ldots j_{r-1}), k)$.

Note that we obtain the (simple) appearance record if we discard the order in the state-lists and delete the h-value. We shall show that the LAR memory structure over Q will suffice for realizing winning strategies in Muller games over Q. Again, it only remains to supply the output function τ for the automaton $\mathcal{S}$ in order to obtain the complete definition of a Mealy automaton.

We add a historical remark. The paper [12] in which McNaughton introduced the fundamental data structure of LAR is a technical report that was not published as a journal paper (it contained an error in an attempted solution of Muller games). McNaughton used the name "order-vector"; the term "latest appearance record" (LAR) was introduced by Gurevich and Harrington in their landmark paper [10] on automata over infinite trees.

4.2   Game Simulations and Parity Conditions

The two versions of appearance record introduced in the previous section allow to reformulate the winning conditions (weak Muller condition, strong Muller condition) in a form that makes the solutions of the corresponding games much easier.

First let us consider weak Muller games. For a play ϱ, consider the sequence of associated appearance records, as assumed in the run of the Mealy automaton that realizes the necessary updates. The set of visited states increases weakly monotonically during the play and finally reaches the value $\text{Occ}(\varrho)$ on which it stays fixed. Similarly the cardinality of the set of visited states increases until it reaches the value $|\text{Occ}(\varrho)|$. This observation enables us to express the weak Muller winning condition "$\text{Occ}(\varrho) \in \mathcal{F}$" in different way. We associate a number $c(R)$ with each subset R of Q, also called its color, which conveys two informations: the size of R, and whether R belongs to $\mathcal{F}$ or not. In the first case, we take the even color $2 \cdot |R|$, otherwise the odd color $2 \cdot |R| - 1$ (assuming $R \neq \emptyset$):

$$c(R) := \begin{cases} 2 \cdot |R| & \text{if } R \in \mathcal{F} \\ 2 \cdot |R| - 1 & \text{for } R \notin \mathcal{F} \end{cases}$$

For $R = \emptyset$ let $c(R) = 0$. – The following claim is then obvious:

Remark 3. Let ϱ be a play and $R_0, R_1, R_2, \ldots$ be the sequence of the associated appearance records. Then $\mathrm{Occ}(\varrho) \in \mathcal{F}$ iff the maximal color in the sequence $c(R_0)c(R_1)c(R_2)\ldots$ is even.

This remark motivates a new winning condition over game graphs $G = (Q, Q_A, E)$ that are equipped with a coloring $c : Q \to \{0, \ldots, k\}$. The *weak parity condition* with respect to coloring c says: Player B wins the play $\varrho = r_0 r_1 r_2 \ldots$ iff the maximum color in the sequence $c(r_0)c(r_1)c(r_2)\ldots$ is even. Given a game graph G and a coloring c with the weak parity winning condition, we speak of the *weak parity game* (G, c).

Using this, one transforms a weak Muller game $(G, \mathcal{F})$ into a weak parity game (G', c): Given $G = (Q, Q_A, E)$ let $G' = (2^Q \times Q, 2^Q \times Q_A, E')$ where $((P, p), (R, r)) \in E'$ iff $(p, r) \in E$ and $R = P \cup \{p\}$, and let $c(R, r) := 2 \cdot |R|$ if $R \in \mathcal{F}$, otherwise $2 \cdot |R| - 1$. Each play $\varrho = r_0 r_1 \ldots$ in G induces the play $\varrho' = (\emptyset, r_0)(\{r_0\}, r_1) \ldots$ in G', which is built up according to the definition of E'. We have by construction that ϱ satisfies the weak Muller condition w.r.t. $\mathcal{F}$ iff ϱ' satisfies the weak parity condition w.r.t. c.

This transformation of $(G, \mathcal{F})$ into (G', c) (with a change of the winning condition) is a "game simulation". (We skip a general definition since we only apply it for the present case and the case of Muller games.)

The simulation has an interesting consequence when the latter game (the weak parity game) allows memoryless winning strategies. Namely, a memoryless strategy over G' immediately determines the output function for the Mealy automaton that computes the appearance records: If the memoryless strategy (say for player B) requires to proceed from position (R, q) (where $q \in Q_B$) to (R', q'), then the output function value $\tau(R, q)$ of the Mealy automaton is set to be q' (and the new state is $R' = R \cup \{q\}$). Also the decision whether a state q of G belongs to the winning region of B is provided by the analysis of the corresponding weak parity game over G', since for each state of a weak parity game we shall determine the winner. Applying this to the state $(\emptyset, q)$ of G' we obtain the answer also for q in the graph G.

In the next section we shall show that weak parity games can indeed be solved with memoryless winning strategies. Using the previous remark this completes the solution of weak Muller games in Theorem 1.

Let us turn to the case of Muller games. We proceed as before, now using the latest appearance record structure LAR in place of the appearance record. Consider a play ϱ over Q and the associated sequence ϱ' of LAR's. We collect the entries of a LAR $((i_1 \ldots i_r), h)$ up to position h as the *hit set* $\{i_1, \ldots, i_h\}$ of the LAR. If h is the maximal hit assumed infinitely often in ϱ', we may pick a position (time instance) in ϱ' where no unlisted state enters any more later in the play and where only hit values $\leq h$ occur afterwards. From that point onwards the states listed after position h stay fixed, and thus also the hit set for the hit value h stays fixed. We call this set *the hit set for the maximal hit occurring infinitely often in ϱ'*. The following statement is now easily verified:

Remark 4. Let ϱ be a sequence over Q and ϱ' be the associated sequence of LAR's. The set $\mathrm{Inf}(\varrho)$ coincides with the hit set H for the maximal hit h occurring infinitely often in ϱ'.

For the proof, consider the point in ϱ from where no new states will occur and where all visits of states that are visited only finitely often are completed. After a further visit of all the states in $\mathrm{Inf}(\varrho)$, these states will stay at the head of the LAR's (in various orders), and the hit values will be $\leq k := |\mathrm{Inf}(\varrho)|$. It remains to show that the hit value in ϱ' reaches k again and again (so that k is the maximal hit occurring infinitely often in ϱ'). If the hit was $< k$ from some point onwards, the state q listed on position k would not be visited later and thus not be in $\mathrm{Inf}(\varrho)$.

Remark 4 allows to transform the Muller winning condition for a play ϱ into a different winning condition applied to the associated play ϱ'. By Remark 4 we know that the Muller winning condition holds for the play ϱ iff *the hit set for the maximal hit occurring infinitely often in ϱ' belongs to $\mathcal{F}$*. This allows us to extract two data from the LAR's which are sufficient to decide whether the play ϱ satisfies the Muller condition: the hit value and the information whether the corresponding hit set belongs to $\mathcal{F}$. We combine these two data in the definition of a coloring of the LAR's. Define

$$c(((i_1 \ldots i_r), h)) := \begin{cases} 2h & \text{if } \{i_1, \ldots, i_h\} \in \mathcal{F} \\ 2h-1 & \text{if } \{i_1, \ldots, i_h\} \notin \mathcal{F} \end{cases}$$

for $h > 0$ and let $c(((i_1 \ldots i_r), 0)) = 0$. Then the Muller condition $\mathrm{Inf}(\varrho) \in \mathcal{F}$ is satisfied iff the maximal color occurring infinitely often in $c(\varrho'(0))c(\varrho'(1)) \ldots$ is even. This is a "parity condition" (as introduced by Mostowski [16] and Emerson and Jutla [9][2]). The only difference to the weak parity condition is the reference to colors occurring infinitely often rather than to those which occur at all.

In general, the parity condition refers to a coloring $c : Q \to \{0, \ldots, k\}$ of a game graph G; it is the following requirement on a play ϱ:

$$\bigvee_{j \text{ even}} (\exists^\omega i : c(\varrho(i)) = j \ \wedge \neg \exists^\omega i : c(\varrho(i)) > j)$$

The pair (G, c) with this convention for the winning condition for player B is called a *parity game*.

In complete analogy to the case of weak Muller games, one can set up a game simulation of a Muller game $(G, \mathcal{F})$ by a parity game (G', c). A state of G' is a pair consisting of a LAR ℓ and a state q from Q. An edge is introduced from (ℓ, q)

---

[2] Other names appearing in the literature are "Mostowski condition" and "Rabin chain condition".

to (ℓ', q') if the edge (q, q') exists in G and ℓ' results from ℓ by the LAR-update that lists q at the head. A play ϱ over G then corresponds to a play ϱ' over G'. The coloring c is defined as above.

We shall show that parity games can be solved with memoryless winning strategies. As for the case of weak Muller and weak parity games, a memoryless winning strategy of player B in the parity game over G' yields a finite-state winning strategy of player B in the Muller game over G. The decision whether for a state q of G such a winning strategy exists is done by testing whether player B wins from position $(((\ ), 0), q)$ in the parity game over G'.

5   Solving Weak and Strong Parity Games

A central difficulty in the solution of weak Muller games and Muller games is the possibly complicated structure of the system $\mathcal{F}$ of "winning state-sets". The first proof of the Büchi-Landweber Theorem [4] involves an intriguing analysis of the partial order (by set inclusion) of the power set of the set Q of states of the game graph. The transformation to a game with a (weak or strong) parity condition stratifies the winning condition by introducing the total order of colors. As we shall see, this order can be exploited for an inductive construction, again for both the weak parity games and the parity games. In both cases, an iterated application of attractor computations suffices; thus, the game solution ultimately rests on simple reachability tests.

Theorem 5. *A weak parity game (G, c) is determined, and one can compute the winning regions W_A, W_B and also construct corresponding memoryless winning strategies for the players A and B.*

It may be noted that we suppressed the initial states q when speaking about memoryless winning strategies. In the proof we shall see that – as for reachability games – the strategies can be defined independently of the start state (as long as it belongs to the winning region of the respective player).

Proof. Let $G = (Q, Q_A, E)$ be a game graph (we do not refer to the special graph G' above), $c : Q \to \{0, \ldots, k\}$ a coloring (w.l.o.g. k even, otherwise switch players). Set $C_i = \{q \in Q \mid c(q) = i\}$.

We first compute the attractor for B of the states with maximal color, which is even. When player B reaches such a state the play is won whatever happens later. So $A_k := \mathrm{Attr}_B(C_k)$ is a part of the winning region of player B.

The remaining vertices form the set $Q \setminus A_k$; the subgraph induced by $Q \setminus A_k$ in G is again a game graph. (Note that from each state q in $Q \setminus A_k$ there is at least one edge back to $Q \setminus A_k$, otherwise – as seen by case distinction whether $q \in Q_A$ or $q \in Q_B$ – q would belong to $A_k = \mathrm{Attr}_B(C_k)$.)

Now in the subgame over $Q \setminus A_k$ we compute $A_{k-1} := \mathrm{Attr}_A(C_{k-1} \setminus A_k)$; from these vertices player A can reach the highest odd color $k - 1$ and guarantee to stay away from A_k, in the same way as explained above for reachability games (see Sect. 4.1).

In both sets we can choose memoryless winning strategies, over A_k for B, and over A_{k-1} for A. In this way we continue to adjoin "slices" of the game graph, taking B- and A-attractors in alternation, in order to obtain the winning regions of B and A. The next set A_{k-2} is the set of all states $q \in Q \setminus (A_{k-1} \cup A_k)$ from which player B can force the play to $C_{k-2} \setminus (A_{k-1} \cup A_k)$. We denote this set by $\mathrm{Attr}_B^{Q \setminus (A_{k-1} \cup A_k)}(C_{k-2} \setminus (A_{k-1} \cup A_k))$. The exponent indicates the (domain of) the game graph in which the attractor computation takes place. In order to facilitate the notation for the general case, set $Q_i := Q \setminus (A_{i+1} \cup \ldots \cup A_k)$.

So we compute the sets $A_k, A_{k-1}, \ldots, A_0$ inductively as follows:

$$A_k := \mathrm{Attr}_B(C_k)$$
$$A_{k-1} := \mathrm{Attr}_A^{Q_{k-1}}(C_{k-1} \setminus A_k)$$

and for $i = k-2, \ldots, 0$:

$$A_i := \begin{cases} \mathrm{Attr}_B^{Q_i}(C_i \setminus (A_{i+1} \cup \ldots \cup A_k)) & \text{if } i \text{ even} \\ \mathrm{Attr}_A^{Q_i}(C_i \setminus (A_{i+1} \cup \ldots \cup A_k)) & \text{if } i \text{ odd} \end{cases}$$

The memoryless strategies for A and B are chosen as explained for the initial cases A_k, A_{k-1}. Now we have

$$W_B = \bigcup_{i \text{ even}} A_i \quad \text{and} \quad W_A = \bigcup_{i \text{ odd}} A_i$$

For the correctness, one verifies by induction on $j = 0, \ldots, k$:

$$\bigcup_{\substack{i=k-j \\ i \text{ even}}}^{k} A_i \subseteq W_B \qquad \bigcup_{\substack{i=k-j \\ i \text{ odd}}}^{k} A_i \subseteq W_A$$

Returning to the solution of weak Muller games, we note that a finite-state winning strategy can be realized with 2^n memory states over a game graph with n states, due to the introduction of appearance records.

Let us turn to the case of parity games, following a proof of McNaughton [14].

Theorem 6. *A parity game (G, c) is determined, and one can compute the winning regions W_A, W_B and also construct corresponding memoryless winning strategies for the players A and B.*

Proof. Given $G = (Q, Q_A, E)$ with coloring $c : Q \to \{0, \ldots, k\}$ we proceed by induction on $|Q|$, the number of states of G.

The induction start (Q is a singleton) is trivial. In the induction step assume that the maximal color k is even (otherwise switch the roles of players A and B). Let q be a state of the highest (even) color k and define $A_0 = \mathrm{Attr}_B(\{q\})$. As the complement of an attractor, the set $Q \setminus A_0$ defines a subgame. The induction hypothesis applied to the game over the subgraph induced by $Q \setminus A_0$ ensures a partition of $Q \setminus A_0$ into the winning regions U_A, U_B of the two players (with corresponding memoryless winning strategies) in the game over $Q \setminus A_0$.

We now distinguish two cases:

1. From q, player B can ensure to be in $U_B \cup A_0$ in the next step,
2. From q, player A can ensure to be in U_A in the next step.

Let us first verify that one of the two cases applies (which gives a kind of local determinacy). Assume Case 1 fails. If $q \in Q_B$, then all transitions from q have to go to U_A, otherwise we would be in Case 1. By the same reason, if $q \in Q_A$, then some transition from q goes to U_A; so Case 2 applies.

In Case 1, one shows that the winning region W_B of B in G is $U_B \cup \text{Attr}_B(\{q\})$ and that $W_A = U_A$. For player B, the memoryless winning strategy is composed of the memoryless strategy over U_B by induction hypothesis in the game over $Q \setminus A_0$, of the attractor strategy over $\text{Attr}_B(\{q\})$, and possibly of the edge choice in q according to "Case 1"; for player A just the memoryless strategy over U_A is taken. For the claim $U_B \cup \text{Attr}_B(\{q\}) \subseteq W_B$ note that a play from a state in $U_B \cup \text{Attr}_B(\{q\})$ either remains in U_B from some point onwards, whence Player B wins by induction hypothesis, or it visits (due to moves of player A) the attractor A_0 and hence q again and again, so that player B wins by seeing the highest color (even!) repeatedly. The claim $U_A \subseteq W_A$ is clear by induction hypothesis.

We turn to Case 2. From our analysis above we know that $q \in \text{Attr}_A(U_A)$. We consider the set $A_1 = \text{Attr}_A(U_A \cup \{q\})$, clearly of cardinality ≥ 1. So we can apply the induction hypothesis to the domain $Q \setminus A_1$. We obtain a partition of this domain into winning regions V_A, V_B for A and B in the subgame over $Q \setminus A_1$, with corresponding memoryless winning strategies. Now it is easy to verify $W_B = V_B$ and $W_A = V_A \cup A_1$; memoryless winning strategies for B, respectively A, are provided by the induction hypothesis and by the attractor strategy over A_1.

Finally we note that the inductive construction can be turned into a recursive procedure which produces, given G and the coloring c, the desired winning regions and memoryless strategies.

The recursive procedure appearing in this proof involves a nested call of the inductive hypothesis, which means that for each induction step the computational effort doubles, resulting in an overall exponential runtime. It is known that the problem "Given a parity game (G, c) and a state q, does q belong to the winning region of B?" is in the complexity class NP $\cap$ co-NP. Whether this problem is decidable in polynomial time is one of the major open problems in the algorithmic theory of infinite games.

For the memory size of Mealy automata that realize winning strategies, we obtain a higher bound than for weak Muller games. Over a graph with n states the bound $n! \cdot n$ states suffices. This bound can be met by simplifying the LAR construction introduced above, in the sense that only state lists of length n (and no shorter lists) are used. It is easy to adapt our construction to this format. That the factorial function also provides a lower bound for the memory size was shown in [7].

6 Conclusion

We have presented an approach to Church's Problem that involves three basic ingredients, namely determinization, the stratification of Muller and weak Muller games by different versions of appearance records, and an iterated application of simple reachability tests in the solution of games.

Today, we see that Church's Problem was the starting point for a highly active area of research in computer science, in the last 20 years even with a great influence in practical verification and program synthesis. Thus the "old" parts of automata theory for infinite computations, as addressed by Boaz Trakhtenbrot in the Preface of the book [28], turned out extremely fruitful. It seems certain that the vision of "new trends" as proposed in [28] already decades ago will lead to many more results that share both beauty and an even wider range of applicability.

Acknowledgment

I thank the editors, in particular Alex Rabinovich, for their encouragement and patience.

References

1. Büchi, J.R.: Weak second-order arithmetic and finite automata. Z. Math. Logik Grundlagen Math. 6, 66–92 (1960)
2. Büchi, J.R.: On a decision method in restricted second order arithmetic. In: Nagel, E., et al. (eds.) Proc. 1960 International Congress on Logic, Methodology and Philosophy of Science, pp. 1–11. Stanford University Press (1962)
3. Büchi, J.R., Elgot, C.C.: Decision problems of weak second-order arithmetics and finite automata, Abstract 553-112, Notices Amer. Math. Soc. 5, 834 (1958)
4. Büchi, J.R., Landweber, L.H.: Solving sequential conditions by finite-state strategies, Trans. Trans. Amer. Math. Soc 138, 367–378 (1969)
5. Church, A.: Applications of recursive arithmetic to the problem of circuit synthesis. In: Summaries of the Summer Institute of Symbolic Logic, vol. I, pp. 3–50. Cornell Univ, Ithaca, N.Y (1957)
6. Church, A.: Logic, arithmetic, and automata. In: Proc. Int. Congr. Math. 1962, Inst. Mittag-Leffler, Djursholm, Sweden, pp. 23–35 (1963)
7. Dziembowski, S., Jurdziński, M., Walukiewicz, I.: How much memory is needed to win infinite games? In: Proc. 12th IEEE Symp. on Logic in Computer Science, pp. 99–110. IEEE Computer Society Press, Los Alamitos (1997)
8. Elgot, C.C.: Decision problems of finite automata design and related arithmetics. Trans. Amer. Math. Soc. 98, 21–52 (1961)
9. Emerson, E.A., Jutla, C.S.: Tree automata, mu-calculus, and determinacy. In: Proc. 32nd FoCS 1991, pp. 368–377. IEEE Comp. Soc. Press, Los Alamitos (1991)
10. Gurevich, Y., Harrington, L.: Trees, automata, and games. In: Proc. 14th ACM Symp. on the Theory of Computing, pp. 60–65. ACM Press, New York (1982)
11. Hopcroft, J.E., Ullman, J.D.: Formal Languages and Their Relation to Automata. Addison-Wesley, Boston (1969)

12. McNaughton, R.: Finite-state infinite games, Project MAC Rep. MIT, Cambridge (1965)
13. McNaughton, R.: Testing and generating infinite sequences by a finite automaton. Inf. Contr. 9, 521–530 (1966)
14. McNaughton, R.: Infinite games played on finite graphs. Ann. Pure Appl. Logic 65, 149–184 (1993)
15. Moore, E.F. (ed.): Sequential Machines – Selected Papers. Addison-Wesley, Reading, Mass (1963)
16. Mostowski, A.W.: Regular expressions for infinite trees and a standard form of automata. In: Skowron, A. (ed.) SCT 1984. LNCS, vol. 208, pp. 157–168. Springer, Heidelberg (1985)
17. Muller, D.E.: Infinite sequences and finite machines. In: Proc. 4th IEEE Ann. Symp. on Switching Circuit Theory and Logical Design, pp. 3–16. IEEE Press, Los Alamitos (1963)
18. Muller, D.E., Schupp, P.E.: Simulating alternating tree automata by nondeterministic automata: New results and new proofs of the results of Rabin, McNaughton, and Safra. Theor. Comput. Sci. 141, 69–107 (1995)
19. Myhill, J.: Finite automata and the representation of events, WADC Tech. Rep. 57-624, pp. 112-137 (1957)
20. Rabin, M.O.: Decidability of second-order theories and automata on infinite trees. Trans. Amer. Math. Soc. 141, 1–35 (1969)
21. Rabin, M.O.: Automata on infinite objects and Church's Problem, Amer. Math. Soc., Providence RI (1972)
22. Rabin, M.O., Scott, D.: Finite automata and their decision problems. IBM J. Res. Develop. 3, 114-125 (1959)
23. Safra, S.: On the complexity of omega-automata. In: Proc. 29th Ann. Symp.on Foundations of Computer Science, White Plains, New York, pp. 319–327. IEEE Computer Society Press, Los Alamitos (1988)
24. Thomas, W.: On the synthesis of strategies in infinite games. In: Mayr, E.W., Puech, C. (eds.) STACS 1995. LNCS, vol. 900, pp. 1–13. Springer, Heidelberg (1995)
25. Thomas, W.: Solution of Church's Problem: A tutorial. In: Apt, K., van Rooij, R. (eds.) New Perspectives on Games and Interaction, vol. 5, Amsterdam Univ. Press, Texts on Logic and Games (to appear)
26. Trakhtenbrot, B.A.: On operators realizable in logical nets. Dokl. Akad. Naut. SSSR 112, 1005–1007 (1957) (in Russian)
27. Trakhtenbrot, B.A.: Synthesis of logical nets whose operators are described of monadic predicates. Dokl. Akad. Naut. SSSR 118, 646–649 (1958) (in Russian)
28. Trakhtenbrot, B.A., Barzdin, Ya.M.: Finite Automata. Behavior and Synthesis. North-Holland, Amsterdam (1973)

From Monadic Logic to PSL*

Moshe Y. Vardi**

Rice University, Department of Computer Science, Rice University,
Houston, TX 77251-1892, U.S.A.
vardi@cs.rice.edu
http://www.cs.rice.edu/~vardi

Two major themes of my research have been finite model theory and the automata theoretic approach. Boaz Trakhtenbrot laid the foundations in both areas. In 1950, he proved the undecidability of the satisfiability in the finite problem for first-order logic. His contributions to the automata-theoretic approach are described in this paper. I met Boaz in a seminar in 1981, when I was a doctoral student. Little did I know then that his work would have such a profound impact on my future research.

Abstract. One of the surprising developments in the area of program verification is how ideas introduced originally by logicians in the 1950s ended up yielding by 2003 an industrial-standard property-specification language called PSL. This development was enabled by the equally unlikely transformation of the mathematical machinery of automata on infinite words, introduced in the early 1960s for second-order arithmetics, into effective algorithms for model-checking tools. This paper attempts to trace the tangled threads of this development.

1 Thread I: Classical Logic of Time

1.1 Logic and Automata

Classical logic views logic as a declarative formalism, aimed at the specification of properties of mathematical objects. For example, the sentence

$$(\forall x, y, x)(mult(x, y, z) \leftrightarrow mult(y, x, z))$$

expressed the commutativity of multiplication. Starting in the 1930s, a different branch of logic focused on formalisms for describing computations, starting with the introduction of Turing machines in the 1930s, and continuing with the development of the theory of finite-state machines in the 1950s. A surprising,

---

* A shorter version of this paper, under the title "From Church and Prior to PSL", appeared in the *Proc. 2006 Workshop on 25 Years of Model Checking, Lecture Notes in Computer Science, Springer.*

** Supported in part by NSF grants CCR-9988322, CCR-0124077, CCR-0311326, and ANI-0216467, by BSF grant 9800096, and by a gift from the Intel Corporation. The "Y" in the author's middle name stands for "Ya'akov".

A. Avron et al. (Eds.): Trakhtenbrot/Festschrift, LNCS 4800, pp. 656–681, 2008.

intimate, connection between these two paradigms of logic emerged in the late 1950s.

A *nondeterministic finite automaton on words* (NFW) $A = (\Sigma, S, S_0, \rho, F)$ consists of a finite input alphabet Σ, a finite state set S, an initial state set $S_0 \subseteq S$, a transition relation $\rho \subseteq S \times \Sigma \times S$, and an accepting state set $F \subseteq S$. An NFW runs over an finite input word $w = a_0, \ldots, a_{n-1} \in \Sigma^*$. A *run* of A on w is a finite sequence $r = s_0, \ldots, s_n$ of states in S such that $s_0 \in S_0$, and $(s_i, a_i, s_{i+1}) \in \rho$, for $0 \le i < n$. The run r is *accepting* if $s_n \in F$. The word w is *accepted* by A if A has an accepting run on w. The *language* of A, denoted $L(A)$, is the set of words accepted by A. The class of languages accepted by NFWs forms the class of *regular* languages, which are defined in terms of regular expressions. This class is extremely robust and has numerous equivalent representations [70].

Example 1. We describe graphically below an NFW that accepts all words over the alphabet $\{0, 1\}$ that end with an occurrence of 1. The arrow on the left designates the initial state, and the circle on the right designates an accepting state.

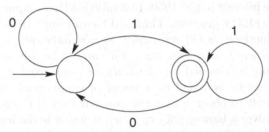

We now view a finite word $w = a_0, \ldots, a_{n-1}$ over an alphabet Σ as a relational structure M_w, with the domain of $0, \ldots, n-1$ ordered by the binary relation $<$, and the unary relations $\{P_a : a \in \Sigma\}$, with the interpretation that $P_a(i)$ holds precisely when $a_i = a$. We refer to such structures as *word structures*. We now use first-order logic (FO) to talk about such words. For example, the sentence

$$(\exists x)((\forall y)(\neg(x < y)) \wedge P_a(x))$$

says that the last letter of the word is a. We say that such a sentence is over the alphabet Σ.

Going beyond FO, we obtain *monadic second-order logic* (MSO), in which we can have monadic second-order quantifiers of the form $\exists Q$, ranging over subsets of the domain, and giving rise to new atomic formulas of the form $Q(x)$. Given a sentence φ in MSO, its set of models models(φ) is a set of words.

The fundamental connection between logic and automata is now given by the following theorem, discovered independently by Julius Richard Büchi, 1924–1984, Calvin Creston Elgot, 1922-1980, and Boris (Boaz) Trakhtenbrot.

Theorem 1. [15,17,43,123,124,125] *Given an MSO sentence φ over alphabet Σ, one can construct an NFW A_φ with alphabet Σ such that a word w in Σ^* is accepted by A_φ iff φ holds in the word structure M_w. Conversely, given an NFW A with alphabet Σ, one can construct an MSO sentence φ_A over Σ such that φ_A holds in a word structure M_w iff w is accepted by A.*

Thus, the class of languages defined by MSO sentences is precisely the class of regular languages.

To decide whether a sentence φ is *satisfiable*, that is, whether models$(\varphi) \neq \emptyset$, we need to check that $L(A_\varphi) \neq \emptyset$. This turns out to be an easy problem. Let $A = (\Sigma, S, S_0, \rho, F)$ be an NFW. Construct a directed graph $G_A = (S, E_A)$, with S as the set of nodes, and $E_A = \{(s,t) : (s,a,t) \in \rho \text{ for some } a \in \Sigma\}$. The following lemma is implicit in [15,17,43,123] and more explicit in [107].

Lemma 1. $L(A) \neq \emptyset$ *iff there are states* $s_0 \in S_0$ *and* $t \in F$ *such that in* G_A *there is a path from* s_0 *to* t.

We thus obtain an algorithm for the SATISFIABILITY problem of MSO over word structures: given an MSO sentence φ, construct the NFW A_φ and check whether $L(A) \neq \emptyset$ by finding a path from an initial state to an accepting state. This approach to satisfiability checking is referred to as the *automata-theoretic approach*, since the decision procedure proceeds by first going from logic to automata, and then searching for a path in the constructed automaton.

There was little interest in the 1950s in analyzing the computational complexity of the SATISFIABILITY problem. That had to wait until 1974. Define the function $exp(k,n)$ inductively as follows: $exp(0,n) = n$ and $exp(k+1,n) = 2^{exp(k,n)}$. We say that a problem is *nonelementary* if it can not be solved by an algorithm whose running time is bounded by $exp(k,n)$ for some fixed $k \geq 0$; that is, the running time cannot be bounded by a tower of exponentials of a fixed height. It is not too difficult to observe that the construction of the automaton A_φ in [15,17,43,123] involves a blow-up of $exp(n,n)$, where n is the length of the MSO sentence being decided. It was shown in [88,116] that the SATISFIABILITY problem for MSO is nonelementary. In fact, the problem is already nonelementary for FO [116].

1.2   Reasoning about Sequential Circuits

The field of hardware verification seems to have been started in a little known 1957 paper by Alonzo Church, 1903–1995, in which he described the use of logic to specify *sequential circuits* [24]. A sequential circuit is a switching circuit whose output depends not only upon its input, but also on what its input has been in the past. A sequential circuit is a particular type of finite-state machine, which became a subject of study in mathematical logic and computer science in the 1950s.

Formally, a sequential circuit $C = (I, O, R, f, g, \mathbf{r}_0)$ consists of a finite set I of Boolean input signals, a finite set O of Boolean output signals, a finite set R of Boolean sequential elements, a transition function $f : 2^I \times 2^R \to 2^R$, an output function $g : 2^R \to 2^O$, and an initial state $\mathbf{r}_0 \in 2^R$. (We refer to elements of $I \cup O \cup R$ as *circuit elements*, and assume that I, O, and R are disjoint.) Intuitively, a state of the circuit is a Boolean assignment to the sequential elements. The initial state is $\mathbf{r}_0$. In a state $\mathbf{r} \in 2^R$, the Boolean assignment to the output signals is $g(\mathbf{r})$. When the circuit is in state $\mathbf{r} \in 2^R$ and it reads an input assignment $\mathbf{i} \in 2^I$, it changes its state to $f(\mathbf{i}, \mathbf{r})$.

A *trace* over a set V of Boolean variables is an infinite word over the alphabet 2^V, i.e., an element of $(2^V)^\omega$. A trace of the sequential circuit C is a trace over $I \cup O \cup R$ that satisfies some conditions. Specifically, a sequence $\tau = (\mathbf{i}_0, \mathbf{r}_0, \mathbf{o}_0)$, $(\mathbf{i}_1, \mathbf{r}_1, \mathbf{o}_1), \ldots$, where $\mathbf{i}_j \in 2^I$, $\mathbf{o}_j \in 2^O$, and $\mathbf{r}_j \in 2^R$, is a trace of C if $\mathbf{r}_{j+1} = f(\mathbf{i}_j, \mathbf{r}_j)$ and $\mathbf{o}_j = g(\mathbf{r}_j)$, for $j \geq 0$. Thus, in modern terminology, Church was following the *linear-time* approach [82] (see discussion in Section 2.1). The set of traces of C is denoted by $\mathrm{traces}(C)$.

We saw earlier how to associate relational structures with words. We can similarly associate with an infinite word $w = a_0, a_1, \ldots$ over an alphabet 2^V, a relational structure $M_w = (\mathbf{N}, \leq, V)$, with the naturals $\mathbf{N}$ as the domain, ordered by $<$, and extended by the set V of unary predicates, where $j \in p$, for $p \in V$, precisely when p *holds* (i.e., is assigned 1) in a_i.[1] We refer to such structures as *infinite word structures*. When we refer to the *vocabulary* of such a structure, we refer explicitly only to V, taking $<$ for granted.

We can now specify traces using First-Order Logic (FO) sentences constructed from atomic formulas of the form $x = y$, $x < y$, and $p(x)$ for $p \in V = I \cup R \cup O$.[2] For example, the FO sentence

$$(\forall x)(\exists y)(x < y \land p(y))$$

says that p holds infinitely often in the trace. In a follow-up paper in 1963 [25], Church considered also specifying traces using monadic second-order logic (MSO), where in addition to first-order quantifiers, which range over the elements of $\mathbf{N}$, we allow also monadic second-order quantifiers, ranging over subsets of $\mathbf{N}$, and atomic formulas of the form $Q(x)$, where Q is a monadic predicate variable. (This logic is also called *S1S*, the "second-order theory of one successor function".) For example, the MSO sentence,

$$(\exists P)(\forall x)(\forall y)((((P(x) \land y = x + 1) \to (\neg P(y))) \land$$
$$(((\neg P(x)) \land y = x + 1) \to P(y))) \land$$
$$(x = 0 \to P(x)) \land (P(x) \to q(x))),$$

where $x = 0$ is an abbrevaition for $(\neg(\exists z)(z < x))$ and $y = x + 1$ is an abbreviation for $(y > x \land \neg(\exists z)(x < z \land z < y))$, says that q holds at every even point on the trace. In effect, Church was proposing to use classical logic (FO or MSO) as a logic of time, by focusing on infinite word structures. The set of infinite models of an FO or MSO sentence φ is denoted by $\mathrm{models}_\omega(\varphi)$.

Church posed two problems related to sequential circuits [24]:

- The DECISION problem: Given circuit C and a sentence φ, does φ hold in all traces of C? That is, does $\mathrm{traces}(C) \subseteq \mathrm{models}(\varphi)$ hold?
- The SYNTHESIS problem: Given sets I and O of input and output signals, and a sentence φ over the vocabulary $I \cup O$, construct, if possible, a sequential circuit C with input signals I and output signals O such that φ holds in all traces of C. That is, construct C such that $\mathrm{traces}(C) \subseteq \mathrm{models}(\varphi)$ holds.

---

[1] We overload notation here and treat p as both a Boolean variable and a predicate.

[2] We overload notation here and treat p as both a circuit element and a predicate symbol.

In modern terminology, Church's DECISION problem is the MODEL-CHECKING problem in the linear-time approach (see Section 2.2). This problem did not receive much attention after [24,25], until the introduction of model checking in the early 1980s. In contrast, the SYNTHESIS problem has remained a subject of ongoing research; see [18,76,78,106,122]. One reason that the DECISION problem did not remain a subject of study, is the easy observation in [25] that the DECISION problem can be reduced to the VALIDITY problem in the underlying logic (FO or MSO). Given a sequential circuit C, we can easily generate an FO sentence α_C that holds in precisely all structures associated with traces of C. Intuitively, the sentence α_C simply has to encode the transition and output functions of C, which are Boolean functions. Then φ holds in all traces of C precisely when $\alpha_C \to \varphi$ holds in all word structures (of the appropriate vocabulary). Thus, to solve the DECISION problem we need to solve the VALIDITY problem over word structures. As we see next, this problem was solved in 1962.

1.3   Reasoning about Infinite Words

Church's DECISION problem was essentially solved in 1962 by Büchi who showed that the VALIDITY problem over infinite word structures is decidable [16]. Actually, Büchi showed the decidability of the dual problem, which is the SATISFIABILITY problem for MSO over infinite word structures. Büchi's approach consisted of extending the automata-theoretic approach, see Theorem 1, which was introduced a few years earlier for word structures, to infinite word structures. To that end, Büchi extended automata theory to automata on infinite words.

A *nondeterministic Büchi automaton on words* (NBW) $A = (\Sigma, S, S_0, \rho, F)$ consists of a finite input alphabet Σ, a finite state set S, an initial state set $S_0 \subseteq S$, a transition relation $\rho \subseteq S \times \Sigma \times S$, and an accepting state set $F \subseteq S$. An NBW runs over an infinite input word $w = a_0, a_1, \ldots \in \Sigma^\omega$. A *run* of A on w is an infinite sequence $r = s_0, s_1, \ldots$ of states in S such that $s_0 \in S_0$, and $(s_i, a_i, s_{i+1}) \in \rho$, for $i \geq 0$. The run r is *accepting* if F is visited by r infinitely often; that is, $s_i \in F$ for infinitely many i's. The word w is *accepted* by A if A has an accepting run on w. The *infinitary language* of A, denoted $L_\omega(A)$, is the set of infinite words accepted by A. The class of languages accepted by NBWs forms the class of ω-*regular* languages, which are defined in terms of regular expressions augmented with the ω-power operator (e^ω denotes an infinitary iteration of e) [16].

Example 2. We describe graphically an NBW that accepts all words over the alphabet $\{0,1\}$ that contain infinitely many occurrences of 1. The arrow on the left designates the initial state, and the circle on the right designates an accepting state. Note that this NBW looks exactly like the NFW in Example 1. The only difference is that in Example 1 we considered finite input words and here we are considering infinite input words.

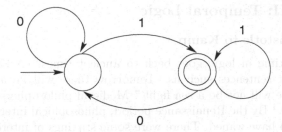

As we saw earlier, the paradigmatic idea of the automata-theoretic approach is that we can compile high-level logical specifications into an equivalent low-level finite-state formalism.

Theorem 2. [16] *Given an MSO sentence φ with vocabulary V, one can construct an NBW A_φ with alphabet 2^V such that a word w in $(2^V)^\omega$ is accepted by A_φ iff φ holds in the word structure M_w. Conversely, given an NBW A with alphabet 2^V, one can construct an MSO sentence φ_A with vocabulary V such that φ_A holds in an infinite word structure M_w iff w is accepted by A.*

Thus, the class of languages defined by MSO sentences is precisely the class of ω-regular languages.

To decide whether sentence φ is satisfiable over infinite words, that is, whether $\text{models}_\omega(\varphi) \neq \emptyset$, we need to check that $L_\omega(A_\varphi) \neq \emptyset$. Let $A = (\Sigma, S, S_0, \rho, F)$ be an NBW. As with NFWs, construct a directed graph $G_A = (S, E_A)$, with S as the set of nodes, and $E_A = \{(s,t) : (s,a,t) \in \rho \text{ for some } a \in \Sigma\}$. The following lemma is implicit in [16] and more explicit in [126].

Lemma 2. $L_\omega(A) \neq \emptyset$ iff there are states $s_0 \in S^0$ and $t \in F$ such that in G_A there is a path from s_0 to t and a path from t to itself.

We thus obtain an algorithm for the SATISFIABILITY problem of MSO over infinite word structures: given an MSO sentence φ, construct the NBW A_φ and check whether $L_\omega(A) \neq \emptyset$ by finding a path from an initial state to an accepting state and a cycle through that accepting state. Since the DECISION problem can be reduced to the SATISFIABILITY problem, this also solves the DECISION problem.

Neither Büchi nor Church analyzed the complexity of the DECISION problem. The non-elementary lower bound mentioned earlier for MSO over words can be easily extended to infinite words. The upper bound here is a bit more subtle. For both finite and infinite words, the construction of A_φ proceeds by induction on the structure of φ, with complementation being the difficult step. For NFW, complementation uses the *subset construction*, which involves a blow-up of 2^n [107,109]. Complementation for NBW is significantly more involved, see [127]. The blow-up of complementation is $2^{\Theta(n \log n)}$, but there is still a gap between the known upper and lower bounds. At any rate, this yields a blow-up of $exp(n, n \log n)$ for the translation from MSO to NBW.

2   Thread II: Temporal Logic

2.1   From Aristotle to Kamp

The history of time in logic goes back to ancient times.[3] Aristotle pondered how to interpret sentences such as "Tomorrow there will be a sea fight," or "Tomorrow there will not be a sea fight." Medieval philosophers also pondered the issue of time.[4] By the Renaissance period, philosophical interest in the logic of time seems to have waned. There were some stirrings of interest in the 19th century, by Boole and Peirce. Peirce wrote:

> "Time has usually been considered by logicians to be what is called 'extra-logical' matter. I have never shared this opinion. But I have thought that logic had not yet reached the state of development at which the introduction of temporal modifications of its forms would not result in great confusion; and I am much of that way of thinking yet."

There were also some stirrings of interest in the first half of the 20th century, but the birth of modern temporal logic is unquestionably credited to Arthur Norman Prior, 1914-1969. Prior was a philosopher, who was interested in theological and ethical issues. His own religious path was somewhat convoluted; he was born a Methodist, converted to Presbytarianism, became an atheist, and ended up an agnostic. In 1949, he published a book titled *"Logic and The Basis of Ethics"*. He was particularly interested in the conflict between the assumption of *free will* ("the future is to some extent, even if it is only a very small extent, something we can make for ourselves"), *foredestination* ("of what will be, it has now been the case that it will be"), and *foreknowledge* ("there is a deity who infallibly knows the entire future"). He was also interested in modal logic [103]. This confluence of interests led Prior to the development of *temporal logic*.[5] His wife, Mary Prior, recalled after his death:

> "I remember his waking me one night [in 1953], coming and sitting on my bed, ..., and saying he thought one could make a formalised tense logic."

---

[3] For a detailed history of temporal logic from ancient times to the modern period, see [92].

[4] For example, William of Ockham, 1288–1348, wrote (rather obscurely for the modern reader): "Wherefore the difference between present tense propositions and past and future tense propositions is that the predicate in a present tense proposition stands in the same way as the subject, unless something added to it stops this; but in a past tense and a future tense proposition it varies, for the predicate does not merely stand for those things concerning which it is truly predicated in the past and future tense propositions, because in order for such a proposition to be true, it is not sufficient that that thing of which the predicate is truly predicated (whether by a verb in the present tense or in the future tense) is that which the subject denotes, although it is required that the very same predicate is truly predicated of that which the subject denotes, by means of what is asserted by such a proposition."

[5] An earlier term was *tense logic*; the term *temporal logic* was introduced in [91]. The technical distinction between the two terms seems fuzzy.

Prior lectured on his new work when he was the John Locke Lecturer at the University of Oxford in 1955–6, and published his book "*Time and Modality*" in 1957 [101].[6] In this book, he presented a temporal logic that is propositional logic extended with two temporal connectives, F and P, corresponding to "sometime in the future" and "sometime in the past". A crucial feature of this logic is that it has an implicit notion of "now", which is treated as an *indexical*, that is, it depends on the context of utterance for its meaning. Both future and past are defined with respect to this implicit "now".

It is interesting to note that the *linear* vs. *branching* time dichotomy, which has been a subject of some controversy in the computer science literature since 1980 (see [132]), has been present from the very beginning of temporal-logic development. In Prior's early work on temporal logic, he assumed that time was linear. In 1958, he received a letter from Saul Kripke,[7] who wrote

"In an indetermined system, we perhaps should not regard time as a linear series, as you have done. Given the present moment, there are several possibilities for what the next moment may be like – and for each possible next moment, there are several possibilities for the moment after that. Thus the situation takes the form, not of a linear sequence, but of a 'tree'."

Prior immediately saw the merit of Kripke's suggestion: "the determinist sees time as a line, and the indeterminist sees times as a system of forking paths." He went on to develop two theories of branching time, which he called "Ockhamist" and "Peircean". (Prior did not use path quantifiers; those were introduced later, in the 1980s. See Section 3.2.)

While the introduction of branching time seems quite reasonable in the context of trying to formalize free will, it is far from being simple philosophically. Prior argued that the nature of the course of time is branching, while the nature of a course of events is linear [102]. In contrast, it was argued in [91] that the nature of time is linear, but the nature of the course of events is branching: "We have 'branching *in* time,' not 'branching *of* time'."[8]

During the 1960s, the development of temporal logic continued through both the linear-time approach and the branching-time approach. There was little connection, however, between research on temporal logic and research on classical logics, as described in Section 1. That changed in 1968, when Johan Anthony Willem (Hans) Kamp tied together the two threads in his doctoral dissertation.

---

[6] Due to the arcane infix notation of the time, the book may not be too accessible to modern readers, who may have difficulties parsing formulas such as $CKMpMqAMKpMqMKqMp$.

[7] Kripke was a high-school student, not quite 18, in Omaha, Nebraska. Kripke's interest in modal logic was inspired by a paper by Prior on this subject [104]. Prior turned out to be the referee of Kripke's first paper [75].

[8] One is reminded of St. Augustin, who said in his *Confessions*: "What, then, is time? If no one asks me, I know; but if I wish to explain it to some who should ask me, I do not know."

Theorem 3. [71] *Linear temporal logic with past and binary temporal connectives ("strict until" and "strict since") has precisely the expressive power of FO over the ordered naturals (with monadic vocabularies).*

It should be noted that Kamp's Theorem is actually more general and asserts expressive equivalence of FO and temporal logic over all "Dedekind-closed orders". The introduction of binary temporal connectives by Kamp was necessary for reaching the expressive power of FO; *unary* linear temporal logic, which has only unary temporal connectives, is weaker than FO [51]. The theorem refers to FO formulas with one free variable, which are satisfied at an element of a structure, analogously to temporal logic formulas, which are satisfied at a point of time.

It should be noted that one direction of Kamp's Theorem, the translation from temporal logic to FO, is quite straightforward; the hard direction is the translation from FO to temporal logic. Both directions are algorithmically effective; translating from temporal logic to FO involves a linear blowup, but translation in the other direction involves a nonelementary blowup.

If we focus on FO sentences rather than FO formulas, then they define sets of traces (a sentence φ defines models(φ)). A characterization of of the expressiveness of FO sentences over the naturals, in terms of their ability to define sets of traces, was obtained in 1979.

Theorem 4. [121] *FO sentences over naturals have the expressive power of $*$-free ω-regular expressions.*

Recall that MSO defines the class of ω-regular languages. It was already shown in [44] that FO over the naturals is weaker expressively than MSO over the naturals. Theorem 4 was inspired by an analogous theorem in [87] for finite words.

2.2 The Temporal Logic of Programs

There were some early observations that temporal logic can be applied to programs. Prior stated: "There are practical gains to be had from this study too, for example, in the representation of time-delay in computer circuits" [102]. Also, a discussion of the application of temporal logic to processes, which are defined as "programmed sequences of states, deterministic or stochastic" appeared in [91].

The "big bang" for the application of temporal logic to program correctness occurred with Amir Pnueli's 1977 paper [94]. In this paper, Pnueli, inspired by [91], advocated using future linear temporal logic (LTL) as a logic for the specification of non-terminating programs; see overview in [59].

LTL is a temporal logic with two temporal connectives, "next" and "until".[9] In LTL, formulas are constructed from a set *Prop* of atomic propositions using the usual Boolean connectives as well as the unary temporal connective X ("next"),

---

[9] Unlike Kamp's "strict until" ("p strict until q" requires q to hold in the strict future), Pnueli's "until" is not strict ("p until q" can be satisfied by q holding now), which is why the "next" connective is required.

and the binary temporal connective U ("until"). Additional unary temporal connectives F ("eventually"), and G ("always") can be defined in terms of U. Note that all temporal connectives refer to the future here, in contrast to Kamp's "strict since" operator, which refers to the past. Thus, LTL is a *future temporal logic*. For extensions with past temporal connectives, see [84,85,129].

LTL is interpreted over traces over the set *Prop* of atomic propositions. For a trace τ and a point $i \in \mathbf{N}$, the notation $\tau, i \models \varphi$ indicates that the formula φ holds at the point i of the trace τ. Thus, the point i is the implicit "now" with respect to which the formula is interpreted. We have that

- $\tau, i \models p$ if p holds at $\tau(i)$,
- $\tau, i \models X\varphi$ if $\tau, i + 1 \models \varphi$, and
- $\tau, i \models \varphi U \psi$ if for some $j \geq i$, we have $\tau, j \models \psi$ and for all k, $i \leq k < j$, we have $\tau, k \models \varphi$.

The temporal connectives F and G can be defined in terms of the temporal connective U; $F\varphi$ is defined as **true** $U\varphi$, and $G\varphi$ is defined as $\neg F\neg\varphi$. We say that τ *satisfies* a formula φ, denoted $\tau \models \varphi$, iff $\tau, 0 \models \varphi$. We denote by models(φ) the set of traces satisfying φ.

As an example, the LTL formula $G(request \rightarrow F\ grant)$, which refers to the atomic propositions *request* and *grant*, is true in a trace precisely when every state in the trace in which *request* holds is followed by some state in the (non-strict) future in which *grant* holds. Also, the LTL formula $G(request \rightarrow (request\ U\ grant))$ is true in a trace precisely if, whenever *request* holds in a state of the trace, it holds until a state in which *grant* holds is reached.

The focus on satisfaction at 0, called *initial semantics*, is motivated by the desire to specify computations at their starting point. It enables an alternative version of Kamp's Theorem, which does not require past temporal connectives, but focuses on initial semantics.

Theorem 5. [56] *LTL has precisely the expressive power of FO over the ordered naturals (with monadic vocabularies) with respect to initial semantics.*

As we saw earlier, FO has the expressive power of star-free ω-regular expressions over the naturals. Thus, LTL has the expressive power of star-free ω-regular expressions (see [96]), and is strictly weaker than MSO. An interesting outcome of the above theorem is that it lead to the following assertion regarding LTL [89]: "The corollary due to Meyer – I have to get in my controversial remark – is that that [Theorem 5] makes it theoretically uninteresting." Developments since 1980 have proven this assertion to be overly pessimistic on the merits of LTL.

Pnueli also discussed the analog of Church's DECISION problem: given a finite-state program P and an LTL formula φ, decide if φ holds in all traces of P. Just like Church, Pnueli observed that this problem can be solved by reduction to MSO. Rather than focus on sequential circuits, Pnueli focused on programs, modeled as (labeled) *transition systems* [72]. A transition system $M = (W, W_0, R, V)$ consists of a set W of states that the system can be in, a set $W_0 \subseteq W$ of initial states, a transition relation $R \subseteq W^2$ that indicates the allowable state transitions of the system, and an assignment $V : W \rightarrow 2^{Prop}$ of truth values to the atomic propositions in each state of the system. (A transition system is essentially a Kripke structure [10].) A *path* in M that *starts at* u is a possible infinite behavior of the system starting at u, i.e., it is an infinite sequence $u_0, u_1 \ldots$ of states in W such that $u_0 = u$, and $(u_i, u_{i+1}) \in R$ for all $i \geq 0$. The sequence $V(u_0), V(u_1) \ldots$ is a *trace* of M that *starts at* u. It is the sequence of truth assignments visited by the path. The *language* of M, denoted $L(M)$, consists of all traces of M that start at a state in W_0. Note that $L(M)$ is a language of infinite words over the alphabet 2^{Prop}. The language $L(M)$ can be viewed as an abstract description of the system M, describing all possible traces. We say that M *satisfies* an LTL formula φ if all traces in $L(M)$ satisfy φ, that is, if $L(M) \subseteq \text{models}(\varphi)$. When W is finite, we have a finite-state system, and can apply algorithmic techniques.

What about the complexity of LTL reasoning? Recall from Section 1 that satisfiability of FO over trace structures is nonelementary. In contrast, it was shown in [61,62,111,112,113,138,139] that LTL SATISFIABILITY is elementary; in fact, it is PSPACE-complete. It was also shown that the DECISION problem for LTL with respect to finite transition systems is PSPACE-complete [111,112,113]. The basic technique for proving these elementary upper bounds is the *tableau* technique, which was adapted from *dynamic logics* [99] (see Section 3.1). Thus, even though FO and LTL are expressively equivalent, they have dramatically different computational properties, as LTL reasoning is in PSPACE, while FO reasoning is nonelementary.

The second "big bang" in the application of temporal logic to program correctness was the introduction of *model checking* by Edmund Melson Clarke and Ernest Allen Emerson [28] and by Jean-Pierre Queille and Joseph Sifakis [105]. The two papers used two different branching-time logics. Clarke and Emerson used CTL (inspired by the branching-time logic UB of [9]), which extends LTL with existential and universal path quantifiers E and A. Queille and Sifakis used a logic introduced by Leslie Lamport [82], which extends propositional logic with the temporal connectives POT (which corresponds to the CTL operator EF) and $INEV$ (which corresponds to the CTL operator AF). The focus in both papers was on model checking, which is essentially what Church called the DECISION problem: does a given finite-state program, viewed as a finite transition system, satisfy its given temporal specification. In particular, Clarke and Emerson showed that model checking transition systems of size m with respect to formulas of size n can be done in time polynomial in m and n. This was refined later to $O(mn)$ (even in the presence of *fairness* constraints, which restrict attention to certain infinite paths in the underlying transition system) [29,30]. We drop the term "DECISION problem" from now on, and replace it with the term "MODEL-CHECKING problem".[10]

---

[10] The model-checking problem is analogous to database query evaluation, where we check the truth of a logical formula, representing a query, with respect to a database, viewed as a finite relational structure. Interestingly, the study of the complexity of database query evaluation started about the same time as that of model checking [128].

It should be noted that the linear complexity of model checking refers to the size of the transition system, rather than the size of the program that gave rise to that system. For sequential circuits, transition-system size is essentially exponential in the size of the description of the circuit (say, in some Hardware Description Language). This is referred to as the "state-explosion problem" [31]. In spite of the state-explosion problem, in the first few years after the publication of the first model-checking papers in 1981-2, Clarke and his students demonstrated that model checking is a highly successful technique for automated program verification [13,33]. By the late 1980s, automated verification had become a recognized research area. Also by the late 1980s, *symbolic* model checking was developed [19,20], and the SMV tool, developed at CMU by Kenneth Laughlin McMillan [86], was starting to have an industrial impact. See [27] for more details.

The detailed complexity analysis in [29] inspired a similar detailed analysis of linear time model checking. It was shown in [83] that model checking transition systems of size m with respect to LTL formulas of size n can be done in time $m2^{O(n)}$. (This again was shown using a tableau-based technique.) While the bound here is exponential in n, the argument was that n is typically rather small, and therefore an exponential bound is acceptable.

2.3   Back to Automata

Since LTL can be translated to FO, and FO can be translated to NBW, it is clear that LTL can be translated to NBW. Going through FO, however, would incur, in general, a nonelementary blowup. In 1983, Pierre Wolper, Aravinda Prasad Sistla, and I showed that this nonelementary blowup can be avoided.

Theorem 6. [136,140] *Given an LTL formula φ of size n, one can construct an NBW A_φ of size $2^{O(n)}$ such that a trace σ satisfies φ if and only if σ is accepted by A_φ.*

It now follows that we can obtain a PSPACE algorithm for LTL SATISFIABILITY: given an LTL formula φ, we construct A_φ and check that $A_\varphi \neq \emptyset$ using the graph-theoretic approach described earlier. We can avoid using exponential space, by constructing the automaton *on the fly* [136,140].

What about model checking? We know that a transition system M satisfies an LTL formula φ if $L(M) \subseteq \text{models}(\varphi)$. It was then observed in [135] that the following are equivalent:

- M satisfies φ
- $L(M) \subseteq \text{models}(\varphi)$
- $L(M) \subseteq L(A_\varphi)$
- $L(M) \cap ((2^{Prop})^\omega - L(A_\varphi)) = \emptyset$
- $L(M) \cap L(A_{\neg\varphi}) = \emptyset$
- $L(M \times A_{\neg\varphi}) = \emptyset$

Thus, rather than complementing A_φ using an exponential complementation construction [16,77,115], we complement the LTL property using logical negation. It is easy to see that we can now get the same bound as in [83]: model checking programs of size m with respect to LTL formulas of size n can be done in time $m2^{O(n)}$. Thus, the optimal bounds for LTL satisfiability and model checking can be obtained without resorting to ad-hoc tableau-based techniques; the key is the exponential translation of LTL to NBW.

One may wonder whether this theory is practical. Reduction to practice took over a decade of further research, which saw the development of

- an optimized search algorithm for explicit-state model checking [36,37],
- a symbolic, BDD-based[11] algorithm for NBW nonemptiness [19,20,49],
- symbolic algorithms for LTL to NBW translation [19,20,32], and
- an optimized explicit algorithm for LTL to NBW translation [58].

By 1995, there were two model-checking tools that implemented LTL model checking via the automata-theoretic approach: Spin [69] is an explicit-state LTL model checker, and Cadence's SMV is a symbolic LTL model checker.[12] See [133] for a description of algorithmic developments since the mid 1990s. Additional tools today are *VIS* [12], *NuSMV* [26], and *SPOT* [38].

It should be noted that Robert Kurshan developed the automata-theoretic approach independently, also going back to the 1980s [1,2,79]. In his approach (as also in [108,140]), one uses automata to represent both the system and its specification [80].[13] The first implementation of COSPAN, a model-checking tool that is based on this approach [63], also goes back to the 1980s; see [81].

2.4 Enhancing Expressiveness

Can the development of LTL model checking [83,135] be viewed as a satisfactory solution to Church's DECISION problem? Almost, but not quite, since, as we observed earlier, LTL is not as expressive as MSO, which means that LTL is expressively weaker than NBW. Why do we need the expressive power of NBWs? First, note that once we add fairness to transitions systems (sse [29,30]), they can be viewed as variants of NBWs. Second, there are good reasons to expect the specification language to be as expressive as the underlying model of programs [95]. Thus, achieving the expressive power of NBWs, which we refer to as ω-regularity, is a desirable goal. This motivated efforts since the early 1980s to extend LTL.

The first attempt along this line was made by Wolper [138,139], who defined ETL (for *Extended Temporal Logic*), which is LTL extended with grammar operators. He showed that ETL is more expressive than LTL, while its SATISFIABILITY problem can still be solved in exponential time (and even PSPACE [111,112,113]). Then, Sistla, Wolper and I showed how to extend LTL with automata connectives, reaching ω-regularity, without losing the PSPACE upper

---

[11] To be precise, one should use the acronym ROBDD, for Reduced Ordered Binary Decision Diagrams [14].

[12] Cadence's SMV is also a CTL model checker. See
www.cadence.com/webforms/cbl_software/index.aspx.

[13] The connection to automata is somewhat difficult to discern in the early papers [1,2].

bound for the SATISFIABILITY problem [136,140]. Actually, three syntactical variations, denoted ETL_f, ETL_l, and ETL_r were shown to be expressively equivalent and have these properties [136,140].

Two other ways to achieve ω-regularity were discovered in the 1980s. The first is to enhance LTL with monadic second-order quantifiers as in MSO, which yields a logic, QPTL, with a nonelementary SATISFIABILITY problem [114,115]. The second is to enhance LTL with least and greatest fixpoints [6,130], which yields a logic, μLTL, that achieves ω-regularity, and has a PSPACE upper bound on its SATISFIABILITY and MODEL-CHECKING problems [130]. For example, the (not too readable) formula

$$(\nu P)(\mu Q)(P \wedge X(p \vee Q)),$$

where ν and μ denote greatest and least fixpoint operators, respectively, is equivalent to the LTL formula GFp, which says that p holds infinitely often.

3   Thread III: Dynamic and Branching-Time Logics

3.1   Dynamic Logics

In 1976, a year before Pnueli proposed using LTL to specify programs, Vaughan Ronald Pratt proposed using *dynamic logic*, an extension of modal logic, to specify programs [97].[14] In modal logic $\Box\varphi$ means that φ holds in all worlds that are possible with respect to the current world [10]. Thus, $\Box\varphi$ can be taken to mean that φ holds after an execution of a program step, taking the transition relation of the program to be the possibility relation of a Kripke structure. Pratt proposed the addition of dynamic modalities $[e]\varphi$, where e is a program, which asserts that φ holds in all states reachable by an execution of the program e. Dynamic logic can then be viewed as an extension of Hoare logic, since $\psi \rightarrow [e]\varphi$ corresponds to the Hoare triple $\{\psi\}e\{\varphi\}$ (see [3]). See [65] for an extensive coverage of dynamic logic.

In 1977, a propositional version of Pratt's dynamic logic, called PDL, was proposed, in which programs are regular expressions over atomic programs [52,53]. It was shown there that the SATISFIABILITY problem for PDL is in NEXPTIME and EXPTIME-hard. Pratt then proved an EXPTIME upper bound, adapting tableau techniques from modal logic [98,99]. (We saw earlier that Wolper then adapted these techniques to linear-time logic.)

Pratt's dynamic logic was designed for terminating programs, while Pnueli was interested in nonterminating programs. This motivated various extensions of dynamic logic to nonterminating programs [68,118,117,119]. Nevertheless, these logics are much less natural for the specification of ongoing behavior than temporal logic. They inspired, however, the introduction of the *(modal) μ-calculus* by Dexter Kozen [73,74]. The μ-calculus is an extension of modal logic with least and greatest fixpoints. It subsumes expressively essentially all dynamic and

---

[14] See discussion of precursor and related developments, such as [21,34,50,110], in [65].

temporal logics [11]. Kozen's paper was inspired by previous papers that showed the usefulness of fixpoints in characterizing correctness properties of programs [45,93] (see also [100]). In turn, the μ-calculus inspired the introduction of μLTL, mentioned earlier. The μ-calculus also played an important role in the development of symbolic model checking [19,20,49].

3.2   Branching-Time Logics

Dynamic logic provided a branching-time approach to reasoning about programs, in contrast to Pnueli's linear-time approach. Lamport was the first to study the dichotomy between linear and branching time in the context of program correctness [82]. This was followed by the introduction of the branching-time logic UB, which extends unary LTL (LTL without the temporal connective "until") with the existential and universal path quantifiers, E and A [9]. Path quantifiers enable us to quantify over different future behavior of the system. By adapting Pratt's tableau-based method for PDL to UB, it was shown that its SATIS-FIABILITY problem is in EXPTIME [9]. Clarke and Emerson then added the temporal connective "until" to UB and obtained CTL [28]. (They did not focus on the SATISFIABILITY problem for CTL, but, as we saw earlier, on its MODEL-CHECKING problem; the SATISFIABILITY problem was shown later to be solvable in EXPTIME [47].) Finally, it was shown that LTL and CTL have incomparable expressive power, leading to the introduction of the branching-time logic CTL*, which unifies LTL and CTL [46,48].

The key feature of branching-time logics in the 1980s was the introduction of explicit path quantifiers in [9]. This was an idea that was not discovered by Prior and his followers in the 1960s and 1970s. Most likely, Prior would have found CTL* satisfactory for his philosophical applications and would have seen no need to introduce the "Ockhamist" and "Peircean" approaches.

3.3   Combining Dynamic and Temporal Logics

By the early 1980s it became clear that temporal logics and dynamic logics provide two distinct perspectives for specifying programs: the first is *state* based, while the second is *action* based. Various efforts have been made to combine the two approaches. These include the introduction of *Process Logic* [64] (branching time), *Yet Another Process Logic* [134] (branching time), *Regular Process Logic* [67] (linear time), *Dynamic LTL* [60] (linear time), and *RCTL* [8] (branching time), which ultimately evolved into *Sugar* [7]. RCTL/Sugar is unique among these logics in that it did not attempt to borrow the action-based part of dynamic logic. It is a state-based branching-time logic with no notion of actions. Rather, what it borrowed from dynamic logic was the use of regular-expression-based dynamic modalities. Unlike dynamic logic, which uses regular expressions over program statements, RCTL/Sugar uses regular expressions over state predicates, analogously to the automata of ETL [136,140], which run over sequences of formulas.

4   Thread IV: From LTL to ForSpec and PSL

In the late 1990s and early 2000s, model checking was having an increasing industrial impact. That led to the development of two industrial temporal logics based on LTL: *ForSpec*, developed by Intel, and *PSL*, developed by an industrial standards committee.

4.1   From LTL to ForSpec

Intel's involvement with model checking started in 1990, when Kurshan, spending a sabbatical year in Israel, conducted a successful feasibility study at the Intel Design Center (IDC) in Haifa, using COSPAN, which at that point was a prototype tool; see [81]. In 1992, IDC started a pilot project using SMV. By 1995, model checking was used by several design projects at Intel, using an internally developed model checker based on SMV. Intel users have found CTL to be lacking in expressive power and the Design Technology group at Intel developed its own specification language, FSL. The FSL language was a linear-time logic, and it was model checked using the automata-theoretic approach, but its design was rather ad-hoc, and its expressive power was unclear; see [54].

In 1997, Intel's Design Technology group at IDC embarked on the development of a second-generation model-checking technology. The goal was to develop a model-checking engine from scratch, as well as a new specification language. A BDD-based model checker was released in 1999 [55], and a SAT-based model checker was released in 2000 [35].

I got involved in the design of the second-generation specification language in 1997. That language, ForSpec, was released in 2000 [5]. The first issue to be decided was whether the language should be linear or branching. This led to an in-depth examination of this issue [132], and the decision was to pursue a linear-time language. An obvious candidate was LTL; we saw that by the mid 1990s there were both explicit-state and symbolic model checkers for LTL, so there was no question of feasibility. I had numerous conversations with Limor Fix, Michael Hadash, Yonit Kesten, and Moshe Sananes on this issue. The conclusion was that LTL is not expressive enough for industrial usage. In particular, many properties that are expressible in FSL are not expressible in LTL. Thus, it turned out that the theoretical considerations regarding the expressiveness of LTL, i.e., its lack of ω-regularity, had practical significance. I offered two extensions of LTL; as we saw earlier both ETL and μLTL achieve ω-regularity and have the same complexity as LTL. Neither of these proposals was accepted, due to the perceived difficulty of usage of such logics by Intel validation engineers, who typically have only basic familiarity with automata theory and logic.

These conversations continued in 1998, now with Avner Landver. Avner also argued that Intel validation engineers would not be receptive to the automata-based formalism of ETL. Being familiar with RCTL/Sugar and its dynamic modalities [7,8], he asked me about regular expressions, and my answer was that regular expressions are equivalent to automata [70], so the automata of ETL_f, which extends LTL with automata on *finite* words, can be replaced by regular expressions over state predicates. This lead to the development of *RELTL*, which is LTL augmented by the dynamic regular modalities of dynamic logic (interpreted linearly, as in ETL). Instead of the dynamic-logic notation $[e]\varphi$, ForSpec uses the more readable (to engineers) (e triggers φ), where e is a regular expression over state predicates (e.g., $(p \vee q)^*, (p \wedge q)$), and φ is a formula. Semantically, $\tau, i \models (e$ triggers $\varphi)$ if, for all $j \geq i$, if $\tau[i,j]$ (that is, the finite word $\tau(i), \ldots, \tau(j)$) "matches" e (in the intuitive formal sense), then $\tau, j \models \varphi$; see [22]. Using the ω-regularity of ETL_f, it is now easy to show that RELTL also achieves ω-regularity [5].

While the addition of dynamic modalities to LTL is sufficient to achieve ω-regularity, we decided to also offer direct support to two specification modes often used by verification engineers at Intel: *clocks* and *resets*. Both clocks and resets are features that are needed to address the fact that modern semiconductor designs consist of interacting parallel modules. While clocks and resets have a simple underlying intuition, defining their semantics formally is quite nontrivial. ForSpec is essentially RELTL, augmented with features corresponding to clocks and resets, as we now explain.

Today's semiconductor designs are still dominated by synchronous circuits. In synchronous circuits, clock signals synchronize the sequential logic, providing the designer with a simple operational model. While the asynchronous approach holds the promise of greater speed (see [23]), designing asynchronous circuits is significantly harder than designing synchronous circuits. Current design methodology attempts to strike a compromise between the two approaches by using multiple clocks. This results in architectures that are globally asynchronous but locally synchronous. The temporal-logic literature mostly ignores the issue of explicitly supporting clocks. ForSpec supports multiple clocks via the notion of *current clock*. Specifically, ForSpec has a construct change_on $c\ \varphi$, which states that the temporal formula φ is to be evaluated with respect to the clock c; that is, the formula φ is to be evaluated in the trace defined by the high phases of the clock c. The key feature of clocks in ForSpec is that each subformula may advance according to a different clock [5].

Another feature of modern designs' consisting of interacting parallel modules is the fact that a process running on one module can be reset by a signal coming from another module. As noted in [120], reset control has long been a critical aspect of embedded control design. ForSpec directly supports reset signals. The formula accept_on $a\ \varphi$ states that the property φ should be checked only until the arrival of the reset signal a, at which point the check is considered to have *succeeded*. In contrast, reject_on $r\ \varphi$ states that the property φ should be checked only until the arrival of the reset signal r, at which point the check is considered to have *failed*. The key feature of resets in ForSpec is that each subformula may be reset (positively or negatively) by a different reset signal; for a longer discussion see [5].

ForSpec is an industrial property-specification language that supports hardware-oriented constructs as well as uniform semantics for formal and dynamic validation, while at the same time it has a well understood expressiveness (ω-regularity) and computational complexity (SATISFIABILITY and

MODEL-CHECKING problems have the same complexity for ForSpec as for LTL) [5]. The design effort strove to find an acceptable compromise, with trade-offs clarified by theory, between conflicting demands, such as expressiveness, usability, and implementability. Clocks and resets, both important to hardware designers, have a clear intuitive semantics, but formalizing this semantics is nontrivial. The rigorous semantics, however, not only enabled mechanical verification of various theorems about the language, but also served as a reference document for the implementors. The implementation of model checking for ForSpec followed the automata-theoretic approach, using *alternating* automata as advocated in [131] (see [57]).

4.2 From ForSpec to PSL

In 2000, the Electronic Design Automation Association instituted a standardization body called *Accellera*.[15] Accellera's mission is to drive worldwide development and use of standards required by systems, semiconductor and design tools companies. Accellera decided that the development of a standard specification language is a requirement for formal verification to become an industrial reality (see [81]). Since the focus was on specifying properties of designs rather than designs themselves, the chosen term was "property specification language" (PSL). The PSL standard committee solicited industrial contributions and received four language contributions: *CBV*, from Motorola, ForSpec, from Intel, *Temporal e*, from Verisity [90], and Sugar, from IBM.

The committee's discussions were quite fierce.[16] Ultimately, it became clear that while technical considerations play an important role, industrial committees' decisions are ultimately made for business considerations. In that contention, IBM had the upper hand, and Accellera chose Sugar as the base language for PSL in 2003. At the same time, the technical merits of ForSpec were accepted and PSL adopted all the main features of ForSpec. In essence, PSL (the current version 1.1) is LTL, extended with dynamic modalities (referred to as the *regular layer*), clocks, and resets (called *aborts*). PSL did inherit the syntax of Sugar, and does include a branching-time extension as an acknowledgment to Sugar.[17]

There was some evolution of PSL with respect to ForSpec. After some debate on the proper way to define resets [4], ForSpec's approach was essentially accepted after some reformulation [41]. ForSpec's fundamental approach to clocks, which is semantic, was accepted, but modified in some important details [42]. In addition to the dynamic modalities, borrowed from dynamic logic, PSL also has weak dynamic modalities [40], which are reminiscent of "looping" modalities in dynamic logic [68,66]. Today PSL 1.1 is an IEEE Standard 1850–2005, and continues to be refined by the IEEE P1850 PSL Working Group.[18]

---

[15] See http://www.accellera.org/.

[16] See http://www.eda-stds.org/vfv/.

[17] See [39] and language reference manual at http://www.eda.org/vfv/docs/PSL-v1.1.pdf.

[18] See http://www.eda.org/ieee-1850/.

Practical use of ForSpec and PSL has shown that the regular layer (that is, the dynamic modalities), is highly popular with verification engineers. Another standardized property specification language, called *SVA* (for SystemVerilog Assertions), is based, in essence, on that regular layer [137].

5   Contemplation

The evolution of ideas, from Church and Prior to PSL, seems to be an amazing development. It reminds me of the medieval period, when building a cathedral spanned more than a mason's lifetime. Many masons spend their whole lives working on a cathedral, never seeing it to completion. We are fortunate to see the completion of this particular "cathedral". Just like the medieval masons, our contributions are often smaller than we'd like to consider them, but even small contributions can have a major impact. Unlike the medieval cathedrals, the scientific cathedral has no architect; the construction is driven by a complex process, whose outcome is unpredictable. Much that has been discovered is forgotten and has to be rediscovered. It is hard to fathom what our particular "cathedral" will look like in 50 years.

Acknowledgments

I am grateful to E. Clarke, A. Emerson, R. Goldblatt, A. Pnueli, P. Sistla, P. Wolper for helping me trace the many threads of this story, to D. Fisman, C. Eisner, J. Halpern, D. Harel and T. Wilke for their many useful comments on earlier drafts of this paper, and to S. Nain, K. Rozier, and D. Tabakov for proofreading earlier drafts. I'd also like to thank K. Rozier for her help with graphics.

References

1. Aggarwal, S., Kurshan, R.P.: Automated implementation from formal specification. In: Proc. 4th Int'l Workshop on Protocol Specification, Testing and Verification, pp. 127–136. North-Holland, Amsterdam (1984)
2. Aggarwal, S., Kurshan, R.P., Sharma, D.: A language for the specification and analysis of protocols. In: Proc. 3rd Int'l Workshop on Protocol Specification, Testing, and Verification, pp. 35–50. North-Holland, Amsterdam (1983)
3. Apt, K., Olderog, E.R.: Verification of Sequential and Concurrent Programs. Springer, Heidelberg (2006)
4. Armoni, R., et al.: Resets vs. aborts in linear temporal logic. In: Garavel, H., Hatcliff, J. (eds.) ETAPS 2003 and TACAS 2003. LNCS, vol. 2619, pp. 65–80. Springer, Heidelberg (2003)
5. Armoni, R., et al.: The ForSpec temporal logic: A new temporal property-specification logic. In: Katoen, J.-P., Stevens, P. (eds.) ETAPS 2002 and TACAS 2002. LNCS, vol. 2280, pp. 211–296. Springer, Heidelberg (2002)

6. Banieqbal, B., Barringer, H.: Temporal logic with fixed points. In: Banieqbal, B., Pnueli, A., Barringer, H. (eds.) Temporal Logic in Specification. LNCS, vol. 398, pp. 62–74. Springer, Heidelberg (1989)
7. Beer, I., et al.: The temporal logic Sugar. In: Berry, G., Comon, H., Finkel, A. (eds.) CAV 2001. LNCS, vol. 2102, pp. 363–367. Springer, Heidelberg (2001)
8. Beer, I., Ben-David, S., Landver, A.: On-the-fly model checking of RCTL formulas. In: Y. Vardi, M. (ed.) CAV 1998. LNCS, vol. 1427, pp. 184–194. Springer, Heidelberg (1998)
9. Ben-Ari, M., Manna, Z., Pnueli, A.: The logic of nexttime. In: Proc. 8th ACM Symp. on Principles of Programming Languages, pp. 164–176 (1981)
10. Blackburn, P., de Rijke, M., Venema, Y.: Modal Logic. Cambridge University Press, Cambridge (2002)
11. Bradfield, J., Stirling, C.: PDL and modal μ-calculus. In: Blackburn, P., van Benthem, J., Wolter, F. (eds.) Handbook of Modal Logic, Elsevier, Amsterdam (2006)
12. Brayton, R.K., et al.: VIS: a system for verification and synthesis. In: Alur, R., Henzinger, T.A. (eds.) CAV 1996. LNCS, vol. 1102, pp. 428–432. Springer, Heidelberg (1996)
13. Browne, M.C., et al.: Automatic verification of sequential circuits using temporal logic. IEEE Transactions on Computing C-35, 1035–1044 (1986)
14. Bryant, R.E.: Graph-based algorithms for Boolean-function manipulation. IEEE Transactions on Computing C-35(8), 677–691 (1986)
15. Büchi, J.R.: Weak second-order arithmetic and finite automata. Zeit. Math. Logik und Grundl. Math. 6, 66–92 (1960)
16. Büchi, J.R.: On a decision method in restricted second order arithmetic. In: Proc. Int. Congress on Logic, Method, and Philosophy of Science. 1960, pp. 1–12. Stanford University Press (1962)
17. Büchi, J.R., Elgot, C.C., Wright, J.B.: The non-existence of certain algorithms for finite automata theory (abstract). Notices Amer. Math. Soc. 5, 98 (1958)
18. Büchi, J.R., Landweber, L.H.: Solving sequential conditions by finite-state strategies. Trans. AMS 138, 295–311 (1969)
19. Burch, J.R., et al.: Symbolic model checking: 10^{20} states and beyond. In: Proc. 5th IEEE Symp. on Logic in Computer Science, pp. 428–439 (1990)
20. Burch, J.R., et al.: Symbolic model checking: 10^{20} states and beyond. Information and Computation 98(2), 142–170 (1992)
21. Burstall, R.M.: Program proving as hand simulation with a little induction. In: Information Processing 1974, Stockholm, Sweden, pp. 308–312. North-Holland, Amsterdam (1974)
22. Bustan, D., et al.: Regular vacuity. In: Borrione, D., Paul, W. (eds.) CHARME 2005. LNCS, vol. 3725, pp. 191–206. Springer, Heidelberg (2005)
23. Nowick, S.M., van Berkel, C.H., Josephs, M.B.: Applications of asynchronous circuits. Proceedings of the IEEE 87(2), 223–233 (1999)
24. Church, A.: Applicaton of recursive arithmetics to the problem of circuit synthesis. In: Summaries of Talks Presented at The Summer Institute for Symbolic Logic, pp. 3–50. Communications Research Division, Institute for Defense Analysis (1957)
25. Church, A.: Logic, arithmetics, and automata. In: Proc. Int. Congress of Mathematicians, 1962, Institut Mittag-Leffler, pp. 23–35 (1963)
26. Cimatti, A., et al.: Nusmv 2: An opensource tool for symbolic model checking. In: Brinksma, E., Larsen, K.G. (eds.) CAV 2002. LNCS, vol. 2404, pp. 359–364. Springer, Heidelberg (2002)

27. Clarke, E.M.: The birth of model checking. In: Avron, A., et al. (eds.) Trakhten-brot/Festschrift. LNCS, vol. 4800 (2007)
28. Clarke, E.M., Emerson, E.A.: Design and synthesis of synchronization skeletons using branching time temporal logic. In: Kozen, D. (ed.) Logic of Programs 1981. LNCS, vol. 131, pp. 52–71. Springer, Heidelberg (1982)
29. Clarke, E.M., Emerson, E.A., Sistla, A.P.: Automatic verification of finite state concurrent systems using temporal logic specifications: A practical approach. In: Proc. 10th ACM Symp. on Principles of Programming Languages, pp. 117–126 (1983)
30. Clarke, E.M., Emerson, E.A., Sistla, A.P.: Automatic verification of finite-state concurrent systems using temporal logic specifications. ACM Transactions on Programming Languagues and Systems 8(2), 244–263 (1986)
31. Clarke, E.M., Grumberg, O.: Avoiding the state explosion problem in temporal logic model-checking algorithms. In: Proc. 16th ACM Symp. on Principles of Distributed Computing, pp. 294–303 (1987)
32. Clarke, E.M., Grumberg, O., Hamaguchi, K.: Another look at LTL model checking. In: Dill, D.L. (ed.) CAV 1994. LNCS, vol. 818, pp. 415–427. Springer, Heidelberg (1994)
33. Clarke, E.M., Mishra, B.: Hierarchical verification of asynchronous circuits using temporal logic. Theoretical Computer Science 38, 269–291 (1985)
34. Constable, R.L.: On the theory of programming logics. In: Proc. 9th ACM Symp. on Theory of Computing, pp. 269–285 (1977)
35. Copty, F., et al.: Benefits of bounded model checking at an industrial setting. In: Berry, G., Comon, H., Finkel, A. (eds.) CAV 2001. LNCS, vol. 2102, pp. 436–453. Springer, Heidelberg (2001)
36. Courcoubetis, C., et al.: Memory efficient algorithms for the verification of temporal properties. In: Clarke, E., Kurshan, R.P. (eds.) CAV 1990. LNCS, vol. 531, pp. 233–242. Springer, Heidelberg (1991)
37. Courcoubetis, C., et al.: Memory efficient algorithms for the verification of temporal properties. Formal Methods in System Design 1, 275–288 (1992)
38. Duret-Lutz, A., Poitrenaud, D.: SPOT: An extensible model checking library using transition-based generalized büchi automata. In: Proc. 12th Int'l Workshop on Modeling, Analysis, and Simulation of Computer and Telecommunication Systems, pp. 76–83. IEEE Computer Society, Los Alamitos (2004)
39. Eisner, C., Fisman, D.: A Practical Introduction to PSL. Springer, Heidelberg (2006)
40. Eisner, C., Fisman, D., Havlicek, J.: A topological characterization of weakness. In: Proc. 24th ACM Symp. on Principles of Distributed Computing, pp. 1–8 (2005)
41. Eisner, C., et al.: Reasoning with temporal logic on truncated paths. In: Hunt Jr., W.A., Somenzi, F. (eds.) CAV 2003. LNCS, vol. 2725, pp. 27–39. Springer, Heidelberg (2003)
42. Eisner, C., et al.: The definition of a temporal clock operator. In: Baeten, J.C.M., et al. (eds.) ICALP 2003. LNCS, vol. 2719, pp. 857–870. Springer, Heidelberg (2003)
43. Elgot, C.: Decision problems of finite-automata design and related arithmetics. Trans. Amer. Math. Soc. 98, 21–51 (1961)
44. Elgot, C.C., Wright, J.: Quantifier elimination in a problem of logical design. Michigan Math. J. 6, 65–69 (1959)
45. Emerson, E.A., Clarke, E.M.: Characterizing correctness properties of parallel programs using fixpoints. In: Proc. 7th Int. Colloq. on Automata, Languages, and Programming, pp. 169–181 (1980)

46. Emerson, E.A., Halpern, J.Y.: "Sometimes" and "not never" revisited: On branching versus linear time. In: Proc. 10th ACM Symp. on Principles of Programming Languages, pp. 127–140 (1983)
47. Emerson, E.A., Halpern, J.Y.: Decision procedures and expressiveness in the temporal logic of branching time. Journal of Computer and Systems Science 30, 1–24 (1985)
48. Emerson, E.A., Halpern, J.Y.: Sometimes and not never revisited: On branching versus linear time. Journal of the ACM 33(1), 151–178 (1986)
49. Emerson, E.A., Lei, C.-L.: Efficient model checking in fragments of the propositional μ-calculus. In: Proc. 1st IEEE Symp. on Logic in Computer Science, pp. 267–278 (1986)
50. Engeler, E.: Algorithmic properties of structures. Math. Syst. Theory 1, 183–195 (1967)
51. Etessami, K., Vardi, M.Y., Wilke, T.: First-order logic with two variables and unary temporal logic. Inf. Comput. 179(2), 279–295 (2002)
52. Fischer, M.J., Ladner, R.E.: Propositional modal logic of programs (extended abstract). In: Proc. 9th ACM Symp. on Theory of Computing, pp. 286–294 (1977)
53. Fischer, M.J., Ladner, R.E.: Propositional dynamic logic of regular programs. Journal of Computer and Systems Science 18, 194–211 (1979)
54. Fix, L.: Fifteen years of formal property verification at Intel. In: Proc. 2006 Workshop on 25 Years of Model Checking. LNCS, Springer, Heidelberg (2007)
55. Fix, L., Kamhi, G.: Adaptive variable reordering for symbolic model checking. In: Proc. ACM/IEEE Int'l Conf. on Computer Aided Design, pp. 359–365 (1998)
56. Gabbay, D., et al.: On the temporal analysis of fairness. In: Proc. 7th ACM Symp. on Principles of Programming Languages, pp. 163–173 (1980)
57. Gastin, P., Oddoux, D.: Fast LTL to Büchi automata translation. In: Berry, G., Comon, H., Finkel, A. (eds.) CAV 2001. LNCS, vol. 2102, pp. 53–65. Springer, Heidelberg (2001)
58. Gerth, R., et al.: Simple on-the-fly automatic verification of linear temporal logic. In: Dembiski, P., Sredniawa, M. (eds.) Protocol Specification, Testing, and Verification, pp. 3–18. Chapman and Hall, Boca Raton (1995)
59. Goldblatt, R.: Logic of time and computation. Technical report, CSLI Lecture Notes, no.7, Stanford University (1987)
60. Hafer, T., Thomas, W.: Computation tree logic CTL* and path quantifiers in the monadic theory of the binary tree. In: Ottmann, T. (ed.) ICALP 1987. LNCS, vol. 267, pp. 269–279. Springer, Heidelberg (1987)
61. Halpern, J., Reif, J.H.: The propositional dynamic logic of deterministic, well-structured programs (extended abstract). In: Proc. 22nd IEEE Symp. on Foundations of Computer Science, pp. 322–334 (1981)
62. Halpern, J.Y., Reif, J.H.: The propositional dynamic logic of deterministic, well-structured programs. Theor. Comput. Sci. 27, 127–165 (1983)
63. Hardin, R.H., Har'el, Z., Kurshan, R.P.: COSPAN. In: Alur, R., Henzinger, T.A. (eds.) CAV 1996. LNCS, vol. 1102, pp. 423–427. Springer, Heidelberg (1996)
64. Harel, D., Kozen, D., Parikh, R.: Process logic: Expressiveness, decidability, completeness. J. Comput. Syst. Sci. 25(2), 144–170 (1982)
65. Harel, D., Kozen, D., Tiuryn, J.: Dynamic Logic. MIT Press, Cambridge (2000)
66. Harel, D., Peleg, D.: More on looping vs. repeating in dynamic logic. Inf. Process. Lett. 20(2), 87–90 (1985)
67. Harel, D., Peleg, D.: Process logic with regular formulas. Theoreti. Comp. Sci. 38(2–3), 307–322 (1985)

68. Harel, D., Sherman, R.: Looping vs. repeating in dynamic logic. Inf. Comput. 55(1–3), 175–192 (1982)
69. Holzmann, G.J.: The model checker SPIN. IEEE Transactions on Software Engineering 23(5), 279–295 (1997)
70. Hopcroft, J.E., Ullman, J.D.: Introduction to Automata Theory, Languages, and Computation. Addison-Wesley, Reading (1979)
71. Kamp, J.A.W.: Tense Logic and the Theory of Order. PhD thesis, UCLA (1968)
72. Keller, R.M.: Formal verification of parallel programs. Communications of the ACM 19, 371–384 (1976)
73. Kozen, D.: Results on the propositional μ-calculus. In: Nielsen, M., Schmidt, E.M. (eds.) ICALP 1982. LNCS, vol. 140, pp. 348–359. Springer, Heidelberg (1982)
74. Kozen, D.: Results on the propositional μ-calculus. Theoretical Computer Science 27, 333–354 (1983)
75. Kripke, S.: A completeness theorem in modal logic. Journal of Symbolic Logic 24, 1–14 (1959)
76. Kupferman, O., Piterman, N., Vardi, M.Y.: Safraless compositional synthesis. In: Ball, T., Jones, R.B. (eds.) CAV 2006. LNCS, vol. 4144, pp. 31–44. Springer, Heidelberg (2006)
77. Kupferman, O., Vardi, M.Y.: Weak alternating automata are not that weak. ACM Transactions on Computational Logic 2(2), 408–429 (2001)
78. Kupferman, O., Vardi, M.Y.: Safraless decision procedures. In: Proc. 46th IEEE Symp. on Foundations of Computer Science, pp. 531–540 (2005)
79. Kurshan, R.P.: Analysis of discrete event coordination. In: de Bakker, J.W., de Roever, W.-P., Rozenberg, G. (eds.) REX 1989. LNCS, vol. 430, pp. 414–453. Springer, Heidelberg (1990)
80. Kurshan, R.P.: Computer Aided Verification of Coordinating Processes. Princeton University Press, Princeton (1994)
81. Kurshan, R.P.: Verification technology transfer. In: Proc. 2006 Workshop on 25 Years of Model Checking. LNCS, Springer, Heidelberg (2007)
82. Lamport, L.: "Sometimes" is sometimes "not never" - on the temporal logic of programs. In: Proc. 7th ACM Symp. on Principles of Programming Languages, pp. 174–185 (1980)
83. Lichtenstein, O., Pnueli, A.: Checking that finite state concurrent programs satisfy their linear specification. In: Proc. 12th ACM Symp. on Principles of Programming Languages, pp. 97–107 (1985)
84. Lichtenstein, O., Pnueli, A., Zuck, L.: The glory of the past. In: Parikh, R. (ed.) Logic of Programs 1985. LNCS, vol. 193, pp. 196–218. Springer, Heidelberg (1985)
85. Markey, N.: Temporal logic with past is exponentially more succinct. EATCS Bulletin 79, 122–128 (2003)
86. McMillan, K.L.: Symbolic Model Checking. Kluwer Academic Publishers, Dordrecht (1993)
87. McNaughton, R., Papert, S.: Counter-Free Automata. MIT Press, Cambridge (1971)
88. Meyer, A.R.: Weak monadic second order theory of successor is not elementary recursive. In: Proc. Logic Colloquium. Lecture Notes in Mathematics, vol. 453, pp. 132–154. Springer, Heidelberg (1975)
89. Meyer, A.R.: Ten thousand and one logics of programming. Technical report, MIT, MIT-LCS-TM-150 (1980)
90. Morley, M.J.: Semantics of temporal e. In: Melham, T.F., Moller, F.G. (eds.) Banff 1999 Higher Order Workshop (Formal Methods in Computation), University of Glasgow, Department of Computing Science Technical Report (1999)

91. Urquhart, A., Rescher, N.: Temporal Logic. Springer, Heidelberg (1971)
92. Øhrstrøm, P., Hasle, P.F.V.: Temporal Logic: from Ancient Times to Artificial Intelligence. Studies in Linguistics and Philosophy, vol. 57. Kluwer Academic Publishers, Dordrecht (1995)
93. Park, D.: Finiteness is μ-ineffable. Theoretical Computer Science 3, 173–181 (1976)
94. Pnueli, A.: The temporal logic of programs. In: Proc. 18th IEEE Symp. on Foundations of Computer Science, pp. 46–57 (1977)
95. Pnueli, A.: Linear and branching structures in the semantics and logics of reactive systems. In: Brauer, W. (ed.) ICALP 1985. LNCS, vol. 194, pp. 15–32. Springer, Heidelberg (1985)
96. Pnueli, A., Zuck, L.: In and out of temporal logic. In: Proc. 8th IEEE Symp. on Logic in Computer Science, pp. 124–135 (1993)
97. Pratt, V.R.: Semantical considerations on Floyd-Hoare logic. In: Proc. 17th IEEE Symp. on Foundations of Computer Science, pp. 109–121 (1976)
98. Pratt, V.R.: A practical decision method for propositional dynamic logic: Preliminary report. In: Proc. 10th Annual ACM Symposium on Theory of Computing, pp. 326–337 (1978)
99. Pratt, V.R.: A near-optimal method for reasoning about action. Journal of Computer and Systems Science 20(2), 231–254 (1980)
100. Pratt, V.R.: A decidable μ-calculus: preliminary report. In: Proc. 22nd IEEE Symp. on Foundations of Computer Science, pp. 421–427 (1981)
101. Prior, A.: Time and Modality. Oxford University Press, Oxford (1957)
102. Prior, A.: Past, Present, and Future. Clarendon Press (1967)
103. Prior, A.N.: Modality de dicto and modality de re. Theoria 18, 174–180 (1952)
104. Prior, A.N.: Modality and quantification in s5. J. Symbolic Logic 21, 60–62 (1956)
105. Queille, J.P., Sifakis, J.: Specification and verification of concurrent systems in Cesar. In: Dezani-Ciancaglini, M., Montanari, U. (eds.) Programming 1982. LNCS, vol. 137, pp. 337–351. Springer, Heidelberg (1982)
106. Rabin, M.O.: Automata on infinite objects and Church's problem. Amer. Mathematical Society (1972)
107. Rabin, M.O., Scott, D.: Finite automata and their decision problems. IBM Journal of Research and Development 3, 115–125 (1959)
108. Sabnani, K., Wolper, P., Lapone, A.: An algorithmic technique for protocol verification. In: Proc. Globecom (1985)
109. Sakoda, W., Sipser, M.: Non-determinism and the size of two-way automata. In: Proc. 10th ACM Symp. on Theory of Computing, pp. 275–286 (1978)
110. Salwicki, A.: Algorithmic logic: a tool for investigations of programs. In: Butts, R.E., Hintikka, J. (eds.) Logic Foundations of Mathematics and Computability Theory, pp. 281–295. Reidel (1977)
111. Sistla, A.P.: Theoretical issues in the design of distributed and concurrent systems. PhD thesis, Harvard University (1983)
112. Sistla, A.P., Clarke, E.M.: The complexity of propositional linear temporal logics. In: Proc. 14th Annual ACM Symposium on Theory of Computing, pp. 159–168 (1982)
113. Sistla, A.P., Clarke, E.M.: The complexity of propositional linear temporal logic. Journal of the ACM 32, 733–749 (1985)
114. Sistla, A.P., Vardi, M.Y., Wolper, P.: The complementation problem for Büchi automata with applications to temporal logic. In: Brauer, W. (ed.) ICALP 1985. LNCS, vol. 194, pp. 465–474. Springer, Heidelberg (1985)

115. Sistla, A.P., Vardi, M.Y., Wolper, P.: The complementation problem for Büchi automata with applications to temporal logic. Theoretical Computer Science 49, 217–237 (1987)

116. Stockmeyer, L.J.: The complexity of decision procedures in Automata Theory and Logic. PhD thesis, MIT, Project MAC Technical Report TR-133 (1974)

117. Street, R.S.: Propositional dynamic logic of looping and converse. In: Proc. 13th ACM Symp. on Theory of Computing, pp. 375–383 (1981)

118. Streett, R.S.: A propositional dynamic logic for reasoning about program divergence. PhD thesis, M.Sc. Thesis, MIT (1980)

119. Streett, R.S.: Propositional dynamic logic of looping and converse. Information and Control 54, 121–141 (1982)

120. A comparison of reset control methods: Application note 11. Summit Microelectronics, Inc. (1999),
http://www.summitmicro.com/tech_support/notes/note11.htm

121. Thomas, W.: Star-free regular sets of ω-sequences. Information and Control 42(2), 148–156 (1979)

122. Thomas, W.: On the synthesis of strategies in infinite games. In: Mayr, E.W., Puech, C. (eds.) STACS 1995. LNCS, vol. 900, pp. 1–13. Springer, Heidelberg (1995)

123. Trakhtenbrot, B.: The synthesis of logical nets whose operators are described in terms of one-place predicate calculus. Doklady Akad. Nauk SSSR 118(4), 646–649 (1958)

124. Trakhtenbrot, B.: Certain constructions in the logic of one-place predicates. Doklady Akad. Nauk SSSR 138, 320–321 (1961)

125. Trakhtenbrot, B.A.: Finite automata and monadic second order logic. Siberian Math. J. 3, 101–131 (1962) Russian; English translation in: AMS Transl. 59, 23–55 (1966)

126. Trakhtenbrot, B.A., Barzdin, Y.M.: Finite Automata. North-Holland, Amsterdam (1973)

127. Vardi, M.Y.: The büchi complementation saga. In: Thomas, W., Weil, P. (eds.) STACS 2007. LNCS, vol. 4393, pp. 12–22. Springer, Heidelberg (2007)

128. Vardi, M.Y.: The complexity of relational query languages. In: Proc. 14th ACM Symp. on Theory of Computing, pp. 137–146 (1982)

129. Vardi, M.Y.: A temporal fixpoint calculus. In: Proc. 15th ACM Symp. on Principles of Programming Languages, pp. 250–259 (1988)

130. Vardi, M.Y.: Unified verification theory. In: Banieqbal, B., Pnueli, A., Barringer, H. (eds.) Temporal Logic in Specification. LNCS, vol. 398, pp. 202–212. Springer, Heidelberg (1989)

131. Vardi, M.Y.: Nontraditional applications of automata theory. In: Hagiya, M., Mitchell, J.C. (eds.) TACS 1994. LNCS, vol. 789, pp. 575–597. Springer, Heidelberg (1994)

132. Vardi, M.Y.: Branching vs. linear time: Final showdown. In: Margaria, T., Yi, W. (eds.) ETAPS 2001 and TACAS 2001. LNCS, vol. 2031, pp. 1–22. Springer, Heidelberg (2001)

133. Vardi, M.Y.: Automata-theoretic model checking revisited. In: Cook, B., Podelski, A. (eds.) VMCAI 2007. LNCS, vol. 4349, pp. 137–150. Springer, Heidelberg (2007)

134. Vardi, M.Y., Wolper, P.: Yet another process logic. In: Clarke, E., Kozen, D. (eds.) Logic of Programs 1983. LNCS, vol. 164, pp. 501–512. Springer, Heidelberg (1984)

135. Vardi, M.Y., Wolper, P.: An automata-theoretic approach to automatic program verification. In: Proc. 1st IEEE Symp. on Logic in Computer Science, pp. 332–344 (1986)
136. Vardi, M.Y., Wolper, P.: Reasoning about infinite computations. Information and Computation 115(1), 1–37 (1994)
137. Vijayaraghavan, S., Ramanathan, M.: A Practical Guide for SystemVerilog Assertions. Springer, Heidelberg (2005)
138. Wolper, P.: Temporal logic can be more expressive. In: Proc. 22nd IEEE Symp. on Foundations of Computer Science, pp. 340–348 (1981)
139. Wolper, P.: Temporal logic can be more expressive. Information and Control 56(1–2), 72–99 (1983)
140. Wolper, P., Vardi, M.Y., Sistla, A.P.: Reasoning about infinite computation paths. In: Proc. 24th IEEE Symp. on Foundations of Computer Science, pp. 185–194 (1983)

Author Index

Lecture Notes in Computer Science

Sublibrary 1: Theoretical Computer Science and General Issues

For information about Vols. 1– 4599
please contact your bookseller or Springer

Vol. 4719: R. Backhouse, J. Gibbons, R. Hinze, J. Jeuring (Eds.), Datatype-Generic Programming. XI, 369 pages. 2007.

Vol. 4711: C.B. Jones, Z. Liu, J. Woodcock (Eds.), Theoretical Aspects of Computing – ICTAC 2007. XI, 483 pages. 2007.

Vol. 4710: C.W. George, Z. Liu, J. Woodcock (Eds.), Domain Modeling and the Duration Calculus. XI, 237 pages. 2007.

Vol. 4708: L. Kučera, A. Kučera (Eds.), Mathematical Foundations of Computer Science 2007. XVIII, 764 pages. 2007.

Vol. 4707: O. Gervasi, M.L. Gavrilova (Eds.), Computational Science and Its Applications – ICCSA 2007, Part III. XXIV, 1205 pages. 2007.

Vol. 4706: O. Gervasi, M.L. Gavrilova (Eds.), Computational Science and Its Applications – ICCSA 2007, Part II. XXIII, 1129 pages. 2007.

Vol. 4705: O. Gervasi, M.L. Gavrilova (Eds.), Computational Science and Its Applications – ICCSA 2007, Part I. XLIV, 1169 pages. 2007.

Vol. 4703: L. Caires, V.T. Vasconcelos (Eds.), CONCUR 2007 – Concurrency Theory. XIII, 507 pages. 2007.

Vol. 4700: C.B. Jones, Z. Liu, J. Woodcock (Eds.), Formal Methods and Hybrid Real-Time Systems. XVI, 539 pages. 2007.

Vol. 4699: B. Kågström, E. Elmroth, J. Dongarra, J. Waśniewski (Eds.), Applied Parallel Computing. XXIX, 1192 pages. 2007.

Vol. 4698: L. Arge, M. Hoffmann, E. Welzl (Eds.), Algorithms – ESA 2007. XV, 769 pages. 2007.

Vol. 4697: L. Choi, Y. Paek, S. Cho (Eds.), Advances in Computer Systems Architecture. XIII, 400 pages. 2007.

Vol. 4688: K. Li, M. Fei, G.W. Irwin, S. Ma (Eds.), Bio-Inspired Computational Intelligence and Applications. XIX, 805 pages. 2007.

Vol. 4684: L. Kang, Y. Liu, S. Zeng (Eds.), Evolvable Systems: From Biology to Hardware. XIV, 446 pages. 2007.

Vol. 4683: L. Kang, Y. Liu, S. Zeng (Eds.), Advances in Computation and Intelligence. XVII, 663 pages. 2007.

Vol. 4681: D.-S. Huang, L. Heutte, M. Loog (Eds.), Advanced Intelligent Computing Theories and Applications. XXVI, 1379 pages. 2007.

Vol. 4672: K. Li, C. Jesshope, H. Jin, J.-L. Gaudiot (Eds.), Network and Parallel Computing. XVIII, 558 pages. 2007.

Vol. 4671: V.E. Malyshkin (Ed.), Parallel Computing Technologies. XIV, 635 pages. 2007.

Vol. 4669: J.M. de Sá, L.A. Alexandre, W. Duch, D. Mandic (Eds.), Artificial Neural Networks – ICANN 2007, Part II. XXXI, 990 pages. 2007.

Vol. 4668: J.M. de Sá, L.A. Alexandre, W. Duch, D. Mandic (Eds.), Artificial Neural Networks – ICANN 2007, Part I. XXXI, 978 pages. 2007.

Vol. 4666: M.E. Davies, C.J. James, S.A. Abdallah, M.D. Plumbley (Eds.), Independent Component Analysis and Signal Separation. XIX, 847 pages. 2007.

Vol. 4665: J. Hromkovič, R. Královič, M. Nunkesser, P. Widmayer (Eds.), Stochastic Algorithms: Foundations and Applications. X, 167 pages. 2007.

Vol. 4664: J. Durand-Lose, M. Margenstern (Eds.), Machines, Computations, and Universality. X, 325 pages. 2007.

Vol. 4661: U. Montanari, D. Sannella, R. Bruni (Eds.), Trustworthy Global Computing. X, 339 pages. 2007.

Vol. 4649: V. Diekert, M.V. Volkov, A. Voronkov (Eds.), Computer Science – Theory and Applications. XIII, 420 pages. 2007.

Vol. 4647: R. Martin, M.A. Sabin, J.R. Winkler (Eds.), Mathematics of Surfaces XII. IX, 509 pages. 2007.

Vol. 4646: J. Duparc, T.A. Henzinger (Eds.), Computer Science Logic. XIV, 600 pages. 2007.

Vol. 4644: N. Azémard, L. Svensson (Eds.), Integrated Circuit and System Design. XIV, 583 pages. 2007.

Vol. 4641: A.-M. Kermarrec, L. Bougé, T. Priol (Eds.), Euro-Par 2007 Parallel Processing. XXVII, 974 pages. 2007.

Vol. 4639: E. Csuhaj-Varjú, Z. Ésik (Eds.), Fundamentals of Computation Theory. XIV, 508 pages. 2007.

Vol. 4638: T. Stützle, M. Birattari, H. H. Hoos (Eds.), Engineering Stochastic Local Search Algorithms. X, 223 pages. 2007.

Vol. 4630: H.J. van den Herik, P. Ciancarini, H.H.L.M.(J.) Donkers (Eds.), Computers and Games. XII, 283 pages. 2007.

Vol. 4628: L.N. de Castro, F.J. Von Zuben, H. Knidel (Eds.), Artificial Immune Systems. XII, 438 pages. 2007.

Vol. 4627: M. Charikar, K. Jansen, O. Reingold, J.D.P. Rolim (Eds.), Approximation, Randomization, and Combinatorial Optimization. XII, 626 pages. 2007.

Vol. 4624: T. Mossakowski, U. Montanari, M. Haveraaen (Eds.), Algebra and Coalgebra in Computer Science. XI, 463 pages. 2007.

Vol. 4623: M. Collard (Ed.), Ontologies-Based Databases and Information Systems. X, 153 pages. 2007.

Vol. 4621: D. Wagner, R. Wattenhofer (Eds.), Algorithms for Sensor and Ad Hoc Networks. XIII, 415 pages. 2007.

Vol. 4619: F. Dehne, J.-R. Sack, N. Zeh (Eds.), Algorithms and Data Structures. XVI, 662 pages. 2007.

Vol. 4618: S.G. Akl, C.S. Calude, M.J. Dinneen, G. Rozenberg, H.T. Wareham (Eds.), Unconventional Computation. X, 243 pages. 2007.

Vol. 4616: A.W.M. Dress, Y. Xu, B. Zhu (Eds.), Combinatorial Optimization and Applications. XI, 390 pages. 2007.

Vol. 4614: B. Chen, M. Paterson, G. Zhang (Eds.), Combinatorics, Algorithms, Probabilistic and Experimental Methodologies. XII, 530 pages. 2007.

Vol. 4613: F.P. Preparata, Q. Fang (Eds.), Frontiers in Algorithmics. XI, 348 pages. 2007.

Vol. 4600: H. Comon-Lundh, C. Kirchner, H. Kirchner (Eds.), Rewriting, Computation and Proof. XVI, 273 pages. 2007.